HANDBOOK OF PETROLEUM
PRODUCT ANALYSIS

CHEMICAL ANALYSIS

A SERIES OF MONOGRAPHS ON ANALYTICAL CHEMISTRY AND ITS APPLICATIONS

Series Editor
MARK F. VITHA

Volume 182

A complete list of the titles in this series appears at the end of this volume.

HANDBOOK OF PETROLEUM PRODUCT ANALYSIS

2nd Edition

JAMES G. SPEIGHT, PhD, DSc
CD & W Inc.,
Laramie, WY, USA

Library of Congress Cataloging-in-Publication Data:

Speight, James G.
  Handbook of petroleum product analysis / James G. Speight, PhD, DSc. – 2nd edition.
    pages cm
  Includes index.
  ISBN 978-1-118-36926-5 (cloth)
1. Petroleum products–Analysis. I. Title.
  TP691.S689 2015
  665.5'38–dc23

                                                     2014020571

Printed in the United States of America

10 9 8 7 6 5 4 3 2 1

CONTENTS

PREFACE

The success of the first edition of this text has been the primary factor in the decision to publish a second edition. During the period (2002–2014) between editions, petroleum products have continued to be produced and used for many different purposes with widely differing requirements leading to criteria for quality which are numerous and complex.

In addition, the demand for petroleum products, particularly liquid fuels (gasoline and diesel fuel) and petrochemical feedstocks (such as aromatics and olefins), is increasing throughout the world. Traditional markets such as North America and Europe are experiencing a steady increase in demand whereas emerging Asian markets, such as India and China, are witnessing a rapid surge in demand for liquid fuels. This has resulted in a tendency for the evolution in product specifications caused by various environmental regulations. In many countries, especially in the United States and Europe, gasoline and diesel fuel specifications have changed radically in the past decade (since the publication of the first edition of this book) and will continue to do so in the future. Currently, reducing the sulfur levels of liquid fuels is the dominant objective of many refiners. This is enhancing the need for accurate analysis of petroleum.

Refineries must, and indeed are eager to, adapt to changing circumstances and are amenable to trying new technologies that are radically different in character. Currently, refineries are also looking to exploit heavy (more viscous) crude oils and tar sand bitumen (sometimes referred to as extra heavy crude oil) provided they have the refinery technology capable of handling such feedstocks. Transforming the higher boiling constituents of these feedstocks components into liquid fuels is becoming a necessity. It is no longer a simple issue of mixing the heavy feedstock with conventional petroleum to make up a blended refinery feedstock. Incompatibility issues arise that can, if not anticipated, close down a refinery or, at best, a major section of the refinery. Therefore handling such feedstocks requires technological change, including more effective and innovative use of hydrogen within the refinery. Heavier crude oil could also be contaminated with sulfur and metal particles that must be detected and removed to meet quality standards.

Thus, this book will deal with the various aspects of petroleum product analysis and will provide a detailed explanation of the necessary standard tests and procedures that are applicable to products in order to help predefine predictability of petroleum behavior during refining. In addition, the application of new methods for determining instability and incompatibility as well as analytical methods related to environmental regulations will be described.

Each chapter is written as a stand-alone chapter that has necessitated some repetition. Repetition is considered necessary for the reader to have all of the relevant information at hand especially where there are tests that can be applied to several products. Where this was not possible, cross-references to the pertinent chapter are included. Several general references are listed for the reader to consult and obtain a more detailed description of petroleum products. No attempt has been made to be exhaustive in the citations of such works. Thereafter, the focus is to cite the relevant test methods that are applied to petroleum products.

The reader might also be surprised at the number of older references that are included. The purpose of this is to remind the reader that there is much valuable work cited in the older literature. Work which is still of value and, even though in some cases, there has been similar work performed with advanced equipment, the older work has stood the test of time. However, the text still maintains its initial premise that is to introduce the reader to the analytical science of petroleum and petroleum products—the standard test methods are up to date and any test methods abandoned or declared obsolete since the publication of the first edition are no longer included. In addition, throughout the chapters, no preference is given to any particular tests. To this end, all lists of tests are ordered alphabetically in the References Section and a newly created Appendix (Tables A01–A029 that are organized by function) contains a more comprehensive list of the various standard test methods.

Thus, it is the purpose of this book to identify quality criteria appropriate analysis and testing. In addition, the book has been adjusted, polished, and improved for the benefit of new readers as well as for the benefit of readers of the first edition.

Dr. James G. Speight
Laramie, Wyoming, USA

1

PETROLEUM AND PETROLEUM PRODUCTS

1.1 INTRODUCTION

Petroleum (also called *crude oil*) is the term used to describe a wide variety of naturally occurring hydrocarbon-rich fluids that has accumulated in subterranean reservoirs and which exhibits considerably simple properties such as specific gravity/API gravity) and the amount of residuum (Table 1.1). More detailed inspections show considerable variations in color, odor, and flow properties that reflect the diversity of the origin of petroleum. From further inspections, variations also occur in the molecular types present in crude oil, which include compounds of nitrogen, oxygen, sulfur, metals (particularly nickel and vanadium), as well as other elements (ASTM D4175) (Speight, 2012a). Consequently, it is not surprising that petroleum can exhibit wide variations in refining behavior, product yields, and product properties (Speight, 2014a).

Over the past four decades, the petroleum being processed in refineries has becoming increasingly heavier (higher amounts of residuum) and higher sulfur content (Speight, 2000, 2014a; Speight and Ozum, 2002; Hsu and Robinson, 2006; Gary et al., 2007). Market demand (*market pull*) dictates that *residua* must be upgraded to higher-value products (Speight and Ozum, 2002; Hsu and Robinson, 2006; Gary et al., 2007; Speight, 2014a). In short, the value of petroleum depends upon its quality for refining and whether or not the product slate and product yields can be obtained to fit market demand.

Thus, process units in a refinery require analytical test methods that can adequately evaluate feedstocks and monitor product quality (Drews, 1998; Nadkarni, 2000, 2011; Rand, 2003; Totten, 2003). In addition, the high sulfur content of petroleum and regulations limiting the maximum sulfur content of fuels makes sulfur removal a priority in refinery processing. Here again, analytical methodology is key to the successful determination of the sulfur compound types present and their subsequent removal.

Upgrading residua involves processing (usually conversion) into a more salable, higher-valued product. Improved characterization methods are necessary for process design, crude oil evaluation, and operational control. Definition of the boiling range and the hydrocarbon-type distribution in heavy distillates and in residua is increasingly important. Feedstock analysis to provide a quantitative boiling range distribution (that accounts for non-eluting components) as well as the distribution of hydrocarbon types in gas oil and higher-boiling materials is important in evaluating feedstocks for further processing.

Sulfur reduction processes are sensitive to both amount and structure of the sulfur compounds being removed. Tests that can provide information about both are becoming increasingly important, and analytical tests that provide information about other constituents of interest (e.g., nitrogen, organometallic constituents) are also valuable and being used for characterization.

But before emerging into the detailed aspects of petroleum product analysis, it is necessary to understand the nature of petroleum as well as the refinery processes required to produce petroleum products. This will present to the reader the background that is necessary to understand petroleum and the processes used to convert it to products. The details of the chemistry are not presented here and can be found elsewhere (Speight, 2000, 2014a; Speight and Ozum, 2002; Hsu and Robinson, 2006; Gary et al., 2007).

Handbook of Petroleum Product Analysis, Second Edition. James G. Speight.
© 2015 John Wiley & Sons, Inc. Published 2015 by John Wiley & Sons, Inc.

TABLE 1.1 Illustration of the variation in petroleum properties—specific gravity/API gravity) and the amount of residuum

Petroleum	Specific gravity	API gravity	Residuum >1050°F (% w/w)
Agbami (Africa)	0.790	48.1	2.5
Alaska North Slope (US)	0.869	31.4	18.3
Alba (North Sea)	0.936	19.5	32.7
Alvheim Blend (North Sea)	0.850	34.9	13.1
Azeri BTC (Asia)	0.843	36.4	13.2
Badak (Indonesia)	0.830	38.9	2.0
Bahrain (Bahrain)	0.861	32.8	26.4
California (US)	0.858	33.4	23.0
Calypso (Trinidad and Tobago)	0.971	30.8	11.6
Dalia (Africa)	0.915	23.1	27.7
Dansk Underground Consortium (DUC) (Denmark)	0.860	33.5	18.2
Draugen (Europe)	0.826	39.9	6.4
Gimboa (Africa)	0.912	25.3	24.0
Grane (North Sea)	0.940	19.0	30.3
Hibernia Blend (Canada)	0.850	35.0	17.2
Iranian Light (Iran)	0.836	37.8	20.8
Iraq Light (Iraq)	0.844	36.2	23.8
Kearl (Canada)	0.918	22.6	31.9
Kutubu Bland (New Guinea)	0.802	44.8	12.0
Kuwaiti Light (Kuwait)	0.860	33.0	31.9
Marib Light (Yemen)	0.809	43.3	7.7
Medanito (Argentina)	0.860	33.0	20.6
Mondo (Africa)	0.877	29.9	22.1
Oklahoma (US)	0.816	41.9	20.0
Oman (Oman)	0.873	30.5	30.5
Pennsylvania (US)	0.800	45.4	2.0
Peregrino (Brazil)	0.974	13.7	40.5
Saudi Arabia	0.840	37.0	27.5
Saxi Batuque Blend (Africa)	0.856	33.9	14.6
Terra Nova (Canada)	0.859	0.9	16.0
Texas (US)	0.827	39.6	15.0
Texas (US)	0.864	32.3	27.9
Venezuela	0.950	17.4	33.6
Zakhum Lower (Abu Dhabi)	0.822	40.5	14.3

1.2 PERSPECTIVES

The following sections are included to introduce the reader to the distant historical and recent historical aspects of petroleum analysis and to show the glimmerings of how it has evolved during the twentieth century and into the twenty-first century. Indeed, in spite of the historical use of petroleum and related materials, the petroleum industry is a modern industry having come into being in 1859. From these comparatively recent beginnings, petroleum analysis has arisen as a dedicated science.

1.2.1 Historical Perspectives

Petroleum is perhaps the most important substance consumed in modern society. The word *petroleum*, derived from the Latin *petra* and *oleum*, means literally *rock oil* and refers to hydrocarbons that occur widely in the sedimentary rocks in the form of gases, liquids, semisolids, or solids. Petroleum provides not only raw materials for the ubiquitous plastics and other products, but also fuel for energy, industry, heating, and transportation.

The *history* of any subject is the means by which the subject is studied in the hopes that much can be learned from the events of the past. In the current context, the occurrence and use of petroleum, petroleum derivatives (naphtha), heavy oil, and bitumen are not new. The use of petroleum and its derivatives was practiced in pre-Christian times and is known largely through historical use in many of the older civilizations (Henry, 1873; Abraham, 1945; Forbes, 1958a, 1958b, 1959, 1964; James and Thorpe, 1994). Thus, the use of petroleum and the development of related technology are not such a modern subject as we are inclined to believe. However, the petroleum industry is essentially a twentieth-century industry, but to understand the evolution of the industry, it is essential to have a brief understanding of the first uses of petroleum.

Briefly, petroleum and bitumen have been used for millennia. For example, the Tigris–Euphrates valley, in what is now Iraq, was inhabited as early as 4000 B.C. by the people known as the Sumerians, who established one of the first great cultures of the civilized world. The Sumerians devised the cuneiform script, built the temple towers known as ziggurats, had an impressive law, as well as a wide and varied collection of literature. As the culture developed, *bitumen* (sometimes referred to as *natural-occurring asphalt*) was frequently used in construction and in ornamental works. Although it is possible, on this basis, to differentiate between the words *bitumen* and *asphalt* in modern use (Speight, 2014a), the occurrence of these words in older texts offers no such possibility. It is significant that the early use of bitumen was in the nature of cement for securing or joining together various objects, and it thus seems likely that the name itself was expressive of this application.

Early references to petroleum and its derivatives occur in the Bible, although by the time the various books of the Bible were written, the use of petroleum and bitumen was established. Investigations at historic sites have confirmed the use of petroleum and bitumen in antiquity for construction, mummification, decorative jewelry, waterproofing, as well as for medicinal use (Speight, 2014a). Many other references to bitumen occur throughout the Greek and Roman empires, and from then to the Middle Ages, early scientists (alchemists) frequently referred to the use of bitumen. In the late fifteenth and early sixteenth centuries, both Christopher Columbus and Sir Walter Raleigh have been credited with the discovery of the asphalt deposit on the island of Trinidad and apparently used the material to caulk their ships. There was also an interest in the thermal product of petroleum (nafta; naphtha) when it was discovered that this material could be used as an illuminant and as a supplement to asphalt incendiaries in warfare.

To continue such references is beyond the scope of this book, although they do give a flavor of the developing interest in petroleum. However, it is sufficient to note that there are many other references to the occurrence and use of bitumen or petroleum derivatives up to the beginning of the modern petroleum industry (Speight, 2014a). However, what is obvious by its absence is any reference to the analysis of the bitumen that was used variously through history. It can only be assumed that there was a correlation between the bitumen character and its behavior. This would be the determining factor(s) in its use as a sealant, a binder, or as a medicine. In this sense, documented history has not been kind to the scientist or engineer.

Thus, the history of analysis of petroleum and its products (as recognized by the modern scientist and engineer) can only be suggested to have started during the second half of the nineteenth century. Further developments of the analytical chemistry of petroleum continued throughout the twentieth century, and it is only through chemical and physical analysis that petroleum can be dealt with logically.

1.2.2 Modern Perspectives

The modern petroleum industry began in 1859 with the discovery and subsequent commercialization of petroleum in Pennsylvania (Speight, 2014a). During the 6000 years of its use, the importance of petroleum has progressed from the relatively simple use of asphalt from Mesopotamian seepage sites to the present-day refining operations that yield a wide variety of products and petrochemicals (Speight, 2014a). However, what is more pertinent to the industry is that throughout the millennia in which petroleum has been known and used, it is only in the twentieth century that attempts have been made to formulate and standardize petroleum analysis.

As the twentieth century matured, there was increased emphasis and reliance on instrumental approaches to petroleum analysis. In particular, *spectroscopic methods* have risen to a level of importance that is perhaps the dreams of those who first applied such methodology to petroleum analysis. There are also potentiometric titration methods that evolved, and the procedures have found favor in the identification of functional types in petroleum and its fractions.

Spectrophotometers came into widespread use—approximately beginning in 1940—and this led to wide acquisition in petroleum analysis (Chapter 2). *Ultraviolet absorption spectroscopy*, *infrared spectroscopy*, *mass spectrometry*, *emission spectroscopy*, and *nuclear magnetic resonance spectroscopy* continue to make major contributions to petroleum analysis (Nadkarni, 2011; Totten, 2003).

Chromatography is another method that is utilized for the most part in the separation of complex mixtures and has found wide use in petroleum analysis (Chapter 2). *Ion exchange* materials, long known in the form of naturally occurring silicates, were used in the earliest types of regenerative water softeners. *Gas chromatography, or vapor-phase chromatography*, found ready applications in the identification of the individual constituents of petroleum. It is still extremely valuable in the analysis of hydrocarbon mixtures of high volatility and has become an important analytical tool in the petroleum industry. With the development of high-temperature columns, the technique has been extended to mixtures of low volatility, such as gas oils and some residua.

In fact, in the petroleum refining industry, boiling range distribution data (for example ASTM D3710) are used (i) to assess petroleum crude quality before purchase, (ii) to monitor petroleum quality during transportation, (iii) to evaluate petroleum for refining, and (iv) to provide information for the optimization of refinery processes. Traditionally, boiling range distributions of the various fractions have been determined by distillation. Yield-on-crude data are still widely reported in the petroleum assay literature, providing information on the yield of specific fractions obtained by distillation (ASTM D86, ASTM D1160). However, to some extent in the laboratory, atmospheric and vacuum distillation techniques have largely been replaced by *simulated* distillation methods, which use low-resolution gas chromatography and correlate retention times to hydrocarbon boiling points (ASTM D2887, ASTM), which typically use external standards such as *n*-alkanes.

1.3 DEFINITIONS

Terminology is the means by which various subjects are named so that reference can be made in conversations and in writings and so that the meaning is passed on.

Definitions are the means by which scientists and engineers communicate the nature of a material to each other and to the world, through either the spoken or the written word. Thus, the definition of a material can be extremely important and have a profound influence on how the technical community and the public perceive that material.

For the purposes of this book, petroleum products and those products that are isolated from petroleum during recovery (such as natural gas, natural gas liquids, and natural gasoline) as well as refined products—petrochemical products—are excluded from this text.

Furthermore, it is necessary to state for the purposes of this text that on the basis of being *chemically correct*, it must be recognized that hydrocarbon molecules (hydrocarbon oils) contain carbon atoms and hydrogen atoms *only*. The presence of atoms (such as nitrogen, oxygen, sulfur, and metals) other than carbon and hydrogen leads to the definition and characterization of such materials as hydrocarbonaceous oils. Also, for the purposes of terminology, it is often convenient to subdivide petroleum and related materials into three major groups (Table 1.2) (Speight, 2014a): (i) materials that are of natural origin, (ii) materials that are manufactured, and (iii) materials that are integral fractions derived from the natural or manufactured products (Speight and Ozum, 2002; Hsu and Robinson, 2006; Gary et al., 2007; Speight, 2014a).

1.3.1 Petroleum

Petroleum is a naturally occurring mixture of hydrocarbons, generally in a liquid state, which may also include compounds of sulfur, nitrogen, oxygen, metals, and other elements (ASTM D4175) (Speight, 2000, 2014a). Although petroleum and fractions thereof have been known since ancient time (Henry, 1873; Abraham, 1945; Forbes, 1958a, b, 1959, 1964; James and Thorpe, 1994; Speight, 2014a),

TABLE 1.2 Subdivision of fossil fuels into various subgroups

Natural Materials	Derived Materials	Manufactured Materials
Natural gas	Saturates	Synthetic crude oil
Petroleum	Aromatics	Distillates
Heavy oil	Resins	Lubricating oils
Bitumen*	Asphaltenes	Wax
Asphaltite	Carbenes[†]	Residuum
Asphaltoid	Carboids[†]	Asphalt
Ozocerite		Coke
(natural wax)		
Kerogen		
Coal		

*Bitumen from tar sand deposits.
[†]Products of petroleum processing.

the current era of petroleum and petroleum product analysis might be assigned to commence in the early-to-mid nineteenth century (Silliman, Sr., 1833. Silliman, Jr., 1860, 1865, 1867, 1871) and continued thereafter. Historically, physical properties such as boiling point, density (gravity), odor, and viscosity have been used to describe crude oil (Speight, 2014a). Petroleum may be called *light* or *heavy* in reference to the amount of low-boiling constituents and the relative density (specific gravity). Likewise, odor is used to distinguish between *sweet* (low-sulfur) and *sour* (high-sulfur) crude oil. Viscosity indicates the ease of (or more correctly the resistance to) flow.

Briefly, the measurement of density is not a pro-forma (i.e., nice-to-have) piece of data as it is often used in combination with other test results to predict crude oil quality. Density or relative density (specific gravity) is used whenever conversions must be made between mass (weight) and volume measurements. Various ASTM procedures for measuring density or specific gravity are also generally applicable to heavy (viscous) oil. In the test methods, heavy oils generally do not create problems because of sample volatility, *but* the test methods are sensitive to the presence of gas bubbles in the heavy oil, and particular care must be taken to exclude or remove gas bubbles before measurement. In addition, heavy oils (with the exception of the more viscous petroleum products such as lubricating oil and white oil) are typically dark-colored samples, and it may be difficult to ascertain whether or not all air bubbles have been eliminated from the sample.

However, there is the need for a thorough understanding of petroleum and the associated technologies; it is essential that the definitions and the terminology of petroleum science and technology be given prime consideration (Speight, 2014a). This presents a better understanding of petroleum, its constituents, and its various fractions. Of the many forms of terminology that have been used, not all have survived, but the more commonly used are illustrated here. Particularly troublesome, and more confusing, are those terms that are applied to the more viscous materials, for example, the use of the terms *bitumen* and *asphalt*. This part of the text attempts to alleviate much of the confusion that exists, but it must be remembered that the terminology of petroleum is still open to personal choice and historical usage.

Conventional (light) petroleum is composed of hydrocarbons together with smaller amounts of organic compounds of nitrogen, oxygen, and sulfur and still smaller amounts of compounds containing metallic constituents, particularly vanadium, nickel, iron, and copper. The processes by which petroleum was formed dictate that petroleum composition will vary and be *site specific* thus leading to a wide variety of compositional differences. By using the term *site specific*, it is intended to convey that petroleum composition will be dependent upon regional and local variations in the proportion of the various precursors that went into the formation of the *protopetroleum* as well as variations in temperature and pressure to which the precursors were subjected.

The active principle is that petroleum is a continuum and has natural product origins (Speight, 2014a). As such, it might be anticipated that there is a continuum of different molecular systems throughout petroleum that might differ from volatile to nonvolatile fractions but which, in fact, are based on natural product systems. It might also be argued that substitution patterns on the aromatic nucleus that are identified in the volatile fractions, or in any natural product counterparts, may also apply to the substitution patterns on the aromatic nucleus of aromatic systems in the nonvolatile fractions.

Because of the complexity of the precursor mix that leads to the intermediate that is often referred to as *protopetroleum* and which eventually to petroleum, the end product contains an extreme range of organic functionality and molecular size. In fact, the large variety of the molecular constituents of petroleum makes it unlikely that a complete compound-by-compound description for even a single crude oil would be possible. Those who propose such molecular identification projects may be in for a very substantial surprise, especially when dealing with heavy oil, extra heavy oil, and tar sand bitumen. At the same time, it must be wondered how such a project, if successful, will help the refiner.

On the other hand, the molecular composition of petroleum can be described in terms of three classes of compounds: saturates, aromatics, and compounds bearing heteroatoms (nitrogen, oxygen, sulfur, and/or metals). Within each class, there are several families of related compounds. The distribution and characteristics of these molecular species account for the rich variety of crude oils. This is the type of information with some modification, but without the need for full molecular identification, that refiners have used for decades with considerable success.

There is no doubt of the need for the application of analytical techniques to petroleum-related issues—refining and environmental—and, accordingly, interest in petroleum analysis has increased over the past four decades because of the change in feedstock composition and feedstock type because of the higher demand for liquid fuels and the increased amounts of the heavier feedstocks that are now used as blendstocks in many refineries. Prior to the energy crises of the 1970s, the heavier feedstocks were used infrequently as sources of liquid fuels and were used to produce asphalt, but, now, these feedstocks have increased in value as sources of liquid fuels.

In conventional (light, sweet) petroleum, the content of pure hydrocarbons (i.e., molecules composed of carbon and hydrogen only) may be as high as 80% w/w for paraffinic petroleum and less than 50% w/w for heavy crude oil and much lower for tar sand bitumen. The non-hydrocarbon constituents are usually concentrated in the higher-boiling portions of the crude oil. The carbon and hydrogen contents are approximately constant from crude oil to crude oil even though the amounts of the various hydrocarbon types and of the individual isomers may vary widely. Thus, the carbon content of various types of petroleum is usually between 83 and 87% by weight, and the hydrogen content is in the range of 11–14% by weight.

The near-constancy of carbon content and the hydrogen content is explained by the fact that variation in the amounts of each series of hydrocarbons does not have a profound effect on overall composition (Speight, 2014a). However, within any petroleum or heavy oil, the atomic ratio of hydrogen to carbon increases from the low- to the high-molecular-weight fractions. This is attributable to an increase in the content of polynuclear aromatics and multi-ring cyclo-paraffins that are molecular constituents of the higher-boiling fractions. For higher-boiling feedstocks such as heavy oil and bitumen, the chemical composition becomes so complex and its relationship to performance so difficult to define that direct correlation of atomic ratios is not always straightforward. In any case, simpler tests are required for quality control purposes. Analysis is typically confined to the determination of certain important elements and to the *characterization* of the feedstock in terms of a variety of structural groups that have the potential to interfere with the thermal decomposition and also with catalysts. Thus, for heavy oil, bitumen, and residua, density and viscosity still are of great interest. But for such materials, hydrogen, nitrogen, sulfur, and metal content as well as carbon residue values become even more important (Table 1.1).

General aspects of petroleum *quality* (as a refinery feedstock) are assessed by the measurement of physical properties such as relative density (specific gravity), refractive index, or viscosity, or by empirical tests such as pour point or oxidation stability that are intended to relate to behavior in service. In some cases, the evaluation may include tests in mechanical rigs and engines either in the laboratory or under actual refinery process conditions.

In the crude state, petroleum has minimal value, but when refined, it provides high-value liquid fuels, solvents, lubricants, and many other products (Speight, 2014a and references cited therein). The fuels derived from petroleum contribute approximately one-third to one-half of the total world energy supply and are used not only for transportation fuels (i.e., gasoline, diesel fuel, and aviation fuel, among others) but also to heat buildings. Petroleum products have a wide variety of uses that vary from gaseous and liquid fuels to near-solid machinery lubricants. In addition, the residue of many refinery processes, asphalt—a once-maligned by-product—is now a premium value product for highway surfaces, roofing materials, and miscellaneous waterproofing uses.

Crude petroleum is a mixture of compounds boiling at different temperatures that can be separated into a variety of different generic fractions by distillation (Speight and Ozum, 2002; Hsu and Robinson, 2006; Gary *et al.*, 2007; Speight, 2014a). And the terminology of these fractions has been bound by utility and often bears little relationship to composition.

The molecular boundaries of petroleum cover a wide range of boiling points and carbon numbers of hydrocarbon compounds and other compounds containing nitrogen, oxygen, and sulfur, as well as metallic (porphyrin) constituents. However, the actual boundaries of such a *petroleum map* can only be arbitrarily defined in terms of boiling point and carbon number (Fig. 1.1). In fact, petroleum is so diverse that materials from different sources exhibit different boundary limits, and for this reason—more than for any other reason—it is not surprising that petroleum has been difficult to *map* in a precise manner (Speight, 2014a).

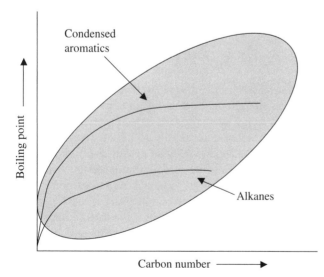

FIGURE 1.1   General boiling point–carbon number profile for petroleum.

Since there is a wide variation in the properties of crude petroleum, the proportions in which the different constituents occur vary with origin (Speight, 2014a). Thus, some crude oils have higher proportions of the lower-boiling components and others (such as heavy oil and bitumen) have higher proportions of higher-boiling components (asphaltic components and residuum).

There are several other definitions that also need to be included in any text on petroleum analysis, in particular since this text also focuses on the analysis of heavy oil and bitumen. These definitions are included because of the increased reliance on the development of these resources and the appearance of the materials in refineries.

Because of the wide range of chemical and physical properties, a wide range of tests have been (and continue to be) developed to provide an indication of the means by which a particular feedstock should be processed. Initial inspection of the nature of the petroleum will provide deductions about the most logical means of refining or correlation of various properties to structural types present and hence attempted classification of the petroleum. Proper interpretation of the data resulting from the inspection of crude oil requires an understanding of their significance.

In terms of the definition of petroleum, there are two formulas that can serve to further defiling petroleum and its products: (i) the correlation index and (ii) the characterization factor—both of which are a means of estimating the character and behavior of crude oil. Both methods rely upon various analytical methods to derive data upon which the outcomes are based.

1.3.1.1   Correlation Index

The *correlation index* is based on the plot of specific gravity versus the reciprocal of the boiling point in degrees Kelvin ($^{\circ}K = ^{\circ}C + 273$). For pure hydrocarbons, the line described by the constants of the individual members of the normal paraffin series is given a value of CI=0 and a parallel line passing through the point for the values of benzene is given as CI=100; thus,

$$CI = 473.7d - 456.8 + 48,640/K$$

In this equation, d is the specific gravity and K is the average boiling point of the petroleum fraction as determined by the standard distillation method.

Values for the index between 0 and 15 indicate a predominance of paraffinic hydrocarbons in the fraction. A value from 15 to 50 indicates predominance of either naphthenes or of mixtures of paraffins, naphthenes, and aromatics. An index value above 50 indicates a predominance of aromatic species. However, it cannot be forgotten that the data used to determine the correlation index are *average* for the fraction of feedstock under study and may not truly represent all constituents of the feedstock, especially those at both ends of a range of physical and chemical properties.

Thus, because of the use of *average data* and the output of a value that falls within a broad range, it is questionable whether or not this correlation index offers realistic or reliable information. As the complexity of feedstocks increase from petroleum to heavy oil and beyond to tar sand bitumen, especially with the considerable overlap of compound types, there must be serious questions about the reliability of the number derived by this method.

1.3.1.2 Characterization Factor

Another derived number, the characterization factor (sometime referred to as the *UOP characterization factor* or the *Watson characterization factor*), is also a widely used method for defining petroleum, and it is derived from the following formula, which is a relationship between boiling point and specific gravity:

$$K = \sqrt[3]{T_B/d}$$

In this equation, T_B is the average boiling point in degrees Rankine (°F+460), and d is the specific gravity (60°/60°F). This factor has been shown to be additive on a weight basis. It was originally devised to show the thermal cracking characteristics of heavy oil. Thus, highly paraffinic oils have K=ca. 12.5–13.0, and cyclic (naphthenic) oils have K=ca. 10.5–12.5.

Again, because of the use of *average data* and the output of a value that falls (in this case) within a narrow range, it is questionable whether or not the data offer realistic or reliable information. Determining whether or not a feedstock is paraffinic is one issue, but one must ask if there is a real difference between feedstocks when the characterization factor is 12.4 or 12.5 or even between feedstocks having characterization factors of 12.4 and 13.0. As the complexity of feedstocks increases from petroleum to heavy oil and beyond to extra heavy oil and tar sand bitumen, especially with the considerable overlap of compound types, there must be serious questions about the reliability of the number derived by this method.

1.3.1.3 Character and Behavior

The data derived from any one or more of the analytical techniques give an indication of the characteristics of petroleum and an indication of the methods of feedstock processing as well as for the predictability of product yields and properties (Dolbear *et al.*, 1987; Speight, 2000, 2014a and references cited therein).

The most promising means of predictability of feedstock behavior during processing and predictability of product yields and properties have arisen from the concept of feedstock mapping (Long and Speight, 1998; Speight, 2014a). In such procedures, properties of feedstock are mapped to show

characteristics that are in visual form rather than in tabular form. In this manner, the visual characteristics of the feedstock are used to evaluate and predict the behavior of the feedstock in various refining scenarios. Whether or not such methods will replace the simpler form of property correlations remains to be determined. It is more than likely that both will continue to be used in a complimentary fashion for some time to come. However, there is also the need to recognize that what is adequate for one refinery and one feedstock (or feedstock blend provided that the blend composition does not change significantly) will not be suitable for a different refinery with a different feedstock (or feedstock blend).

One of the most effective means of feedstock mapping has arisen through the use of a multidisciplinary approach that involves use of all of the necessary properties of a feedstock. However, it must be recognized that such *maps* do not give any indication of the complex interactions that occur between, for example, such fractions as the asphaltene constituents and resins as well as the chemical transformations and interactions that occur during processing (Koots and Speight, 1975; Speight, 1994; Ancheyta *et al.*, 2010), but it does allow predictions of feedstock behavior. It must also be recognized that such a representation varies for different feedstocks. More recent work related to feedstock mapping has involved the development of a different type of compositional map using the molecular weight distribution and the molecular type distribution as coordinates. Such a map can provide insights into many separation and conversion processes used in petroleum refining (Long and Speight, 1998; Speight, 2014a).

Thus, a feedstock map can be used to show where a particular physical or chemical property tends to concentrate on the map. For example, the coke-forming propensity, that is, the amount of the carbon residue, can be illustrated for various regions on the map for a sample of atmospheric residuum (Long and Speight, 1998; Speight, 2014a). In addition, a feedstock map can be extremely useful for predicting the effectiveness of various types of separation (and other refinery) processes as applied to petroleum (Long and Speight, 1998; Speight, 2014a).

In contrast to the cut lines generated by separation processes, conversion processes move materials in the composition from one molecular type to another. For example, reforming converts saturates to aromatics and hydrogenation converts aromatic molecules to saturated molecules and polar aromatic molecules to either aromatic molecules or saturated molecules (Speight, 2014a). Hydrotreating removes nitrogen and sulfur compounds from polar aromatics without much change in molecular weight, while hydrocracking converts polar species to aromatics while at the same time reducing molecular weight. Visbreaking and heat soaking primarily lower or raise the molecular weight of the polar species in the composition map. Thus, visbreaking is used to lower the viscosity of heavy oils, whereas heat soaking is a coking method. Thus, conversion processes can change the shape and size of the composition map.

Thus, the data derived from any one, or more, of the analytical methods described in this chapter can be combined to give an indication of the characteristics of the feedstock as well as options for feedstock processing as well as for the prediction of product properties. Indeed, the use of physical properties for feedstock evaluation has continued in refineries and in process research laboratories to the present time and will continue for some time. It is, of course, a matter of choosing the relevant and meaningful properties to meet the nature of the task. What is certain is that the use of one single property cannot accurately portray the character and behavior of petroleum.

1.3.1.4 Bulk Fractions

While not truly a petroleum product in the refining sense, the bulk fractions produced from petroleum during laboratory fractionation studies can also be designated as derived materials and, thence, petroleum products. The data derived from the analysis of these fractions can be used to predict the refinability of the crude oil and to formulate refining procedures.

Briefly, in addition to distillation, petroleum can be subdivided into bulk fractions by a variety of precipitation/adsorption procedures: (i) asphaltene constituents, (ii) resin constituents, (iii) aromatic constituents, and (iv) saturated constituents (Fig. 1.2) (Speight, 2014a). However, the fractionation methods available to the petroleum industry allow a reasonably effective degree of separation of hydrocarbon mixtures. However, the problems are separating the petroleum

constituents without alteration of their molecular structure and obtaining these constituents in a substantially pure state. Thus, the general procedure is to employ techniques that segregate the constituents according to molecular size and molecular type. Furthermore, the names given to the fraction (i.e., asphaltene constituents, resin constituents, aromatic constituents, and saturated constituents) are based on separation procedures rather than on an accurate account of the molecular constituents of the fractions.

These investigations of the character of petroleum have been focused on the influence of the bulk makeup of petroleum on refining operations and the nature of the products that will be produced. However, the fractional composition of petroleum varies markedly with the method of isolation or separation, thereby leading to potential complications (especially in the case of the heavier feedstocks) in the choice of suitable processing schemes for these feedstocks. Because of this, the application of analytical techniques to these *other* petroleum products should also be applied assiduously and the data interpreted accordingly.

1.3.2 Natural Gas

Natural gas is the gaseous mixture associated with petroleum reservoirs and is predominantly methane, but does contain other combustible hydrocarbon compounds as well as non-hydrocarbon compounds (Mokhatab *et al.*, 2006, Speight 2014a). In fact, associated natural gas is believed to be the most economical form of ethane.

The gas occurs in the porous rock of the earth's crust either alone or with accumulations of petroleum. In the latter

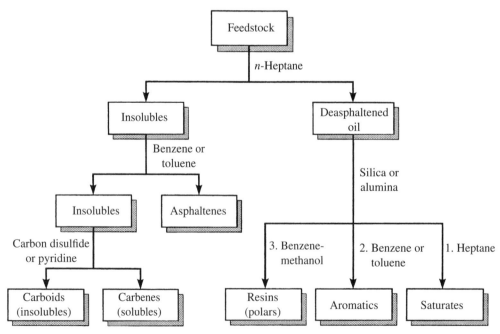

FIGURE 1.2 Feedstock fractionation.

case, the gas forms the gas cap, which is the mass of gas trapped between the liquid petroleum and the impervious cap rock of the petroleum reservoir. When the pressure in the reservoir is sufficiently high, the natural gas may be dissolved in the petroleum and is released upon penetration of the reservoir as a result of drilling operations.

Natural gas is also associated with shale formations, and such gas is commonly referred to as *shale gas*—to define the origin of the gas rather than the character and properties (Speight, 2013b). Chemically, shale gas is typically a dry gas composed primarily of methane (60–95% v/v), but some formations do produce wet gas—in the United States, the Antrim and New Albany plays have typically produced water and gas. Gas shale formations are organic-rich shale formations that were previously regarded only as source rocks and seals for gas accumulating in the strata near sandstone and carbonate reservoirs of traditional onshore gas development. Analysis of shale gas follows the methods of analysis for natural gas.

The principal types of gaseous fuels are oil (distillation) gas, reformed natural gas, and reformed propane or liquefied petroleum gas (LPG). *Mixed gas* is a gas prepared by adding natural gas or LPG to a manufactured gas, giving a product of better utility and higher heat content or Btu value.

The principal constituent of natural gas is methane (CH_4). Other constituents are paraffinic hydrocarbons such as ethane (CH_3CH_3), propane, and the butanes. Many natural gases contain nitrogen (N_2) as well as carbon dioxide (CO_2) and hydrogen sulfide (H_2S). Trace quantities of argon, hydrogen, and helium may also be present. Generally, the hydrocarbons having a higher molecular weight than methane, carbon dioxide, and hydrogen sulfide are removed from natural gas prior to its use as a fuel. Gases produced in a refinery contain methane, ethane, ethylene, propylene, hydrogen, carbon monoxide, carbon dioxide, and nitrogen, with low concentrations of water vapor, oxygen, and other gases.

Types of natural gas vary according to composition. There is *dry gas* or *lean gas*, which is mostly methane, and *wet gas*, which contains considerable amounts of higher-molecular-weight and higher-boiling hydrocarbons (Mokhatab *et al.*, 2006; Speight, 2007, 2014a). *Sour gas* contains high proportions of hydrogen sulfide, whereas *sweet gas* contains little or no hydrogen sulfide. *Residue gas* is the gas remaining (mostly methane) after the higher-molecular-weight paraffins has been extracted. *Casinghead gas* is the gas derived from an oil well by extraction at the surface. Natural gas has no distinct odor and the main use is for fuel, but it can also be used to make chemicals and LPG.

Some natural gas wells also produce helium, which can occur in commercial quantities; nitrogen and carbon dioxide are also found in some natural gases. Gas is usually separated at as high a pressure as possible, reducing compression costs when the gas is to be used for gas lift or delivered to a pipeline. After gas removal, lighter hydrocarbons and

hydrogen sulfide are removed as necessary to obtain petroleum of suitable vapor pressure for transport yet retaining most of the natural gasoline constituents.

In addition to composition and thermal content (Btu/scf, Btu/ft^3), natural gas can also be characterized on the basis of the mode of the natural gas found in reservoirs where there is no, or at best only minimal amounts of, petroleum.

Thus, there is *nonassociated* natural gas, which is found in reservoirs in which there is no, or at best only minimal amounts of, petroleum. Nonassociated gas is usually richer in methane but is markedly leaner in terms of the higher-molecular-weight hydrocarbons and condensate. Conversely, there is also *associated* natural gas (*dissolved* natural gas) that occurs either as free gas or as gas in solution in the petroleum. Gas that occurs as a solution with the crude petroleum is *dissolved gas*, whereas the gas that exists in contact with the crude petroleum (*gas cap*) is *associated gas*. Associated gas is usually leaner in methane than the nonassociated gas but is richer in the higher-molecular-weight constituents.

The most preferred type of natural gas is the nonassociated gas. Such gas can be produced at high pressure, whereas associated, or dissolved, gas must be separated from petroleum at lower separator pressures, which usually involves increased expenditure for compression. Thus, it is not surprising that such gas (under conditions that are not economically favorable) is often flared or vented.

As with petroleum, natural gas from different wells varies widely in composition and analyses (Mokhatab *et al.*, 2006; Speight, 2014a and references cited therein), and the proportion of non-hydrocarbon constituents can vary over a very wide range. The non-hydrocarbon constituents of natural gas can be classified as two types of materials: (i) diluents, such as nitrogen, carbon dioxide, and water vapors; and (ii) contaminants, such as hydrogen sulfide and/or other sulfur compounds. Thus, a particular natural gas field could require production, processing, and handling protocols different from those used for gas from another field.

The diluents are noncombustible gases that reduce the heating value of the gas and are on occasion used as *fillers* when it is necessary to reduce the heat content of the gas. On the other hand, the contaminants are detrimental to production and transportation equipment in addition to being obnoxious pollutants. Thus, the primary reason for gas refining is to remove the unwanted constituents of natural gas and to separate the gas into its various constituents. The processes are analogous to the distillation unit in a refinery where the feedstock is separated into its various constituent fractions before further processing to products.

The major diluents or contaminants of natural gas are (i) acid gas, which is predominantly hydrogen sulfide although carbon dioxide does occur to a lesser extent; (ii) water, which includes all entrained free water or water in condensed forms; (iii) liquids in the gas, such as higher-boiling hydrocarbons as well as pump lubricating oil, scrubber oil, and, on occasion, methanol; and (iv) any solid matter that may be present, such as fine silica (sand) and scaling from the pipe.

1.3.3 Natural Gas Liquids and Natural Gasoline

Natural gas liquids are products other than methane from natural gas: ethane, butane, *iso*-butane, and propane. Natural gasoline may also be included in this group.

Natural gas liquids are, in fact, separate and distinct hydrocarbons contained within some streams of natural gas. Streams that contain commercial quantities of natural gas liquids are called *wet gas*, and those with little or no liquids present are known as *dry gas* (see earlier text).

Chemical manufacturers use ethane in making *ethylene,* an important petrochemical. Butane and propane, and mixtures of the two, are classified as LPG that is used chiefly as a heating fuel in industry and homes. Pentane, hexane, and heptane are collectively referred to as *gas condensate* (*natural gasoline, casinghead gasoline, natural gas gasoline*). However, at high pressures, such as those existing in the deeper fields, the density of the gas increases and the density of the oil decreases until they form a single phase in the reservoir.

Wet natural gas contains natural gasoline in vapor form. The wet gas, also known as *casinghead gas*, is chiefly a mixture of methane, ethane, and the volatile hydrocarbons propane, butane, pentane (C_5H_{12}), hexane (C_6H_{14}), and heptane (C_7H_{16}). The latter three hydrocarbons form the main constituents of natural gasoline, which is recovered in refineries in liquid form mainly by absorption or compression processes. Pentane, hexane, and heptane are liquids under normal atmospheric conditions and are the chief components of ordinary refinery gasoline. Natural gasoline is used as blending stock for refinery gasoline and may be *cracked* to produce lower-boiling products, such as ethylene, propylene, and butylene. Caution should be taken not to confuse *natural gasoline* with *straight-run gasoline* (often also incorrectly referred to as natural gasoline), which is the gasoline distilled unchanged from petroleum.

The various tests that are applied to specifications for this group of low-boiling liquids will be referenced in the chapters dealing with LPG (Chapter 4) and gasoline (Chapter 5).

1.3.4 Opportunity Crudes

There is also the need for a refinery to be configured to accommodate *opportunity crude oils* and/or *high-acid crude oils*, which, for many purposes, are often included with heavy feedstocks.

Opportunity crude oils are either new crude oils with unknown or poorly understood processing issues or are existing crude oils with well-known processing concerns.

Opportunity crude oils are often, but not always, heavy crude oils but in either case are more difficult to desalt, most commonly due to high solid content, high levels of acidity, viscosity, electrical conductivity, or contaminants. They may also be oils that are incompatible, causing excessive equipment fouling when processed either in blends or separately.

Typically, opportunity crude oils are often dirty and need cleaning before refining by removal of undesirable constituents such as high-sulfur, high-nitrogen, and high-aromatics (such as polynuclear aromatic) components (Speight, 2014a, b). A controlled visbreaking treatment would *clean up* such crude oils by removing these undesirable constituents (which, if not removed, would cause problems further down the refinery sequence) as coke or sediment.

In addition to taking preventative measure for the refinery to process these feedstocks without serious deleterious effects on the equipment, refiners will need to develop programs for detailed and immediate feedstock evaluation so that they can understand the qualities of a crude oil very quickly, and it can be valued appropriately, and management of the crude processing can be planned meticulously.

Compatibility of opportunity crudes with other opportunity crudes and with conventional crude oil and heavy oil is a very important property to consider when making decisions regarding which crude to purchase. Blending crudes that are incompatible can lead to extensive fouling and processing difficulties due to unstable asphaltene constituents (Speight, 2014a). These problems can quickly reduce the benefits of purchasing the opportunity crude in the first place. For example, extensive fouling in the crude preheat train may occur resulting in decreased energy efficiency, increased emissions of carbon dioxide, and increased frequency at which heat exchangers need to be cleaned. In a worst-case scenario, crude throughput may be reduced leading to significant financial losses.

Opportunity crude oils, while offering initial pricing advantages, may have composition problems that can cause severe problems at the refinery, harming infrastructure, yield, and profitability. Before refining, there is the need for comprehensive evaluations of opportunity crudes, giving the potential buyer and seller the needed data to make informed decisions regarding fair pricing and the suitability of a particular opportunity crude oil for a refinery. This will assist the refiner to manage the ever-changing crude oil quality input to a refinery—including quality and quantity requirements and situations, crude oil variations, contractual specifications, and risks associated with such opportunity crudes.

1.3.5 High-Acid Crudes

Acidity in crude oils is typically caused by the presence of naphthenic acids, which are natural constituents of petroleum, where they evolve through the oxidation of naphthenes (cycloalkanes). Initially, the presence of these acidic species was suggested due to process artifacts formed during refining processes—and this may still be the case in some instances. However, it was shown that only a small quantity of acids was produced during these processes (Costantinides and Arich, 1967). Currently, it is generally assumed that acids may have been incorporated into the oil from three different sources: (i) acidic compounds found in source rocks, derived from the original organic matter that created the crude oil (plants and animals); (ii) neo-formed acids during biodegradation (although the high-acid concentration in biodegraded oils is believed to be related principally to the removal of nonacidic compounds, leading to a relative increase in the acid concentration levels); and (iii) acids that are derived from the bacteria themselves, for example, from cell walls that the organisms leave behind when their life cycle is completed (Mackenzie *et al.*, 1981; Thorn and Aiken, 1998; Meredith et al., 2000; Tomczyk et al., 2001; Watson *et al.*, 2002; Wilkes *et al.*, 2003; Barth *et al.*, 2004; Kim *et al.*, 2005; Fafet *et al.*, 2008).

The naphthenic acid subclass of the oxygen-containing species known as *naphthenic acids* and the term *naphthenic acids* is commonly used to describe an isomeric mixture of carboxylic acids (*predominantly* monocarboxylic acids) containing one or several saturated fused alicyclic rings (Hell and Medinger, 1874; Lochte, 1952; Ney *et al.*, 1943; Tomczyk *et al.*, 2001; Rodgers *et al.*, 2002; Barrow *et al.*, 2003; Clemente *et al.*, 2003a, b; Zhao *et al.*, 2012). However, in petroleum terminology, it has become customary to use this term to describe the whole range of organic acids found in crude oils; species such as phenols and other acidic species are often included in the naphthenic acid category (Speight, 2014b).

High-acid crude oils (Speight, 2014a, b) cause corrosion in the refinery—corrosion is predominant at temperatures in excess of 180°C (355°F) (Speight, 2014c)—and occurs particularly in the atmospheric distillation unit (the first point of entry of the high-acid crude oil) and also in the vacuum distillation units. In addition, overhead corrosion is caused by the mineral salts, magnesium, calcium, and sodium chloride, which are hydrolyzed to produce volatile hydrochloric acid, causing a highly corrosive condition in the overhead exchangers. Therefore, these salts present a significant contamination in opportunity crude oils. Other contaminants in opportunity crude oils that are shown to accelerate the hydrolysis reactions are inorganic clays and organic acids.

1.3.6 Foamy Oil

Foamy oil is oil-continuous foam that contains dispersed gas bubbles produced at the wellhead from heavy oil reservoirs under solution gas drive. The nature of the gas dispersions in oil distinguishes foamy oil behavior from conventional heavy oil. The gas that comes out of solution in the reservoir does not coalesce into large gas bubbles or into a continuous flowing gas phase. Instead, it remains as small bubbles entrained in the crude oil, keeping the effective oil viscosity low while providing expansive energy that helps drive the oil toward producing. Foamy oil accounts for unusually high production in heavy oil reservoirs under solution gas drive.

Thus, foamy oil is formed in solution gas drive reservoirs when gas is released from solution as reservoir pressure declines. It has been noted that the oil at the wellhead of these heavy-oil reservoirs resembles the form of foam, hence the term *foamy oil*. The gas initially exists in the form of small bubbles within individual pores in the rock. As time passes and pressure continues to decline, the bubbles grow to fill the pores. With further declines in pressure, the bubbles created in different locations become large enough to coalesce into a continuous gas phase. Once the gas phase becomes continuous (i.e., when gas saturation exceeds the critical level—the minimum saturation at which a continuous gas phase exists in porous media) traditional two-phase (oil and gas) flow with classical relative permeability occurs. As a result, the production gas–oil ratio increases rapidly after the critical.

Before analysis, the foam should be dismissed either by use of an appropriate separator vessel or by use of antifoaming agents. However, modification of the separator design may not always be feasible because of the limited space at many wellhead facilities, especially offshore platforms. Therefore, chemical additives (antifoaming agents, foam inhibitors) are employed to prevent or break up the foam. In the case of antifoaming agents, analysts must (i) determine the chemical nature of the agents, (ii) remove the agents prior to commencing analysis, and (iii) if item 2 is not possible or difficult, ensure that the presence of these agents does not interfere with the test method result.

1.3.7 Oil from Shale

One of the newest terms in the petroleum lexicon arbitrarily named (even erroneously named) *shale oil* is the crude oil that is produced from tight shale formation and should not be confused with *shale oil*, which is the oil produced by the thermal treatment of oil shale and the decomposition of kerogen contained therein (Speight, 2012b). The tight shale formations are those same formations that produce gas (*tight gas*) (Speight, 2013b). The introduction of the term *shale oil* to define crude oil from tight shale formations is the latest term to add confusion to the system of nomenclature of petroleum–heavy oil–bitumen materials. The term has been used without any consideration of the original term shale oil produced by the thermal decomposition of kerogen in oil shale. It is not quite analogous, but is certainly similarly confusing, to the term *black oil* that has been used to define petroleum by color rather than by any meaningful properties.

Typical of the oil from tight shale formations is the Bakken crude oil, which is a light crude oil. Briefly, Bakken crude oil is a light sweet (low-sulfur) crude oil that has a relatively high proportion of volatile constituents. The production of the oil yields not only petroleum but also a significant amount of volatile gases (including propane and butane) and low-boiling liquids (such as pentane and natural gasoline), which are often referred to collectively as (low-boiling or light) naphtha. By definition, natural gasoline (sometime also referred to as *gas condensate*) is a mixture of low-boiling liquid hydrocarbons isolated from petroleum and natural gas wells suitable for blending with light naphtha and/or refinery gasoline (Mokhatab *et al.*, 2006; Speight, 2007, 2014a). Because of the presence of low-boiling hydrocarbons, light naphtha can become extremely explosive, even at relatively low ambient temperatures. Some of these gases may be burned off (flared) at the field wellhead, but others remain in the liquid products extracted from the well (Speight, 2014a).

The liquid stream produced from the Bakken formation will include the crude oil, the low-boiling liquids, and gases that were not flared, along with the materials and by-products of the fracking process. These products are then mechanically separated into three streams: (i) produced salt water, often referred to as brine; (ii) gases; and (iii) petroleum liquids, which include condensates, natural gas liquids, and light oil. Depending on the effectiveness and appropriate calibration of the separation equipment, which is controlled by the oil producers, varying quantities of gases remain dissolved and/or mixed in the liquids, and the whole is then transported from the separation equipment to the well-pad storage tanks, where emissions of volatile hydrocarbons have been detected as emanating from the oil.

Bakken crude oil is considered to be a low-sulfur (*sweet*) crude oil, and there have been increasing observations of elevated levels of hydrogen sulfide (H_2S) in the oil. Hydrogen sulfide is a toxic, highly flammable, corrosive, explosive gas (hydrogen sulfide), and there have been increasing observations of elevated levels of hydrogen sulfide in Bakken oil.

1.3.8 Heavy Oil

Heavy oil (*heavy crude oil*) is more viscous than conventional crude oil and has a lower mobility in the reservoir but can be recovered through a well from the reservoir by the application of secondary or enhanced recovery methods (Speight, 2009, 2013a, 2014a). The term *heavy oil* has also been arbitrarily used to describe both the heavy oils that require thermal stimulation of recovery from the reservoir and (incorrectly) to the bitumen in bituminous sand (tar sand, *q.v.*) formations from which the heavy bituminous material is recovered by a recovery operation other than the recognized enhanced oil recovery methods (Speight, 2009, 2014a).

When petroleum occurs in a reservoir that allows the crude material to be recovered by pumping operations as a free-flowing dark- to light-colored liquid, it is often referred to as *conventional petroleum*. Heavy oil is a *type* of petroleum that is different from the conventional petroleum insofar as they are much more difficult to recover from the subsurface reservoir. These materials have a much higher viscosity (and lower API gravity) than conventional petroleum, and primary recovery of these petroleum types usually requires thermal stimulation of the reservoir (Speight, 2009, 2013a, b). Heavy oils are more difficult to recover from the subsurface reservoir than light oils. The definition of heavy oils is usually based on the API gravity or viscosity, and the definition is quite arbitrary although there have been attempts to rationalize the definition based upon viscosity, API gravity, and density.

For many years, petroleum and heavy oil were very generally defined in terms of physical properties. For example, heavy oils were considered to be those crude oils that had gravity somewhat less than 20° API with the heavy oils falling into the API gravity range 10–15°. For example, Cold Lake heavy crude oil has an API gravity equal to 12°, and extra heavy oils, such as tar sand bitumen, usually have an API gravity in the range 5–10° (Athabasca bitumen: 8° API). Residua would vary depending upon the temperature at which distillation was terminated but usually vacuum residua are in the range 2–8° API (Speight, 2000 and references cited therein; Gary *et al.*, 2007; Hsu and Robinson, 2006; Speight, 2014a; Speight and Ozum, 2002).

Heavy oil may also be called *viscous oil*—but the latter term has also been generally applied to petroleum products such a lubricating oil. More typically, and for the purposes of this book, viscous oils are those petroleum fractions and derived products that have higher-boiling points than distillate fuels and are liquid at, or slightly above, room temperature. The molecular constituents of these product oils contain 20 to ≥50 carbon atoms and distill at temperatures above 260°C (500°F). Examples include refinery streams such as gas oils, residua, cracked residua, and viscous oils as well as finished products such as white oil, lubricating oil, and other process oils.

The analysis of heavy oils is more complex than the analysis of hydrocarbon gases and lower-molecular-weight volatile liquids. In addition, the analysis of heavy oil may be complicated by issues related to handling—the higher viscosity of these oils makes them more difficult to sample, and the dark color can cause problems with some test methods.

Moreover, the types of molecular species present increases rapidly as the number of carbon atoms per molecule increases, and in addition, hydrocarbons in heavy oils are much more complex. Characterization does not (and should not) focus on identifying specific molecular structures but on classes of molecules (such as paraffins, naphthenes, aromatics, polycyclic compounds, and polar compounds). In addition to carbon and hydrogen, heavy oils contain oxygen, sulfur, and nitrogen compounds, as well as trace quantities of metals (such as nickel and vanadium) may also be present. Determining the chemical form present for these elements provides additional important information.

Test methods of interest for the analysis of heavy oil (extra heavy oil and tar sand bitumen can be included here) include tests that measure physical properties such as density, refractive index, molecular weight, and boiling range; those methods that are used to measure chemical composition such as elemental and molecular structure analysis; and derivative methods that correlate measured properties with aspects of chemical composition (Speight, 2001, 2014a).

Test methods for conventional crude oil are generally applicable to heavy oil. However, depending upon the properties of the heavy oil, modification may be necessary as illustrated for tar sand bitumen (Section 1.3.10) (see also Wallace, 1988; Wallace *et al.*, 1988).

1.3.9 Extra Heavy Oil

Briefly, extra heavy oil is a material that occurs in the solid or near-solid state and generally has mobility under reservoir conditions (Speight, 2009, 2013a, 2014a). *extra heavy oil* is a recently evolved term that is of questionable scientific meaning and is subject to much verbal and written variation. While this material may resemble tar sand bitumen and does not flow easily, extra heavy oil can generally be recognized (for the purposes of this text) as being more viscous than heavy oil but having mobility in the reservoir—tar sand bitumen is typically incapable of mobility (free flow) under the conditions in the tar sand deposit.

For example, the tar sand bitumen located in Alberta Canada is not mobile in the deposit and requires extreme methods of recovery to recover the bitumen. On the other hand, much of the extra heavy oil located in the Orinoco belt of Venezuela requires recovery methods that are less extreme because of the mobility of the material in the reservoir. Whether the mobility of extra heavy oil is due to a high reservoir temperature (that is higher than the pour point of the extra heavy oil) or due to other factors is variable and subject to localized conditions in the reservoir.

Test methods for conventional crude oil are not always applicable to extra heavy oil, and modification of the test method(s) may be necessary—as illustrated for tar sand bitumen (Section 1.3.10) (Wallace, 1988; Wallace *et al.*, 1988).

1.3.10 Tar Sand Bitumen

Tar sand is the several rock types that contain an extremely viscous hydrocarbon that is not recoverable in its natural state by conventional oil well production methods including currently used enhanced recovery techniques (Speight, 2009, 2013a, 2014a).

More descriptively, *tar sand* is unconsolidated-to-consolidated sandstone or a porous carbonate rock, impregnated with bitumen. In simple terms, an unconsolidated rock approximates the consistency of dry or moist sand, and a consolidated rock may approximate the consistency of set concrete. Alternative names, such as *bituminous sand* or (in Canada) *oil sand*, are gradually finding usage, with the former name (*bituminous sand*) more technically correct. The term *oil sand* is also used in the same way as the term *tar sand*, and the terms are used interchangeably. The term *oil sand* is analogous to the term *oil shale*. Neither material contains oil, but oil is produced therefrom by application of thermal decomposition methods. It is important to understand that tar sand and the bitumen contained therein are different components of the deposit. The recovery of the bitumen, a hydrocarbonaceous material that can be converted into *synthetic crude oil* (Speight, 2014a), depends to a large degree on the composition and construction of the sands.

The molecular constituents found in tar sand bitumen also contain nitrogen, oxygen, sulfur, and metals (particularly nickel and vanadium) chemically bound in their molecular structures. Thus, it is chemically correct to refer to bitumen as a *hydrocarbonaceous* material, that is, a material that is composed predominantly of carbon and hydrogen but recognizing the presence of other atoms.

The term *bitumen* (also, on occasion, referred to as *native asphalt*, and *extra heavy oil*) includes a wide variety of reddish brown to black materials of semisolid, viscous to brittle character that can exist in nature with no mineral impurity or with mineral matter contents that exceed 50% by weight. *Bitumen* is frequently found filling pores and crevices of sandstone, limestone, or argillaceous sediments, in which case the organic and associated mineral matrix is known as *rock asphalt*.

On the basis of the definition of tar sand (earlier text), *bitumen*, also on occasion referred to as *native asphalt* and *extra heavy oil*, is a naturally occurring hydrocarbonaceous material that has little or no mobility under reservoir conditions and which cannot be recovered through a well by conventional oil well production methods including currently used enhanced recovery techniques; current methods involve mining for bitumen recovery (Speight, 2014a).

Because of the immobility of the bitumen, the permeability of the deposit is low, and passage of fluids through the deposit is prevented. Bitumen is a high-boiling material with little, if any, material boiling below 350°C (660°F), and the boiling range approximates the boiling range of an atmospheric residuum and has a much lower proportion of volatile constituents than a conventional crude oil (Speight and Ozum, 2002; Hsu and Robinson, 2006; Gary *et al.*, 2007; Speight, 2014a).

Synthetic crude oil is the *hydrocarbon* liquid that is produced from *bitumen*, by a variety of processes that involve thermal decomposition. *Synthetic crude oil* (also referred to as *syncrude*) is a marketable and transportable product that resembles conventional crude oil. *Synthetic crude oil*, although it may be produced from one of the less conventional fossil fuel sources, can be accepted into and refined by the usual refinery system.

Many analytical test methods developed for conventional crude oil have been applied to the analysis of tar sand bitumen. However, there may be a need to adapt the published test method to accommodate the differences in behavior that are apparent when conventional crude oil and bitumen are compared. For example, the removal of an asphaltene fraction from conventional crude oil requires addition of an excess of heptane to the oil whereupon the asphaltene constituents are separated as a precipitate. The procedure for bitumen requires modification. The addition of heptane to bitumen proceeds not as a precipitation process but as an extraction process (which is diffusion-controlled) and is subject to more limitations than the precipitation process. As a consequence, some workers prefer to add a solvent (typically toluene) to the bitumen, allow dissolution of the bitumen to occur, and then add heptane.

1.4 PETROLEUM REFINING

Petroleum is rarely used in its raw form and contains many thousands of different compounds that vary in molecular weight from methane (CH_4, molecular weight: 16) to more than 2000 (Speight, 2000, 2014a; Speight and Ozum, 2002; Parkash, 2003; Hsu and Robinson, 2006; Gary et al., 2007). This broad range in molecular weights results in boiling points that range from −160°C (−288°F) to temperatures in excess of 1100°C (2000°F) (Speight and Ozum, 2002; Hsu and Robinson, 2006; Gary et al., 2007; Speight, 2014a).

Thus, a refinery is, of necessity, a complex network of integrated unit processes for the purpose of producing a variety of products from petroleum (Table 1.3) (Speight and Ozum, 2002; Hsu and Robinson, 2006; Gary et al., 2007; Speight, 2014b). Each refinery has its own range of preferred petroleum feedstock from which a desired distribution of products is obtained. Nevertheless, refinery processes can be divided into three major types: (i) *separation*—division of petroleum into various streams (or fractions) depending on the nature of the crude material; (ii) *conversion*—production of salable materials from petroleum, usually by skeletal alteration, or even by alteration of the chemical type, of the petroleum constituents; and (iii) *finishing*—purification of various product streams by a variety of processes that essentially remove impurities from the product; for convenience, processes that accomplish molecular alteration, such as *reforming*, are also included in this category.

The *separation* and *finishing* processes may involve distillation or even treatment with a *wash* solution, either to remove impurities or, in the case of distillation, to produce a material boiling over a narrower range, and the chemistry of these processes is quite simple. *Conversion processes* are, in essence, processes that change the number of carbon atoms per molecule, alter the molecular hydrogen-to-carbon ratio, or change the molecular structure of the material without affecting the number of carbon atoms per molecule. These latter processes (*isomerization processes*) essentially change the shape of the molecule(s) and are used to improve the quality of the product (Speight and Ozum, 2002; Hsu and Robinson, 2006; Gary et al., 2007; Speight, 2014a).

Thermal cracking processes are commonly used to convert petroleum residua into distillable liquid products, and examples of thermal cracking processes currently in use are *visbreaking* and *coking* (*delayed coking*, *fluid coking*, and *flexicoking*). In all of these processes, the simultaneous formation of sediment or coke limits the conversion to usable liquid products (Speight and Ozum 2002; Hsu and Robinson, 2006; Gary et al., 2007; Speight, 2014a).

1.4.1 Visbreaking

The visbreaking process is used primarily as a means of reducing the viscosity of heavy feedstocks by *controlled thermal decomposition* insofar as the hot products are

TABLE 1.3 Separation processes and conversion processes

Process	Action	Method	Purpose	Feedstock(s)	Product(s)
Separation processes					
Atmospheric distillation	Separation	Thermal	Separate feedstock	Desalted crude oil	Gas, naphtha, gas oil, resid
Vacuum distillation	Separation	Thermal	Separate without cracking	Atmospheric resid	Gas oil, lube stock, residual
Conversion processes					
Catalytic cracking	Alteration	Catalytic	Upgrading	Gas oil	Naphtha, petrochemical feedstock
Coking	Polymerize	Thermal	Convert resid	Gas oil, coke distillate	Naphtha, petrochemical feedstock
Hydrocracking	Hydrogenate	Catalytic	Convert to lower-boiling products	Gas oil, resid	Lower-boiling products
Visbreaking	Decompose	Thermal	Reduce viscosity	Atmospheric resid	Distillate, fuel oil

quenched before complete conversion can occur (Speight and Ozum, 2002; Hsu and Robinson, 2006; Gary *et al.*, 2007; Speight, 2014a). However, the process is often plagued by sediment formation in the products. This sediment, or sludge, must be removed if the products are to meet fuel oil specifications.

The process uses the mild thermal cracking (*partial conversion*) as a relatively low-cost and low-severity approach to improving the viscosity characteristics of the residue without attempting significant conversion to distillates. Low residence times are required to avoid polymerization and coking reactions, although additives can help to suppress coke deposits on the tubes of the furnace.

A visbreaking unit consists of a reaction furnace, followed by quenching with a recycled oil, and fractionation of the product mixture (Speight, 2000, 2014a; Speight and Ozum, 2002; Parkash, 2003; Hsu and Robinson, 2006; Gary *et al.*, 2007). All of the reaction in this process occurs as the oil flows through the tubes of the reaction furnace. The severity of the process is controlled by the flow rate through the furnace and the temperature; typical conditions are 475–500°C (885–930°F) at the furnace exit with a residence time of 1–3min, with operation for 3–6 months on stream (continuous use) is possible before the furnace tubes cleaned and the coke removed. The operating pressure in the furnace tubes can range from 0.7 to 5MPa depending on the degree of vaporization and the residence time desired. For a given furnace tube volume, a lower operating pressure will reduce the actual residence time of the liquid phase.

1.4.2 Coking

Coking, as the term is used in the petroleum industry, is a process for converting non-distillable fractions (residua) of petroleum to lower-boiling products and coke. Coking is often used in preference to catalytic cracking because of the presence of metals and nitrogen components that poison catalysts (Speight and Ozum, 2002; Hsu and Robinson, 2006; Gary *et al.*, 2007; Speight, 2014a).

There are several coking processes: *delayed coking*, *fluid coking*, and *flexicoking* as well as several other variations.

Delayed coking is the oldest, most widely used process and has changed very little in the five or more decades in which it has been on stream in refineries.

In the semi-continuous process, the residuum or other heavy feedstock is heated to the cracking/coking temperature (>350°C, >660°F); but usually at temperatures on the order of 480°C (895°F) and the hot liquid is charged, usually by upflow, to the coke drum where the coking reactions occur. Liquid and gaseous products pass to the fractionator for separation and coke deposits in the drum. The coke drums are arranged in pairs, one on stream and the other off stream, and used alternately to allow continuous processing. The process can be operated on a cycle, typically 24–48h.

The overhead oil is fractionated into fuel gas (ethane and lower-molecular-weight gases), propane–propylene, butane–butene, naphtha, light gas oil, and heavy gas oil. Yields and product quality vary widely due to the broad range of feedstock types charged to the delayed coking process. The function of the coke drum is to provide the residence time required for the coking reactions and to accumulate the coke. Hydraulic cutters are used to remove coke from the drum.

Fluid coking is a continuous fluidized solid process that cracks feed thermally over heated coke particles in a reactor vessel to gas, liquid products, and coke. Heat for the process is supplied by partial combustion of the coke, with the remaining coke being drawn as product. The new coke is deposited in a thin fresh layer on the outside surface of the circulating coke particle.

Small particles of coke made in the process circulate in a fluidized state between the vessels and are the heat transfer medium. Thus, the process requires no high-temperature pre-heat furnace. Fluid coking is carried out at essentially atmospheric pressure and temperatures in excess of 485°C (900°F) with residence times of the order of 15–30s. The longer residence time is in direct contrast to the delayed coking process, in which the coking reactions are allowed to proceed to completion. This is evident from the somewhat higher liquid yields observed in many fluid coking processes. However, the products from a fluid coker may contain more olefin species and be less desirable for downstream processing.

The *flexicoking* process is a modification of the fluid coking process that includes a gasifier adjoining the burner/regenerator to convert excess coke to a clean fuel gas with a heating value of about 90Btu/ft³. The coke gasification can be controlled to burn about 95% of the coke to maximize the production of coke gas or to operate at a reduced level to produce both gas and a coke. This flexibility permits adjustment for coke market conditions over a considerable range of feedstock properties.

The *liquid products* from the coker can, following cleanup via commercially available hydrodesulfurization technology (Speight, 2000, 2014a; Ancheyta and Speight, 2007), provide low-sulfur liquid fuels (<0.2% w/w sulfur). Coker naphtha has a boiling range up to 220°C (430°F), contains olefins, and must be upgraded by hydrogen processing for removal of olefins, sulfur, and nitrogen. They are then used conventionally for reforming to gasoline or chemical feedstock. Middle distillates, boiling in the range of 220–360°C (430–680°F), are also hydrogen treated for improved storage stability, sulfur removal, and nitrogen reduction. They can then be used as precursors to gasoline, diesel fuel, or fuel oil. The gas oil boiling up to 510°C (950°F) is usually low in metals and may be used as the feedstock for fluid catalytic cracking (Occelli, 2010).

Another major application for the coking processes is in upgrading heavy (high-viscosity) low-value petroleum into lighter products.

Petroleum *coke* is used principally as a fuel or, after calcining, for carbon electrodes. The feedstock from which the coke is produced controls the coke properties, especially sulfur, nitrogen, and metal content. A concentration effect tends to deposit the majority of the sulfur, nitrogen, and metals in the coke. Cokes exceeding about 2.5% sulfur content and 200ppm vanadium are mainly used, environmental regulations permitting for fuel or fuel additives. The properties of coke for non-fuel use include low sulfur, metals, and ash content as well as a definable physical structure.

1.4.3   Hydroprocessing

Hydroprocessing is the conversion of various feedstocks using the physical aspects of temperature, residence time, and the presence of hydrogen under pressure. Hydroprocessing is more conveniently subdivided into *hydrotreating* and *hydrocracking*.

Hydrotreating is defined as the lower-temperature removal of heteroatomic species by treatment of a feedstock or product in the presence of hydrogen. On the other hand, *hydrocracking* is the thermal decomposition (in the presence of hydrogen) of a feedstock in which carbon–carbon bonds are cleaved in addition to the removal of heteroatomic species (nitrogen oxygen, and sulfur) as the respective hydrogenated analogs (ammonia, NH_3; water, H_2O; and hydrogen sulfide, H_2S); in reality, hydrotreating and hydrocracking may occur simultaneously.

In contrast to the visbreaking process, in which the general principle is the production of products for use as fuel oil, the hydroprocessing is employed to produce a slate of products for use as liquid fuels.

1.5   PETROLEUM PRODUCTS

Petroleum products are any petroleum-based products that can be obtained by refining (Speight and Ozum, 2002; Hsu and Robinson, 2006; Gary *et al.*, 2007; Speight, 2014a) and comprise refinery gas, ethane, LPG, naphtha, gasoline, aviation fuel, marine fuel, kerosene, diesel oil fuel, distillate fuel oil, residual fuel oil, gas oil, lubricants, white oil, grease, wax, asphalt, as well as coke. Petrochemical products are not included here.

Petroleum products are highly complex chemicals, and considerable effort is required to characterize their chemical and physical properties with a high degree of precision and accuracy. Indeed, the analysis of petroleum products is necessary to determine the properties that can assist in resolving a process problem as well as the properties that indicate the function and performance of the product in service.

Petroleum and the products obtained from refinery operations contain a variety of compounds, usually but not always hydrocarbons, and are not always immediately suitable for commercial use (Speight and Ozum, 2002; Hsu and Robinson, 2006; Gary *et al.*, 2007; Speight, 2014a). As the number of carbon atoms in, for example, the paraffin series increases, the complexity of petroleum mixtures also rapidly increases. Consequently, detailed analysis of the individual constituents of the higher-boiling fractions becomes increasingly difficult, if not impossible.

Additionally, *classes* (or *types*) of hydrocarbons were, and still are, determined based on the capability to isolate them by separation techniques. The four fractional types into which petroleum is subdivided are paraffins, olefins, naphthenes, and aromatics. Paraffinic hydrocarbons include both normal and branched alkanes, whereas olefins refer to normal and branched alkenes that contain one or more double or triple carbon–carbon bonds. *Naphthene* (not to be confused with *naphthalene*) is a term specific to the petroleum industry that refers to the *saturated cyclic hydrocarbons* (*cycloalkanes*). Finally, the term *aromatics* includes all hydrocarbons containing one or more rings of the benzene-type structure.

These general definitions of the different fractions are subject to the many combinations of the hydrocarbon types (Speight and Ozum, 2002; Hsu and Robinson, 2006; Gary *et al.*, 2007; Speight, 2014a) and the action of the adsorbent or the solvent used in the separation procedure. For example, a compound containing 1 benzene-type ring (6 aromatic carbon atoms) that has 12 non-aromatic carbon atoms in alkyl side chains can be separated as an aromatic compound depending upon the adsorbent employed.

Although not directly derived from composition, the terms *light* or *heavy* or *sweet* and *sour* provide convenient terms for use in descriptions. For example, *light petroleum* (often referred to as *conventional petroleum*) is usually rich in low-boiling constituents and waxy molecules, while *heavy petroleum* contains greater proportions of higher-boiling, more aromatic, and heteroatom-containing (N-, O-, S-, and metal-containing) constituents. *Heavy oil* is more viscous than conventional petroleum and requires enhanced methods for recovery. *Bitumen* is *near solid* or *solid* and cannot be recovered by enhanced oil recovery methods (Speight, 2009, 2014a).

1.5.1   Types

The term *petroleum products* encompasses a wide range of chemical mixtures with a variable assortment of properties that vary from gases to liquids to solids. In order to understand the analytical method for petroleum products, a brief survey of the types of products and their general properties is warranted. More recently, the popularity of products from the Fischer–Tropsch process has increased ion prominence as has bio-oil—the product from the thermal decomposition of biomass (Speight, 2008, 2011b). Thus, for the sake of completeness, these definitions are also included.

1.5.1.1 Gases

Hydrocarbon products with four or less carbon atoms have boiling points that are lower than room temperature, and these products are gases at ambient temperature and pressure (Chapter 4). For example, *natural gas* (predominantly *methane*, CH_4) is the lowest boiling and least complex of all hydrocarbons. Higher-boiling hydrocarbons up to *n*-octane (C_8H_{18}) may also be present in which case the gas is referred to as *wet gas*, which is usually processed (refined) to remove the high-boiling hydrocarbons (when isolated, the higher-boiling hydrocarbon mixture is often referred to as *natural gas condensate*). In fact, there are wells from which the predominant product is *condensate*.

Still gas is a term reserved for low-boiling hydrocarbon mixtures that represent the lowest-boiling fraction isolated from an atmospheric distillation unit (*still*) in the refinery. Typically, the still gas will contain methane, ethane (CH_3CH_3), propane ($CH_3CH_2CH_3$), butane ($CH_3CH_2CH_2CH_3$), and their respective isomers. If cracked products have been recycled to the atmospheric distillation unit, the still gas may also contain volatile olefins such as ethylene ($CH_2=CH_2$). The term *fuel gas* is often used interchangeably with *still gas*, but the term *fuel gas* is intended to denote the destination of the gaseous mixture, such as fuel for boilers, furnaces, or heaters, whereas the term *still gas* may be indicative of the use of the mixture for petrochemical production.

LPG is composed of propane (C_3H_8) and butane (C_4H_{10})—the ratio of the two gases is variable and depends upon the ultimate use of the gas and the conditions of use—and is stored under pressure in order to keep the two hydrocarbons liquefied within the container. Before use, LPG passes through a pressure relief valve that causes a reduction in pressure, and the liquid vaporizes (gasifies). Winter-grade LPG is mostly propane, the lower boiling of the two gases that is easier to vaporize at lower temperatures. Summer-grade LPG contains higher amounts of butane.

Mixtures of these hydrocarbons are commonly encountered in material testing, and the composition varies depending upon the source and intended use of the material. Other non-hydrocarbon constituents of these mixtures are important analytes since they may be useful products or may be undesirable as a source of processing problems. Some of these constituents are helium-, hydrogen-, argon-, oxygen-, nitrogen-, carbon monoxide-, carbon dioxide-, sulfur-, and nitrogen-containing compounds, as well as higher-molecular-weight hydrocarbons. Desired testing of these hydrocarbon mixtures usually involves the determination of bulk physical or chemical properties and component speciation and quantitation.

1.5.1.2 Naphtha, Solvents, and Gasoline

Naphtha is the general term that is applied to refined, partly refined, or unrefined low-boiling liquid petroleum products (Chapter 5). Naphtha is prepared by any one of several methods including (i) fractionation of distillates or even crude petroleum, (ii) fluid catalytic cracking, (iii) solvent extraction, (iv) hydrogenation of distillates, (v) polymerization of unsaturated (olefin) compounds, and (vi) alkylation processes. Naphtha may also be a combination of product streams from more than one of these processes.

The term aliphatic naphtha refers to naphtha containing less than 0.1% benzene and with carbon numbers from C_3 through C_{16}. Aromatic naphtha has constituents composed of C_6 through C_{16} hydrocarbons and contains significant quantities of aromatic hydrocarbons such as benzene (>0.1%), toluene, and xylene. The final gasoline product as a transport fuel is a carefully blended mixture having a predetermined octane value. Thus, gasoline is a complex mixture of hydrocarbons that boils below 200°C (390°F). The hydrocarbon constituents in this boiling range are those that have 4–12 carbon atoms.

The uses of petroleum naphtha include the following: (i) precursor to gasoline and other liquid fuels, (ii) solvents (diluents) for paints, (iii) dry-cleaning solvents, (iv) solvents for cutback asphalts, (v) solvents in rubber industry, and (vi) solvents for industrial extraction processes. Turpentine, the older and more conventional solvent for paints, has now been almost completely replaced by the cheaper and more abundant petroleum naphtha.

Petroleum solvents (also called *naphtha*) are valuable because of their good dissolving power. The wide range of naphtha available and the varying degree of volatility possible offer products suitable for many uses.

Stoddard solvent is a petroleum distillate widely used as a dry-cleaning solvent and as a general cleaner and degreaser. It may also be used as paint thinner, as a solvent in some types of photocopier toners, in some types of printing inks, and in some adhesives. Stoddard solvent is considered to be a form of mineral spirits, white spirits, and naphtha; however, not all forms of mineral spirits, white spirits, and naphtha are considered to be Stoddard solvent. The solvent consists of linear and branched alkanes (30–50% v/v), cycloalkanes (30–40% v/v), and aromatic hydrocarbons (10–20% v/v).

Gasoline varies widely in composition, and even those with the same octane number may be quite different (Chapter 6). The variation in aromatic content as well as the variation in the content of normal paraffins, branched paraffins, cyclopentane derivatives, and cyclohexane derivatives all involves characteristics of any one individual crude oil and influence the octane number of the gasoline. However, in spite of the varied nomenclature of the refining industry, there is no one refinery process that produces gasoline–gasoline is a blend of several refinery streams that is sold as gasoline after the inclusion of the necessary additives (Chapter 6) (Speight and Ozum, 2002; Hsu and Robinson, 2006; Gary *et al.*, 2007; Speight, 2014a).

Automotive gasoline is a mixture of low-boiling hydrocarbon compounds suitable for use in spark-ignited internal combustion engines and having an octane rating of at least 60. Automotive gasoline contains 150 or more different chemical compounds, and the relative concentrations of the compounds vary considerably depending on the source of crude oil, refinery process, and product specifications. Typical hydrocarbon constituents are alkanes (4–8% v/v), alkenes (2–5% v/v), iso-alkanes (25–40% v/v), cycloalkanes (3–7% v/v), cycloalkenes (l–4% v/v), and aromatics (20–50% v/v). However, these proportions vary greatly.

Additives that have been used in gasoline include improvers for octane number, and other categories of compounds that may be added to gasoline include antiknock agents, antioxidants, metal deactivators, lead scavengers, antirust agents, anti-icing agents, upper-cylinder lubricants, detergents, and dyes.

Aviation fuel occurs as two variations: (i) *aviation gasoline* and (ii) *jet fuel* (Chapter 7). Aviation gasoline, now usually found in use in light aircraft and older civil aircraft, has narrower boiling ranges than conventional (automobile) gasoline, that is, 38–170°C (100–340°F), compared to −1 to 200°C (30–390°F) for automobile gasoline. The narrower boiling range ensures better distribution of the vaporized fuel through the more complicated induction systems of aircraft engines. Since aircraft operate at altitudes where the prevailing pressure is relatively low—the pressure at 17,500ft is 7.5psi (0.5 atmosphere) compared to 14.8psi (1.0 atmosphere) at the surface of the earth—the vapor pressure of aviation gasoline must be limited to reduce boiling in the tanks, fuel lines, and carburetors.

Aviation gasoline consists primarily of straight and branched alkanes and cycloalkanes. Aromatic hydrocarbons are limited to 20–25% v/v of the total mixture because of the smoke produced when they are burned. *Jet fuel* is classified as *aviation turbine fuel*, and, in the specifications, ratings relative to octane number are replaced with properties concerned with the ability of the fuel to burn cleanly.

Jet fuel is composed of distillate fractions and comprises both gasoline-type and kerosene-type jet fuels that meet specifications for use in aviation turbine units. *Gasoline-type jet fuel* includes naphtha constituents that distil between 100 and 250°C (212 and 480°F). This fuel is obtained by blending kerosene and gasoline or naphtha in such a way that the aromatic content does not exceed 25% v/v, with limits on the range of vapor pressure. *Kerosene-type jet fuel* is a middle distillate fraction used for aviation turbine power units. It has the same distillation characteristics and flash point as kerosene—between 150 and 300°C (300 and 570°F) but not generally above 250°C (480°F).

Volatility is an important property of all types of gasoline since it is related to performance and requires sufficient low-boiling hydrocarbons to vaporize easily in cold weather. The gasoline must also contain sufficient high-boiling hydrocarbons to remain a liquid in an engine's fuel supply system during hotter periods.

1.5.1.3   Kerosene and Diesel Fuel

In the early days of the refining industry, before the onset of the *automobile age*, *kerosene* (*kerosine*) (Chapter 8) was the major refinery product but now is considered one of several other petroleum products after gasoline (Speight and Ozum, 2002; Hsu and Robinson, 2006; Gary *et al.*, 2007; Speight, 2011a, 2014a). Kerosene originated as a straight-run (distilled without molecular change) petroleum fraction that boiled between approximately 150 and 350°C (300 and 660°F). In the early days of petroleum refining, some crude oils contained kerosene fractions of very high quality, but other crude oils, such as those having a high proportion of asphaltic materials, had to be carefully refined to remove aromatics and sulfur compounds before a satisfactory kerosene fraction can be obtained.

Diesel fuel (Chapter 9) also forms part of the kerosene boiling range (or middle distillate group of products). Diesel fuels come in two broad groups: for high-speed engines in cars and trucks requiring a high-quality product, and lower-quality heavier diesel fuel for slower engines, such as in marine engines or for stationary power plants. However, like gasoline, in spite of the varied nomenclature of the refining industry, there is no one refinery process that produces diesel fuel and the diesel fuel for sale is actually a blend of several refinery streams that is then sold as diesel fuel after the inclusion of the necessary additives (Chapter 9) (Speight and Ozum, 2002; Hsu and Robinson, 2006; Gary *et al.*, 2007; Speight, 2014a).

The quality of diesel fuel is measured using the cetane number, which is a measure of the tendency of a diesel fuel to knock in a diesel engine, and the scale, from which the cetane number is derived, is based upon the ignition characteristics of two hydrocarbons: (i) *n*-hexadecane (cetane) and (ii) 2,3,4,5,6,7,8-heptamethylnonane.

1.5.1.4   Fuel Oil

Fuel oil is generally subdivided into two main types: (i) *distillate fuel oil* (Chapter 10) and (ii) *residual fuel oil* (Chapter 11). However, the terms distillate fuel oil and residual fuel oil have lost some of their significance because fuel oils are now made for specific uses and may be distillates, residuals, or mixtures of the two. The terms *domestic fuel oil*, *diesel fuel oil*, and *heavy fuel oil* are more indicative of the uses of fuel oils. More often than not, fuel oil is prepared by using a visbreaker unit to perform mild thermal cracking on a residuum or a high-boiling distillate so that the product meets specifications (Speight and Ozum, 2002; Hsu and Robinson, 2006; Gary *et al.*, 2007; Speight, 2014a).

Distillate fuel oil is produced by distillation, has a definite boiling range, and does not contain constituents boiling above the specified distillation limit and does not contain asphaltic constituents. A fuel oil that contains any amount of the residue from crude distillation or thermal cracking is *residual fuel oil*. *Domestic fuel oil* is fuel oil that is used primarily in the home and includes kerosene, stove oil, and furnace fuel oil. *Diesel fuel oil* is also a distillate fuel oil, while *furnace fuel oil* is similar to diesel fuel, but the proportion of cracked gas oil in diesel fuel is usually less since the high aromatic content of the cracked gas oil reduces the cetane number of the diesel fuel.

Stove oil is a straight-run (distilled without molecular change) fraction from crude oil, whereas other fuel oils are usually blends of two or more fractions that include high-boiling naphtha, light gas oil, heavy gas oil, and residuum. Cracked fractions such as light gas oil, heavy gas oil from catalytic cracking, and fractionator bottoms from catalytic cracking may also be used as blends to meet the specifications of the different fuel oils.

Heavy fuel oil includes a variety of oils ranging from distillates to residual oils that must be heated to 260°C (500°F) or higher before they can be used. In general, heavy fuel oils consist of residual oils blended with distillates to suit specific needs. Included among heavy fuel oils are various industrial oils; when used to fuel ships, heavy fuel oil is called bunker oil.

Fuel oil that is used for heating is graded from No. 1 Fuel Oil to No. 6 Fuel Oil and cover light distillate oils, medium distillate, heavy distillate, a blend of distillate and residue, and residue oil.

No. 1 fuel oil is a relatively low-boiling distillate (straight-run kerosene) consisting primarily of hydrocarbons in the range C_9–C_{16} and is similar in composition to diesel fuel; the primary difference is in the additives. This fuel oil is used in atomizing burners that spray the fuel oil into a combustion chamber where the tiny droplets burn while in suspension. It is also used in asphalt coatings, enamels, paints, thinners, and varnishes.

No. 2 fuel oil is a petroleum distillate that may be referred to as domestic fuel oil or industrial fuel oil. The domestic fuel oil is usually lighter and straight-run refined; it is used primarily for home heating and to produce diesel fuel. Industrial distillate is the cracked type or a blend of both. It is used in smelting furnaces, ceramic kilns, and packaged boilers. It contains hydrocarbons in the C_{11}–C_{20} range, whereas diesel fuel predominantly contains a mixture of C_{10}–C_{19} hydrocarbons but contains less than 5% v/v polynuclear aromatic hydrocarbons (PNAs).

No. 6 fuel oil (also called *Bunker C*) is the residual material left after the light oil, naphtha, No. 1 fuel oil, and No. 2 fuel oil have been distilled from crude oil. No. 6 fuel oil can be blended directly to heavy fuel oil or made into asphalt. It is limited to commercial and industrial uses where sufficient heat is available to fluidize the oil for pumping and combustion. On the other hand, *residual fuel oil* is generally more complex in composition and impurities than distillate fuel oils and Bunker C. Because of the mode of manufacture, PNAs and their alkyl derivatives and metals are persistent constituents of No. 6 fuel oil.

1.5.1.5 White Oil, Insulating Oil, Insecticides

There is also a category of petroleum products known as *white oil* (Chapter 12) that generally falls into two classes: (i) *technical white oil* that is employed for cosmetics, textile lubrication, insecticide vehicles, and paper impregnation, and (ii) *pharmaceutical white oil* that is employed medicinally (e.g., as a laxative) or for the lubrication of food-handling machinery.

Insulating oil (as part of the white oil family) falls into two general classes: (i) oil used in transformers, circuit breakers, and oil-filled cables; and (ii) oil employed for impregnating the paper covering of wrapped cables. The first is highly refined fractions of low viscosity and comparatively high boiling range and resembles heavy burning oils, such as mineral seal oil, or the very light lubricating fractions known as nonviscous neutral oils. The second is usually highly viscous products and is often naphthenic distillate that is not usually highly refined.

Insecticides are derived from petroleum oil that can usually be applied in water-emulsion form and which have marked killing power for certain species of insects. For many applications for which their own effectiveness is too slight, the oils serve as carriers for active poisons, as in the household and livestock sprays.

Medicinal oil is a petroleum product that is used for medicinal purposes and must be ultrapure. For example, liquid paraffin is a clear water-white medicinal oil that is used to be given to miners to lubricate the alimentary tract and prevent coal dust buildup in the tract.

Finally and by way of clarification, *white spirit* is not necessarily the same as *white oil* and is a refined distillate boiling in the naphtha–kerosene range. White spirit has a flash point above 30°C (86°F) and a distillation range of 135–200°C (275–390°F). *Industrial spirit* comprises light oils distilling between 30 and 200°C (86 and 390°F). There are several grades of industrial spirit, depending on the distillation range.

1.5.1.6 Lubricating Oil

Lubricating oil (Chapter 13) is distinguished from other fractions of crude oil by a usually high (>400°C, >750°F) boiling point, as well as high viscosity. Lubricating oil may be divided into many categories according to the types of service it is intended to perform. However, there are two main groups: (i) oils used in intermittent service, such as motor and aviation oils; and (ii) oils designed for continuous service such as turbine oils. Crankcase oil or motor oil may be either petroleum based or synthetic (Speight and Ozum, 2002; Hsu and Robinson, 2006; Gary *et al.*, 2007; Speight, 2014a; Speight and Exall, 2014).

Petroleum-based oils are produced from heavy-end crude oil distillates, which may be treated in several ways, such as (i) vacuum distillation, (ii) solvent treatment, (ii) acid treatment, or (iv) hydrotreating, to produce lubricating oils with the oils with necessary properties. Hydrocarbon types ranging from C_{15} to C_{50} occur in the various types of lubricating oils. The hydrocarbon constituents are predominantly mixtures of straight- and branched-chain hydrocarbons (alkanes), cycloalkanes, and aromatic hydrocarbons. PNAs, alkyl PNAs, and metals are also components of motor oils and crankcase oils, with the used oils typically having higher concentrations than the new unused oils.

1.5.1.7 Grease

Grease (Chapter 14) is a lubricating oil to which a thickening agent has been added for the purpose of holding the oil to surfaces that must be lubricated. The most widely used thickening agents are soaps of various kinds, and grease manufacture is essentially the mixing of soaps with lubricating oils.

Soap is made by chemically combining a metal hydroxide with a fat or fatty acid:

$$R CO_2H + NaOH \rightarrow R CO^-_2 Na^+ + H_2O$$

Fatty acid   Soap

The most common metal hydroxides used for this purpose are calcium hydroxide, sodium/potassium hydroxide (lye), lithium hydroxide, and barium hydroxide. Commonly used fatty acids for grease-making soaps come from cottonseed oil, tallow, and lard. Among the fatty acids used are stearic acid (from tallow), oleic acid (from cottonseed oil), and animal fatty acids (from lard).

1.5.1.8 Wax

Wax (Chapter 15) is of two general types: (i) paraffin wax in petroleum distillates and (ii) microcrystalline wax in petroleum residua.

Paraffin wax is a solid crystalline mixture of straight-chain (normal) hydrocarbons ranging from 20 to 30 carbon atoms per molecule, and even higher.

Paraffin wax is a solid crystalline mixture of straight-chain (normal) hydrocarbons ranging from C_{20} to C_{30} and possibly higher, that is, $CH_3(CH_2)_nCH_3$, where $n \geq 18$. It is distinguished by its solid state at ordinary temperatures (25°C, 77°F) and low viscosity (35–45 SUS at 99°C, 210°F) when melted. However, in contrast to petroleum wax, petrolatum (*petroleum jelly*), although solid at ordinary temperatures, does in fact contain both solid and liquid hydrocarbons. It is essentially a low-melting, ductile, microcrystalline wax.

Microcrystalline waxes form approximately 1–2% w/w of crude oil and are valuable products having numerous applications. These waxes are usually obtained from heavy lube distillates by solvent dewaxing and from tank bottom sludge by acid clay treatment. However, these crude wax products usually contain appreciable quantity (10–20% w/w) of residual oil and, as such, are not suitable for many applications such as paper coating, electrical insulation, textile printing, and polishes.

1.5.1.9 Residua and Asphalt

Residua (*resids*, sing.: *resid*, *residuum*) are the residues obtained from petroleum after nondestructive distillation has removed all the volatile materials (Chapter 16). The temperature of the distillation is usually maintained below 350°C (660°F) since the rate of thermal decomposition of petroleum constituents is minimal below this temperature, but the rate of thermal decomposition of petroleum constituents is substantially above 350°C (660°F). A residuum that has held above this temperature so that thermal decomposition has occurred is known as a *cracked resid*.

Asphalt (sometimes referred to as bitumen in Europe and other parts of the world) is produced from the distillation residuum (Chapter 16). In addition to road asphalt, a variety of asphalt grades for roofing and waterproofing are also produced. Asphalt has complex chemical and physical compositions that usually vary with the source of the crude oil, and it is produced to certain standards of hardness or softness in controlled vacuum distillation processes. There are wide variations in refinery operations and in the types of crude oils so different asphalts will be produced—for example, soft asphalts can be converted into harder asphalts by oxidation (air blowing). Asphalt is also produced by propane deasphalting and can be made softer by blending the hard asphalt with the extract obtained in the solvent treatment of lubricating oils (Speight and Ozum, 2002; Hsu and Robinson, 2006; Gary *et al.*, 2007; Speight, 2014a).

Road oil is liquid asphalt intended for easy application to earth roads and provides a strong base or a hard surface and will maintain satisfactory in-service performance for light traffic loads. *Cutback asphalt* is a mixture of hard asphalt that has been diluted with lower-boiling oil (aromatic naphtha or aromatic kerosene) to permit application as a liquid without drastic heating. *Asphalt emulsions* are typically oil-in-water emulsions that break on application to an earthen or stone surface—the asphaltic oil adheres to the stone, and the water disappears. In addition to their usefulness in road and soil stabilization, asphalt emulsions are also used for paper impregnation and waterproofing.

1.5.1.10 Coke, Carbon Black, and Graphite

Petroleum coke (Chapter 17) is the residue left by the destructive distillation (thermal cracking or coking) of petroleum residua. It consists mainly of carbon (90–95% w/w) and has low mineral matter ash content (measure as mineral ash). It is used as a feedstock in coke ovens for the steel industry, for heating purposes, for electrode manufacture, and for production of chemicals. The two most important types are *green coke* and *calcined coke*. This product category (petroleum coke) also includes *catalyst coke* deposited on the catalyst during refining processes—this type of coke is not recoverable because of adherence to the catalyst and is removed by burning to produce fuel gas that may serve as process fuel gas. The composition of coke varies with the source of the crude oil but, in general, is insoluble in organic solvents and has a honeycomb-type appearance.

With the influx of heavy crude oil into refineries, the use of coke as a refinery fuel must proceed with some caution—the coke from heavy oil typically has a higher content of sulfur and nitrogen than coke from lighter cure oil. Both of these elements will produce unacceptable pollutants—sulfur oxides (SO_x) and nitrogen oxides (NO_x) produced during combustion of the coke are serious environmental contaminates (Speight, 2005).

Carbon black (Chapter 17) is produced from coke as a form of para-crystalline carbon that has a high surface-area-to-volume ratio. It is dissimilar to soot in its much higher surface-area-to-volume ratio and significantly lower content of PNAs. However, carbon black is widely used as a model compound for diesel soot for diesel oxidation experiments, as a filler in tires, and as a color pigment in plastics, paints, and ink.

Graphite (Chapter 17) is produced by treating petroleum coke in an oxygen-free oven at extremely high temperature. Graphite is also a naturally occurring mineral that occurs in two forms: (i) alpha and (ii) beta, which have identical physical properties but different crystal structures. All graphite produced from petroleum sources is of the alpha type.

1.5.1.11 Fischer–Tropsch Liquids and Bio-Oil

While not truly petroleum products in the literal sense, there are two other fuel products that must be included here: (i) Fischer–Tropsch liquids and (ii) bio-oil.

Fischer–Tropsch liquids are products (which vary from naphtha-type liquids to wax) produced by the Fischer–Tropsch synthesis using mixture of hydrogen and carbon monoxide (synthesis gas, syngas) as the feedstock (Davis and Occelli, 2010). The synthesis gas is produced by the gasification conversion of carbonaceous feedstock such as petroleum residua, coal, and biomass, and production of hydrocarbon products can be represented simply as follows:

$$CH_{resid,etc.} + O_2 \rightarrow CO + H_2$$

$$nCO + nH_2 \rightarrow C_nH_{2n+2}$$

However, before conversion of the carbon monoxide and hydrogen to hydrocarbon products, several reactions are employed to adjust the hydrogen/carbon monoxide ratio. Most important is the water gas shift reaction in which additional hydrogen is produced (at the expense of carbon monoxide) to satisfy the hydrogen/carbon monoxide ratio necessary for the production of hydrocarbons:

$$H_2O + CO \rightarrow H_2 + CO_2$$

The boiling range of Fischer–Tropsch typically spans the naphtha and kerogen boiling ranges and is suitable for analysis by application of the standard test methods. With the suitable choice of a catalyst, the preference for products boiling in the naphtha range (<200°C, <390°F) or for product boiling in the diesel range (~150 to 300°C, 300 to 570°F) can be realized.

Bio-oil (*pyrolysis oil*, *bio-crude*) is the liquid product produced by the thermal decomposition (destructive distillation) of biomass at temperatures on the order of 500°C (930°F). The product is a synthetic crude oil and is of interest as a possible complement (eventually a substitute) to petroleum. The product can vary from a light tarry material to a free-flowing liquid—both require further refining to produce specification-grade fuels. Hydrocarbon moieties are predominant in the product, but the presence of varying levels of oxygen (depending upon the character of the feedstock) requires testament (using, e.g., hydrotreating) during refining.

In summary, the Fischer–Tropsch process produces hydrocarbon products of different molecular weights from a gas mixture of hydrogen and carbon monoxide (synthesis gas, syngas). The process uses various catalysts to produce linear hydrocarbons and oxygenates, including unrefined naphtha, diesel, and waxes. Analysis of these products is achieved using the relevant analytical test method presented in the following chapters.

1.5.2 Properties

Measurements of bulk properties are generally easy to perform and, therefore, quick and economical. Several properties may correlate well with certain compositional characteristics and are widely used as a quick and inexpensive means to determine those. The most important properties of a whole crude oil are its boiling-point distribution, its density (or API gravity), and its viscosity. The *boiling-point distribution*, *boiling profile*, or *distillation assay* gives the yield of the various distillation cuts, and selected properties of the fractions are usually determined (Table 1.4). It is a prime property in its own right that indicates how much gasoline and other transportation fuels can be made from petroleum without conversion. Density and viscosity are measured for secondary reasons. The former helps to estimate the paraffinic character of the oil, and the latter permits the assessment of its undesirable residual material that causes resistance to flow. Boiling-point distribution, density, and viscosity are easily measured and give a quick first evaluation of petroleum oil. Sulfur content, another crucial and primary property of a crude oil, is also readily determined.

TABLE 1.4   Distillation profile of petroleum (Leduc, Woodbend, Upper Devonian, Alberta, Canada) and selected properties of the fractions

Whole Bitumen	Boiling Range		Wt %	Wt % Cumulative 100.0	Specific Gravity 1.030	API Gravity 5.9	Sulfur Wt % 5.8	Carbon Residue (Conradson) 19.6
	°C	°F						
Fraction*								
1	0–50	0–122	0.0	0.0				
2	50–75	122–167	0.0	0.0				
3	75–100	167–212	0.0	0.0				
4	100–125	212–257	0.0	0.0				
5	125–150	257–302	0.9	0.9				
6	150–175	302–347	0.8	1.7	0.809	43.4		
7	175–200	347–392	1.1	2.8	0.823	40.4		
8	200–225	392–437	1.1	3.9	0.848	35.4		
9	225–250	437–482	4.1	8.0	0.866	31.8		
10	250–275	482–527	11.9	19.9	0.867	31.7		
11	<200	<392	1.6	21.5	0.878	29.7		
12	200–225	392–437	3.2	24.7	0.929	20.8		
13	225–250	437–482	6.1	30.8	0.947	17.9		
14	250–275	482–527	6.4	37.2	0.958	16.2		
15	275–300	527–572	10.6	47.8	0.972	14.1		
Residuum	>300	>572	49.5	97.3				39.6

*Distillation at 762 mmHg then at 40 mmHg for fractions 11–15.

Certain composite characterization values, calculated from density and mid-boiling point, correlate better with molecular composition than density alone (Speight and Ozum, 2002; Hsu and Robinson, 2006; Gary et al., 2007; Speight, 2014a).

The acceptance of heavy oil and bitumen as refinery feedstocks has meant that the analytical techniques used for the lighter feedstocks have had to evolve to produce meaningful data that can be employed to assist in defining refinery scenarios for processing the feedstocks. In addition, selection of the most appropriate analytical procedures will aid in the predictability of feedstock behavior during refining. This same rationale can also be applied to feedstock behavior during recovery operations. Indeed, bitumen, a source of synthetic crude oil, is different from petroleum (Speight and Moschopedis, 1979; Speight, 2014a), and many of the test methods designed for petroleum may need modification before application to bitumen (Wallace, 1988; Wallace et al., 1988).

Thus, knowledge of the composition of petroleum and the resulting products allows the refiner to optimize the conversion of raw petroleum into high-value products. Petroleum is now the world's main source of energy and petrochemical feedstock. Originally, petroleum was distilled and sold as fractions with desirable physical properties. Today crude oil is sold in the form of gasoline, solvents, diesel and jet fuel, heating oil, lubricant oils, and asphalts, or it is converted to petrochemical feedstocks such as ethylene, propylene, the butenes, butadiene, and isoprene. These feedstocks are important, for they form the basis for, among others, the plastics, elastomers, and artificial-fiber industries. Modern refining uses a sophisticated combination of heat, catalyst, and hydrogen to rearrange the petroleum molecules into these products. Conversion processes include (i) coking, (ii) hydrocracking, and (iii) catalytic cracking to break large molecules into smaller fractions, as well as (iv) hydrotreating to reduce heteroatoms and aromatics creating environmentally acceptable products, and (v) isomerization and reforming to rearrange molecules to those with high value, for example, gasoline with a high octane number.

Also, knowledge of the molecular composition of petroleum allows environmental scientists to consider the biological impact of environmental exposure (Speight, 2005; Speight and Arjoon, 2012). Increasingly, petroleum is being produced and transported from remote areas of the world to refineries located closer to their markets. Although a minuscule fraction of that oil is released into the environment, the sheer volume involved has the potential for environmental exposure. Molecular composition is needed not only to identify the sources of contamination hut also to understand the fate and effects of its potentially hazardous components.

In addition, knowledge of the composition of petroleum allows the geologist to answer questions of precursor–product relationships and conversion mechanisms. Biomarkers—molecules that retain the basic carbon skeletons of biological compounds from living organisms after losing functional groups through the maturation process—play an important role in such studies. The distribution of biomarker isomers can not only serve as fingerprints for oil/oil and oil/source correlation (to relate the source and reservoir) but also give geochemical information on organic source input (marine, lacustrine, or land-based sources), age, maturity, depositional environment (e.g., clay or carbonate, oxygen levels, salinity), and alteration (e.g., water washing, biodegradation) (Speight and Arjoon, 2012).

The need for the application of analytical techniques for petroleum and petroleum products has increased over the past three decades because of the change in feedstock composition. This has arisen because of the increased amounts of the heavier feedstocks that are now used to produce liquid products. Prior to the energy crises of the 1970s, the heavier feedstocks were used infrequently as sources of liquid fuels and were used to produce asphalt. Now these feedstocks have increased in value as sources of liquid fuels.

Because of the wide range of chemical and physical properties, a wide range of tests have been (and continue to be) developed to provide an indication of the means by which a particular feedstock should be processed. Initial inspection of the nature of the petroleum will provide deductions about the most logical means of refining or correlation of various properties to structural types present and hence attempted classification of the petroleum. Proper interpretation of the data resulting from the inspection of crude oil requires an understanding of their significance.

Having decided what characteristics are necessary, it then remains to describe the product in terms of a specification. This entails selecting suitable test methods and setting appropriate limits. Many specifications in widespread use have evolved usually by the addition of extra clauses (rarely is a clause deleted). This has resulted in unnecessary restrictions that, in turn, result in increased cost of the products specified.

REFERENCES

Abraham, H. 1945. Asphalts and Allied Substances. Van Nostrand-Reinhold Company, New York.

Ancheyta, J. and Speight, J.G. 2007. Hydroprocessing of Heavy Oils and Residua. CRC Press, Taylor & Francis Group, Boca Raton, FL.

Ancheyta, J., Trejo, F., and Rana, M.S. 2010. Asphaltenes: Chemical Transformation during Hydroprocessing of Heavy Oils. CRC Press, Taylor & Francis Group, Boca Raton, FL.

ASTM D86. Standard Test Method for Distillation of Petroleum Products at Atmospheric Pressure. Annual Book of Standards. ASTM International, West Conshohocken, PA.

ASTM D2887. Standard Test Method for Boiling Range Distribution of Petroleum Fractions by Gas Chromatography. Annual Book of Standards. ASTM International, West Conshohocken, PA.

ASTM D3710. Standard Test Method for Boiling Range Distribution of Gasoline and Gasoline Fractions by Gas Chromatography. Annual Book of Standards. ASTM International, West Conshohocken, PA.

ASTM D1160. Standard Test Method for Distillation of Petroleum Products at Reduced Pressure. Annual Book of Standards. ASTM International, West Conshohocken, PA.

ASTM D4175. Standard Terminology Relating to Petroleum, Petroleum Products, and Lubricants, Volume 05.02. Annual Book of Standards. ASTM International, West Conshohocken, PA.

Barrow, M.P., McDonnell, L.A., Feng, X., Walker, J., and Derrick, P.J. 2003. Determination of the Nature of Naphthenic Acids Present in Crude Oils Using Nanospray Fourier Transform Ion Cyclotron Resonance Mass Spectrometry: The Continued Battle Against Corrosion. Analytical Chemistry, 75(4): 860–866.

Barth, T., Høiland, S., Fotland, P., Askvik, K.M., Pedersen, B.S., and Borgund. A.E. 2004. Acidic Compounds in Biodegraded Petroleum. Organic Geochemistry, 35(11–12): 1513–1525.

Clemente, J.S., Yen, T.W., and Fedorak, P.M. 2003a. Development of a High Performance Liquid Chromatography Method to Monitor the Biodegradation of Naphthenic Acids. Journal of Environmental Engineering and Science, 2(3):177–186.

Clemente, J.S., Prasad, N.G.N, MacKinnon. M.D., and Fedorak, P.M. 2003b. A Statistical Comparison of Naphthenic Acids Characterized by Gas Chromatography-Mass Spectrometry. Chemosphere, 50: 1265–1267.

Costantinides, G. and Arich, G. 1967. Non-hydrocarbon Compounds in Petroleum. In: Fundamental Aspects of Petroleum Geochemistry. B. Nagy and U. Colombo (Editors). Elsevier Publishing Company, London. p. 109–175.

Davis, B.H. and Occelli, M.L. 2010. Advances in Fischer-Tropsch Synthesis, Catalysts, and Catalysis. CRC Press, Taylor & Francis Group, Boca Raton, FL.

Dolbear, G.E., Tang, A., and Moorehead, E.L. 1987. Upgrading studies with Californian, Mexican and Middle Eastern heavy oils. In Metal Complexes in Fossil Fuels. R.H. Filby and J.F. Branthaver (Editors). Symposium Series No. 344. American Chemical Society, Washington, DC. p. 220–232.

Drews, A.W. (Editor). 1998. Manual on Hydrocarbon Analysis, 6th Edition. ASTM International, West Conshohocken, PA.

Fafet, A., Kergall, F., Da Silva, M., and Behar, F. 2008. Characterization of Acidic Compounds in Biodegraded Oils. Organic Geochemistry, 39(8): 1235–1242.

Forbes, R.J. 1958a. A History of Technology, Oxford University Press, Oxford.

Forbes, R.J. 1958b. Studies in Early Petroleum Chemistry. E. J. Brill, Leiden.

Forbes, R.J. 1959. More Studies in Early Petroleum Chemistry. E.J. Brill, Leiden.

Forbes, R.J. 1964. Studies in Ancient Technology. E. J. Brill, Leiden.

Gary, J.G., Handwerk, G.E., and Kaiser, M.J. 2007. Petroleum Refining: Technology and Economics, 5th Edition. CRC Press, Taylor & Francis Group, Boca Raton, FL.

Hell, C.C., and Medinger, E. 1874. Ueber das Vorkommen und die Zusammensetzung von Säuren im Rohpetroleum. Berichte der deutschen chemischen Gesellschaft, 7(2):1216–1223.

Henry, J.T. 1873. The Early and Later History of Petroleum. Volumes I and II. APRP Co., Philadelphia, PA.

Hsu, C.S., and Robinson, P.R. (Editors) 2006. Practical Advances in Petroleum Processing. Volumes 1 and 2. Springer Science, New York.

James, P., and Thorpe, N. 1994. Ancient Inventions. Ballantine Books, Random House Publishing Group, New York.

Kim, K., Stanford, L.A., Rodgers, R.P., Marshall, A.G., Walters, C.C., Qian, K., Wenger, L.M., and Mankiewicz, P. 2005. Microbial Alteration of the Acidic and Neutral Polar NSO Compounds Revealed by Fourier Transform Ion Cyclotron Resonance Mass Spectrometry. Organic Geochemistry, 36(8):1117–1134.

Koots, J.A., and Speight, J.G. 1975. The Relation of Petroleum Resins to Asphaltenes. Fuel, 54: 179–184.

Lochte, H.L. 1952. Petroleum Acids and Bases. Industrial and Engineering Chemistry, 44(11): 2597–2601.

Long, R.B., and Speight, J.G. 1998. The Composition of Petroleum. In Petroleum Chemistry and Refining. Taylor & Francis Publishers, Washington, DC. p. 1–38.

Mackenzie, A.S., Wolff, G.A., and Maxwell, J.R. 1981. Fatty Acids in Some Biodegraded Petroleums. Possible Origins and Significance. In: Advances in Organic Geochemistry. M. Bjorøy (Editor). John Wiley & Sons Ltd, Chichester. p. 637–643.

Meredith, W., Kelland, S.J., and Jones, D.M. 2000. Influence of Biodegradation on Crude Oil Activity and Carboxylic Acid Composition. Organic Geochemistry, 31(11): 1059–1073.

Mokhatab, S., Poe, W.A., and Speight, J.G. 2006. Handbook of Natural Gas Transmission and Processing. Elsevier, Amsterdam.

Nadkarni, R.A.K. 2000. Guide to the ASTM Test Methods for the Analysis of Petroleum Products and Lubricants. ASTM International, West Conshohocken, PA.

Nadkarni, R.A.K. 2011. Spectroscopic Analysis of Petroleum Products and Lubricants. ASTM International, West Conshohocken, PA.

Ney, W.O., Crouch, W.W., Rannefeld, C.E., and Lochte, H.L. 1943. Petroleum Acids. VI. Naphthenic Acids from California Petroleum. Journal of the American Chemical Society, 65(5): 770–777.

Occelli, M.L. 2010. Advances in Fluid Catalytic Cracking: Testing, Characterization, and Environmental Regulations. CRC Press, Taylor & Francis Group, Boca Raton, FL.

Parkash, S. 2003. Refining Processes Handbook. Gulf Professional Publishing Company, Elsevier, Burlington, MA.

Rand, S. 2003. Significance of Tests for Petroleum Products, 7th Edition. ASTM International, West Conshohocken, PA.

Rodgers, R.P., Hughey, C.A., Hendrickson, C.L., and Marshall, A.G. 2002. Advanced Characterization of Petroleum Crude and Products by High Field Fourier Transform Ion Cyclotron Resonance Mass Spectrometry. Preprints. Div. Fuel Chem., Am Chem. Soc., 47(2): 636–637.

Silliman, B. Sr. 1833. Notice of a Fountain of Petroleum Called the Oil Spring. American Journal, Series 1, Volume XXIII, 97–103.

Silliman, B. Jr. 1860. Oil Wells of Pennsylvania and Ohio. American Journal, Series 2, Volume XXX, 305–306.

Silliman, B. Jr. 1865. Examination of Petroleum from California. American Journal, Series 2, Volume XXXIX, page 341–343.

Silliman, B. Jr. 1867. On Naphtha and Illuminating Oil from Heavy California Tar. American Journal, Series 2, Volume XLIII, page 242–246.

Silliman, B. Jr. 1871. Report on the Rock Oil or Petroleum from Venango County Pennsylvania. Am. Chemist. 2: 18–23.

Speight, J.G., and Moschopedis, S.E. 1979. In The Future of Heavy Crude Oil and Tar Sands. F.R. Meyer and C.T. Steele (Editors). McGraw-Hill, New York. p. 603.

Speight, J.G. 1994. Chemical and Physical Studies of Petroleum Asphaltenes. In Asphaltenes and Asphalts, I. Developments in Petroleum Science, 40. T.F. Yen and G.V. Chilingarian (Editors). Elsevier, Amsterdam, Netherlands. Chapter 2.

Speight, J.G. 2000. The Desulfurization of Heavy Oils and Residua, 2nd Edition. Marcel Dekker Inc., New York.

Speight, J.G. 2001. Handbook of Petroleum Analysis. John Wiley & Sons Inc., Hoboken, NJ.

Speight, J.G., and Ozum, B. 2002. Petroleum Refining Processes. Marcel Dekker Inc., New York.

Speight. J.G. 2005. Environmental Analysis and Technology for the Refining Industry. John Wiley & Sons Inc., Hoboken, NJ.

Speight, J.G. 2007. Natural Gas: A Basic Handbook. GPC Books, Gulf Publishing Company, Houston, TX.

Speight, J.G. 2008. Synthetic Fuels Handbook: Properties, Processes, and Performance. McGraw-Hill, NY.

Speight, J.G. 2009. Enhanced Recovery Methods for Heavy Oil and Tar Sands. Gulf Publishing Company, Houston, TX.

Speight, J.G. 2011a. The Refinery of the Future. Gulf Professional Publishing, Elsevier, Oxford.

Speight, J.G. (Editor) 2011b. The Biofuels Handbook. The Royal Society of Chemistry, London.

Speight, J.G. 2012a. Crude Oil Assay Database. Elsevier-Knovel, New York.

Speight, J.G. 2012b. Shale Oil Production Processes. Gulf Professional Publishing, Elsevier, Oxford.

Speight, J.G., and Arjoon, K.K. 2012. Bioremediation of Petroleum and Petroleum Products. Scrivener Publishing, Salem, MA.

Speight, J.G. 2013a. Heavy Oil Production Processes. Gulf Professional Publishing, Elsevier, Oxford.

Speight, J.G. 2013b. Shale Gas Production Processes. Gulf Professional Publishing, Elsevier, Oxford.

Speight, J.G. 2014a. The Chemistry and Technology of Petroleum, 5th Edition. CRC Press, Taylor & Francis Group, Boca Raton, FL.

Speight, J.G. 2014b. High Acid Crudes. Gulf Professional Publishing, Elsevier, Oxford.

Speight, J.G. 2014c. Oil and Gas Corrosion Prevention. Gulf Professional Publishing, Elsevier, Oxford.

Speight, J.G., and Exall, D.I. 2014. Refining Used Lubricating Oils. CRC Press, Taylor & Francis Group, Boca Raton, FL.

Thorn, K.A., and Aiken. G.R. 1998. Biodegradation of Crude Oil into Nonvolatile Organic Acids in a Contaminated Aquifer near Bemidji, Minnesota. Organic Geochemistry, 29(4): 909–931.

Tomczyk, N.A., Winans, R.E., Shinn, J.H., and Robinson, R.C. 2001. On the Nature and Origin of Acidic Species in Petroleum. 1. Detailed Acidic Type Distribution in a California Crude Oil. Energy and Fuels, 15(6): 1498–1504.

Totten, G.E. (Editor). 2003. Fuels and Lubricants Handbook: Technology, Properties, Performance, and Testing. ASTM International, West Conshohocken, PA.

Wallace, D. (Editor). 1988. A Review of Analytical Methods for Bitumens and Heavy Oils. Alberta Oil Sands Technology and Research Authority, Edmonton, AB.

Wallace, D., Starr, J., Thomas, K.P., and Dorrence, S.M. 1988. Characterization of Oil Sand Resources. Alberta Oil Sands Technology and Research Authority, Edmonton, AB.

Watson, J.S., Jones, D.M., Swannell, R.P.G., and Van Duin, A.C.T. 2002. Formation of Carboxylic Acids during Aerobic Biodegradation of Crude Oil and Evidence of Microbial Oxidation of Hopanes. Organic Geochemistry, 33(10): 1153–1169.

Wilkes, H., Kühner, S., Bolm, C., Fischer, T., Classen, A., Widdel, F., and Rabus, R. 2003. Formation of n-Alkane- and Cycloalkane-Derived Organic Acids during Anaerobic Growth of a Denitrifying Bacterium with Crude Oil. Organic Geochemistry, 34(9): 1313–1323.

Zhao, B., Currie, R., and Mian, H. 2012. Catalogue of Analytical Methods for Naphthenic Acids Related to Oil Sands Operations. OSRIN Report No. TR-21. Oil Sands Research and Information Network, University of Alberta, School of Energy and the Environment, Edmonton, AB.

2

ANALYTICAL METHODS

2.1 INTRODUCTION

The chemical composition of petroleum and petroleum products is complex (Chapter 1), which makes it essential that the most appropriate analytical methods are selected from a comprehensive list of methods and techniques that are used for the analysis of samples (Dean, 1998; Miller, 2000; Budde, 2001; Speight, 2001, 2005, 2014; Speight and Arjoon, 2012). Furthermore, samples may be disturbed during sampling, storage, and pretreatment. Also, most laboratory experiments impose steady environmental conditions, while in an outdoor climate, these conditions show dynamic behavior (Speight and Arjoon, 2012). However, the manner in which the corrections are applied must be quality data (ASTM D6299, ASTM D6792) and must be beyond reproach, or claims of falsification of the data will be the most likely result (Speight and Foote, 2011).

In the early days of petroleum processing, there was not the need to understand the character and behavior of petroleum in detail that is currently required. Refining was relatively simple and involved distillation of the valuable kerosene fraction that was then sold as an illuminant. After the commercialization of the internal combustion engine, the desired product became gasoline, and it was also obtained by distillation. Even when crude oil that contained little natural gasoline was used, cracking (i.e. thermal decomposition with simultaneous removal of distillate) became the modus operandi.

However, with the startling demands on the petroleum industry during, and after, World War II and the emergence of the age of petrochemicals and plastics, the petroleum industry needed to produce materials not even considered as products in the decade before the war. Thus, petroleum refining took on the role of technological innovator, as new and better processes were invented and advances in the use of materials for reactors were developed. In addition, there became a necessity to find out more about petroleum so that the refiner might be able to enjoy the luxury of predictability and plan a product slate that was based on market demand, a difficult task when the character of the crude oil was unknown. The idea that petroleum refining should be a *hit-and-miss* affair was not acceptable.

The processing of petroleum requires not only knowledge of its chemical and physical properties, but also knowledge of its chemical and physical reactivities. The former is dealt with in this chapter; the latter, because of the structure of petroleum, is dealt with elsewhere in this book (Chapter 18). Because petroleum varies markedly in its properties and composition, according to the source, it also varies in its chemical and physical reactivities. Thus, knowledge of petroleum reactivity is required for optimization of the existing processes as well as for the development and design of new processes.

For example, valuable information can be obtained from the true boiling point (TBP) curve, which is a function of percent weight distilled and temperature, that is, a boiling point distribution (Speight, 2001, 2014). However, the boiling range does not convey much detail about the chemical reactivity of crude oil. In addition to the boiling point distribution, it is possible to measure bulk physical properties, such as specific gravity and viscosity that have assisted in the establishment of certain empirical relationships for

Handbook of Petroleum Product Analysis, Second Edition. James G. Speight.
© 2015 John Wiley & Sons, Inc. Published 2015 by John Wiley & Sons, Inc.

petroleum processing from the TBP curve. Many of these relationships include assumptions that are based on experience with a range of feedstocks. However, the chemical aspects of refining feedstocks that contain different proportions of chemical species emphasize the need for more definitive data that would enable more realistic predictions to be made of crude oil behavior in refinery operations.

Thus, this chapter presents some of the methods that are generally applied to study the makeup of the feedstock in terms of chemical structures as well as methods that might be preferred for refining. There are, of course, many analytical methods that can be applied to the analysis of petroleum and petroleum products, but they vary with sample condition and composition. More specifically, this chapter deals with the more common methods used to define the chemical and physical properties of the sample.

Any of the methods described herein might also be applied to the analysis of the sample for environmental purposes. However, methods for analytical purpose are not intended to be the focus of this text and are described elsewhere (Speight, 2005; Speight and Arjoon, 2012).

Finally, there is the need for a comment on the analysis for TPH. The term *total petroleum hydrocarbons*, which is used in the area of environmental analysis, is not covered in this book, but is covered in detail elsewhere (Speight and Arjoon, 2012). The analysis for TPH in a sample as a means of evaluating petroleum-contaminated sites is also an analytical method in common use. The data are used to establish target cleanup levels for soil or water, which is a common approach implemented by regulatory agencies in the United States, and in many other countries. There is a wide variety of methods for measurement of the TPH in a sample, but analytical inconsistencies must be recognized because of the definition of TPH (Speight and Arjoon, 2012). In practice, as in many analyses of petroleum and petroleum products, the term *total petroleum hydrocarbons* is defined by the analytical method, since different methods are designed to extract and measure slightly different subsets of petroleum hydrocarbons and, thus, often give different results.

2.2 CHEMICAL AND PHYSICAL ANALYSES

The importance of chemical and physical properties is dependent upon the purity of the petroleum or petroleum products. In the strictest sense, petroleum and petroleum products are complex chemical mixtures. These mixtures contain hydrocarbons and nonhydrocarbon compounds that confer properties on the mixture that may not be reflected in the composition. Therefore, it is necessary to apply various text methods to petroleum and petroleum products to determine if the material is suitable for processing and (in the case of the products) for sale with a designated use in mind.

The more common tests are introduced in the following subsections—these tests are presented in alphabetical order with no preference given to any partilcuar test method.

2.2.1 Boiling Point Distribution

In the petroleum refining industry, boiling range distribution data are used (i) to assess petroleum crude quality before purchase, (ii) to monitor petroleum quality during transportation, (iii) to evaluate petroleum for refining, and (iv) to provide information for the optimization of refinery processes.

Traditionally, boiling range distributions of the various fractions have been determined by distillation. Yield-on-crude data are still widely reported in the petroleum assay literature, providing information on the yield of specific fractions obtained by distillation.

To some extent, in the laboratory, atmospheric and vacuum distillation techniques have largely been replaced by *simulated* distillation methods, which use low-resolution gas chromatography (GC) and correlate retention times to hydrocarbon boiling points. Two test methods (ASTM D2887, ASTM D3710) use external standards composed of *n*-alkanes, while similar third test method (ASTM D5307) requires two determinations to be made with each sample, one of which uses an internal standard. The amount of material boiling above 538°C (1000°F) (reported as residue) is calculated from the difference between the two determinations.

From the point of view of petroleum and petroleum product analysis for environmental purposes, boiling range distributions provide an indication of volatility and component distribution. In addition, boiling range distribution data is also useful for the development of equations for predicting evaporative loss (Speight, 2005; Speight and Arjoon, 2012).

2.2.2 Density, Specific Gravity, and API Gravity

Density is the mass per unit volume of a substance. It is most often reported for oils in units of g/ml or g/cm^3, and less often in units of kg/m^3. Density is temperature-dependent. It is an important property of petroleum and petroleum products, as it gives the investigator(s) indications of whether or not the contaminant(s) will float on water.

Two density-related properties of petroleum and petroleum products are often used: (i) specific gravity and (ii) American Petroleum Institute (API) gravity. Specific gravity (*relative density*) is the ratio, at a specified temperature, of the oil density to the density of pure water. The API gravity scale (presented as °API) arbitrarily assigns an API gravity of 10° to pure water. Thus,

$$°API = \left[141.5 / \left(\text{specific gravity at } 15.6°C \right) - 131.5 \right].$$

The scale is commercially important for ranking petroleum quality: heavy oil typically has a gravity less than 20°API; medium oils—20–35° API; light oils—35–45°API; liquid petroleum products can have an API gravity up to 65°; Tar sand bitumen typically has an API gravity less than 10°.

Petroleum with low density, and hence low specific gravity, has a high API gravity. The price of crude oil is usually based on its API gravity, with high-gravity oils commanding higher prices (Speight, 2011). API gravity and density or specific gravity, at 15°C (60°F), can be interconverted using petroleum measurements.

Petroleum (unless it is a specific heavy oil or tar sand bitumen) and petroleum products (unless it is a residual fuel oil or asphalt) will float on water if the density of the petroleum or petroleum product is less than that of the water. This behavior is typical of all crude oils and distillate products for both salt and freshwater. Some heavy oils, tar sand bitumen, and residual fuel oils may have densities greater than 1.0 g/ml, and their buoyancy behavior varies depending on the salinity and temperature of the water (Speight, 2009).

2.2.3 Emulsion Formation

A water-in-oil emulsion is a stable dispersion of small droplets of water in oil. When formed from crude oils spilled at sea, these emulsions can have very different characteristics from their parent crude oils. This has important implications for the fate and behavior of the oil and its subsequent cleanup. It is desirable, therefore, to determine if the oil is likely to form an emulsion, and if so, whether that emulsion is stable, and the physical characteristics of the emulsion.

In an older test method, the tendency for a crude oil to form a water-in-oil emulsion was measured using a method based on the rotating flask apparatus (Mackay and Zagorski, 1982). All numerical values (mostly ones or zeroes) based on this method have subsequently been reduced to *yes* or *no*, respectively, and indicate the formation (or not) of an emulsion that remained stable 24 h after settling. In a newer variation, the reproducibility is considerably improved, and several parameters—(i) the water-to-oil ratio, (ii) the fill volume, and (iii) the orientation of the vessels—were found to be important parameters affecting emulsion formation.

However, such effects are not lasting. Emulsion formation and behavior is influenced by the oxidation of petroleum constituents (Speight, 2014). The inclusion of polar functions such as hydroxyl groups ($-OH$) or carbonyl groups ($>C=O$) (a result of the oxidation process) causes an increase in the density of the emulsion (relative to the original unoxidized petroleum), with an increased propensity to form emulsions. As a result, the emulsion sinks to various depths, or even to the seabed, depending on the extent of oxidation and the resulting density. This may give an erroneous appearance (leading to erroneous deductions with catastrophic consequences) that the petroleum spill (as evidenced from the petroleum remaining on the surface of the water) is less than it actually was. The so-called *missing* oil will undergo further chemical changes and eventually reappear on the water surface or on a distant beach.

2.2.4 Evaporation

Evaporation is the removal of the lower-boiling constituents from petroleum or a petroleum product, usually under ambient conditions or, in the current context, under the conditions prevalent at the spill site.

Evaporation rate and loss are of importance for all the volatile constituents of petroleum and petroleum products. While standard test methods such as those designated for distillation and vapor pressure determination are often used to determine evaporation properties (ASTM, 2011), test methods for determining evaporation loss are available for higher-boiling petroleum products (ASTM D972, ASTM D2595). Though not necessarily applicable to petroleum and petroleum products in general, evaporation loss data can be obtained at any temperature in the range from 93 to 316°C (200 to 600°F). Viscous samples can be analyzed using a water vaporizer accessory that heats the sample in the evaporation chamber, and the vaporized water is carried into the Karl Fischer titration cell by a dry inert carrier gas.

Petroleum and petroleum products evaporate at a logarithmic rate with respect to time (Fingas, 1998). This is attributed to the overall logarithmic appearance of many components evaporating at different linear rates. Petroleum products with fewer constituents (such as diesel fuel) evaporate at a rate which is square root with respect to time, which is a result of the number of components evaporating. The evaporation process, as evidenced by petroleum and petroleum products, is not strictly boundary-layer regulated, which is largely a result of the high saturation concentrations of oil components in air and is associated with a high boundary-layer regulated rate. Some volatile crude oils and petroleum products show some effects of boundary-layer regulation at the start of the evaporation process, but after several minutes, evaporation slows because of the loss of the more volatile components, at which point evaporation ceases to be boundary-layer regulated.

It must also be recognized that as evaporation occurs, the density and viscosity of the residual petroleum or residual petroleum products increase, thereby causing behavioral changes in the contaminant. Such changes could be reflected in an increase in adhesion of the contaminant constituents to the soil or rock.

2.2.5 Fire Point and Flash Point

The *fire point* is the lowest temperature, corrected to 1 atmosphere pressure (14.7 psi), at which the application of a test flame to the petroleum or petroleum product sample surface causes the vapor of the oil to ignite and burn for at least 5 s.

At any time after a spill of petroleum or a petroleum product, fire should always be considered an imminent hazard. Related to fire point, the flash point is a measure of the tendency of the petroleum or a petroleum product to form a flammable mixture with air under controlled laboratory conditions. It is only one of a number of properties that should be considered in assessing the overall flammability hazard of a spilled material (ASTM D92).

The ignition temperature (sometimes called the *autoignition temperature*) is the minimum temperature at which the material will ignite without a spark or flame being present. The method of measurement is given by ASTM E659 (Standard Test Method for Autoignition Temperature of Liquid Chemicals).

Also related to fire point, the flammability limit of vapor in air is an expression of the percent concentration in air (by volume), which is given for the lower and upper limits. These values give an indication of relative flammability. The limits are sometimes referred to as *lower explosive limit* (LEL) and *upper explosive limit* (UEL).

The *flash point* of petroleum or a petroleum product is the temperature to which the sample must be heated to produce a vapor/air mixture above the liquid fuel that is ignitable when exposed to an open flame under specified test conditions. In North America, flash point is used as an index of fire hazard.

Flash point is an extremely important factor in relation to the safety of spill cleanup operations. Gasoline and other low-boiling liquid fuels can be ignited under most ambient conditions and therefore pose a serious hazard when spilled. Many freshly spilled crude oils also have low flash points, until the lighter components have evaporated or dispersed.

There are several ASTM methods for measuring flash points (ASTM D56, ASTM D93, IP 34 are among the most commonly used). The minimum flash point that can be determined (ASTM D93, IP 34) is 10°C (50°F). One method (ASTM D56) is intended for liquids with a viscosity less than 9.5 cSt at 25°C (77°F). The flash point and fire point of lubricating oil are determined by a separate method (ASTM D92, IP 36).

2.2.6 Fractionation

Rather than quantifying a complex TPH mixture as a single number, petroleum fractionation methods break the mixture into discrete fractions, thus providing data that can be used in a risk assessment and in characterizing the product type and compositional changes such as may occur during refining and during weathering (oxidation) after a spill (Speight and Arjoon, 2012). The fractionation methods can be used to measure both the volatile constituents and the extractable constituents.

In contrast to the traditional methods for the analysis of petroleum products that report a single concentration number for complex mixtures, the fractionation methods report separate concentrations for discrete aliphatic and aromatic fractions of the product (Speight and Arjoon, 2012; Speight, 2014). The available petroleum fractionation methods are chromatography-based and are thus sensitive to a broad range of hydrocarbon constituents as well as to the constituents containing one or more polar (nonhydrocarbon) functions (Speight, 2014) . Identification and quantification of aliphatic and aromatic fractions allow identification of petroleum products and evaluation of the extent of contamination by the nonhydrocarbon constituents or the extent of participation in product petroleum behavior.

2.2.7 Metal Content

Metal content in crude oils can provide valuable information about the origin of those oils, potentially aiding in identifying the source of oil spills.

Crude oil assays often include nickel and vanadium contents due to the detrimental effects of these metals on catalysts used in cracking and desulfurization processes. In lubricating oils, metal contents can provide information on both the types of additives used in the oil and on the wear history of the equipment being lubricated. ASTM D5185—standard test method for the determination of additive elements, wear metals, and contaminants in used lubricating oils and determination of selected elements in base oils by inductively coupled plasma atomic emission spectrometry (ICP-AES)—can be used to determine over 20 different metals in a variety of petroleum products.

2.2.8 Pour Point

The pour point of petroleum or a petroleum product is the lowest temperature at which the oil will just flow, under standard test conditions (ASTM D97). The failure to flow at the pour point is usually attributed to the separation of waxes from the oil, but can also be due to the effect of viscosity in the case of very viscous oils. Also, particularly in the case of residual fuel oils, pour points may be influenced by the thermal history of the sample, that is, the degree and duration of heating and cooling to which the sample has been exposed.

From a spill response point of view, it must be emphasized that the tendency of the oil to flow will be influenced by the size and shape of the container, the head of the oil, and the physical structure of the solidified oil. The pour point of the oils is therefore an indication, and not an exact measure, of the temperature at which flow ceases (Dyroff, 1993).

2.2.9 Sulfur Content

The sulfur content of a crude oil is important for a number of reasons. Downstream processes such as catalytic cracking and refining will be adversely affected by high sulfur contents. If high-sulfur oils are burned, they produce high levels of sulfur dioxide, which can lead to acid deposition (acid rain) (Speight, 2005; Speight and Arjoon, 2012).

The total sulfur content of oil can be determined by numerous standard techniques. ASTM D129 (standard test method for sulfur in petroleum products) is applicable to petroleum products of low volatility and contains at least 0.1 mass percent sulfur. Sulfur contents are also determined in accordance with ASTM D4294 (standard test method for sulfur in petroleum products by energy-dispersive X-ray fluorescence spectroscopy). This method is applicable to both volatile and nonvolatile petroleum products with sulfur concentrations ranging from 0.05 to 5 mass percent.

2.2.10   Surface Tension and Interfacial Tension

Interfacial tension is the force of attraction between the molecules at the interface of two fluids. At the air/liquid interface, this force is often referred to as surface tension. The SI units for interfacial tension are milli-newtons per meter (mN/m). These are equivalent to the former units of dynes per centimeter (dyne/cm).

The surface tension of petroleum (or a petroleum product), together with its viscosity, affects the rate at which an oil spill spreads. Air/oil and oil/water interfacial tensions can be used to calculate a spreading coefficient, which gives an indication of the tendency for the oil to spread. It is defined as

$$\text{Spreading coefficient} = S_{WA} - S_{OA} - S_{WO},$$

where S_{WA} is water/air interfacial tension, S_{OA} is oil/air interfacial tension, and S_{WO} is water/oil interfacial tension.

Unlike density and viscosity, which show systematic variations with the temperature and degree of evaporation, interfacial tensions of crude oils and oil products show no such correlations.

A single test method (ASTM D971—Standard Test Method for Interfacial Tension of Oil against Water by the Ring Method) is applicable to the measurement of oil/water interfacial tensions. Unlike manually operated ring tensiometers, the maximum deformation of the lamella is detected electronically and occurs before the ring pulls completely through the interface. This results in interfacial tension that is somewhat lower than that measured manually.

2.2.11   Viscosity

Viscosity is a measure of a fluid's resistance to flow; the lower the viscosity of a fluid, the more easily it flows. Like density, viscosity is affected by temperature. As temperature decreases, viscosity increases.

Viscosity is a very important property of oils, because it affects the rate at which spilled oil will spread, the degree to which it will penetrate shoreline substrates, and the selection of mechanical spill countermeasures equipment.

Viscosity measurements may be absolute or relative (sometimes called "apparent"). Absolute viscosities are those measured by a standard method, with the results traceable to fundamental units. Absolute viscosities are distinguished from relative measurements made with instruments that measure viscous drag in a fluid, without known and/or uniform applied shear rates (Schramm, 1992). An important benefit of absolute viscometry is that the test results are independent of the particular type or make of the viscometer used. Absolute viscosity data can be compared easily between laboratories worldwide.

Modern rotational viscometers are capable of making absolute viscosity measurements for both Newtonian and non-Newtonian fluids at a variety of well-controlled, known, and/or uniform shear rates. Unfortunately, no ASTM standard method exists that makes use of these viscometers. Nonetheless, these instruments are in widespread use in many industries.

There are standard test methods for measuring the viscosity of oils, such as ASTM D445 and ASTM D4486, which make use of glass capillary kinematic viscometers and will produce absolute measurements in units of centistokes (cSt), only for oils that exhibit Newtonian flow behavior (viscosity independent of the rate of shear). Though now obsolete, at one time, the petroleum industry relied on measuring kinematic viscosity with the Saybolt viscometer and expressing kinematic viscosity in Saybolt universal seconds (SUS) or Saybolt Furol seconds (SFS). Occasionally, Saybolt viscosities are still reported in the literature, using the official equations relating SUS and SFS to kinematic viscosity (ASTM D2161).

2.2.12   Water Content

Some of the petroleum and petroleum products sampled contain substantial amounts of water. Because any process that would separate the oil and water would also change the composition of the oil, most properties are often determined on the oils as received. Therefore, for those oils with significant water contents (>5% w/w), many of the properties measured will not represent the properties of the *dry* oil. Generally, water content is determined by the Karl Fischer titration (ASTM D6304). The procedure is adopted in many official standards as the standard test method for water determination in petroleum and petroleum products.

Water content has not been conclusively identified as a direct governing environmental factor in the biodegradation of petroleum and petroleum products (Frijer *et al.*, 1996). However, an indirect effect of water content is that it influences the effective gas diffusion coefficient in soils. The rate of production of carbon dioxide is a means of expressing the mineralization rate of petroleum hydrocarbons (Speight, 2005; Speight and Arjoon, 2012).

2.3   CHROMATOGRAPHIC ANALYSES

Petroleum group analyses are conducted to determine the amounts of the petroleum compound classes (e.g. saturates, aromatics, and polars/resins) present in petroleum-contaminated samples. This type of measurement is sometimes used to identify the fuel type or to track plumes. It may be particularly useful for higher-boiling products, such as asphalt. Group-type test methods include multidimensional GC (not often used for environmental samples), high performance liquid chromatography (HPLC), and thin layer chromatography (TLC) (Miller, 2000; Patnaik, 2004; Speight, 2005; Speight and Arjoon, 2012).

Test methods that analyze individual compounds (e.g. benzene–toluene–ethylbenzene–xylene mixtures and polynuclear aromatic hydrocarbons) are generally applied to detect the presence of an additive or to provide the concentration data needed to estimate the environmental and health risks that are associated with individual compounds. Common constituent measurement techniques include GC with second column confirmation, GC with multiple selective detectors, and GC with mass spectrometry detection (GC/MS) (Speight, 2005; Speight and Arjoon, 2012).

2.3.1   Adsorption Chromatography

Adsorption is the bonding of molecules or particles to a surface. On the other hand, *absorption* is the filling of pores in a solid. The bonding to the surface is usually (but not always) weak and reversible. Compounds that contain functional groups are very often strongly adsorbed on activated carbon.

The composition of the higher molecular weight fraction and products (such as residual fuel oil and asphalt) is varied and is often reported in the form of four or five major fractions as deduced by adsorption chromatography—either the standard column method (Fig. 2.1) or a recycle method (Fig. 2.2) (Speight, 2001). In the case of cracked products, thermal decomposition products (such as carbenes and carboids) may also be present (Fig. 2.3) (Speight, 2014).

Column chromatography is used for several hydrocarbon-type analyses that involve fractionation of viscous oils (ASTM D2007, ASTM D2549), including residual fuel oil. The former method (ASTM D2007) advocates the use of adsorption on clay and clay–silica gel, followed by elution of the clay with pentane to separate the saturates, elution of clay with acetone–toluene to separate the polar compounds, and elution of the silica gel fraction with toluene to separate the aromatic compounds. The latter method (ASTM D2549) uses adsorption on a bauxite–silica gel column. Saturates are eluted with pentane; aromatics are eluted with ether, chloroform, and ethanol.

The most common industrial adsorbents are activated clay, carbon, silica gel, and alumina, because they present

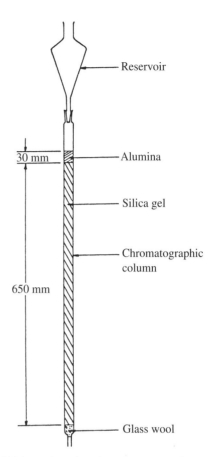

FIGURE 2.1   Adsorption chromatography using a standard column.

enormous surface areas per unit weight. Clay is a naturally occurring mineral. Thermal treatment (*roasting*) of organic material to decompose it to granules of carbon produces carbon black or activated carbon (Chapter 17). Coconut shell, wood, and bone are common sources of activated carbon. Silica gel is a matrix of hydrated silicon dioxide. Alumina is mined or precipitated aluminum oxide and hydroxide.

Temperature effects on adsorption are profound, and the measurements are usually at a constant temperature. Graphs of the data are called *isotherms*. Most steps using adsorbents have little variation in temperature.

2.3.2   Gas Chromatography

Gas chromatography, which is based on the principle of a stationary phase and a mobile phase, remains a primary technique for determining hydrocarbon distribution in petroleum products through the identification of individual hydrocarbons. Though a measure of purity by GC is often sufficient for many purposes, it is not always sufficient for measuring absolute purity—not all possible impurities will pass through the chromatographic column, and not all those that do will be measured by the detector. Absolute purity is best measured by distillation range, or freeze or solidification points. Despite this disadvantage, the technique is still widely used, and is the basis of many current standard test methods for the determination and measurement of hydrocarbons in

petroleum products. When classes of hydrocarbons such as olefins need to be measured, techniques such as bromine index are used (ASTM D1492, ASTM D2710, ASTM D5776).

Briefly, *gas–liquid chromatography* (GLC) is a method for separating the volatile components of various mixtures (Fowlis, 1995; Grob, 1995). It is, in fact, a highly efficient fractionating technique, and it is ideally suited to the quantitative analysis of mixtures when the possible components are known, and the interest lies only in determining the amounts of each present. In this type of application, GC has taken over much of the work previously done by the other techniques; it is now the preferred technique for the analysis of hydrocarbon gases, and gas chromatographic in-line monitors are having increasing application in refinery plant control. Gas–liquid chromatography is also used extensively for individual component identification, as well as percentage composition, in the gasoline boiling range.

The mobile phase is the carrier gas, and the gas selected has a bearing on the resolution. Nitrogen has very poor resolution ability, while helium and/or hydrogen are better choices, with hydrogen being the best carrier gas for resolution. However, hydrogen is reactive and may not be compatible with all sets of target analytes. There is an optimum flow rate for each carrier gas to achieve maximum resolution. As the temperature of the oven increases, the flow rate of the gas changes due to thermal expansion of the gas. Most modern gas chromatographs are equipped with constant flow devices that change the gas valve settings, as the temperature in the oven changes; so, changing flow rates are no longer a concern. Once the flow is optimized at one temperature, it is optimized for all temperatures.

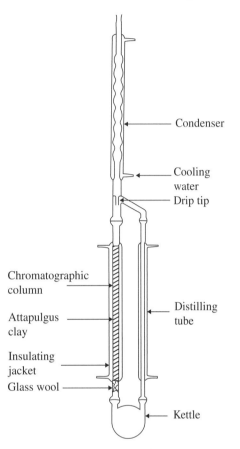

FIGURE 2.2   Adsorption chromatography using a recycle column.

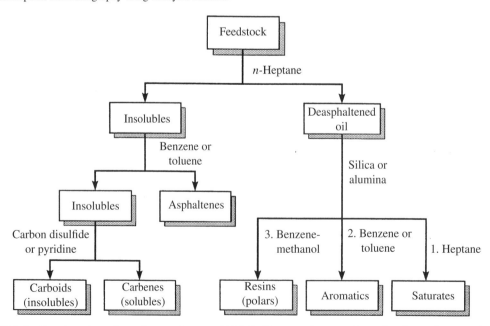

FIGURE 2.3   Fractionation scheme for petroleum and high-boiling petroleum products showing carbenes and carboids.

2.3.3 Gas Chromatography–Mass Spectrometry

A gas chromatography–mass spectrometry system is used to measure concentrations of target volatile and semivolatile petroleum constituents. It is not typically used to measure the amount of TPH. The advantage of the technique is the high selectivity, or the ability to confirm compound identity through retention time and unique spectral pattern. The method is used for identification and quantification of the constituents of petroleum fractions and petroleum products.

To reduce the possibility of false-positives, the intensities of one to three selected ions are compared to the intensity of a unique target ion of the same spectrum. The sample ratios are compared to the ratios of a standard. If the sample ratios fall within a certain range of the standard, and if the retention time matches the standard within specifications, the analyte is considered present. Quantification is performed by integrating the response of the target ion only.

Mass spectrometers are among the most selective detectors, but they are still susceptible to interferences. Isomers have identical spectra, while many other compounds have similar mass spectra. Heavy petroleum products can contain thousands of major components that are not resolved by the gas chromatograph. As a result, multiple compounds are simultaneously entering the mass spectrometer. Different compounds may share many of the same ions, confusing the identification process. The probability of misidentification is high in complex mixtures such as petroleum products.

2.3.4 High-Performance Liquid Chromatography

A HPLC system can be used to measure the concentrations of target semivolatile and nonvolatile petroleum constituents. The system only requires that the sample be dissolved in a solvent compatible with those used in the separation. The detector most often used in petroleum environmental analysis is the fluorescence detector. These detectors are particularly sensitive to aromatic molecules, especially polynuclear aromatic hydrocarbons. An ultraviolet (UV) detector may be used to measure compounds that do not fluoresce.

In the method, polynuclear aromatic hydrocarbons are extracted from the sample matrix with a suitable solvent, which is then injected into the chromatographic system. Usually, the extract must be filtered, because fine particulate matter can collect on the inlet frit of the column, resulting in high back pressures and eventual plugging of the column. For most hydrocarbon analyses, reverse-phase HPLC (i.e. using a nonpolar column packing with a more polar mobile phase) is used. The most common bonded phase is the octadecyl (C18) phase. The mobile phase is commonly aqueous mixtures of either acetonitrile or methanol.

After the chromatographic separation, the analytes flow through the cell of the detector. A fluorescence detector shines light of a particular wavelength (the excitation wavelength) into the cell. Fluorescent compounds absorb light and reemit light of other, higher wavelengths (emission wavelengths). The emission wavelengths of a molecule are mainly determined by its structure. For polynuclear aromatic hydrocarbons, the emission wavelengths are mainly determined by the arrangement of the rings and vary greatly between isomers.

Some of the polynuclear aromatic hydrocarbons (such as phenanthrene, pyrene, and benzo(g,h,i)perylene) are commonly seen in products boiling in the middle to heavy distillate range. In a method for their detection and analysis (EPA 8310), an octadecyl column and an aqueous acetonitrile mobile phase are used. Analytes are excited at 280 nm and detected at emission wavelengths of greater than 389 nm. Naphthalene, acenaphthene, and fluorene must be detected by a less-sensitive UV detector, because they emit light at wavelengths below 389 nm. Acenaphthylene is also detected by UV detector.

The methods using fluorescence detection will measure any compounds that elute in the appropriate retention time range and which fluoresce at the targeted emission wavelength(s) (Falla Sotelo *et al.*, 2008). In the case of one method (EPA 8310), the excitation wavelength excites most aromatic compounds. These include the target compounds and also many aromatics derivatives, such as alkyl aromatics, phenols, anilines, and heterocyclic aromatic compounds containing the pyrrole (indole, carbazole, etc.), pyridine (quinoline, acridine, etc.), furan (benzofuran, naphthofuran, etc.), and thiophene (benzothiophene, naphthothiophene, etc.) structures. In petroleum samples, alkyl polynuclear aromatic hydrocarbons are strong interfering compounds. For example, there are 5 methylphenanthrenes and over 20 dimethylphenanthrenes. The alkyl substitution does not significantly affect either the wavelengths or intensity of the phenanthrene fluorescence. For a very long time after the retention time of phenanthrene, the alkylphenanthrene derivatives will interfere, affecting the measurements of all later-eluting target polynuclear aromatic hydrocarbons.

Interfering compounds will vary considerably from source to source, and samples may require a variety of cleanup steps to reach the required method detection limits. The emission wavelengths used (EPA 8310) are not optimal for the sensitivity of the small ring compounds. With modern electronically controlled monochromators, wavelength programs can be used, which tune the excitation and emission wavelengths to maximize sensitivity and/or selectivity for a specific analyte in its retention time window.

2.3.5 Thin Layer Chromatography

In the environmental field, TLC is best used for screening analyses and characterization of semivolatile and nonvolatile petroleum products. The precision and accuracy of the technique are inferior to other methods (Speight, 2005; Speight and Arjoon, 2012), but when speed and simplicity are desired, TLC may be a suitable alternative. For characterizations of petroleum products such as asphalt, the method has the advantage of separating compounds that are too high boiling to pass through a gas chromatograph. While TLC does not have the resolving power of a gas chromatograph, it is able to separate different classes of compounds. Thin layer chromatography analysis is fairly simple and, since the method does not give highly accurate or precise results, there is no need to perform the highest quality extractions.

In the method, soil samples are extracted by shaking or vortexing with the solvent. Water samples are extracted by shaking in a separatory funnel. If there is the potential for the presence of compounds that interfere with the method and make the data suspect, silica gel can be added to clean the extract. Sample extract aliquots are placed close to the bottom of a glass plate coated with a stationary phase. The most widely used stationary phases are made of an organic hydrocarbon moiety bonded to a silica backbone.

For the analysis of petroleum hydrocarbons, a moderately polar material stationary phase works well. The plate is placed in a sealed chamber with a solvent (mobile phase). The solvent travels up the plate carrying compounds present in the sample. The distance a compound travels is a function of the affinity of the compound to the stationary phase relative to the mobile phase. Compounds with chemical structure and polarity similar to the solvent pass readily into the mobile phase. For example, the saturated hydrocarbons seen in diesel fuel travel readily up a plate in a hexane mobile phase. Polar compounds such as ketones or alcohols travel a smaller distance in hexane than saturated hydrocarbons.

After a plate has been exposed to the mobile phase solvent for the required time, the compounds present can be viewed by several methods. Polynuclear aromatic hydrocarbons, other compounds with conjugated systems, and compounds containing heteroatoms (nitrogen, oxygen, or sulfur) can be viewed with long-wave and short-wave UV light. The unaided eye can see other material or plates can be developed using iodine, which has an affinity for most petroleum compounds, including the saturated hydrocarbons, and stains the compounds a reddish/brown color.

The method is considered to be a qualitative and useful tool for rapid sample screening. Limitations of the method center on its moderate reproducibility, detection limits, and resolving capabilities. Variability between operators can be as high as 30%. Detection limits (without any concentration of the sample extract) are near 50 ppm (mg/kg) for most petroleum products in soils. When the aromatic content of a sample is high, as with bunker C fuel oil, the detection limit can be near 100 ppm. It is often not possible to distinguish between similar products such as diesel and jet fuel. As with all chemical analyses, quality assurance tests should be run to verify the accuracy and precision of the method.

2.4   SPECTROSCOPIC ANALYSES

The chemical composition of a feedstock has always been considered to be a valuable indicator of refining behavior. Whether or not it is the ultimate indicator of refining behavior remains to be seen and is audience-dependent. More than likely, chemical composition studies truly complement physical property and physical behavior studies, and the true picture is a combination of all of these studies.

However, the chemical composition of a feedstock is represented in terms of compound types and/or in terms of generic compound classes, thus allowing the analytical chemist, the process chemist, the process engineer, and the refiner to determine the nature of the reactions. Hence, chemical composition can play a large part in determining the nature of the products that arise from the refining operations. It can also play a role in determining the means by which a particular feedstock should be processed (Wallace et al., 1988; Speight, 2000). However, proper interpretation of the data resulting from the composition studies requires an understanding of chemical structures, their significance, and an open mind.

The physical and chemical characteristics of crude oils and the yields and properties of products or factions prepared from them vary considerably and are dependent on the concentration of the various types of hydrocarbons and minor constituents present. Some types of petroleum have economic advantages as sources of fuels and lubricants with highly restrictive characteristics, because they require less specialized processing than that needed for the production of the same products from many types of crude oil. Others may contain unusually low concentrations of components that are desirable fuel or lubricant constituents, and the production of these products from such crude oils may not be economically feasible.

Spectroscopic studies have played an important role in the evaluation of petroleum and of petroleum products for the last three decades, and many of the methods are now used as standard methods of analysis for refinery feedstocks and products. Application of these methods to feedstocks and products is a natural consequence for the refiner.

The methods include the use of *mass spectrometry* to determine the (i) hydrocarbon types in middle distillates (ASTM D2425); (ii) hydrocarbon types of gas oil saturate fractions (ASTM D2786); (iii) hydrocarbon types in low-olefin gasoline (ASTM D2789); and (iv) aromatic types of gas oil aromatic fractions (ASTM D3239). *Nuclear magnetic resonance spectroscopy* (NMR) has been developed as a standard method for the determination of hydrogen types in aviation turbine fuels (ASTM D3701). *X-ray fluorescence spectrometry* has been applied to the determination of the analysis of selected elements (nitrogen, sulfur, nickel, and vanadium) as well as to the determination of sulfur in various petroleum products (ASTM D2622, ASTM D4294).

Infrared spectroscopy is used for the determination of benzene in motor and/or aviation gasoline, while UV spectroscopy is employed for the evaluation of mineral oils (ASTM D2269) and for determining the naphthalene content of aviation turbine fuels (ASTM D1840).

Other techniques include the use of *flame emission spectroscopy* for determining trace metals in gas turbine fuels (ASTM D3605) and the use of *absorption spectrophotometry* for the determination of the alkyl nitrate content of diesel fuel (ASTM D4046). *Atomic absorption* has been employed as a means of the analysis of metals (ASTM D1971, ASTM D4698, ASTM D5056), for the manganese content of gasoline (ASTM D3831), as well as for determining the barium, calcium, magnesium, and zinc contents of lubricating oils (ASTM D4628). *Flame photometry* has been employed as a means of measuring the lithium/sodium content of lubricating greases and the sodium content of residual fuel oil (ASTM D1318).

Nowhere is the contribution of spectroscopic studies more emphatic than in the application to the delineation of structural types in the heavier feedstocks. This has been necessary because of the unknown nature of these feedstocks by refiners. One particular example is the *n-d-M method* (ASTM D3238) that is designed for the carbon distribution and structural group analysis of petroleum oils. Later investigators have taken structural group analysis several steps further than the n-d-M method.

It is also appropriate at this point to give a brief description of the other methods that are used for the identification of the constituents of petroleum. Though useful information about the composition of high–molecular weight petroleum fractions is provided by the right combination of separation and analytical methods, spectroscopic methods can also be applied to the problems of characterization and identification. For example, information about the nature of polar functional groups or the elucidation of the way vanadium and nickel are bound to the molecules is available from a variety of spectroscopy techniques, as well as by the use of a variety of chemical techniques (Speight, 2001, 2014).

It is not intended to convey here that any one of these methods can be used for complete characterization and identification of the high–molecular weight petroleum constituents. Even though any one of these methods may fall short of complete acceptability as a method for the characterization of individual constituents of feedstocks, they can be used as methods by which an overall evaluation of the feedstock may be obtained in terms of molecular types. This is especially true when the methods are used in conjunction with each other.

2.4.1 Infrared Spectroscopy

Infrared spectroscopy is a well-established method that was developed for comparative, semiquantitative analysis, leading to the present quantitative analysis (ASTM E168,

ASTM E204, ASTM E334, ASTM E1252, IP 429). Infrared spectra are displayed either as percent transmittance or as absorptivity versus frequency (cm^{-1}). Transmittance, T, defined as the ratio of transmitted light over incident light, or percent transmittance, usually shows more detail over the entire range, and is generally the preferred display. Absorbance, A, on the other hand, is proportional to the concentration and is, therefore, used for quantitative measurements.

Infrared spectroscopy is a simple procedure, and is one of several techniques that can provide quick information about the distribution of several structural and functional groups (Drews, 1998; Nadkarni, 2000; Rand, 2003; Totten 2003). In combination with NMR spectroscopy, it will provide quick, yet fairly, detail on the distribution of carbon–hydrogen groups. However, in the context of high–molecular weight petroleum fractions, conventional infrared spectroscopy yields information about the functional features of various petroleum constituents. For example, infrared spectroscopy will aid in the identification of imino functions ($=N-H$) and hydroxyl ($-O-H$) functions, as well as the nature of the various carbonyl ($-C=O$) functions.

In the older, infrared spectroscopy, light is refracted by a prism or a grating and scanned by a moving slit that takes several minutes for one measurement. In Fourier transform infrared (FTIR) spectroscopy, the entire spectrum is obtained by an interferometer in a fraction of a second. Thus, several hundred measurements can be taken in a matter of minutes and averaged by computer. This multiplexing leads to greatly increased (about 100-fold) sensitivity and precision over those achievable with dispersive instruments.

Thus, with the recent progress of *FTIR spectroscopy*, quantitative estimates of the various functional groups can also be made. This is particularly important for application to the higher molecular weight solid constituents of petroleum (i.e. the asphaltene fraction) and for group-type analysis.

It is also possible to derive structural parameters from the infrared spectroscopic data, and these are (i) saturated hydrogen to saturated carbon ratio; (ii) paraffinic character; (iii) naphthenic character; (iv) methyl group content; and (v) paraffin chain length.

In conjunction with proton magnetic resonance (PMR; see next section), structural parameters such as the fraction of paraffinic methyl groups to aromatic methyl groups can be obtained.

The newer diffuse reflectance infrared (DRIR) techniques seem to give equally good spectra as obtained in conventional ways from solutions in cells or from potassium bromide (KBr) pellets. Another synonym for this technique is DRIFT (diffuse reflectance infrared Fourier transform).

The sample is deposited from a solution onto finely ground KBr in a small cup that is placed into a diffuse reflectance accessory, after removal of the solvent, in a vacuum oven. A related technique, variable angle specular reflectance, allows rotation of the sample holder for optimization.

Increased resolution by band-narrowing techniques in conjunction with reference spectra allows the distinction of methylene (CH_2) groups next to other groups such as alkyls, aromatic rings, carbonyls, or alkoxy groups. Carbon–hydrogen (CH_n) groups and aromatic carbon can be identified, measured, and compared. The ratio of the band at $1602\,cm^{-1}$ (aromatic C—C stretching) to that at $2920\,cm^{-1}$ (aliphatic hydrogen–carbon stretching) in methylene groups (β or further away from the aromatic ring) is a good relative measure for the aromaticity of a sample.

Fourier transform infrared spectroscopy is sufficiently sensitive that it can be used for detection in HPLC and even in GC and supercritical fluid chromatography (SFC).

Band assignments (Table 7.1) for infrared spectra were established several decades ago. Resolution enhancement by Fourier self-deconvolution (band-narrowing) led to the recognition and distinction of bands that could previously not be separated. Examples of such new assignments are methyl (CH_3) groups next to aromatic rings ($2948\,cm^{-1}$) and next to alkyl groups ($2960\,cm^{-1}$); methylene (CH_2) groups (together with some methyl groups) next to aromatic rings ($2916\,cm^{-1}$) and in aliphatic chains ($2926\,cm^{-1}$); methylene groups next to an alkoxy function ($2928\,cm^{-1}$) and next to a carbonyl group ($2933\,cm^{-1}$); and methane (CH) groups in unspecified environments (2905 and $1897\,cm^{-1}$). The rocking vibration band of methylene groups in straight aliphatic chains varies slightly with their length. For *n*-pentyl groups, it is located at $726\,cm^{-1}$, and with longer chains, it shifts to lower frequencies until it stabilizes at $720\,cm^{-1}$ for chains with more than nine carbon atoms.

In the field of the analysis of the high–molecular weight petroleum constituents, infrared spectroscopy has been mostly employed for measuring oxygen- and nitrogen-containing groups and for evaluating shifts in certain bands due to aggregation or other interactions such as carboxylic acid derivatives, phenol derivatives, carbazole derivatives, cyclic amides, as well as pyridine and acridine derivatives among the bases.

In the lower-boiling petroleum fractions, parts of the hydrocarbon skeleton can be assessed by infrared spectroscopy. Specifically, the alkyl substitution of aromatic rings can be determined from the out-of-plane carbon–hydrogen deformation bands.

2.4.2   Mass Spectrometry

Mass spectrometry furnishes the molecular weight and chemical formula of compounds and their relative amounts in a mixture and offers nondestructive examination of the sample (ASTM D2425, ASTM D2650, ASTM D2786, ASTM D2789, ASTM D3239, ASTM E1316). The technique can also provide important information on their molecular structure. The earliest and the most common type of mass spectrometry, electron impact mass spectrometry (EI-MS), gives a fragmentation pattern displaying both parent ion peaks and fragment ion peaks, characteristic of each molecular type.

Fragmentation is frequently employed to differentiate between isomers of pure compounds and of molecules in fairly simple mixtures. However, it is usually avoided with such complex samples as high–molecular weight petroleum fractions, which are composed of such a multitude of closely related compounds that their fragmentation patterns are nondistinctive and cannot be readily interpreted.

Mass spectrometry can play a key role in the identification of the constituents of feedstocks and products, either in the laboratory or online (ASTM D2425; ASTM D2786; ASTM D2789; ASTM D3239). The principal advantages of mass spectrometric methods are (i) high reproducibility of quantitative analyses; (ii) the potential for obtaining detailed data on the individual components and/or carbon number homologues in complex mixtures; and (iii) a minimal sample size required for analysis. The ability of mass spectrometry to identify individual components in complex mixtures is unmatched by any modern analytical technique. Perhaps, the exception is GC.

However, there are disadvantages arising from the use of mass spectrometry, and these are (i) the limitation of the method to organic materials that are volatile and stable at temperatures up to 300°C (570°F); and (ii) the difficulty in separating isomers for absolute identification. The sample is usually destroyed, but this is seldom a disadvantage.

Nevertheless, in spite of these limitations, mass spectrometry does furnish useful information about the composition of feedstocks and products, even if this information is not as exhaustive as might be required. There are structural similarities that might hinder identification of individual components. Consequently, identification by type or by homologue will be more meaningful, since similar structural types may be presumed to behave similarly in processing situations. Knowledge of the individual isomeric distribution may add only a little to an understanding of the relationships between composition and processing parameters.

Mass spectrometry should be used discriminately, where a maximum amount of information can be expected. The heavier nonvolatile feedstocks are for practical purposes, beyond the useful range of routine mass spectrometry. At the elevated temperatures necessary to encourage volatility, thermal decomposition will occur in the inlet, and any subsequent analysis would be biased to the low–molecular weight end and to the lower molecular products produced by the thermal decomposition.

High-voltage electron impact mass spectrometry (HVEI-MS or EI-MS) can cause repeated fragmentation of daughter ions. Considering the repeated fragmentation of even the simplest hydrocarbons and the fact that the higher molecular weight fractions of petroleum contain such a broad range of molecular weight species (Speight, 2001, 2014), the patterns obtained from petroleum fractions are so complex as to almost defy interpretation. Thus, nonfragmenting mass spectrometric (NF-MS) methods are preferred.

For high–molecular weight petroleum fractions, the use of NF-MS methods is now preferred. These methods, also called *soft ionization* methods, produce predominantly parent ion (molecular ion) peaks and, thus, much simpler spectra than the methods producing fragments. By rendering the molecular weight of each compound in a sample, and reasonably well its abundance, NF-MS can also be used to determine the molecular weight distribution, that is, the molar mass profile of a sample.

Indeed, the great advantage of *NF-MS* is the relative simplicity of the spectra. A disadvantage is the relatively low signal intensity of the ions. With *low-voltage electron impact ionization*, the number of parent ions formed increases rapidly as the ionizing voltage increases above the ionizing potential of the molecules in the sample. Thus, higher voltage gives a more intense signal (up to about 20–40 eV, depending on the compound type). However, the higher ionizing voltage transfers more energy to the sample molecules and causes fragmentation to increase. The surplus energy transferred to the molecule in excess of the ionizing potential is equilibrated and dissipated in various ways, for example, by increasing the internal energy of the molecule or by breaking the atomic bonds.

Thus, the challenge in *NF-MS* is to maximize the number of parent ions and, at the same time, maintain the number of fragment ions at acceptable levels. Paraffins are highly susceptible to fragmentation; an ionizing voltage high enough to generate a good parent ion spectrum also breaks many (paraffinic) carbon–carbon bonds, producing a significant number of fragment ions.

Among the most important NF-MS techniques are field ionization mass spectrometry (FI-MS), field desorption mass spectrometry (FD-MS), chemical ionization mass spectrometry (CI-MS), and low-voltage (10–20 eV) electron impact mass spectrometry (LVEI-MS). A particularly powerful version of electron impact mass spectrometry is high-resolution electron impact mass spectrometry (HR-LEVI-MS or LVRH-MS). All of these are designed to generate *cold* ions of such low excess energy that they do not undergo fragmentation to any great extent, in contrast to the *hot* ions generated by conventional electron impact ionization (70–100 eV). With such ionization methods, a compromise must be found between low fragmentation and sufficiently high sensitivity. Unfortunately, samples rich in aliphatic hydrocarbons are hard to keep completely from fragmenting, but most other compound types, especially aromatic compounds, yield clean parent ion spectra.

Field ionization mass spectrometry shows extremely low fragmentation, because the ionization process imparts little

excess energy to the formed ion. This is crucial when ionizing saturates. Both low-voltage electron impact (LVEI) and chemical ionization (CI) transfer more excess energy than FI-MS, leading to higher levels of fragmentation with saturates. As long as the analyte is volatile, the FI-MS experiment is not sample-limited. That is, additional sample can be vaporized as needed to obtain the required signal-to-noise ratio. Field desorption mass spectrometry can be considered a special form of EI-MS, where the sample material is placed on the ionizing surface of the probe before insertion into the sample chamber of the mass spectrometer. It is, however, sample-limited and subject to experimental artifacts not normally present in EI-MS.

In the EI-MS technique, the emitting surface is a cathode, consisting of an array of sharp tips on a support such as a wire, a grid, or a razor blade. This surface is located very close to the anode. The combination of an extremely small radius of curvature of the tips, the short distance, and the high electric field creates a very high field gradient. The sample is introduced to this field gradient by evaporation and diffusion from a heated surface nearby.

Field desorption mass spectrometry is a special form of FI-MS. The cathode is the same, but here it is coated with the sample before it is placed in the mass spectrometer source. This eliminates the need to evaporate the sample and any losses on transport to the cathode. Instead, the ions are desorbed directly from the solid sample. The emitter is heated to the point where the sample melts and can be drawn by the surface tension to the tip where the molecules are converted to ions. As the ions are ejected from the surface into the mass spectrometer, more sample molecules can replace them at the tip. In this way, labile molecules can be handled without significant fragmentation, and even conventionally nonvolatile samples can be run. Thus, the method is especially suitable for the high–molecular weight fractions of petroleum residua. Because of the low thermal energy to which the sample molecules are exposed, the total excess energy in this process is minimal. Thus, FD-MS is an even milder method than FI-MS, and produces less fragmentation.

Chemical ionization mass spectrometry (CI-MS), another *soft* method, has been used only occasionally for petroleum samples. In the method, a relatively high-pressure (0.1–1 Torr) reagent gas (e.g. methane, *n*-butane, *iso*-butane) is ionized and allowed to contact the sample. At such pressures, the reagent ions undergo many collisions that help them lose excess energy left after ionization. The ions also exchange charges during collision with one another and with the sample molecules. The latter is the desired reaction, and the method works best when the sample molecules retain the charge better than the reagent molecules.

Fast atom bombardment mass spectrometry (FAB-MS) is a method in which the sample is dissolved in a high-boiling polar liquid matrix (such as glycerol or triethanolamine) and spread on a plate. Rare gas molecules (e.g. xenon or argon) are ionized, accelerated, focused at the sample, and stripped of their charge before they hit their target. The impact of uncharged atoms sputters small amounts of the matrix and sample off the sample plate. The liquid matrix evaporates, liberating charged sample ions, which are now propelled by the accelerating voltage into the mass spectrometer.

The method produces substantial amounts of fragments and rarely works for neutral compounds—two disadvantages for our purpose. Other problems and interferences with FAB-MS are that the matrix material is ionized along with the sample, and that both the sample and the matrix material may form cluster ions that complicate the spectrum.

For high-resolution mass spectrometers, precise standardization of operating conditions is even more important for quantitative measurements than for regular mass spectrometers. Generally, the higher the resolution, the weaker is the signal and the longer is the data acquisition time. Special mass standards, mostly chlorinated hydrocarbon petroleum fractions, are employed for mass calibration. The parent petroleum fractions guarantee a complete mass sequence that is needed for the large number of hetero-compound peaks. Even with such a standard, there will still be several hetero-compound peaks between each pair of hydrocarbon peaks. The chloro-compounds are preferred because of their higher sensitivities at the low ionization voltage, compared to those of the fluorinated standards, common with the regular EI-MS.

Gas chromatography coupled with mass spectrometry (GC-MS) is a powerful method for petroleum distillates. Heavy petroleum fractions (345–450°C, 650–850°F) are usually first separated by liquid chromatography into compound class fractions before the application of GC that separates the sample by boiling point. Mass spectrometry, used as a detection method, gives the molecular weight and, with fragmenting ionization, the compound type of each peak.

Petroleum samples, boiling higher than about 450°C (850°C), other than well-separated narrow fractions, are too complex and high-boiling for GC. In these cases, NF-MS should be combined with liquid chromatography, instead of GC. However, higher-boiling samples (>650°C, >1200°F) may not be seen with the same sensitivity as lower-boiling constituents by most mass spectrometric methods. For these cases, mass spectrometric methods, not limited by volatility, are now available. For example, FD-MS, laser desorption (LS), fast atom bombardment (FAB), and electrospray (ES) techniques can be applied to very large (polar) molecules, even of several thousand molecular weight. However, because FAB and ES work only with highly polar, easily-to-ionize compounds, their application to petroleum fractions may be limited.

One of the applications of mass spectrometry is group-type mass spectrometry. The method uses fragment patterns and sets of empirical relations to sort the contributions from different molecules to certain selected peaks and to

reconstruct the molecular distribution from these peak areas. One test method (ASTM D2786) measures the concentration of eight compound types—paraffins, naphthenes with one to six rings, and mono-aromatics in saturated fractions. The mass spectrometric group-type method (ASTM D3239) is used to determine the aromatic compound types present in chromatographic fractions of aromatics separated from vacuum gas oils. Other ASTM group-type methods determine the various aliphatic and aromatic compound types in middle distillates (ASTM D2425).

Mass spectrometric group-type methods are fast, relatively cheap, and require only a few milligrams of sample. However, they have their limits. They are restricted to samples, free of olefins and, ordinarily, with less than 2–5% w/w sulfur, nitrogen, or oxygen compounds, and in some cases, with much more stringent limits. Other restrictions, for example, to certain boiling ranges (and molecular weight ranges) also apply, depending on the method. Furthermore, only those compound types are considered in the calculation, for which results are given in the output. If any other compound types are present in the sample to any large degree, which may easily be the case, major errors may result.

The mass spectrometric group-type analysis makes use of average fragmentation patterns specific to the different compound types, that is, to certain groups of molecular structures. But, the masses, as well as the abundances, of fragment ions are characteristic of the specific parent molecule. The sum of all of the peaks gives the average distribution of the parent molecules. Generally, a limited set of peaks of a molecule is selected for evaluation. Calibration ensures that the system is operating in a reproducible fashion and the response factor matrix approach is used for quantitative measurement and to compensate for overlaps in fragmentation.

Selected fragment peak intensities within the homologous series of compound types are added together, and the sums are evaluated with a matrix of response factors to give the concentration of each compound type in the sample. Mass spectrometric group-type analyses can be expected to have errors in the order of up to 20%. However, *relative* changes are ordinarily reported with better accuracy.

2.4.3 Nuclear Magnetic Resonance

Nuclear magnetic resonance has frequently been employed for general studies of hydrogen types in petroleum and petroleum products (ASTM D4808, IP 392) as well as for the structural studies of petroleum constituents (ASTM E386; see also Speight, 1994). The technique has recently been adapted to measure the hydrogen content of fuels and other petroleum products (ASTM D3701, ASTM D4808). In fact, PMR studies (along with infrared spectroscopic studies) were, perhaps, the first studies of the modern era that allowed structural inferences to be made about the polynuclear aromatic systems that occur in the high–molecular weight constituents of petroleum.

Thus, NMR methods have gained a prominent place in the compositional and structural analyses of petroleum fractions (Speight, 1994). In its basic applications, NMR is fast and relatively inexpensive. Because of its convenience, speed, and greater wealth of detailed information, particularly from ^{13}C-NMR, it has displaced the n-d-M and related methods in most laboratories (Speight, 2001, 2014). Nuclear magnetic resonance presents a direct measure of the aromatic and aliphatic carbon as well as hydrogen distribution. Beyond these results, both carbon and hydrogen in various structural groupings in a molecule can be determined. Thus, proton (^{1}H) and carbon-13 (^{13}C) nuclei are the most common ones used in NMR spectroscopy; nitrogen (^{15}N and ^{14}N) and sulfur (^{33}S) have been employed on occasion with petroleum fractions for special applications.

Proton magnetic resonance has been widely employed in the structural analysis of petroleum fractions. It is a relatively inexpensive technique that allows measurement of hydrogen atoms in aromatic and aliphatic groups, even allowing differentiation between hydrogen attached next to an aromatic ring (α-position) and those farther removed from the ring. Atoms in single-ring and multi-ring aromatic compounds as well as those in olefin locations can also be identified.

Only a small amount (<10 mg) of the sample is required, dissolved in a solvent such as deuterochloroform, contained in a glass tube of 5 mm diameter, and placed in a highly homogeneous magnetic field, where it is surrounded by one or more coils. The coils serve to subject the sample to a weak radio-frequency (RF) field. The hydrogen nuclei of the sample can be visualized as magnets, and when the RF is equal to the processing frequency, resonance occurs between the two, and the spin resonance is detected by a receiver coil. The position of a sample resonance with that of tetramethylsilane (TMS) as a reference difference is reported as *chemical shift*, δ, which is a dimensional number that is expressed in terms of parts per million (ppm) difference from the reference (TMS).

Quantitative accuracy of PMR for aromatic and aliphatic hydrogen is about 1% for distillates and 2–3% for residua; that for the distinction of aliphatic hydrogen atoms attached to an aromatic ring from those β and farther away from the ring is somewhat lower. Methyl (CH_3), methylene (CH_2), and methine (CH) hydrogen can ordinarily not be distinguished, except for methyl hydrogen γ or those farther away from the aromatic rings. Even this methyl peak is sometimes difficult to quantify, because of the interference by naphthenic methine and methylene hydrogen.

Protons attached to single-ring and multi-ring aromatic compounds can usually be distinguished with reasonable accuracy, especially when the sample concentration is 2% or less. At such low concentrations, the dividing line between these protons is at 7.25 ppm. In the spectra of samples with high boiling point, much of the detail is lost, and differentiation between group-types is difficult.

Thus, in general, the proton (hydrogen) types in petroleum fractions can be subdivided into three types: (1) aromatic ring hydrogen, (2), aliphatic hydrogen adjacent to an aromatic ring, and (3) aliphatic hydrogen, remote from an aromatic ring. In other cases, five types of hydrogen locations are identified: (1) aromatic hydrogen, (2) substituted hydrogen next to an aromatic ring, (3) naphthene hydrogen, (4) methylene hydrogen, and (5) terminal methyl hydrogen, remote from an aromatic ring. Other ratios are also derived from which a series of structural parameters can be calculated.

However, it must be remembered that the structural details of the carbon backbone obtained from proton spectra are derived by inference, but it must be recognized that protons at peripheral positions can be obscured by intermolecular interactions. This, of course, can cause errors in the ratios that can have a substantial influence on the outcome of the calculations (Ebert *et al.*, 1987; Ebert, 1990).

It is in this regard that *carbon-13 magnetic resonance* (CMR) can play a useful role. Since carbon magnetic resonance deals with analyzing the carbon distribution types, the obvious structural parameter to be determined is the aromaticity, f_a. A direct determination from the various carbon-type environments is one of the better methods for the determination of aromaticity. Thus, through a combination of proton and carbon magnetic resonance techniques, refinements can be made on the structural parameters and, for the solid-state high-resolution CMR technique, additional structural parameters can be obtained.

The basic instrumentation for carbon-13 magnetic resonance is the same as that for PMR, except that there are two RF fields orthogonal (at right angles) to the main magnetic field, one for observing the carbon-13 nuclei and the other for decoupling the proton nuclei. The low abundance of carbon-13 isotopes (1.1%) and the lower gyromagnetic ratio of the carbon-13 nucleus make the signal weaker by more than two orders of magnitude, and, moreover, the nuclei have longer relaxation times. The effect is that, even with Fourier transform (FT) data acquisition, carbon-13 magnetic resonance measurements can take hours to perform.

In contrast to PMR spectroscopy, the peak areas arising from the carbon-13 nuclei in different molecular positions ordinarily are not proportional to their concentration. Quantitative measurements require that two effects must be overcome: (1) the different relaxation times of the carbon-13 nuclei in different chemical groups, and (2) the nuclear Overhauser enhancement (NOE). The latter effect refers to the rise in signal intensity when C–H coupled protons are saturated by the decoupling field.

One way to do this is to add a small amount of a paramagnetic relaxation reagent, such as trisacetylacetonatochromium

(III), $Cr(AcAc)_3$, which changes the dominant relaxation mechanism into one involving the interaction between unpaired electrons and ^{13}C nuclei. It also reduces the long relaxation times of some carbons.

In its simplest form, carbon-13 magnetic resonance can distinguish between aliphatic and aromatic carbons. In the aliphatic region of petroleum ^{13}C-NMR spectra, several sharp peaks stand out and are used for quantitative measurements. The most prominent peak is usually that at 29.7 ppm. It is attributed to the methylene (CH_2) carbon atoms in long alkyl chains, positioned four or more carbons ($>\gamma$) away from an aromatic ring and from the terminal methyl (CH_3) groups.

Normally, the absorption of 29.5–30.3 ppm is an estimation of the amount of carbon atoms in long alkyl chains ($>C_5$). Because this band represents the methylene (CH_2) groups two or more carbons away from an aromatic ring and a terminal group, there must be four more carbons per chain than indicated by the area under these peaks. The *number of long chains* (n_{CH2}) can be estimated from the peaks at 14.2 ppm (ω-CH_3) and at 28.1 ppm (CH_2), next to a terminal branch point, that is, a methane (CH) group. On the one hand, the peak at 14.2 ppm gives results too high for this purpose, because it also indicates CH_3 groups from chain branches. Subtracting half of the 37.6-ppm peak area (CH_2 next to CH groups inside a chain) corrects for this feature. On the other hand, the peak at 14.2 ppm does not cover twin CH_3 groups (as in an isopropyl group,). This is why the 28.1-ppm peak (CH next to two terminal methyl groups) is needed. Thus,

$$n_{CH_2 \text{ long chains}} = A(14.2) - 1/2A(37.6) + 1/3A(28.1).$$

The number of methylene groups in long chains is the sum of the peak at 29.7 ppm, the number of methylene groups at the two ends, that is, six times the number of long chains for each of the two ends (Fig. 5.8), and the number close to the branch points inside the chain, namely, those next to them and one carbon atom removed, on both sides of the CH group. Thus, the total number is

$$n_{CH_2} = C(29.7) + 6n_{CH_2 \text{ long chains}} + 4C(37.6).$$

The number of methine (CH) groups in long chains can be estimated from the absorbance at 37.6 ppm and at 39.5 ppm. The peak at 37.6 ppm represents methylene groups next to methane carbon and thus does not represent a direct measurement of the methane groups. Thus,

$$n_{CH \text{ long chains}} = 1/2C(37.6) + C(39.5).$$

The methyl groups give rise to at least four peaks in carbon-13 spectra, namely, the peaks at 11.5, 14.2, 19.5, and 22.7 ppm. The peak 22.7 ppm represents twin methyl groups as in an isopropyl group, and it also has a contribution from a methylene group next to a methyl group and does not need to be represented again by a contribution to the equation from the peak at 14.2 ppm. Thus,

$$n_{CH_3 \text{ long chains}} = C(11.5) + C(19.5) + C(22.7).$$

The final estimate for carbon atoms in long chains is then the sum of the three types. The average length of long chains is derived from dividing the number of carbon atoms in long chains by the number of long chains. Remembering, of course, that averages derived from magnetic resonance spectra can be very misleading if the data are interpreted too literally (Speight, 2014).

The methyl group (CH_3) gives rise to several peaks, depending on its position in the molecule. A peak at 14.2 ppm signals such a group positioned at the end of an unbranched chain segment of at least two or three methylene (CH_2) groups. A methyl group next to a branch point (methine group, CH) at the end of a chain with at least two methylene groups produces a peak at 22.7 ppm. Farther away from the end of the chain, such a group gives a peak at or near 19.8 ppm. The methyl group at the end is also affected by these branch points (Fig. 5.8).

Naphthenic carbon in high–molecular weight petroleum fractions usually occupies so many slightly different positions that the peaks are unresolved and form a broad hump in the range 25–60 ppm, under the generally well-resolved paraffinic peaks.

The evaluation of the hump is the only direct method for the determination of naphthenic methylene and methane (CH_n) groups. However, its measurement may not always be reliable. In very high-boiling petroleum samples, such as asphaltene constituents, the paraffin-type resonance may be only partly resolved. Though this leads mainly to broader peaks, the overlap may, in some cases, add to the hump and, thus, cause erroneously high results for naphthenic carbon. The aromatic region of the ^{13}C-NMR spectrum can be evaluated by conventional integration of the peaks due to the main aromatic group types.

Thus, proton and carbon-13 magnetic resonance spectroscopic techniques offer the potential information about the molecular types in the nonvolatile fractions of petroleum. The techniques, by the application of the estimation of peak areas and further application of mathematical methods, have been used to obtain information about *structural parameters* that are then converted to *average structures*.

In most cases, the average structure of such complex mixtures as petroleum fractions is not the same as the representative structure. As already noted, average structures derived from magnetic resonance spectra can be very misleading if the data are interpreted too literally (Speight, 2014). Even though the average structures may always be questions, one must also treat with some caution the structural parameters, since they have been derived using assumptions that themselves are subject to debate.

2.4.4 Ultraviolet Spectroscopy

The ultraviolet–visible (UV–Vis) spectrum, though not as specific for chemical group types as infrared spectroscopy and NMR spectroscopy, can distinguish between aromatic compounds with different ring numbers and configurations (Fig. 9.4) (e.g. ASTM D1840, ASTM D2269). The patterns are not distinct enough to recognize or distinguish these compounds in complex mixtures, but they can be useful for their identification in narrow fractions.

Ultraviolet–visible spectroscopy (ASTM E169) can be employed as a detector for the fractionation of petroleum samples, for example, for the chromatographic separation and/or identification of aromatics by ring number, especially in combination with a technique such as liquid chromatography (Speight, 1986). Thus, UV–Vis spectroscopy lends itself to studies of refining processes as an online detector.

2.4.5 X-Ray Diffraction

X-ray diffraction had been used for the determination of the fraction of aromatic carbon (f_a) in petroleum constituents. This ratio can also be conveniently and precisely obtained by carbon-13 NMR spectroscopy as well as by infrared spectroscopy. The determination of the fraction of carbon that is aromatic may be in error (Ebert, 1990), since X-ray diffraction data can be very misleading (Ebert et al., 1984), especially when the data are used to translate geometric data—measurements of (aromatic) sheet diameter—into structural information. On the one hand, not all aromatic atoms contribute to the stack diameter, seen by X-ray diffraction, whereas, on the other hand, nonaromatic atoms such as hydroaromatic carbons and other substituents on aromatic rings may contribute to the diffraction pattern. Thus, the interpretation of these measurements is quite arbitrary.

Extended X-ray absorption fine structure (EXAFS) and *X-ray absorption near-edge structure* (XANES) spectroscopy are the tools for the investigation of the immediate chemical environment of X-ray absorbing elements such as metals and sulfur. X-ray absorption near-edge structure and X-ray photoelectron spectroscopy (XPS) been applied to the determination of sulfur compounds as well as nickel and vanadium in petroleum samples.

2.5   MOLECULAR WEIGHT

The molecular weight (formula weight) of a compound is the sum of the atomic weights of all the atoms in a molecule and can be determined by a variety of methods. Petroleum, being a complex mixture of (at least) several thousand constituents, requires qualification of the molecular weight as either (i) number-average molecular weight or (ii) weight-average molecular weight.

The *number-average molecular weight* is the ordinary arithmetic mean or the average of the molecular weights of the individual constituents. It is determined by measuring the molecular weight of n molecules, summing the weights, and dividing by n. The *weight-average molecular weight* is a way of describing the molecular weight of a complex mixture such as petroleum, even if the molecular constituents are not of the same type and exist in different sizes. Molecular weight is often used in refineries to provide mass-average or number-average measurements. Consequently, a number of methods are available to measure the molecular weight of petroleum and petroleum products.

Molecular weight may be calculated from viscosity data (ASTM D2502). The test method requires viscosity data from different temperatures, typically at 37.8 and 98.9°C (100 and 210°F). The method is generally applicable to a variety of petroleum fractions and products, but the number is an average number and applicable to those fractions or products with molecular weights in the range 250–700. Samples with unusual composition, such as aromatic-free white mineral oils, or oils with very narrow boiling range, may give atypical or questionable results. For samples with higher molecular weight (up to 3000 or more), with unusual composition, or for polymers, another test method is recommended (ASTM D2503). This method uses a vapor pressure osmometer to determine the molecular weight of the sample. Low-boiling samples may not be suitable—the vapor pressure of the constituents of the sample can interfere with the method.

A third method (ASTM D2878) developed for measuring the molecular weight of lubricating oil provides a procedure to calculate these properties from test data on evaporation. The procedure is based on the test method for measuring the evaporation loss of lubricating greases and other high-boiling petroleum products (ASTM D972). In the procedure, the sample is partly evaporated at a temperature of 250–500°C (480–930°F). However, fluids that are unstable in this temperature range are not suitable for submission to this test method.

If the molecular weight determination involves use of a solvent, it is recommended that, to negate concentration effects and temperature effects, the molecular weight determination be carried out at three different concentrations at three different temperatures. The data for each temperature are then extrapolated to zero concentration, and the zero concentration data at each temperature are then extrapolated to room temperature (Speight, 2001, 2014). Furthermore, each method of molecular weight determination has proponents and opponents because of the assumptions made in the use of the method or because of the mere complexity of the sample and the nature of the intermolecular and intramolecular interactions.

Methods for molecular weight measurement are also included in other more comprehensive standards (e.g. ASTM D128, ASTM D3712), and there are several indirect methods that have been proposed for the estimation of molecular weight by correlation with other, more readily measured physical properties. They are satisfactory when dealing with the conventional type of crude oils or their fractions and products when approximate values are desired.

2.6 INSTABILITY AND INCOMPATIBILITY

The problem of instability in petroleum and petroleum products may manifest itself in changes to the liquid, such as (i) asphaltene separation that occurs when the paraffinic nature of the liquid medium increases—as is often the case in residual fuel oil, (ii) wax separation that occurs when there is a drop in temperature or the aromaticity of the liquid medium increases, as may be the case in lubricating oil, (iii) sludge/sediment formation in fuel products that occurs because of the interplay of several chemical and physical factors (Mushrush and Speight, 1995; Speight, 2000, 2001, 2014). In addition to the phase-separation phenomenon, there may also be a general darkening of the color of the liquid product (ASTM D1500, IP 17).

Asphaltene-type deposition may, however, result from the mixing of fuels of different origin and treatment, each of which may be perfectly satisfactory when used alone. For example, straight-run fuel oils from the same crude oil are normally stable and mutually compatible, whereas fuel oils produced from thermal cracking and visbreaking operations may be stable, but can be unstable or incompatible if blended with straight-run fuels and vice versa. In fact, problems of thermal stability and incompatibility in residual fuel oils are associated with those fuels used in oil-fired naval or marine vessels, where the fuel is usually passed through a preheater before being fed to the burner system. In earlier days, this preheating, with some fuels, could result in the deposition of asphaltic matter culminating, in the extreme case, in the blockage of preheaters, pipelines, and even complete combustion failure.

Paraffin precipitation and deposition in petroleum pipelines and in petroleum product pipelines is an issue that decreases the cross-sectional area of the pipeline, thereby restricting operating capacities, and places additional strain on the pumping equipment. On the other hand, asphaltene deposition can lead to fouling of preheaters on heating the fuel to elevated temperatures (Park and Mansoori, 1988; Mushrush and Speight, 1995; Mushrush and Speight, 1998a, 1998b; Mushrush et al., 1999; Speight, 2014).

To assess the possibility of gum formation of a product during storage, in other words, the gum stability of gasoline, a test (ASTM D525, IP 40) is employed, which determines, in a pressure vessel, the *induction period* or time of heating at 100°C (212°F) with oxygen at an initial pressure of 100 pounds per square inch, which elapses before the oxygen pressure begins to fall, due to the oxidation of the sample, and the formation of gum therein indicates the onset of oxidative instability (ASTM D2893, ASTM D4636, ASTM D5483, IP 40, IP 48, IP 138, IP 142, IP 157, IP 229, IP 280, IP 306, IP 323, IP 388). The figure for oxidation stability, or *breakdown time* as it is sometimes called, is thus regarded as a measure of the stability of the fuel.

However, due to the multiplicity of types and conditions of storage, it is impossible to equate induction period with safe storage time, but it has been found by long experience that a minimum figure of 240 min induction period usually ensures a satisfactory level of gum stability for most normal marketing and distributing purposes. Induction period is also a useful control test for determining the amount of gum inhibitor to be added to gasoline, provided that the storage stability of the combination of the gasoline and inhibitor has been established by practical storage experiments.

Sludge (or sediment) formation takes one of the following forms: (i) material dissolved in the liquid; (ii) precipitated material; and (iii) material emulsified in the liquid. Under favorable conditions, sludge or sediment will dissolve in the crude oil or product with the potential of increasing the viscosity. Sludge or sediment, which is not soluble in the crude oil (ASTM D473, ASTM D1796, ASTM D2273, ASTM D4007, ASTM D4807, ASTM D4870), may either settle at the bottom of the storage tanks or remain in the crude oil as an emulsion. In most of the cases, the smaller part of the sludge/sediment will settle satisfactorily, and the larger part will stay in the crude oil as emulsions. In any case, there is a need of breaking the emulsion, whether it is a water-in-oil emulsion or whether it is the sludge itself, which has to be separated into the oily phase and the aqueous phase. The oily phase can be then processed with the crude oil, and the aqueous phase can be drained out of the system.

Emulsion breaking, whether the emulsions are due to crude oil–sludge emulsions, crude oil–water emulsions, or breaking of the sludge themselves into their oily and inorganic components, is of major importance from operational as well as commercial aspects. With some heavy fuel oil products and heavy crude oils, phase-separation difficulties often arise (Mushrush and Speight, 1995). Also, some crude oil emulsions may be stabilized by naturally occurring substances in the crude oil. Many of these polar particles accumulate at the oil–water interface, with the polar groups directed toward the water and the hydrocarbon groups toward the oil. A stable interfacial skin may be so formed; particles of clay or similar impurities, as well as wax crystals present in the oil may be embedded in this skin and make the emulsion very difficult to break (see Schramm, 1992, and references cited therein).

Chemical and electrical methods for sludge removal and for water removal, often combined with chemical additives, have to be used for breaking such emulsions. Each emulsion has its own structure and characteristics: water-in-oil emulsions, where the oil is the major component, or oil in water emulsions, where the water is the major component. The chemical nature and physical nature of the components of the emulsion play a major role in their susceptibility to the various surface-active agents used for breaking them.

Therefore, appropriate emulsion-breaking agents have to be chosen very carefully, usually with the help of previous laboratory evaluations. Water- or oil-soluble demulsifiers, the latter being often nonionic surface-active alkylene oxide adducts, are used for this purpose. But, as had been said in the foregoing, the most suitable demulsifier has to be chosen for each case from a large number of such substances in the market, by a prior laboratory evaluation.

In addition to the stability of the crude oil system, there are many crude oil products that are unstable under service conditions. For example, many products that are manufactured by cracking processes contain unsaturated components that may oxidize during storage and form undesirable oxidation products. If storage for considerable time before use is anticipated, it is essential that the product should not undergo any deleterious change under storage conditions and should remain stable.

Heteroatoms (particularly nitrogen, sulfur, and trace metals) are present in petroleum and might be also expected to be present in liquid fuels and other products from petroleum (Speight and Ozum, 2002; Hsu and Robinson, 2006; Gary et al., 2007; Speight, 2014). And, indeed, this is often the case, although there might have been some skeletal changes induced by the refining process(es). Oxygen is much more difficult to define in petroleum and liquid fuels. However, it must be stressed that instability/incompatibility is not directly related to the total nitrogen, oxygen, or sulfur content. The formation of color/sludge/sediment is a result of several factors. Perhaps, the main factor is the location and nature of the heteroatom that, in turn, determines reactivity (Mushrush and Speight, 1995).

Compatibility in distillate products is important to commercial and consumers as well as to the producer. Distillate products that are made from the refining process based on straight-run distillation show very little incompatibility problems. However, at present and in the future, problems for refiners will continue to increase as the quality of the available crude decreases worldwide. This decrease, coupled with the inevitable future use of liquid fuels from bio-sources, coal, and shale sources, will exacerbate the present problems for the producers.

When various stocks are blended at the refinery, incompatibility can be explained by the onset of acid–base catalyzed condensation reactions of the various organo-nitrogen compounds in the individual blending stock themselves. These are usually very rapid reactions with practically no observed induction time period (Mushrush and Speight, 1995).

When the product is transferred to a storage tank or some other holding tank, incompatibility can occur due to the free-radical hydroperoxide induced polymerization of active olefins. This is a relatively slow reaction, because the observed increase in hydroperoxide concentration is dependent on the dissolved oxygen content. Another incompatibility mechanism involves degradation when the product is stored for prolonged periods, as might occur when the fuel is stockpiled for military use. This incompatibility process involves (i) the buildup of hydroperoxide moieties after the gum reactions; (ii) a free-radical reaction with the various organo-sulfur compounds present (such as mercaptan sulfur, R-SH) (ASTM D3227, ASTM D5305), which can be oxidized to sulfonic acids; and (iii) reactions such as condensations between organo-sulfur and nitrogen compounds and esterification reactions.

2.7 THE FUTURE

In the petroleum industry, as in other chemical industries, the importance of feedstock and product analyses continues to grow. Instrumental and automated methods are replacing chemical and physical methods in the laboratories.

More stringent product requirements, related not only to product performance but also environmental issues, advanced catalytic processing techniques, and improved feedstock purification for specific downstream processes are driving the limits of impurities into the less than parts-per-million range, and in some cases, into the parts-per-billion range. Efforts to provide quantitative analyses at this level continue. As refinery feedstocks, product distributions, and methodologies change, efforts to improve the analytical methods based on the current instrumental technology will continue to go hand-in-hand with refinery evolution.

With the passage of time since the 1960s, the crude oils being processed in refineries are on average becoming increasingly heavier (more residuum) and more sour (higher sulfur content). In addition, refinery economics dictate that the *bottom of the barrel* (residuum) must be upgraded to higher value products. To produce a viable product slate with these crudes, refiners must add to or expand the existing treatment and processing options. The high sulfur content of crude, coupled with the government regulations limiting the maximum sulfur content of fuels, makes sulfur removal a priority in refinery processing. This is not the only answer—fuels containing aromatic constituents can, and often do, emit black fumes when traveling under a load or moving up an incline. New treatment and process units in the refinery usually translate into a need for new analytical test methods that can adequately evaluate feedstocks and monitor product quality.

Desulfurization processes (hydrodesulfurization processes) in a refinery use catalysts that are sensitive to (i) the amount of sulfur and (ii) the structure of the sulfur compounds being removed, and test methods that can provide data about both of the sulfur-related issues are continuing and will continue to increase in importance. Extension of these types of analyses are combining the separation power of GC with sulfur-selective detectors to provide data on the boiling range distribution of the sulfur compounds and the molecular types of sulfur compounds that occur within a specific boiling range. Work on extending this type of analysis to the higher boiling ranges is also being used for characterization.

Specific to the *bottom-of-the-barrel* upgrading, the production of quality products involves taking more (all, if possible) of the residuum and producing more salable, higher value, on-specification products. As this form of upgrading expands, improved test methods and characterization techniques are necessary for (i) feedstock evaluation, (ii) process design, and (iii) predictability of the yields of the product and the character of the product (see Tables A1–A29, Appendix).

In particular, there is a need to continue development of text methods that define the boiling range and distribution of the molecular types. The boiling range distribution of heavy distillates and residua are, for example, increasingly being carried out by high-temperature simulated distillation (HTSD) using GC as the operational technique. The distributions of hydrocarbon types in gas oil and heavier materials are important in evaluating them as feedstocks for further processing.

The goal for any such method is automated, instrumental analyses, as the option of choice when developing new methods and the trend to automation appears to be increasing. Reduction in analytical time as well as improving the quality of test results (eliminating dependency on the manual skills of the analyst) is needed to fulfill the required levels of accuracy and precision. Moved ahead by advances in technology, it is necessary that the analytical challenges of the refining industry are addressed.

REFERENCES

ASTM D56. Standard Test Method for Flash Point by Tag Closed Tester. Annual Book of Standards. ASTM International, West Conshohocken, PA.

ASTM D92. Standard Test Method for Flash and Fire Points by Cleveland Open Cup. Annual Book of Standards. ASTM International, West Conshohocken, PA.

ASTM D93. Standard Test Methods for Flash Point by Pensky-Martens Closed Tester. Annual Book of Standards. ASTM International, West Conshohocken, PA.

ASTM D97. Standard Test Method for Pour Point of Petroleum Products. Annual Book of Standards. ASTM International, West Conshohocken, PA.

ASTM D129. Standard Test Method for Sulfur in Petroleum Products. Annual Book of Standards. ASTM International, West Conshohocken, PA.

ASTM D323. Standard Test Method for Vapor Pressure of Petroleum Products. Annual Book of Standards. ASTM International, West Conshohocken, PA.

ASTM D445. Standard Test Method for Kinematic Viscosity of Transparent and Opaque Liquids. Annual Book of Standards. ASTM International, West Conshohocken, PA.

ASTM D473. Standard Test Method for Sediment in Crude Oils and Fuel Oils by the Extraction Method. Annual Book of Standards. ASTM International, West Conshohocken, PA.

ASTM D525. Standard Test Method for Oxidation Stability of Gasoline (Induction Period Method). Annual Book of Standards. ASTM International, West Conshohocken, PA.

ASTM D873. Standard Test Method for Oxidation Stability of Aviation Fuels (Potential Residue Method). Annual Book of Standards. ASTM International, West Conshohocken, PA.

ASTM D942. Standard Test Method for Oxidation Stability of Lubricating Greases by the Oxygen Pressure Vessel Method. Annual Book of Standards. ASTM International, West Conshohocken, PA.

ASTM D971. Standard Test Method for Interfacial Tension of Oil against Water by the Ring Method. Annual Book of Standards. ASTM International, West Conshohocken, PA.

ASTM D972. Standard Test Method for Evaporation Loss of Lubricating Greases and Oils. Annual Book of Standards. ASTM International, West Conshohocken, PA.

ASTM D1318. Standard Test Method for Sodium in Residual Fuel Oil (Flame Photometric Method). Annual Book of Standards. ASTM International, West Conshohocken, PA.

ASTM D1492. Standard Test Method for Bromine Index of Aromatic Hydrocarbons by Coulometric Titration. Annual Book of Standards. ASTM International, West Conshohocken, PA.

ASTM D1500. Standard Test Method for ASTM Color of Petroleum Products (ASTM Color Scale). Annual Book of Standards. ASTM International, West Conshohocken, PA.

ASTM D1796. Standard Test Method for Water and Sediment in Fuel Oils by the Centrifuge Method (Laboratory Procedure). Annual Book of Standards. ASTM International, West Conshohocken, PA.

ASTM D1840. Standard Test Method for Naphthalene Hydrocarbons in Aviation Turbine Fuels by Ultraviolet Spectrophotometry. Annual Book of Standards. ASTM International, West Conshohocken, PA.

ASTM D1971. Standard Practices for Digestion of Water Samples for Determination of Metals by Flame Atomic Absorption, Graphite Furnace Atomic Absorption, Plasma Emission Spectroscopy, or Plasma Mass Spectrometry. Annual Book of Standards. ASTM International, West Conshohocken, PA.

ASTM D2007. Standard Test Method for Characteristic Groups in Rubber Extender and Processing Oils and Other Petroleum-Derived Oils by the Clay-Gel Absorption Chromatographic Method. Annual Book of Standards. ASTM International, West Conshohocken, PA.

ASTM D2161. Standard Practice for Conversion of Kinematic Viscosity to Saybolt Universal Viscosity or to Saybolt Furol Viscosity. Annual Book of Standards. ASTM International, West Conshohocken, PA.

ASTM D2269. Standard Test Method for Evaluation of White Mineral Oils by Ultraviolet Absorption. Annual Book of Standards. ASTM International, West Conshohocken, PA.

ASTM D2273. Standard Test Method for Trace Sediment in Lubricating Oils. Annual Book of Standards. ASTM International, West Conshohocken, PA.

ASTM D2425. Standard Test Method for Hydrocarbon Types in Middle Distillates by Mass Spectrometry. Annual Book of Standards. ASTM International, West Conshohocken, PA.

ASTM D2502. Standard Test Method for Estimation of Mean Relative Molecular Mass of Petroleum Oils from Viscosity Measurements. Annual Book of Standards. ASTM International, West Conshohocken, PA.

ASTM D2503. Standard Test Method for Relative Molecular Mass (Molecular Weight) of Hydrocarbons by Thermoelectric Measurement of Vapor Pressure. Annual Book of Standards. ASTM International, West Conshohocken, PA.

ASTM D2549. Standard Test Method for Separation of Representative Aromatics and Nonaromatics Fractions of High-Boiling Oils by Elution Chromatography. Annual Book of Standards. ASTM International, West Conshohocken, PA.

ASTM D2595. Standard Test Method for Evaporation Loss of Lubricating Greases Over Wide-Temperature Range). Annual Book of Standards. ASTM International, West Conshohocken, PA.

ASTM D2622. Standard Test Method for Sulfur in Petroleum Products by Wavelength Dispersive X-ray Fluorescence Spectrometry. Annual Book of Standards. ASTM International, West Conshohocken, PA.

ASTM D2625. Standard Test Method for Endurance (Wear) Life and Load-Carrying Capacity of Solid Film Lubricants (Falex Pin and Vee Method). Annual Book of Standards. ASTM International, West Conshohocken, PA.

ASTM D2650. Standard Test Method for Chemical Composition of Gases by Mass Spectrometry. Annual Book of Standards. ASTM International, West Conshohocken, PA.

ASTM D2710. Standard Test Method for Bromine Index of Petroleum Hydrocarbons by Electrometric Titration. Annual Book of Standards. ASTM International, West Conshohocken, PA.

ASTM D2786. Standard Test Method for Hydrocarbon Types Analysis of Gas-Oil Saturates Fractions by High Ionizing Voltage Mass Spectrometry. Annual Book of Standards. ASTM International, West Conshohocken, PA.

ASTM D2789. Standard Test Method for Hydrocarbon Types in Low Olefinic Gasoline by Mass Spectrometry. Annual Book of Standards. ASTM International, West Conshohocken, PA.

ASTM D2878. Method for Estimating Apparent Vapor Pressures and Molecular Weights of Lubricating Oils. Annual Book of Standards. ASTM International, West Conshohocken, PA.

ASTM D2887. Standard Test Method for Boiling Range Distribution of Petroleum Fractions by Gas Chromatography. Annual Book of Standards. ASTM International, West Conshohocken, PA.

ASTM D2893. Standard Test Methods for Oxidation Characteristics of Extreme-Pressure Lubrication Oils. Annual Book of Standards. ASTM International, West Conshohocken, PA.

ASTM D3227. Standard Test Method for Lead in Gasoline by Atomic Absorption Spectroscopy. Annual Book of Standards. ASTM International, West Conshohocken, PA.

ASTM D3238. Standard Test Method for Calculation of Carbon Distribution and Structural Group Analysis of Petroleum Oils by the n-d-M Method. Annual Book of Standards. ASTM International, West Conshohocken, PA.

ASTM D3239. Standard Test Method for Aromatic Types Analysis of Gas-Oil Aromatic Fractions by High Ionizing Voltage Mass Spectrometry. Annual Book of Standards. ASTM International, West Conshohocken, PA.

ASTM D3241. Standard Test Method for Thermal Oxidation Stability of Aviation Turbine Fuels. Annual Book of Standards. ASTM International, West Conshohocken, PA.

ASTM D3605. Standard Test Method for Trace Metals in Gas Turbine Fuels by Atomic Absorption and Flame Emission Spectroscopy. Annual Book of Standards. ASTM International, West Conshohocken, PA.

ASTM D3701. Standard Test Method for Hydrogen Content of Aviation Turbine Fuels by Low Resolution Nuclear Magnetic Resonance Spectrometry. Annual Book of Standards. ASTM International, West Conshohocken, PA.

ASTM D3710. Standard Test Method for Boiling Range Distribution of Gasoline and Gasoline Fractions by Gas Chromatography. Annual Book of Standards. ASTM International, West Conshohocken, PA.

ASTM D3831. Standard Test Method for Manganese in Gasoline By Atomic Absorption Spectroscopy. Annual Book of Standards. ASTM International, West Conshohocken, PA.

ASTM D4007. Standard Test Method for Water and Sediment in Crude Oil by the Centrifuge Method (Laboratory Procedure). Annual Book of Standards. ASTM International, West Conshohocken, PA.

ASTM D4046. Standard Test Method for Alkyl Nitrate in Diesel Fuels by Spectrophotometry. Annual Book of Standards. ASTM International, West Conshohocken, PA.

ASTM D4294. Standard Test Method for Sulfur in Petroleum and Petroleum Products by Energy Dispersive X-ray Fluorescence Spectrometry. Annual Book of Standards. ASTM International, West Conshohocken, PA.

ASTM D4486. Standard Test Method for Kinematic Viscosity of Volatile and Reactive Liquids. Annual Book of Standards. ASTM International, West Conshohocken, PA.

ASTM D4628. Standard Test Method for Analysis of Barium, Calcium, Magnesium, and Zinc in Unused Lubricating Oils by Atomic Absorption Spectrometry. Annual Book of Standards. ASTM International, West Conshohocken, PA.

ASTM D4636. Standard Test Method for Corrosiveness and Oxidation Stability of Hydraulic Oils, Aircraft Turbine Engine Lubricants, and Other Highly Refined Oils. Annual Book of Standards. ASTM International, West Conshohocken, PA.

ASTM D4698. Standard Practice for Total Digestion of Sediment Samples for Chemical Analysis of Various Metals.

Annual Book of Standards. ASTM International, West Conshohocken, PA.

ASTM D4807. Standard Test Method for Sediment in Crude Oil by Membrane Filtration. Annual Book of Standards. ASTM International, West Conshohocken, PA.

ASTM D4808. Standard Test Methods for Hydrogen Content of Light Distillates, Middle Distillates, Gas Oils, and Residua by Low-Resolution Nuclear Magnetic Resonance Spectroscopy. Annual Book of Standards. ASTM International, West Conshohocken, PA.

ASTM D4870. Standard Test Method for Determination of Total Sediment in Residual Fuels. Annual Book of Standards. ASTM International, West Conshohocken, PA.

ASTM D5056. Standard Test Method for Trace Metals in Petroleum Coke by Atomic Absorption. Annual Book of Standards. ASTM International, West Conshohocken, PA.

ASTM D5185. Standard Test Method for Determination of Additive Elements, Wear Metals, and Contaminants in Used Lubricating Oils and Determination of Selected Elements in Base Oils by Inductively Coupled Plasma Atomic Emission Spectrometry (ICP-AES). Annual Book of Standards. ASTM International, West Conshohocken, PA.

ASTM D5305. Standard Test Method for Determination of Ethyl Mercaptan in LP-Gas Vapor. Annual Book of Standards. ASTM International, West Conshohocken, PA.

ASTM D5307. Standard Test Method for Determination of Boiling Range Distribution of Crude Petroleum by Gas Chromatography. Annual Book of Standards. ASTM International, West Conshohocken, PA.

ASTM D5483. Standard Test Method for Oxidation Induction Time of Lubricating Greases by Pressure Differential Scanning Calorimetry. Annual Book of Standards. ASTM International, West Conshohocken, PA.

ASTM D5776. Standard Test Method for Bromine Index of Aromatic Hydrocarbons by Electrometric Titration. Annual Book of Standards. ASTM International, West Conshohocken, PA.

ASTM D6299. Standard Practice for Applying Statistical Quality Assurance and Control Charting Techniques to Evaluate Analytical Measurement System Performance. Annual Book of Standards. ASTM International, West Conshohocken, PA.

ASTM D6792. Standard Practice for Quality System in Petroleum Products and Lubricants Testing Laboratories. Annual Book of Standards. ASTM International, West Conshohocken, PA.

ASTM E168. Standard Practices for General Techniques of Infrared Quantitative Analysis. Annual Book of Standards. ASTM International, West Conshohocken, PA.

ASTM E169. Standard Practices for General Techniques of Ultraviolet-Visible Quantitative Analysis. Annual Book of Standards. ASTM International, West Conshohocken, PA.

ASTM E204. Standard Practices for Identification of Material by Infrared Absorption Spectroscopy, Using the ASTM Coded Band and Chemical Classification Index. Annual Book of Standards. ASTM International, West Conshohocken, PA.

ASTM E334. Standard Practice for General Techniques of Infrared Microanalysis. Annual Book of Standards. ASTM International, West Conshohocken, PA.

ASTM E386. Standard Practice for Data Presentation Relating to High-Resolution Nuclear Magnetic Resonance (NMR) Spectroscopy. Annual Book of Standards. ASTM International, West Conshohocken, PA.

ASTM E659. Standard Test Method for Autoignition Temperature of Liquid Chemicals. Annual Book of Standards. ASTM International, West Conshohocken, PA.

ASTM E1252. Standard Practice for General Techniques for Obtaining Infrared Spectra for Qualitative Analysis. Annual Book of Standards. ASTM International, West Conshohocken, PA.

ASTM E1316. Standard Terminology for Nondestructive Examinations. Annual Book of Standards. ASTM International, West Conshohocken, PA.

Budde, W.L. 2001. The Manual of Manuals. Office of Research and Development, Environmental Protection Agency, Washington, DC.

Dean, J.R. 1998. Extraction Methods for Environmental Analysis. John Wiley & Sons, Inc., New York.

Drews, A.W. (Editor). 1998. Manual on Hydrocarbon Analysis, 6th Edition. ASTM International, West Conshohocken, PA.

Dyroff, G.V. (Editor). 1993. Manual on Significance of Tests for Petroleum Products, 6th Edition. American Society for Testing and Materials, West Conshohocken, PA.

Ebert, L.B. 1990. Fuel Science and Technology International, 8: 563.

Ebert, L.B., Mills, D.R., and Scanlon, J.C. 1987. Preprints, American Chemical Society, Division of Petroleum Chemistry, 32(2): 419.

Ebert, L.B., Scanlon, J.C., and Mills, D.R. 1984. Liquid Fuels Technology, 2: 257.

Falla Sotelo, F., Araujo Pantoja, P., López-Gejo, J., Le Roux, J.G.A.C., Quina, F.H., and Nascimento, C.A.O. 2008. Application of fluorescence spectroscopy for spectral discrimination of crude oil samples. Brazilian Journal of Petroleum and Gas, 2(2): 63–71.

Fingas, M.F. 1998. Studies on the evaporation of crude oil and petroleum products. ii. boundary layer regulation. Journal of Hazardous Materials, 57(1–3): 41–58.

Fowlis, I.A. 1995. Gas Chromatography, 2nd Edition. John Wiley & Sons Inc., New York.

Frijer, J.I., De Jonge, H., Bounten, W., and Verstraten, J.M. 1996. Assessing mineralization rates of petroleum hydrocarbons in soils in relation to environmental factors and experimental scale. Biodegradation, 7: 487–500.

Gary, J.G., Handwerk, G.E., and Kaiser, M.J. 2007. Petroleum Refining: Technology and Economics, 5th Edition. CRC Press, Taylor & Francis Group, Boca Raton, FL.

Grob, R.L. 1995. Modern Practice of Gas Chromatography, 3rd Edition. John Wiley & Sons Inc., New York.

Hsu, C.S., and Robinson, P.R. (Editors) 2006. Practical Advances in Petroleum Processing, Volumes 1 and 2. Springer Science, New York.

IP 17. Determination of Color – Lovibond Tintometer Method. IP Standard Methods 2013. The Energy Institute, London.

IP 34. Determination of Flash Point – Pensky-Martens Closed Cup Method. IP Standard Methods 2013. The Energy Institute, London.

IP 36. Determination of Open Flash and Fire Point – Cleveland Method. IP Standard Methods 2013. The Energy Institute, London.

IP 40 (ASTM D525). Petroleum Products – Determination of Oxidation Stability of Gasoline – Induction Period Method. IP Standard Methods 2013. The Energy Institute, London.

IP 48. Determination of Oxidation Characteristics of Lubricating Oil. IP Standard Methods 2013. The Energy Institute, London.

IP 138 (ASTM D873). Determination of Oxidation Stability of Aviation Fuel – Potential Residue Method. IP Standard Methods 2013. The Energy Institute, London.

IP 142 (ASTM D942). Determination of Oxidation Stability of Lubricating Grease – Oxygen Bomb Method. IP Standard Methods 2013. The Energy Institute, London.

IP 157. Determination of the Oxidation Stability of Inhibited Mineral Oils (TOST Test). IP Standard Methods 2013. The Energy Institute, London.

IP 229. Determination of the Relative Oxidation Stability by Rotating Bomb of Mineral Turbine Oil. IP Standard Methods 2013. The Energy Institute, London.

IP 280. Petroleum Products and Lubricants – Inhibited Mineral Turbine Oils – Determination of Oxidation Stability. IP Standard Methods 2013. The Energy Institute, London.

IP 306. Determination of Oxidation Stability of Straight Mineral Oil. IP Standard Methods 2013. The Energy Institute, London.

IP 323 (ASTM D3241). Petroleum Products – Determination of Thermal Oxidation Stability of Gas Turbine Fuels – JFTOT Method. IP Standard Methods 2013. The Energy Institute, London.

IP 388 (ASTM D2274). Petroleum Products – Determination of the Oxidation Stability of Middle-Distillate Fuels. IP Standard Methods 2013. The Energy Institute, London.

IP 392. Aromatic Hydrogen and Carbon Contents – High Resolution Nuclear Magnetic Resonance Spectroscopy Method. IP Standard Methods 2013. The Energy Institute, London.

IP 429. Liquid Petroleum Products – Petrol – Determination of the Benzene Content by Infrared Spectrometry. IP Standard Methods 2013. The Energy Institute, London.

Mackay, D., and Zagorski, W. 1982. Studies of Water-in-Oil Emulsions. Report No. EE-34. Environment Canada, Ottawa, ON.

Miller, M. (Editor). 2000. Encyclopedia of Analytical Chemistry. John Wiley & Sons Inc., New York.

Mushrush, G.W., and Speight, J.G. 1995. Petroleum Products: Instability and Incompatibility. Taylor & Francis, New York.

Mushrush, G.W., and Speight, J.G. 1998a. Instability and Incompatibility of Petroleum Products. In Petroleum Chemistry and Refining. J.G. Speight (Editor). Taylor & Francis, New York.

Mushrush, G.W., and Speight, J.G. 1998b. The chemistry of the incompatibility process in middle distillate fuels. Reviews in Process Chemistry and Engineering, 1: 5.

Mushrush, G.W., Speight, J.G., Beal, E.J., and Hardy, D.R. 1999. Instability chemistry of middle distillate fuel oils. American Chemical Society, Division of Petroleum Chemistry, 44(2): 175.

Nadkarni, R.A.K. 2011. Spectroscopic Analysis of Petroleum Products and Lubricants. ASTM International, West Conshohocken, PA.

Park, S.J., and Mansoori, G.A. 1988. Aggregation and deposition of heavy organics in petroleum crudes. Energy Sources, 10: 109–125.

Patnaik, P. (Editor). 2004. Dean's Analytical Chemistry Handbook, 2nd Edition. McGraw-Hill, New York.

Rand, S. 2003. Significance of Tests for Petroleum Products, 7th Edition. ASTM International, West Conshohocken, PA.

Schramm, L.L. (Editor). 1992. Emulsions. Fundamentals and Applications in the Petroleum Industry. American Chemical Society, Washington, DC.

Speight, J.G. 1986. Polynuclear Aromatic Systems in Petroleum. Preprints, American Chemical Society, Division of Petroleum Chemistry, 31(4): 818.

Speight, J.G. 1994. Application of spectroscopic techniques to the structural analysis of petroleum. Applied Spectroscopy Reviews, 29: 269.

Speight, J.G. 2000. The Desulfurization of Heavy Oils and Residua, 2nd Edition. Marcel Dekker, New York.

Speight, J.G. 2001. Handbook of Petroleum Analysis. John Wiley & Sons Inc., New York.

Speight, J.G. 2005. Environmental Analysis and Technology for the Refining Industry. John Wiley & Sons Inc., Hoboken, NJ.

Speight, J.G. 2009. Enhanced Recovery Methods for Heavy Oil and Tar Sands. Gulf Publishing Company, Houston, TX.

Speight, J.G. 2011. An Introduction to Petroleum Technology, Economics, and Politics. Scrivener Publishing, Salem, MA.

Speight, J.G. 2014. The Chemistry and Technology of Petroleum, 4th Edition. CRC Press, Taylor & Francis Group, Boca Raton, FL.

Speight, J.G., and Arjoon, K.K. 2012. Bioremediation of Petroleum and Petroleum Products. Scrivener Publishing, Salem, MA.

Speight, J.G., and Foote, R. 2011. Ethics in Science and Engineering. Scrivener Publishing, Salem, MA.

Speight, J.G., and Ozum, B. 2002. Petroleum Refining Processes. Marcel Dekker Inc., New York.

Totten, G.E. (Editor). 2003. Fuels and Lubricants Handbook: Technology, Properties, Performance, and Testing. ASTM International, West Conshohocken, PA.

Wallace, D., Starr, J., Thomas, K.P., and Dorrence. S.M. 1988. Characterization of Oil Sand Resources. Alberta Oil Sands Technology and Research Authority, Edmonton, AB.

3

SAMPLING AND MEASUREMENT

3.1 INTRODUCTION

Petroleum exhibits a wide range of physical properties, and several relationships can be made between various physical properties (Chapter 1) (Speight and Ozum, 2002; Hsu and Robinson, 2006; Gary et al., 2007; Speight, 2014a). Whereas the properties such as viscosity, density, boiling point, and color of petroleum may vary widely, the elemental analysis (ultimate analysis) varies, as already noted, over a narrow range for a large number of petroleum samples (Speight, 2014a). The carbon content is relatively constant, while the hydrogen and heteroatom contents are responsible for the major differences between petroleum. Coupled with the changes brought about to the feedstock constituents by refinery operations, it is not surprising that petroleum characterization is a monumental task.

The chemical and physical properties of crude oil and the yields, and properties of products or fractions produced therefrom vary considerably and are dependent on the concentration of the hydrocarbon constituents and non-hydrocarbon constituents present in the crude oil. Some types of crude oils have pronounced advantages as sources of fuels and lubricants because they require less specialized processing than that needed for production of the same products from other types of crude oil. Other crude oils may contain unusually low concentrations of components that are desirable fuel or lubricant constituents or precursors to these products, and the production of these products from such crude oils may require considerable refining.

Evaluation of petroleum for use as a feedstock usually involves an examination of one or more of the physical properties of the material (Speight, 2001, 2014a). By this means, a set of basic characteristics can be obtained that can be correlated with utility. To satisfy specific needs with regard to the type of petroleum to be processed, as well as to the nature of the product, various standards organizations, such as the American Society for Testing and Materials (ASTM, 2013) in North America and the Energy Institute in the United Kingdom (formerly the Institute of Petroleum) (Energy Institute, 2013), have devoted considerable time and effort to the correlation and standardization of methods for the inspection and evaluation of petroleum and petroleum products. Many other countries—too numerous to mention here—also have similar organizations for developing standard test methods or industrial products. For the purposes of this text, it is appropriate that in any discussion of the physical properties of petroleum and petroleum products, reference be made to the corresponding test methods (designated by the letters ASTM and IP along with a number), and accordingly, the various test numbers have been included in the text.

In North America and the United Kingdom, these test methods are prime assets for the evaluation and assessment of the physical, mechanical, rheological, thermal, and chemical properties of crude oils, lubricating grease, automobile and aviation gasoline, hydrocarbons, and other naturally occurring energy resources used for various industrial applications. This is emphasized by the series of text used for crude oils collected in the US Strategic Petroleum reserve. These crudes and their respective distillation products (Fig. 3.1) are tested for their composition, purity, density, thermal stability, and a series of other tests. The data

Handbook of Petroleum Product Analysis, Second Edition. James G. Speight.
© 2015 John Wiley & Sons, Inc. Published 2015 by John Wiley & Sons, Inc.

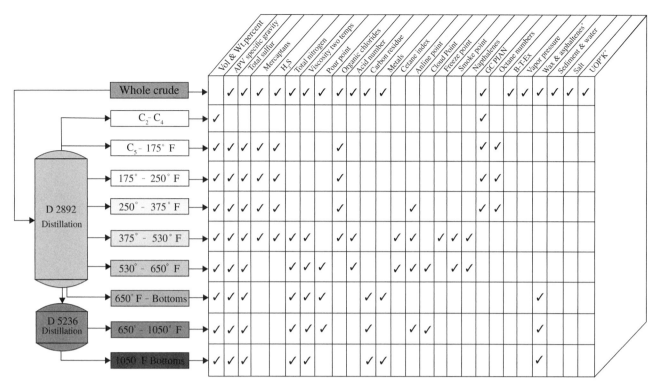

FIGURE 3.1 Crude oil analysis scheme for the strategic petroleum reserve (*Source*: US DOE, 2002).

for the text methods allow refineries to examine and process these crude oils efficiently and to ensure the quality of the products for use.

3.2 SAMPLING

The reliability of a sampling method is the degree of perfection with which the identical composition and properties of the entire whole petroleum or petroleum product are obtained (preserved completely without alteration) in the sample—such a sample is referred to as a *representative sample*. The reliability of the storage procedure is the degree to which the sample remains unchanged thereby guaranteeing the accuracy and usefulness of the analytical data. At this point, a review of the sampling methods applied to petroleum and petroleum products is worth of inclusion.

Since the value of any product is judged by the characteristics of the sample as determined by laboratory tests, the ability to collect and preserve a sample that is representative of the site is a critically important step. Furthermore, obtaining a representative sample of petroleum or a petroleum product is always a challenge due to the heterogeneity of the materials. Additional difficulties are encountered due to the ranges in volatility, solubility, and adsorption potential of individual constituents. And the procedures used for sample collection and preparation must be analytically reproducible with a high degree of accuracy (Weisman, 1998; Speight, 2001,

2014a; Dean, 2003; Patnaik, 2004). The sample to be used for the test methods must be representative of the bulk material, or data will be produced that are not a true representation of the material and will, to be blunt, be incorrect no matter how accurate or precise the test method and no matter how much *fine-tuning* is applied by the submitter of the sample. In addition, the type and cleanliness of sample containers are important; if the container is contaminated or is made of material that either reacts with the product or is a catalyst, the test results may be wrong.

3.2.1 Sampling Protocol

The importance of the correct sampling of any product destined for analysis should always be overemphasized. Incorrect sampling protocols can lead to erroneous analytical data from which performance of the product in service cannot be accurately deduced. For example, properties such as specific gravity, distillation yield, vapor pressure, hydrogen sulfide content, and octane number of the gasoline are affected by the content of low-boiling hydrocarbons so that suitable cooling or pressure sampling methods have to be used and care taken during the subsequent handling of the sample in order to avoid the loss of any volatile components. In addition, adequate records of the circumstances and conditions during sampling have to be made; for example, in sampling from storage tanks, the temperatures and pressures of the separation plant and the atmospheric temperature would be noted.

At the other end of the volatility scale, products that contain, or are composed of, high-molecular-weight paraffin hydrocarbons (wax) that are also in a solid state may require judicious heating (to dissolve the wax) and agitation (homogenized, to ensure thorough mixing) before sampling. If room temperature sampling is the *modus operandi* and product cooling causes wax to precipitate, homogenization to ensure correct sampling is also necessary.

Representative samples are prerequisite for the laboratory evaluation of any type of product, and many precautions are required in obtaining and handling representative samples (ASTM D1265). The precautions depend upon the sampling procedure, the characteristics (low-boiling or high-boiling constituents) of the product being sampled, and the storage tank, container, or tank carrier from which the sample is obtained. In addition, the sample container must be clean, and the type to be used depends not only on the product but also on the data to be produced.

3.2.2 Representative Sample

Representative samples are always a prerequisite for the laboratory evaluation of petroleum and petroleum products (Speight, 2001), and precautions are required in (i) obtaining and (ii) handling representative samples. The commercial value of any petroleum product (as well as the parent petroleum) is assessed by the characteristics of the sample as determined by standard test methods. In summary, the sample must be representative of the lot, or a wrong evaluation will be made, no matter how accurately the sample may be tested. In addition, each type of petroleum or petroleum products (gas, liquid, or solid) will more than likely require application of a different sampling method to assure that the sample is truly a representative sample.

Whether or not a sample is a representative sample depends on the sampling procedure, the type of petroleum or petroleum product being sampled, and the storage tank, container, or tank carrier from which the sample is obtained. The sample container must be clean, and the type to be used depends not only on the product but also on the laboratory data desired. Thus, the objective of each procedure is to obtain a sample or a composite of several samples in such manner and from such location in the tank or other container that the sample or composite will be representative of the entire bulk material. The sample container—which must be clean and free of any possible source of sample contamination and should not be made of material that either reacts with the sample or is a catalyst—should be labeled immediately by the sampler, to indicate at least (i) the nature/name of the sample, (ii) time of sampling, (iii) location of the sampling point, and any other information necessary for the sample identification.

Thus, the basic objective of each sampling procedure is to obtain a truly representative sample or, more often, a composite of several samples that can be considered to be a representative sample. In some cases, because of the size of the storage tank and the lack of suitable methods of agitation, several samples are taken from large storage tanks in such a manner that the samples represent the properties of the bulk material from different locations in the tank, and thus, the composite sample will be representative of the entire lot being sampled. This procedure allows for differences in sample that might be due to the stratification of the bulk material due to tank size or temperature at different levels of the storage tank. Solid samples require a different protocol that might involve melting (liquefying) the bulk material (assuming that thermal decomposition is not induced) followed by homogenization. On the other hand, the protocol used for coal or coke sampling (ASTM D346, ASTM D2013) might also be applied to sampling petroleum products, such as coke, that are solid and for which accurate analysis is required before sales.

Once the sampling procedure is accomplished, the sample container should be labeled immediately, to indicate the product, time of sampling, location of the sampling point, and any other information necessary for the sample identification. And if the samples were taken from different levels of the storage tank, the levels from which the samples were taken and the amounts taken and mixed into the composite should be indicated on the sample documentation.

While the earlier text as focused on the acquisitions of samples from storage tanks, sampling records for any procedure must be complete and should include, but is not restricted to items, information such as (i) the precise (geographic or other) location (or site or refinery or process) from which the sample was obtained, (ii) the identification of the location (or site or refinery or process) by name, (iii) the character of the bulk material (solid, liquid, or gas) at the time of sampling, (iv) the means by which the sample was obtained, (v) the means and protocols that were used to obtain the sample, (vi) the date, the amount of sample that was originally placed into storage, (vii) any chemical analyses (elemental analyses, fractionation by adsorbents or by liquids, functional type analyses) that have been determined to date, (viii) any physical analyses (API gravity, viscosity, distillation profile) that have been determined to date, (ix) the date of any such analyses included in items v and vi, (x) the methods used for analyses that were employed in v and vi, (xi) the analysts who carried out the work in v and vi, and (xii) a log sheet showing the names of the persons (with the date and the reason for the removal of an aliquot) who removed the samples from storage and the amount of each sample (aliquot) that was removed for testing.

In summary, there must be a means to accurately track and identify the sample history so that each sample is tracked and defined in terms of source, activity, and the personnel involved in any of the earlier stages. Thus, the accuracy of the data from any subsequent procedures and tests for which the sample is used will be placed beyond a *reasonable doubt* and would stand the test of time in court should legal issues arise.

3.2.3 Sampling Error

When a property of the sample (which exists as a large volume of material) is to be measured, there usually will be differences between the analytical data derived from application of the test methods to a *gross lot* or *gross consignment* and the data from the *sample lot*. This difference (the *sampling error*) has a frequency distribution with a mean value and a variance. *Variance* is a statistical term defined as the mean square of errors; the square root of the variance is more generally known as the *standard deviation* or the *standard error of sampling*.

Analytical test methods applied in the field—and such logic can be applied to all sampling protocols—must take into consideration the various site conditions comprising local heterogeneity, as well as the changing ambient conditions. Furthermore, the analysts must take into account the potential for samples to be disturbed during sampling, storage, and pretreatment. Rather than attempt to rationalize the application of corrections because of such disturbances, it is better that steps be taken to avoid sample disturbance. If corrections must be applied, the manner in which the corrections are applied must be beyond reproach or claims that the data will not stand up (when needed) to legal scrutiny, and falsification of the data will be the most likely result (Speight and Foote, 2011).

3.3 VOLUME MEASUREMENT

As part of the analytical procedure for petroleum and petroleum products, refiners also need to know not only the bulk properties of the material but also the amount of materials held in storage tanks—as part of, say, the refinery inventory. Once the petroleum or petroleum product is sampled, there is the need to calculate the total volume of the material in storage. Following this, the amount of petroleum or petroleum product involved in commercial transactions can be calculated with a high degree of accuracy.

The complex nature of petroleum and petroleum products requires several procedures for calculating volume or density of liquid products in commercial transactions. In addition, temperature affects the volume and density of petroleum and its products. Measured volumes must be converted to net quantities at 15.6°C (60°F), which is the accepted base temperature used widely in commercial transactions. Again, obtaining a representative sample of the material that is held in storage or carrier tanks is essential for converting measured quantities to the standard volume.

To assist with volume calculations, petroleum measurement tables are available for use in the calculation of quantities of crude petroleum and petroleum products as they exist under designated conditions (ASTM D1250, IP 200). The tables provide data that cover typical operating ranges for the reduction of gravity and volume to standard states, for calculation of weight–volume relationships, and interconversion of a wide variety of commercially useful units.

Following from this, the issues that always face analytical laboratory personnel include the need to provide higher-quality results. In addition, environmental regulations may influence the method of choice. Nevertheless, the method of choice still depends to a large extent on the boiling range (or carbon number) of the sample to be analyzed. For example, there is a large variation in the carbon number range and boiling points (of normal paraffins) for some of the more common petroleum products and thus a variation in the methods that may be applied to these products (Speight, 2001).

Testing for suspended water and sediment is used primarily with fuel oils, where appreciable amounts of water and sediment may cause fouling of facilities for handling the oil and give trouble in burner mechanisms. Three standard methods are available for this determination: (i) the centrifuge method, which gives the total water and sediment content of the sample by volume; (ii) the distillation method, which gives the volumetric amount of water; and (iii) the extraction method, which gives the solid sediment in percentage by weight.

Sediment usually consists of finely divided solids that may be dispersed in the oil or carried in water droplets. The solids may be drilling mud or sand or scale picked up during the transport of the oil, or may consist of chlorides derived from the evaporation of brine droplets in the oil. In any event, the sediment can lead to serious plugging of the equipment, corrosion due to chloride decomposition, and a lowering of residual fuel quality.

Water may be found in the crude either in an emulsified form or in large droplets. The quantity is generally limited by pipeline companies and by refiners, and steps are normally taken at the wellhead to reduce the water content as low as possible. Prior to analyses, it is often necessary to separate the water from a crude oil sample, and this is usually carried out by one of the procedures described in the distillation of crude petroleum (ASTM D4006) or by the coulometric Karl Fischer titration method (ASTM D4928).

The determination of density of specific gravity (ASTM D287, ASTM D1298) in the measurement and calculation of volume of petroleum products is important since gravity is an index of the weight of a measured volume of the product. Two scales are in use in the petroleum industry, specific gravity and API gravity, the determination being made in each case by means of a hydrometer of constant weight displacing a variable volume of oil. The reading obtained depends upon both the gravity and the temperature of the oil.

Gaging petroleum products involves the use of procedures for determining the liquid contents of tanks, ships and barges, tank cars, and tank trucks. Depth of liquid is determined by gaging through specified hatches or by reading gage glasses or other devices. There are two basic types of gages, innage and outage. The procedures used depend upon the type of tank to be gaged, its equipment, and the gaging apparatus.

An innage gage is the depth of liquid in a tank measured from the surface of the liquid to the tank bottom or to a datum plate attached to the shell or bottom. The innage gage is used directly with the tank calibration table and temperature of the product to calculate the volume of product (ASTM D1250). On the other hand, an outage gage is the distance between the surface of the product in the tank and the reference point above the surface, which is usually located in the gaging hatch. The outage gage is used either directly or indirectly with the tank calibration table and the temperature of the product to calculate the volume of product. The amount of any free water and sediment in the bottom of the tank is also gaged so that corrections can be made when calculating the net volume of the crude oil or petroleum product.

There are also procedures for determining the temperatures of petroleum and its products when in a liquid state. Temperatures are determined at specified locations in tanks, ships and barges, tank cars, and tank trucks. For a non-pressure tank, a temperature is obtained by lowering a tank thermometer of proper range through the gaging hatch to the specified liquid level. After the entire thermometer assembly has had time to attain the temperature of the product, the thermometer is withdrawn and read quickly. This procedure is also used for low-pressure tanks equipped with gaging hatches or standpipes, and for any pressure tank that has a pressure lock. For tanks equipped with thermometer wells, temperatures are obtained by reading thermometers placed in the wells with their bulbs at the desired tank levels. If more than one temperature is determined, the average temperature of the product is calculated from the observed temperatures. Electrical-resistance thermometers are sometimes used to determine both average and spot temperatures.

In general, the volume received or delivered is calculated from the observed gage readings. Corrections are made for any *free* water and sediment as determined by the gage of the water level in the tank. The resultant volume is then corrected to the equivalent volume at 15.6°C (60°F) by use of the observed average temperature and the appropriate volume correction table (ASTM D1250). When necessary, a further correction is made for any suspended water and sediment that may be present in materials such as crude petroleum and heavy fuel oils.

For the measurement of other petroleum products, a wide variety of tests is available. In fact, there are approximately 350 tests (ASTM, 2013) that are used to determine the different properties of petroleum products. Each test has its own limits of accuracy and precision that must be adhered to if the data are to be accepted.

3.4   METHOD VALIDATION

Method validation is the process of demonstrating that an analytical method is acceptable for its intended purpose. In general, methods for product specifications and regulatory submission must include studies on specificity, linearity, detection limit, and quantitation limit as well as accuracy and precision.

The process of method development and validation covers all aspects of the analytical procedure, and the best way to minimize method problems is to perform validation experiments during development. However, method validation cannot mitigate out all of the potential problems; the process of method development and validation should address the most common (anticipated) problems.

3.4.1   Requirements

In order to perform validation studies, the approach should be viewed with the understanding that validation requirements are continually changing and vary widely, depending on the type of product under test and compliance with any necessary regulatory group. Furthermore, issues arising from the use of analytical methods increase as additional (new) people unfamiliar with the method, different laboratories, and equipment are used to perform the analytical investigation. For example, a small adjustment (no adjustment is really *small*) can usually be made to make the method work, but the flexibility to change or reverse the adjustment on an as-needed basis is lost once the method is transferred to other analysts or to other analytical laboratories.

The first step in the method development and validation cycle should be to set minimum requirements, which are essentially acceptance specifications for the method. A complete list of criteria should be agreed on during method development with the end users before the method is developed so that expectations are clear. Once the validation studies are complete, the method developers should be confident in the ability of the method to provide good quantitation in their own laboratories. The remaining studies should provide greater assurance that the method will work well in other laboratories, where different operators, instruments, and reagents are involved and where it will be used over much longer periods of time.

The remaining precision studies comprise much of what is often referred to as *ruggedness*. *Intermediate precision* is the precision obtained when an assay is performed by multiple analysts using several instruments on different days in one laboratory. Intermediate precision results are used to identify which of the earlier factors contribute significant variability to the final result.

The last type of precision study is *reproducibility (q.v.)* that is determined by testing homogeneous samples in multiple laboratories, often as part of inter-laboratory crossover studies. The evaluation of reproducibility results often focuses more on measuring bias in results than on determining differences in precision alone. Statistical equivalence is often used as a measure of acceptable inter-laboratory results. An alternative, more practical approach is the use of *analytical equivalence* in which a range of acceptable results is chosen prior to the study and used to judge the acceptability of the results obtained from different laboratories.

Performing a thorough method validation can be a tedious process, but the quality of data generated with the method is directly linked to the quality of this process. Time constraints often do not allow for sufficient method validation. Many researchers have experienced the consequences of invalid methods and realized that the amount of time and resources required to solve problems discovered later exceeds what would have been expended initially if the validation studies had been performed properly. *Putting in time and effort up front* will help any analysts to find a way through the method validation maze and eliminate many of the problems common to inadequately validated analytical methods.

Once the validation studies are complete, there should be a high degree of confidence in the ability of the method to provide good quantitation. Any remaining studies should provide greater assurance that the method will work well in other laboratories, where different operators, instruments, and reagents are involved and where it will be used over much longer periods of time.

3.4.2 Detection Limit

The method detection limit (MDL) is the smallest quantity or concentration of a substance that the instrument can measure (Patnaik, 2004). It is related to the instrument detection limit (IDL) that depends on the type of instrument and its sensitivity, and on the physical and chemical properties of the test substance.

The MDL is, in reality, a statistical concept that is applicable only in trace analysis of certain type of substances, such as organic pollutants by gas chromatographic methods. The MDL measures the minimum detection limit of the method and involves all analytical steps including sample extraction, concentration, and determination by an analytical instrument. Unlike the IDL, the MDL is not confined only to the detection limit of the instrument.

In the environmental analysis of organic pollutants, the MDL is the minimum concentration of a substance that can be measured and reported with 99% confidence that the analyte concentration is greater than zero and is determined from the analysis of a sample in a given matrix containing the analyte. For the determination of the MDL, several replicate analyses are performed at the concentration level of the IDL or at a level equivalent to two to five times the background noise level. The standard deviation of the

replicate tests is found. The MDL is determined by multiplying the standard derivation by the *t-factor*.

In environmental analysis, however, periodic determination of the MDL (e.g., once per year or with any change in personnel, location, or instrument) is part of the quality control (QC) requirement.

3.4.3 Accuracy

The *accuracy* of a test is a measure of how close the test result will be to the true value of the property being measured. As such, the accuracy can be expressed as the *bias* between the test result and the true value. However, the *absolute accuracy* can only be established if the true value is known.

In the simplest sense, a convenient method to determine a relationship between two measured properties is to plot one against the other (Fig. 3.2). Such an exercise will provide either a line fit of the points or a spread that may or may not be within the limits of experimental error. The data can then be used to determine the approximate accuracy of one or more points employed in the plot. For example, a point that lies outside the limits of experimental error (a *flyer*) will indicate an issue of accuracy with that test and the need for a repeat determination.

However, the graphical approach is not appropriate for finding the absolute accuracy between more than two properties. The well-established statistical technique of regression analysis is more pertinent to determining the accuracy of points derived from one property and any number of other properties. There are many instances in which relationships of this sort enable properties to be predicted from other measured properties with as good precision as they can be measured by a single test. It would be possible to examine in this way the relationships between all the specified properties of a product and to establish certain key properties from which the remainder could be predicted, but this would be a tedious task.

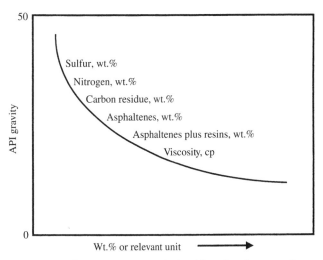

FIGURE 3.2 Inter-property relationships of various samples.

An alternative approach to that of picking out the essential tests in a specification using regression analysis is to take a look at the specification as a whole and extract the essential features (termed *principal components analysis*).

Principal components analysis involves an examination of set of data as points in *n*-dimensional space (corresponding to *n* original tests) and determines (first) the direction that accounts for the biggest variability in the data (*first principal component*). The process is repeated until *n* principal components are evaluated, but not all of these are of practical importance since some may be attributable purely to experimental error. The number of significant principal components shows the number of independent properties being measured by the tests considered.

Following from this, it is necessary to establish the number of independent properties that are necessary to predict product performance in service with the goals of rendering any specification more meaningful and allowing a high degree of predictability of product behavior. On a long-term approach, it might be possible to obtain new tests of a fundamental nature to replace, or certainly to supplement, existing tests. In the short term, selecting the best of the existing tests to define product quality is the most beneficial route to predictability.

3.4.4 Precision

The precision of a test method is the variability between test results obtained on the same material, using a specific test method. The precision of a test is usually unrelated to its accuracy. The results may be precise, but not necessarily accurate. In fact, the precision of an analytical method is the amount of scatter in the results obtained from multiple analyses of a homogeneous sample. To be meaningful, the precision study must be performed using the exact sample and standard preparation procedures that will be used in the final method. Precision is expressed as repeatability and reproducibility.

The *intra-laboratory precision* or the *within-laboratory precision* refers to the precision of a test method when the results are obtained by the same operator in the same laboratory using the same apparatus. In some cases, the precision is applied to data gathered by a different operator in the same laboratory using the same apparatus. Thus, intra-laboratory precision has an expanded meaning insofar as it can be applied to laboratory precision.

Repeatability or repeatability interval of a test (*r*) is the maximum permissible difference due to test error between two results obtained on the same material in the same laboratory.

$$r = 2.77 \times \text{standard deviation of test}$$

The repeatability interval (*r*) is, statistically, the 95% probability level, or the differences between two test results are

unlikely to exceed this repeatability interval more than five times in a hundred.

The *inter-laboratory precision* or the *between-laboratory precision* is defined in terms of the variability between test results obtained on the aliquots of the same homogeneous material in different laboratories using the same test method.

The term *reproducibility* or *reproducibility interval* (*R*) is analogous to the term repeatability, but it is the maximum permissible difference between two results obtained on the same material but now in different laboratories. Therefore, differences between two or more laboratories should not exceed the reproducibility interval more than five times in a hundred.

$$R = 2.77 \times \text{standard deviation of test}$$

The repeatability value and the reproducibility value have important implications for quality (Ralli et al., 2003). As the demand for clear product specifications, and hence control over product consistency, grows, it is meaningless to establish product specifications that are more restrictive than the reproducibility/repeatability values of the specification test methods.

3.5 QUALITY CONTROL AND QUALITY ASSURANCE

Quality control (QC) and quality assurance (QA) programs are key components of all analytical protocols in all areas of analysis, including environmental, pharmaceutical, and forensic testing, among others (Patnaik, 2004). These programs mandate that the laboratories follow a set of well-defined guidelines to achieve valid analytical results to a high degree of reliability and accuracy within an acceptable range. Although such programs may vary depending on the regulatory authority, certain key features of these programs are more or less the same.

However, there is often confusion between the terms *quality assurance* and *quality control*, perhaps because there is also considerable overlap between certain aspects of QA and QC programs.

3.5.1 Quality Control

QCs are single procedures that are performed in conjunction with the analysis to help assess in a quantitative manner the success of the individual analysis. Examples of QCs are blanks, calibration, calibration verification, surrogate additions, matrix spikes, laboratory control samples, performance evaluation samples, and determination of detection limits. The success of the QC is evaluated against an acceptance limit. The actual generation of the acceptance limit is a function of QA; it would not be termed a QC.

Unlike QA plans that mostly address regulatory requirements involving comprehensive documentation, QC programs are science-based, the components of which may be defined statistically. The two most important components of QC are (i) determination of precision of analysis and (ii) determination of accuracy of measurement.

Whereas *precision* is a measure of the reproducibility of data from replicate analyses, the *accuracy* of a test estimates how accurate are the data, that is, how close the data would fall to probable true values or how accurate is the analytical procedure to give results that may be close to true values. Both the precision and accuracy are measured on one or more samples selected at random from a given batch of samples for analysis. The precision of analysis is usually determined by running duplicate or replicate tests on one of the samples in a given batch of samples. It is expressed statistically as standard deviation, relative standard deviation (RSD), coefficient of variance, standard error of the mean (*M*), and the relative percent difference (RPD).

The standard deviation in measurements, however, can vary with the concentrations of the analytes. On the other hand, the RSD, which is expressed as the ratio of standard deviation to the arithmetic mean of replicate analyses and is given as a percent, does not have this problem and is a more rational way of expressing precision:

$$\text{RSD} = -\left(\frac{\text{Standard deviation}}{\text{Arithmetic mean of replicate analysis}} \right) \times 100\%$$

The standard error of the mean, *M*, is the ratio of the standard deviation and the square root of the number of measurements (*n*):

$$M = \frac{\text{Standard deviation}}{n}$$

This scale, too, will vary in the same proportion as standard deviation with the size of the analyte in the sample.

In routine testing, many repeat analyses of a sample aliquot may not be possible. Alternatively, therefore, the precision of a test may be determined from duplicate analyses of sample aliquots and expressed as the RPD:

$$\text{RPD} = \frac{a_1 - a_2}{(a_1 + a_2)/2} \times 100\%$$

In another iteration,

$$\text{RPD} = \frac{a_2 - a_1}{(a_1 + a_2)/2} \times 100\%$$

In this equation, a_1 and a_2 are the results of duplicate analyses of a sample. Since only two tests are performed on a selected sample, the RPD may not be as accurate a measure of precision as the RSD, and since the RSD does not vary with sample size, it should be used whenever possible to estimate precision of analysis from replicate tests.

The accuracy of an analysis can be determined by several procedures. One common method is to analyze a "known" sample, such as a standard solution or a QC check standard solution that may be commercially available or a laboratory-prepared standard solution made from a neat compound, and compare the test results with the true values (theoretically expected values). Such samples must be subjected to all analytical steps, including sample extraction, digestion, or concentration, similar to regular samples. Alternatively, accuracy may be estimated from the recovery of a known standard solution *spiked* or added into the sample in which a known amount of the same substance that is to be tested is added to an aliquot of the sample, usually as a solution, prior to the analysis. The concentration of the analyte in the spiked solution of the sample is then measured. The percent spike recovery is then calculated. A correction for the bias in the analytical procedure can then be made, based on the percent spike recovery. However, in most routine analysis, such bias correction is not required. Percent spike recovery then may be calculated as follows:

$$\text{Recovery}\,(\%) = \left(\frac{\text{Measured concentration}}{\text{Theoretical concentration}} \right) \times 100\%$$

The percent spike recovery to measure the accuracy of analysis may also be determined by the US Environmental Protection Agency method often used in environmental analysis:

$$\text{Recovery}\,(\%) = \frac{100\left(X_s - X_u\right)}{K}$$

X_s is the measured value for the spiked sample, X_u is the measured value for the unspiked sample adjusted for the dilution of the spike, and *K* is the known value of spike in the sample.

3.5.2 Quality Assurance

QA is an umbrella term that is correctly applied to everything that the laboratory does to assure product reliability. As the product of a laboratory is information, anything that is done to improve the reliability of the generated information falls under QA.

QA includes all the QCs, the generation of expectations (acceptance limits) from the QCs, plus a great number of other activities such as (i) analyst training and certification; (ii) data review and evaluation; (iii) preparation of final reports of analysis; (iv) information given to clients about tests that are needed to fulfill regulatory requirements; (v) use of the appropriate tests in the laboratory; (vi) obtaining and maintaining laboratory certifications/accreditations; (vii) conducting internal and external audits; (viii) preparing responses to the audit results; (ix) the receipt, storage, and tracking of samples; and (x) tracking the acquisition of standards and reagents.

Thus, the objective of QA (usually in the form of a *QA plan*) is to obtain reliable and accurate analytical results that may be stated with a high level of confidence (statistically), so that such results are legally defensible. The key features of any plan involve essentially documentation and record keeping. In short, QA program involves documentation of sample collection for testing, the receipt of samples in the laboratory, and their transfer to the individuals who perform the analyses. The information is recorded on chain-of-custody forms stating the dates and times along with the names and signatures of individuals who carry out these tasks. Also, other pertinent information is recorded, such as any preservatives added to the sample to prevent degradation of test analytes, the temperature at which the sample is stored, the temperature to which the sample is brought prior to its analysis, the nature of the container (which may affect the stability of the sample), and its holding time prior to testing.

In fact, the performance of QC is just one small aspect of the QA program. The functions of QA are embodied in the terms *analytically valid* and (in these days of perpetual litigation) *legally defensible.*

Analytically valid means that the target analyte has been (i) correctly identified and (ii) quantified using fully calibrated tests. In addition, the sensitivity of the test (MDL) has been established. And the analysts have demonstrated that they are capable of performing the test. The accuracy and precision of the test on the particular sample must also have been determined, and the possibility of false-positive and false-negative results has been evaluated through performance of blanks and other test-specific interference procedures.

3.6 ASSAY AND SPECIFICATIONS

Petroleum exhibits wide variations in composition and properties, and these occur not only in petroleum from different fields but also in oils taken from different production depths in the same well. Historically, physical properties such as boiling point, density (gravity), and viscosity have been used to describe petroleum, but the needs are even more extensive as increased amounts of heavy feedstocks are introduced into refineries (Table 3.1).

Petroleum analysis involves not only determining the composition of the material under investigations but, more appropriately, determining the suitability of the petroleum for refining or the product for use. In this sense, the end

TABLE 3.1 Analytical inspections for petroleum, heavy oil, extra heavy oil, tar sand bitumen (oil sand bitumen), residua

Petroleum	Heavy feedstocks
Density (specific gravity)	Density (specific gravity)
API gravity	API gravity
Carbon (wt.%)	Carbon (wt.%)
Hydrogen (wt.%)	Hydrogen (wt.%)
Nitrogen (wt.%)	Nitrogen (wt.%)
Sulfur (wt.%)	Sulfur (wt.%)
	Nickel (ppm)
	Vanadium (ppm)
	Iron (ppm)
Pour point	Pour point
Wax content	
Wax appearance temperature	
Viscosity (various temperatures)	Viscosity (various temperatures and at reservoir temperature)
Carbon residue of residuum	Carbon residue*
	Ash (wt.%)
Distillation profile	Fractional composition:
All fractions plus vacuum residue	Asphaltenes (wt.%)
	Resins (wt.%)
	Aromatics (wt.%)
	Saturates (wt.%)

*Conradson carbon residue or microcarbon residue.

product of petroleum analysis (or testing) is a series of data that allow the investigator to *specify* the character and quality of the material under investigation. Thus, a series of specifications are determined for petroleum and its products.

Because of the differences in petroleum composition, the importance of the correct sampling of crude oil that contains light hydrocarbons cannot be overestimated. Properties such as specific gravity, distillation profile, vapor pressure, hydrogen sulfide content, and octane number of gasoline are affected by the light hydrocarbon content so that suitable cooling or pressure sampling methods have to be used and care taken during the subsequent handling of the oil in order to avoid the loss of any volatile constituents. In addition, adequate records of the circumstances and conditions during sampling have to be made. For example, sampling from oilfield separators, the temperatures and pressures of the separation plant, and the atmospheric temperature should be noted.

Hence, the production of data focuses on (i) measurement, (ii) accuracy, (iii) precision, and (iv) method validation, all of which depend upon the sampling protocols that were used to obtain the sample. Without strict sampling protocols, variation and loss of accuracy (or precision) must be anticipated. For example, correct sampling of the product in storage or carrier tanks is important in order to obtain a representative sample for the laboratory tests that are essential in converting measured quantities to the standard volume.

Elemental analyses of petroleum show that it contains mainly carbon and hydrogen. Nitrogen, oxygen, and sulfur (heteroelements) are present in smaller amounts and also trace elements such as vanadium, nickel, etc. Of the heteroelements, sulfur is the most important. The mixture of hydrocarbons is highly complex. Paraffinic, naphthenic, and aromatic structures can occur in the same molecule, and the complexity increases with boiling range. The attempted classification of crude oils in terms of these three main structural types has proved inadequate.

3.6.1 Assay

An assay, as the term is used in the petroleum industry, is the application of a series of test to petroleum, heavy oil, extra heavy oil, or tar sand bitumen that present data upon which the character and processability of the feedstock can be estimated (Table 3.1). Since petroleum is a naturally occurring mixture of hydrocarbons, generally in a liquid state, that may also include compounds of sulfur, nitrogen, oxygen, metals, and other elements (ASTM D4175). Consequently, it is not surprising that petroleum can vary in composition properties and produce wide variations in refining behavior as well as product properties. Thus, a feedstock assay is a necessary preliminary step before refining.

The petroleum being processed in refineries is becoming increasingly heavier (higher amounts of residuum and higher sulfur content) (Speight, 2002, 2014a and references cited therein). Market demand (*market pull*) dictates that *residua* must be upgraded to higher-value products (Speight and Ozum, 2002; Hsu and Robinson, 2006; Gary et al., 2007; Speight, 2014a). In short, the value of petroleum depends upon its quality for refining and whether or not a product slate can be obtained to fit market demand.

Thus, process units in a refinery require analytical test methods that can adequately evaluate feedstocks and monitor product quality. In addition, the high sulfur content of petroleum and regulations limiting the maximum sulfur content of fuels make sulfur removal a priority in refinery processing. Here again, analytical methodology is key to the successful determination of the sulfur compound types present and their subsequent removal.

Upgrading residua involves processing (usually conversion) into a more salable, higher-valued product. Improved characterization methods are necessary for process design, crude oil evaluation, and operational control. Definition of the boiling-range and the hydrocarbon-type distribution in heavy distillates and in residua is increasingly important. Feedstock analysis to provide a quantitative boiling range distribution (that accounts for non-eluting components) as well as the distribution of hydrocarbon types in gas oil and higher-boiling materials is important in evaluating feedstocks for further processing.

Sulfur reduction processes are sensitive to both amount and structure of the sulfur compounds being removed. Tests that can provide information about both are becoming increasingly important, and analytical tests that provide information about other constituents of interest (e.g., nitrogen, organometallic constituents) are also valuable and being used for characterization.

But before emerging into the detailed aspects of petroleum product analysis, it is necessary to understand the nature and character of petroleum as well as the methods used to produce petroleum products. This will present to the reader the background that is necessary to understand petroleum and the processes used to convert it to products. The details of the chemistry are not presented here and can be found elsewhere (Speight, 2014a; Speight and Ozum, 2002).

The underlying premise for these methods of definition or classification is uniformity of the molecular nature of the feedstocks. This is not the case. And when blends are employed as refinery feedstocks, the methods do not take into account any potential interactions between the constituents of each member of the blend.

The most adequate definitions of petroleum come from legal documents where petroleum is defined directly or by inference.

In all of these attempts at a definition or classification of petroleum, it must be remembered that petroleum exhibits wide variations in composition and properties, and these variations occur not only in petroleum from different fields but may also be manifested in petroleum taken from different production depths in the same well. The mixture of hydrocarbons is highly complex. Paraffinic, naphthenic, and aromatic structures can occur in the same molecule, and the complexity increases with the boiling range of the petroleum fraction. In addition, petroleum varies in physical appearance from a light-colored liquid to the more viscous heavy oil. The near-solid or solid bitumen that occurs in tar sand deposits is different from petroleum and heavy oil, as evidenced by the respective methods of recovery.

Elemental analyses of petroleum show that the major constituents are carbon and hydrogen with smaller amounts of sulfur (0.1–8% w/w), nitrogen (0.1–1.0% w/w), and oxygen (0.1–3% w/w) and trace elements such as vanadium, nickel, iron, and copper present at the part-per-million (ppm) level. Of the non-hydrocarbon (heteroelements) elements, sulfur is the most abundant and often considered the most important by refiners. However, nitrogen and the trace metals also have deleterious effects on refinery catalysts and should not be discounted because of relative abundance. Process units that have, for example, a capacity of 50,000 bbl/day and which are in operation continuously can soon reflect the presence of the trace elements. The effects of oxygen, which also has an effect on refining catalysts, has received somewhat less study than the other heteroelements but remains equally important in refining.

Petroleum suitability for refining (to produce a slate of predetermined products) is determined by application of a series of analytical methods (Speight, 2014a) that provide information that is sufficient to assess the potential quality of the petroleum as a feedstock and also to indicate whether any difficulties might arise in handling, refining, or transportation. Such information may be obtained either by (i) a preliminary assay of petroleum or by (ii) a full assay of petroleum that involves presentation of a true boiling point (TBP) curve and the analysis of fractions throughout the full range of petroleum.

Moreover, the suitability of a petroleum product for a specific use cannot be assumed, and the suitability must be determined by a sequence of analytical test data that must be in agreement with (or within the range of) the data presented as part of the specifications. Thus, an efficient assay is derived from a series of test data that give an accurate description of petroleum quality and petroleum product quality and allow an indication of its behavior during refining or service. The first step is, of course, to assure adequate (correct) sampling by use of the prescribed protocols (ASTM D4057).

Thus, analyses are performed to determine whether each batch of crude oil received at the refinery or each batch of the product is suitable for refining purposes or for the designated use. The tests are also applied to determine if there has been any contamination during storage or transportation that may increase the processing difficulty (cost). The information required is generally specific to a particular refinery and is also a function of refinery operations and desired product slate.

To obtain the necessary information, two different analytical schemes are commonly used and these are (i) an inspection assay and (ii) a comprehensive assay.

Inspection assays usually involve determination of several key bulk properties of the sample (such as API gravity, sulfur content, pour point, and distillation range) as a means of determining if *major* changes in characteristics have occurred since the last comprehensive assay was performed. For example, a more detailed inspection assay might consist of the following tests: API gravity (or density or relative density), sulfur content, pour point, viscosity, salt content, water and sediment content, trace metals (or organic halides). The results from these tests—as well as the archived data from a comprehensive assay—provide an estimate of any changes that have occurred in the crude oil that may be critical to refinery operations. Inspection assays are routinely performed on all crude oils received at a refinery.

On the other hand, the comprehensive (or full) assay is more complex (as well as time consuming and costly) and is usually performed only when a new field comes on stream, or when the inspection assay indicates that significant changes in the composition of the crude oil have occurred. Except for these circumstances, a comprehensive assay of a particular crude oil stream may not (unfortunately) be updated for several years.

3.6.2   Specifications

A *feedstock specification* or *product specification* is the data that give adequate control of feedstock behavior in a refinery or product quality. More accurately, the specifications are derived from the set of tests and data limits applicable to the crude oil or to a finished product in order to ensure that every batch is of satisfactory and consistent quality at release for sales. The specifications should include all critical parameters in which variations would be likely to affect the safety and in-service use of the product—they are, in fact, part of the assay.

Often, the specifications against which a finished product is tested before release for sale are referred to as the *batch release specifications*. These are the product data against which the finished product is tested to ensure satisfactory quality. If the product should be in service before a certain date, the purchases will be advised (by law) through a note on the product of the *expiry specifications* and the product, if tested at any time within the functional life period, must comply with the requirements in the expiry specifications.

In terms of the whole crude oil, a specification offers the luxury of *predictability* of feedstock behavior in a refinery or *predictability* of product quality (therefore, product behavior) relative to market demand. Ultimately, feedstock behavior and/or product quality are judged by an assessment of performance. And it is *performance* that is the ultimate criterion of quality. It is therefore necessary to determine those properties, the values of which can be established precisely and relatively simply by inspection tests in a control laboratory that correlate closely with the important performance properties.

The value of crude oil to a refiner depends upon its quality and whether he can economically obtain a satisfactory product pattern that matches market demand (*market pull*). In the main, the refiner is not concerned with the actual chemical nature of the material but with methods of analysis that would provide information sufficient to assess the potential quality of the oil, to supply preliminary engineering data, and also to indicate whether any difficulties might arise in handling, refining, or transporting petroleum or its products. Such information may be obtained in one of two ways: (i) inspection data from a preliminary assay, and (ii) inspection data from a full assay, which involves the derivation of a TBP curve and the analysis of fractions and product blends throughout the full range of the crude oil.

The *preliminary assay* provides general data on the oil and is based on simple tests such as distillation range, water content, specific gravity, and sulfur content that enable desirable or undesirable features to be noted. This form of assay requires only a small quantity of sample and is therefore particularly useful for the characterization of oilfield samples produced from cores, drill stem tests, or seepages.

The tests in the preliminary assay are relatively simple and can be completed in a short time and generally on a routine basis. This assay gives a useful general picture of the quality of petroleum, but it does not cover the work necessary to provide adequate data, for example, for the design of refinery equipment, nor does it produce a sufficient quantity of the various products from the crude so that they can be examined for quality. A *full assay* of petroleum is based on a TBP distillation of the crude, and sufficient data are obtained to assess the yields and properties of the straight-run products, covering light hydrocarbons; light, middle, and heavy distillate; lubricants; residual fuel oil; and residuum. Often, the middle-ground is reached between the preliminary assay and the full assay, but the requirements may also be feedstock dependent (Table 3.1).

Sometimes the inspection tests attempt to measure these properties, for example, the carbon residue of a feedstock that is an approximation of the amount of the thermal coke that will be formed during refining. Or, a research octane number test that was devised to measure performance of motor fuel. In other cases, the behavior must be determined indirectly from a series of test results.

The carbon residue of petroleum and petroleum products serve as an indication of the propensity of the sample to form carbonaceous deposits (thermal coke) under the influence of heat.

Tests for Conradson carbon residue (ASTM D189, IP 13), the Ramsbottom carbon residue (ASTM D524, IP 14), the microcarbon carbon residue (ASTM D4530, IP 398), and asphaltene content (ASTM D893, ASTM D2007, ASTM D3279, ASTM D4124, ASTM D6560, IP 143) are sometimes included in inspection data on petroleum. The data give an indication of the amount of coke that will be formed during thermal processes as well as an indication of the amount of high-boiling constituents in petroleum.

The determination of the *carbon residue* of petroleum or a petroleum product is applicable to relatively nonvolatile samples that decompose on distillation at atmospheric pressure. Samples that contain ash-forming constituents will have an erroneously high carbon residue, depending upon the amount of ash formed. All three methods are applicable to relatively nonvolatile petroleum products that partially decompose on distillation at atmospheric pressure. Crude oils having a low carbon residue may be distilled to a specified residue and the carbon residue test of choice then applied to the residue.

In the Conradson carbon residue test (ASTM D189, IP 13), a weighed quantity of sample is placed in a crucible and subjected to destructive distillation for a fixed period of severe heating. At the end of the specified heating period, the test crucible containing the carbonaceous residue is cooled in a desiccator and weighed, and the residue is reported as a percentage (% w/w) of the original sample (Conradson carbon residue).

In the Ramsbottom carbon residue test (ASTM D524, IP 14), the sample is weighed into a glass bulb that has a capillary opening and is placed into a furnace (at 550°C, 1020°F). The volatile matter is distilled from the bulb, and the nonvolatile matter that remains in the bulb cracks to form thermal coke. After a specified heating period, the bulb is removed from the bath, cooled in a desiccator, and weighed to report the residue (Ramsbottom carbon residue) as a percentage (% w/w) of the original sample.

In the microcarbon residue test (ASTM D4530, IP 398), a weighed quantity of the sample placed in a glass vial is heated to 500°C (930°F) under an inert (nitrogen) atmosphere in a controlled manner for a specific time, and the carbonaceous residue [*carbon residue (micro)*] is reported as a percentage (% w/w) of the original sample.

The data produced by the microcarbon test (ASTM D4530, IP 398) are equivalent to those by Conradson Carbon method (ASTM D189, IP 13). However, this microcarbon test method offers better control of test conditions and requires a smaller sample. Up to 12 samples can be run simultaneously. This test method is applicable to petroleum and to petroleum products that partially decompose on distillation at atmospheric pressure and is applicable to a variety of samples that generate a range of yields (0.01–30% w/w) of thermal coke.

As noted, in any of the carbon residue tests, ash-forming constituents (ASTM D482) or nonvolatile additives present in the sample will be included in the total carbon residue reported leading to higher carbon residue values and erroneous conclusions about the coke-forming propensity of the sample.

The asphaltene fraction (ASTM D893, ASTM D2007, ASTM D3279, ASTM D4124, ASTM D6560, IP 143) is the highest-molecular-weight and most complex fraction in petroleum. The yield of the asphaltene fraction gives an indication of the amount of coke that can be expected during processing (Speight, 2001, Speight and Ozum, 2002).

In any of the methods for the determination of the asphaltene content, the crude oil or product (such as asphalt) is mixed with a large excess of (usually >30 volumes of hydrocarbon per volume of sample) low-boiling hydrocarbon such as *n*-pentane or *n*-heptane. For an extremely viscous sample, a solvent such as toluene may be used prior to the addition of the low-boiling hydrocarbon, but an additional amount of the hydrocarbon (usually >30 volumes of hydrocarbon per volume of solvent) must be added to compensate for the presence of the solvent. After a specified time, the insoluble material (the asphaltene fraction) is separated (by filtration) and dried. The yield is reported as percentage (% w/w) of the original sample.

It must be recognized that, in any of these tests, different hydrocarbons (such as *n*-pentane or *n*-heptane) will give different yields of the asphaltene fraction, and if the presence of the solvent is not compensated by the use of additional hydrocarbon, the yield will be erroneous. In addition, if the hydrocarbon is not present in large excess, the yields of the asphaltene fraction will vary and will be erroneous.

The *precipitation number* is often equated to the asphaltene content, but there are several issues that remain obvious in its rejection for this purpose. For example, the method to determine the precipitation number (ASTM D91) advocates the use of naphtha for use with black oil or lubricating oil, and the amount of insoluble material (as a % v/v of the sample) is the precipitating number. In the test, 10 ml of sample is mixed with 90 ml of ASTM-specified precipitation naphtha (that may or may not have a constant chemical composition) in a graduated centrifuge cone and centrifuged for 10 min at 600–700 rpm. The volume of material on the bottom of the centrifuge cone is noted until repeat centrifugation gives a value within 0.1 ml (the precipitation number). Obviously, this can be substantially different from the yield of the asphaltene fraction.

For clarification, it is necessary to understand the basic definitions that are used to define the density of crude oil. Thus, the *density* is the mass of liquid per unit volume at 15°C, whereas the *relative density* is the ratio of the mass of a given volume of liquid at 15°C to the mass of an equal volume of pure water at the same temperature, and the *specific gravity* is the same as the relative density, and the terms are used interchangeably.

Density (ASTM D1298, IP 160) is an important property of petroleum products since petroleum and especially petroleum products are usually bought and sold on that basis or if on volume basis then converted to mass basis via density measurements. This property is almost synonymously termed as density, relative density, gravity, and specific gravity, all terms related to each other. Usually a hydrometer, pycnometer, or more modern digital density meter is used for the determination of density or specific gravity (ASTM, 2013; Speight, 2014a).

In the most commonly used method (ASTM D1298, IP 160), the sample is brought to the prescribed temperature and transferred to a cylinder at approximately the same temperature. The appropriate hydrometer is lowered into the sample and allowed to settle, and after temperature equilibrium has been reached, the hydrometer scale is read and the temperature of the sample is noted.

Although there are many methods for the determination of density due to the different nature of petroleum itself and the different products, one test method (ASTM D5002) is used for the determination of the density or relative density of petroleum that can be handled in a normal fashion as liquids at test temperatures between 15 and 35°C (59 and 95°F). This test method applies to petroleum oils with high vapor pressures provided appropriate precautions are taken to prevent vapor loss during transfer of the sample to the density analyzer. In the method, approximately 0.7 ml of crude oil sample is introduced into an oscillating sample tube, and the change in oscillating frequency caused by the change in mass of the tube is used in conjunction with calibration data to determine the density of the sample.

Another test determines density and specific gravity by means of a digital densimeter (ASTM D4052, IP 365). In the test, a small volume (~0.7 ml) of liquid sample is introduced into an oscillating sample tube, and the change in oscillating frequency caused by the change in the mass of the tube is used in conjunction with calibration data to determine the density of the sample. The test is usually applied to petroleum, petroleum distillates, and petroleum products that are liquids at temperatures between 15 and 35°C (59 and 95°F) that have vapor pressures below 600 mm Hg and viscosities below about 15,000 cSt at the temperature of test. However, the method should not be applied to samples so dark in color that the absence of air bubbles in the sample cell cannot be established with certainty.

Accurate determination of the density or specific gravity of crude oil is necessary for the conversion of measured volumes to volumes at the standard temperature of 15.56°C (60°F) (ASTM D1250, IP 200). The specific gravity is also a factor reflecting the quality of crude oils.

The accurate determination of the API gravity of petroleum and its products is necessary for the conversion of measured volumes to volumes at the standard temperature of 60°F (15.56°C). Gravity is a factor governing the quality of crude oils. However, the gravity of a petroleum product is an uncertain indication of its quality. Correlated with other properties, gravity can be used to give approximate hydrocarbon composition and heat of combustion. This is usually accomplished through use of the API gravity that is derived from the specific gravity:

$$\text{API gravity}\left(\text{API}\right) = \frac{141.5}{\text{sp gr} 60/60°\text{F}} - 131.5$$

The API gravity is also an important measure for estimating the quality and refinability of petroleum.

API gravity or density or relative density can be determined using one of two hydrometer methods (ASTM D287, ASTM D1298). The use of a digital analyzer (ASTM D5002) is finding increasing popularity for the measurement of density and specific gravity.

In the method (ASTM D287), the API gravity is determined using a glass hydrometer for petroleum and petroleum products that are normally handled as liquids and that have a Reid vapor pressure of 26 psi (180 kPa) or less. The API gravity is determined at 15.6°C (60°F), or converted to values at 60°F, by means of standard tables. These tables are not applicable to non-hydrocarbons or essentially pure hydrocarbons such as the aromatics.

This test method is based on the principle that the gravity of a liquid varies directly with the depth of immersion of a body floating in it. The API gravity is determined using a hydrometer by observing the freely floating API hydrometer and noting the graduation nearest to the apparent intersection of the horizontal plane surface of the liquid with the vertical scale of the hydrometer, after temperature equilibrium has been reached. The temperature of the sample is determined using a standard test thermometer that is immersed in the sample or from the thermometer that is an integral part of the hydrometer (thermohydrometer).

3.6.2.1 Distillation

The distillation tests give an indication of the types of products and the quality of the products that can be obtained from the petroleum, and the tests are used to compare different petroleum through the yield and quality of the 300°C (570°F) residuum fraction. For example, the waxiness or viscosity of this fraction gives an indication of the amount, types, and quality of the residual fuel that can be obtained from the petroleum. In this respect, the determination of the aniline point (ASTM D611, IP 2) can be used to determine the aromatic or aliphatic character of petroleum. Although not necessarily the same as the wax content, correlative relationships can be derived from the data.

The basic method of distillation (ASTM D86) is one of the oldest methods in use since the distillation characteristics of hydrocarbons have an important effect on safety and performance, especially in the case of fuels and solvents. The boiling range gives information on the composition, the properties, and the behavior of petroleum and derived products during storage and use. Volatility is the major determinant of the tendency of a hydrocarbon mixture to produce potentially explosive vapors. Several methods are available to define the distillation characteristics of petroleum and its various petroleum products. In addition to these physical methods, other test methods based on gas chromatography are also used to derive the boiling point distribution of a sample (ASTM D2887, ASTM D3710, ASTM D6352).

In the preliminary assay of petroleum, the method of distillation is often used to give a rough indication of the boiling range of the crude (ASTM D2892, IP 123). The test method is used for the distillation of stabilized (gases-removed) crude oil to a final-cut temperature of up to 400°C (750°F) *atmospheric equivalent temperature* (AET). The crude oil is heated and separated by the distillation column into lower-boiling products such as naphtha and kerosene. The distillate and the residuum can be further examined by testing, for example, specific gravity (ASTM D1298, IP 160), sulfur content (ASTM D129, IP 61) and viscosity (ASTM D445, IP 71). In fact, using a method (ASTM D2569) developed for determining the distillation characteristics of pitch allows further detailed examination of residua.

In addition to the whole crude oil tests performed as part of the inspection assay, a comprehensive or full assay requires that the crude be fractionally distilled and the fractions characterized by the relevant tests. Fractionation of the crude oil begins with a TBP distillation employing a fractionating column having an efficiency of 14–18 theoretical plates and operated at a reflux ratio of 5:1 (ASTM D2892). The TBP distillation may be used for all fractions up to a maximum cut point of approximately 350°C AET, but low residence time in the still (or reduced pressure) is needed to minimize cracking.

It is often useful to extend the boiling point data to higher temperatures than are possible in the fractionating distillation method previously described, and for this purpose, a vacuum distillation in a simple still, with no fractionating column (ASTM D1160), can be carried out. This distillation, which is done under fractionating conditions equivalent to one theoretical plate, allows the boiling point data to be extended to about 600°C (1112°F) with many crude oils. This method gives useful comparative and reproducible results that are often accurate enough for refinery purposes, provided significant cracking does not occur.

Usually seven fractions provide the basis for a reasonably thorough evaluation of the distillation properties of the feedstock: (i) gas, boiling range: <15.6°C/60°F; (ii) gasoline (light naphtha), boiling range: 15.6–150°C/60–300°F; (iii) kerosene (medium naphtha), boiling range: 150–230°C/300–450°F; (iv) gas oil, boiling range: 230–345°C/450–650°F; (v) light vacuum gas oil boiling range: 345–370°C/650–700°F; (vi) heavy vacuum gas oil, boiling range: 370–565°C/700–1050°F; and (vii) residuum, boiling range: >565°C/1050°F. From 5 to 50 l of crude oil are necessary to complete a full assay, depending on the number of cuts to be taken and the tests to be performed on the fractions.

A more recent test method (ASTM D5236) is seeing increasingly more use and appears to be the method of choice for crude assay vacuum distillations. The method employs a vacuum pot still with a low pressure drop entrainment separator operated under total takeoff conditions. The reduced pressure allows volatilization at a lower temperature than under atmospheric conditions, thus allowing temperatures up to 565°C (1050°F) for most samples and which avoids thermal decomposition (cracking) of the oil (caused by prolonged exposure to temperatures in excess of above 350°C, 650°F). The test method applies to the higher-boiling fractions of petroleum that are in the gas oil and lubricating oil range as well as heavy crude oils and residua.

Wiped-wall or thin-film molecular stills can also be used to separate the higher-boiling fractions under conditions that minimize cracking. In these units, however, cut points cannot be directly selected, because vapor temperature in the distillation column cannot be measured accurately under operating conditions. Instead, the wall (film) temperature, pressure, and feed rate that will produce a fraction with a given end point are determined from in-house correlations developed by matching yields between the wiped-wall distillation and the conventional distillation (ASTM D1160, ASTM D5236). And wiped-wall stills are often used because they allow higher end points and can easily provide sufficient quantities of the fractions for characterization purposes.

3.6.2.2 Low-Boiling Hydrocarbons

The amount of the individual light hydrocarbons in petroleum (methane to butane or pentane) is often included as part of the preliminary assay. For safety reasons, it is essential to know the content of low-boiling hydrocarbons in crude oil prior to handling and shipping the crude oil.

For example, Bakken crude oil is a light sweet (low-sulfur) crude oil that has a relatively high proportion of volatile constituents. The production of the oil yields not only petroleum but also a significant amount of volatile gases (including propane and butane) and low-boiling liquids (such as pentane and natural gasoline), which are often referred to collectively as (low-boiling or light) naphtha. By definition, natural gasoline (sometimes also referred to as *gas condensate*) is a mixture of low-boiling liquid hydrocarbons isolated from petroleum and natural gas wells suitable for blending with light naphtha and/or refinery gasoline (Mokhatab et al., 2006; Speight, 2007, 2014a). Because of the presence of low-boiling hydrocarbons, light naphtha can become extremely explosive, even at relatively low ambient temperatures. Some of these gases may be burned off (flared) at the field wellhead, but others remain in the liquid products extracted from the well (Speight, 2014a).

Although one of the more conventional distillation procedures might be employed, the determination of light hydrocarbons in petroleum is best carried out using a gas chromatographic method (ASTM D2427).

3.6.2.3 Metallic Constituents

Petroleum, as recovered from the reservoir, contains metallic constituents but also picks up metallic constituents during recovery, transportation, and storage. Even trace amounts of these metals can be deleterious to refining processes, especially processes where catalysts are used. Trace components, such as metallic constituents, can also produce adverse effects in refining either (i) by causing corrosion or (ii) by affecting the quality of refined products.

Hence, it is important to have test methods that can determine metals, both at trace levels and at major concentrations. Thus, test methods have evolved that are used for the determination of specific metals, as well as for the multielement methods, using techniques such as atomic absorption (AA) spectrometry, inductively coupled plasma atomic emission spectrometry (ICP-AES), and X-ray fluorescence spectroscopy.

Nickel and vanadium along with iron and sodium (from the brine) are the major metallic constituents of crude oil.

These metals can be determined by AA spectrophotometric methods (ASTM D5863, IP 285, IP 465), wave length–dispersive X-ray fluorescence spectrometry (IP 433), and ICP-AES. Several other analytical methods are available for the routine determination of trace elements in crude oil, some of which allow direct aspiration of the samples (diluted in a solvent) instead of the time-consuming sample preparation procedures such as wet ashing (acid decomposition) or flame or dry ashing (removal of volatile/combustible constituents) (ASTM D5863). Among the techniques used for trace element determinations are conductivity (IP 265), flameless and flame AA spectrophotometry (ASTM D5863) and ICP spectrophotometry (ASTM D5708).

ICP-AES (ASTM D5708) has an advantage over AA spectrophotometry (ASTM D4628, ASTM D5863) because it can provide more complete elemental composition data than the AA method. Flame emission spectroscopy is often used successfully in conjunction with AA spectrophotometry (ASTM D3605). X-ray fluorescence spectrophotometry (ASTM D4927, ASTM D6443) is also sometimes used, but matrix effects can be a problem.

The method to be used for the determination of metallic constituents in petroleum is often a matter of individual preference.

3.6.2.4 Salt Content

The salt content of crude oil is highly variable and results principally from production practices used in the field and, to a lesser extent, from its handling aboard tankers bringing it to terminals. The bulk of the salt present will be dissolved in coexisting water and can be removed in desalters, but small amounts of salt may be dissolved in the crude oil itself. Salt may be derived from reservoir or formation waters, or from other waters used in secondary recovery operations. Aboard tankers, ballast water of varying salinity may also be a source of salt contamination.

Salt in crude oil may be deleterious in several ways. Even in small concentrations, salts will accumulate in stills, heaters, and exchangers leading to fouling that requires expensive cleanup. More importantly, during flash vaporization of crude oil, certain metallic salts can be hydrolyzed to hydrochloric acid according to the following reactions:

$$2NaCl + H_2O \rightarrow 2HCl + Na_2O$$
$$MgCl_2 + H_2O \rightarrow 2HCl + MgO$$

The hydrochloric acid evolved is extremely corrosive, necessitating the injection of a basic compound, such as ammonia, into the overhead lines to minimize corrosion damage. Salts and evolved acids can also contaminate both overhead and residual products, and certain metallic salts can deactivate catalysts.

Thus, knowledge of the content of salt in crude oil is important in deciding whether or not and to what extent the crude oil needs desalting.

The salt content is determined by potentiometric titration in a nonaqueous solution in which the conductivity of a solution of crude oil in a polar solvent is compared to that of a series of standard salt solutions in the same solvent (ASTM D3230). In the method, the sample is dissolved in a mixed solvent and placed in a test cell consisting of a beaker and two parallel stainless steel plates. An alternating voltage is passed through the plates, and the salt content is obtained by reference to a calibration curve of the relationship of salt content of known mixtures to the current.

It is necessary, however, to employ other methods, such as AA, ICP-AES, and ion chromatography to determine the composition of the salts present. A method involving application of extraction and volumetric titration is also used (IP 77).

3.6.2.5 *Sulfur Content*

Sulfur is present in petroleum as sulfides, thiophene derivatives, benzothiophene derivatives, and dibenzothiophene derivatives. In most cases, the presence of sulfur is detrimental to the processing since sulfur can act as catalytic poisons during processing.

The sulfur content of petroleum is an important property and varies widely within the rough limits 0.1–3.0% w/w, and a sulfur content up to 8.0% w/w has been noted for tar sand bitumen. Compounds containing this element are among the most undesirable constituents of petroleum since they can give rise to plant corrosion and atmospheric pollution. Petroleum can evolve hydrogen sulfide during distillation as well as low-boiling sulfur compounds.

Hydrogen sulfide may be evolved during the distillation process either from free hydrogen sulfide in the feedstocks or because of low-temperature thermal decomposition, of sulfur compounds; the latter is less likely than the former. Generally, however, the sulfur compounds concentrate in the distillation residue, the volatile sulfur compounds in the distillates being removed by such processes as hydrofining and caustic washing. The sulfur content of fuels obtained from petroleum residua and the atmospheric pollution arising from the use of these fuels are important factors in petroleum utilization, so that the increasing insistence on a low-sulfur-content fuel oil has increased the value of low-sulfur petroleum.

Sulfur compounds contribute to corrosion of refinery equipment and poisoning of catalysts, cause corrosiveness in refined products, and contribute to environmental pollution as a result of the combustion of fuel products. Sulfur compounds may be present throughout the boiling range of crude oils although, as a rule, they are more abundant in the higher-boiling fractions. In some crude oils, thermally labile sulfur compounds can decompose on heating to produce hydrogen sulfide that is corrosive and toxic.

A considerable number of tests are available to estimate the sulfur in petroleum or to study its effect on various products. Hydrogen sulfide dissolved in petroleum is normally determined by absorption of the hydrogen sulfide in a suitable solution that is subsequently analyzed chemically (Doctor method) (ASTM D4952, IP 30) or by the formation of cadmium sulfate (IP 103).

The Doctor test measures the amount of sulfur available to react with metallic surfaces at the temperature of the test. The rates of reaction are metal type, temperature, and time dependent. In the test, a sample is treated with copper powder at 150°C (300°F). The copper powder is filtered from the mixture. Active sulfur is calculated from the difference between the sulfur contents of the sample (ASTM D129) before and after treatment with copper.

Sulfur that is chemically combined as an organic constituent of crude is usually estimated by oxidizing a sample in a bomb and converting the sulfur compounds to barium sulfate that is determined gravimetrically (ASTM D129, IP 61). This method is applicable to any sample of sufficiently low volatility (e.g., a residuum or tar sand bitumen) that can be weighed accurately in an open sample boat and that contains at least 0.1% sulfur. In the method, the sample is oxidized by combustion in a pressure vessel (bomb) containing oxygen under pressure. The sulfur in the sample is converted to sulfate and from the bomb washings is gravimetrically determined as barium sulfate. However, the method is not applicable to samples containing elements that give residues, other than barium sulfate, which are insoluble in dilute hydrochloric acid and would interfere in the precipitation step. In addition, the method is also subject to inaccuracies that arise from interference by the sediment inherently present in petroleum.

Until recently, one of the most widely used methods for the determination of total sulfur content has been the combustion of a sample in oxygen to convert the sulfur to sulfur dioxide, which is collected and subsequently titrated iodometrically or detected by nondispersive infrared (ASTM D1552). This method is particularly applicable to heavier oil and fractions such as residua that boil above 177°C (350°F) and contain more than 0.06% w/w sulfur. In addition, the sulfur content of petroleum coke containing up to 8% w/w sulfur can be determined.

In the iodate detection system, the sample is bummed in a stream of oxygen at a sufficiently high temperature to convert the sulfur to sulfur dioxide. The combustion products are passed into an absorber that contains an acidic solution of potassium iodide and starch indicator. A faint blue color is developed in the absorber solution by the addition of standard potassium iodate solution, and as combustion proceeds, bleaching the blue color, more iodate is added. From the amount of standard iodate consumed during the combustion, the sulfur content of the sample is calculated.

In the infrared detection system, the sample is weighed into a special ceramic boat that is then placed into a combustion furnace at 1370°C (2500°F) in an oxygen atmosphere. Moisture and dust are removed using traps, and the sulfur dioxide is measured with an infrared detector.

Other methods such as the lamp combustion method (ASTM D1266, IP 107) and the Wickbold combustion method (IP 243) are used for the determination of sulfur in petroleum and as trace quantities of total sulfur in petroleum products and is related to various other methods (ASTM D2384, ASTM D2784, ASTM D4045).

In the lamp method (ASTM D1266, IP 107), a sample is burned in a closed system using a suitable lamp and an artificial atmosphere composed of 70% carbon dioxide and 30% oxygen to prevent formation of nitrogen oxides. The sulfur oxides are absorbed and oxidized to sulfuric acid (H_2SO_4) by means of hydrogen peroxide (H_2O_2) solution that is then flushed with air to remove dissolved carbon dioxide. Sulfur as sulfate in the absorbent is determined acidimetrically by titration with standard sodium hydroxide (NaOH) solution. Alternatively, the sample can be burned in air, and the sulfur as sulfate in the absorbent be determined gravimetrically by barium sulfate ($BaSO_4$) after precipitation. If the sulfur content of the sample is less than 0.01% w/w, it is necessary to determine sulfur in the absorber solution turbidimetrically as barium sulfate.

The older, classical techniques for sulfur determination are being supplanted by two instrumental methods (ASTM D2622, ASTM D4294, IP 447).

In the former method (ASTM D2622), the sample is placed in an X-ray beam, and the peak intensity of the sulfur Kα line at 5.373Å is measured. The background intensity, measured at 5.190Å, is subtracted from the peak intensity, and resultant net counting rate is then compared to a previously prepared calibration curve or equation to obtain the sulfur concentration in % w/w.

The latter method (ASTM D4294, IP 447) employs energy-dispersive X-ray fluorescence spectroscopy and has slightly better repeatability and reproducibility than the high-temperature method and is adaptable to field applications but can be affected by some commonly present interferences such as halides. In the method, the sample is placed in a beam emitted from an X-ray source. The resultant excited characteristic X radiation is measured, and the accumulated count is compared with counts from previously prepared calibration standard to obtain the sulfur concentration. Two groups of calibration standards are required to span the concentration range, one standard ranges from 0.015 to 0.1% w/w sulfur, and the other from 0.1 to 5.0% w/w sulfur.

3.6.2.6 Viscosity and Pour Point

Viscosity and pour point determinations are performed principally to ascertain the handling (flow) characteristics of petroleum at low temperatures. There are, however, some general relationships about crude oil composition that can be derived from pour point and viscosity data. Commonly, the lower the pour point of a crude oil, the more aromatic it is, and the higher the pour point, the more paraffinic it is.

Viscosity is usually determined at different temperatures (e.g., 25°C, 77°F, and 100°C, 212°F) by measuring the time for a volume of liquid to flow under gravity through a calibrated glass capillary viscometer (ASTM D445).

In the test, the time for a fixed volume of liquid to flow under gravity through the capillary of a calibrated viscometer under a reproducible driving head and at a closely controlled temperature is measured in seconds. The kinematic viscosity is the product of the measured flow time and the calibration constant of the viscometer—tables and formula are available for the conversion of the kinematic viscosity in centistokes (cSt) at any temperature to Saybolt Universal viscosity in Saybolt Universal seconds at the same temperature and for converting kinematic viscosity in centistokes at 122 and 210°F to Saybolt Furol viscosity in Saybolt Furol seconds at the same temperature (ASTM D2161).

The *viscosity index* (ASTM D2270, IP 226) is a widely used measure of the variation in kinematic viscosity due to changes in the temperature of petroleum between 40 and 100°C (104 and 212°F). For crude oils of similar kinematic viscosity, the higher the viscosity index, the smaller is the effect of temperature on its kinematic viscosity. The accuracy of the calculated viscosity index is dependent only on the accuracy of the original viscosity determination.

The *pour point* of petroleum is an index of the lowest temperature at which the crude oil will flow under specified conditions. The maximum and minimum pour point temperatures provide a temperature window where petroleum, depending on its thermal history, might appear in the liquid as well as the solid state. The pour point data can be used to supplement other measurements of cold flow behavior, and the data are particularly useful for the screening of the effect of wax interaction modifiers on the flow behavior of petroleum.

In the original (and still widely used) test for pour point (ASTM D97, IP 15), a sample is cooled at a specified rate and examined at intervals of 3°C (5.4°F) for flow characteristics. The lowest temperature at which the movement of the oil is observed is recorded as the pour point.

A later test method (ASTM D5853) covers two procedures for the determination of the pour point of crude oils down to −36°C. One method provides a measure of the maximum (upper) pour point temperature. The second method measures the minimum (lower) pour point temperature. In these methods, the test specimen is cooled (after preliminary heating) at a specified rate and examined at intervals of 3°C (5.4°F) for flow characteristics. Again, the lowest temperature at which movement of the test specimen is observed is recorded as the pour point.

In any determination of the pour point, petroleum that contains wax produces an irregular flow behavior when the wax begins to separate. Such petroleum possesses viscosity relationships that are difficult to predict in pipeline operation. In addition, some waxy petroleum is sensitive to heat treatment that can also affect the viscosity characteristics. This complex behavior limits the value of viscosity and pour point tests on waxy petroleum. However, laboratory pumpability tests (ASTM D3829, ASTM D7528) are available that give an estimate of minimum handling temperature and minimum line or storage temperature.

3.6.2.7 Water and Sediment

Considerable importance is attached to the presence of water or sediment in petroleum for they lead to difficulties in the refinery, for example, corrosion of equipment, uneven running on the distillation unit, blockages in heat exchangers and adverse effects on product quality.

The water and sediment content of crude oil, like salt, results from production and transportation practices. Water, with its dissolved salts, may occur as easily removable suspended droplets or as an emulsion. The sediment dispersed in crude oil may be comprised of inorganic minerals from the production horizon or from drilling fluids, and scale and rust from pipelines and tanks used for oil transportation and storage. Usually water is present in far greater amounts than sediment, but collectively, it is unusual for them to exceed 1% of the crude oil on a delivered basis. Like salt, water and sediment can foul heaters, stills, and exchangers and can contribute to corrosion and to deleterious product quality. Also, water and sediment are principal components of the sludge that accumulates in storage tanks and must be disposed of periodically in an environmentally acceptable manner. Knowledge of the water and sediment content is also important in accurately determining net volumes of crude oil in sales, taxation, exchanges, and custody transfers.

The sediment consists of finely divided solids that may be drilling mud or sand or scale picked up during the transport of the oil or may consist of chlorides derived from evaporation of brine droplets in the oil. The solids may be dispersed in the oil or carried in water droplets. Sediment in petroleum can lead to serious plugging of the equipment, corrosion due to chloride decomposition, and a lowering of residual fuel quality.

Water may be found in the crude either in an emulsified form or in large droplets and can cause flooding of distillation units and excessive accumulation of sludge in tanks. Refiners generally limit the quantity, and although steps are normally taken at the oilfield to reduce the water content as low as possible, water may be later introduced during shipment. In any form, water and sediment are highly undesirable in a refinery feedstock, and the relevant tests involving distillation (ASTM D95, ASTM D4006, IP 74, IP 358),

centrifuging (ASTM D4007), extraction (ASTM D473, IP 53), and the Karl Fischer titration (ASTM D4377, ASTM D4928, IP 356, IP 386, IP 438, IP 439) are regarded as important in petroleum quality examinations.

Prior to the assay, it is sometimes necessary to separate the water from a petroleum sample. Certain types of petroleum, notably heavy oil, often form persistent emulsions that are difficult to separate. On the other hand, testing wax-bearing petroleum for sediment and water care has to be taken to ensure that wax suspended in the sample is brought into solution prior to the test; otherwise, it will be recorded as sediment.

The Karl Fischer test method (ASTM D1364, ASTM D6304) covers the direct determination of water in petroleum. In the test, the sample injection in the titration vessel can be performed on a volumetric or gravimetric basis. Viscous samples can be analyzed using a water vaporizer accessory that heats the sample in the evaporation chamber, and the vaporized water is carried into the Karl Fischer titration cell by a dry inert carrier gas.

Water and sediment in petroleum can be determined simultaneously (ASTM D4007) by the centrifuge method. Known volumes of petroleum and solvent are placed in a centrifuge tube and heated to 60°C (140°F). After centrifugation, the volume of the sediment-and-water layer at the bottom of the tube is read. For petroleum that contains wax, a temperature of 71°C (160°F) or higher may be required to completely melt the wax crystals so that they are not measured as sediment.

Sediment is also determined by an extraction method (ASTM D473, IP 53) or by membrane filtration (ASTM D4807). In the former method (ASTM D473, IP 53), an oil sample contained in a refractory thimble is extracted with hot toluene until the residue reaches a constant mass. In the latter test, the sample is dissolved in hot toluene and filtered under vacuum through a 0.45 μm porosity membrane filter. The filter with residue is washed, dried, and weighed.

3.6.3 Other Tests

The inspection assay tests discussed earlier are not exhaustive but are the ones most commonly used and provide data on the impurities present as well as a general idea of the products that may be recoverable. Other properties that are determined on an as-needed basis include, but are not limited to, the following: (i) wax content; (ii) vapor pressure—Reid method; (iii) total acid number, which is rapidly becoming essential in terms of the data relating to the content of acidic species in crude oil and crude oil products (Speight, 2014b); (iv) chloride content; and as the occasion demands, (v) the *aniline point* (or *mixed aniline point*.

Not every type of petroleum contains significant amounts of wax constituents. However, petroleum with high *wax content* presents difficulties in handling and pumping as well as producing distillate and residual fuels of high pour point and lubricating oils that are costly to dewax.

All the standard methods for the determination of the wax involve precipitating the wax from solvents such as methylene chloride or acetone under specified conditions of solvent/oil ratio and temperature. Measurements such as these give comparative results that are often useful in characterizing the wax content of petroleum or for investigating factors involved in flow problems.

On the other hand, the cloud point (ASTM D2500, ASTM D5772, ASTM D7397, ASTM D7689), which is often used to indicate the temperature at which wax deposits from oil, may be determined by cooling of a sample under prescribed conditions with stirring. The temperature at which the wax first appears is the wax appearance point.

The Reid vapor pressure test method (ASTM D323, IP 69) measures the vapor pressure of volatile petroleum. The Reid vapor pressure differs from the true vapor pressure of the sample due to some small sample vaporization and the presence of water vapor and air in the confined space.

The *acid number* is the quantity of base, expressed in milligrams of potassium hydroxide per gram of sample that is required to titrate a sample in this solvent to a green/green-brown end point, using *p*-naphtholbenzein indicator solution. The *strong acid number* is the quantity of base, expressed as milligrams of potassium hydroxide per gram of sample, required to titrate a sample in the solvent from its initial meter reading to a meter reading corresponding to a freshly prepared nonaqueous acidic buffer solution or a well-defined inflection point as specified in the test method (ASTM D664, IP 177).

To determine the acid number by the color indicator method (ASTM D974, IP 139), the sample is dissolved in a mixture of toluene and isopropyl alcohol containing a small amount of water, and the resulting single-phase solution is titrated at room temperature with standard alcoholic base or alcoholic acid solution, respectively, to the end point indicated by the color change of the added *p*-naphtholbenzein solution (orange in acid and green-brown in base). To determine the strong acid number, a separate portion of the sample is extracted with hot water and the aqueous extract is titrated with potassium hydroxide solution, using methyl orange as an indicator.

To determine the acid number by the potentiometric titration method (ASTM D664, IP 177), the sample is dissolved in a mixture of toluene and isopropyl alcohol containing a small amount of water and titrated potentiometrically with alcoholic potassium hydroxide using a glass indicating electrode and a calomel reference electrode. The meter readings are plotted manually or automatically against the respective volumes of titrating solution, and the end points are taken only at well-defined inflections in the resulting curve. When no definite inflections are obtained, end points are taken at

meter readings corresponding to those found for freshly prepared nonaqueous acidic and basic buffer solutions.

The acid numbers obtained by this color indicator test method (ASTM D974, IP 139) may or may not be numerically the same as those obtained by potentiometric titration method (ASTM D664, IP 177). In addition, the color of the crude oil sample can interfere with observation of the end point when the color indicator method is used. Determination of the acid number is more appropriate for high-acid crude oils, heavy oils, and tar sand bitumen (which are often high in acid content), and various petroleum products (Speight, 2014b).

The test method for the determination of the acid number by the color indicator titration method (ASTM D3339, IP 431) measures the acid number of oils obtained from laboratory oxidation test (ASTM D943) using smaller amounts of samples than those used in other acid number tests (ASTM D664, ASTM D974, IP 139, IP 177).

In this test, the sample is dissolved in a solvent mixture of toluene, isopropyl alcohol, and a small amount of water, and the solution is titrated at room temperature under a nitrogen atmosphere with standard potassium hydroxide (KOH) in isopropyl alcohol to the stable green color of the added indicator *p*-naptholbenzein. Dark-colored crude oils (and crude oil products) are more difficult to analyze by this method because of the difficulty in detecting color change. In such cases, the potentiometric titration method (ASTM D664, IP 177) may be used if sufficient sample is available.

The acid numbers will not provide the data essential to determining whether a single crude oil (or a blend of the crude oil petroleum with other crude oils) will yield the desired product slate. Such data can only be generated when a comprehensive assay of the crude oil (and its partners in the blending) is performed and the data from several tests are taken in relation to each other.

The chloride content of crude oil (ASTM D4929) is, like the total acid content, also a must-have—especially at the wellhead and after the desalting operation. Just as *high-acid crude oils* cause corrosion in the refinery, overhead corrosion is caused by the mineral salts, magnesium, calcium, and sodium chloride, which are hydrolyzed to produce volatile hydrochloric acid, causing a highly corrosive condition in the overhead exchangers (Speight, 2014c). Therefore, these salts present a significant contamination in opportunity crude oils.

Finally, the *aniline point* (or *mixed aniline point*) (ASTM D611, IP 2) has been used for the characterization of crude oil although it is more applicable to pure hydrocarbons and in their mixtures and is used to estimate the aromatic content of mixtures. Aromatics exhibit the lowest aniline points and paraffins the highest aniline points. Cycloparaffins and olefins exhibit values between these two extremes. In any hydrocarbon homologous series, the aniline point increases with increasing molecular weight.

REFERENCES

ASTM 2013. Annual Book of ASTM Standards. American Society for Testing and Materials, West Conshohocken, PA.

ASTM D86. Standard Test Method for Distillation of Petroleum Products at Atmospheric Pressure. Annual Book of Standards. ASTM International, West Conshohocken, PA.

ASTM D91. Standard Test Method for Precipitation Number of Lubricating Oils. Annual Book of Standards. ASTM International, West Conshohocken, PA.

ASTM D95. Standard Test Method for Water in Petroleum Products and Bituminous Materials by Distillation. Annual Book of Standards. ASTM International, West Conshohocken, PA.

ASTM D97. Standard Test Method for Pour Point of Petroleum Products. Annual Book of Standards. ASTM International, West Conshohocken, PA.

ASTM D129. Standard Test Method for Sulfur in Petroleum Products (General High Pressure Decomposition Device Method). Annual Book of Standards. ASTM International, West Conshohocken, PA.

ASTM D189. Standard Test Method for Conradson Carbon Residue of Petroleum Products. Annual Book of Standards. ASTM International, West Conshohocken, PA.

ASTM D287. Standard Test Method for API Gravity of Crude Petroleum and Petroleum Products (Hydrometer Method). Annual Book of Standards. ASTM International, West Conshohocken, PA.

ASTM D323. Standard Test Method for Vapor Pressure of Petroleum Products (Reid Method). Annual Book of Standards. ASTM International, West Conshohocken, PA.

ASTM D346. Standard Practice for Collection and Preparation of Coke Samples for Laboratory Analysis. ASTM International, West Conshohocken, PA.

ASTM D445. Standard Test Method for Kinematic Viscosity of Transparent and Opaque Liquids (and Calculation of Dynamic Viscosity). Annual Book of Standards. ASTM International, West Conshohocken, PA.

ASTM D473. Standard Test Method for Sediment in Crude Oils and Fuel Oils by the Extraction Method. Annual Book of Standards. ASTM International, West Conshohocken, PA.

ASTM D482. Standard Test Method for Ash from Petroleum Products. Annual Book of Standards. ASTM International, West Conshohocken, PA.

ASTM D524. Standard Test Method for Ramsbottom Carbon Residue of Petroleum Products. Annual Book of Standards. ASTM International, West Conshohocken, PA.

ASTM D611. Standard Test Methods for Aniline Point and Mixed Aniline Point of Petroleum Products and Hydrocarbon Solvents. Annual Book of Standards. ASTM International, West Conshohocken, PA.

ASTM D664. Standard Test Method for Acid Number of Petroleum Products by Potentiometric Titration. Annual Book of Standards. ASTM International, West Conshohocken, PA.

ASTM D893. Standard Test Method for Insolubles in Used Lubricating Oils. Annual Book of Standards. ASTM International, West Conshohocken, PA.

ASTM D943. Standard Test Method for Oxidation Characteristics of Inhibited Mineral Oils. Annual Book of Standards. ASTM International, West Conshohocken, PA.

ASTM D974. Standard Test Method for Acid and Base Number by Color-Indicator Titration. Annual Book of Standards. ASTM International, West Conshohocken, PA.

ASTM D1160. Standard Test Method for Distillation of Petroleum Products at Reduced Pressure. Annual Book of Standards. ASTM International, West Conshohocken, PA.

ASTM D1250. Standard Guide for Use of the Petroleum Measurement Tables. ASTM International, West Conshohocken, PA.

ASTM D1265. Standard Practice for Sampling Liquefied Petroleum (LP) Gases, Manual Method. ASTM International, West Conshohocken, PA.

ASTM D1266. Standard Test Method for Sulfur in Petroleum Products (Lamp Method). Annual Book of Standards. ASTM International, West Conshohocken, PA.

ASTM D1298. Standard Test Method for Density, Relative Density, or API Gravity of Crude Petroleum and Liquid Petroleum Products by Hydrometer Method. ASTM International, West Conshohocken, PA.

ASTM D1364. Standard Test Method for Water in Volatile Solvents (Karl Fischer Reagent Titration Method). Annual Book of Standards. ASTM International, West Conshohocken, PA.

ASTM D1552. Standard Test Method for Sulfur in Petroleum Products (High-Temperature Method). Annual Book of Standards. ASTM International, West Conshohocken, PA.

ASTM D2007. Standard Test Method for Characteristic Groups in Rubber Extender and Processing Oils and Other Petroleum-Derived Oils by the Clay-Gel Absorption Chromatographic Method. Annual Book of Standards. ASTM International, West Conshohocken, PA.

ASTM D2013. Standard Practice for Preparing Coal Samples for Analysis. ASTM International, West Conshohocken, PA.

ASTM D2161. Standard Practice for Conversion of Kinematic Viscosity to Saybolt Universal Viscosity or to Saybolt Furol Viscosity. Annual Book of Standards. ASTM International, West Conshohocken, PA.

ASTM D2270. Standard Practice for Calculating Viscosity Index from Kinematic Viscosity at 40 and 100°C. Annual Book of Standards. ASTM International, West Conshohocken, PA.

ASTM D2384. Standard Test Methods for Traces of Volatile Chlorides in Butane–Butene Mixtures. Annual Book of Standards. ASTM International, West Conshohocken, PA.

ASTM D2427. Standard Test Method for Determination of C2 through C5 Hydrocarbons in Gasolines by Gas Chromatography. Annual Book of Standards. ASTM International, West Conshohocken, PA.

ASTM D2500. Standard Test Method for Cloud Point of Petroleum Products. Annual Book of Standards. ASTM International, West Conshohocken, PA.

ASTM D2569. Standard Test Method for Distillation of Pitch (Withdrawn 2006 but still in use in some laboratories). Annual Book of Standards. ASTM International, West Conshohocken, PA.

ASTM D2622. Standard Test Method for Sulfur in Petroleum Products by Wavelength Dispersive X-ray Fluorescence

Spectrometry. Annual Book of Standards. ASTM International, West Conshohocken, PA.

ASTM D2784. Standard Test Method for Sulfur in Liquefied Petroleum Gases (Oxy-Hydrogen Burner or Lamp). Annual Book of Standards. ASTM International, West Conshohocken, PA.

ASTM D2887. Standard Test Method for Boiling Range Distribution of Petroleum Fractions by Gas Chromatography. Annual Book of Standards. ASTM International, West Conshohocken, PA.

ASTM D2892. Standard Test Method for Distillation of Crude Petroleum (15-Theoretical Plate Column). Annual Book of Standards. ASTM International, West Conshohocken, PA.

ASTM D3230. Standard Test Method for Salts in Crude Oil (Electrometric Method). Annual Book of Standards. ASTM International, West Conshohocken, PA.

ASTM D3279. Standard Test Method for n-Heptane Insolubles. Annual Book of Standards. ASTM International, West Conshohocken, PA.

ASTM D3339. Standard Test Method for Acid Number of Petroleum Products by Semi-Micro Color Indicator Titration. Annual Book of Standards. ASTM International, West Conshohocken, PA.

ASTM D3605. Standard Test Method for Trace Metals in Gas Turbine Fuels by Atomic Absorption and Flame Emission Spectroscopy. Annual Book of Standards. ASTM International, West Conshohocken, PA.

ASTM D3710. Standard Test Method for Boiling Range Distribution of Gasoline and Gasoline Fractions by Gas Chromatography. Annual Book of Standards. ASTM International, West Conshohocken, PA.

ASTM D3829. Standard Test Method for Predicting the Borderline Pumping Temperature of Engine Oil. Annual Book of Standards. ASTM International, West Conshohocken, PA.

ASTM D4006. Standard Test Method for Water in Crude Oil by Distillation. Annual Book of Standards. ASTM International, West Conshohocken, PA.

ASTM D4007. Standard Test Method for Water and Sediment in Crude Oil by the Centrifuge Method (Laboratory Procedure). Annual Book of Standards. ASTM International, West Conshohocken, PA.

ASTM D4045. Standard Test Method for Sulfur in Petroleum Products by Hydrogenolysis and Rateometric Colorimetry. Annual Book of Standards. ASTM International, West Conshohocken, PA.

ASTM D4052. Standard Test Method for Density, Relative Density, and API Gravity of Liquids by Digital Density Meter. Annual Book of Standards. ASTM International, West Conshohocken, PA.

ASTM D4057. Standard Practice for Manual Sampling of Petroleum and Petroleum Products. Annual Book of Standards. ASTM International, West Conshohocken, PA.

ASTM D4124. Standard Test Method for Separation of Asphalt into Four Fractions. Annual Book of Standards. ASTM International, West Conshohocken, PA.

ASTM D4175. Standard Terminology Relating to Petroleum, Petroleum Products, and Lubricants. ASTM International, West Conshohocken, PA.

ASTM D4294. Standard Test Method for Sulfur in Petroleum and Petroleum Products by Energy Dispersive X-ray Fluorescence Spectrometry. Annual Book of Standards. ASTM International, West Conshohocken, PA.

ASTM D4377. Standard Test Method for Water in Crude Oils by Potentiometric Karl Fischer Titration. Annual Book of Standards. ASTM International, West Conshohocken, PA.

ASTM D4530. Standard Test Method for Determination of Carbon Residue (Micro Method). Annual Book of Standards. ASTM International, West Conshohocken, PA.

ASTM D4628. Standard Test Method for Analysis of Barium, Calcium, Magnesium, and Zinc in Unused Lubricating Oils by Atomic Absorption Spectrometry. Annual Book of Standards. ASTM International, West Conshohocken, PA.

ASTM D4807. Standard Test Method for Sediment in Crude Oil by Membrane Filtration. Annual Book of Standards. ASTM International, West Conshohocken, PA.

ASTM D4927. Standard Test Methods for Elemental Analysis of Lubricant and Additive Components – Barium, Calcium, Phosphorus, Sulfur, and Zinc by Wavelength-Dispersive X-Ray Fluorescence Spectroscopy. Annual Book of Standards. ASTM International, West Conshohocken, PA.

ASTM D4928. Standard Test Method for Water in Crude Oils by Coulometric Karl Fischer Titration. Annual Book of Standards. ASTM International, West Conshohocken, PA.

ASTM D4929. Standard Test Methods for Determination of Organic Chloride Content in Crude Oil. Annual Book of Standards. ASTM International, West Conshohocken, PA.

ASTM D4952. Standard Test Method for Qualitative Analysis for Active Sulfur Species in Fuels and Solvents (Doctor Test). Annual Book of Standards. ASTM International, West Conshohocken, PA.

ASTM D5002. Standard Test Method for Density and Relative Density of Crude Oils by Digital Density Analyzer. Annual Book of Standards. ASTM International, West Conshohocken, PA.

ASTM D5236. Standard Test Method for Distillation of Heavy Hydrocarbon Mixtures (Vacuum Potstill Method). Annual Book of Standards. ASTM International, West Conshohocken, PA.

ASTM D5708. Standard Test Methods for Determination of Nickel, Vanadium, and Iron in Crude Oils and Residual Fuels by Inductively Coupled Plasma (ICP) Atomic Emission Spectrometry. Annual Book of Standards. ASTM International, West Conshohocken, PA.

ASTM D5772. Standard Test Method for Cloud Point of Petroleum Products (Linear Cooling Rate Method. Annual Book of Standards. ASTM International, West Conshohocken, PA.

ASTM D5853. Standard Test Method for Pour Point of Crude Oils. Annual Book of Standards. ASTM International, West Conshohocken, PA.

ASTM D5863. Standard Test Methods for Determination of Nickel, Vanadium, Iron, and Sodium in Crude Oils and Residual Fuels by Flame Atomic Absorption Spectrometry. Annual Book of Standards. ASTM International, West Conshohocken, PA.

ASTM D6304. Standard Test Method for Determination of Water in Petroleum Products, Lubricating Oils, and Additives by

Coulometric Karl Fischer Titration. Annual Book of Standards. ASTM International, West Conshohocken, PA.

ASTM D6352. Standard Test Method for Boiling Range Distribution of Petroleum Distillates in Boiling Range from 174 to 700°C by Gas Chromatography. Annual Book of Standards. ASTM International, West Conshohocken, PA.

ASTM D6443. Standard Test Method for Determination of Calcium, Chlorine, Copper, Magnesium, Phosphorus, Sulfur, and Zinc in Unused Lubricating Oils and Additives by Wavelength Dispersive X-ray Fluorescence Spectrometry (Mathematical Correction Procedure). Annual Book of Standards. ASTM International, West Conshohocken, PA.

ASTM D6560. Standard Test Method for Determination of Asphaltenes (Heptane Insolubles) in Crude Petroleum and Petroleum Products. Annual Book of Standards. ASTM International, West Conshohocken, PA.

ASTM D7397. Standard Test Method for Cloud Point of Petroleum Products (Miniaturized Optical Method). Annual Book of Standards. ASTM International, West Conshohocken, PA.

ASTM D7528. Standard Test Method for Bench Oxidation of Engine Oils by ROBO Apparatus. Annual Book of Standards. ASTM International, West Conshohocken, PA.

ASTM D7689. Standard Test Method for Cloud Point of Petroleum Products (Mini Method). Annual Book of Standards. ASTM International, West Conshohocken, PA.

Dean, J.R. 2003. Methods for Environmental Trace Analysis. John Wiley & Sons Inc., Hoboken, NJ.

Energy Institute. 2013. IP Standard Methods 2013. The Institute of Petroleum, London.

Gary, J.G., Handwerk, G.E., and Kaiser, M.J. 2007. Petroleum Refining: Technology and Economics, 5th Edition. CRC Press, Taylor & Francis Group, Boca Raton, FL.

Hsu, C.S. and Robinson, P.R. (Editors) 2006. Practical Advances in Petroleum Processing (Vols. 1 and 2). Springer Science, New York.

IP 2 (ASTM D611). Petroleum Products and Hydrocarbon Solvents – Determination of Aniline and Mixed Aniline Point. IP Standard Methods 2013. The Energy Institute, London.

IP 13 (ASTM D189). Petroleum Products – Determination of Carbon Residue – Conradson Method. IP Standard Methods 2013. The Energy Institute, London.

IP 14 (ASTM D524). Petroleum Products – Determination of Carbon Residue – Ramsbottom Method. IP Standard Methods 2013. The Energy Institute, London.

IP 15 (ASTM D97). Petroleum Products – Determination of Pour Point. IP Standard Methods 2013. The Energy Institute, London.

IP 30. Detection of Mercaptans, Hydrogen Sulfide, Elemental Sulfur and Peroxides – Doctor Test Method. IP Standard Methods 2013. The Energy Institute, London.

IP 53 (ASTM D473). Crude Petroleum and Fuel Oils – Determination of Sediment – Extraction Method. IP Standard Methods 2013. The Energy Institute, London.

IP 61 (ASTM D129). Determination of Sulfur – High Pressure Combustion Method. IP Standard Methods 2013. The Energy Institute, London.

IP 69. Determination of Vapor Pressure – Reid Method. IP Standard Methods 2013. The Energy Institute, London.

IP 71 (ASTM D445). Petroleum Products – Transparent and Opaque Liquids – Determination of Kinematic Viscosity and Calculation of Dynamic Viscosity. IP Standard Methods 2013. The Energy Institute, London.

IP 74 (ASTM D95). Petroleum Products and Bituminous Materials – Determination of Water – Distillation Method. IP Standard Methods 2013. The Energy Institute, London.

IP 77. Determination of Salt Content – Extraction and Volumetric Titration Method. IP Standard Methods 2013. The Energy Institute, London.

IP 103. Hydrogen Sulfide – Calcium Sulfate Method. Energy Institute, London.

IP 107 (ASTM D1266). Determination of Sulfur – Lamp Combustion Method. IP Standard Methods 2013. The Energy Institute, London.

IP 123. Petroleum Products – Determination of Distillation Characteristics at Atmospheric Pressure. IP Standard Methods 2013. The Energy Institute, London.

IP 139 (ASTM D974). Petroleum Products and Lubricants – Determination of Acid or Base Number – Color Indicator Titration Method. IP Standard Methods 2013. The Energy Institute, London.

IP 143 (ASTM D6560). Determination of Asphaltenes (Heptane Insolubles) in Crude Petroleum and Petroleum Products. IP Standard Methods 2013. The Energy Institute, London.

IP 160 (ASTM D1298). Crude Petroleum and Liquid Petroleum Products – Laboratory Determination of Density – Hydrometer Method. IP Standard Methods 2013. The Energy Institute, London.

IP 177 (ASTM D664). Determination of Weak and Strong Acid Number – Potentiometric Titration Method. IP Standard Methods 2013. The Energy Institute, London.

IP 200 (ASTM D1250). Guidelines for the Use of the Petroleum Measurement Tables. IP Standard Methods 2013. The Energy Institute, London.

IP 226 (ASTM D2270). Petroleum Products – Calculation of Viscosity Index From Kinematic Viscosity. IP Standard Methods 2013. The Energy Institute, London.

IP 243. Petroleum Products and Hydrocarbons – Determination of Sulfur Content – Wickbold Combustion Method. IP Standard Methods 2013. The Energy Institute, London.

IP 265. Determination of Total Salts Content of Crude Oil – Conductivity Method. IP Standard Methods 2013. The Energy Institute, London.

IP 285. Determination of Nickel and Vanadium – Spectrophotometric Method. IP Standard Methods 2013. The Energy Institute, London.

IP 356. Crude Petroleum – Determination of Water – Potentiometric Karl Fischer Titration Method. IP Standard Methods 2013. The Energy Institute, London.

IP 358. Crude Petroleum – Determination of Water – Distillation Method. IP Standard Methods 2013. The Energy Institute, London.

IP 365. Crude Petroleum and Petroleum Products – Determination of Density – Oscillating U-Tube Method. IP Standard Methods 2013. The Energy Institute, London.

IP 386 (ASTM D4928). Crude Petroleum – Determination of Water – Coulometric Karl Fischer Titration Method. IP Standard Methods 2013. The Energy Institute, London.

IP 398. Petroleum Products – Determination of Carbon Residue – Micro Method. IP Standard Methods 2013. The Energy Institute, London.

IP 431. Petroleum Products – Determination of Acid Number – Semi-Micro Color-Indicator Titration Method. IP Standard Methods 2013. The Energy Institute, London.

IP 433. Petroleum Products – Determination of Vanadium and Nickel Content – Wavelength-Dispersive X-Ray Fluorescence Spectrometry. IP Standard Methods 2013. The Energy Institute, London.

IP 438. Water Content by Coulometric Karl Fischer Titration Method. IP Standard Methods 2013. The Energy Institute, London.

IP 439. Petroleum Products – Determination of Water – Potentiometric Karl Fischer Titration Method. IP Standard Methods 2013. The Energy Institute, London.

IP 447. Petroleum Products – Determination of Sulfur Content – Wavelength Dispersive X-Ray Fluorescence Spectrometry. IP Standard Methods 2013. The Energy Institute, London.

IP 465. Liquid Petroleum Products – Determination of Nickel and Vanadium Content – Atomic Absorption Spectrometric Method. IP Standard Methods 2013. The Energy Institute, London.

Mokhatab, S., William, W.A., and Speight, J.G. 2006. Handbook of Natural Gas Transmission and Processing. Elsevier, Amsterdam.

Patnaik, P. (Editor) 2004. Dean's Analytical Chemistry Handbook. 2nd Edition. McGraw-Hill, New York.

Ralli, D.K., Pandey, S.C., Saxena, A.K., and Alamkhan, W.K. 2003. Impact of a Quality Management System on Product Quality at ONGC, Uran. Journal of Scientific & Industrial Research, 62: 1001–1007.

Speight, J.G. 2001. Handbook of Petroleum Analysis. John Wiley & Sons Inc., Hoboken, NJ.

Speight, J.G. 2007. Natural Gas: A Basic Handbook. GPC Books, Gulf Publishing Company, Houston, TX.

Speight, J.G. 2014a. The Chemistry and Technology of Petroleum. 5th Edition. CRC Press, Taylor & Francis Group, Boca Raton, FL.

Speight, J.G. 2014b. High Acid Crudes. Gulf Professional Publishing, Elsevier, Oxford.

Speight, J.G. 2014c. Oil and Gas Corrosion Prevention. Gulf Professional Publishing, Elsevier, Oxford.

Speight, J.G. and Ozum, B. 2002. Petroleum Refining Processes. Marcel Dekker Inc., New York.

Speight, J.G. and Foote, R. 2011. Ethics in Science and Engineering. Scrivener Publishing, Salem, MA.

United States Department of Energy (US DOE). 2002. Strategic Petroleum Reserve Crude Oil Assay Manual. 2nd Edition. US DOE, Washington, DC. Revision 2 – November.

Weisman, W. 1998. Analysis of Petroleum Hydrocarbons in Environmental Media. Total Petroleum Hydrocarbons Criteria Working Group Series. Volume 1. Amherst Scientific Publishers, Amherst, MA. (See also: Volume 2: Composition of Petroleum Mixtures, 1998; Volume 3: Selection of Representation Total Petroleum Hydrocarbons Fractions Based on Fate and Transport Considerations, 1997; Volume 4: Development of Fraction-Specific, 1997; and Volume 5: Human Health Risk-Based Evaluation of Petroleum Contaminated Sites – Implementation of the Working Group Approach, 1999).

4

GASES

4.1 INTRODUCTION

Petroleum-related gases (including natural gas) and refinery gases (process gases), as well as product gases produced from petroleum refining, upgrading, or natural gas-processing facilities are a category of saturated and unsaturated gaseous hydrocarbons, predominantly C_1–C_6. Some gases may also contain inorganic compounds, such as hydrogen, nitrogen, hydrogen sulfide, carbon monoxide, and carbon dioxide. As such, petroleum and refinery gases (unless produced as a salable product that must meet specifications prior to sale) are considered to be of unknown or variable composition and toxic (API, 2009). The site-restricted petroleum and refinery gases (i.e. those not produced for sale) often serve as fuels consumed on-site, as intermediates for purification and recovery of various gaseous products, or as feedstocks for isomerization and alkylation processes within a facility.

In addition to the gases obtained by distillation of crude petroleum, more highly volatile products result from the subsequent processing of naphtha and middle distillate to produce gasoline, from desulfurization processes involving hydrogen treatment of naphtha, distillate, and residual fuel; and from the coking or similar thermal treatments of vacuum gas oils and residual fuels. The most common processing step in the production of gasoline is the catalytic reforming of hydrocarbon fractions in the heptane (C_7) to decane (C_{10}) range.

Like any other refinery product, the gas must be processed to prepare it for final use and to ascertain the extent of contaminants that could cause environmental damage.

Furthermore, gas processing is a complex industrial process designed to clean raw (dirty, contaminated) gas by separating impurities and various nonmethane and fluids to produce what is known as *pipeline-quality* dry natural gas.

Natural-gas processing typically begins at the wellhead, where the composition of the raw natural gas extracted from producing wells depends on the type, depth, as well as the geology and location of the underground reservoir (Speight, 2014). Thus, gas-processing plants purify raw gas to remove common contaminates such as water, carbon dioxide, and hydrogen. However, some of the substances which contaminate gas have economic value and are further processed for sale or for use in a refinery petrochemical plant. On the other hand, the gases produced within the refinery from the refinery processes are, in a sense, much more complex, and often, the composition cannot be predicted. Such gases contain a wider variety of useful hydrocarbons as well as higher amounts of hydrogen sulfide (from the hydrodesulfurization units).

There are many variables in treating natural gas and refinery gas. The precise area of application of a given process is difficult to define. Several factors must be considered:

1. the types of contaminants in the gas;
2. the concentrations of contaminants in the gas;
3. the degree of contaminant removal desired;
4. the selectivity of acid gas removal required;
5. the temperature of the gas to be processed;
6. the pressure of the gas to be processed;

Handbook of Petroleum Product Analysis, Second Edition. James G. Speight.
© 2015 John Wiley & Sons, Inc. Published 2015 by John Wiley & Sons, Inc.

7. the volume of the gas to be processed;
8. the composition of the gas to be processed;
9. the carbon dioxide–hydrogen sulfide ratio in the gas;
10. the desirability of sulfur recovery due to process economics or environmental issues.

Thus, in many cases, the process complexities arise because of the need for the recovery of the materials used to remove the contaminants or even recovery of the contaminants in the original, or altered, form. In any case, whatever be the source of the gas, careful analysis before processing is necessary (Speight, 2001, 2007, 2014; Mokhatab *et al.*, 2006).

4.2 TYPES OF GASES

The gaseous products that occur in a refinery comprise mixtures that vary from natural gas to gases produced during refining (*refinery gas, process gas*). The constituents of each type of gas may be similar (with the exception of the olefin-type gases produced during thermal processes), but the variations of the amounts of these constituents cover wide ranges. Thus, the gas products of a refinery are (i) *liquefied petroleum gas* (*LPG*), (ii) *natural gas,* (iii) *shale gas*, and (iv) *refinery gas*, which also include *still gas* and *process gas*.

Each type of gas may be analyzed by similar methods, though the presence of high-boiling hydrocarbons and non-hydrocarbon species such as carbon dioxide and hydrogen sulfide may require slight modifications to the analytical test methods.

4.2.1 Liquefied Petroleum Gas

Liquefied petroleum gas is the term applied to certain specific hydrocarbons and their mixtures, which exist in the gaseous state under atmospheric ambient conditions but can be converted to the liquid state under conditions of moderate pressure at ambient temperature. Typically, fuel gases with four or less carbon atoms in the hydrogen–carbon combination have boiling points that are lower than room temperature, and these products are gases at ambient temperature and pressure.

The constituents of LPG are produced during natural gas refining, petroleum stabilization, and petroleum refining (Speight and Ozum, 2002; Hsu and Robinson, 2006; Gary *et al.*, 2007; Speight, 2014).

Thus, LPG is a hydrocarbon mixture containing propane ($CH_3CH_2CH_3$; boiling point: $-42°C$, $-44°F$), butane ($CH_3CH_2CH_2CH_3$; boiling point: $0°C$, $32°F$), *iso*-butane ($CH_3CH(CH_3)CH_3$; boiling point: $-7°C$, $-20°F$) and to a lesser extent propylene ($CH_3CH=CH_2$, boiling point: $-47°C$, $-54°F$), or butylene ($CH_3CH_2CH=CH_2$, boiling point: $-6°C$, $20°F$). The most common commercial products are propane,

butane, or some mixture of the two, and are generally extracted from natural gas or crude petroleum. Propylene and butylene isomers result from cracking other hydrocarbons in a petroleum refinery and are two important chemical feedstocks.

Propane and butane can be derived from natural gas or from refinery operations, but, in this latter case, substantial proportions of the corresponding olefins will be present and need to be separated. The hydrocarbons are normally liquefied under pressure for transportation and storage.

The presence of propylene and butylene in LPG used as fuel gas is not critical. The vapor pressures of these olefins are slightly higher than those of propane and butane, and the flame speed is substantially higher, but this may be an advantage, since the flame speeds of propane and butane are slow. However, one issue that often limits the amount of olefins in LPG is the propensity of the olefins to form soot.

In addition, LPG is usually available in different grades (usually specified as commercial propane, commercial butane, commercial propane–butane (p–b) mixtures, and special duty propane). During the use of LPG, the gas must vaporize completely and burn satisfactorily in the appliance, without causing any corrosion or producing any deposits in the system.

Commercial Propane consists predominantly of propane and/or propylene, while Commercial Butane is mainly composed of butanes and/or butylenes. Both must be free from harmful amounts of toxic constituents (API, 2009) and free from mechanic ally entrained water (that may be further limited by specifications) (ASTM D1835). Analysis by gas chromatography is possible (IP 405).

Commercial Propane–Butane mixtures are produced to meet particular requirements, such as volatility, vapor pressure, specific gravity, hydrocarbon composition, sulfur and its compounds, corrosion of copper, residues, and water content. These mixtures are used as fuels in areas and at times, where low ambient temperatures are less frequently encountered. Analysis by gas chromatography is possible (ASTM D5504, ASTM D6228, IP 405).

Special Duty Propane is intended for use in spark-ignition engines, and the specification includes a minimum *motor octane number* to ensure satisfactory antiknock performance. Propylene ($CH_3CH=CH_2$) has a significantly lower octane number than propane; so, there is a limit to the amount of this component that can be tolerated in the mixture. Analysis by gas chromatography is possible (ASTM D5504, ASTM D6228, IP 405).

Liquefied petroleum gas and liquefied natural gas can share the facility of being stored and transported as a liquid and then vaporized and used as a gas. To achieve this, LPG must be maintained at a moderate pressure, but at ambient temperature. The liquefied natural gas can be at ambient pressure, but must be maintained at a temperature of roughly -1 to $60°C$ (30 to $140°F$). In fact, in some applications, it is actually economical and convenient to use LPG in the liquid phase. In such cases, certain aspects of gas composition (or quality such as the ratio of propane to butane and the presence of traces of heavier hydrocarbons, water, and other extraneous materials) may be of lesser importance compared to the use of the gas in the vapor phase.

For normal (gaseous) use, the contaminants of LPG are controlled at a level at which they do not corrode fittings and appliances or impede the flow of the gas. For example, hydrogen sulfide (H_2S) and carbonyl sulfide (COS) should be absent. Organic sulfur, to the level required for adequate odorization (ASTM D5305), or *stenching*, is a normal requirement in LPG; dimethyl sulfide (CH_3SCH_3) and ethyl mercaptan (C_2H_5SH) are commonly used at a concentration of up to 50 ppm. Natural gas is similarly treated, possibly with a wider range of volatile sulfur compounds.

The presence of water in LPG (or in natural gas) is undesirable, since it can produce hydrates that will cause, for example, line blockage due to the formation of hydrates under conditions where the water *dew point* is attained (ASTM D1142). If the amount of water is above acceptable levels, the addition of a small methanol will counteract any such effect.

In addition to other gases, LPG may also be contaminated by higher boiling constituents, such as the constituents of middle distillates to lubricating oil. These contaminants become included in the gas during handling and must be prevented from reaching unacceptable levels. Olefins and especially diolefins are prone to polymerization and should be removed.

4.2.2 Natural Gas

Natural gas (predominantly *methane*), denoted by the chemical structure CH_4, is the lowest boiling and least complex of all hydrocarbons. Natural gas from an underground reservoir, when brought to the surface, can contain other higher boiling hydrocarbons and is often referred to as *wet gas*. Wet gas is usually processed to remove the entrained hydrocarbons that are higher boiling than methane and, when isolated, the higher boiling hydrocarbons sometimes liquefy and are called *natural gas condensate*.

Natural gas is found in petroleum reservoirs as free gas (*associated gas*) or in solution with petroleum in the reservoir (*dissolved gas*) or in reservoirs that contain only gaseous constituents and no (or little) petroleum (*unassociated gas*) (Cranmore and Stanton, 2000; Speight, 2014). The hydrocarbon content varies from mixtures of methane and ethane with very few other constituents (*dry* gas) to mixtures containing all of the hydrocarbons from methane to pentane, and even hexane (C_6H_{14}) and heptane (C_7H_{16}) (*wet* gas). In both cases, some carbon dioxide (CO_2) and inert gases, including helium (He), are present together with hydrogen sulfide (H_2S) and a small quantity of organic sulfur.

The term *petroleum gas(es)* in this context is also used to describe the gaseous phase and liquid phase mixtures comprised mainly of methane to butane (C_1–C_4 hydrocarbons) that are dissolved in the crude oil and natural gas, as well as gases produced during thermal processes in which the crude oil is converted to other products. It is necessary, however, to acknowledge that in addition to the hydrocarbons, gases such as carbon dioxide, hydrogen sulfide, and ammonia are also produced during petroleum refining and will be the constituents of refinery gas that have to be removed. Olefins are also present in the gas streams of various processes and are not included in LPG, but are removed for use in petrochemical operations (Crawford *et al.*, 1993).

Raw natural gas varies greatly in composition (Table 4.1), and the constituents can be several group of hydrocarbons (Table 4.2) and nonhydrocarbons. The treatment required to prepare natural gas for distribution as an industrial or household fuel is specified in terms of the use and environmental regulations.

Briefly, natural gas contains hydrocarbons and nonhydrocarbon gases. Hydrocarbon gases are methane (CH_4), ethane (C_2H_6), propane (C_3H_8), butane (C_4H_{10}), pentane (C_5H_{12}), hexane (C_6H_{14}), heptane (C_7H_{16}), and sometimes trace amounts of octane (C_8H_{18}), and higher molecular weight hydrocarbons. Some aromatics [BTX—benzene (C_6H_6), toluene ($C_6H_5CH_3$), and the xylenes ($CH_3C_6H_4CH_3$)] can also be present, raising safety issues due to their toxicity. The nonhydrocarbon gas portion of the natural gas contains nitrogen (N_2), carbon dioxide (CO_2), helium (He), hydrogen sulfide (H_2S), water vapor (H_2O), and other sulfur compounds (such as carbonyl sulfide (COS)) and mercaptans (e.g. methyl mercaptan, CH_3SH), and trace amounts of other gases. Carbon

TABLE 4.1 Composition of associated natural gas from a petroleum well

Category	Component	Amount (%)
Paraffinic	Methane (CH_4)	70–98
	Ethane (C_2H_6)	1–10
	Propane (C_3H_8)	Trace–5
	Butane (C_4H_{10})	Trace–2
	Pentane (C_5H_{12})	Trace–1
	Hexane (C_6H_{14})	Trace–0.5
	Heptane and higher (C_7^+)	None–trace
Cyclic	Cyclopropane (C_3H_6)	Traces
	Cyclohexane (C_6H_{12})	Traces
Aromatic	Benzene (C_6H_6), others	Traces
Non-hydrocarbon	Nitrogen (N_2)	Trace–15
	Carbon dioxide (CO_2)	Trace–1
	Hydrogen sulfide (H_2S)	Trace occasionally
	Helium (He)	Trace–5
	Other sulfur and nitrogen compounds	Trace occasionally
	Water (H_2O)	Trace–5

TABLE 4.2 Possible constituents of natural gas and refinery gas

Gas	Molecular weight	Boiling point 1 atm °C (°F)	Density at 60 °F (15.6 °C), 1 atm	
			g/l	Relative to air = 1
Methane	16.043	−161.5 (−258.7)	0.6786	0.5547
Ethylene	28.054	−103.7 (−154.7)	1.1949	0.9768
Ethane	30.068	−88.6 (−127.5)	1.2795	1.0460
Propylene	42.081	−47.7 (−53.9)	1.8052	1.4757
Propane	44.097	−42.1 (−43.8)	1.8917	1.5464
1,2-Butadiene	54.088	10.9 (51.6)	2.3451	1.9172
1,3-Butadiene	54.088	−4.4 (24.1)	2.3491	1.9203
1-Butene	56.108	−6.3 (20.7)	2.4442	1.9981
cis-2-Butene	56.108	3.7 (38.7)	2.4543	2.0063
trans-2-Butene	56.108	0.9 (33.6)	2.4543	2.0063
iso-Butene	56.104	−6.9 (19.6)	2.4442	1.9981
n-Butane	58.124	−0.5 (31.1)	2.5320	2.0698
iso-Butane	58.124	−11.7 (10.9)	2.5268	2.0656

dioxide and hydrogen sulfide are commonly referred to as *acid gases*, since they form corrosive compounds in the presence of water. Nitrogen, helium, and carbon dioxide are also referred to as *diluents*, since none of these burn, and thus they have no heating value. Mercury can also be present, either as a metal in vapor phase or as an organo-metallic compound in liquid fractions. Concentration levels are generally very small, but even at very small concentration levels, mercury can be detrimental due its toxicity and its corrosive properties (reaction with aluminum alloys).

A natural gas stream traditionally has high proportions of *natural gas liquids* (NGLs) and is referred to as *rich gas*. Natural gas liquids are constituents, such as ethane, propane, butane, and pentane, and higher molecular weight hydrocarbon constituents. The higher molecular weight constituents (i.e. the C_{5+} product) are commonly referred to as *natural gasoline*. Rich gas will have a high heating value and a high hydrocarbon dew point. When referring to natural gas liquids in the gas stream, the term *gallon per thousand cubic feet is* used as a measure of high–molecular weight hydrocarbon content. On the other hand, the composition of non-associated gas (sometimes called *well gas*) is deficient in natural gas liquids. The gas is produced from geological formations that typically do not contain much, if any, hydrocarbon liquids.

Carbon dioxide (ASTM D1945, ASTM D4984) in excess of 3% is normally removed for reasons of corrosion prevention (ASTM D1838). Hydrogen sulfide (ASTM D2420, ASTM D4084, ASTM D4810, IP 272) is also removed, and the odor of the gas must not be objectionable (ASTM D6273); so, mercaptan content (ASTM D1988, IP 272) is important. A simple lead acetate test (ASTM D2420, ASTM D4084) is available for detecting the presence of hydrogen sulfide and is an additional safeguard that hydrogen sulfide not be present (ASTM D1835). The odor of the gases must not be objectionable. Methyl mercaptan, if present, produces

a transitory yellow stain on the lead acetate paper that fades completely in less than 5 min. Other sulfur compounds (ASTM D5504, ASTM D6228) present in LPG do not interfere.

In the lead acetate test (ASTM D2420), the vaporized gas is passed over moist lead acetate paper under controlled conditions. Hydrogen sulfide reacts with lead acetate to form lead sulfide, resulting in a stain on the paper, varying in color from yellow to black, depending on the amount of hydrogen sulfide present. Other pollutants can be determined by gas chromatography (ASTM D5504, ASTM D6228).

The total sulfur content (ASTM D1072, ASTM D2784) is normally acceptably low, and frequently so low that it needs augmenting by means of alkyl sulfides, mercaptan derivatives, or thiophene derivatives to maintain an acceptable safe level of odor.

The hydrocarbon dew point is reduced to such a level that retrograde condensation, that is, condensation resulting from pressure drop, cannot occur under the worst conditions likely to be experienced in the gas transmission system. Similarly, the water dew point is reduced to a level sufficient to preclude formation of C_1–C_4 hydrates in the system.

The natural gas after appropriate treatment for acid gas reduction, odorization, and hydrocarbon and moisture dew-point adjustment (ASTM D1142), would then be sold within prescribed limits of pressure, calorific value, and possibly Wobbe Index (cv/(sp. gr.)).

4.2.3 Shale Gas

Shale gas is natural gas produced from shale formations that typically function as both the reservoir and source rocks for the natural gas (Speight, 2013b). In terms of chemical makeup, shale gas is typically a dry gas composed primarily of methane (60%–95% v/v), but some formations do produce wet gas. The Antrim and New Albany plays have typically produced water and gas. Gas shale formations are organic-rich shale formations that were previously regarded only as source rocks and seals for gas accumulating in the strata near sandstone and carbonate reservoirs of traditional onshore gas.

Gas production from unconventional shale gas reservoirs (such as tight shale formations) has become more common in the past decade. Produced shale gas observed to date has shown a broad variation in compositional makeup, with some having wider component ranges, a wider span of minimum and maximum heating values, and higher levels of water vapor and other substances than pipeline tariffs or purchase contracts may typically allow. Indeed, because of these variations in gas composition, each shale gas formation can have unique processing requirements for the produced shale gas to be marketable.

Ethane can be removed by cryogenic extraction, while carbon dioxide can be removed through a scrubbing process. However, it is not always necessary (or practical) to process shale gas to make its composition identical to *conventional* transmission-quality gases. Instead, the gas should be interchangeable with other sources of natural gas now provided to end-users. The interchangeability of shale gas with conventional gases is crucial to its acceptability and eventual widespread use in the United States.

Though not highly sour in the usual sense of having high hydrogen sulfide content, and with considerable variation from play to resource to resource, and even from well to well within the same resource (due to extremely low permeability of the shale even after fracturing) (Speight, 2013b), shale gas often contains varying amounts of hydrogen sulfide with wide variability in the carbon dioxide content. The gas is not ready for pipelining immediately after it has exited the shale formation.

The challenge in treating such gases is the low (or differing) hydrogen sulfide/carbon dioxide ratio and the need to meet pipeline specifications. In a traditional gas-processing plant, the choice of olamine content for hydrogen sulfide removal is *N*-methyldiethanolamine (MDEA) (Mokhatab *et al.*, 2006; Speight, 2007, 2014), but whether or not this olamine will suffice to remove the hydrogen sulfide without removal of excessive amounts of carbon dioxide is another issue.

Gas treatment may begin at the wellhead—condensates and free water usually are separated at the wellhead, using mechanical separators, he observes. Gas, condensate, and water are separated in the field separator and are directed to separate storage tanks, and the gas flows to a gathering system. After the free water has been removed, the gas is still saturated with water vapor, and depending on the temperature and pressure of the gas stream, it may need to be dehydrated or treated with methanol to prevent hydrates, as the temperature drops. But, this may not be always the case in actual practice.

4.2.4 Refinery Gas

The terms *petroleum gas* or *refinery gas* are often used to identify LPG or even gas that emanates from the top of a refinery distillation column. For the purpose of this text, petroleum gas not only describes LPG, but also natural gas and refinery gas (Speight and Ozum, 2002; Hsu and Robinson, 2006; Mokhatab *et al.*, 2006; Gary *et al.*, 2007; Speight, 2014). In this chapter, each gas is, in turn, referenced by its name rather than the generic term *petroleum gas* (ASTM D4150). However, the composition of each gas varies (Table 4.3), and recognition of this is essential before testing protocols are applied.

TABLE 4.3 General summary of product types and distillation range

Product	Lower carbon limit	Upper carbon limit	Lower boiling point (°C)	Upper boiling point (°C)	Lower boiling point (°F)	Upper boiling point (°F)
Refinery gas	C_1	C_4	−161	−1	−259	31
Liquefied petroleum gas	C_3	C_4	−42	−1	−44	31
Naphtha	C_5	C_{17}	36	302	97	575
Gasoline	C_4	C_{12}	−1	216	31	421
Kerosene/diesel fuel	C_8	C_{18}	126	258	302	575
Aviation turbine fuel	C_8	C_{16}	126	287	302	548
Fuel oil	C_{12}	>C_{20}	216	421	>343	>649
Lubricating oil	>C_{20}		>343		>649	
Wax	C_{17}	>C_{20}	302	>343	575	>649
Asphalt	>C_{20}		>343		>649	
Coke	>C_{50}[a]		>1000[a]		>1832[a]	

[a]Carbon number and boiling point difficult to assess; inserted for illustrative purposes only.

Refinery gas varies in composition and volume, depending on the crude origin and on any additions to the crude made at the loading point—in this sense, it is probably the most difficult gas to analyze. Furthermore, it is not uncommon to reinject low–molecular weight hydrocarbons such as propane and butane into the crude before dispatch by tanker or pipeline. This results in a higher vapor pressure of the crude, but it allows one to increase the quantity of light products obtained at the refinery. Since light ends in most petroleum markets command a premium, while in the oil field itself propane and butane may have to be reinjected or flared, the practice of *spiking* crude oil with LPG is becoming fairly common.

Thus, *refinery gas (fuel gas)* is the noncondensable gas that is obtained during distillation of crude oil or treatment (cracking, thermal decomposition) of petroleum (Robinson and Faulkner, 2000; Speight, 2014). Refinery gas is produced in considerable quantities during the different refining processes and is used as fuel for the refinery itself and as an important feedstock for the production of petrochemicals. It consists mainly of hydrogen (H_2), methane (CH_4), ethane (C_2H_6), propane (C_3H_8), butane (C_4H_{10}), and olefins (RCH=CHR[1], where R and R[1] can be hydrogen or a methyl group) and may also include off-gases from petrochemical processes (Table 4.2). Olefins such as ethylene (CH_2=CH_2, boiling point: $-104°C$, $-155°F$), propene (propylene, CH_3CH=CH_2, boiling point: $-47°C$, $-53°F$), butene (butene-1, CH_3CH_2CH=CH_2, boiling point: $-5°C$, $23°F$), *iso*-butylene $(CH_3)_2C$=CH_2, $-6°C$, $21°F$), *cis*- and *trans*-butene-2 (CH_3CH=$CHCH_3$, boiling point: *ca.* $1°C$, $30°F$), and butadiene (CH_2=$CHCH$=CH_2, boiling point: $-4°C$, $24°F$), as well as higher boiling olefins are produced by various refining processes.

Still gas is a broad terminology for low-boiling hydrocarbon mixtures and is the lowest boiling fraction isolated from a distillation (*still*) unit in the refinery. If the distillation unit is separating light hydrocarbon fractions, the still gas will be almost entirely methane, with only traces of ethane (CH_3CH_3) and ethylene (CH_2=CH_2). If the distillation unit is handling higher boiling fractions, the still gas might also contain propane ($CH_3CH_2CH_3$), butane ($CH_3CH_2CH_2CH_3$), and their respective isomers. *Fuel gas* and still gas are terms that are often used interchangeably, but the term *fuel gas* is intended to denote the product's destination, to be used as a fuel for boilers, furnaces, or heaters.

In a series of processes commercialized under the names Platforming, Powerforming, Catforming, and Ultraforming, paraffinic and naphthenic (cyclic nonaromatic) hydrocarbons, in the presence of hydrogen and a catalyst, are converted into aromatics, or isomerized to more highly branched hydrocarbons. Catalytic reforming processes, thus not only result in the formation of a liquid product of higher octane number, but also produce substantial quantities of gases. The latter are rich in hydrogen, but also contain hydrocarbons from methane to

butanes, with a preponderance of propane ($CH_3CH_2CH_3$), *n*-butane ($CH_3CH_2CH_2CH_3$), and isobutane [$(CH_3)_3CH$]. The composition of these gases varies in accordance with reforming severity and reformer feedstock. Since all catalytic reforming processes require substantial recycling of a hydrogen stream, it is normal to separate reformer gas into a propane ($CH_3CH_2CH_3$) and/or a butane [$CH_3CH_2CH_2CH_3/(CH_3)_3CH$] stream, which becomes part of the refinery LPG production, and a lighter gas fraction, part of which is recycled. In view of the excess of hydrogen in the gas, all products of catalytic reforming are saturated, and there are usually no olefin-type gases present in either gas stream.

A second group of refining operations that contributes to gas production is that of the catalytic cracking processes, Thermofor catalytic cracking, and other variants in which heavy gas oils are converted into cracked gas, LPG, catalytic naphtha, fuel oil, and coke by contacting the heavy hydrocarbon with the hot catalyst (Speight and Ozum, 2002; Hsu and Robinson, 2006; Gary *et al.*, 2007; Speight, 2014). Both catalytic and thermal cracking processes, the latter being now largely used for the production of chemical raw materials, result in the formation of unsaturated hydrocarbons, particularly ethylene (CH_2=CH_2), and also propylene (propene, CH_3CH=CH_2), isobutylene [isobutene, $(CH_3)_2C$=CH_2], and the *n*-butenes (CH_3CH_2CH=CH_2, and CH_3CH=$CHCH_3$), in addition to hydrogen (H_2), methane (CH_4) and smaller quantities of ethane (CH_3CH_3), propane ($CH_3CH_2CH_3$), and butane isomers [$CH_3CH_2CH_2CH_3$, $(CH_3)_3CH$]. Diolefins such as butadiene (CH_2=$CHCH$=CH_2) are also present.

Additional gases are produced in refineries with coking or visbreaking facilities for the processing of their heaviest crude fractions. In the visbreaking process, fuel oil is passed through externally fired tubes, where it undergoes liquid-phase cracking reactions, which result in the formation of lighter fuel oil components. Oil viscosity is thereby reduced, and some gases, mainly hydrogen, methane, and ethane, are formed. Substantial quantities of both gas and carbon are also formed in coking (both fluid coking and delayed coking), in addition to the middle distillate and naphtha. When coking a residual fuel oil or heavy gas oil, the feedstock is preheated and contacted with hot carbon (coke), which causes extensive cracking of the feedstock constituents of higher molecular weight to produce lower–molecular weight products, ranging from methane, via liquefied petroleum gas(es) and naphtha, to gas oil and heating oil. Products from coking processes tend to be unsaturated, and olefin-type components predominate in the tail gases from coking processes.

A further source of refinery gas is hydrocracking, a catalytic high-pressure pyrolysis process in the presence of fresh and recycled hydrogen. The feedstock is again heavy gas oil or residual fuel oil, and the process is mainly directed at the production of additional middle distillates and gasoline. Since hydrogen is to be recycled, the gases produced in this process again have to be separated into lighter and heavier streams; any surplus recycle gas and the LPG from the hydrocracking process are both saturated.

Both hydrocracker and catalytic reformer tail gases are commonly used in catalytic desulfurization processes. In the latter, feedstocks ranging from light to vacuum gas oils are passed at pressures of 500–1000 psi ($3.5–7.0 \times 10^3$ kPa), with hydrogen over a hydrofining catalyst. This results mainly in the conversion of organic sulfur compounds to hydrogen sulfide (Speight, 2014):

$$[\text{S}]_{\text{feedstock}} + \text{H}_2 \rightarrow \text{H}_2\text{S} + \text{hydrocarbons}$$

The process also produces some light hydrocarbons by hydrocracking.

The first and most important aspect of gaseous testing is the measurement of the volume of gas (ASTM D1071). In this test method, several techniques are described and may be employed for any purpose where it is necessary to know the quantity of gaseous fuel. In addition, the thermophysical properties of methane (ASTM D3956), ethane (ASTM D3984), propane (ASTM D4362), n-butane (ASTM D4650), and iso-butane (ASTM D4651) should be available for use and consultation (Stephenson and Malanowski, 1987).

In all cases, it is the composition of the gas in terms of *hydrocarbon type* that is more important in the context of the application. For example, in petrochemical applications, the presence of propylene and butylene above 10% v/v can have an adverse effect on hydrodesulfurization, prior to steam-reforming. On the other hand, petrochemical processes, such as the production of *iso*-octane from *iso*-butane and butylene, can require the exclusion of the saturated hydrocarbons.

Refinery gas specifications will vary according to the gas quality available and the end use (Johansen, 1998). For fuel uses, gas, as specified earlier, presents little difficulty, used as supplied. Alternatively, a gas of constant Wobbe Index, say for gas turbine use, could readily be produced by the user. Part of the combustion air would be diverted into the gas stream by a Wobbe Index controller. This would be set to supply gas at the lowest Wobbe Index of the undiluted gas.

Residual refinery gases, usually in more than one stream, which allows a degree of quality control, are treated for hydrogen sulfide removal, and gas sales are usually on a thermal content (calorific value, heating value) basis, with some adjustment for variation in the calorific value and hydrocarbon type (Cranmore and Stanton, 2000; Speight, 2007, 2014).

4.3 SAMPLING

One of the more critical aspects for the analysis of low-boiling hydrocarbons is the question of volumetric measurement (ASTM D1071) and sampling (ASTM D1265). However, sampling LPG from a liquid storage system is complicated by the existence of two phases (gas and liquid), and the composition of the supernatant vapor phase will, most probably, differ from the composition of the liquid phase. Furthermore, the compositions of both phases will vary as a sample (or sample) is removed from one or both phases. An accurate check of composition can only be made if samples are taken during filling of the tank or from a fully charged tank.

In general, the sampling of gaseous constituents and of liquefied gases is the subject of a variety of sampling methods (ASTM D5503), such as the manual method (ASTM D1265, ASTM D4057), the floating piston cylinder method (ASTM D3700), and the automatic sampling method (ASTM D4177, ASTM D5287). Methods for the preparation of gaseous and liquid blends are also available (ASTM D4051, ASTM D4307), including the sampling and handling of fuels for volatility measurements (ASTM D5842).

Sampling methane (CH_4) and ethane (C_2H_6) hydrocarbons is usually achieved using stainless steel cylinders, either lined or unlined. However, other containers may also be employed, dependent upon particular situations. For example, glass cylinder containers or polyvinyl fluoride (PVF) sampling bags may also be used, but, obviously, cannot be subjected to pressures that are far in excess of ambient pressure. The preferred method for sampling propane (C_3H_8) and butane (C_4H_{10}) hydrocarbons is by the use of piston cylinders (ASTM D3700), though sampling these materials as gases is also acceptable in many cases. The sampling of propane and higher boiling hydrocarbons is dependent upon the vapor pressure of the sample (IP 410). Piston cylinders or pressurized steel cylinders are recommended for high vapor pressure sampling, where significant amounts of low-boiling gases are present, while atmospheric sampling may be used for samples having a low vapor pressure.

4.4 STORAGE

Gas storage is principally used to meet use and load variations. Gas injected into storage during periods of low demand and withdrawn from storage during periods of peak demand should also be analyzed carefully. It is also used for a variety of secondary purposes, such as leveling production over periods of fluctuating demand, which involves storing any gas that is not immediately marketable, typically over the summer when the demand is low and deliver it in the winter months when the demand is high. In addition, storage of natural gas can be used to offset changes in gas demands, such as storing sufficient gas meet the winter peak demand as well as the summer peak demand, required by electricity generation via gas-fired power plants.

The most important type of gas storage is in underground reservoirs. There are three principal types—depleted gas reservoirs, aquifer reservoirs, and salt cavern reservoirs. Each of these types has distinct physical and economic characteristics, which govern the suitability of a particular type of storage type for a given application.

4.4.1 Depleted Gas Reservoirs

Depleted gas reservoirs are the most prominent and common form of underground storage. They are the reservoir formations of natural gas fields that have produced all their economically recoverable gas. The depleted reservoir formation is readily capable of holding injected natural gas. Using such a facility is economically attractive, because it allows the reuse, with suitable modification, of the extraction and distribution infrastructure, remaining from the productive life of the gas field which reduces the start-up costs. Depleted reservoirs are also attractive, because their geological and physical characteristics are usually well-known and are generally the cheapest and easiest to develop, operate, and maintain, of the three types of underground storage.

4.4.2 Aquifer Reservoirs

Aquifers are underground, porous, and permeable rock formations that act as natural water reservoirs. In some cases, they can be used for natural gas storage. Usually, these facilities are operated on a single annual cycle, as with depleted reservoirs. The geological and physical characteristics of aquifer formation are not always known ahead of time, and a significant investment has to go into investigating these and evaluating the aquifer's suitability for natural gas storage.

If the aquifer is suitable, all of the associated infrastructure must be developed from scratch, increasing the development costs compared to depleted reservoirs. This includes installation of wells, extraction equipment, pipelines, dehydration facilities, and possibly, compression equipment. Since the aquifer initially contains water, there is little or no naturally occurring gas in the formation, and, of the gas injected, some will be physically unrecoverable. As a result, aquifer storage typically requires significantly more cushion gas than depleted reservoirs, up to 80% of the total gas volume. Most aquifer storage facilities were developed when the price of natural gas was low, meaning this cushion gas was inexpensive to sacrifice. With rising gas prices, aquifer storage becomes more expensive to develop.

4.4.3 Salt Formations

Underground salt formations are well suited to natural gas storage. Salt caverns allow very little of the injected natural gas to escape from storage, unless specifically extracted. The walls of a salt cavern are strong and impervious to gas over the life span of the storage facility.

Once a suitable salt feature is discovered and found to be appropriate for the development of a gas storage facility, a cavern is created within the salt feature. This is done by the process of solution-mining. Freshwater is pumped down a borehole into the salt. Some of the salt is dissolved, leaving a void, and the water, now saline, is pumped back to the surface. The process continues until the cavern is created of the desired size. Once created, a salt cavern offers an underground natural gas storage vessel with very high deliverability.

4.4.4 Gasholders

Gas can be stored above ground in gasholders (gasometers), largely for balancing, not long-term storage, and this has been done since Victorian times, especially for the storage of coal gas (Speight, 2013a). These store gas at district pressure, meaning that they can provide extra gas very quickly at peak times. There are two kinds of gasholders: (1) column-guided, which are guided up by a large frame that is always visible, regardless of the position of the holder, and (2) spiral-guided, which have no frame and are guided up by concentric runners in the previous lift.

4.5 TEST METHODS

Test methods for gaseous fuels have been developed over many years, extending back into the 1930s. Bulk physical property tests, such as density and heating value, as well as some compositional tests, such as the Orsat analysis and the mercuric nitrate method for the determination of unsaturation, were widely used. More recently, mass spectrometry has become a popular method of choice for compositional analysis of low molecular weight and has replaced several older methods (ASTM D2421, ASTM D2650). Gas chromatography (ASTM D1945) is another method of choice for hydrocarbon identification in gases.

In fact, gas chromatography, in one form or another, will continue to be the method of choice for the characterization of low–molecular weight hydrocarbons. Developments in high-speed test methods using gas chromatographic instrumentation and data-processing are already leading to new and revised test methods. Other detection techniques such as chemiluminescence, atomic emission, and mass spectrometry will undoubtedly enhance selectivity, detection limits, and analytical productivity. Furthermore, laboratory automation using autosampling and data handling will provide improved precision and productivity, accompanied by simplified method operation.

Specifications for petroleum gases typically focus on LPG, propane, and butane, and the specifications generally define the physical properties and characteristics of LPG, which make them suitable for private, commercial, or industrial applications. The specifications do not purport to specifically define all possible requirements to meet all possible applications, and the user is cautioned to exercise judgment in formulating final specifications for specific applications.

The Gas Processors Association, its management, and supporting companies claim no specific knowledge of how manufacturers and users will produce, handle, store, transfer, or consume the products defined herein and, therefore, are not responsible for any claims, causes of action, liabilities, losses, or expenses resulting from injury or death of persons and/or damage to property arising directly or indirectly from the use of LPG or these specifications relating to LPG (GPA, 1997).

Liquefied petroleum gas is composed of hydrocarbon compounds, predominately propane and butane, produced during the processing of natural gas and also in the conventional processing of crude oil (Mokhatab *et al.*, 2006; Speight, 2007, 2014). The composition of LPG may vary depending on the source and the relative content of propane and butane content. These hydrocarbons exist as gases at atmospheric pressure and ambient temperatures, but are readily liquefied under moderate pressures for transportation and utilization.

There are many uses for LPG, the major ones being as (i) petrochemical, synthetic rubber, and motor gasoline feedstocks, and as (ii) commercial, domestic, and industrial fuel. The following may be accepted as a general guide for the common uses for the four fuel types covered by these specifications:

1. Commercial Propane is the preferred fuel type for domestic, commercial, and industrial fuels; it is also a suitable fuel for low-severity internal combustion engines.
2. Commercial Butane is used principally as feedstock for petrochemicals, synthetic rubber, and as blending stocks or feedstocks in the manufacture of motor gasoline; use as a fuel is generally limited to industrial applications where vaporization problems are not encountered; however, small quantities are used as domestic fuel.
3. Commercial butane–propane mixtures cover a broad range of mixtures, which permits the tailoring of fuels or feedstocks to specific needs.
4. Propane is less variable in composition and combustion characteristics than other products covered by these specifications; it is also suitable as a fuel for internal combustion engines operating at moderate to high engine severity.

Hydrocarbon gases are amenable to analytical techniques, and there has been the tendency, and it remains, for the determination of both major constituents and trace constituents than is the case with the heavier hydrocarbons. The complexity of the mixtures that is evident as the boiling point of petroleum fractions and petroleum products increases makes identification of many of the individual constituents difficult, if not impossible. In addition, methods have been developed for the determination of physical characteristics such as calorific value, specific gravity, and enthalpy from the analyses of mixed hydrocarbon gases, but the accuracy does suffer when compared to the data produced by methods for the direct determination of these properties.

Bulk physical property tests, such as density and heating value, as well as some compositional tests, such as the Orsat analysis (the measurement of the quantitative amounts by selective absorption of carbon dioxide, oxygen, and carbon monoxide) and the mercuric nitrate method for the determination of unsaturation, are still used.

However, the choice of a particular test rests upon the decision of the analyst that, then, depends upon the nature of the gas under study. For example, judgment by the analyst is necessary, whether or not a test that is applied to LPG is suitable for natural gas insofar as inference from the nonhydrocarbon constituents will be minimal.

4.5.1 Calorific Value

The calorific value (heat of combustion) gives an indication of the satisfactory combustion of hydrocarbon gases, which depends upon the matching of burner and appliance design with certain gas characteristics. Various types of test methods are available for the direct determination of calorific value (ASTM D1826, ASTM D3588).

The most important of these are the Wobbe Index (or Wobbe Number = calorific value/(specific gravity) and the flame speed, usually expressed as a factor or an arbitrary scale on which that of hydrogen is 100. This factor can be calculated from the gas analysis. In fact, calorific value and specific gravity can be calculated from compositional analysis (ASTM D3588).

The Wobbe Number gives a measure of the heat input to an appliance through a given aperture at a given gas pressure. Using this as a vertical coordinate and the flame speed factor as the horizontal coordinate, a combustion diagram can be constructed for an appliance, or a whole range of appliances, with the aid of appropriate test gases. This diagram shows the area within which variations in the Wobbe index of gases may occur for the given range of appliances without resulting in either incomplete combustion, flame lift, or the lighting back of pre-aerated flames. This method of prediction of combustion characteristics is not sufficiently accurate to eliminate entirely the need for the practical testing of new gases.

Another important combustion criterion is the Gas Modulus, $M = P/W$, where P is the gas pressure and W the Wobbe Number of the gas. This must remain constant if a given degree of aeration is to be maintained in a pre-aerated burner using air at atmospheric pressure.

4.5.2 Composition

Liquefied petroleum gas, natural gas, and refinery gas are mixtures of products or naturally occurring materials and, fortunately, are relative simple mixtures, which do not suffer the complexities of the isomeric variations of the higher molecular weight hydrocarbons (Table 4.4) (Drews, 1998; Speight, 2014).

Thus, because of the lower molecular weight constituents of these gases and their volatility, gas chromatography has been the technique of choice for fixed gas and hydrocarbon speciation, and mass spectrometry is also a method of choice for compositional analysis of low–molecular weight hydrocarbons (ASTM D2421, ASTM D2650). More recently, piggyback methods (such as gas chromatography/mass spectrometry and other double-technique methods) have been developed for the identification of gaseous and low-boiling liquid constituents of mixtures. The hydrocarbon composition is limited to set to the total amount of ethane, butane, or pentane, as well as ethylene and total dienes.

By limiting the amount of hydrocarbons that are lower boiling than the main component, the vapor pressure control is reinforced. Tests are available for vapor pressure at 38°C (100°F) (ASTM D1267) and at 45°C (113°F) (IP 161). The limitation on the amount of higher boiling hydrocarbons supports the volatility clause. The vapor pressure and volatility specifications will often be met automatically if the hydrocarbon composition is correct.

The amount of ethylene is limited, because it is necessary to restrict the amount of unsaturated components so as to avoid the formation of deposits caused by the polymerization of the olefin(s). In addition, ethylene (boiling point: −104°C, −155°F)

is more volatile than ethane (boiling point: −88°C, −127°F), and therefore a product with a substantial proportion of ethylene will have a higher vapor pressure and volatility than one that is predominantly ethane. Butadiene is also undesirable, because it may also produce polymeric products that form deposits and cause blockage of lines.

Currently, the preferred method for the analysis of LPG, and indeed for most petroleum-related gases, is gas chromatography (ASTM D2163, IP 264). This technique can be used for the identification and measurement of both main constituents and trace constituents. However, there may be some accuracy issues that arise in the measurement of the higher boiling constituents due to relative volatility under the conditions in which the sample is held.

Capillary column gas chromatography is an even quicker and equally accurate alternative. Mass spectrometry is also suitable for the analysis of petroleum gases. Of the other spectroscopic techniques, infrared and ultraviolet absorption may be applied to petroleum gas analysis for some specialized applications. Gas chromatography has also largely supplanted chemical absorption methods of analysis, but again, these may have some limited specialized applications.

Once the composition of a mixture has been determined, it is possible to calculate various properties such as specific gravity, vapor pressure, calorific value, and dew point.

Simple evaporation tests in conjunction with vapor pressure measurement give a further guide to composition. In these tests, a LPG sample is allowed to evaporate naturally from an open graduated vessel. Results are recorded on the basis of volume/temperature changes, such as the temperature recorded when 95% has evaporated, or volume left at a particular temperature (ASTM D1837).

Since dew point can be calculated from composition, direct determination of dew point for a particular LPG sample is a measure of composition. It is, of course, of more direct practical value, and if there are small quantities of higher molecular weight material present, it is preferable to use a direct measurement.

Specific gravity again can be calculated; but, if it is necessary to measure it, several pieces of apparatus are available. For determining the density or specific gravity of LPG in its liquid state, there are two methods, using a metal pressure pycnometer. A pressure hydrometer may be used (ASTM D1267) for measuring the relative density that may also be calculated from compositional analysis (ASTM D2598). Various procedures, manual and recording, for specific gravity or density in the gaseous state are given in two methods (ASTM D1070). Calculation of the density is also possible using any one of four models, depending upon the composition of the gas (ASTM D4784).

Gases such as ethane that are destined for use as petrochemical feedstocks must adhere to stringent composition controls that are dependent upon the process. For example, moisture content (ASTM D1142), oxygen content (ASTM

TABLE 4.4 **Number of isomers for the selected hydrocarbons**

Carbon atoms	Number of isomers
1	1
2	1
3	1
4	2
5	3
6	5
7	9
8	18
9	35
10	75
15	4,347
20	366,319
25	36,797,588
30	4,111,846,763
40	62,491,178,805,831

D1945), carbon dioxide content (ASTM D1945), and sulfur content (ASTM D1072) must be monitored, as they all interfere with catalyst performance in petrochemical processes.

The hydrocarbon composition of natural gasoline (though not specifically a gas) for petrochemical use must undergo a compositional analysis (ASTM D2427) and a test for total sulfur (ASTM D1266, IP 107, IP 191).

An issue that arises during the characterization of LPG relates to the accurate determination of heavy residues (i.e. higher molecular weight hydrocarbons and even oils) in the gas. Test methods using procedures similar to those employed in gas chromatographic simulated distillation are becoming available. In fact, the presence of any component substantially less volatile than the main constituents of LPG will give rise to unsatisfactory performance. It is difficult to set limits to the amount and nature of the *residue*, which will make a product unsatisfactory. Obviously, small amounts of oil-like material can block regulators and valves. In liquid vaporizer feed systems, even gasoline-type materials could cause difficulty. The residue as determined by the *end-point index* (EPI) (ASTM D2158) is a measure of the concentration of contaminants boiling above 37.8°C (100°F) that may be present in the gas. Other methods are available which measure the residue more directly, and for particular applications, it may be possible to relate the values obtained to the performance required and so set satisfactory limits.

Analytical methods are available in standard form for determining the volatile sulfur content and certain specific corrosive sulfur compounds that are likely to be present. Volatile sulfur determination is made by a combustion procedure (ASTM D126, IP 107) that uses a modification of the standard wick-fed lamp. Many laboratories use rapid combustion techniques with an oxyhydrogen flame in a Wickbold or Martin-Floret burner (ASTM D2784, IP 243).

This test method (ASTM D2784, IP 243) is valid for sulfur levels of greater than 1 µg/g of sulfur in LPG, but the samples should not contain more than 100 µg/g of chlorine. In the test, the sample is burned in an oxyhydrogen burner or in a lamp, in a closed system in a carbon dioxide–oxygen atmosphere. The latter is not recommended for trace quantities of sulfur due to the inordinately long combustion times needed. The sulfur oxides produced are absorbed and oxidized to sulfuric acid in a hydrogen peroxide solution. The sulfate ions are then determined by either titrating with barium perchlorate solution using a thorin–methylene blue mixed indicator, or by precipitating as barium sulfate and measuring the turbidity of the precipitate with a photometer.

Trace hydrocarbons that may be regarded as contaminants may be determined by the gas chromatographic methods already discussed. Heavier hydrocarbons in small amounts may not be completely removed from the column. If accurate information is required about the nature and amount of heavy ends, then temperature programming or a concentration procedure may be used.

Analytical methods for determining traces of various other impurities, such as chlorides (ASTM D2384), are known to be in use. The presence of acetylenes in refinery gases, though unlikely, must still be considered. Acetylenes can be determined using a chemical test method, while carbonyls are determined by the classical hydroxylamine hydrochloride reaction.

The determination of traces of higher boiling hydrocarbons and oily matter involves use of a method for residue that involves a preliminary weathering. The residue, after weathering, is dissolved in a solvent, and the solution is applied to a filter paper. The presence of residue is indicated by the formation of an oil stain. The procedure is taken further by combining the oil stain observation with the other observed values to calculate an End Point Index (ASTM D2158). The method is not very precise, and work is proceeding in several laboratories to develop a better method for the determination of residue in the form of oily matter.

In LPG, where the composition is such that the hydrocarbon dew point is known to be low, a dew-point method will detect the presence of traces of water (ASTM D1142).

The odor of LPG has to be detectable to avoid the risk of explosion. Odor is a very subjective matter, and no standard method is available. It is desirable to set up some system in which the concentration of gas can be measured in relation to its explosive limits and in which some variables can be standardized, for example, flow rate and orifice size. This will ensure that in any one location LPG is always being assessed under similar conditions from day to day.

Propane, *iso*-butane (boiling point: −12°C, 11°F), and butane generally constitute this sample type and are used for heating, as motor fuels, and as chemical feedstocks (ASTM D2504, ASTM D2505, ASTM D2597).

Procedures for the determination of hydrogen, helium, oxygen, nitrogen, carbon monoxide, carbon dioxide, methane, ethane, ethylene, propane, butanes, pentanes, and hexanes, as well as higher molecular weight hydrocarbons in natural and reformed gases by packed column gas chromatography are available (ASTM D1945, ASTM D1946). These compositional analyses are used to calculate many other properties of gases, such as density, heating value, and compressibility. The first five components listed are determined using a molecular sieve column (argon-carrier gas), while the remaining components are determined using polydimethylsiloxane partition or porous polymer columns. The hexanes-plus analysis is accomplished by backflushing the column after the elution of pentane or by the use of a back-flushed precolumn.

Important constituents of natural gas not accounted for in these analyses are moisture (water) and hydrogen sulfide, as well as other sulfur compounds (ASTM D1142, ASTM D1988, ASTM D4888, ASTM D5454, ASTM D5504, ASTM D6228).

Olefins (ethylene, propylene, butylenes, and pentylenes) that occur in refinery (process) gas have specific characteristics and require specific testing protocols (ASTM D5234, ASTM D5273, ASTM D5274).

Thus, hydrocarbon analysis of ethylene is accomplished by two methods (ASTM D2505, ASTM D6159) one of which (ASTM D6159) uses wide-bore (0.53 mm) capillary columns, including an alumina–potassium chloride (Al_2O_3/KCl) PLOT column. Another method (ASTM D2504) is recommended for the determination of noncondensable gases, and yet another (ASTM D2505) is used for the determination of carbon dioxide.

Hydrocarbon impurities in propylene can be determined by gas chromatographic methods (ASTM D2163, ASTM D2712), and another test is available for the determination of traces of methanol in propylene (ASTM D4864). A gas chromatographic method (ASTM D5303) is available for the determination of trace amounts of carbonyl sulfide in propylene, using a flame-photometric detector. Also, sulfur in petroleum gas can be determined by oxidative microcoulometry (ASTM D3246).

Commercial butylenes, high-purity butylenes, and butane–butylene mixtures are analyzed for hydrocarbon constituents (ASTM D4424), and hydrocarbon impurities in 1,3-butadiene can also be determined by gas chromatography (ASTM D2593). The presence of butadiene dimer and styrene are determined in butadiene by gas chromatography (ASTM D2426).

Carbonyls in C4 hydrocarbons are determined by a titrimetric technique (ASTM D4423) and by use of a peroxide method (ASTM D5799).

In general, gas chromatography will undoubtedly continue to be the method of choice for the characterization of light hydrocarbon materials. New and improved detection devices and techniques, such as chemiluminescence, atomic emission, and mass spectroscopy, will enhance selectivity, detection limits, and analytical productivity. Laboratory automation through autosampling, computer control, and data handling will provide improved precision and productivity, as well as simplified method operation.

Compositional analysis can be used to calculate calorific value, specific gravity, and compressibility factor (ASTM D3588).

Mercury in natural gas is also measured by atomic fluorescence spectroscopy (ASTM D6350) and by atomic absorption spectroscopy (ASTM D5954).

4.5.3 Density

The density of light hydrocarbons can be determined by several methods (ASTM D1070), including a hydrometer method (ASTM D1298), or by a pressure hydrometer method (ASTM D1657, IP 235). The specific gravity (relative density) (ASTM D1070, ASTM D1657) by itself has little significance compared to its use for higher molecular weight liquid petroleum products) and can only give an indication of quality characteristics when combined with values for volatility and vapor pressure. It is important for stock quantity calculations and is used in connection with transport and storage.

4.5.4 Relative Density

A statement is often made that *natural gas is lighter than air*. This statement often arises because of the continued insistence by engineers and scientists that the properties of a mixture are determined by the mathematical average of the properties of the individual constituents of the mixture. Such mathematical bravado and inconsistency of thought are *detrimental to safety* and need to be qualified.

The *relative density* (*specific gravity*) is the ratio of the density (mass of a unit volume) of a substance to the density of a given reference material. Specific gravity usually means relative density (of a liquid) with respect to water.

$$\text{Relative density} = \left[\rho \left(\text{substance} \right) \right] / \left[\rho \left(\text{reference} \right) \right]$$

As it pertains to gases, particularly in relation to safety considerations at commercial and industrial facilities in the United States, the relative density of a gas is usually defined with respect to air, in which air is assigned a *vapor density* of 1 (unity). With this definition, the vapor density indicates whether a gas is denser (>1) or less dense (<1) than air. The vapor density has implications for container storage and personnel safety—if a container can release a dense gas, its vapor could sink and, if flammable, collect until it is at a concentration sufficient for ignition. Even if not flammable, it could collect in the lower floor or the level of a confined space and displace air, possibly presenting a smothering hazard to individuals entering the lower part of that space.

Gases can be divided into two groups, based upon their vapor density: (1) gases which are heavier than air, and (2) gases which are as light as air or lighter than air. Gases that have a vapor density greater than 1 will be found in the bottom of storage containers and will tend to migrate downhill and accumulate in low-lying areas. Gases that have a vapor density which is the same or less than the vapor density of air will disperse readily into the surrounding environment. Additionally, chemicals that have the same vapor density as air (1.0) tend to disperse uniformly into the surrounding air when contained and, when released into the open air, chemicals that are lighter than air will travel up and away from the ground.

Methane is the only hydrocarbon constituent of natural gas that is lighter than air (Table 4.5). The higher molecular weight hydrocarbons have a higher vapor density than air and are likely, after a release, to accumulate in low-lying areas

TABLE 4.5 Relative density (specific gravity) of natural gas hydrocarbons, relative to air

Gas	Specific gravity
Air	1.000
Methane, CH_4	0.5537
Ethane, C_2H_6	1.0378
Propane, C_3H_8	1.5219
Butane, C_4H_{10}	2.0061
Pentane, C_5H_{12}	2.487
Hexane, C_6H_{14}	2.973

and represent a danger to the investigator (of the release). However, the other hydrocarbon constituents of unrefined natural gas (i.e. ethane, propane, butane, etc.) are denser than air. Therefore, should a natural gas leak occur in field operations, especially where the natural gas contains constituents other than methane, only methane dissipates readily into the air, whereas the other hydrocarbon constituents that are heavier than air do not readily dissipate into the atmosphere. This poses considerable risk if these constituents of natural gas accumulate or pool at ground level when it has been erroneously assumed that natural gas is lighter than air.

4.5.5 Sulfur Content

The manufacturing processes for LPG are designed so that the majority, if not all, of the sulfur compounds are removed. The total sulfur level is therefore considerably lower than for other petroleum fuels, and a maximum limit for sulfur content helps to define the product more completely. The sulfur compounds that are mainly responsible for corrosion are hydrogen sulfide, carbonyl sulfide, and, sometimes, elemental sulfur. Hydrogen sulfide and mercaptans have distinctive unpleasant odors.

A control of the total sulfur content, hydrogen sulfide, and mercaptans ensures that the product is not corrosive or nauseating. Stipulating a satisfactory copper strip test further ensures the control of the corrosion.

Total sulfur in gas can be determined by combustion (ASTM D1072), by the lamp method (ASTM D1266), or by hydrogenation (ASTM D4468). Trace total organically-bound nitrogen is determined (ASTM D4629). The current test method for heavy residues in LPG (ASTM D2158) involves evaporation of a LPG sample, measuring the volume of residue, and observing the residue for oil stain on a piece of filter paper.

Corrosive sulfur compounds can be detected by their effect on copper and the form in which the general copper strip corrosion test (ASTM D1838) for petroleum products is applied to LPG. Hydrogen sulfide can be detected by its action on moist lead acetate paper, and the procedure is also used as a measure of sulfur compounds. The method follows the principle of the standard Doctor test.

4.5.6 Volatility and Vapor Pressure

The vaporization and combustion characteristics of LPG are defined for normal applications by volatility, vapor pressure, and to a lesser extent, specific gravity.

Volatility is expressed in terms of the temperature at which 95% of the sample is evaporated and presents a measure of the least volatile component present (ASTM D1837). Vapor pressure (IP 410) is, therefore, a measure of the most extreme low-temperature conditions under which initial vaporization can take place. By setting limits to vapor pressure and volatility jointly, the specification serves to ensure essentially single-component products for the butane and propane grades (ASTM D1267, ASTM D2598, IP 410). By combining vapor pressure/volatility limits with specific gravity for propane–butane mixtures, essentially two-component systems are ensured.

The residue (ASTM D1025, ASTM D2158, IP 317), that is, nonvolatile matter, is a measure of the concentration of contaminants boiling above 37.8°C (100°F) that may be present in the gas.

For natural gasoline, the primary criteria are volatility (vapor pressure and knock performance. Determination of the vapor pressure (ASTM D323, ASTM D4953, ASTM D5191) and distillation profile (ASTM D216, IP 191) is essential. Knock performance is determined by rating in knock test engines by both the motor method (ASTM D2700, IP 236) and the research method (ASTM D2699, IP 237). The knock characteristics of LPG can also be determined.

Other considerations for natural gasoline are copper corrosion (ASTM D130, IP 154, IP 411) and specific gravity (ASTM D1298, IP 160), the latter determination being necessary for measurement and transportation.

4.5.7 Water

It is a fundamental requirement that LPG should not contain free water (ASTM D2713). Dissolved water may give trouble by forming hydrates and giving moisture vapor in the gas phase. Both of these will lead to blockages. Therefore, test methods are available to determine the presence of water, using electronic moisture analyzers (ASTM D5454), dew-point temperature (ASTM D1142), and length-of-stain detector tubes (ASTM D4888).

REFERENCES

API. 2009. Refinery Gases Category Analysis and Hazard Characterization. Submitted to the EPA by the American Petroleum Institute, Petroleum HPV Testing Group. HPV Consortium Registration # 1100997 United States Environmental Protection Agency, Washington, DC.

ASTM D130. Standard Test Method for Corrosiveness to Copper from Petroleum Products by Copper Strip Test. Annual Book of Standards. ASTM International, West Conshohocken, PA.

ASTM D216. Standard Practice for Conversion of Kinematic Viscosity to Saybolt Universal Viscosity or to Saybolt Furol Viscosity. Annual Book of Standards. ASTM International, West Conshohocken, PA.

ASTM D323. Standard Test Method for Vapor Pressure of Petroleum Products (Reid Method). Annual Book of Standards. ASTM International, West Conshohocken, PA.

ASTM D1025. Standard Test Method for Nonvolatile Residue of Polymerization-Grade Butadiene. Annual Book of Standards. ASTM International, West Conshohocken, PA.

ASTM D1070. Standard Test Methods for Relative Density of Gaseous Fuels. Annual Book of Standards. ASTM International, West Conshohocken, PA.

ASTM D1071. Standard Test Methods for Volumetric Measurement of Gaseous Fuel Samples. Annual Book of Standards. ASTM International, West Conshohocken, PA.

ASTM D1072. Standard Test Method for Total Sulfur in Fuel Gases by Combustion and Barium Chloride Titration. Annual Book of Standards. ASTM International, West Conshohocken, PA.

ASTM D1142. Standard Test Method for Water Vapor Content of Gaseous Fuels by Measurement of Dew-Point Temperature. Annual Book of Standards. ASTM International, West Conshohocken, PA.

ASTM D1265. Standard Practice for Sampling Liquefied Petroleum (LP) Gases, Manual Method. Annual Book of Standards. ASTM International, West Conshohocken, PA.

ASTM D1266. Standard Test Method for Sulfur in Petroleum Products (Lamp Method). Annual Book of Standards. ASTM International, West Conshohocken, PA.

ASTM D1267. Standard Test Method for Gage Vapor Pressure of Liquefied Petroleum (LP) Gases (LP-Gas Method). Annual Book of Standards. ASTM International, West Conshohocken, PA.

ASTM D1298. Standard Test Method for Density, Relative Density, or API Gravity of Crude Petroleum and Liquid Petroleum Products by Hydrometer Method. Annual Book of Standards. ASTM International, West Conshohocken, PA.

ASTM D1657. Standard Test Method for Density or Relative Density of Light Hydrocarbons by Pressure Hydrometer. Annual Book of Standards. ASTM International, West Conshohocken, PA.

ASTM D1826. Standard Test Method for Calorific (Heating) Value of Gases in Natural Gas Range by Continuous Recording Calorimeter. Annual Book of Standards. ASTM International, West Conshohocken, PA.

ASTM D1835. Standard Specification for Liquefied Petroleum (LP) Gases. Annual Book of Standards. ASTM International, West Conshohocken, PA.

ASTM D1837. Standard Test Method for Volatility of Liquefied Petroleum (LP) Gases. Annual Book of Standards. ASTM International, West Conshohocken, PA.

ASTM D1838. Standard Test Method for Copper Strip Corrosion by Liquefied Petroleum (LP) Gases. Annual Book of Standards. ASTM International, West Conshohocken, PA.

ASTM D1945. Standard Test Method for Analysis of Natural Gas by Gas Chromatography. Annual Book of Standards. ASTM International, West Conshohocken, PA.

ASTM D1946. Standard Practice for Analysis of Reformed Gas by Gas Chromatography. Annual Book of Standards. ASTM International, West Conshohocken, PA.

ASTM D1988. Standard Test Method for Mercaptans in Natural Gas Using Length-of-Stain Detector Tubes. Annual Book of Standards. ASTM International, West Conshohocken, PA.

ASTM D2158. Standard Test Method for Residues in Liquefied Petroleum (LP) Gases. Annual Book of Standards. ASTM International, West Conshohocken, PA.

ASTM D2163. Standard Test Method for Determination of Hydrocarbons in Liquefied Petroleum (LP) Gases and Propane/Propene Mixtures by Gas Chromatography. Annual Book of Standards. ASTM International, West Conshohocken, PA.

ASTM D2384. Standard Test Methods for Traces of Volatile Chlorides in Butane-Butene Mixtures. Annual Book of Standards. ASTM International, West Conshohocken, PA.

ASTM D2420. Standard Test Method for Hydrogen Sulfide in Liquefied Petroleum (LP) Gases (Lead Acetate Method). Annual Book of Standards. ASTM International, West Conshohocken, PA.

ASTM D2421. Standard Practice for Interconversion of Analysis of C5 and Lighter Hydrocarbons to Gas-Volume, Liquid-Volume, or Mass Basis. Annual Book of Standards. ASTM International, West Conshohocken, PA.

ASTM D2426. Standard Test Method for Butadiene Dimer and Styrene in Butadiene Concentrates by Gas Chromatography. Annual Book of Standards. ASTM International, West Conshohocken, PA.

ASTM D2427. Standard Test Method for Determination of C2 through C5 Hydrocarbons in Gasolines by Gas Chromatography. Annual Book of Standards. ASTM International, West Conshohocken, PA.

ASTM D2504. Standard Test Method for Noncondensable Gases in C2 and Lighter Hydrocarbon Products by Gas Chromatography. Annual Book of Standards. ASTM International, West Conshohocken, PA.

ASTM D2505. Standard Test Method for Ethylene, Other Hydrocarbons, and Carbon Dioxide in High-Purity Ethylene by Gas Chromatography. Annual Book of Standards. ASTM International, West Conshohocken, PA.

ASTM D2593. Standard Test Method for Butadiene Purity and Hydrocarbon Impurities by Gas Chromatography. Annual Book of Standards. ASTM International, West Conshohocken, PA.

ASTM D2597. Standard Test Method for Analysis of Demethanized Hydrocarbon Liquid Mixtures Containing Nitrogen and Carbon Dioxide by Gas Chromatography. Annual Book of Standards. ASTM International, West Conshohocken, PA.

ASTM D2598. Standard Practice for Calculation of Certain Physical Properties of Liquefied Petroleum (LP) Gases from Compositional Analysis. Annual Book of Standards. ASTM International, West Conshohocken, PA.

ASTM D2650. Standard Test Method for Chemical Composition of Gases by Mass Spectrometry. Annual Book of Standards. ASTM International, West Conshohocken, PA.

ASTM D2699. Standard Test Method for Research Octane Number of Spark-Ignition Engine Fuel. Annual Book of Standards. ASTM International, West Conshohocken, PA.

ASTM D2700. Standard Test Method for Motor Octane Number of Spark-Ignition Engine Fuel. Annual Book of Standards. ASTM International, West Conshohocken, PA.

ASTM D2712. Standard Test Method for Hydrocarbon Traces in Propylene Concentrates By Gas Chromatography. Annual Book of Standards. ASTM International, West Conshohocken, PA.

ASTM D2713. Standard Test Method for Dryness of Propane (Valve Freeze Method). Annual Book of Standards. ASTM International, West Conshohocken, PA.

ASTM D2784. Standard Test Method for Sulfur in Liquefied Petroleum Gases (Oxy-Hydrogen Burner or Lamp). Annual Book of Standards. ASTM International, West Conshohocken, PA.

ASTM D3246. Standard Test Method for Sulfur in Petroleum Gas by Oxidative Microcoulometry. Annual Book of Standards. ASTM International, West Conshohocken, PA.

ASTM D3588. Standard Practice for Calculating Heat Value, Compressibility Factor, and Relative Density of Gaseous Fuels. Annual Book of Standards. ASTM International, West Conshohocken, PA.

ASTM D3700. Standard Practice for Obtaining LPG Samples Using a Floating Piston Cylinder. Annual Book of Standards. ASTM International, West Conshohocken, PA.

ASTM D3956. Standard Specification for Methane Thermophysical Property Tables. Annual Book of Standards. ASTM International, West Conshohocken, PA.

ASTM D3984. Standard Specification for Ethane Thermophysical Property Tables. Annual Book of Standards. ASTM International, West Conshohocken, PA.

ASTM D4051. Standard Practice for Preparation of Low-Pressure Gas Blends. Annual Book of Standards. ASTM International, West Conshohocken, PA.

ASTM D4057. Standard Practice for Manual Sampling of Petroleum and Petroleum Products. Annual Book of Standards. ASTM International, West Conshohocken, PA.

ASTM D4084. Standard Test Method for Analysis of Hydrogen Sulfide in Gaseous Fuels (Lead Acetate Reaction Rate Method). Annual Book of Standards. ASTM International, West Conshohocken, PA.

ASTM D4150. Standard Terminology Relating to Gaseous Fuels. Annual Book of Standards. ASTM International, West Conshohocken, PA.

ASTM D4177. Standard Practice for Automatic Sampling of Petroleum and Petroleum Products. Annual Book of Standards. ASTM International, West Conshohocken, PA.

ASTM D4307. Standard Practice for Preparation of Liquid Blends for Use as Analytical Standards. Annual Book of Standards. ASTM International, West Conshohocken, PA.

ASTM D4362. Standard Specification for Propane Thermophysical Property Tables. Annual Book of Standards. ASTM International, West Conshohocken, PA.

ASTM D4424. Standard Test Method for Butylene Analysis by Gas Chromatography. Annual Book of Standards. ASTM International, West Conshohocken, PA.

ASTM D4468. Standard Test Method for Total Sulfur in Gaseous Fuels by Hydrogenolysis and Rateometric Colorimetry. Annual Book of Standards. ASTM International, West Conshohocken, PA.

ASTM D4629. Standard Test Method for Trace Nitrogen in Liquid Petroleum Hydrocarbons by Syringe/Inlet Oxidative Combustion and Chemiluminescence Detection. Annual Book of Standards. ASTM International, West Conshohocken, PA.

ASTM D4650. Standard Specification for Normal Butane Thermophysical Property Tables. Annual Book of Standards. ASTM International, West Conshohocken, PA.

ASTM D4651. Standard Specification for Isobutane Thermophysical Property Tables. Annual Book of Standards. ASTM International, West Conshohocken, PA.

ASTM D4784. Standard for LNG Density Calculation Models. Annual Book of Standards. ASTM International, West Conshohocken, PA.

ASTM D4810. Standard Test Method for Hydrogen Sulfide in Natural Gas Using Length-of-Stain Detector Tubes. Annual Book of Standards. ASTM International, West Conshohocken, PA.

ASTM D4864. Standard Test Method for Determination of Traces of Methanol in Propylene Concentrates by Gas Chromatography. Annual Book of Standards. ASTM International, West Conshohocken, PA.

ASTM D4888. Standard Test Method for Water Vapor in Natural Gas Using Length-of-Stain Detector Tubes. Annual Book of Standards. ASTM International, West Conshohocken, PA.

ASTM D4953. Standard Test Method for Vapor Pressure of Gasoline and Gasoline-Oxygenate Blends (Dry Method). Annual Book of Standards. ASTM International, West Conshohocken, PA.

ASTM D4984. Standard Test Method for Carbon Dioxide in Natural Gas Using Length-of-Stain Detector Tubes. Annual Book of Standards. ASTM International, West Conshohocken, PA.

ASTM D5191. Standard Test Method for Vapor Pressure of Petroleum Products (Mini Method). Annual Book of Standards. ASTM International, West Conshohocken, PA.

ASTM D5234. Standard Guide for Analysis of Ethylene Product. Annual Book of Standards. ASTM International, West Conshohocken, PA.

ASTM D5273. Standard Guide for Analysis of Propylene Concentrates. Annual Book of Standards. ASTM International, West Conshohocken, PA.

ASTM D5274. Standard Guide for Analysis of 1,3-Butadiene Product. Annual Book of Standards. ASTM International, West Conshohocken, PA.

ASTM D5287. Standard Practice for Automatic Sampling of Gaseous Fuels. Annual Book of Standards. ASTM International, West Conshohocken, PA.

ASTM D5303. Standard Test Method for Trace Carbonyl Sulfide in Propylene by Gas Chromatography. Annual Book of Standards. ASTM International, West Conshohocken, PA.

ASTM D5305. Standard Test Method for Determination of Ethyl Mercaptan in LP-Gas Vapor. Annual Book of Standards. ASTM International, West Conshohocken, PA.

ASTM D5454. Standard Test Method for Water Vapor Content of Gaseous Fuels Using Electronic Moisture Analyzers. Annual Book of Standards. ASTM International, West Conshohocken, PA.

ASTM D5503. Standard Practice for Natural Gas Sample-Handling and Conditioning Systems for Pipeline Instrumentation. Annual Book of Standards. ASTM International, West Conshohocken, PA.

ASTM D5504. Standard Test Method for Determination of Sulfur Compounds in Natural Gas and Gaseous Fuels by Gas Chromatography and Chemiluminescence. Annual Book of Standards. ASTM International, West Conshohocken, PA.

ASTM D5799. Standard Test Method for Determination of Peroxides in Butadiene. Annual Book of Standards. ASTM International, West Conshohocken, PA.

ASTM D5842. Standard Practice for Sampling and Handling of Fuels for Volatility Measurement. Annual Book of Standards. ASTM International, West Conshohocken, PA.

ASTM D5954. Standard Test Method for Mercury Sampling and Measurement in Natural Gas by Atomic Absorption Spectroscopy. Annual Book of Standards. ASTM International, West Conshohocken, PA.

ASTM D6159. Standard Test Method for Determination of Hydrocarbon Impurities in Ethylene by Gas Chromatography. Annual Book of Standards. ASTM International, West Conshohocken, PA.

ASTM D6228. Standard Test Method for Determination of Sulfur Compounds in Natural Gas and Gaseous Fuels by Gas Chromatography and Flame Photometric Detection. Annual Book of Standards. ASTM International, West Conshohocken, PA.

ASTM D6273. Standard Test Methods for Natural Gas Odor Intensity. Annual Book of Standards. ASTM International, West Conshohocken, PA.

ASTM D6350. Standard Test Method for Mercury Sampling and Analysis in Natural Gas by Atomic Fluorescence Spectroscopy. Annual Book of Standards. ASTM International, West Conshohocken, PA.

Cranmore, R.E., and Stanton, E. 2000. Natural Gas. In Modern Petroleum Technology. Volume 1: Upstream. R.A. Dawe (Editor). John Wiley & Sons Inc., New York, pp. 337–82.

Crawford, D.B., Durr, C.A., Finneran, J.A., and Turner, W. 1993. Chemicals from Natural Gas. In Chemical Processing Handbook. J.J. McKetta (Editor). Marcel Dekker Inc., New York. p. 2.

Drews, A.W. 1998. Introduction. In Manual on Hydrocarbon Analysis. 6th Edition. A.W. Drews (Editor). American Society for Testing and Materials, West Conshohocken, PA.

Gary, J.G., Handwerk, G.E., and Kaiser, M.J. 2007. Petroleum Refining: Technology and Economics. 5th Edition. CRC Press, Taylor & Francis Group, Boca Raton, FL.

GPA. 1997. Liquefied Petroleum Gas Specifications and Test Methods. GPA Standard 2140. Gas Processors Association, Tulsa, OK.

Hsu, C.S., and Robinson, P.R. (Editors) 2006. Practical Advances in Petroleum Processing, Volumes 1 and 2. Springer Science, New York.

IP 107 (ASTM D1266). Determination of Sulfur – Lamp Combustion Method. IP Standard Methods 2013. The Energy Institute, London.

IP 154 (ASTM D1838). Petroleum Products – Corrosiveness to Copper – Copper Strip Test. IP Standard Methods 2013. The Energy Institute, London.

IP 160 (ASTM D1298). Crude Petroleum and Liquid Petroleum Products – Laboratory determination of density – Hydrometer method. IP Standard Methods 2013. The Energy Institute, London.

IP 161. Liquefied Petroleum Gases – Determination of Gauge Vapor Pressure – LPG Method (superseded by IP 410 but still used in some laboratories). IP Standard Methods 2013. The Energy Institute, London.

IP 191. Determination of Distillation Characteristics of Natural Gasoline (incorporated into IP 123 but still used in some laboratories). IP Standard Methods 2013. The Energy Institute, London.

IP 235. Determination of Density of Light Hydrocarbons – Pressure Hydrometer Method. IP Standard Methods 2013. The Energy Institute, London.

IP 236 (ASTM D2700). Motor and Aviation-Type Fuels – Determination of Knock Characteristics – Motor Method. IP Standard Methods 2013. The Energy Institute, London.

IP 237 (ASTM D2699). Petroleum Products – Determination of Knock Characteristics of Motor Fuels – Research Method (Research Octane Number, RON). IP Standard Methods 2013. The Energy Institute, London.

IP 243. Petroleum Products and Hydrocarbons – Determination of Sulfur Content – Wickbold Combustion Method. IP Standard Methods 2013. The Energy Institute, London.

IP 264. Determination of Composition of LPG and Propylene Concentrates – Gas Chromatography Method. IP Standard Methods 2013. The Energy Institute, London.

IP 272. Determination of Mercaptan Sulfur and Hydrogen Sulfide Content of Liquefied Petroleum Gases (LPG) – Electrometric Titration Method. IP Standard Methods 2013. The Energy Institute, London.

IP 317 (ASTM D2158). Determination of Residues in Liquefied Petroleum Gases (LPG) – Low Temperature Evaporation Method. IP Standard Methods 2013. The Energy Institute, London.

IP 405. Commercial Propane and Butane – Analysis by Gas Chromatography. IP Standard Methods 2013. The Energy Institute, London.

IP 410. Liquefied Petroleum Products – Determination of Gauge Vapor Pressure – LPG Method. IP Standard Methods 2013. The Energy Institute, London.

IP 411. Liquefied Petroleum Gases – Corrosiveness to Copper – Copper Strip Test. IP Standard Methods 2013. The Energy Institute, London.

Johansen, N.G. 1998. In Manual on Hydrocarbon Analysis. 6th Edition. A.W. Drews (Editor). American Society for Testing and Materials, West Conshohocken, PA.

Mokhatab, S., Poe, W.A., and Speight, J.G. 2006. Handbook of Natural Gas Transmission and Processing. Elsevier, Amsterdam.

Robinson, J.D., and Faulkner, R.P. 2000. The Oil Refinery: Types, Structure and Configuration. In Modern Petroleum technology. Volume 2: Downstream. A.G. Lucas (Editor). John Wiley & Sons Inc., New York.

Speight, J.G. 2001. Handbook of Petroleum Analysis. John Wiley & Sons Inc., New York.

Speight, J.G. 2007. Natural Gas: A Basic Handbook. GPC Books, Gulf Publishing Company, Houston, TX.

Speight, J.G. 2013a. The Chemistry and Technology of Coal. 5th Edition. CRC Press, Taylor & Francis Group, Boca Raton, FL.

Speight, J.G. 2013b. Shale Gas Production Processes. Gulf Professional Publishing, Elsevier, Oxford.

Speight, J.G. 2014. The Chemistry and Technology of Petroleum. 5th Edition. CRC Press, Taylor & Francis Group, Boca Raton, FL.

Stephenson, R.M., and Malanowski, S. 1987. Handbook of the Thermodynamics of Organic Compounds. Elsevier, New York.

Speight, J.G., and Ozum, B. 2002. Petroleum Refining Processes. Marcel Dekker Inc., New York.

5

NAPHTHA AND SOLVENTS

5.1 INTRODUCTION

Naphtha (*petroleum naphtha*) is a generic term applied to refined, partly refined, or unrefined petroleum products and liquid products of natural gas, which distill below 240°C (465°F) and is the volatile fraction of the petroleum, which is used as a solvent or as a precursor to gasoline. In fact, not less than 10% v/v of material should distill below 75°C (167°F); not less than 95% v/v of the material should distill below 240°C (465°F) under standard distillation conditions, though there are different grades of naphtha within this extensive boiling range that have different boiling ranges (Hori, 2000; Speight and Ozum, 2002; Pandey *et al.*, 2004; Hsu and Robinson, 2006; Gary *et al.*, 2007; Speight, 2014a). The term *petroleum solvent* is often used synonymously with *naphtha*. However, naphtha may also be produced from tar sand bitumen, coal tar, and oil shale kerogen, as well as by the destructive distillation of wood and from synthesis gas (mixtures of carbon monoxide and hydrogen produced by the gasification of coal and/or biomass or other feedstocks), which is then converted to liquid products by the Fischer–Tropsch process (Davis and Occelli, 2010; Chadeesingh, 2011; Speight, 2011, 2013, 2014a, 2014b). For that reason and to remain within the context of this book, this chapter deals only with naphtha produced by the processing of crude oil in petroleum refineries.

In the refinery, naphtha is an unrefined or refined low-boiling distillate fraction, which usually boils below 250°C (480°F), but often with a fairly wide boiling range, depending upon the crude oil from which the naphtha is produced as well as the process which generates the naphtha. More specifically, there is a range of special-purpose hydrocarbon fractions that can be described as *naphtha*. For example, the 0–100°C (32–212°F) fraction from the distillation of crude oil is referred to as *light virgin naphtha* and the 100–200°C (212–390°F) fraction is referred to as *heavy virgin naphtha*. The product stream from the fluid catalytic cracker is often split into three fractions: (1) the fraction boiling <105°C/<220°F is *light FCC naphtha*; (2) the fraction boiling from 105 to 160°C (220 to 320°F) is *intermediate FCC naphtha*; and (3) the fraction boiling from 160 to 200°C (320 to 390°F) is referred to as *heavy FCC naphtha* (Occelli, 2010). These boiling ranges can vary from refinery to refinery and even within a refinery, when the crude oil feedstock changes or blends of crude oil are used as refinery feedstock.

More generally, naphtha is an intermediate hydrocarbon liquid stream derived from crude oil and is usually desulfurized and then catalytically reformed to produce high-octane naphtha before blending into the streams that make up gasoline. Because of the variations in crude oil composition and quality, as well as differences in refinery operations, it is difficult (if not impossible) to provide a definitive, single definition of the word naphtha, since each refinery produces a site-specific naphtha—often with a unique boiling range (unique initial point and final boiling point) as well as other physical and compositional characteristics.

On a chemical basis, (petroleum) naphtha is difficult to define precisely, because it can contain varying amounts of the constituents (paraffins, naphthenes, aromatics, and olefins) in different proportions, in addition to the potential isomers of the paraffins that exist in the naphtha boiling

Handbook of Petroleum Product Analysis, Second Edition. James G. Speight.
© 2015 John Wiley & Sons, Inc. Published 2015 by John Wiley & Sons, Inc.

TABLE 5.1   General summary of product types and distillation range

Product	Lower carbon limit	Upper carbon limit	Lower boiling point (°C)	Upper boiling point (°C)	Lower boiling point (°F)	Upper boiling point (°F)
Refinery gas	C_1	C_4	−161	−1	−259	31
Liquefied petroleum gas	C_3	C_4	−42	−1	−44	31
Naphtha	C_5	C_{17}	36	302	97	575
Gasoline	C_4	C_{12}	−1	216	31	421
Kerosene/diesel fuel	C_8	C_{18}	126	258	302	575
Aviation turbine fuel	C_8	C_{16}	126	287	302	548
Fuel oil	C_{12}	$>C_{20}$	216	421	>343	>649
Lubricating oil	$>C_{20}$		>343		>649	
Wax	C_{17}	$>C_{20}$	302	>343	575	>649
Asphalt	$>C_{20}$		>343		>649	
Coke	$>C_{50}{}^a$		$>1000^a$		$>1832^a$	

^{a} Carbon number and boiling point difficult to assess; inserted for illustrative purposes only.

TABLE 5.2   Increase in the number of isomers with carbon number

Carbon atoms	Number of isomers
1	1
2	1
3	1
4	2
5	3
6	5
7	9
8	18
9	35
10	75
15	4,347
20	366,319
25	36,797,588
30	4,111,846,763
40	62,491,178,805,831

range (Tables 5.1 and 5.2). Naphtha is also represented as having the boiling range and carbon number similar to gasoline (Fig. 5.1), being a precursor to gasoline.

The so-called *petroleum ether* solvents are specific boiling range naphtha as is *ligroin*. Thus, the term *petroleum solvent* describes a special liquid hydrocarbon fraction obtained from naphtha and used in industrial processes and formulations (Speight, 2014a). These fractions are also referred to as *industrial naphtha*. Other solvents include *white spirit* that is subdivided into *industrial spirit* (distilling between 30 and 200°C, 86 and 392°F) and *white spirit* (light oil with a distillation range of 135–200°C (275–392°F). The special value of naphtha as a solvent lies in its stability and purity.

In this chapter, references to the various test methods dedicated to the determination of the amounts of carbon, hydrogen, and nitrogen (ASTM D5291), as well as to the determination of oxygen, sulfur, metals, and elements such

as chlorine (ASTM D808) are not included. Such tests might be deemed necessary, but, in the light of the various tests available for composition, are presumed to be at the discretion of the analyst.

5.2   PRODUCTION AND PROPERTIES

Naphtha is produced by any one of several methods that include (i) fractionation of straight-run, cracked, and reforming distillates, or even fractionation of crude petroleum; (ii) solvent extraction; (iii) hydrogenation of cracked distillates; (iv) polymerization of unsaturated compounds (olefins); and (v) alkylation processes. In fact, naphtha may be a combination of product streams from more than one of these processes. On occasion, depending upon the availability, gas condensate, natural gas liquids (NGL), and natural gasoline are combined with the naphtha streams to complement the composition and volatility requirements.

Briefly and by way of explanation, *gas condensate* is a mixture of light hydrocarbon liquids obtained by condensation of hydrocarbon vapors: predominately butane, propane, and pentane with some heavier hydrocarbons and relatively little methane or ethane. *Natural gas liquids* are the hydrocarbon liquids that condense during the processing of hydrocarbon gases that are produced from oil or gas reservoir, and *natural gasoline* is a mixture of liquid hydrocarbons extracted from natural gas that is suitable for blending with the various refinery streams that constitute the final sales gasoline (Speight, 2014a).

The more common method of naphtha preparation is distillation. Depending on the design of the distillation unit, either one or two naphtha streams may be produced: (1) a single naphtha with an end point of about 205°C (400°F) and similar to straight-run gasoline, or (2) this same fraction divided into light naphtha and heavy naphtha. The end point of the light naphtha is varied to suit the subsequent

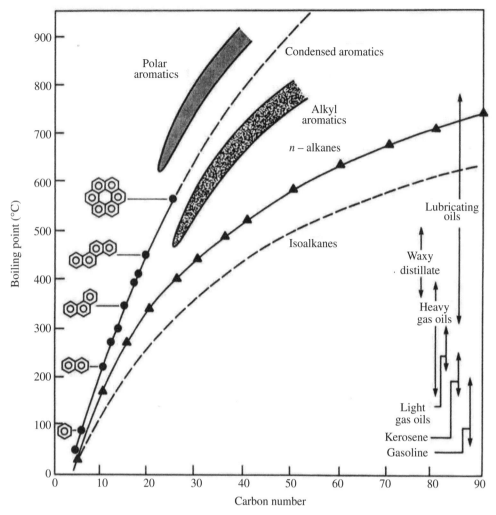

FIGURE 5.1   Boiling point and carbon number for various hydrocarbons and petroleum products.

subdivision of the naphtha into narrower boiling fractions, and may be of the order of 120°C (250°F).

Sulfur compounds are most commonly removed or converted to a harmless form by chemical treatment with lye, doctor solution, copper chloride, or similar treating agents (Speight, 1999). Hydrorefining processes (Speight, 1999) are also often used in place of chemical treatment. When used as a solvent, naphtha is selected for low sulfur content, and the usual treatment processes remove only the sulfur compounds. Naphtha, with a small aromatic content, has a slight odor, but the aromatics increase the solvent power of the naphtha, and there is no need to remove the aromatics, unless odor-free naphtha is specified.

Generally, naphtha is comprised of shorter chain hydrocarbons (C_5–C_8), can be aromatic (sweet smelling and carcinogenic), and can contain sulfuric compounds or other impurities. It has the lowest boiling point of all the petroleum-derived liquids, which gives it the highest vapor pressure. As a result, naphtha produces flammable vapors whenever it is not contained in a vapor-seal container, but those vapors will disperse quickly if well-ventilated. Odor is particularly important, since, unlike most other petroleum liquids, many of the manufactured products containing naphtha are used in confined spaces, in factory workshops, and in the home.

Naphtha is readily flammable, will evaporate quickly from most surfaces, and must be very carefully contained at all times. Aromatic naphtha contains a high percentage of aromatic constituents, and it can also be smoky, toxic, and carcinogenic. Some naphtha-based fuels have a reduced aromatic content, but many are naturally high or augmented in aromatics.

The term aliphatic naphtha refers to naphtha containing less than 0.1% benzene and with carbon numbers from C_5 through C_{16}. Aromatic naphtha has carbon numbers from C_6 through C_{16} and contains significant quantities of aromatic hydrocarbons such as benzene (>0.1%), toluene, and xylene (ASTM D5580, ASTM D5769). The final gasoline product as a transport fuel is a carefully blended mixture having a predetermined octane value. Thus, gasoline is a complex mixture of hydrocarbons that boils below 200°C (390°F).

The hydrocarbon constituents in this boiling range are those that have 4–12 carbon atoms in their molecular structure.

To meet the demands of a variety of uses, certain basic naphtha grades are produced, which are identified by boiling range. The complete range of naphtha solvents may be divided, for convenience, into four general categories: (1) special boiling point spirits having overall distillation range within the limits 30–l65°C (86–329°F), (2) pure aromatic compounds such as benzene, toluene, xylenes, or mixtures (BTX), thereof, (3) white spirit, also known as mineral spirit and naphtha, usually boiling within 150–210°C (302–410°F), and (4) high-boiling petroleum fractions, boiling within the limits 160–325°C (320617°F).

Since the end use dictates the required composition of naphtha, most grades are available in both high- and low-solvency categories, and the various text methods can have major significance in some applications and lesser significance in others. Hence, the application and significance of tests must be considered in the light of the proposed end use.

Naphtha contains varying amounts of its constituents, viz., paraffins, naphthenes, aromatics, and olefins in different proportions, in addition to the potential isomers of paraffin that exist in the naphtha boiling range. Naphtha resembles gasoline in terms of boiling range and carbon number, being a precursor to gasoline. Naphtha is used as an automotive fuel, engine fuel, and jet-B (naphtha-type). Broadly, naphtha is classified as *light naphtha* and *heavy naphtha*. Light naphtha is used as a rubber solvent, lacquer diluent, while heavy naphtha finds its application as a varnish solvent, dyer's naphtha, and cleaner's naphtha.

Volatility, solvent properties (dissolving power), purity, and odor determine the suitability of naphtha for a particular use. The use of naphtha as an incendiary device in warfare and as an illuminant dates back to 1200 A.D. Naphtha is characterized as lean (high paraffin content) or rich (low paraffin content). The rich naphtha with higher proportion of naphthene content is easier to process in the Platforming unit (Speight and Ozum, 2002; Hsu and Robinson, 2006; Gary *et al.*, 2007; Speight, 2014a).

Naphtha solvents may belong to categories such as special boiling spirits, having distillation range 30–165°C (86–330°F), white spirit (mineral spirit), boiling within 150–210°C (300–410°F), and high-boiling petroleum fractions (l60–325°C; 320–615°F). In aromatic complexes, naphtha is converted into basic petrochemical intermediates: BTX. Petroleum naphtha is by far the most popular feedstock for aromatics production.

If spilled or discharged in the environment, naphtha represents a threat of the toxicity of the constituents to land and/or to aquatic organisms. A significant spill may cause long-term adverse effects in the aquatic environment. The constituents of naphtha predominantly fall in the C_5–C_{16} carbon range: alkanes, cycloalkanes, aromatics, and, if they are subject to a cracking process, alkenes as well. Naphtha

may also contain a preponderance of aromatic constituents (up to 65%); others contain up to 40% alkenes, while all of the others are aliphatic in composition, up to 100%.

Water solubility ranges from very low solubility for the longest chain alkanes to high solubility for the simplest monoaromatic constituents. Generally, the aromatic compounds are more soluble than the same-sized alkanes, iso-alkanes, and cycloalkanes. This indicates that the components likely to remain in water are the one- and two-ring aromatics (C_6–C_{12}). The C_9–C_{16} alkanes, iso-alkanes, and one- and two-ring cycloalkanes are likely to be attracted to the sediments, based on their low water solubility and moderate-to-high octanol–water partition coefficient (log K_{ow}) and organic carbon–water partition coefficient (log K_{oc}) values.

Naphtha (especially the low-boiling naphtha) contains volatile organic compounds (VOCs) that are rapidly degraded in air, water, and soil. Considerable measures must be taken to prevent the release of naphtha (constituents) to the atmosphere and minimize any exposure to the environment from the activities in which naphtha is manufactured and used.

Constituents of naphtha can be carcinogenic, and, frequently, products sold as naphtha contain some impurities, which may also have harmful properties of their own. The method of manufacture means that there is a range of distinct chemicals in naphtha, which makes rigorous comparisons and identification of specific carcinogens difficult, and is further complicated by exposure to a significant range of other known and potential carcinogens.

The variety of applications emphasizes the versatility of naphtha. For example, naphtha is used in the paint, printing ink, and polish manufactures; the rubber and adhesive industries; as well as in the preparation of edible oils, perfumes, glues, and fats. Further uses are found in the dry-cleaning, leather, and fur industries, and also in the pesticide field. The characteristics that determine the suitability of naphtha for a particular use are volatility, solvent properties (dissolving power), purity, and odor (generally lack thereof).

The main uses of petroleum naphtha fall into the general areas of (i) precursor to gasoline and other liquid fuels, (ii) solvents (diluents) for paints, (iii) dry-cleaning solvents, (iv) solvents for cutback asphalts, (v) solvents in rubber industry, and (vi) solvents for industrial extraction processes. Turpentine, the older and more conventional solvent for paints, has now been almost completely replaced by the cheaper and more abundant petroleum naphtha. Test methods (Section 5.3) are based on defining the suitability of naphtha for any of the aforementioned uses.

5.3   TEST METHODS

Because of the high standards set for naphtha (McCann, 1998), it is essential to employ the correct techniques when taking samples for testing (ASTM D4057). Mishandling of the sample or the slightest trace of contaminant in the sample can give rise to misleading results. Special care is necessary to ensure that containers are scrupulously clean and free from odor. Samples should be taken with the minimum of disturbance so as to avoid loss of volatile components; in the case of low-boiling naphtha, it may be necessary to chill the sample. And, while awaiting examination, samples should be kept in a cool dark place so as to ensure that they do not lose volatile constituents or discolor and develop odors due to oxidation.

The physical properties of naphtha depend on the hydrocarbon types present, in general, the aromatic hydrocarbons having the highest solvent power and the straight-chain aliphatic compounds the lowest. The solvent properties can be assessed by estimating the amount of the various hydrocarbon types present. This method provides an indication of the solvent power of the naphtha on the basis that aromatic constituents and naphthenic constituents provide dissolving ability that paraffinic constituents do not. Another method for assessing the solvent properties of naphtha measures the performance of the fraction when used as a solvent under specified conditions, such as, for example, by the Kauri–Butanol test method (ASTM D1133). Another method involves measurement of the surface tension from which the solubility parameter is calculated and then provides an indication of the dissolving power and compatibility. Such calculations have been used to determine the yield of asphaltene fraction from petroleum by use of various solvents (Mitchell and Speight, 1973; Speight, 2001, 2014a). A similar principle is applied to determine the amount of insoluble material in lubricating oil using *n*-pentane (ASTM D893, ASTM D4055).

Insoluble constituents in lubricating oil can cause wear that can lead to equipment failure. Pentane-insoluble materials can include oil-insoluble materials and some oil-insoluble resinous matter originating from oil or additive degradation, or both. Toluene-insoluble constituents arise from external contamination, fuel carbon, and highly carbonized materials from degradation of fuel, oil, and additives, or engine wear and corrosion materials. A significant change in pentane-insoluble or toluene-insoluble constituents indicates a change in oil properties that could lead to machinery failure. The insoluble constituents measured can also assist in evaluating the performance characteristics of used oil or in determining the cause of equipment failure (Chapter 13) (Speight and Exall, 2014a).

Thus, one test (ASTM D893) covers the determination of pentane-insoluble and toluene-insoluble constituents in lubricating oils, using pentane dilution and centrifugation as the method of separation. The other test (ASTM D4055) uses pentane dilution followed by membrane filtration to remove insoluble constituents that have a size greater than 0.8μm.

5.3.1 Aniline Point and Mixed Aniline Point

The test method for the determination of aniline point and mixed aniline point of hydrocarbon solvents (ASTM D611, IP 2) is a means of determining the solvent power of naphtha by estimating the relative amounts of the various hydrocarbon constituents. It is a more precise technique than the method for Kauri–Butanol number (ASTM D1133).

The aniline (or mixed aniline) (ASTM D611, IP 2) point helps in the characterization of pure hydrocarbons and their mixtures, and is most often used to estimate the aromatic content of naphtha. Aromatic compounds exhibit the lowest aniline points, and paraffin compounds have the highest aniline points, with cycloparaffins (naphthenes) and olefins having aniline points between the two extremes. In any homologous series, the aniline point increases with increasing molecular weight.

There are five submethods in the test (ASTM D611, IP 2) for the determination of the aniline point:

1. Method A is used for transparent samples with an initial boiling point above room temperature and where the aniline point is below the bubble point and above the solidification point of the aniline–sample mixture.
2. Method B, a thin film method, is suitable for samples too dark for testing by method A.
3. Methods C and D are employed when there is the potential for sample vaporization at the aniline point.
4. Method D is particularly suitable, where only small quantities of sample are available.
5. Method E uses an automatic apparatus suitable for the range covered by methods A and B.

The results obtained by the Kauri–Butanol test (ASTM D1133) depend upon factors other than solvent power and are specific to the solute employed. For this reason, the aniline point is often preferred to the Kauri–Butanol number.

5.3.2 Composition

The number of potential hydrocarbon isomers in the naphtha boiling range (Tables 5.1 and 5.2) renders complete speciation of individual hydrocarbons impossible for the naphtha distillation range, and methods are used that identify the hydrocarbon types as chemical groups rather than as individual constituents.

The data from the density (specific gravity) test method (ASTM D1298, IP 160) provide a means of identification of a grade of naphtha, but is not a guarantee of composition and can only be used to indicate and evaluate product composition or quality when used in conjunction with the data from other test methods. Density data are used primarily to convert naphtha volume to a weight basis, a requirement in many of the industries concerned. For the necessary temperature corrections and also for volume corrections, appropriate sections of the petroleum measurement tables (ASTM D1250, IP 200) are used.

The first level of compositional information is group-type totals, as deduced by adsorption chromatography (ASTM D1319, IP 156), to give volume percent saturates, olefins, and aromatics in materials that boil below 315°C (600°F).

In this test method, a small amount of sample is introduced into a glass adsorption column packed with activated silica gel, of which a small layer contains a mixture of fluorescent dyes. When the sample has been adsorbed on the gel, alcohol is added to desorb the sample down the column, and the hydrocarbon constituents are separated according to their affinities into three types (aromatics, olefins, and saturates). The fluorescent dyes also react selectively with the hydrocarbon types and make the boundary zones visible under ultraviolet light. The volume percentage of each hydrocarbon type is calculated from the length of each zone in the column.

There are other test methods available. Benzene content and other aromatics may be estimated by spectrophotometric analysis and also by gas–liquid chromatography. However, these two test methods, based on the adsorption concept (ASTM D2007, ASTM D2549), are used for classifying oil samples of initial boiling point of at least 200°C (392°F) into the hydrocarbon types of polar compounds, aromatics, and saturates, and for the recovery of representative fractions of these types. Such methods are unsuitable for the majority of naphtha samples because of volatility constraints.

An indication of naphtha composition may also be obtained from the determination of aniline point (IP 2), freezing point (ASTM D852, ASTM D1015) (Fig. 5.2), cloud point (ASTM D2500) (Fig. 5.3), and the solidification point. And, though refinery treatment should ensure no alkalinity and acidity (ASTM D847, ASTM D1093, ASTM D1613, ASTM D2896, IP 1) and no olefins present, the relevant tests using bromine number (ASTM D1159, IP 130), bromine index (ASTM D2710), and flame ionization absorption (ASTM D1319, IP 156) are necessary to insure low levels (at the maximum) of hydrogen sulfide as well as the sulfur compounds in general (ASTM D130, ASTM D849, ASTM D1266, ASTM D3120, ASTM D4045, IP 107, IP 154), and especially corrosive sulfur compounds, such as that are determined by the doctor test method (ASTM D4952, IP 30).

Aromatic content is a key property of low-boiling distillates such as naphtha and gasoline, because the aromatic constituents influence a variety of properties, including boiling range (ASTM D86, IP 123), viscosity (ASTM D88, ASTM D445, ASTM D2161, IP 71), stability (ASTM D525, IP 40), and compatibility (ASTM D1133), with a variety of solutes. Existing methods use physical measurements and need suitable standards. Tests such as aniline point

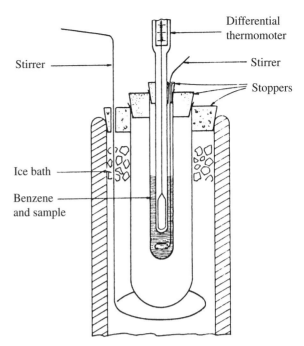

FIGURE 5.2   Freezing point apparatus for use in the depression of the freezing point of benzene test.

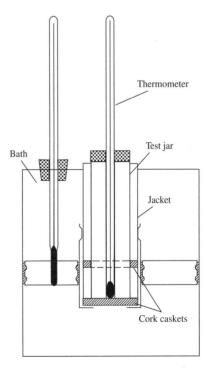

FIGURE 5.3   Apparatus for the determination of cloud point and pour point.

(ASTM D611) and Kauri–Butanol number (ASTM D1133) are of a somewhat empirical nature and can serve a useful function as control tests. Naphtha composition, however, is monitored mainly by gas chromatography (GC), and though most of the methods might have been developed for gasoline

(ASTM D2427, ASTM D6296), the applicability of the test methods to naphtha is in order.

A multidimensional GC method (ASTM D5443) provides for the determination of paraffins, naphthenes, and aromatics by carbon number in low olefin-type hydrocarbon streams having final boiling points lower than 200°C (392°F). In the method, the sample is injected into a GC system that contains a series of columns and switching values. First, a polar column retains polar aromatic compounds, binaphthenes, and high-boiling paraffins and naphthenes. The eluate from this column goes through a platinum column that hydrogenates olefins, and then to a molecular sieve column that performs a carbon number separation based on the molecular structure, that is, naphthenes and paraffins. The fraction remaining on the polar column is further divided into three separate fractions that are then separated on a nonpolar column by boiling point. A flame ionization detector detects the eluting compounds.

Other methods (ASTM D3257, ASTM D7576) for the determination of the amount of aromatic constituents include various types of detectors and offer alternate routes to determining aromatics in naphtha. Hydrocarbon composition is also determined by mass spectrometry—a technique that has seen wide use for hydrocarbon-type analysis of naphtha and gasoline (ASTM D2789), as well as to the identification of hydrocarbon constituents in higher boiling naphtha fractions (ASTM D2425).

One method (ASTM D6379, IP 436) is used to determine the monoaromatic and diaromatic hydrocarbon contents in distillates boiling in the range from 50 to 300°C (122 to 572°F). In the method, the sample is diluted with an equal volume of hydrocarbon, such as heptane, and a fixed volume of this solution is injected into a high-performance liquid chromatograph fitted with a polar column, where separation of the aromatic hydrocarbons from the nonaromatic hydrocarbons occurs. The separation of the aromatic constituents appears as distinct bands according to the ring structure, and a refractive index detector is used to identify the components as they elute from the column. The peak areas of the aromatic constituents are compared with those obtained from the previously run calibration standards to calculate the %w/w monoaromatic hydrocarbon constituents and diaromatic hydrocarbon constituents in the sample.

Compounds containing sulfur, nitrogen, and oxygen could possibly interfere with the performance of the test. Monoalkenes do not interfere, but conjugated di- and polyalkenes, if present, may interfere with the test performance.

Another method (ASTM D2425) provides more compositional detail (in terms of molecular species) than the chromatographic analysis, and the hydrocarbon types are classified in terms of a Z-series, in which Z (in the empirical formula C_nH_{2n+z}) is a measure of the hydrogen deficiency of the compound. This method requires that the sample be separated into saturated and aromatic fractions before mass

spectrometric analysis (ASTM D2549), and the separation is applicable only to some fractions, but not others. For example, the method is applicable to high-boiling naphtha, but not to the low-boiling naphtha, since it is impossible to evaporate the solvent used in the separation without also losing the lower boiling constituents of the naphtha under investigation.

The percentage of aromatic hydrogen atoms and aromatic carbon atoms can be determined by high-resolution nuclear magnetic resonance spectroscopy (ASTM D5292) that gives the mol percent of aromatic hydrogen or carbon atoms. Proton (hydrogen) magnetic resonance spectra are obtained on sample solutions in either chloroform or carbon tetrachloride, using a continuous-wave or pulse-Fourier transform high-resolution magnetic resonance spectrometer. Carbon magnetic resonance spectra are obtained on the sample solution in chloroform-d, using a pulse-Fourier transform high-resolution magnetic resonance.

The determination of hydrocarbon components in petroleum naphtha is of great importance to the petrochemical industry and for process control of reforming processes, as well as for regulatory purposes. The data obtained by nuclear magnetic resonance (ASTM D5292) can be used to evaluate changes in aromatic contents in naphtha as well as kerosene, gas oil, mineral oil, and lubricating oil. However, the results from this test are not equivalent to the mass–percent or volume–percent aromatics determined by the chromatographic methods, because the chromatographic methods determine the percent by weight or percent by volume of molecules that have one or more aromatic rings, and the alkyl substituents on the rings will contribute to the percentage of aromatics determined by chromatographic techniques.

Low-resolution nuclear magnetic resonance spectroscopy can also be used to determine percent by weight hydrogen in jet fuel (ASTM D3701), and in light distillate, middle distillate, and gas oil (ASTM D4808). As noted earlier, chromatographic methods are not applicable to naphtha, where losses can occur by evaporation.

The nature of the uses found for naphtha demands compatibility with the many other materials employed in formulation, with waxes, pigments, resins, and so on; thus, the solvent properties of a given fraction must be carefully measured and controlled. For most purposes, volatility is important, and, because of the wide use of naphtha in industrial and recovery plants, information on some other fundamental characteristics is required for plant design.

Though the focus of many tests is analysis of the hydrocarbon constituents of naphtha and other petroleum fractions, heteroatom compounds that contain sulfur and nitrogen atoms cannot be ignored, and methods for their determination are available. The combination of GC with element-selective detection gives information about the distribution of the element. In addition, many individual heteroatomic compounds can be determined.

Nitrogen compounds in middle distillates can be selectively detected by chemiluminescence. Individual nitrogen compounds can be detected down to 100ppb nitrogen. Gas chromatography with either sulfur chemiluminescence detection or atomic emission detection has been used for sulfur-selective detection.

Estimates of the purity of these products were determined in laboratories, using a variety of procedures such as freezing point, flame ionization absorbance, ultraviolet absorbance, GC, and capillary gas chromatography (ASTM D850, ASTM D852, ASTM D848, ASTM D849, ASTM D1015, ASTM D1016, ASTM D1078, ASTM D1319, ASTM D2008, ASTM D2360, ASTM D5134, ASTM D5917, IP 156).

Gas chromatography has become a primary technique for determining hydrocarbon impurities in individual aromatic hydrocarbons and the composition of mixed aromatic hydrocarbons. Though the measure of purity by GC is often sufficient, GC is not capable of measuring absolute purity; not all possible impurities will pass through the GC column, and not all those that do will be measured by the detector. Despite some shortcomings, GC is a standard, widely used technique, and is the basis of many current test methods for aromatic hydrocarbons (ASTM D2360, ASTM D3797, ASTM D4492, ASTM D4735, ASTM D5060, ASTM D5135, ASTM D5713, ASTM D5917, ASTM D6144).

When classes of hydrocarbons, such as olefins, need to be measured, techniques such as bromine index are used (ASTM D1492, ASTM D5776).

Impurities other than hydrocarbons are of concern in the petroleum industry. For example, many catalytic processes are sensitive to sulfur contaminants. Consequently, there is also a series of methods to determine trace concentrations of sulfur-containing compounds (ASTM D4045, ASTM D4735).

Chloride-containing impurities are determined by various test methods (ASTM D5194, ASTM D5808, ASTM D6069) that have sensitivity to 1mg/kg, reflecting the needs of the industry to determine very low levels of these contaminants.

Water is a contaminant in naphtha and should be measured using the Karl Fischer method (ASTM D1364, ASTM D1744, ASTM D4377, ASTM D4928, ASTM D6304, ASTM E203), by distillation (ASTM D4006), or by centrifuging, and excluded by relevant drying methods.

Tests should also be carried out for sediments, if the naphtha has been subjected to events (such as oxidation) that could lead to sediment formation and instability of the naphtha and the resulting products. Test methods are available for the determination of sediment by extraction (ASTM D473, IP 285) or by membrane filtration (ASTM D4807) and for the determination of sediment simultaneously with water by centrifugation (ASTM D1796, ASTM D2709, ASTM D4007, IP 373).

5.3.3 Correlative Methods

Correlative methods have long been used as a way of dealing with the complexity of various petroleum fractions, including naphtha. Relatively easy-to-measure physical properties such as density (or specific gravity) (ASTM D3505, ASTM D4052) are also required. Viscosity (ASTM D88, ASTM D445, ASTM D2161, IP 71), density (ASTM D287, ASTM D891, ASTM D1217, ASTM D1298, ASTM D1555, ASTM D1657, ASTM D4052, ASTM D5002, IP 160, IP 235, IP 365), and refractive index (ASTM D1218) have been correlated to hydrocarbon composition (Tables 5.3 and 5.4).

5.3.4 Density

Density (the mass of liquid per unit volume at 15°C) and the related terms, *specific gravity* (the ratio of the mass of a given volume of liquid at 15°C to the mass of an equal volume of pure water at the same temperature) and *relative density* (same as *specific gravity*), are important properties of petroleum products as they are part of product sales specifications, though they only play a minor role in the studies of product composition. Usually, a hydrometer, pycnometer, or digital density meter is used for the determination of all these standards.

Density is an important parameter for naphtha and solvents, and the determination of density (specific gravity) (ASTM D287, ASTM D891, ASTM D1217, ASTM D1298, ASTM D1555, ASTM D1657, ASTM D4052, ASTM D5002, IP 160, IP 235, IP 365) (Fig. 5.4) provides a check

TABLE 5.3   Refractive index of selected hydrocarbons

Compound	Refractive index n_D^{20}
n-Pentane	1.3578
n-Hexane	1.3750
n-Hexadecane	1.4340
Cyclopentane	1.4065
Cyclopentene	1.4224
Pentene-1	1.3714
1,3-Pentadiene	1.4309
Benzene	1.5011
cis-Decahydronaphthalene	1.4814
Methylnaphthalenes	1.6150

TABLE 5.4   Physical properties of selected petroleum products

Refractive index n_D^{20}	Specific gravity 60°/60°F	Viscosity (cSt)		Molecular weight
		100°F	210°F	
1.5185	0.9250	26.35	3.87	300
1.4637	0.8406	14.55	3.28	353
1.5276	0.9367	1955.00	49.50	646
1.4799	0.8724	3597.00	27.35	822

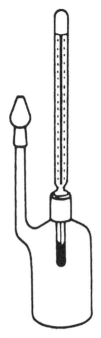

FIGURE 5.4 Density-weighing bottle.

on the uniformity of the naphtha and permits calculation of the weight per gallon. The temperature at which the determination is carried out and for which the calculations are to be made should also be known. However, the methods are subject to vapor pressure constraints and are used with appropriate precautions to prevent vapor loss during sample-handling and density measurement. In addition, some test methods should not be applied if the samples are so dark in color, because the absence of air bubbles in the sample cell cannot be established with certainty. The presence of such bubbles can have serious consequences for the reliability of the test data.

5.3.5 Evaporation Rate

The vaporizing tendencies of petroleum products are the basis for the general characterization of liquid petroleum fuels, such as liquefied petroleum gas, natural gasoline, motor and aviation gasoline, naphtha, kerosene, gas oil, diesel fuel, and fuel oil. Standard test methods (ASTM D6, ASTM D20, ASTM D2715, IP 506) are available to determine the volatility of higher boiling products, and can, with some cautious modifications, be adapted for lower boiling products.

For some purposes, it is necessary to have information on the initial stage of vaporization and the potential hazards that such a property can cause. To supply this need, flash and fire, vapor pressure, and evaporation methods are available. The data from the early stages of the several distillation methods are also useful. For other uses, it is important to know the tendency of a product to partially vaporize or to completely vaporize and, in some cases, to know if small quantities of high-boiling components are present. For such purposes, chief reliance is placed on the distillation methods.

Nevertheless, the evaporation rate is an important property of naphtha, and though there is a significant relation between distillation range and evaporation rate, the relationship is not straightforward.

A simple procedure for determining the evaporation rate involves use of at least a pair of weighed shallow containers, each containing a weighed amount of naphtha. The cover-free containers are placed in a temperature-controlled and humidity-controlled draft-free area. The containers are reweighed at intervals, until the samples have completely evaporated or left a residue that does not evaporate further (ASTM D381, ASTM D1353, IP 131).

The evaporation rate can be derived either (i) by a plot of time versus weight, using a solvent having a known evaporation rate for comparison or (ii) from the distillation profile (ASTM D86, IP 123).

5.3.6 Flash Point

The flash point is the lowest temperature at atmospheric pressure (760mm Hg, 101.3 kPa), at which application of a test flame will cause the vapor of a sample to ignite under specified test conditions. The sample is deemed to have reached the flash point when a large flame appears and instantaneously propagates itself over the surface of the sample. The flash point data is used in shipping and safety regulations to define *flammable* and *combustible* materials. Flash point data can also indicate the possible presence of highly volatile and flammable constituents in a relatively nonvolatile or nonflammable material.

Of the available test methods, the most common method of determining the flash point confines the vapor (closed cup), until the instant flame is applied (ASTM D56, ASTM D93, ASTM D3828, ASTM D6450, IP 34, IP 523) (Fig. 5.5). An alternate method that does not confine the vapor (open cup method (ASTM D92, ASTM D1310, IP 36) gives slightly higher values of the flash point.

Erroneously high flash points can be obtained when precautions are not taken to avoid the loss of volatile material. Samples should not be stored in plastic bottles, since the volatile material may diffuse through the walls of the container. The containers should not be opened unnecessarily. The samples should not be transferred between containers, unless the sample temperature is at least 20°F (11°C) below the expected flash point.

Another test (ASTM E659) is available that can be used as a complement to the flash point test and involves determination of the autoignition temperature. However, the flash point should not be confused with the autoignition temperature that measures spontaneous combustion with no external source of ignition.

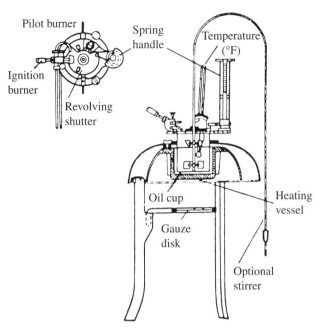

FIGURE 5.5   The Pensky–Marten flash point apparatus (ASTM D56).

5.3.7   Kauri–Butanol Value

The Kauri–Butanol value (ASTM D1133) is the number of milliliters of the solvent, at 15°C (77°F), required to produce a defined degree of turbidity when added to 20g of a standard solution of gum kauri resin in *n*-butyl alcohol. The Kauri–Butanol value of naphtha is used to determine relative solvent power.

For Kauri–Butanol values of 60 and higher, the standard is toluene, which has an assigned value of 105, whereas for Kauri–Butanol values less than 60, the standard is a blend of 75% *n*-heptane and 25% toluene, which has an assigned value of 40. The Kauri–Butanol values of products that are classified as regular mineral spirits normally vary between 34 and 44; xylene is 93; and aromatic naphtha falls in the range 55–108.

However, the data obtained by the Kauri–Butanol test depend upon the factors other than solvent power and are specific to the solute employed. For this reason, the aniline point is often preferred to the Kauri–Butanol number.

5.3.8   Odor and Color

The degree of purity of naphtha is an important aspect of naphtha properties, and strict segregation of all distribution equipment is maintained to ensure strict and uniform specification for the product handled. Naphtha is refined to a low level of odor to meet the specifications for use. Purified naphtha is required to have a low level of odor to meet the specifications for use.

In general, the paraffinic hydrocarbons possess the mildest odor and the aromatics the strongest, the odor level

(ASTM D268, ASTM D1296) being related to the chemical character and volatility of the constituents. Odors due to the presence of sulfur compounds or unsaturated constituents are excluded by specification. And, apart from certain high-boiling aromatic fractions (that are usually excluded by volatility from the majority of the naphtha fractions) which may be pale yellow in color, naphtha is usually colorless (water white).

Usually, naphtha is colorless (water white), but naphtha containing higher amounts of aromatic constituents may be pale yellow. Measurement of color (ASTM D156, ASTM D848, ASTM D1209, ASTM D1555, ASTM D5386, IP 17) provides a rapid method of checking the degree of freedom from contamination. Observation of the test for residue on evaporation (ASTM D381, ASTM D1353, IP 131) provides a further guard against adventitious contamination.

5.3.9   Volatility

Distillation, as a means of determining the boiling range (hence the *volatility*) of petroleum and petroleum products, has been in use since the beginning of the petroleum industry and is an important aspect of product specifications.

Thus, one of the most important physical parameters is the boiling-range distribution (ASTM D86, ASTM D1078, ASTM D2268, ASTM D2887, ASTM D2892, IP 123). The significance of the distillation test is the indication of volatility that dictates the evaporation rate, which is an important property for naphtha used in coatings and similar applications, where the premise is that the naphtha evaporates over time, leaving the coating applied to the surface.

In the basic test method (ASTM D86, IP 123), a 100-ml sample is distilled (manually or automatically) under prescribed conditions.. Temperatures and volumes of condensate are recorded at regular intervals from which the boiling profile is derived.

The determination of the boiling-range distribution of distillates such as naphtha and gasoline by GC (ASTM D3710) not only helps identify the constituents but also facilitates on-line controls at the refinery. This test method is designed to measure the entire boiling range of naphtha with either high or low Reid vapor pressure (ASTM D323, IP 69). In the method, the sample is injected into a GC column that separates hydrocarbons in boiling-point order. The column temperature is raised at a reproducible rate, and the area under the chromatogram is recorded throughout the run. Calibration is performed using a known mixture of hydrocarbons covering the expected boiling of the sample.

Another method is described as a method for determining the carbon number distribution (ASTM D2887), and the data derived by this test method are essentially equivalent to that obtained by TBP distillation (ASTM D2892). The sample is introduced into a GC column that separates hydrocarbons in boiling-point order. The column temperature is raised at a reproducible rate, and the area under the chromatogram is recorded throughout the run. Boiling temperatures are assigned to the time axis from a calibration curve, obtained under the same conditions by running a known mixture of hydrocarbons covering the boiling range expected in the sample. From these data, the boiling-range distribution may be obtained. However, this test method is limited to samples having a boiling range greater than 55°C (l00°F) and having a vapor pressure (ASTM D323, ASTM D4953, ASTM D5191, ASTM D5482, ASTM D6377, ASTM D6378, IP 69) sufficiently low to permit sampling at ambient temperature without contamination.

Naphtha grades are often referred to by a boiling range, which is the defined temperature range in which the fraction is distilled. The ranges are determined by standard methods (ASTM D86, IP 123, IP 195); it is being especially necessary to use a recognized method, since the initial and final boiling points which ensure conformity with volatility requirements and absence of *heavy ends* are affected by the testing procedure.

A simple test for the evaporation properties of naphtha is available (ASTM D381, IP 131), but the volatility of naphtha is generally considered a measure of its drying time in use. And, the temperature of use obviously governs the choice of naphtha. A high-boiling narrow distillation fraction of gas oil may be required for a heatset ink, where the operating temperature may be as high as 316°C (600°F). However, the need for vacuum distillation (ASTM D1160) as a product specification in the boiling range of naphtha is not necessary. By definition, naphtha very rarely has such a high boiling point.

While pure hydrocarbons such as pentane, hexane, heptane, and BTX, which are now largely of petroleum origin may be characterized by a fixed boiling point, naphtha is a mixture of many hydrocarbons and cannot be so identified. The distillation test does however give a useful indication of their volatility. The data obtained should include the initial and final temperatures of distillation, together with sufficient temperature and volume observations, to permit a characteristic distillation curve to be drawn.

This information is especially important when a formulation includes other volatile liquids, since the performance of the product will be affected by the relative volatility of the constituents. An illustration of the importance of this aspect is found in the use of specifically defined boiling point naphtha in cellulose lacquers, where a mixture with ester, alcohols, and other solvents may be employed. The naphtha does not act as a solvent for the cellulose ester, but is incorporated as a diluent to control the viscosity and flow properties of the mixture. If the solvent evaporates too rapidly, then blistering of the surface coating may result, while if the solvent evaporates unevenly, leaving behind a higher proportion of the naphtha, precipitation of the cellulose may occur leading to a milky opaqueness known as blushing.

Though much dependence is placed on the assessment of volatility by distillation methods, some specifications include measurement of drying time by evaporation from a filter paper or dish. Laboratory measurements are expressed as *evaporation rate*, either by reference to a pure compound evaporated under similar conditions as the sample under test, or by constructing a time–weight loss curve under standard conditions. Though the results obtained on naphtha provide a useful guide, it is, wherever possible, better to carry out a performance test on the final product when assessing formulations.

In choosing naphtha for a particular purpose, it is necessary to relate volatility to the fire hazard associated with its use, storage, and transport, and also with the handling of the products arising from the process. This is normally based on the characterization of the solvent by flash point limits (ASTM D56, ASTM D93, IP 34, IP 170)

5.4 STORAGE

Naphtha can (as do other hydrocarbon liquids) act as a non-conductive flammable liquid (or static accumulators), and may form ignitable vapor–air mixtures in storage tanks or other containers.

Precautions to prevent static-initiated fire or explosion during transfer, storage, or handling, include, but are not limited to, these examples:

1. Ground and bond containers during product transfers. Grounding and bonding may not provide adequate protection to prevent ignition or explosion of hydrocarbon liquids and vapors that are static accumulators.

2. Special slow-load procedures for *switch loading* must be followed to avoid the static ignition hazard that can exist when higher flash point material (such as fuel oil or diesel) is loaded into tanks previously containing low flash point products (such as gasoline or naphtha).

3. Storage tank level floats must be effectively bonded (NFPA 77, API Recommended Practice, 2003; Recommended Practice on Static Electricity, 2007; Protection Against Ignitions Arising Out of Static, Lightning, and Stray Currents, 2008).

REFERENCES

ASTM D6. Standard Test Method for Loss on Heating of Oil and Asphaltic Compounds. Annual Book of Standards. ASTM International, West Conshohocken, PA.

ASTM D20. Standard Test Method for Distillation of Road Tars. Annual Book of Standards. ASTM International, West Conshohocken, PA.

ASTM D56. Standard Test Method for Flash Point by Tag Closed Cup Tester. Annual Book of Standards. ASTM International, West Conshohocken, PA.

ASTM D86. Standard Test Method for Distillation of Petroleum Products at Atmospheric Pressure. Annual Book of Standards. ASTM International, West Conshohocken, PA.

ASTM D88. Standard Test Method for Saybolt Viscosity. Annual Book of Standards. ASTM International, West Conshohocken, PA.

ASTM D92. Standard Test Method for Flash and Fire Points by Cleveland Open Cup Tester. Annual Book of Standards. ASTM International, West Conshohocken, PA.

ASTM D93. Standard Test Methods for Flash Point by Pensky-Martens Closed Cup Tester. Annual Book of Standards. ASTM International, West Conshohocken, PA.

ASTM D130. Standard Test Method for Corrosiveness to Copper from Petroleum Products by Copper Strip Test. Annual Book of Standards. ASTM International, West Conshohocken, PA.

ASTM D156. Standard Test Method for Saybolt Color of Petroleum Products (Saybolt Chromometer Method). Annual Book of Standards. ASTM International, West Conshohocken, PA.

ASTM D268. Standard Guide for Sampling and Testing Volatile Solvents and Chemical Intermediates for Use in Paint and Related Coatings and Materials. Annual Book of Standards. ASTM International, West Conshohocken, PA.

ASTM D287. Standard Test Method for API Gravity of Crude Petroleum and Petroleum Products (Hydrometer Method). Annual Book of Standards. ASTM International, West Conshohocken, PA.

ASTM D323. Standard Test Method for Vapor Pressure of Petroleum Products (Reid Method). Annual Book of Standards. ASTM International, West Conshohocken, PA.

ASTM D381. Standard Test Method for Gum Content in Fuels by Jet Evaporation. Annual Book of Standards. ASTM International, West Conshohocken, PA.

ASTM D445. Standard Test Method for Kinematic Viscosity of Transparent and Opaque Liquids (and Calculation of Dynamic Viscosity). Annual Book of Standards. ASTM International, West Conshohocken, PA.

ASTM D473. Standard Test Method for Sediment in Crude Oils and Fuel Oils by the Extraction Method. Annual Book of Standards. ASTM International, West Conshohocken, PA.

ASTM D525. Standard Test Method for Oxidation Stability of Gasoline (Induction Period Method). Annual Book of Standards. ASTM International, West Conshohocken, PA.

ASTM D611. Standard Test Methods for Aniline Point and Mixed Aniline Point of Petroleum Products and Hydrocarbon Solvents. Annual Book of Standards. ASTM International, West Conshohocken, PA.

ASTM D808. Standard Test Method for Chlorine in New and Used Petroleum Products (High Pressure Decomposition Device Method). Annual Book of Standards. ASTM International, West Conshohocken, PA.

ASTM D847. Standard Test Method for Acidity of Benzene, Toluene, Xylenes, Solvent Naphthas, and Similar Industrial Aromatic Hydrocarbons. Annual Book of Standards. ASTM International, West Conshohocken, PA.

ASTM D848. Standard Test Method for Acid Wash Color of Industrial Aromatic Hydrocarbons. Annual Book of Standards. ASTM International, West Conshohocken, PA.

ASTM D849. Standard Test Method for Copper Strip Corrosion by Industrial Aromatic Hydrocarbons. Annual Book of Standards. ASTM International, West Conshohocken, PA.

ASTM D850. Standard Test Method for Distillation of Industrial Aromatic Hydrocarbons and Related Materials. Annual Book of Standards. ASTM International, West Conshohocken, PA.

ASTM D852. Standard Test Method for Solidification Point of Benzene. Annual Book of Standards. ASTM International, West Conshohocken, PA.

ASTM D891. Standard Test Methods for Specific Gravity, Apparent, of Liquid Industrial Chemicals. Annual Book of Standards. ASTM International, West Conshohocken, PA.

ASTM D893. Standard Test Method for Insolubles in Used Lubricating Oils. Annual Book of Standards. ASTM International, West Conshohocken, PA.

ASTM D1015. Standard Test Method for Freezing Points of High-Purity Hydrocarbons. Annual Book of Standards. ASTM International, West Conshohocken, PA.

ASTM D1016. Standard Test Method for Purity of Hydrocarbons from Freezing Points. Annual Book of Standards. ASTM International, West Conshohocken, PA.

ASTM D1078. Standard Test Method for Hydrogen in Petroleum Fractions. Annual Book of Standards. ASTM International, West Conshohocken, PA.

ASTM D1093. Standard Test Method for Acidity of Hydrocarbon Liquids and Their Distillation Residues. Annual Book of Standards. ASTM International, West Conshohocken, PA.

ASTM D1133. Standard Test Method for Kauri-Butanol Value of Hydrocarbon Solvents. Annual Book of Standards. ASTM International, West Conshohocken, PA.

ASTM D1159. Standard Test Method for Bromine Numbers of Petroleum Distillates and Commercial Aliphatic Olefins by Electrometric Titration. Annual Book of Standards. ASTM International, West Conshohocken, PA.

ASTM D1160. Standard Test Method for Distillation of Petroleum Products at Reduced Pressure. Annual Book of Standards. ASTM International, West Conshohocken, PA.

ASTM D1209. Standard Test Method for Color of Clear Liquids (Platinum-Cobalt Scale). Annual Book of Standards. ASTM International, West Conshohocken, PA.

ASTM D1217. Standard Test Method for Density and Relative Density (Specific Gravity) of Liquids by Bingham Pycnometer. Annual Book of Standards. ASTM International, West Conshohocken, PA.

ASTM D1218. Standard Test Method for Refractive Index and Refractive Dispersion of Hydrocarbon Liquids. Annual Book of Standards. ASTM International, West Conshohocken, PA.

ASTM D1250. Standard Guide for Use of the Petroleum Measurement Tables. Annual Book of Standards. ASTM International, West Conshohocken, PA.

ASTM D1266. Standard Test Method for Sulfur in Petroleum Products (Lamp Method). Annual Book of Standards. ASTM International, West Conshohocken, PA.

ASTM D1296. Standard Test Method for Odor of Volatile Solvents and Diluents. Annual Book of Standards. ASTM International, West Conshohocken, PA.

ASTM D1298. Standard Test Method for Density, Relative Density, or API Gravity of Crude Petroleum and Liquid Petroleum Products by Hydrometer Method. Annual Book of Standards. ASTM International, West Conshohocken, PA.

ASTM D1310. Standard Test Method for Flash Point and Fire Point of Liquids by Tag Open-Cup Apparatus. Annual Book of Standards. ASTM International, West Conshohocken, PA.

ASTM D1319. Standard Test Method for Hydrocarbon Types in Liquid Petroleum Products by Fluorescent Indicator Adsorption. Annual Book of Standards. ASTM International, West Conshohocken, PA.

ASTM D1353. Standard Test Method for Nonvolatile Matter in Volatile Solvents for Use in Paint, Varnish, Lacquer, and Related Products. Annual Book of Standards. ASTM International, West Conshohocken, PA.

ASTM D1364. Standard Test Method for Water in Volatile Solvents (Karl Fischer Reagent Titration Method). Annual Book of Standards. ASTM International, West Conshohocken, PA.

ASTM D1492. Standard Test Method for Bromine Index of Aromatic Hydrocarbons by Coulometric Titration. Annual Book of Standards. ASTM International, West Conshohocken, PA.

ASTM D1555. Standard Test Method for Calculation of Volume and Weight of Industrial Aromatic Hydrocarbons and Cyclohexane [Metric]. Annual Book of Standards. ASTM International, West Conshohocken, PA.

ASTM D1613. Standard Test Method for Acidity in Volatile Solvents and Chemical Intermediates Used in Paint, Varnish, Lacquer, and Related Products. Annual Book of Standards. ASTM International, West Conshohocken, PA.

ASTM D1657. Standard Test Method for Density or Relative Density of Light Hydrocarbons by Pressure Hydrometer. Annual Book of Standards. ASTM International, West Conshohocken, PA.

ASTM D1744. Standard Test Method for Determination of Water in Liquid Petroleum Products by Karl Fischer Reagent. Annual Book of Standards. ASTM International, West Conshohocken, PA.

ASTM D1796. Standard Test Method for Water and Sediment in Fuel Oils by the Centrifuge Method (Laboratory Procedure). Annual Book of Standards. ASTM International, West Conshohocken, PA.

ASTM D2007. Standard Test Method for Characteristic Groups in Rubber Extender and Processing Oils and Other Petroleum-Derived Oils by the Clay-Gel Absorption Chromatographic Method. Annual Book of Standards. ASTM International, West Conshohocken, PA.

ASTM D2008. Standard Test Method for Ultraviolet Absorbance and Absorptivity of Petroleum Products. Annual Book of Standards. ASTM International, West Conshohocken, PA.

ASTM D2161. Standard Practice for Conversion of Kinematic Viscosity to Saybolt Universal Viscosity or to Saybolt Furol Viscosity. Annual Book of Standards. ASTM International, West Conshohocken, PA.

ASTM D2268. Standard Test Method for Analysis of High-Purity n-Heptane and Isooctane by Capillary Gas Chromatograph. Annual Book of Standards. ASTM International, West Conshohocken, PA.

ASTM D2360. Standard Test Method for Trace Impurities in Monocyclic Aromatic Hydrocarbons by Gas Chromatography. Annual Book of Standards. ASTM International, West Conshohocken, PA.

ASTM D2425. Standard Test Method for Hydrocarbon Types in Middle Distillates by Mass Spectrometry. Annual Book of Standards. ASTM International, West Conshohocken, PA.

ASTM D2427. Standard Test Method for Determination of C2 through C5 Hydrocarbons in Gasolines by Gas Chromatography. Annual Book of Standards. ASTM International, West Conshohocken, PA.

ASTM D2500. Standard Test Method for Cloud Point of Petroleum Products. Annual Book of Standards. ASTM International, West Conshohocken, PA.

ASTM D2549. Standard Test Method for Separation of Representative Aromatics and Nonaromatics Fractions of High-Boiling Oils by Elution Chromatography. Annual Book of Standards. ASTM International, West Conshohocken, PA.

ASTM D2709. Standard Test Method for Water and Sediment in Middle Distillate Fuels by Centrifuge. Annual Book of Standards. ASTM International, West Conshohocken, PA.

ASTM D2710. Standard Test Method for Bromine Index of Petroleum Hydrocarbons by Electrometric Titration. Annual Book of Standards. ASTM International, West Conshohocken, PA.

ASTM D2715. Standard Test Method for Volatilization Rates of Lubricants in Vacuum. Annual Book of Standards. ASTM International, West Conshohocken, PA.

ASTM D2789. Standard Test Method for Hydrocarbon Types in Low Olefinic Gasoline by Mass Spectrometry. Annual Book of Standards. ASTM International, West Conshohocken, PA.

ASTM D2887. Standard Test Method for Boiling Range Distribution of Petroleum Fractions by Gas Chromatography. Annual Book of Standards. ASTM International, West Conshohocken, PA.

ASTM D2892. Standard Test Method for Distillation of Crude Petroleum (15-Theoretical Plate Column). Annual Book of Standards. ASTM International, West Conshohocken, PA.

ASTM D2896. Standard Test Method for Base Number of Petroleum Products by Potentiometric Perchloric Acid Titration. Annual Book of Standards. ASTM International, West Conshohocken, PA.

ASTM D3120. Standard Test Method for Trace Quantities of Sulfur in Light Liquid Petroleum Hydrocarbons by Oxidative Microcoulometry. Annual Book of Standards. ASTM International, West Conshohocken, PA.

ASTM D3257. Standard Test Methods for Aromatics in Mineral Spirits by Gas Chromatography. Annual Book of Standards. ASTM International, West Conshohocken, PA.

ASTM D3505. Standard Test Method for Density or Relative Density of Pure Liquid Chemicals. Annual Book of Standards. ASTM International, West Conshohocken, PA.

ASTM D3701. Standard Test Method for Hydrogen Content of Aviation Turbine Fuels by Low Resolution Nuclear Magnetic Resonance Spectrometry. Annual Book of Standards. ASTM International, West Conshohocken, PA.

ASTM D3710. Standard Test Method for Boiling Range Distribution of Gasoline and Gasoline Fractions by Gas Chromatography. Annual Book of Standards. ASTM International, West Conshohocken, PA.

ASTM D3797. Standard Test Method for Analysis of o-Xylene by Gas Chromatography. Annual Book of Standards. ASTM International, West Conshohocken, PA.

ASTM D3828. Standard Test Method for pH of Activated Carbon. Annual Book of Standards. ASTM International, West Conshohocken, PA.

ASTM D4006. Standard Test Method for Water in Crude Oil by Distillation. Annual Book of Standards. ASTM International, West Conshohocken, PA.

ASTM D4007. Standard Test Method for Water and Sediment in Crude Oil by the Centrifuge Method (Laboratory Procedure). Annual Book of Standards. ASTM International, West Conshohocken, PA.

ASTM D4045. Standard Test Method for Sulfur in Petroleum Products by Hydrogenolysis and Rateometric Colorimetry. Annual Book of Standards. ASTM International, West Conshohocken, PA.

ASTM D4052. Standard Test Method for Density, Relative Density, and API Gravity of Liquids by Digital Density Meter. Annual Book of Standards. ASTM International, West Conshohocken, PA.

ASTM D4055. Standard Test Method for Pentane Insolubles by Membrane Filtration. Annual Book of Standards. ASTM International, West Conshohocken, PA.

ASTM D4057. Standard Practice for Manual Sampling of Petroleum and Petroleum Products. Annual Book of Standards. ASTM International, West Conshohocken, PA.

ASTM D4377. Standard Test Method for Water in Crude Oils by Potentiometric Karl Fischer Titration. Annual Book of Standards. ASTM International, West Conshohocken, PA.

ASTM D4492. Standard Test Method for Analysis of Benzene by Gas Chromatography. Annual Book of Standards. ASTM International, West Conshohocken, PA.

ASTM D4735. Standard Test Method for Determination of Trace Thiophene in Refined Benzene by Gas Chromatography. Annual Book of Standards. ASTM International, West Conshohocken, PA.

ASTM D4807. Standard Test Method for Sediment in Crude Oil by Membrane Filtration. Annual Book of Standards. ASTM International, West Conshohocken, PA.

ASTM D4808. Standard Test Methods for Hydrogen Content of Light Distillates, Middle Distillates, Gas Oils, and Residua by Low-Resolution Nuclear Magnetic Resonance Spectroscopy. Annual Book of Standards. ASTM International, West Conshohocken, PA.

ASTM D4928. Standard Test Method for Water in Crude Oils by Coulometric Karl Fischer Titration. Annual Book of Standards. ASTM International, West Conshohocken, PA.

ASTM D4952. Standard Test Method for Qualitative Analysis for Active Sulfur Species in Fuels and Solvents (Doctor Test). Annual Book of Standards. ASTM International, West Conshohocken, PA.

ASTM D4953. Standard Test Method for Vapor Pressure of Gasoline and Gasoline-Oxygenate Blends (Dry Method). Annual Book of Standards. ASTM International, West Conshohocken, PA.

ASTM D5002. Standard Test Method for Density and Relative Density of Crude Oils by Digital Density Analyzer. Annual Book of Standards. ASTM International, West Conshohocken, PA.

ASTM D5060. Standard Test Method for Determining Impurities in High-Purity Ethylbenzene by Gas Chromatography. Annual Book of Standards. ASTM International, West Conshohocken, PA.

ASTM D5134. Standard Test Method for Detailed Analysis of Petroleum Naphthas through n-Nonane by Capillary Gas Chromatography. Annual Book of Standards. ASTM International, West Conshohocken, PA.

ASTM D5135. Standard Test Method for Analysis of Styrene by Capillary Gas Chromatography. Annual Book of Standards. ASTM International, West Conshohocken, PA.

ASTM D5191. Standard Test Method for Vapor Pressure of Petroleum Products (Mini Method). Annual Book of Standards. ASTM International, West Conshohocken, PA.

ASTM D5194. Standard Test Method for Trace Chloride in Liquid Aromatic Hydrocarbons. Annual Book of Standards. ASTM International, West Conshohocken, PA.

ASTM D5291. Standard Test Methods for Instrumental Determination of Carbon, Hydrogen, and Nitrogen in Petroleum Products and Lubricants. Annual Book of Standards. ASTM International, West Conshohocken, PA.

ASTM D5292. Standard Test Method for Aromatic Carbon Contents of Hydrocarbon Oils by High Resolution Nuclear Magnetic Resonance Spectroscopy. Annual Book of Standards. ASTM International, West Conshohocken, PA.

ASTM D5386. Standard Test Method for Color of Liquids Using Tristimulus Colorimetry. Annual Book of Standards. ASTM International, West Conshohocken, PA.

ASTM D5443. Standard Test Method for Paraffin, Naphthene, and Aromatic Hydrocarbon Type Analysis in Petroleum Distillates through 200°C by Multi-Dimensional Gas Chromatography. Annual Book of Standards. ASTM International, West Conshohocken, PA.

ASTM D5482. Standard Test Method for Vapor Pressure of Petroleum Products (Mini Method—Atmospheric). Annual Book of Standards. ASTM International, West Conshohocken, PA.

ASTM D5580. Standard Test Method for Determination of Benzene, Toluene, Ethylbenzene, p/m-Xylene, o-Xylene, C9 and Heavier Aromatics, and Total Aromatics in Finished Gasoline by Gas Chromatography. Annual Book of Standards. ASTM International, West Conshohocken, PA.

ASTM D5713. Standard Test Method for Analysis of High Purity Benzene for Cyclohexane Feedstock by Capillary Gas Chromatography. Annual Book of Standards. ASTM International, West Conshohocken, PA.

ASTM D5769. Standard Test Method for Determination of Benzene, Toluene, and Total Aromatics in Finished Gasolines by Gas Chromatography/Mass Spectrometry. Annual Book of Standards. ASTM International, West Conshohocken, PA.

ASTM D5776. Standard Test Method for Bromine Index of Aromatic Hydrocarbons by Electrometric Titration. Annual Book of Standards. ASTM International, West Conshohocken, PA.

ASTM D5808. Standard Test Method for Determining Chloride in Aromatic Hydrocarbons and Related Chemicals by Microcoulometry. Annual Book of Standards. ASTM International, West Conshohocken, PA.

ASTM D5917. Standard Test Method for Trace Impurities in Monocyclic Aromatic Hydrocarbons by Gas Chromatography and External Calibration. Annual Book of Standards. ASTM International, West Conshohocken, PA.

ASTM D6069. Standard Test Method for Trace Nitrogen in Aromatic Hydrocarbons by Oxidative Combustion and Reduced Pressure Chemiluminescence Detection. Annual Book of Standards. ASTM International, West Conshohocken, PA.

ASTM D6144. Standard Test Method for Analysis of AMS (α-Methylstyrene) by Capillary Gas Chromatography. Annual Book of Standards. ASTM International, West Conshohocken, PA.

ASTM D6296. Standard Test Method for Total Olefins in Spark-ignition Engine Fuels by Multidimensional Gas Chromatography. Annual Book of Standards. ASTM International, West Conshohocken, PA.

ASTM D6304. Standard Test Method for Determination of Water in Petroleum Products, Lubricating Oils, and Additives by Coulometric Karl Fischer Titration. Annual Book of Standards. ASTM International, West Conshohocken, PA.

ASTM D6377. Standard Test Method for Determination of Vapor Pressure of Crude Oil: VPCRx (Expansion Method). Annual Book of Standards. ASTM International, West Conshohocken, PA.

ASTM D6378. Standard Test Method for Determination of Vapor Pressure (VPX) of Petroleum Products, Hydrocarbons, and Hydrocarbon-Oxygenate Mixtures (Triple Expansion Method). Annual Book of Standards. ASTM International, West Conshohocken, PA.

ASTM D6379. Standard Test Method for Determination of Aromatic Hydrocarbon Types in Aviation Fuels and Petroleum Distillates—High Performance Liquid Chromatography Method with Refractive Index Detection. Annual Book of Standards. ASTM International, West Conshohocken, PA.

ASTM D6450. Standard Test Method for Flash Point by Continuously Closed Cup (CCCFP) Tester. Annual Book of Standards. ASTM International, West Conshohocken, PA.

ASTM D7576. Standard Test Method for Determination of Benzene and Total Aromatics in Denatured Fuel Ethanol by Gas Chromatography. Annual Book of Standards. ASTM International, West Conshohocken, PA.

ASTM E203. Standard Test Method for Water Using Volumetric Karl Fischer Titration. Annual Book of Standards. ASTM International, West Conshohocken, PA.

ASTM E659. Standard Test Method for Autoignition Temperature of Liquid Chemicals. Annual Book of Standards. ASTM International, West Conshohocken, PA.

Chadeesingh, R. 2011. The Fischer-Tropsch Process. In The Biofuels Handbook. J.G. Speight (Editor). The Royal Society of Chemistry, London. Part 3, p. 476–517.

Davis, B.H., and Occelli, M.L. 2010. Advances in Fischer-Tropsch Synthesis, Catalysts, and Catalysis. CRC Press, Taylor & Francis Group, Boca Raton, FL.

Gary, J.G., Handwerk, G.E., and Kaiser, M.J. 2007. Petroleum Refining: Technology and Economics, 5th Edition. CRC Press, Taylor & Francis Group, Boca Raton, FL.

Hori, Y. 2000. Crude Oil Processing. In Modern Petroleum Technology. Volume 2: Downstream. A.G. Lucas (Editor). John Wiley & Sons Inc., New York.

Hsu, C.S., and Robinson, P.R. (Editors) 2006. Practical Advances in Petroleum Processing, Volumes 1 and 2. Springer Science, New York.

IP 1. Determination of Acidity, Neutralization Value – Color Indicator Titration Method. IP Standard Methods 2013. The Energy Institute, London.

IP 2 (ASTM D611). Petroleum Products and Hydrocarbon Solvents – Determination of Aniline and Mixed Aniline Point. IP Standard Methods 2013. The Energy Institute, London.

IP 17. Determination of Color – Lovibond Tintometer Method. IP Standard Methods 2013. The Energy Institute, London.

IP 30. Detection of Mercaptans, Hydrogen Sulfide, Elemental Sulfur and Peroxides – Doctor Test Method. IP Standard Methods 2013. The Energy Institute, London.

IP 34 (ASTM D93). Determination of Flash Point – Pensky-Martens Closed Cup Method. IP Standard Methods 2013. The Energy Institute, London.

IP 36. Determination of Open Flash and Fire Point – Cleveland Method. IP Standard Methods 2013. The Energy Institute, London.

IP 40 (ASTM D525). Petroleum Products – Determination of Oxidation Stability of Gasoline – Induction Period Method. IP Standard Methods 2013. The Energy Institute, London.

IP 69. Determination of Vapor Pressure – Reid Method. IP Standard Methods 2013. The Energy Institute, London.

IP 71 (ASTM D445). Petroleum Products – Transparent and Opaque Liquids – Determination of Kinematic Viscosity and Calculation of Dynamic Viscosity. IP Standard Methods 2013. The Energy Institute, London.

IP 107 (ASTM D1266). Determination of Sulfur – Lamp Combustion Method. IP Standard Methods 2013. The Energy Institute, London.

IP 123. Petroleum Products – Determination of Distillation Characteristics at Atmospheric Pressure. IP Standard Methods 2013. The Energy Institute, London.

IP 130 (ASTM D1159). Petroleum Products – Determination of Bromine Number of Distillates and Aliphatic Olefins – Electrometric Method. IP Standard Methods 2013. The Energy Institute, London.

IP 131. Petroleum Products – Gum Content of Light and Middle Distillates – Jet Evaporation Method. IP Standard Methods 2013. The Energy Institute, London.

IP 154. Petroleum Products – Corrosiveness to Copper – Copper Strip Test. IP Standard Methods 2013. The Energy Institute, London.

IP 156. Petroleum Products and Related Materials – Determination of Hydrocarbon Types – Fluorescent Indicator Adsorption Method. IP Standard Methods 2013. The Energy Institute, London.

IP 160 (ASTM D1298). Crude Petroleum and Liquid Petroleum Products – Laboratory Determination of Density – Hydrometer Method. IP Standard Methods 2013. The Energy Institute, London.

IP 170. Petroleum Products and Other Liquids – Determination of Flash Point – Abel Closed Cup Method. IP Standard Methods 2013. The Energy Institute, London.

IP 195. Determination of Distillation Characteristics of Volatile Organic Liquids. IP Standard Methods 2013. The Energy Institute, London.

IP 200 (ASTM D1250). Guidelines for the Use of the Petroleum Measurement Tables. IP Standard Methods 2013. The Energy Institute, London.

IP 235. Determination of Density of Light Hydrocarbons – Pressure Hydrometer Method. IP Standard Methods 2013. The Energy Institute, London.

IP 264. Determination of Composition of LPG and Propylene Concentrates – Gas chromatography Method. IP Standard Methods 2013. The Energy Institute, London.

IP 285. Determination of Nickel and Vanadium – Spectrophotometric Method. IP Standard Methods 2013. The Energy Institute, London.

IP 365. Crude Petroleum and Petroleum Products – Determination of Density – Oscillating U-Tube Method. IP Standard Methods 2013. The Energy Institute, London.

IP 373. Sulfur Content – Microcoulometry (Oxidative) Method. IP Standard Methods 2013. The Energy Institute, London.

IP 436 (ASTM D6379). Aromatic Hydrocarbon Types in Aviation Fuels and Petroleum Distillates – High Performance Liquid Chromatography Method with Refractive Index Detection. IP Standard Methods 2013. The Energy Institute, London.

IP 506. Loss on Heating Bitumen. IP Standard Methods 2013. The Energy Institute, London.

IP 523. Determination of Flash Point – Rapid Equilibrium Closed Cup Method. IP Standard Methods 2013. The Energy Institute, London.

McCann, J.M. 1998. Analysis of Gasoline and Other Light Distillate Fuels. In Manual on Hydrocarbon Analysis, 6th Edition. A.W. Drews (Editor). American Society for Testing and Materials, West Conshohocken, PA.

Mitchell, D.L., and Speight, J.G. 1973. The Solubility of Asphaltenes in Hydrocarbon Solvents. Fuel, 52: 149.

Occelli, M.L. 2010. Advances in Fluid Catalytic Cracking: Testing, Characterization, and Environmental Regulations. CRC Press, Taylor & Francis Group, Boca Raton, FL.

Pandey, S.C., Ralli, D.K., Saxena, A.K., and Alamkhan, W.K. 2004. Physicochemical Characterization and Application of Naphtha. Journal of Scientific & Industrial Research, 63: 276–282.

Speight, J.G. 2001. Handbook of Petroleum Analysis. John Wiley & Sons Inc., New York.

Speight, J.G. (Editor). 2011. The Biofuels Handbook. Royal Society of Chemistry, London, London.

Speight, J.G. 2013. The Chemistry and Technology of Coal, 3rd Edition. CRC Press, Taylor & Francis Group, Boca Raton, FL.

Speight, J.G. 2014a. The Chemistry and Technology of Petroleum, 5th Edition. CRC Press, Taylor & Francis Group, Boca Raton, FL.

Speight, J.G. 2014b. Gasification of Unconventional Feedstocks. Gulf Professional Publishing, Elsevier, Oxford, London.

Speight, J.G., and Exall, D.I. 2014. Refining Used Lubricating Oils. CRC Press, Taylor & Francis Group, Boca Raton, FL.

Speight, J.G., and Ozum, B. 2002. Petroleum Refining Processes. Marcel Dekker Inc., New York.

6

GASOLINE

6.1 INTRODUCTION

Gasoline (also referred to as *motor gasoline, petrol* in Britain, *benzine* in Europe) is mixture of volatile, flammable liquid hydrocarbons derived from petroleum that is used as fuel for internal-combustion engines such as occur in motor vehicles, excluding aircraft (Gary et al., 2007; Hsu and Robinson, 2006; Speight, 2014; Speight and Ozum, 2002).

The boiling range of motor gasoline falls between −1°C (30°F) and 215°C (420°F) and has the potential to contain several hundred isomers of the various hydrocarbons (Tables 6.1 and 6.2)—a potential that may be theoretical and never realized in practice (Speight, 2001, 2014). The hydrocarbon constituents in this boiling range are those that have 4–12 carbon atoms in their molecular structure and fall into three general types: (i) paraffins (including the cycloparaffins and branched materials), (ii) olefins, and (iii) aromatics. Gasoline boils at about the same range as naphtha (a precursor to gasoline) but below kerosene (Fig. 6.1).

The various test methods dedicated to the determination of the amounts of carbon, hydrogen, and nitrogen (ASTM D5291) as well as the determination of oxygen, sulfur, metals, and chlorine (ASTM D808) are not included. Although necessary, the various tests available for composition (Chapter 2) are left to the discretion of the analyst. In addition, test methods recommended for naphtha (Chapter 5) may also be applied, in many circumstances to gasoline.

6.2 PRODUCTION AND PROPERTIES

Gasoline was at first produced by distillation, simply separating the volatile, more valuable fractions of crude petroleum, and was composed of the naturally occurring constituents of petroleum. Later processes, designed to raise the yield of gasoline from crude oil, split higher-molecular-weight constituents into lower-molecular-weight products by processes known as *cracking*.

The final gasoline product as a transport fuel is a carefully blended mixture having a predetermined octane value. Thus, gasoline is a complex mixture of hydrocarbons that boils below 200°C (390°F). The hydrocarbon constituents in this boiling range are those that have 4–12 carbon atoms in their molecular structure. Like typical gasoline, several processes (Table 6.3) produce the blending stocks for gasoline.

By way of definition of some of these processes, *polymerization* is the conversion of gaseous olefins, such as propylene and butylene, into larger molecules in the gasoline range.

$$CH_3CH=CH_2+CH_3CH_2CH$$
$$=CH_2(+H_2)\rightarrow CH_3CH_2CH_2CH_2CH(CH_3)CH_3$$

Alkylation is a process combining an olefin and a paraffin (such as *iso*-butane).

$$CH_3CH=CH_2+(CH_3)_3CH\rightarrow CH_3CHCH_2CH(CH_3)_2CH_3$$

Handbook of Petroleum Product Analysis, Second Edition. James G. Speight.
© 2015 John Wiley & Sons, Inc. Published 2015 by John Wiley & Sons, Inc.

TABLE 6.1 General summary of product types and distillation range

Product	Lower carbon limit	Upper carbon limit	Lower boiling point (°C)	Upper boiling point (°C)	Lower boiling point (°F)	Upper boiling point (°F)
Refinery gas	C_1	C_4	−161	−1	−259	31
Liquefied petroleum gas	C_3	C_4	−42	−1	−44	31
Naphtha	C_5	C_{17}	36	302	97	575
Gasoline	C_4	C_{12}	−1	216	31	421
Kerosene/diesel fuel	C_8	C_{18}	126	258	302	575
Aviation turbine fuel	C_8	C_{16}	126	287	302	548
Fuel oil	C_{12}	>C_{20}	216	421	>343	>649
Lubricating oil	>C_{20}		>343		>649	
Wax	C_{17}	>C_{20}	302	>343	575	>649
Asphalt	>C_{20}		>343		>649	
Coke	>C_{50}^{a}		>1000^a		>1832^1	

^{a} Carbon number and boiling point difficult to assess; inserted for illustrative purposes only.

TABLE 6.2 Increase in the number of isomers with carbon number

Carbon atoms	Number of isomers
1	1
2	1
3	1
4	2
5	3
6	5
7	9
8	18
9	35
10	75
15	4,347
20	366,319
25	36,797,588
30	4,111,846,763
40	62,491,178,805,831

Isomerization is the conversion of straight-chain hydrocarbons to branched-chain hydrocarbons.

$$CH_3CH_2CH_2CH_2CH_2CH_2CH_2CH_3$$
$$\rightarrow CH_3CH_2CH_2CH_2CH_2CH(CH_3)CH_3$$

Reforming is the use of either heat or a catalyst to rearrange the molecular structure. Selection of the components and their proportions in a blend is the most complex problem in a refinery.

$$C_6H_{12} \rightarrow C_6H_6$$

Thus, gasoline is a mixture of hydrocarbons that boils below 180°C (355°F) or, at most, below 200°C (390°F). The hydrocarbon constituents in this boiling range are those that have 4–12 carbon atoms in their molecular structure.

TABLE 6.3 Component streams and processes for gasoline production

Stream	Producing process	Boiling range (°C)	Boiling range (°F)
Paraffinic			
Butane	Distillation Conversion	0	32
Isopentane	Distillation Conversion Isomerization	27	81
Alkylate	Alkylation	40–150	105–300
Isomerate	Isomerization	40–70	105–160
Straight-run naphtha	Distillation	30–100	85–212
Hydrocrackate	Hydrocracking	40–200	105–390
Olefinic			
Catalytic naphtha	Catalytic cracking	40–200	105–390
Steam-cracked naphtha	Steam cracking	40–200	105–390
Polymer	Polymerization	60–200	140–390
Aromatic			
Catalytic reformate	Catalytic reforming	40–200	105–390

The hydrocarbons of which gasoline is composed fall into three general types: paraffins (including the cycloparaffins and branched materials), olefins, and aromatics.

The hydrocarbons produced by modern refining techniques (distillation, cracking, reforming, alkylation, isomerization, and polymerization) provide blending components for gasoline production (Gary et al., 2007; Hsu and Robinson, 2006; Speight, 2014; Speight and Ozum, 2002).

Gasoline consists of a very large number of different hydrocarbons, and the individual hydrocarbons in gasoline

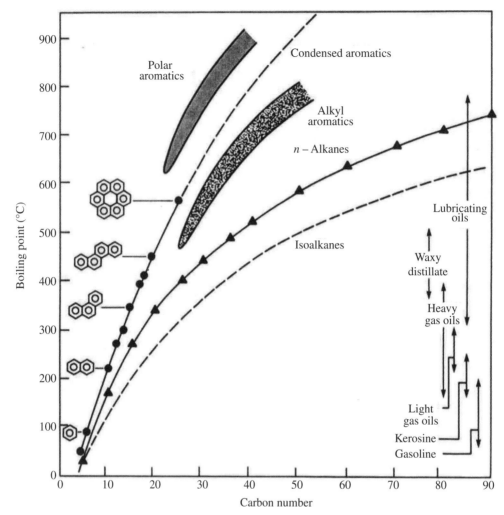

FIGURE 6.1   Boiling point and carbon number for various hydrocarbons and petroleum products.

cannot be conveniently used to describe gasoline. The composition of gasoline is best expressed in terms of hydrocarbon types (saturates, olefins, and aromatics) that enable inferences to be made about the behavior in service.

6.3   VOLATILITY REQUIREMENTS AND OTHER PROPERTIES

The volatility of a liquid is the tendency of the liquid to change from the liquid to the vapor or gaseous state. Thus, volatility is a primary and necessary characteristic of most liquid gasoline. The distillation profile is also a measure of the relative amounts of the gasoline constituents in petroleum.

The volatility of gasoline affects the performance of the engine in a number of ways, the chief of which are ease of starting, rate of warm-up, vapor lock, carburetor icing, and

crankcase dilution (the dilution of the engine lubricating oil with the higher-boiling constituents of the gasoline). The gasoline must be sufficiently volatile to give easy starting, rapid warm-up, and adequate vaporization for proper distribution between the cylinders. Conversely, it must not be so volatile that vapor losses from the gasoline tank are excessive or that vapor is formed in the gasoline line causing vapor lock that may impede the flow of gasoline to the carburetor.

Thus, one of the most important physical parameters of gasoline is the boiling range distribution (ASTM D86; ASTM D1078; ASTM D2887; ASTM D2892; IP 123). The significance of the distillation test is the indication of volatility that dictates the evaporation rate, which is an important property for gasoline used in coatings and similar application where the premise is that the gasoline evaporates over time, leaving the coating applied to the surface. In the basic test method (ASTM D86; IP 123) a 100-ml sample is distilled (manually or automatically) under prescribed conditions. Temperatures and volumes of condensate are recorded at regular intervals from which the boiling profile is derived.

The determination of the boiling range distribution of gasoline by gas chromatography (GC; ASTM D3710) not only helps to identify the constituents but also facilitates online controls at the refinery. This test method is designed to measure the entire boiling range of gasoline with either high or low Reid vapor pressure (ASTM D323; IP 69). In the method, the sample is injected into a gas chromatographic column that separates hydrocarbons in boiling point order. The column temperature is raised at a reproducible rate, and the area under the chromatogram is recorded throughout the run. Calibration is performed using a known mixture of hydrocarbons covering the expected boiling of the sample.

Another method is described as a method for determining the carbon number distribution (ASTM D2887), and the data derived by this test method are essentially equivalent to that obtained by TBP distillation (ASTM D2892). The sample is introduced into a gas chromatographic column that separates hydrocarbons in boiling point order. The column temperature is raised at a reproducible rate, and the area under the chromatogram is recorded throughout the run. Boiling temperatures are assigned to the time axis from a calibration curve, obtained under the same conditions by running a known mixture of hydrocarbons covering the boiling range expected in the sample. From these data, the boiling range distribution may be obtained. However, this test method is limited to samples having a boiling range greater than 55°C (l00°F) and having a vapor pressure (ASTM D323; ASTM D4953; ASTM D5191; ASTM D5482; ASTM D6377; ASTM D6378; IP 69; IP 394) sufficiently low to permit sampling at ambient temperature.

The volatility of petroleum and petroleum products is an important aspect of safety and quality. It would be unsafe to attempt to store highly volatile materials in the open sunlight or in an enclosed space where temperature can rise to be in excess of 37.8°C (100°F). However, without any indications of when the material might vaporize and spontaneously ignite, there is no way of even considering the correct storage and handling conditions.

The *boiling points* of petroleum fractions are rarely, if ever, distinct temperatures, and it is, in fact, more correct to refer to the *boiling range* of a particular fraction. To determine these ranges, the petroleum is tested in various methods of distillation, either at atmospheric pressure or at reduced pressure. A general estimate is that the limiting molecular weight range for distillation at atmospheric pressure without thermal degradation is 200–250, whereas the limiting molecular weight range for conventional vacuum distillation is 500–600.

In each homologous series of hydrocarbons, the *boiling point* increases with molecular weight. Structure also has a marked influence, and it is a general rule that branched paraffin isomers have lower boiling points than the corresponding *n*-alkane. However, the most dramatic illustration of the variation in boiling point with carbon number is an actual plot for different hydrocarbons (Fig. 6.1). In any given series, steric effects notwithstanding, there is an increase in boiling point with an increase in carbon number of the alkyl side chain (Fig. 6.2). This particularly applies to alkyl aromatic compounds where alkyl-substituted aromatic compounds can have higher boiling points than polycondensed aromatic systems. And this fact is very meaningful when attempts are made to develop hypothetical structures for the higher-molecular-weight constituents of petroleum (Gary et al., 2007; Hsu and Robinson, 2006; Speight, 2001, 2014; Speight and Ozum, 2002).

As an early part of characterization studies, a correlation was observed between the quality of petroleum products and their hydrogen content since gasoline, kerosene, diesel gasoline, and lubricating oil are made up of hydrocarbon constituents containing high proportions of hydrogen. Thus, it is not surprising that tests to determine the volatility of petroleum and petroleum products were among the first to be defined. Indeed, volatility is one of the major tests for petroleum products, and it is inevitable that all products will, at some stage of their history, be tested for volatility characteristics.

The boiling range of motor gasoline falls in the range 30°C–210°C (86°F–410°F), and blending of the available refinery components can maintain balance between different volatility requirements. The volatility of gasoline is normally assessed by the distillation test (ASTM D86; IP 123). Determination of the boiling range distribution of gasoline and gasoline fractions by GC (ASTM D3710) can be used for determining the boiling point properties of oxygenate-free gasoline distillates, and this test has the advantage that it uses a smaller sample size and can be more easily automated but the data from the two tests may not be directly equivalent.

The apparatus employed for the test provides little in the way of fractionation, and although the thermometer is accurately standardized at total immersion, it is used at partial immersion, and no temperature corrections are made for emergent stem. However, provision is made in the test for correcting the thermometer readings for variations in atmospheric pressure whenever the barometric reading is sufficiently far from standard atmospheric pressure to make corrections desirable. Therefore, while the temperatures are highly reproducible, they are not true vapor temperatures.

The *vapor pressure* of petroleum or a petroleum product is the force exerted on the walls of a closed container by the vaporized portion of a liquid. Conversely, it is the force that must be applied to the liquid to prevent it vaporizing further. The vapor pressure increases with increase in

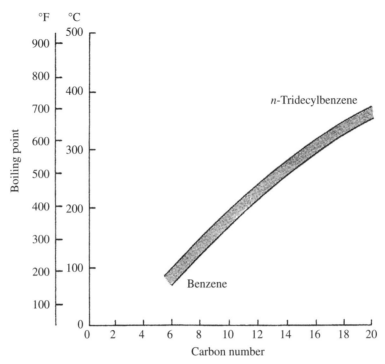

FIGURE 6.2 Effect of increasing length of the alkyl chain on the boiling point.

temperature and is variously expressed in terms of millimeters of mercury, pounds per square inch, or other equivalent units of pressure depending on common usage. Gasoline vapor pressure depends critically on its butane content, and in the refinery, the final adjustment of vapor pressure of a gasoline to meet the specification is often made by butane injection.

The *Reid vapor pressure* (ASTM D323; IP 69) is a measure of the vapor pressure of petroleum or a petroleum product at 37.8°C (100°F) expressed as millimeters of mercury. The apparatus used to determine the Reid vapor pressure consists of a metal cylinder, or *bomb*, fitted with an accurate dial pressure gauge, or a mercury manometer. The bomb consists of two parts: an upper expansion chamber and a lower liquid chamber. The sample is cooled and poured into the lower chamber until full. The temperature of the air in the upper chamber is taken, and the two chambers are connected together in a gas-tight manner. The bomb is immersed upright in a water bath at 37.8°C (100°F) and shaken repeatedly until a constant pressure reading is obtained. This is corrected, from tables, for initial air temperature and pressure.

Another method (ASTM D5191) is now most commonly referenced in gasoline regulations, and the method requires a lesser amount of the sample and is much easier and faster to run. Other methods for determining the vapor pressure of gasoline are also available (ASTM D4953; ASTM D5482).

Thus, petroleum can be subdivided by distillation into a variety of fractions of different *boiling ranges* (*cut points*) using a variety of standard methods specifically designed for this task.

Distillation involves the general procedure of vaporizing the petroleum liquid in a suitable flask either at *atmospheric pressure* (ASTM D86; ASTM D2892; ASTM D5236; IP 123; IP 191) or at *reduced pressure* (ASTM D1160). Most of the methods specify an upper atmospheric equivalent temperature (AET) limit of 360°C (680°F) to mitigate the onset of thermal decomposition. For higher-boiling fractions, the spinning band distillation method and distillation using a flash still are available.

In the simplest case, the distillation method involves using a standard round-bottom distillation flask of 250-ml capacity attached to a water-cooled condenser. The thermometer bulb is placed at the opening to the side arm of the flask. Sample of 100 ml is placed in the flask and heated by a small gas flame so as to produce 10 ml of distillate every 4 or 5 min. The temperature of initial distillation is recorded; the temperature at which each further 10 ml distils and the final boiling point are also recorded.

Samples containing light components must be debutanized in a preliminary step. In the method, a weighed sample of 1–10 l is distilled in a fractionating column having an efficiency at total reflux of 14–18 theoretical plates. A reflux ratio of 5:1 is maintained throughout except at the minimum pressure of 2 mm Hg where a ratio of 2:1 can be used. The mass and density of each fraction are used to prepare distillation curves either by mass or volume. This method is often referred to as the *15-5 method*.

One method (ASTM D1160) is available to determine the boiling ranges of petroleum products to a maximum liquid temperature of 400°C (750°F) at pressures as low as 1 mm Hg and is adequate for gasoline. In the method, a 200-ml sample is weighed to the nearest 0.1 g in a distillation flask. The distillation assembly is evacuated to the desired pressure, and heat is applied to the flask as rapidly as possible using a 750-W heater. When refluxing liquid appears, the rate of heating is adjusted so that the distillate is recovered at 4–8 ml/min until the distillation is complete.

Generally, the distillation tests are planned so that the data are reported in terms of one or more of the following items:

1. *Initial boiling point* is the thermometer reading in the neck of the distillation flask when the first drop of distillate leaves the tip of the condenser tube. This reading is materially affected by a number of test conditions, namely, room temperature, rate of heating, and condenser temperature.

2. *Distillation temperatures* are usually observed when the level of the distillate reaches each 10% mark on the graduated receiver, with the temperatures for the 5% and 95% marks often included. Conversely, the volume of the distillate in the receiver, that is, the percentage recovered, is often observed at specified thermometer readings.

3. *Endpoint* or *maximum temperature* is the highest thermometer reading observed during distillation. In most cases, it is reached when all of the sample has been vaporized. If a liquid residue remains in the flask after the maximum permissible adjustments are made in heating rate, this is recorded as indicative of the presence of very-high-boiling compounds.

4. *Dry point* is the thermometer reading at the instant the flask becomes dry and is for special purposes, such as for solvents and for relatively pure hydrocarbons. For these purposes, dry point is considered more indicative of the final boiling point than the end point or maximum temperature.

5. *Recovery* is the total volume of distillate recovered in the graduated receiver, and *residue* is the liquid material, mostly condensed vapors, left in the flask after it has been allowed to cool at the end of distillation. The residue is measured by transferring it to an appropriate small graduated cylinder. Low or abnormally high residues indicate the absence or presence, respectively, of high-boiling components.

6. *Total recovery* is the sum of the liquid recovery and residue; *distillation loss* is determined by subtracting the total recovery from 100%. It is, of course, the measure of the portion of the vaporized sample that does not condense under the conditions of the test. Like the initial boiling point, distillation loss is affected materially by a number of test conditions, namely, condenser temperature, sampling and receiving temperatures, barometric pressure, heating rate in the early part of the distillation, and others. Provisions are made for correcting high distillation losses for the effect of low barometric pressure because of the practice of including distillation loss as one of the items in some specifications for motor gasoline.

7. *Percentage evaporated* is the percentage recovered at a specific thermometer reading or other distillation temperatures, or the converse. The amounts that have been evaporated are usually obtained by plotting observed thermometer readings against the corresponding observed recoveries plus, in each case, the distillation loss. The initial boiling point is plotted with the distillation loss as the percentage evaporated. Distillation data are reproducible, particularly for the more volatile products.

The capillary method can be used as one of two standardized methods (ASTM D2887; ASTM D3710) that are available for the boiling point determination of petroleum fractions and gasoline, respectively. The former method (ASTM D2887) utilizes nonpolar, packed gas chromatographic columns in conjunction with flame ionization detection. The upper limit of the boiling range covered by this method is to approximately 540°C (1000°F) atmospheric equivalent boiling point. Recent efforts in which high-temperature GC was used have focused on extending the scope of the method for higher-boiling petroleum materials to 800°C (1470°F) atmospheric equivalent boiling point (Speight, 2014).

The measurement of pressure at the *top* of the distillation column is critical to valid distillation results because the observed vapor temperature must be corrected to the AET at standard pressure conditions (760 mm Hg). There is a general belief that the minimum pressure should be 2 mm Hg or greater for reasonably accurate measurements and correction to the AET. At pressures below 2 mm Hg, the pressure measurement is too inaccurate, and a discontinuity can arise in the AET distillation curve from atmospheric to vacuum distillation.

The precision and accuracy of the single internal standard method depend on the accuracy of estimating the area of the internal standard. In addition to the problems in determining the percentage of sample that elutes from a gas chromatographic column, errors can occur because the components of a sample can elute at a different time than the standards of the same boiling point. Another aspect of volatility, that is, the tendency for gasoline to vaporize, can be expressed in terms of vapor-to-liquid (V/L) ratio at temperatures approximating those found in critical parts of the gasoline system. An instrumental method that does not use a confining fluid can be used for both gasoline and gasoline-oxygenate blends (ASTM D5188). A linear equation method and a nomograph method can be used for estimating vapor–liquid equilibria of gasoline from vapor pressure and distillation test results (ASTM D4814). However, these estimation methods are not applicable to gasoline-oxygenate blends.

6.4   OCTANE RATING

Gasoline performance and hence quality of an automobile gasoline are determined by its resistance to knock, for example, *detonation* or *ping* during service. The antiknock quality of the fuel limits the power and economy that an engine using that fuel can produce: the higher the antiknock quality of the fuel, the more the power and efficiency of the engine.

In the early days of automobile engine development, there was a demand for more powerful engines, and knocking was not a problem. This demand for more powerful engines was met at first by adding more and larger pistons to the engines; some were even built with 16 cylinders. However, the practical limit to the size to which an engine could be built meant that other avenues to increased power had to be explored. The obvious method of getting more power from an engine of given size was to increase its compression ratio (a measure of the extent to which the gasoline–air mixture is compressed in the cylinder of an engine). The more the mixture is compressed before ignition, the more power the engine can deliver, but the increased performance was accompanied by an increased tendency for detonation or *knocking*. Eventually, the cause of engine knock was traced to the gasoline. Since some gasoline seemed to cause more knock than another, and since there was no suitable testing procedure, it was difficult to determine the relative antiknock characteristics of gasoline. It appeared, however, that cracked gasoline caused less knock than straight-run gasoline, and hence, the use of cracked gasoline as a fuel increased.

In 1922, tetraethyl lead (TEL) was discovered to be an excellent antiknock material when added in small quantities to gasoline, and gasoline containing TEL became widely available. However, the problem of how to increase the antiknock characteristics of cracked gasoline became acute in the 1930s. One feature of the problem concerned the need to measure the antiknock characteristics of gasoline accurately. This was solved in 1933 by the general use of a single-cylinder test engine, which allowed comparisons of the antiknock characteristics of gasoline to be made in terms of octane numbers. The octane numbers formed a scale ranging from 0 to 100: the higher the number, the greater the antiknock characteristics. In 1939, a second and less severe test procedure using the same test engine was developed, and

results obtained by this test were also expressed in octane numbers.

Octane numbers are obtained by two test procedures; those obtained by the first method are called *motor octane numbers* (indicative of high-speed performance) (ASTM D2700). Those obtained by the second method are called *research octane numbers* (indicative of normal road performance) (ASTM D2699). Octane numbers quoted are usually, unless stated otherwise, research octane numbers.

In the test methods used to determine the antiknock properties of gasoline, comparisons are made with blends of two pure hydrocarbons, *n*-heptane and *iso*-octane (2,2,4-trimethylpentane). *Iso*-octane has an octane number of 100 and is high in its resistance to knocking; *n*-heptane is quite low (with an octane number of 0) in its resistance to knocking.

Extensive studies of the octane numbers of individual hydrocarbons have brought to light some general rules. For example, normal paraffins have the least desirable knocking characteristics, and these become progressively worse as the molecular weight increases. *Iso*-paraffins have higher octane numbers than the corresponding normal isomers, and the octane number increases as the degree of branching of the chain is increased. Olefins have markedly higher octane numbers than the related paraffins; naphthenes are usually better than the corresponding normal paraffins but rarely have very high octane numbers; aromatics usually have quite high octane numbers.

Blends of *n*-heptane and *iso*-octane thus serve as a reference system for gasoline and provide a wide range of quality used as an antiknock scale. The exact blend, which matches identically the antiknock resistance of the fuel under test, is found, and the percentage of *iso*-octane in that blend is termed the *octane number* of the gasoline. For example, gasoline with a knocking ability that matches that of a blend of 90% *iso*-octane and 10% *n*-heptane has an octane number of 90. However, many pure hydrocarbons and even commercial gasoline have antiknock quality above an octane number of 100. In this range, it is common practice to extend the reference values by the use of varying amounts of TEL in pure *iso*-octane.

With an accurate and reliable means of measuring octane numbers, it was possible to determine the cracking conditions—temperature, cracking time, and pressure—that caused increases in the antiknock characteristics of cracked gasoline. In general, it was found that higher cracking temperatures and lower pressures produced higher octane gasoline, but unfortunately, more gas, cracked residua, and coke were formed at the expense of the volume of cracked gasoline.

To produce higher-octane gasoline, cracking coil temperatures were pushed up to 510°C (950°F), and pressures dropped from 1000 to 350 psi. This was the limit of thermal cracking units, for at temperatures over 510°C (950°F), coke formed so rapidly in the cracking coil that the unit became inoperative after only a short time on stream. Hence, it was at this stage that the nature of the gasoline-producing process was reexamined, leading to the development of other processes, such as reforming, polymerization, and alkylation (Gary et al., 2007; Hsu and Robinson, 2006; Speight, 2014; Speight and Ozum, 2002) for the production of gasoline components having suitably high octane numbers.

It is worthy of note here that the continued decline in petroleum reserves and the issue of environmental protection have emerged as of extreme importance in the search for alternatives to petroleum. In this light, oxygenated derivatives, either neat or as additives to fuels, appear to be the principal alternative fuel candidates beyond the petroleum refinery.

6.5 ADDITIVES

Additives are chemical compounds intended to improve some specific properties of gasoline or other petroleum products and can be monofunctional or multifunctional (Table 6.4) (ASTM D2669). Different additives, even when added for identical purposes, may be incompatible with each other, for example, react and form new compounds. Consequently, a blend of two or more gasoline, containing different additives, may form a system in which the additives react with each other and so deprive the blend of their beneficial effect.

Thus, certain substances added to gasoline, notably the lead alkyls, have a profound effect on antiknock properties and inhibit the precombustion oxidation chain that is known to promote knocking. For a considerable period, TEL was the preferred compound, but more recently, tetramethyl lead (TML) has been shown to have advantages with certain modern types of gasoline because of its lower boiling point [110°C (230°F) as against 200°C (392°F) for TEL] and

therefore its higher vapor pressure that enables it to be more evenly distributed among the engine cylinders with the more volatile components of the gasoline.

Some gasoline may still contain TML and TEL, while others contain compounds prepared by a chemical reaction between TML and TEL in the presence of a catalyst. These chemically reacted compounds contain various proportions of TML and TEL and their intermediates, trimethyl ethyl lead, dimethyl diethyl lead, and methyl triethyl lead, and thus provide antiknock compounds with a boiling range of 110°C–200°C (230°F–392°F). The lead compounds, if used alone, would cause an excessive accumulation of lead compounds in the combustion chambers of the engine and on sparking plugs and valves. Therefore, *scavengers* such as dibromoethane, alone or in admixture with dichloroethane, are added to the lead alkyl and combine with the lead during the combustion process to form volatile compounds that pass harmlessly from the engine.

The amount of lead alkyl compounds used in gasoline is normally expressed in terms of equivalent grams of metallic lead per gallon or per liter. The maximum concentration of lead permitted in gasoline varies from country to country according to governmental legislation or accepted commercial practice, and it is a subject that is currently under discussion in many countries due to the attention being paid to reduction of exhaust emissions from the spark ignition engine. Metals (including lead) in gasoline may be determined by atomic absorption spectrometry (ASTM D3237; IP 428), by the iodine chloride method (ASTM D3341; IP 270), by inductively coupled plasma atomic emission spectrometry (ASTM D5185), and by X-ray fluorescence (ASTM D5059). When it is desired to estimate TEL, a method is available, while for the separate determination of TML and TEL, recourse can be made to separate methods (ASTM D1949).

TABLE 6.4   Additives for gasoline

Class and function	Additive type
Oxidation inhibitors—minimize oxidation and gum formation	Aromatic amines and hindered phenols
Corrosion Inhibitors—inhibit ferrous corrosion in pipelines, storage tanks, and vehicle fuel systems	Carboxylic acids and carboxylates
Metal deactivators—inhibit oxidation and gum formation catalyzed by ions of copper and other metals	Chelating agent
Carburetor/injector detergents—prevent and remove deposits in carburetors and port fuel injectors	Amines, amides, and amine carboxylates
Deposit control additives—remove and prevent deposits throughout fuel injectors, carburetors, intake ports and valves, and intake manifold	Polybutene amines and polyether amines
Demulsifiers—minimize emulsion formation by improving water separation	Polyglycol derivatives
Anti-Icing Additives—minimize engine stalling and starting problems by preventing ice formation in the carburetor and fuel system	Surfactants, alcohols, and glycols
Antiknock Compounds—improve octane quality of gasoline	Lead alkyls and methylcyclopentadienylmanganese tricarbonyl
Dyes—Identification of gasoline	Oil-soluble solid and liquid dyes

Other additives used in gasoline include antioxidants and metal deactivators for inhibiting gum formation, surface-active agents and freezing point depressants for preventing carburetor icing, deposit modifiers for reducing spark plug fouling and surface ignition, and rust inhibitors (ASTM D665; IP 135) for preventing the rusting of steel tanks and pipe work by the traces of water carried in gasoline. For their estimation, specialized procedures involving chemical tests and physical techniques such as spectroscopy and chromatography have been used successfully.

6.6   TEST METHODS

The test protocols used for gasoline are similar to the protocols used for naphtha. The similarity of the two liquids requires the application of similar test methods. However, knocking properties are emphasized for gasoline, and there are several other differences that must be recognized. But, all in all, consultation of the test methods used for the analysis of naphtha (Chapter 5) can assist in developing protocols for gasoline.

The properties of gasoline are quite diverse, and the principal properties affecting the performance of gasoline are volatility and combustion characteristics. These properties are adjusted according to the topography and climate of the country in which the gasoline is to be used. For example, mountainous regions will require gasoline with volatility and knock characteristics somewhat different from those that are satisfactory in flat or undulating country only a little above sea level. Similarly, areas that exhibit extremes of climatic temperature, such as the northern provinces in Canada where temperatures of 30°C (86°F) in the summer are often accompanied by temperatures as low as −40°C (−40°F) in the winter, necessitate special consideration, particularly with regard to volatility.

Because of the high standards set for gasoline, as with naphtha, it is essential to employ the correct techniques when taking samples for testing (ASTM D4057). Mishandling, or the slightest trace of contaminants, can give rise to misleading results. Special care is necessary to ensure that containers are scrupulously clean and free from odor. Samples should be taken with the minimum of disturbance so as to avoid loss of volatile components; in the case of the lightest solvents, it may be necessary to chill the sample.

While awaiting examination, samples should be kept in a cool dark place so as to ensure that they do not discolor or develop odors.

Test methods have been developed to measure ethers and alcohols in gasoline-range hydrocarbons, because oxygenated components such as *methyl t-butyl ether* and ethanol are common blending components in gasoline (ASTM D4814; ASTM D4815; ASTM D5441; ASTM D5599; ASTM D5622; ASTM D5845; ASTM D5986).

Another type of gasoline sometimes referred to as *vaporizing oil* or *power kerosene* is primarily intended as a gasoline for agricultural tractors and is, in effect, a low-volatility (higher-boiling) gasoline. For reliable operation, such a gasoline must not be prone to deposit formation (sediment or gum), and the residue on evaporation (ASTM D381; IP 131) must, therefore, be low. The volatility of vaporizing oil is, as for regular gasoline, assessed by the distillation test (ASTM D86; IP 123), the requirement normally being controlled by the percentages boiling at 160°C and 200°C (320°F and 392°F).

Because of the lower volatility of vaporizing oil compared with that of gasoline, a relatively high proportion of aromatics may be necessary to maintain the octane number although unsaturated hydrocarbons may also be used in proportions compatible with stability requirements. However, the presence of unsaturated constituents must be carefully monitored because of the potential for incompatibility through the formation of sediment and gum. Other tests include flash point (closed-cup method: ASTM D56; ASTM D93; ASTM D3828; ASTM D6450; IP 34; IP 523; open-cup method: ASTM D92; ASTM D1310; IP 36), sulfur content (ASTM D1266; IP 107), corrosion (ASTM D130; ASTM D849; IP 154), octane number (ASTM D2699; ASTM D2700; ASTM D2885; IP 236; IP 237), and residue on evaporation (ASTM D381; ASTM D1353; IP 131).

Although not always present as additives, tests must be stringently followed for the presence of trace elements such as lead (ASTM D3237; ASTM D5059; ASTM D5185; IP 228; IP 270), manganese (ASTM D3831), and phosphorus (ASTM D3231) since the presence of these elements can have an adverse effect on gasoline performance or on the catalytic converter.

6.6.1   Combustion Characteristics

Combustion in the spark ignition engine depends chiefly on engine design and gasoline quality. Under ideal conditions, the flame initiated at the sparking plug spreads evenly across the combustion space until all the gasoline has been burned. The increase in temperature caused by the spreading of the flame results in an increase in pressure in the end gas zone, which is that part of the gasoline–air mixture that the flame has not yet reached. The increase in temperature and pressure in the end gas zone causes the gasoline to undergo preflame reactions. Among the main preflame products are the highly temperature-sensitive peroxides, and if these exceed a certain critical threshold concentration, the end gas will spontaneously ignite before the arrival of the flame front emanating from the sparking plug: this causes detonation or *knocking*. On the other hand, if the flame front reaches the end gas zone before the buildup of the critical threshold peroxide concentration, the combustion of the gasoline–air mixture will be without *knock*.

TABLE 6.5 Octane numbers of selected hydrocarbons

Hydrocarbons	Octane number	
	Research	Motor
Normal paraffins		
Pentane	61.7	61.9
Hexane	24.8	26.0
Heptane	0.0	0.0
Octane	−19.0	−15.0
Nonane	−17.0	−20.0
Isoparaffins		
2-Methylbutane (isopentane)	92.3	90.3
2-Methylhexane (isoheptane)	42.4	46.4
2-Methylheptane (isooctane)	21.7	23.8
2,4-Dimethylhexene	65.2	69.9
2,2,4-Trimethylpentane ("isooctane")	100.0	100.0
Olefins		
1-Pentene	90.9	77.1
1-Octene	28.7	34.7
3-Octene	72.5	68.1
4-Methyl-l-pentene	95.7	80.9
Aromatics		
Benzene		114.8
Toluene	120.1	103.5

The various types of hydrocarbons in gasoline behave differently in their preflame reactions and, thus, their tendency to knock. It is difficult to find any precise relationship between chemical structure and antiknock performance in an engine. Members of the same hydrocarbon series may show very different *antiknock* effects. For example, normal heptane and normal pentane, both paraffins, have antiknock ratings (octane numbers) of 0 and 61.9, respectively (Table 6.5). Very generally, aromatic hydrocarbons (e.g., benzene and toluene), highly branched iso-paraffins (e.g., iso-octane), and olefins (e.g., di-isobutylene) have high antiknock values. In an intermediate position are iso-paraffins with little branching and naphthenic hydrocarbons (e.g., cyclohexane), while of low antiknock value are the normal paraffins (e.g., normal heptane).

The knock rating of a gasoline is expressed as *octane number* and is the percentage by volume of iso-octane (octane number: 100, by definition) in admixture with normal heptane (octane number 0, by definition), which has the same knock characteristics as the gasoline being assessed.

Gasoline is normally rated using two sets of conditions of differing severity. One, known as the Research Method (ASTM D2699; IP 237) gives a rating applicable to mild operating conditions, that is, low inlet mixture temperature and relatively low engine loading such as would be experienced generally in passenger cars and light duty commercial vehicles. The other is the Motor Method (ASTM D2700; ASTM D2885; IP 236), which represents

more severe operating conditions, that is, relatively high inlet mixture temperature and high engine loading such as would be experienced during full throttle operation at high speed.

Research octane numbers are generally higher than those obtained by the Motor Method, and the difference between the two ratings is known as the *sensitivity* of the gasoline. The sensitivity of low-octane-number gasoline is usually small, but with high-octane gasoline, it varies greatly according to gasoline composition, and for most commercial blends, it is between 7 and 12 in the 90–100 Research octane number range. The actual performance of a gasoline on the road (the *road octane number*) usually falls between the Research and Motor values, and depends on the engine used and also on the gasoline composition and choice of anti-knock compounds.

To avoid the confusion arising from the use of two separate octane number scales, one below and one above 100, an arbitrary extension of the octane number scale was selected so that the value in terms of automotive engine performance of each unit between 100 and 103 was similar to those between 97 and 100. The relationship between the octane number scale above 100 and the performance number scale is

$$\text{Octane number} = 100 + (\text{performance number} - 100)/3$$

The relationship between TEL concentration in iso-octane and octane number is given in various test methods (ASTM D2699; IP 236; IP 237).

The *heat of combustion* (ASTM D240; IP 12) is a direct measure of gasoline energy content and is determined as the quantity of heat liberated by the combustion of a unit quantity of gasoline with oxygen in a standard bomb calorimeter. This gasoline property affects the economics of engine performance, and the specified minimum value is a compromise between the conflicting requirements of maximum gasoline availability and consumption characteristics.

An alternative criterion of energy content is the *aniline gravity product*, which is fairly accurately related to calorific value but more easily determined. It is the product of the gravity at 15.6°C (60°F) (expressed in degrees API) and the aniline point of the gasoline in °F (ASTM D611; IP 2). The aniline point is the lowest temperature at which the gasoline is miscible with an equal volume of aniline and is inversely proportional to the aromatic content and is related to the calorific value (ASTM D1405).

6.6.2 Composition

As with naphtha, the number of potential hydrocarbon isomers in the gasoline boiling range (Table 6.2) renders complete speciation of individual hydrocarbons impossible for the gasoline distillation range, and methods are used that identify the hydrocarbon types as chemical groups rather than as individual constituents.

In terms of hydrocarbon components, several procedures have been devised for the determination of hydrocarbon type, and the method based on fluorescent indicator adsorption (ASTM D1319; IP 156) is the most widely employed. Furthermore, since aromatic content is a key property of low-boiling distillates such as gasoline because the aromatic constituents influence a variety of properties including boiling range (ASTM D86; IP 123), viscosity (ASTM D88; ASTM D445; ASTM D2161; IP 71), and stability (ASTM D525; IP 40). Existing methods use physical measurements and need suitable standards. Tests such as aniline point (ASTM D611) and Kauri-butanol number (ASTM D1133) are of a somewhat empirical nature and can serve a useful function as control tests. However, gasoline composition is monitored mainly by GC (ASTM D2427; ASTM D6296).

A multidimensional gas chromatographic method (ASTM D5443) provides for the determination of paraffins, naphthenes, and aromatics by carbon number in low-olefin hydrocarbon streams having final boiling points lower than 200°C (392°F). In the method, the sample is injected into a gas chromatographic system that contains a series of columns and switching values. First a polar column retains polar aromatic compounds, bi-naphthenes, and high-boiling paraffins and naphthenes. The eluate from this column goes through a platinum column that hydrogenates olefins, and then to a molecular sieve column that performs a carbon number separation based on the molecular structure, that is, naphthenes and paraffins. The fraction remaining on the polar column is further divided into three separate fractions that are then separated on a nonpolar column by boiling point. A flame ionization detector detects eluting compounds.

Methods for the determination of aromatics in gasoline include a method (ASTM D5580) using a flame ionization detector and methods in which there is a combination of gas chromatography and Fourier transform infrared spectroscopy (GC–FTIR) (ASTM D5986) and gas chromatography and mass spectrometry (GC–MS) (ASTM D5769). Many of the methods applied to the determination of aromatics in naphtha (Chapter 5) are applicable to gasoline. The accurate measurements of benzene, and/or toluene, and total aromatics in gasoline are regulated test parameters in gasoline (ASTM D3606; ASTM D5580; ASTM D5769; ASTM D5986). The precision and accuracy of some of these tests are diminished in gasoline containing ethanol or methanol, since these components often do not completely separate from the benzene peak.

Benzene, toluene, ethylbenzene, the xylene isomers, as well as C_9 aromatics and higher-boiling aromatics are determined by GC, and the test (ASTM D5580) was developed to include gasoline containing commonly encountered alcohols and ethers. This test is the designated test for determining benzene and total aromatics in gasoline and includes testing gasoline containing oxygenates and uses a flame ionization detector. Another method that employs the flame ionization technique (ASTM D1319; IP 156) is widely used for measuring total olefins in gasoline fractions as well as aromatics and saturates, although the results may need correction for the presence of oxygenates. GC is also used for the determination of olefins in gasoline (ASTM D6296).

Benzene in gasoline can also be measured by infrared spectroscopy (ASTM D4053). But additional benefits are derived from *hyphenated* analytical methods such as GC–MS (ASTM D5769) and GC–FTIR (ASTM D5986), which also accurately measure benzene in gasoline. The GC–MS method (ASTM D5769) is based upon the EPA GC/MS procedure for aromatics.

Hydrocarbon composition is also determined by MS—a technique that has seen wide use for hydrocarbon-type analysis of gasoline (ASTM D2789) as well as to the identification of hydrocarbon constituents in higher-boiling gasoline fractions (ASTM D2425).

One method (ASTM D6379; IP 436) is used to determine the mono-aromatic and di-aromatic hydrocarbon contents in distillates boiling in the range from 50°C to 300°C (122°F to 572°F). In this method, the sample is diluted with an equal volume of hydrocarbon, such as heptane, and a fixed volume of this solution is injected into a high-performance liquid chromatograph fitted with a polar column where separation of the aromatic hydrocarbons from the non-aromatic hydrocarbons occurs. The separation of the aromatic constituents appears as distinct bands according to ring structure, and a refractive index detector is used to identify the components as they elute from the column. The peak areas of the aromatic constituents are compared with those obtained from previously run calibration standards to calculate the %w/w mono-aromatic hydrocarbon constituents and di-aromatic hydrocarbon constituents in the sample.

Compounds containing sulfur, nitrogen, and oxygen could possibly interfere with the performance of the test. Mono-alkenes do not interfere, but conjugated di- and poly-alkenes, if present, may interfere with the test performance.

Paraffins, naphthenes, and aromatic hydrocarbons in gasoline and other distillates boiling up to 200°C (392°F) are determined by multidimensional GC (ASTM D5443). Olefins that are present are converted to saturated hydrocarbons and are included in the paraffin and naphthene distribution. However, the scope of this test does not allow it to be applicable to hydrocarbons containing oxygenates. An extended version of the method can be used to determine the amounts of paraffins, olefins, naphthenes, and aromatics in gasoline-range hydrocarbon fractions (Speight, 2001, 2014).

A titration procedure (ASTM D1159), which determined the bromine number of petroleum distillates and aliphatic olefins by electrometric titration, can be used to provide an approximation of olefin content in a sample. Another related method (ASTM D2710) is used to determine the bromine index of petroleum hydrocarbons by electrometric titration and is valuable for determining trace levels of olefins in gasoline. Obviously, these methods use an indirect route to determine the total olefins, and the type of olefin derivative present affects the results since the results depend upon the ability of the olefins to react with bromine. Steric factors can prove to have an adverse effect on the experimental data.

The compositional analysis of gasoline, up to and including *n*-nonane, can be achieved using by capillary GC (ASTM D5134). Higher-resolution GC capillary column techniques provide a detailed analysis of most of the individual hydrocarbons in gasoline, including many of the oxygenated blending components. Capillary gas chromatographic techniques can be combined with MS to enhance the identification of the individual components and hydrocarbon types.

The presence of pentane and lighter hydrocarbons in gasoline interferes in the determination of hydrocarbon types (ASTM D1319; ASTM D2789). Pentane and lighter hydrocarbons are separated by this test method so that the depentanized residue can be analyzed, and pentane and lighter hydrocarbons can be analyzed by other methods, if desired. Typically, about 2% by volume of pentane and lower-boiling hydrocarbons remain in the bottoms, and hexane and higher-boiling hydrocarbons carry over to the overhead. In the test (ASTM D2001), a 50-ml sample is distilled into an *overhead* (pentane and lower-boiling hydrocarbons) fraction and a *bottoms* (hexane and higher-boiling hydrocarbons) fraction. The volume of bottoms is measured, and the percentage by volume, based on the original gasoline charged to the unit, is calculated.

Sulfur-containing components exist in gasoline-range hydrocarbons and can be identified using a gas chromatographic capillary column coupled with either a sulfur chemiluminescence detector or atomic emission detector (ASTM D5623). The most widely specified method for total sulfur content uses X-ray spectrometry (ASTM D2622), and other methods are available that use ultraviolet fluorescence spectroscopy (ASTM D5453), and/or hydrogenolysis and rateometric colorimetry (ASTM D4045) are also applicable, particularly when the sulfur level is low.

6.6.3 Corrosiveness

Because a gasoline would be unsuitable for use if it corroded the metallic parts of the gasoline system or the engine, it must be substantially free from corrosive compounds both before and after combustion.

Corrosiveness is usually due to the presence of free sulfur and sulfur compounds that burn to form sulfur dioxide (SO_2) that combines with water vapor formed by the combustion of the gasoline to produce sulfurous acid (H_2SO_3). Sulfurous acid can, in turn, oxidize to sulfuric acid (H_2SO_4), and both acids are corrosive toward iron and steel and would attack the cooler parts of the engine's exhaust system and cylinders as they cool off after the engine is shut down.

The total sulfur content of gasoline is very low, and knowledge of its magnitude is of chief interest to the refiner who must produce a product that conforms to a stringent specification. Various methods are available for the determination of total sulfur content. The one most frequently quoted in specifications is the Lamp Method (ASTM D1266; IP 107), in which the gasoline is burned in a small wick-fed lamp in an artificial atmosphere of carbon dioxide and oxygen; the oxides of sulfur are converted to sulfuric acid, which is then determined either volumetrically or gravimetrically.

A more recent development is the Wickbold Method (IP 243). This is basically similar to the Lamp Method except that the sample is burned in an oxy-hydrogen burner to give much more rapid combustion. An alternative technique, which has the advantage of being nondestructive, is X-ray spectrography (ASTM D2622).

Mercaptan sulfur (R-SH) and hydrogen sulfide (H_2S) are undesirable contaminants because, apart from their corrosive nature, they possess an extremely unpleasant odor. Such compounds should have been removed completely during refining, but their presence and that of free sulfur are detected by application of the *Doctor Test* (ASTM D4952; IP 30). The action on copper of any free or corrosive sulfur present in gasoline may be estimated by a procedure (ASTM D130; ASTM D849; IP 154) in which a strip of polished copper is immersed in the sample that is heated under specified conditions of temperature and time, and any staining of the copper is subsequently compared with the stains on a set of reference copper strips and thus the degree of corrosivity of the test sample determined.

Total sulfur is determined by combustion in a bomb calorimeter (ASTM D129; IP 61) and is often carried out with the determination of calorific value. The contents of the bomb are washed with distilled water into a beaker after which hydrochloric acid is added, and the solution is raised to boiling point. Barium chloride is added drop by drop to the boiling solution to precipitate the sulfuric acid as granular barium sulfate. After cooling, and standing for 24 h, the precipitate is filtered off on an ashless paper, washed, ignited, and weighed as barium sulfate.

$$\text{Sulfur } \%\text{w}/\text{w} = \frac{\text{Wt. of barium sulfate} \times 13.73}{\text{Weight of sample}}$$

As an addition to the test for mercaptan sulfur by potentiometric titration (ASTM D3227; IP 342), a piece of mechanically cleaned copper is also used to determine the amount of *corrosive sulfur* in a sample (ASTM D130; IP 112; IP 154; IP 411). The pure sheet of copper $3.0'' \times 0.5''$ (75×12 mm) is placed in a test tube with 40 ml of the sample, so that the copper is completely immersed. The tube is closed with a vented cork and heated in a boiling-water bath for 3 h.

The copper strip is then compared visually with a new strip of copper for signs of tarnish. The results are recorded as follows:

No change:	result negative
Slight discoloration:	result negative
Brown shade:	some effect
Steel gray:	some effect
Black, not scaled:	result positive, corrosive sulfur present
Black, scaled:	result positive, corrosive sulfur present

Thus, visual observation of the copper strip can present an indication or a conclusion of the presence or absence of corrosive sulfur. There is also a copper strip corrosion method for liquefied petroleum gases (ASTM D1838).

6.6.4   Density

Density (the mass of liquid per unit volume at 15°C) and the related terms *specific gravity* (the ratio of the mass of a given volume of liquid at 15°C to the mass of an equal volume of pure water at the same temperature) and *relative density* (same as *specific gravity*) are important properties of petroleum products (Chapter 2) (Gary et al., 2007; Hsu and Robinson, 2006; Speight, 2014; Speight and Ozum, 2002). Usually, a hydrometer, pycnometer, or digital density meter is used for the determination in all these standards.

The determination of density (specific gravity) (ASTM D287; ASTM D891; ASTM D1217; ASTM D1298; ASTM D1555; ASTM D1657; ASTM D4052; ASTM D5002; IP 160; IP 235; IP 365) provides a check on the uniformity of the gasoline, and it permits calculation of the weight per gallon. The temperature at which the determination is carried out and for which the calculations are to be made should also be known. However, the methods are subject to vapor pressure constraints and are used with appropriate precautions to prevent vapor loss during sample handling and density measurement. In addition, some test methods should not be applied if the samples are so dark in color that the absence of air bubbles in the sample cell cannot be established with certainty. The presence of such bubbles can have serious consequences for the reliability of the test data.

The current specification for automotive gasoline (ASTM D4814) does not set limits on the density of gasoline (ASTM D1298; IP 160). However, the density is fixed by the other chemical and physical properties of the gasoline and is important because gasoline is often bought and sold with reference to a specific temperature, usually 15.6°C (60°F). Since the gasoline is usually not at the specified temperature, volume correction factors based on the change in density with temperature are used to correct the volume to that temperature. Volume correction factors for oxygenates differ somewhat from those for hydrocarbons, and work is in progress to determine precise correction factors for gasoline–oxygenate blends.

6.6.5   Flash Point and Fire Point

For some purposes, it is necessary to have information on the initial stage of vaporization and the potential hazards that such a property can cause. To supply this need, *flash point, fire point, vapor pressure*, and evaporation/distillation methods are available.

The *flash point* is the lowest temperature at atmospheric pressure (760 mm Hg, 101.3 kPa) at which application of a test flame will cause the vapor of a sample to ignite under specified test conditions. The sample is deemed to have reached the flash point when a large flame appears and instantaneously propagates itself over the surface of the sample. The flash point data are used in shipping and safety regulations to define *flammable* and *combustible* materials. Flash point data can also indicate the possible presence of highly volatile and flammable constituents in a relatively nonvolatile or nonflammable material. The flash point of a petroleum product is also used to detect contamination. A substantially lower flash point than expected for a product is a reliable indicator that a product has become contaminated with a more volatile product, such as gasoline. The flash point is also an aid in establishing the identity of a particular petroleum product.

Of the available test methods, the most common method of determining the flash point confines the vapor (closed cup) until the instant the flame is applied (ASTM D56; ASTM D93; ASTM D3828; ASTM D6450; IP 34; IP 523). An alternate method that does not confine the vapor (open-cup method: ASTM D92; ASTM D1310; IP 36) gives slightly higher values of the flash point.

The Pensky–Marten apparatus using a closed or open system (ASTM D93; IP 34; IP 35) is the standard instrument for flash points above 50°C (122°F), and the Abel apparatus (IP 170) is used for more volatile oils, with flash points below 50°C (122°F). The Cleveland Open-Cup method (ASTM D92; IP 36) is also used for the determination of the *fire point* (the temperature at which the sample will ignite and burn for at least 5 s).

The Pensky–Marten apparatus consists of a brass cup, mounted in an air bath and heated by a gas flame. A propeller-type stirrer, operated by a flexible drive, extends from the center of the cover into the cup. The cover has four openings: one for a thermometer, and the others fitted with sliding shutters for the introduction of a pilot flame and for ventilation. The temperature of the oil in the cup is raised at a rate of 5–6°C/min (9–11°F/min). The stirrer is rotated at approximately 60 rpm. When the temperature has risen to approximately 15°C (27°F) above the anticipated flash point, the pilot flame is dipped into the oil vapor for 2 s for every 1°C

(1.8°F) rise in temperature up to 105°C (221°F). Above 105°C (221°F), the flame is introduced for every 2°C (3.6°F) rise in temperature. The flash point is the temperature at which a distinct flash is observed when the pilot flame meets the vapor in the cup.

Erroneously high flash points can be obtained when precautions are not taken to avoid the loss of volatile material. Samples should not be stored in plastic bottles, since the volatile material may diffuse through the walls of the container. The containers should not be opened unnecessarily. The samples should not be transferred between containers unless the sample temperature is at least 20°F (11°C) below the expected flash point.

The Abel *closed-cup* apparatus (IP 170) consists of a brass cup sealed in a small water bath that is immersed in a second water bath. The cover of the brass cup is fitted in a manner similar to that in the Pensky–Marten apparatus. For crude oils and products with flash point higher than 30°C (<86°F), the outer bath is filled with water at 55°C (131°F) and is not heated further. The oil under test is then placed inside the cup. When the temperature reaches 19°C (66°F), the pilot flame is introduced every 0.5°C (1.0°F) until a flash is obtained. For oils with flash points in excess of 30°C (>86°F) and less than 50°C (<122°F), the inner water bath is filled with cold water to a depth of 35 mm. The outer bath is filled with cold water and heated at a rate of 1°C/min (1.8°F/min). The flash point is obtained as before.

From the viewpoint of safety, information about the *flash point* is of most significance at or slightly above the maximum temperatures (30°C–60°C, 86°F–140°F) that may be encountered in storage, transportation, and use of liquid petroleum products, in either closed or open containers. In this temperature range, the relative fire and explosion hazard can be estimated from the flash point. For products with flash point below 40°C (104°F), special precautions are necessary for safe handling. Flash points above 60°C (140°F) gradually lose their safety significance until they become indirect measures of some other quality.

Another test (ASTM E659) is available that can be used as a complement to the flash point test and involves determination of the auto-ignition temperature. However, the flash point should not be confused with auto-ignition temperature that measures spontaneous combustion with no external source of ignition.

The *fire point* is the temperature to which the product must be heated under the prescribed conditions of the method to burn continuously when the mixture of vapor and air is ignited by a specified flame (ASTM D92; IP 36).

6.6.6 Oxygenates

Blends of gasoline with oxygenates are common and, in fact, are required in certain areas. These blends consist primarily of gasoline with substantial amounts of oxygenates, which are oxygen-containing, ashless, organic compounds such as alcohols and ethers. The most common oxygenates are ethanol and methyl-*t*-butyl ether with lesser attention focused on ethyl-*t*-butyl ether, *t*-amyl methyl ether, and di-*iso*-propyl ether.

Some of the test methods originally developed for gasoline can be used for gasoline–oxygenate blends, while certain other test methods for gasoline are not suitable for blends. To avoid the necessity of determining in advance whether a gasoline contains oxygenates, there are test methods that can be used for both gasoline and gasoline-oxygenate blends (e.g., ASTM D4814).

In general, the test methods for determining distillation temperatures, lead content, sulfur content, copper corrosion, existent gum, and oxidation stability can be used for both gasoline and gasoline–oxygenate blends. Some of the test methods for vapor pressure and vapor–liquid ratio are sensitive to the presence of oxygenates in the gasoline and must be acknowledged accordingly.

Whereas gasoline and water are almost entirely immiscible, and will readily separate into two phases, a gasoline–oxygenate blend is able to dissolve a limited amount of water without phase separation. Hence, the use of the term *water tolerance*. Gasoline–oxygenate blends will dissolve some water but will also separate into two phases when contacted with water that is beyond the threshold concentration. Such a phenomenon is a major issue for alcohol-containing blends.

Phase separation can usually be avoided if the gasoline is sufficiently water-free initially and care is taken during distribution to prevent contact with water. Gasoline–oxygenate blends can be tested for water tolerance (ASTM D4814) in which gasoline is cooled under specified conditions to its expected use temperature. Formation of a haze (analogous to the cloud point, ASTM D2500; ASTM D5771; ASTM D5772; ASTM D5773; IP 219) must be carefully distinguished from separation into two distinct phases and is not a reason for the rejection of the gasoline.

6.6.7 Stability and Instability

The study of the analysis of crude oil products is complete without some mention of *instability* and *incompatibility*. Both result in formation and appearance of degradation products or other undesirable changes in the original properties of petroleum products.

In terms of the instability and incompatibility of petroleum and petroleum products, the unsaturated hydrocarbon content and the heteroatom content appear to represent the greatest influence. In fact, the sulfur and nitrogen contents of crude oil are an important parameter in respect of the processing methods that have to be used in order to produce gasoline that meets specifications for sulfur concentration.

There could well be a relation between nitrogen and sulfur contents and crude oil (or product) stability; higher nitrogen and sulfur crude oils often exhibit higher sludge-forming tendencies.

Gasoline manufactured by cracking processes contains unsaturated components that may oxidize during storage and form undesirable oxidation products. Since gasoline is often stored before use, it is essential that any components should not undergo any deleterious change under storage conditions and remain stable during their passage from the gasoline tank of a vehicle to the cylinder of its engine so that no harmful deposits are built up in the tank and gasoline lines and in the inlet system and on the valves.

During storage, gasoline is exposed to the action of air at ambient temperature, and in its path from vehicle tank to engine, it is mixed with air and also subjected to the effects of heat. An unstable gasoline will undergo oxidation and polymerization under such conditions form gum, a resinous material that in the early stages of chain reaction initiated by the formation of peroxides and catalyzed by the presence of metals, particularly copper, which may have been picked up during refining and handling operations.

Instability occurs because of a low resistance of the product to environmental (in use) influences during storage or to its susceptibility to oxidative and/or other degradative processes. In case of incompatibility, degradation products form or changes occur due to an interaction of some chemical groups present in the components of the final blend.

Petroleum product components can be defined as being *incompatible* when sludge, semisolid, or solid particles (for convenience here, these are termed *secondary products* to distinguish them from the actual petroleum product) are formed during and after blending. This phenomenon usually occurs prior to use. If the secondary products are marginally soluble in the blended petroleum product, use might detract from solubility of the secondary products, and they will appear as sludge or sediment that can be separated by centrifuge with water (ASTM D2709) or by extraction (ASTM D4310; IP 53; IP 375). When the secondary products are truly insoluble, they separate and settle out as a semisolid or solid phase floating in the gasoline or are deposited on the walls and floors of containers. In addition, secondary products usually increase the viscosity of the petroleum product. Standing at low temperatures will also cause a viscosity change (ASTM D2532). Usually, the viscosity change might be due to separation of paraffins as might occur when gasoline, but more especially diesel gasoline, is allowed to cool and stand unused overnight in low-temperature climates.

Briefly, the term *incompatibility* refers to the formation of a precipitate (or sediment) or phase separation when two liquids are mixed. The term *instability* is often used in reference to the formation of color, sediment, or gum in the liquid over a period of time. This term may be used to contrast the formation of a precipitate in the near term (almost immediately). However, the terms are often used interchangeably.

The measurement of color (ASTM D156; ASTM D848; ASTM D1209; ASTM D1555; ASTM D5386; IP 17) provides a rapid method of checking the degree of freedom from contamination. Observation of the test for residue on evaporation (ASTM D381; ASTM D1353; IP 131) provides a further guard against adventitious contamination.

Tests should also be carried out for sediment if the gasoline has been subjected to events (such as oxidation) that could lead to sediment formation and instability of the gasoline and the resulting products. Test methods are available for the determination of sediment by extraction (ASTM D473; IP 285) or by membrane filtration (ASTM D4807) and the determination of simultaneously sediment with water by centrifugation (ASTM D1796; ASTM D2709; ASTM D4007; IP 373).

Gum formation (ASTM D525; IP 40) alludes to the formation of soluble organic material, whereas *sediment* is the insoluble organic material. *Storage stability* (or *storage instability*) (ASTM D381; ASTM D4625; IP 131; IP 378) is a term used to describe the ability of the liquid to remain in storage over extended periods of time without appreciable deterioration as measured by gum formation and/or the formation sediment. *Thermal stability* is also defined as the ability of the liquid to withstand relatively high temperatures for short periods of time without the formation of sediment (i.e., carbonaceous deposits and/or coke). *Thermal oxidative stability* is the ability of the liquid to withstand relatively high temperatures for short periods of time in the presence of oxidation and without the formation of sediment or deterioration of properties (ASTM D3241), and there is standard equipment for various oxidation tests (ASTM D4871). *Stability* is also as the ability of the liquid to withstand long periods at temperatures up to 100°C (212°F) without degradation. Determination of the *reaction threshold temperature* for various liquid and solid materials might be beneficial (ASTM D2883).

Existent-gum is the name given to the nonvolatile residue present in the gasoline as received for test (ASTM D381; IP 131). In this test, the sample is evaporated from a beaker maintained at a temperature of 160°C–166°C (320°F–331°F) with the aid of a similarly heated jet of air. This material is distinguished from the *potential gum* that is obtained by aging the sample at an elevated temperature.

Thus, *potential gum* is determined by the *accelerated gum test* (ASTM D873; IP 138) that is used as a safeguard of storage stability and can be used to predict the potential for gum formation during prolonged storage. In this test, the gasoline is heated for 16h with oxygen under pressure in a bomb at 100°C (212°F). After this time, both the gum content and the solid precipitate are measured.

Dry sludge is defined as the material separated from the bulk of an oil by filtration and which is insoluble in heptane. *Existent dry sludge* is the dry sludge in the original sample as received and is distinguished from the accelerated dry sludge obtained after aging the sample by chemical addition or heat. The *existent dry sludge* is distinguished from the *potential dry sludge* that is obtained by aging the sample at an elevated temperature.

The *existent dry sludge* is operationally defined as the material separated from the bulk of a crude oil or crude oil product by filtration and which is insoluble in heptane. The test is used as an indicator of process operability and as a measure of potential downstream fouling.

Attractive as they may be, any tests that involve *accelerated oxidation of* the sample must be used with caution and consideration of the chemistry. Depending on the constituents of the sample, it is quite possible that the higher temperature and extreme conditions (oxygen under pressure) may not be truly representative of the deterioration of the sample under storage conditions. The higher temperature and the oxygen under pressure might change the chemistry of the system and produce products that would not be produced under ambient storage conditions. An assessment of the composition of the gasoline prior to storage and application of the test will assist in this determination.

Because gasoline contains traces of nonvolatile oils and additives, the residue left in the beakers is washed with heptane before the gum-residue is dried and weighed. The existent gum test (ASTM D381; IP 131) is useful as a refinery control but is to some extent unrealistic as a criterion of performance, and therefore engine tests have been developed to determine the tendency toward inlet system deposits.

In general, gasoline instability and incompatibility can be related to the heteroatom containing (i.e., nitrogen-, oxygen-, and sulfur-containing compounds). The degree of unsaturation of the gasoline (i.e., the level of olefin species) also plays a role in determining instability/incompatibility. And, recent investigations have also implicated catalytic levels of various oxidized intermediates and acids as especially deleterious for middle distillate gasoline.

6.6.8 Water and Sediment

However, before any volatility tests are carried out, all water must be removed since the presence of more than 0.5% water in test samples of crude can cause several problems during distillation procedures. Water has a high heat of vaporization, necessitating the application of additional thermal energy to the distillation flask. Water is relatively easily superheated and therefore excessive *bumping* can occur, leading to erroneous readings, and the potential for destruction of the glass equipment is real. Steam formed during distillation can act as a carrier gas, and high-boiling-point components may end up in the distillate (*steam distillation*) (Fig. 6.3).

Centrifugation can be used to remove water (and sediment) if the sample is not a tight emulsion. Other methods that are used to remove water include the following: (i) heating in a pressure vessel to control loss of light ends, (ii) addition of calcium chloride as recommended in ASTM D1160, (iii) addition of an azeotroping agent such as *iso*-propanol or *n*-butanol, (iv) removal of water in a preliminary low-efficiency or flash distillation followed by reblending

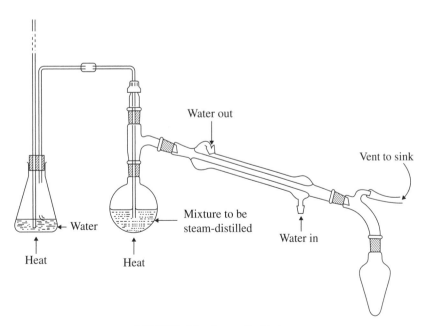

FIGURE 6.3   Steam distillation.

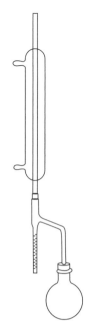

FIGURE 6.4 Dean and Stark adaptor on the bottom of a condenser.

the hydrocarbon which co-distills with the water into the sample (IP 74), and (v) separation of the water from the hydrocarbon distillate by freezing.

Water is a contaminant in gasoline and should be measured using the Karl Fischer method (ASTM D1364; ASTM D1744; ASTM D4377; ASTM D4928; ASTM D6304; ASTM E203), by distillation using a Dean and Stark condenser (ASTM D4006) (Fig. 6.4), or by centrifuging and excluded by relevant drying methods.

Tests should also be carried out for sediment if the gasoline has been subjected to events (such as oxidation) that could lead to sediment formation and instability of the gasoline and the resulting products. Test methods are available for the determination of sediment by extraction (ASTM D473; IP 285) or by membrane filtration (ASTM D4807) and the determination of simultaneously sediment with water by centrifugation (ASTM D1796; ASTM D2709; ASTM D4007; IP 373).

REFERENCES

ASTM D56. Standard Test Method for Flash Point by Tag Closed Cup Tester. Annual Book of Standards. ASTM International, West Conshohocken, Pennsylvania.

ASTM D86. Standard Test Method for Distillation of Petroleum Products at Atmospheric Pressure. Annual Book of Standards. ASTM International, West Conshohocken, Pennsylvania.

ASTM D88. Standard Test Method for Saybolt Viscosity. Annual Book of Standards. ASTM International, West Conshohocken, Pennsylvania.

ASTM D92. Standard Test Method for Flash and Fire Points by Cleveland Open Cup Tester. Annual Book of Standards. ASTM International, West Conshohocken, Pennsylvania.

ASTM D93. Standard Test Methods for Flash Point by Pensky-Martens Closed Cup Tester. Annual Book of Standards. ASTM International, West Conshohocken, Pennsylvania.

ASTM D129. Standard Test Method for Sulfur in Petroleum Products (General High Pressure Decomposition Device Method). Annual Book of Standards. ASTM International, West Conshohocken, Pennsylvania.

ASTM D130. Standard Test Method for Corrosiveness to Copper from Petroleum Products by Copper Strip Test. Annual Book of Standards. ASTM International, West Conshohocken, Pennsylvania.

ASTM D156. Standard Test Method for Saybolt Color of Petroleum Products (Saybolt Chromometer Method). Annual Book of Standards. ASTM International, West Conshohocken, Pennsylvania.

ASTM D240. Standard Test Method for Heat of Combustion of Liquid Hydrocarbon Fuels by Bomb Calorimeter. Annual Book of Standards. ASTM International, West Conshohocken, Pennsylvania.

ASTM D287. Standard Test Method for API Gravity of Crude Petroleum and Petroleum Products (Hydrometer Method). Annual Book of Standards. ASTM International, West Conshohocken, Pennsylvania.

ASTM D323. Standard Test Method for Vapor Pressure of Petroleum Products (Reid Method). Annual Book of Standards. ASTM International, West Conshohocken, Pennsylvania.

ASTM D381. Standard Test Method for Gum Content in Fuels by Jet Evaporation. Annual Book of Standards. ASTM International, West Conshohocken, Pennsylvania.

ASTM D445. Standard Test Method for Kinematic Viscosity of Transparent and Opaque Liquids (and Calculation of Dynamic Viscosity). Annual Book of Standards. ASTM International, West Conshohocken, Pennsylvania.

ASTM D473. Standard Test Method for Sediment in Crude Oils and Fuel Oils by the Extraction Method. Annual Book of Standards. ASTM International, West Conshohocken, Pennsylvania.

ASTM D525. Standard Test Method for Oxidation Stability of Gasoline (Induction Period Method). Annual Book of Standards. ASTM International, West Conshohocken, Pennsylvania.

ASTM D611. Standard Test Methods for Aniline Point and Mixed Aniline Point of Petroleum Products and Hydrocarbon Solvents. Annual Book of Standards. ASTM International, West Conshohocken, Pennsylvania.

ASTM D665. Standard Test Method for Rust-Preventing Characteristics of Inhibited Mineral Oil in the Presence of Water. Annual Book of Standards. ASTM International, West Conshohocken, Pennsylvania.

ASTM D808. Standard Test Method for Chlorine in New and Used Petroleum Products (High Pressure Decomposition Device Method). Annual Book of Standards. ASTM International, West Conshohocken, Pennsylvania.

ASTM D848. Standard Test Method for Acid Wash Color of Industrial Aromatic Hydrocarbons. Annual Book of Standards. ASTM International, West Conshohocken, Pennsylvania.

ASTM D849. Standard Test Method for Copper Strip Corrosion by Industrial Aromatic Hydrocarbons. Annual Book of Standards. ASTM International, West Conshohocken, Pennsylvania.

ASTM D873. Standard Test Method for Oxidation Stability of Aviation Fuels (Potential Residue Method). Annual Book of Standards. ASTM International, West Conshohocken, Pennsylvania.

ASTM D891. Standard Test Methods for Specific Gravity, Apparent, of Liquid Industrial Chemicals. Annual Book of Standards. ASTM International, West Conshohocken, Pennsylvania.

ASTM D1078. Standard Test Method for Distillation Range of Volatile Organic Liquids. Annual Book of Standards. ASTM International, West Conshohocken, Pennsylvania.

ASTM D1133. Standard Test Method for Kauri-Butanol Value of Hydrocarbon Solvents. Annual Book of Standards. ASTM International, West Conshohocken, Pennsylvania.

ASTM D1159. Standard Test Method for Bromine Numbers of Petroleum Distillates and Commercial Aliphatic Olefins by Electrometric Titration. Annual Book of Standards. ASTM International, West Conshohocken, Pennsylvania.

ASTM D1160. Standard Test Method for Distillation of Petroleum Products at Reduced Pressure. Annual Book of Standards. ASTM International, West Conshohocken, Pennsylvania.

ASTM D1209. Standard Test Method for Color of Clear Liquids (Platinum-Cobalt Scale). Annual Book of Standards. ASTM International, West Conshohocken, Pennsylvania.

ASTM D1217. Standard Test Method for Density and Relative Density (Specific Gravity) of Liquids by Bingham Pycnometer. Annual Book of Standards. ASTM International, West Conshohocken, Pennsylvania.

ASTM D1266. Standard Test Method for Sulfur in Petroleum Products (Lamp Method). Annual Book of Standards. ASTM International, West Conshohocken, Pennsylvania.

ASTM D1298. Standard Test Method for Density, Relative Density, or API Gravity of Crude Petroleum and Liquid Petroleum Products by Hydrometer Method. Annual Book of Standards. ASTM International, West Conshohocken, Pennsylvania.

ASTM D1310. Standard Test Method for Flash Point and Fire Point of Liquids by Tag Open-Cup Apparatus. Annual Book of Standards. ASTM International, West Conshohocken, Pennsylvania.

ASTM D1319. Standard Test Method for Hydrocarbon Types in Liquid Petroleum Products by Fluorescent Indicator Adsorption. Annual Book of Standards. ASTM International, West Conshohocken, Pennsylvania.

ASTM D1353. Standard Test Method for Nonvolatile Matter in Volatile Solvents for Use in Paint, Varnish, Lacquer, and Related Products. Annual Book of Standards. ASTM International, West Conshohocken, Pennsylvania.

ASTM D1364. Standard Test Method for Water in Volatile Solvents (Karl Fischer Reagent Titration Method). Annual Book of Standards. ASTM International, West Conshohocken, Pennsylvania.

ASTM D1405. Standard Test Method for Estimation of Net Heat of Combustion of Aviation Fuels. Annual Book of Standards. ASTM International, West Conshohocken, Pennsylvania.

ASTM D1555. Standard Test Method for Calculation of Volume and Weight of Industrial Aromatic Hydrocarbons and Cyclohexane [Metric]. Annual Book of Standards. ASTM International, West Conshohocken, Pennsylvania.

ASTM D1657. Standard Test Method for Density or Relative Density of Light Hydrocarbons by Pressure Hydrometer. Annual Book of Standards. ASTM International, West Conshohocken, Pennsylvania.

ASTM D1744. Standard Test Method for Determination of Water in Liquid Petroleum Products by Karl Fischer Reagent. Annual Book of Standards. ASTM International, West Conshohocken, Pennsylvania.

ASTM D1796. Standard Test Method for Water and Sediment in Fuel Oils by the Centrifuge Method (Laboratory Procedure). Annual Book of Standards. ASTM International, West Conshohocken, Pennsylvania.

ASTM D1838. Standard Test Method for Copper Strip Corrosion by Liquefied Petroleum (LP) Gases. Annual Book of Standards. ASTM International, West Conshohocken, Pennsylvania.

ASTM D1949. Separation of Tetraethyl Lead and Tetramethyl Lead in Gasoline. Annual Book of Standards. ASTM International, West Conshohocken, PA.

ASTM D2001. Standard Test Method for Depentanization of Gasoline and Naphtha. Annual Book of Standards. ASTM International, West Conshohocken, Pennsylvania.

ASTM D2161. Standard Practice for Conversion of Kinematic Viscosity to Saybolt Universal Viscosity or to Saybolt Furol Viscosity. Annual Book of Standards. ASTM International, West Conshohocken, Pennsylvania.

ASTM D2425. Standard Test Method for Hydrocarbon Types in Middle Distillates by Mass Spectrometry. Annual Book of Standards. ASTM International, West Conshohocken, Pennsylvania.

ASTM D2427. Standard Test Method for Determination of C2 through C5 Hydrocarbons in Gasolines by Gas Chromatography. Annual Book of Standards. ASTM International, West Conshohocken, Pennsylvania.

ASTM D2500. Standard Test Method for Cloud Point of Petroleum Products. Annual Book of Standards. ASTM International, West Conshohocken, Pennsylvania.

ASTM D2532. Standard Test Method for Viscosity and Viscosity Change after Standing at Low Temperature of Aircraft Turbine Lubricants. Annual Book of Standards. ASTM International, West Conshohocken, Pennsylvania.

ASTM D2622. Standard Test Method for Sulfur in Petroleum Products by Wavelength Dispersive X-ray Fluorescence Spectrometry. Annual Book of Standards. ASTM International, West Conshohocken, Pennsylvania.

ASTM D2669. Standard Test Method for Apparent Viscosity of Petroleum Waxes Compounded with Additives (Hot Melts). Annual Book of Standards. ASTM International, West Conshohocken, Pennsylvania.

ASTM D2699. Standard Test Method for Research Octane Number of Spark-Ignition Engine Fuel. Annual Book of Standards. ASTM International, West Conshohocken, Pennsylvania.

ASTM D2700. Standard Test Method for Motor Octane Number of Spark-Ignition Engine Fuel. Annual Book of Standards. ASTM International, West Conshohocken, Pennsylvania.

ASTM D2709. Standard Test Method for Water and Sediment in Middle Distillate Fuels by Centrifuge. Annual Book of Standards. ASTM International, West Conshohocken, Pennsylvania.

ASTM D2710. Standard Test Method for Bromine Index of Petroleum Hydrocarbons by Electrometric Titration. Annual Book of Standards. ASTM International, West Conshohocken, Pennsylvania.

ASTM D2789. Standard Test Method for Hydrocarbon Types in Low Olefinic Gasoline by Mass Spectrometry. Annual Book of Standards. ASTM International, West Conshohocken, Pennsylvania.

ASTM D2883. Standard Test Method for Reaction Threshold Temperature of Liquid and Solid Materials. Annual Book of Standards. ASTM International, West Conshohocken, Pennsylvania.

ASTM D2885. Standard Test Method for Determination of Octane Number of Spark-Ignition Engine Fuels by On-Line Direct Comparison Technique. Annual Book of Standards. ASTM International, West Conshohocken, Pennsylvania.

ASTM D2887. Standard Test Method for Boiling Range Distribution of Petroleum Fractions by Gas Chromatography. Annual Book of Standards. ASTM International, West Conshohocken, Pennsylvania.

ASTM D2892. Standard Test Method for Distillation of Crude Petroleum (15-Theoretical Plate Column). Annual Book of Standards. ASTM International, West Conshohocken, Pennsylvania.

ASTM D3227. Standard Test Method for (Thiol Mercaptan) Sulfur in Gasoline, Kerosene, Aviation Turbine, and Distillate Fuels (Potentiometric Method). Annual Book of Standards. ASTM International, West Conshohocken, Pennsylvania.

ASTM D3231. Standard Test Method for Phosphorus in Gasoline. Annual Book of Standards. ASTM International, West Conshohocken, Pennsylvania.

ASTM D3237. Standard Test Method for Lead in Gasoline by Atomic Absorption Spectroscopy. Annual Book of Standards. ASTM International, West Conshohocken, Pennsylvania.

ASTM D3241. Standard Test Method for Thermal Oxidation Stability of Aviation Turbine Fuels. Annual Book of Standards. ASTM International, West Conshohocken, Pennsylvania.

ASTM D3341. Standard Test Method for Lead in Gasoline—Iodine Monochloride Method. Annual Book of Standards. ASTM International, West Conshohocken, Pennsylvania.

ASTM D3606. Standard Test Method for Determination of Benzene and Toluene in Finished Motor and Aviation Gasoline by Gas Chromatography. Annual Book of Standards. ASTM International, West Conshohocken, Pennsylvania.

ASTM D3710. Standard Test Method for Boiling Range Distribution of Gasoline and Gasoline Fractions by Gas Chromatography. Annual Book of Standards. ASTM International, West Conshohocken, Pennsylvania.

ASTM D3828. Standard Test Methods for Flash Point by Small Scale Closed Cup Tester. Annual Book of Standards. ASTM International, West Conshohocken, Pennsylvania.

ASTM D3831. Standard Test Method for Manganese in Gasoline by Atomic Absorption Spectroscopy. Annual Book of Standards. ASTM International, West Conshohocken, Pennsylvania.

ASTM D4006. Standard Test Method for Water in Crude Oil by Distillation. Annual Book of Standards. ASTM International, West Conshohocken, Pennsylvania.

ASTM D4007. Standard Test Method for Water and Sediment in Crude Oil by the Centrifuge Method (Laboratory Procedure). Annual Book of Standards. ASTM International, West Conshohocken, Pennsylvania.

ASTM D4045. Standard Test Method for Sulfur in Petroleum Products by Hydrogenolysis and Rateometric Colorimetry. Annual Book of Standards. ASTM International, West Conshohocken, Pennsylvania.

ASTM D4052. Standard Test Method for Density, Relative Density, and API Gravity of Liquids by Digital Density Meter. Annual Book of Standards. ASTM International, West Conshohocken, Pennsylvania.

ASTM D4053. Standard Test Method for Benzene in Motor and Aviation Gasoline by Infrared Spectroscopy. Annual Book of Standards. ASTM International, West Conshohocken, Pennsylvania.

ASTM D4057. Standard Practice for Manual Sampling of Petroleum and Petroleum Products. Annual Book of Standards. ASTM International, West Conshohocken, Pennsylvania.

ASTM D4310. Standard Test Method for Determination of Sludging and Corrosion Tendencies of Inhibited Mineral Oils. Annual Book of Standards. ASTM International, West Conshohocken, Pennsylvania.

ASTM D4377. Standard Test Method for Water in Crude Oils by Potentiometric Karl Fischer Titration. Annual Book of Standards. ASTM International, West Conshohocken, Pennsylvania.

ASTM D4625. Standard Test Method for Distillate Fuel Storage Stability at 43°C (110°F). Annual Book of Standards. ASTM International, West Conshohocken, Pennsylvania.

ASTM D4807. Standard Test Method for Sediment in Crude Oil by Membrane Filtration. Annual Book of Standards. ASTM International, West Conshohocken, Pennsylvania.

ASTM D4814. Standard Specification for Automotive Spark-Ignition Engine Fuel. Annual Book of Standards. ASTM International, West Conshohocken, Pennsylvania.

ASTM D4815. Standard Test Method for Determination of MTBE, ETBE, TAME, DIPE, tertiary-Amyl Alcohol and C1 to C4 Alcohols in Gasoline by Gas Chromatography. Annual Book of Standards. ASTM International, West Conshohocken, Pennsylvania.

ASTM D4871. Standard Guide for Universal Oxidation/Thermal Stability Test Apparatus. Annual Book of Standards. ASTM International, West Conshohocken, Pennsylvania.

ASTM D4928. Standard Test Method for Water in Crude Oils by Coulometric Karl Fischer Titration. Annual Book of Standards. ASTM International, West Conshohocken, Pennsylvania.

ASTM D4952. Standard Test Method for Qualitative Analysis for Active Sulfur Species in Fuels and Solvents (Doctor Test).

Annual Book of Standards. ASTM International, West Conshohocken, Pennsylvania.

ASTM D4953. Standard Test Method for Vapor Pressure of Gasoline and Gasoline-Oxygenate Blends (Dry Method). Annual Book of Standards. ASTM International, West Conshohocken, Pennsylvania.

ASTM D5002. Standard Test Method for Density and Relative Density of Crude Oils by Digital Density Analyzer. Annual Book of Standards. ASTM International, West Conshohocken, Pennsylvania.

ASTM D5059. Standard Test Methods for Lead in Gasoline by X-Ray Spectroscopy. Annual Book of Standards. ASTM International, West Conshohocken, Pennsylvania.

ASTM D5134. Standard Test Method for Detailed Analysis of Petroleum Naphthas through n-Nonane by Capillary Gas Chromatography. Annual Book of Standards. ASTM International, West Conshohocken, Pennsylvania.

ASTM D5185. Standard Test Method for Determination of Additive Elements, Wear Metals, and Contaminants in Used Lubricating Oils and Determination of Selected Elements in Base Oils by Inductively Coupled Plasma Atomic Emission Spectrometry (ICP-AES). Annual Book of Standards. ASTM International, West Conshohocken, Pennsylvania.

ASTM D5188. Standard Test Method for Vapor-Liquid Ratio Temperature Determination of Fuels (Evacuated Chamber Method). Annual Book of Standards. ASTM International, West Conshohocken, Pennsylvania.

ASTM D5191. Standard Test Method for Vapor Pressure of Petroleum Products (Mini Method). Annual Book of Standards. ASTM International, West Conshohocken, Pennsylvania.

ASTM D5236. Standard Test Method for Distillation of Heavy Hydrocarbon Mixtures (Vacuum Potstill Method). Annual Book of Standards. ASTM International, West Conshohocken, Pennsylvania.

ASTM D5291. Standard Test Methods for Instrumental Determination of Carbon, Hydrogen, and Nitrogen in Petroleum Products and Lubricants. Annual Book of Standards. ASTM International, West Conshohocken, Pennsylvania.

ASTM D5386. Standard Test Method for Color of Liquids Using Tristimulus Colorimetry. Annual Book of Standards. ASTM International, West Conshohocken, Pennsylvania.

ASTM D5441. Standard Test Method for Analysis of Methyl t-Butyl Ether (MTBE) by Gas Chromatography. Annual Book of Standards. ASTM International, West Conshohocken, Pennsylvania.

ASTM D5443. Standard Test Method for Paraffin, Naphthene, and Aromatic Hydrocarbon Type Analysis in Petroleum Distillates through 200°C by Multi-Dimensional Gas Chromatography. Annual Book of Standards. ASTM International, West Conshohocken, Pennsylvania.

ASTM D5453. Standard Test Method for Determination of Total Sulfur in Light Hydrocarbons, Spark Ignition Engine Fuel, Diesel Engine Fuel, and Engine Oil by Ultraviolet Fluorescence. Annual Book of Standards. ASTM International, West Conshohocken, Pennsylvania.

ASTM D5482. Standard Test Method for Vapor Pressure of Petroleum Products (Mini Method – Atmospheric). Annual

Book of Standards. ASTM International, West Conshohocken, Pennsylvania.

ASTM D5580. Standard Test Method for Determination of Benzene, Toluene, Ethylbenzene, p/m-Xylene, o-Xylene, C9 and Heavier Aromatics, and Total Aromatics in Finished Gasoline by Gas Chromatography. Annual Book of Standards. ASTM International, West Conshohocken, Pennsylvania.

ASTM D5599. Standard Test Method for Determination of Oxygenates in Gasoline by Gas Chromatography and Oxygen Selective Flame Ionization Detection. Annual Book of Standards. ASTM International, West Conshohocken, Pennsylvania.

ASTM D5622. Standard Test Methods for Determination of Total Oxygen in Gasoline and Methanol Fuels by Reductive Pyrolysis. Annual Book of Standards. ASTM International, West Conshohocken, Pennsylvania.

ASTM D5623. Standard Test Method for Sulfur Compounds in Light Petroleum Liquids by Gas Chromatography and Sulfur Selective Detection. Annual Book of Standards. ASTM International, West Conshohocken, Pennsylvania.

ASTM D5769. Standard Test Method for Determination of Benzene, Toluene, and Total Aromatics in Finished Gasolines by Gas Chromatography/Mass Spectrometry. Annual Book of Standards. ASTM International, West Conshohocken, Pennsylvania.

ASTM D5771. Standard Test Method for Cloud Point of Petroleum Products (Optical Detection Stepped Cooling Method). Annual Book of Standards. ASTM International, West Conshohocken, Pennsylvania.

ASTM D5772. Standard Test Method for Cloud Point of Petroleum Products (Linear Cooling Rate Method). Annual Book of Standards. ASTM International, West Conshohocken, Pennsylvania.

ASTM D5773. Standard Test Method for Cloud Point of Petroleum Products (Constant Cooling Rate Method). Annual Book of Standards. ASTM International, West Conshohocken, Pennsylvania.

ASTM D5845. Standard Test Method for Determination of MTBE, ETBE, TAME, DIPE, Methanol, Ethanol and t-Butanol in Gasoline by Infrared Spectroscopy. Annual Book of Standards. ASTM International, West Conshohocken, Pennsylvania.

ASTM D5986. Standard Test Method for Determination of Oxygenates, Benzene, Toluene, C8-C12 Aromatics and Total Aromatics in Finished Gasoline by Gas Chromatography/Fourier Transform Infrared Spectroscopy. Annual Book of Standards. ASTM International, West Conshohocken, Pennsylvania.

ASTM D6296. Standard Test Method for Total Olefins in Spark-ignition Engine Fuels by Multidimensional Gas Chromatography. Annual Book of Standards. ASTM International, West Conshohocken, Pennsylvania.

ASTM D6304. Standard Test Method for Determination of Water in Petroleum Products, Lubricating Oils, and Additives by Coulometric Karl Fischer Titration. Annual Book of Standards. ASTM International, West Conshohocken, Pennsylvania.

ASTM D6377. Standard Test Method for Determination of Vapor Pressure of Crude Oil: VPCRx (Expansion Method). Annual

Book of Standards. ASTM International, West Conshohocken, Pennsylvania.

ASTM D6378. Standard Test Method for Determination of Vapor Pressure (VPX) of Petroleum Products, Hydrocarbons, and Hydrocarbon-Oxygenate Mixtures (Triple Expansion Method). Annual Book of Standards. ASTM International, West Conshohocken, Pennsylvania.

ASTM D6379. Standard Test Method for Determination of Aromatic Hydrocarbon Types in Aviation Fuels and Petroleum Distillates—High Performance Liquid Chromatography Method with Refractive Index Detection. Annual Book of Standards. ASTM International, West Conshohocken, Pennsylvania.

ASTM D6450. Standard Test Method for Flash Point by Continuously Closed Cup (CCCFP) Tester. Annual Book of Standards. ASTM International, West Conshohocken, Pennsylvania.

ASTM E203. Standard Test Method for Water Using Volumetric Karl Fischer Titration. Annual Book of Standards. ASTM International, West Conshohocken, Pennsylvania.

ASTM E659. Standard Test Method for Autoignition Temperature of Liquid Chemicals. Annual Book of Standards. ASTM International, West Conshohocken, Pennsylvania.

Energy Institute. 2013. IP Standard Methods 2013. The Institute of Petroleum, London, United Kingdom.

Gary, J.G., Handwerk, G.E., and Kaiser, M.J. 2007. Petroleum Refining: Technology and Economics, 5th Edition. CRC Press, Taylor & Francis Group, Boca Raton, Florida.

Hsu, C.S. and Robinson, P.R. (Editors). 2006. Practical Advances in Petroleum Processing, Volume 1 and Volume 2. Springer Science, New York.

IP 2 (ASTM D611). Petroleum Products and Hydrocarbon Solvents – Determination of Aniline and Mixed Aniline Point. IP Standard Methods 2013. The Energy Institute, United Kingdom.

IP 12. Determination of Specific Energy. IP Standard Methods 2013. The Energy Institute, United Kingdom.

IP 17. Determination of Color – Lovibond Tintometer Method. IP Standard Methods 2013. The Energy Institute, United Kingdom.

IP 30. Detection of Mercaptans, Hydrogen Sulfide, Elemental Sulfur and Peroxides – Doctor Test Method. IP Standard Methods 2013. The Energy Institute, United Kingdom.

IP 34 (ASTM D93). Determination of Flash Point – Pensky-Martens Closed Cup Method. IP Standard Methods 2013. The Energy Institute, United Kingdom.

IP 35. Determination of Open Flash and Fire Point – Pensky-Martens Method. IP Standard Methods 2013. The Energy Institute, United Kingdom.

IP 36. Determination of Open Flash and Fire Point – Cleveland Method. IP Standard Methods 2013. The Energy Institute, United Kingdom.

IP 40 (ASTM D525). Petroleum Products – Determination of Oxidation Stability of Gasoline – Induction Period Method. IP Standard Methods 2013. The Energy Institute, United Kingdom.

IP 53 (ASTM D473). Crude Petroleum and Fuel Oils – Determination of Sediment – Extraction Method. IP Standard Methods 2013. The Energy Institute, United Kingdom.

IP 61 (ASTM D129). Determination of Sulfur – High Pressure Combustion Method. IP Standard Methods 2013. The Energy Institute, United Kingdom.

IP 69. Determination of Vapor Pressure – Reid Method. IP Standard Methods 2013. The Energy Institute, United Kingdom.

IP 71 (ASTM D445). Petroleum Products – Transparent and Opaque Liquids – Determination of Kinematic Viscosity and Calculation of Dynamic Viscosity. IP Standard Methods 2013. The Energy Institute, United Kingdom.

IP 74. Water Content of Petroleum Products – Distillation Method. IP Standard Methods 2013. The Energy Institute, United Kingdom.

IP 107 (ASTM D1266). Determination of Sulfur – Lamp Combustion Method. IP Standard Methods 2013. The Energy Institute, United Kingdom.

IP 112. Determination of Corrosiveness to Copper of Lubricating Grease – Copper Strip Method. IP Standard Methods 2013. The Energy Institute, United Kingdom.

IP 123. Petroleum Products – Determination of Distillation Characteristics at Atmospheric Pressure. IP Standard Methods 2013. The Energy Institute, United Kingdom.

IP 131. Petroleum Products – Gum Content of Light and Middle Distillates – Jet Evaporation Method. IP Standard Methods 2013. The Energy Institute, United Kingdom.

IP 135 (ASTM D665). Determination of Rust-Preventing Characteristics of Steam-Turbine Oil in the Presence of Water. IP Standard Methods 2013. The Energy Institute, United Kingdom.

IP 138 (ASTM D873). Determination of Oxidation Stability of Aviation Fuel – Potential Residue Method. IP Standard Methods 2013. The Energy Institute, United Kingdom.

IP 154. Petroleum Products – Corrosiveness to Copper – Copper Strip Test. IP Standard Methods 2013. The Energy Institute, United Kingdom.

IP 156. Petroleum Products and Related Materials – Determination of Hydrocarbon Types – Fluorescent Indicator Adsorption Method. IP Standard Methods 2013. The Energy Institute, United Kingdom.

IP 160 (ASTM D1298). Crude Petroleum and Liquid Petroleum Products – Laboratory Determination of Density – Hydrometer Method. IP Standard Methods 2013. The Energy Institute, United Kingdom.

IP 170. Petroleum Products and Other Liquids – Determination of Flash Point – Abel Closed Cup Method. IP Standard Methods 2013. The Energy Institute, United Kingdom.

IP 191. Determination of Distillation Characteristics of Natural Gasoline (Incorporated into IP 123). IP Standard Methods 2013. The Energy Institute, United Kingdom.

IP 219. Petroleum Products – Determination of Cloud Point. IP Standard Methods 2013. The Energy Institute, United Kingdom.

IP 228. Determination of Lead Content of Gasoline – X-ray Spectrometric Method. IP Standard Methods 2013. The Energy Institute, United Kingdom.

IP 235. Determination of Density of Light Hydrocarbons – Pressure Hydrometer Method. IP Standard Methods 2013. The Energy Institute, United Kingdom.

IP 236 (ASTM D2700). Motor and Aviation-Type Fuels – Determination of Knock Characteristics – Motor Method. IP Standard Methods 2013. The Energy Institute, United Kingdom.

IP 237 (ASTM D2699). Petroleum Products – Determination of Knock Characteristics of Motor Fuels – Research Method – Research Octane Number (RON). IP Standard Methods 2013. The Energy Institute, United Kingdom.

IP 243. Petroleum Products and Hydrocarbons – Determination of Sulfur Content – Wickbold Combustion Method. IP Standard Methods 2013. The Energy Institute, United Kingdom.

IP 270. Petroleum Products – Determination of Lead Content of Gasoline – Iodine Monochloride Method. IP Standard Methods 2013. The Energy Institute, United Kingdom.

IP 285. Determination of Nickel and Vanadium – Spectrophotometric Method. IP Standard Methods 2013. The Energy Institute, United Kingdom.

IP 342 (ASTM D3227). Determination of Thiol (Mercaptan) Sulfur in Light and Middle Distillate Fuels – Potentiometric Method. IP Standard Methods 2013. The Energy Institute, United Kingdom.

IP 365. Crude Petroleum and Petroleum Products – Determination of Density – Oscillating U-Tube Method. IP Standard Methods 2013. The Energy Institute, United Kingdom.

IP 373. Sulfur Content – Microcoulometry (Oxidative) Method. IP Standard Methods 2013. The Energy Institute, United Kingdom.

IP 375 (ASTM D4870). Petroleum Products – Total Sediment in Residual Fuel Oils – Part 1: Determination by Hot Filtration. IP Standard Methods 2013. The Energy Institute, United Kingdom.

IP 378 (ASTM D4625). Storage Stability at 43°C of Distillate Fuel. IP Standard Methods 2013. The Energy Institute, United Kingdom.

IP 394. Determination of Air Saturated Vapor Pressure (ASVP). IP Standard Methods 2013. The Energy Institute, United Kingdom.

IP 411. Liquefied Petroleum Gases – Corrosiveness to Copper – Copper Strip Test. IP Standard Methods 2013. The Energy Institute, United Kingdom.

IP 428. Determination of the Oil Content of Effluent Water – Extraction and Infrared Spectrometric Method. IP Standard Methods 2013. The Energy Institute, United Kingdom.

IP 436 (ASTM D6379). Aromatic Hydrocarbon Types in Aviation Fuels and Petroleum Distillates – High Performance Liquid Chromatography Method with Refractive Index Detection. IP Standard Methods 2013. The Energy Institute, United Kingdom.

IP 523. Determination of Flash Point – Rapid Equilibrium Closed Cup Method. IP Standard Methods 2013. The Energy Institute, United Kingdom.

Speight, J.G. 2001. Handbook of Petroleum Analysis. John Wiley & Sons Inc., New York.

Speight, J.G. 2014. The Chemistry and Technology of Petroleum, 5th Edition. CRC Press, Taylor & Francis Group, Boca Raton, Florida.

Speight, J.G. and Ozum, B. 2002. Petroleum Refining Processes. Marcel Dekker Inc., New York.

7

AVIATION AND MARINE FUELS

7.1 INTRODUCTION

The term *aviation fuel*, as used in this text, is a collective term that includes *aviation gasoline* and *aviation gas turbine fuel* as well as various types of *jet fuel* (Wolveridge, 2000; Speight and Ozum, 2002; Hsu and Robinson, 2006; Gary *et al.*, 2007; Speight 2014). Aviation fuels consist of hydrocarbons and sulfur-containing as well as oxygen-containing impurities that are limited strictly by specification. Composition specifications usually state that aviation fuel must consist entirely of hydrocarbons except for trace amounts of *approved* additives.

The two basic types of jet fuels that are in general use are based on kerosene (kerosene-type jet fuel) and gasoline (naphtha) (gasoline-type jet fuel). The kerosene-type jet fuel is a modified development of the illuminating kerosene originally used in gas–turbine engines (ASTM D2880). The gasoline-type jet fuel has a wider boiling range and includes some gasoline fractions. In addition, a number of specialized fuel grades are required for military use in high-performance military aircraft.

Kerosene-type jet fuel is medium distillate used for aviation turbine power units and usually has the same distillation characteristics and flash point as kerosene (between 150 and 300°C but not generally above 250°C). In addition, this fuel has particular specifications (such as freezing point) which are established by the International Air Transport Association (IATA). On the other hand, aircraft gas turbine engines require a fuel with different properties from those required for aviation gasoline (ASTM D1655, IP 528). The major difference is that aircraft turbine engines require a fuel with good combustion characteristics and high energy content. However, as engine and fuel system designs have become more complicated, the fuel specifications have become more varied and restrictive.

The first aviation gas turbine engines were regarded as having non-critical fuel requirements. Ordinary illuminating kerosene was the original development fuel but the increased complexity in design of the engine has required fuel specification tests to be more complicated and numerous. Demands for improved performance, economy, and overhaul life will indirectly continue the trend toward additional tests.

7.2 PRODUCTION AND PROPERTIES

7.2.1 Aviation Fuels

Aviation fuels have a narrower distillation range than motor gasoline, and the octane ratings of aviation gasoline and motor gasoline are not comparable due to the different test methods used to rate the two types of fuels. In addition, motor gasoline has a shorter storage stability lifetime than aviation gasoline and can form gum deposits that can induce poor mixture distribution and valve sticking.

Furthermore, the higher aromatics content and the possible presence of oxygenates in motor gasoline can induce solvent characteristics that are unsuitable for seals, gaskets, fuel lines, and some fuel tank materials in air craft. Motor gasoline may also contain additives that could be incompatible with certain in-service aviation turbine fuel (ASTM D4054, ASTM D4307). For example, alcohols or other

Handbook of Petroleum Product Analysis, Second Edition. James G. Speight.
© 2015 John Wiley & Sons, Inc. Published 2015 by John Wiley & Sons, Inc.

oxygenates can increase the tendency for the fuel to hold water either in solution or in suspension.

The manufacture of *aviation gasoline*, for aviation piston engines, is produced from petroleum distillation fractions containing lower boiling hydrocarbons that are usually found in straight-run naphtha. These fractions have high contents of *iso*-pentanes and *iso*-hexanes and provide needed volatility, as well as high octane numbers (ON). Higher boiling *iso*-paraffins are provided by aviation alkylate, which consists mostly of branched octanes. Aromatics, such as benzene, toluene, and xylene, are obtained from processes such as catalytic reforming.

To increase the proportion of higher boiling octane components, such as aviation alkylate and xylenes, the proportion of lower boiling components must also be increased to maintain the proper volatility. *Iso*-pentane and, to some extent, *iso*-hexanes are the lower boiling components used and can be separated from naphtha by superfractionators or synthesized from the normal hydrocarbons by isomerization. In general, most aviation gasoline is made by blending a selected straight-run naphtha fraction (aviation base stock) with *iso*-pentane and aviation alkylate.

Aviation gasoline has an ON suited to the engine, a freezing point of −60°C, and a distillation range usually within the limits of 30°C (86°F) and 180°C (356°F). Aviation gasoline specifications generally contain three main sections covering suitability, composition, and chemical and physical requirements. In addition *gasoline-type jet fuel* includes all light hydrocarbon fractions for use in aviation turbine power units and distills between 100°C (212°F) and 250°C (482°F). It is obtained by blending kerosene and gasoline or naphtha in such a way that the aromatic content does not exceed 25% v/v and the vapor pressure is between 13.7 kPa (2 psi) and 20.6 kPa (3 psi).

Aviation turbine fuels are manufactured predominantly from straight-run kerosene or kerosene/naphtha blends in the case of wide-cut fuel that are produced from the atmospheric distillation of crude oil. Straight-run kerosene from low-sulfur (sweet) crude oil will meet all the requirements of the jet fuel specification without further refinery processing, but for the majority of feedstocks, the kerosene fraction will contain trace constituents that have to be removed by hydrotreating (hydrofining) or by a chemical sweetening process (Speight, 2000).

Traditionally, *jet fuel* has been manufactured only from straight-run components, but in recent years, however, hydrocracking processes (Speight and Ozum, 2002, Hsu and Robinson, 2006; Gary *et al.*, 2007; Speight, 2014) have been introduced that produce high-quality kerosene fractions ideal for jet fuel blending. As aviation activities are of international standard, the technical requirements of all the western specifications are virtually identical, and only a minor difference exists between various specifications (ASTM D910).

7.2.2 Marine Fuels

The residual fuel oil that remains after the distillation of crude oil is mainly used as fuel for the larger two stroke engines. These large two stroke engines are mainly used in marine vessels and, to a lesser extent, power plants; therefore residual fuel oil is also known as marine or bunker fuel.

Marine fuel quality can significantly affect the performance, operation, and maintenance of the diesel engine. Whereas a marine diesel engine is a very efficient power plant, it does have a higher degree of sensitivity to specific properties of the fuel properties and contaminants than does the steam boiler. As is the case with most fuels, additives may also be included in marine fuels to improve performance, which is usually achieved by preventing or suppressing undesirable fuel behavior. The effectiveness of additives is due to their chemical nature and the resulting interactions with constituents of the fuel (usually trace constituents) and the additives must not produce adverse side effects on fuel performance in an engine or on its behavior in marine engines (possibly by interfering with the action of other additives present).

7.3 TEST METHODS

The requirements for jet fuels stress a different combination of properties and tests than those required for aviation gasoline (ASTM D1655, IP 528). The same basic controls are needed for such properties as storage stability and corrosivity, but the gasoline antiknock tests are replaced by tests directly and indirectly controlling energy content and combustion characteristics. However, as with other petroleum products, application of sampling protocols (ASTM D3700, ASTM D4057, ASTM D4177, ASTM D4306, ASTM D5842) are of prime importance.

7.3.1 Acidity

Acidity is a property usually found in lubricating oil (ASTM D664, ASTM D974, ASTM D3339, ASTM D5770, IP 139, IP 177, IP 431); acidic compounds can also be present in aviation turbine fuels due either to the acid treatment during the refining process or to naturally occurring organic acids. Acidity is an undesirable property because of the possibility of metal corrosion and impairment of water separation characteristics of the fuel.

In the test method for the determination of the acidity in an aviation turbine fuel (ASTM D3242, IP 354), a sample is dissolved in a solvent mixture (toluene plus isopropyl alcohol and a small amount of water) and under a stream of nitrogen is titrated with standard alcoholic potassium hydroxide; on addition of indicator *p*-naptholbenzein solution, a change in color is observed from orange in acid to green in base.

7.3.2 Additives

Additives may be included in aviation fuels for a number of reasons; while their purpose is generally to improve certain aspects of fuel performance, they usually achieve the desired effect by preventing or suppressing some undesirable fuel behavior, such as corrosion, icing, oxidation, and detonation. The effectiveness of additives is due to their chemical nature and the resulting interactions with constituents of the fuel (usually trace constituents). It is important when approving additives not only to establish that they achieve the desired results and are fully compatible with all materials likely to be contacted, but to ensure also that they do not react in other ways to produce adverse side effects on fuel performance in an engine or on its behavior in aircraft or ground handling systems (possibly by interfering with the action of other additives present).

The various approved additives for jet fuels include oxidation inhibitors to improve storage stability, copper deactivators to neutralize the known adverse effect of copper on fuel stability, and corrosion inhibitors intended for the protection of storage tanks and pipelines. An anti-icing additive (fuel system icing inhibitor [FSII]) is called for in many military fuels and a static dissipator additive (antistatic additive) may be required to minimize fire and explosion risks due to electrostatic discharges in installations and equipment during pumping operations. Details of the various approved additives (mandatory or optional) are included in the individual specifications and, moreover, the additives must be compatible with the fuel (ASTM D4054).

Only a limited number of additives are permitted in aviation fuels and for each fuel grade the type and concentration are closely controlled by the appropriate fuel specifications. Additives may be included for a variety of reasons, but in every case, the specifications define the requirements as follows:

1. Mandatory: must be present between minimum and maximum limits.
2. Permitted: may be added up to a maximum limit.
3. Optional: may be added only within specified limits.
4. Not allowed: additives not listed in the specifications.

While the type and amount of each additive permitted in aviation fuels are strictly limited to color dye, antioxidant, metal deactivator, corrosion inhibitor, FSII, static dissipator, and lubricity, additive test methods for checking the concentration present are not specified in every case. In some cases, tests to determine the additive content (or its effect) are called for, but in other cases, a written statement of its original addition (e.g., at the refinery) is accepted as adequate evidence of its presence.

After the specified amounts of color dyes have been added to aviation gasoline, the color is normally checked only visually (by inspection), although the Saybolt method is recommended (ASTM D156). In the past jet fuel, color has been checked by Lovibond method (IP 17), the method currently used to check the color of dyed aviation gasoline, although color *by inspection* might, but not always, be considered adequate.

After the required amounts of antioxidant, metal deactivator, or corrosion inhibitor have been added to aviation fuels, it is not normal to carry out any checks on the concentrations and no test methods are included in specifications for this purpose. Occasionally a need arises to determine the amount of corrosion inhibitor remaining in a fuel and several analytical methods have been developed, none of which has yet been standardized.

FSII used in jet fuels can be lost by evaporation and is also lost rapidly into any water that may contact the fuel during transportation. Routine checks have therefore to be made on the icing inhibitor content of the fuel, right up to the point of delivery to aircraft in some instances, but for routine test purposes, a simpler colorimetric version of this test is commonly used.

Many fuel specifications require the use of static dissipator additive to improve safety in fuel handling. In such cases, the specification defines both minimum and maximum electrical conductivity; the minimum level ensures adequate charge relaxation while the maximum prevents too high a conductivity, since this can upset some capacitance-type *fuel* gauges in aircraft (ASTM D2624, ASTM D4308, IP 274). The standard test methods (ASTM D2624, IP 274) employ an immersible conductivity cell and field meter intended for measuring the conductivity of fuel in storage tanks.

As a valuable step toward rationalizing the approval procedure for aviation fuel additives, guidelines are available (ASTM D1655, ASTM D4054, IP 528). Tests are available for measuring or specifying additives such as color dyes (ASTM D156, ASTM D2392, ASTM D5386, IP 17), corrosion inhibitors (often measured by the corrosivity of the fuel—ASTM D130, ASTM D5968, IP 154), lubricity (ASTM D5001), FSIIs (ASTM D910, ASTM D4171, ASTM D5006), and static dissipator additives (ASTM D2624, ASTM D4865, IP 274).

7.3.3 Calorific Value

The calorific value (heat of combustion; ASTM D240, ASTM D1405, ASTM D3338, ASTM D4529, ASTM D4809, IP 12) is a direct measure of fuel energy content and is determined as the quantity of heat liberated by the combustion of a unit quantity of fuel with oxygen in a standard bomb calorimeter. This fuel property affects the economics of engine performance, and the specified minimum value is a compromise between the conflicting requirements of maximum fuel availability and good fuel consumption characteristics.

As a general guideline, the heat of combustion of petroleum is on the order of 18,000–21,000 Btu/lb (10,000–11,600 cal/g), gasoline is on the order of 20,000–20,700 Btu/lb (11,100–11,500 cal/g), kerosene and similar fuels are on the order of 19,000–20,200 Btu/lb (10,500–11,200 cal/g), and fuel oil is on the order of 17,300–20,200 Btu/lb (9,600–11,200 cal/g).

When an experimental determination of heat of combustion is not available and cannot be made conveniently, an estimate might be considered satisfactory. In this test method (ASTM D3338), the net heat of combustion is calculated from the density, sulfur, and hydrogen content, but this calculation is justifiable only when the fuel belongs to a well-defined class for which a relationship between these quantities has been derived from accurate experimental measurements on representative samples. Thus, the hydrogen content (ASTM D1217, ASTM D1298, ASTM D3701, ASTM D4052, ASTM D4808, ASTM D5291, IP 160, IP 365), density (ASTM D129, ASTM D1250, ASTM D1266, ASTM D2622, ASTM D3120, IP 61, IP 107), and sulfur content (ASTM D2622, ASTM D3120, ASTM D3246, ASTM D4294, ASTM D5453, ASTM D5623, IP 336, IP 373) of the sample are determined by experimental test methods, and the net heat of combustion is calculated using the values obtained by these test methods based on reported correlations.

A simple equation for calculating the heat of combustion is:

$$Q = 12,400 - 2,100d^2$$

where Q is the heat of combustion and d is the specific gravity. However, the accuracy of any method used to calculate such a property is not guaranteed and can only be used as a guide or approximation to the measured value.

An alternative criterion of energy content is the *aniline gravity product* (AGP) that is related to calorific value (ASTM D1405). The *AGP* is the product of the API gravity (ASTM D287, ASTM D1298) and the aniline point of the fuel (ASTM D611, IP 2). The aniline point is the lowest temperature at which the fuel is miscible with an equal volume of aniline and is inversely proportional to the aromatic content. The relationship between the AGP and calorific value is given in method. In another method (ASTM D3338), the heat of combustion is calculated from the fuel density, the 10, 50, and 90% distillation temperatures, and the aromatic content. However, neither method is legally acceptable, and other methods such as ASTM D240, ASTM D1655, and ASTM D4809 are preferred.

Jet fuels of the same class can vary widely in their burning quality as measured by carbon deposition, smoke formation, and flame radiation. This is a function of hydrocarbon composition—paraffins have excellent burning properties in contrast to those of the aromatics (particularly the polynuclear aromatic hydrocarbons). As a control measure, the smoke point test (ASTM D1322, IP 57) gives the maximum smokeless flame height in millimeters at which the fuel will burn in a wick-fed lamp under prescribed conditions. The combustion performance of wide-cut fuels correlates well with smoke point when a fuel volatility factor is included, since carbon formation tends to increase with boiling point. A minimum smoke volatility index (SVI) value is specified and is defined as:

$$SVI = \text{smoke point} + 0.42$$
$$\left(\text{percent distilled below } 204°C : 400°F\right)$$

However, the smoke point test is not always a reliable criterion of combustion performance and should be used in conjunction with other properties. Various alternative laboratory test methods have previously been specified such as the lamp burning test (ASTM D187, IP 10) and a limit on the polynuclear aromatic content (ASTM D1840), as well as by the luminometer number (LN) (ASTM D1740). The test for the LN was developed because certain designs of jet engine have the potential for a shortened combustion chamber life due to high liner temperatures caused by radiant heat from luminous flames. The test apparatus is a smoke point lamp modified to include a photoelectric cell for flame radiation measurement and a thermocouple to measure temperature rise across the flame. The fuel LN is expressed on an arbitrary scale on which values of 0–100 are given to reference fuels tetralin and isooctane, respectively. However, the text was withdrawn in 2006 but may still be in use in some laboratories.

7.3.4 Composition

The first level of compositional information is group-type totals as deduced by adsorption chromatography (ASTM D1319, IP 156). This method is applied to data related to the volume percent saturates, olefins, and aromatics in materials that boil below 315°C (600°F). This temperature range includes jet fuel (but not all diesel fuel) most of which have an end point above 315°C.

Aviation gasoline consists substantially of hydrocarbons (Table 7.1). Sulfur-containing and oxygen-containing impurities are strictly limited by specification and only certain additives are permitted. Straight-run gasoline from crude oil, containing varying proportions of paraffins, naphthenes, and aromatics, invariably lacks the high proportion of branch-chain paraffins (iso-paraffins) required to produce the higher quality aviation fuels.

Unsaturated hydrocarbons (olefins) are relatively unstable and give rise to excessive gum formation. Only the lower grades of fuel include a proportion of straight-run gasoline and the higher grades consist mainly of iso-paraffins with a small amount of aromatic material to improve the rich

TABLE 7.1 General summary of product types and distillation range

Product	Lower carbon limit	Upper carbon limit	Lower boiling point (°C)	Upper boiling point (°C)	Lower boiling point (°F)	Upper boiling point (°F)
Refinery gas	C_1	C_4	−161	−1	−259	31
Liquefied petroleum gas	C_3	C_4	−42	−1	−44	31
Naphtha	C_5	C_{17}	36	302	97	575
Gasoline	C_4	C_{12}	−1	216	31	421
Kerosene/diesel fuel	C_8	C_{18}	126	258	302	575
Aviation turbine fuel	C_8	C_{16}	126	287	302	548
Fuel oil	C_{12}	$>C_{20}$	216	421	>343	>649
Lubricating oil	$>C_{20}$		>343		>649	
Wax	C_{17}	$>C_{20}$	302	>343	575	>649
Asphalt	$>C_{20}$		>343		>649	
Coke	$>C_{50}$[a]		>1000[a]		>1832[a]	

[a]Carbon number and boiling point difficult to assess; inserted for illustrative purposes only.

mixture antiknock performance. The main component of these high-grade fuels is *iso*-octane produced in the alkylation process by reaction of refinery butenes with iso-butane over the acid catalysts. In order to meet the volatility requirements of the final blend, a small proportion of *iso*-pentane is added, obtained by superfractionation of light straight-run gasoline. The aromatic component required to improve rich mixture rating is now usually a catalytic reformate and the amount added is indirectly limited by the gravimetric calorific value requirement.

Jet fuels consist entirely of hydrocarbons except for trace quantities of sulfur compounds and approved additives. Jet fuels are produced, for example, by blending straight run distillate components, and olefins are limited by specification (ASTM D1319, IP 156) or by the bromine number (ASTM D1159, ASTM D2710, IP 130).

The bromine number is the grams of bromine that will react with 100g of the sample under the test conditions. The magnitude of bromine number is an indication of the quantity of bromine-reactive constituents and is not an identification of constituents. It is used as a measure of aliphatic unsaturation in petroleum samples, and as percentage of olefins in petroleum distillates boiling up to approximately 315°C (600°F). In the test, a known weight of the sample dissolved in a specified solvent maintained at 0–5°C (32–41°F) is titrated with standard bromide–bromate solution. Determination of the end point is method dependent.

Because the aromatic hydrocarbon content of aviation turbine fuels affects their combustion characteristics and smoke-forming tendencies, the amounts of aromatics (ASTM D1319, IP 156) are limited. Aromatic constituents also increase the intensity of the combustion flame, which can have an adverse effect on the in-service life of the combustion chamber.

The aromatics content of aviation turbine fuel is included in the aviation turbine fuel specification (ASTM D1655, IP 528). Another test method for aromatics content

(ASTM D5186) involves the injection of a small aliquot of the fuel sample onto a packed silica adsorption column and eluted using supercritical carbon dioxide as the mobile phase. Mono- and polynuclear aromatics in the sample are separated from non-aromatics and detected using a flame ionization detector. The chromatographic areas corresponding to the mono-, polynuclear, and non-aromatic components are determined and the mass percent content of each of these groups is calculated by area normalization. The results obtained by this method are at least statistically more precise than those obtained by other test methods (ASTM D1319, ASTM D2425).

Although the boiling range of aviation gasoline will differ from that of automobile gasoline, many of the tests designated for automotive gasoline (Chapter 6) can also be applied to the determination of the aromatic constituents of aviation gasoline (ASTM D86, ASTM D1319, ASTM D5443, ASTM D5580, ASTM D5769, ASTM D5986, IP 123; IP 156).

The percentage of aromatic hydrogen atoms and aromatic carbon atoms can be determined by low-resolution magnetic resonance spectroscopy (ASTM D3701, ASTM D4808) and by high-resolution nuclear magnetic resonance spectroscopy (ASTM D5292). The data produced by magnetic resonance spectroscopic methods are not equivalent to mass- or volume-percent aromatics determined by the chromatographic methods since these methods determine the mass- or volume percentage of molecules that have one or more aromatic rings. Chromatographic methods can also include alkyl side chains (on aromatic rings) within the aromatics fraction. Naphthalene content is an important quality parameter of jet fuel and is determined by ultraviolet spectrophotometry (ASTM D1840).

As with other fuels, heteroatoms, mainly sulfur and nitrogen compounds, cannot be ignored, and well established methods are available for determining the concentration of these elements. The combination of gas chromatography with element selective detection gives information about the distribution of the element. In addition, many individual heteroatomic compounds can be determined.

The principal non-hydrocarbon components are sulfur compounds that vary with the source of the crude oil. The sulfur content of a feedstock or fuel is determined by burning a sample of the fuel and determining the amount of sulfur oxides that are formed (ASTM D126, IP 107). Generally, current desulfurization technologies (Speight, 2001, 2014) are capable of reducing sulfur to the desired levels (ASTM D1266, ASTM D1552, ASTM D2622, ASTM D4294, IP 107). High levels of sulfur compounds adversely affect the fuel performance in the combustion chamber, and the presence of large amounts of oxides of sulfur in the combustion gases is undesirable because of possible corrosion.

Some sulfur compounds can also have a corroding action on the various metals of the engine system, varying according to the chemical type of sulfur compound present. Fuel corrosivity is assessed by its action on copper and is controlled by the copper strip test (ASTM D130, IP 154) which specifies that not more than a slight stain shall be observed when the polished strip is immersed in fuel heated for 2h in a bomb at 100°C (2l2°F). This particular method is not always capable of reflecting fuel corrosivity toward other fuel system metals. In addition, the mercaptan sulfur content (ASTM D3227, IP 342) of jet fuels is limited because of objectionable odor, adverse effect on certain fuel system elastomers, and corrosiveness toward fuel system metals. As an alternative to determining the mercaptan content, a negative result by the Doctor Test (ASTM D4952, IP 30) is usually acceptable.

Oxygenated constituents present as acidic compounds such as phenols and naphthenic acids are controlled in different specifications by a variety of acidity tests. The total acidity (ASTM D974, IP 139) is still widely used but has been found to be insufficiently sensitive to detect trace acidic materials that can adversely affect the water separating properties of fuel. Oxygen-containing impurities in the form of gum are limited by the existent gum method (ASTM D381, IP 131) and potential gum method (ASTM D873, IP 138). With respect to aviation turbine fuels, large quantities of gum are indicative of contamination of fuel by higher boiling oils (ASTM D86, IP 123) or by particulate matter (ASTM D2276, ASTM D5452, ASTM D6217, IP 216, IP 415). In the existent gum test for aviation fuel, measured quantity of fuel is evaporated under controlled conditions of temperature and flow of air or steam. The residue is weighed and reported.

Control of dirt and other particles involved is by means of membrane filtration method (ASTM D2276, IP 216) in which the dirt retained by filtration of a sample through a cellulose membrane is expressed as weight per unit volume of the fuel. This test provides field quality control of dirt content and can be supplemented by a visual assessment of membrane appearance of the color. However, no direct relationship exists between particulate content weight and membrane color, and

field experience is required to assess the results by either method. Nevertheless, jet fuel is tested for being *clear and bright* by visual examination of a sample (ASTM D4176).

Another contamination problem is that of microbiological growth activity that can give rise to various types of service troubles. This problem can generally be avoided by the adoption of good housekeeping techniques by all concerned, but major incidents in recent years have led to the development of several microbiological monitoring tests for aviation fuel. In one of these tests, fuel is filtered through a sterile membrane that is subsequently cultured for microbiological growths; other tests employ various techniques to detect the presence of viable microbiological matter but none of the tests have yet been standardized.

Correlative methods are also available for application to aviation fuels. Such methods include the use of viscosity-temperature charts (ASTM D341), calculation of the cetane index (ASTM D976, ASTM D4737), calculation of the viscosity index (ASTM D2270), calculation of the viscosity gravity constant (ASTM D2501), calculation of the true vapor pressure (ASTM D2889), and estimation of the heat of combustion (ASTM D3338).

7.3.5 Density

The density (specific gravity) of a fuel is a measure of the mass per unit volume and can be determined directly using calibrated glass hydrometers (Chapters 2, 5, and 6).

Density (specific gravity) (ASTM D1298, IP 160) is an important property of aviation fuel as it is an indicator of the total energy content of a fuel uplift on a weight and/or volume basis. Variation in density is controlled within broad limits to ensure engine control. Both fuel specific gravity and calorific value vary somewhat according to crude source, paraffinic fuels having a slightly lower specific gravity but higher gravimetric calorific value than those from naphthenic crude oils.

Density is used in fuel load calculations, since weight or volume fuel limitations (or both) may be necessary according to the type of aircraft and flight pattern involved. In most cases, the volume of fuel that can be carried is limited by tank capacity, and to achieve maximum range, a high-density fuel is preferred, as this will provide the greatest heating value per gallon (liter) of fuel. The calorific (heating) value per unit weight kilogram of fuel decreases with increasing density, and when the weight of fuel that can be carried is limited, it may be advantageous to use a lower density fuel, provided adequate tank capacity is available.

In the United States, it is more common to specify fuel density in terms of the API Gravity (ASTM D287):

$$API\ gravity, degrees = \left[141.5 / \left(specific\ gravity\ 60° / 60°F \right) \right] - 131.5$$

Aviation fuel might be expected to have an API gravity in the range 57–35 (specific gravity: 0.75–0.85, respectively).

The importance of density (specific gravity) relative to diesel engine operation lies in the fact that the standard fuel/water separating techniques are based upon the difference in density between the two substances. Therefore, as the specific gravity of fuel approaches 1.0, centrifuging becomes less effective and, since diesel engine fuels should be free of water and the salts normally dissolved therein, extra centrifuging capacity will be required for high gravity fuel.

High specific gravity of the fuel oil indicates the presence of heavily cracked, aromatic constituents with poor combustion qualities, which can cause abnormal engine wear. Heating the fuel oil prior to centrifuging assist the separation process—the density of fuel oil changes more rapidly with temperature than does the density of water.

7.3.6 Flash Point

The flashpoint test is a guide to the fire hazard associated with the use of the fuel and can be determined by several test methods and the results are not always strictly comparable.

The minimum flash point is usually defined by the Abel method (IP 170) but except for high-flash kerosene where the Pensky–Martens method (ASTM D93, IP 34) is specified. The TAG method (ASTM D56) is used for both the minimum and maximum limits, while certain military specifications also give minimum limits by the Pensky–Martens method (ASTM D93, IP 34). The Abel method (IP 170) can give results up to 2–3°C (3–5°F) lower than the TAG method (ASTM D56).

Similarly, for jet fuel the flash point is a guide to the fire hazard associated with the fuel and can be determined by the same test methods as noted earlier (ASTM D56, ASTM D93, ASTM D3828, IP 34, IP 170, IP 303), except for high-flash kerosene where the method (ASTM D93, IP 34) is specified.

It should be noted that the various flash point methods can yield different numerical results, and in the case of the two most commonly used methods (Abel and TAG), it has been found that the former (IP 170) can give results up to 1–2°C lower than the latter method (ASTM D56). Setaflash (D3828/IP 303) results are generally very close to Abel values.

7.3.7 Freezing Point

The freezing of aviation fuel is an index of the lowest temperature of its utility for the specified applications. Aviation fuels must have acceptable freezing point and low temperature pumpability characteristics so that adequate fuel flow to the engine is maintained at high altitude and is a requirement of aviation specifications (ASTM D910, ASTM D1655, IP 528). Maximum freezing point values are specified for all types of aviation fuel as a guide to the lowest temperature at which the fuel can be used without risk of the separation of solid hydrocarbons. The solidified hydrocarbons could lead to clogging of fuel lines or fuel filters and loss in available fuel load due to retention of solidified fuel in the tanks. The freezing point of the fuel (typically in the range −40 to −65°C, −40 to −85°F) must always be lower than the minimum operational fuel temperature. The freezing point specification is retained as a specification property in order to predict and safeguard high altitude performance.

There are three test methods available for determination of the freezing point. All three methods have been found to give equivalent results. However, when a specification calls for a specific test, only that test must be used.

In this test (ASTM D2386, IP 16), a measured fuel sample is placed in a jacketed sample tube that also holds a thermometer and a stirrer. The tube is placed in a vacuum flask containing the cooling medium. Various coolants used are acetone, methyl alcohol, ethyl alcohol, or iso-propyl alcohol, solid carbon dioxide, or liquid nitrogen. As the sample cools, it is continuously stirred. The temperature at which the hydrocarbon crystals appear is recorded. The jacketed sample is removed from the coolant and allowed to warm, stirring it continuously. The temperature at which the crystals completely disappear is recorded. In another test method (IP 434), an automated optical method is used for the temperature range as to −70°C (−94°F). In this method, a 25-min portion of the fuel is placed in a test chamber that is cooled while continuously stirred and monitored by an optical system. The temperature of the specimen is measured with an electronic measuring device, and the temperatures when crystals first appear, and then on warming disappear, are recorded.

In the third method (ASTM D5972, IP 435), an automated phase-transition method is used in the temperature range −80 to 20°C (−112 to 68°F). In this test, a specimen is cooled at a rate of 15±5°C/min while continuously being illuminated by a light source. The specimen is continuously monitored by an array of optical detectors for the first formation of solid hydrocarbon crystals. After that the specimen is warmed at the rate of 10±0.5°C/min until all crystals return to the liquid phase, and that temperature is also recorded.

The cold flow test is also available but may not give an adequate safety margin for the behavior of the fuel in service.

7.3.8 Knock and Antiknock Properties

The various fuel grades are classified by their *antiknock* quality characteristics as determined in single cylinder laboratory engines. Knock, or detonation, in an engine is a form of abnormal combustion where the air/fuel charge in the cylinder ignites spontaneously in a localized area instead of being consumed progressively by the spark-initiated flame front. Such knocking combustion can damage the engine and give serious power loss if allowed to persist and the various grades are designed to guarantee knock-free operation for a range of engines from those used in light aircraft up to high powered transport and military types.

The antiknock ratings of aviation gasoline are determined in standard laboratory engines by matching their performance against reference blends of pure isooctane and *n*-heptane. Fuel rating is expressed as an ON that is defined as the percentage of isooctane in the matching reference blend. Fuels of higher performance than *iso*-octane (100 ON) are tested against blends of isooctane with various amounts of antiknock additive. The rating of such fuel is expressed as a performance number (PN), defined as the maximum knock-free power output obtained from the fuel expressed as a percentage of the power obtainable on isooctane.

The antiknock rating of fuel varies according to the air/fuel mixture strength employed and this fact is used in defining the performance requirements of the higher grade aviation fuels. As mixture strength is increased (richened), the additional fuel acts as an internal coolant and suppresses knocking combustion thus permitting a higher power rating to be obtained. Since maximum power output is the prime requirement of an engine under rich take-off conditions, the *rich mixture performance* of a fuel is determined in a special supercharged single cylinder engine (ASTM D909, IP 119); *weak mixture performance* is also determined (ASTM D2700, IP 236).

The higher grades of fuel are thus classified by their specified antiknock ratings under both sets of test conditions. For example, 100/130 grade fuel has an antiknock quality of 100 minimum by the weak mixture test procedure and 130 minimum by the rich mixture procedure. ONs are used to specify ratings of 100 and below while PNs are used above 100.

7.3.9 Pour Point

The *pour point* of a petroleum product is an index of the lowest temperature at which the product will flow under specified conditions. Pour point data can be used to supplement other measurements of cold flow behavior (such as the freezing point).

In the original (and still widely used) test for pour point (ASTM D97, IP 15), a sample is cooled at a specified rate and examined at intervals of 3°C (5.4°F) for flow characteristics. The lowest temperature at which the movement of the oil is observed is recorded as the pour point. A later test method (ASTM D5853) covers two procedures for the determination of the pour point of crude oils down to −36°C. One method provides a measure of the maximum (upper) pour point temperature. The second method measures the minimum (lower) pour point temperature. In these methods, the test specimen is cooled (after preliminary heating) at a specified rate and examined at intervals of 3°C (5.4°F) for flow characteristics. Again, the lowest temperature at which movement of the test specimen is observed is recorded as the pour point.

In any determination of the pour point, a petroleum product that contains wax produces an irregular flow behavior when the wax begins to separate. This type of product petroleum possesses viscosity relationships that are difficult to predict in operating conditions. This complex behavior may limit the value of pour point data but, with use of the freezing point test method (ASTM D2386, ASTM D5972, IP 16, IP 434, IP 435), the data give an estimate of minimum handling temperature and minimum line or storage temperature.

7.3.10 Storage Stability

Aviation fuel must retain its required properties for long periods of storage in all kinds of climates. Unstable fuels oxidize and form oxidation products that remain as a resinous solid or *gum* on induction manifolds, carburetors, and valves as the fuel is evaporated. Hence, there is a limitation for olefins in the fuel; they are extremely reactive and form resinous products readily. Thus, formation of this undesirable gum is strictly limited and is assessed by the existent and accelerated (or potential) gum tests.

The existent gum value (ASTM D381, IP 131) is the gum actually present in the fuel at the time of test and is measured as the weight of residue obtained after controlled evaporation of a standard volume of fuel. The accelerated gum test (ASTM D873, IP 138) is a safeguard of storage stability and predicts the possibility of gum forming during protracted storage and decomposition of the antiknock additive. In this test, the fuel is heated for 16h with oxygen under pressure in a bomb at 100°C (212°F) and then both the gum content and amount of precipitate are measured.

Another test used for determining the extent of oxidation of aviation fuels is the determination of the hydroperoxide number (ASTM D6447) and the peroxide number (ASTM D3703). Deterioration of aviation fuel results in the formation of the peroxides as well as other oxygen-containing compounds and these numbers are indications of the quantity of oxidizing constituents present in the sample as determined by the measurement of the compounds that will oxidize potassium iodide.

The determination of hydroperoxide number is significant because of the adverse effect of hydroperoxides upon certain elastomers in the fuel systems. The method (ASTM D6447) measures the same peroxide species, primarily the hydroperoxides in aviation fuels. This test method does not use the ozone depleting substance 1,1,2-trichloro-1,2,2-trifluoroethane (ASTM D3703) and is applicable to any water-insoluble, organic fluid, particularly diesel fuels, gasoline, and kerosene. In this method, a quantity of sample is contacted with aqueous potassium iodide (KI) solution in the presence of an acid.

The hydroperoxides present are reduced by potassium iodide liberating an equivalent amount of iodine that is quantified by voltametric analysis.

The determination of peroxide number of aviation turbine fuel is important because of the adverse effects of peroxides upon certain elastomers in the fuel system. In the test, the sample is dissolved (unlike ASTM D6447) in 1,1,2-tri-chloro-l,2,2-trifluoroethane and is contacted within an aqueous potassium iodide solution. The peroxides present are reduced by the potassium iodide whereupon an equivalent amount of iodine is released that is titrated with standard sodium thiosulfate solution using a starch indicator.

Other tests for storage stability include determination of color formation and sediment (ASTM D4625, ASTM D5304) in which reactivity to oxygen at high temperatures is determined by the amount of sediment formation as well as any color changes.

7.3.11   Thermal Stability

Although the conventional (storage) stability of aviation fuel has long been defined and controlled by the existent and accelerated gum tests, another test is required to measure the stability of a fuel to the thermal stresses which can arise during sustained supersonic flight and in some high-subsonic applications.

In high-speed flight, the fuel is subjected to considerable heat input due to kinetic heating of the airframe and also to the use of the bulk fuel as a coolant for engine oil, hydraulic, and air conditioning equipment, etc. Consequently, fuel for supersonic flight must perform satisfactorily at temperatures up to about 250°C (480°F) without formation of lacquer and deposits that can adversely affect the efficiency of heat exchangers, metering devices, fuel filters, and injector nozzles. The initial problem was that of reduced overhaul life in military engines due to the high temperatures in the fuel system upstream of the injector nozzles giving rise to deposit formation.

Hence the application of a test is to determine the coking index, which is a measure of the tendency of jet fuels to deposit thermal decomposition products in fuel systems. In this test, fuel is pumped through a preheater tube assembly representing fuel/oil heat exchange systems and then through a sintered stainless steel filter that represents nozzles and fine orifices where fuel degradation products could become trapped. Fuel degradation is determined by pressure drop across the filter as well as by visual preheater tube deposit condition and is rated numerically using various color standard tests (ASTM D156, ASTM D848, ASTM D1209, ASTM D1555, ASTM D5386, IP 17).

The fuel coker test suffered from precision problems and has been largely replaced by a test for the thermal oxidation stability of the fuel (ASTM D3241, IP 323) which overcomes the disadvantages of the fuel coker test in fuel specifications.

7.3.12   Viscosity

Heavy fuel oil (such as marine fuel oil) is typically purchased on the basis of a limiting viscosity due to storage, handling, or engine-related restrictions. However, viscosity data do not indicate that fuel quality carry a quality implication, regardless of the fact that many purchasers of marine fuel oils believe this to be the case. Fuel oils are being produced by more intensive secondary processing and the relationship between fuel oil viscosity and fuel oil quality becomes less and less meaningful. Generally, viscosity is still used as a yardstick (often inadequate) which is used to judge the quality of heavy fuel oil (residual fuel oil).

However, the flow properties of the fuel oil can have an important influence on fuel pump service life. Thus, the viscosity of fuel oil at low temperature (ASTM D445, IP 71) is limited to ensure that adequate fuel flow and pressure are maintained under all operating conditions and that fuel injection nozzles and system controls will operate down to design temperature conditions.

7.3.13   Volatility

Fuels must be easily convertible from storage in the liquid form to the vapor phase in the engine to allow formation of the combustible air/fuel vapor mixture. If gasoline fuel volatility were too low, liquid fuel would enter the cylinder and wash lubricating oil from the walls and pistons and hence lead to increased engine wear; a further effect would be to cause dilution of the crankcase oil; poor volatility can also give rise to poor distribution of mixture strength between cylinders. Conversely, if volatility is too high, the fuel can vaporize in the fuel tank and supply lines giving undue venting losses and the possibility of fuel starvation through *vapor lock* in the fuel lines. The cooling effect due to rapid vaporization of excessive amounts of highly volatile materials can also cause ice formation in the carburetor under certain conditions of humidity and air temperature.

One of the most important physical parameters defining these products is their boiling range distribution (ASTM D86, ASTM D1078, ASTM D2887, ASTM D2892, IP 123). However, this method is a low-efficiency, one theoretical plate distillation, and although it has been adequate for product specification purposes, true boiling point (TBP) data are also required (ASTM D2887, ASTM D2892).

In the simplest test method (ASTM D86, IP 123), a 100-ml sample is distilled (manually or automatically) under prescribed conditions. Temperatures and volumes of condensate are recorded at regular intervals from which the boiling profile is derived. Distillation points of 10, 20, 50, and 90% are specified in various ways to ensure that a properly balanced fuel is produced with no undue proportion of light or heavy fractions. The distillation end point excludes any heavy material that would give poor fuel vaporization and ultimately affect engine combustion performance.

The determination of the boiling range distribution of aviation fuel by gas chromatography (ASTM D2887, ASTM D3710) not only helps identify the constituents but also facilitates online controls at the refinery. This test method is designed to measure the entire boiling range of the fuel that has either high or low Reid vapor pressure (ASTM D323, IP 69). In the either method, the sample is injected into a gas chromatographic column that separates hydrocarbons in boiling point order. The column temperature is raised at a reproducible rate and the area under the chromatogram is recorded throughout the run. Calibration is performed using a known mixture of hydrocarbons covering the expected boiling of the sample.

Another method is described as a method for determining the carbon number distribution (ASTM D2887) and the data derived by this test method are essentially equivalent to that obtained by TBP distillation (ASTM D2892). The sample is introduced into a gas chromatographic column that separates hydrocarbons in boiling point order. The column temperature is raised at a reproducible rate and the area under the chromatogram is recorded throughout the run. Boiling temperatures are assigned to the time axis from a calibration curve, obtained under the same conditions by running a known mixture of hydrocarbons covering the boiling range expected in the sample. From these data, the boiling range distribution may be obtained. However, this test method is limited to samples having a boiling range greater than 55°C (100°F) and having a vapor pressure (ASTM D323, ASTM D4953, ASTM D5191, ASTM D5482, ASTM D6377, ASTM D6378, IP 69, IP 394) sufficiently low to permit sampling at ambient temperature.

7.3.14 Water

Because of their higher density and viscosity, jet fuels tend to retain fine particulate matter and water droplets in suspension for a much longer time than aviation gasoline.

Jet fuels can also vary considerably in their tendency to pick up and retain water droplets or to hold fine water hazes in suspension depending on the presence of trace surface-active impurities (surfactants). Some of these materials (such as sulfonic and naphthenic acids and their sodium salts) may originate from the crude source or from certain refinery treating processes, while others may be picked up by contact with other products during transportation to the airfield, particularly in multiproduct pipelines. These latter materials may be natural contaminants from other less highly refined products (e.g., burning oils) or may consist of additives from motor gasoline (such as glycol-type anti-icing agents). It should be noted that some of the additives specified for jet fuel use (e.g., corrosion inhibitors and static dissipator additive) also have surface-active properties.

The presence of surfactants can also impair the performance of the water separating equipment (filter/separators) widely used throughout fuel handling systems to remove the traces of free (undissolved) water, particularly at the later stages prior to delivery to aircraft. Very small traces of free water can adversely affect jet engine and aircraft operations in several ways, and the water retention and separating properties of jet fuels have become a critical quality consideration in recent years.

Free water in jet fuels can be detected by Karl Fischer titration method (ASTM D1744) or by chemical methods, such as by observing color changes when a chemical reacts with aqueous solution (ASTM D3240). The standard water reaction test for jet fuel (ASTM D1094, IP 289) is the same as for aviation gasoline, but the interface and separation ratings are more critically defined. Test assessment is by subjective visual observation and, while quite precise when made by an experienced operator, the test can cause rating difficulties under borderline conditions. As a consequence, a more objective test is now included in many specifications known as the water separometer test (ASTM D3948, ASTM D7224, ASTM D7261).

In this test, fuel is mechanically mixed with a small quantity of water and the resulting emulsion is passed through a miniature water coalescing pad and then through a settling chamber followed by a photoelectric device that measures the clarity of the effluent fuel. A good fuel, which has shed easily the entrained water, has a high rating or water separometer index modified (WSIM) on a numerical scale directly related to the percentage of light transmission.

There is also a *water reaction test* that is used to estimate, and prevent, the addition of high-octane, water-soluble components such as ethyl alcohol to aviation gasoline. The test method involves shaking 80 ml of fuel with 20 ml of water under standard conditions and observing phase volume changes and interface condition.

It is specified that phase volume change shall not exceed 2 ml and that the interface shall be substantially free from bubbles or scum, with sharp separation of the phases without emulsion or precipitate within or upon either layer. The long-established standard test methods for water reaction (ASTM D1094, ASTM D3948, IP 289) cover the volume change and the interface condition, and special clauses have been included in most specifications to cover the phase separation requirements.

In addition to appreciable amounts of water (ASTM D4176, ASTM D4860), sediment can also occur and will cause fouling of the fuel handling facilities and the fuel system. An accumulation of sediment in storage tanks and on filter screens can obstruct the flow of oil from the tank to the combustor. A test method is available to determine the water and sediment in fuels (ASTM D2709). In the test method, a sample of fuel is centrifuged at a rpm of 800 for 10 min at 21–32°C in a centrifuge tube readable to 0.005 ml and measurable to 0.01 ml. After centrifugation, the volume of water and sediment that has settled into the tip of the centrifuge tube is read to the nearest 0.005 ml.

REFERENCES

ASTM D56. Standard test method for flash point by tag closed cup tester. Annual Book of Standards. ASTM International, West Conshohocken, PA.

ASTM D86. Standard test method for distillation of petroleum products at atmospheric pressure. Annual Book of Standards. ASTM International, West Conshohocken, PA.

ASTM D93. Standard test methods for flash point by Pensky–Martens closed cup tester. Annual Book of Standards. ASTM International, West Conshohocken, PA.

ASTM D97. Standard test method for pour point of petroleum products. Annual Book of Standards. ASTM International, West Conshohocken, PA.

ASTM D130. Standard test method for corrosiveness to copper from petroleum products by copper strip test. Annual Book of Standards. ASTM International, West Conshohocken, PA.

ASTM D156. Standard test method for saybolt color of petroleum products (Saybolt chromometer method). Annual Book of Standards. ASTM International, West Conshohocken, PA.

ASTM D187. Standard test method for burning quality of kerosene. Annual Book of Standards. ASTM International, West Conshohocken, PA.

ASTM D240. Standard test method for heat of combustion of liquid hydrocarbon fuels by bomb calorimeter. Annual Book of Standards. ASTM International, West Conshohocken, PA.

ASTM D287. Standard test method for API gravity of crude petroleum and petroleum products (Hydrometer method). Annual Book of Standards. ASTM International, West Conshohocken, PA.

ASTM D323. Standard test method for vapor pressure of petroleum products (Reid method). Annual Book of Standards. ASTM International, West Conshohocken, PA.

ASTM D341. Standard practice for viscosity-temperature charts for liquid petroleum products. Annual Book of Standards. ASTM International, West Conshohocken, PA.

ASTM D381. Standard test method for gum content in fuels by jet evaporation. Annual Book of Standards. ASTM International, West Conshohocken, PA.

ASTM D445. Standard test method for kinematic viscosity of transparent and opaque liquids (and calculation of dynamic viscosity). Annual Book of Standards. ASTM International, West Conshohocken, PA.

ASTM D611. Standard test methods for aniline point and mixed aniline point of petroleum products and hydrocarbon solvents. Annual Book of Standards. ASTM International, West Conshohocken, PA.

ASTM D664. Standard test method for acid number of petroleum products by potentiometric titration. Annual Book of Standards. ASTM International, West Conshohocken, PA.

ASTM D848. Standard test method for acid wash color of industrial aromatic hydrocarbons. Annual Book of Standards. ASTM International, West Conshohocken, PA.

ASTM D873. Standard test method for oxidation stability of aviation fuels (Potential residue method). Annual Book of Standards. ASTM International, West Conshohocken, PA.

ASTM D909. Standard test method for supercharge rating of spark-ignition aviation gasoline. Annual Book of Standards. ASTM International, West Conshohocken, PA.

ASTM D910. Standard specification for aviation gasoline. Annual Book of Standards. ASTM International, West Conshohocken, PA.

ASTM D974. Standard test method for acid and base number by color-indicator titration. Annual Book of Standards. ASTM International, West Conshohocken, PA.

ASTM D976. Standard test method for calculated cetane index of distillate fuels. Annual Book of Standards. ASTM International, West Conshohocken, PA.

ASTM D1078. Standard test method for distillation range of volatile organic liquids. Annual Book of Standards. ASTM International, West Conshohocken, PA.

ASTM D1094. Standard test method for water reaction of aviation fuels. Annual Book of Standards. ASTM International, West Conshohocken, PA.

ASTM D1159. Standard test method for bromine numbers of petroleum distillates and commercial aliphatic olefins by electrometric titration. Annual Book of Standards. ASTM International, West Conshohocken, PA.

ASTM D1209. Standard test method for color of clear liquids (platinum-cobalt scale). Annual Book of Standards. ASTM International, West Conshohocken, PA.

ASTM D1217. Standard test method for density and relative density (specific gravity) of liquids by Bingham pycnometer. Annual Book of Standards. ASTM International, West Conshohocken, PA.

ASTM D1250. Standard guide for use of the petroleum measurement tables. Annual Book of Standards. ASTM International, West Conshohocken, PA.

ASTM D1266. Standard test method for sulfur in petroleum products (Lamp method). Annual Book of Standards. ASTM International, West Conshohocken, PA.

ASTM D1298. Standard test method for density, relative density, or API gravity of crude petroleum and liquid petroleum products by hydrometer method. Annual Book of Standards. ASTM International, West Conshohocken, PA.

ASTM D1319. Standard test method for hydrocarbon types in liquid petroleum products by fluorescent indicator adsorption. Annual Book of Standards. ASTM International, West Conshohocken, PA.

ASTM D1322. Standard test method for smoke point of kerosene and aviation turbine fuel. Annual Book of Standards. ASTM International, West Conshohocken, PA.

ASTM D1405. Standard test method for estimation of net heat of combustion of aviation fuels. Annual Book of Standards. ASTM International, West Conshohocken, PA.

ASTM D1552. Standard test method for sulfur in petroleum products (high-temperature method). Annual Book of Standards. ASTM International, West Conshohocken, PA.

ASTM D1555. Standard test method for calculation of volume and weight of industrial aromatic hydrocarbons and cyclohexane [metric]. Annual Book of Standards. ASTM International, West Conshohocken, PA.

ASTM D1655. Standard specification for aviation turbine fuels. Annual Book of Standards. ASTM International, West Conshohocken, PA.

ASTM D1740. Standard Test Method for Luminometer Numbers of Aviation Turbine Fuels (Withdrawn 2006 but still in use in some laboratories). ASTM International, West Conshohocken, PA.

ASTM D1744. Standard test method for determination of water in liquid petroleum products by Karl Fischer reagent. Annual Book of Standards. ASTM International, West Conshohocken, PA.

ASTM D1840. Standard test method for naphthalene hydrocarbons in aviation turbine fuels by ultraviolet spectrophotometry. Annual Book of Standards. ASTM International, West Conshohocken, PA.

ASTM D2270. Standard practice for calculating viscosity index from kinematic viscosity at 40 and 100°C. Annual Book of Standards. ASTM International, West Conshohocken, PA.

ASTM D2276. Standard test method for particulate contaminant in aviation fuel by line sampling. Annual Book of Standards. ASTM International, West Conshohocken, PA.

ASTM D2386. Standard test method for freezing point of aviation fuels. Annual Book of Standards. ASTM International, West Conshohocken, PA.

ASTM D2392. Standard test method for color of dyed aviation gasoline. Annual Book of Standards. ASTM International, West Conshohocken, PA.

ASTM D2425. Standard test method for hydrocarbon types in middle distillates by mass spectrometry. Annual Book of Standards. ASTM International, West Conshohocken, PA.

ASTM D2501. Standard test method for calculation of viscosity-gravity constant (VGC) of petroleum oils. Annual Book of Standards. ASTM International, West Conshohocken, PA.

ASTM D2622. Standard test method for sulfur in petroleum products by wavelength dispersive x-ray fluorescence spectrometry. Annual Book of Standards. ASTM International, West Conshohocken, PA.

ASTM D2624. Standard test methods for electrical conductivity of aviation and distillate fuels. Annual Book of Standards. ASTM International, West Conshohocken, PA.

ASTM D2700. Standard test method for motor octane number of spark-ignition engine fuel. Annual Book of Standards. ASTM International, West Conshohocken, PA.

ASTM D2709. Standard test method for water and sediment in middle distillate fuels by centrifuge. Annual Book of Standards. ASTM International, West Conshohocken, PA.

ASTM D2710. Standard test method for bromine index of petroleum hydrocarbons by electrometric titration. Annual Book of Standards. ASTM International, West Conshohocken, PA.

ASTM D2880. Standard specification for gas turbine fuel oils. Annual Book of Standards. ASTM International, West Conshohocken, PA.

ASTM D2887. Standard test method for boiling range distribution of petroleum fractions by gas chromatography. Annual Book of Standards. ASTM International, West Conshohocken, PA.

ASTM D2889. Standard test method for calculation of true vapor pressures of petroleum distillate fuels. Annual Book of Standards. ASTM International, West Conshohocken, PA.

ASTM D2892. Standard test method for distillation of crude petroleum (15-theoretical plate column). Annual Book of Standards. ASTM International, West Conshohocken, PA.

ASTM D3120. Standard test method for flash point and fire point of liquids by tag open-cup apparatus. Annual Book of Standards. ASTM International, West Conshohocken, PA.

ASTM D3227. Standard test method for (thiol mercaptan) sulfur in gasoline, kerosene, aviation turbine, and distillate fuels (potentiometric method). Annual Book of Standards. ASTM International, West Conshohocken, PA.

ASTM D3240. Standard test method for undissolved water in aviation turbine fuels. Annual Book of Standards. ASTM International, West Conshohocken, PA.

ASTM D3241. Standard test method for thermal oxidation stability of aviation turbine fuels. Annual Book of Standards. ASTM International, West Conshohocken, PA.

ASTM D3242. Standard test method for acidity in aviation turbine fuel. Annual Book of Standards. ASTM International, West Conshohocken, PA.

ASTM D3246. Standard test method for sulfur in petroleum gas by oxidative microcoulometry. Annual Book of Standards. ASTM International, West Conshohocken, PA.

ASTM D3338. Standard test method for estimation of net heat of combustion of aviation fuels. Annual Book of Standards. ASTM International, West Conshohocken, PA.

ASTM D3339. Standard test method for acid number of petroleum products by semi-micro color indicator titration. Annual Book of Standards. ASTM International, West Conshohocken, PA.

ASTM D3700. Standard practice for obtaining LPG samples using a floating piston cylinder. Annual Book of Standards. ASTM International, West Conshohocken, PA.

ASTM D3701. Standard test method for hydrogen content of aviation turbine fuels by low resolution nuclear magnetic resonance spectrometry. Annual Book of Standards. ASTM International, West Conshohocken, PA.

ASTM D3703. Standard test method for hydroperoxide number of aviation turbine fuels, gasoline and diesel fuels. Annual Book of Standards. ASTM International, West Conshohocken, PA.

ASTM D3710. Standard test method for boiling range distribution of gasoline and gasoline fractions by gas chromatography. Annual Book of Standards. ASTM International, West Conshohocken, PA.

ASTM D3828. Standard test methods for flash point by small scale closed cup tester. Annual Book of Standards. ASTM International, West Conshohocken, PA.

ASTM D3948. Standard test method for determining water separation characteristics of aviation turbine fuels by portable separometer. Annual Book of Standards. ASTM International, West Conshohocken, PA.

ASTM D4052. Standard test method for density, relative density, and API gravity of liquids by digital density meter. Annual Book of Standards. ASTM International, West Conshohocken, PA.

ASTM D4054. Standard Practice for Qualification and Approval of New Aviation Turbine Fuels and Fuel Additives. Annual Book of Standards. ASTM International, West Conshohocken, PA.

ASTM D4057. Standard practice for manual sampling of petroleum and petroleum products. Annual Book of Standards. ASTM International, West Conshohocken, PA.

ASTM D4171. Standard specification for fuel system icing inhibitors. Annual Book of Standards. ASTM International, West Conshohocken, PA.

ASTM D4176. Standard test method for free water and particulate contamination in distillate fuels (visual inspection procedures). Annual Book of Standards. ASTM International, West Conshohocken, PA.

ASTM D4177. Standard practice for automatic sampling of petroleum and petroleum products. Annual Book of Standards. ASTM International, West Conshohocken, PA.

ASTM D4294. Standard test method for sulfur in petroleum and petroleum products by energy dispersive x-ray fluorescence spectrometry. Annual Book of Standards. ASTM International, West Conshohocken, PA.

ASTM D4306. Standard practice for aviation fuel sample containers for tests affected by trace contamination. Annual Book of Standards. ASTM International, West Conshohocken, PA.

ASTM D4307. Standard practice for preparation of liquid blends for use as analytical standards. Annual Book of Standards. ASTM International, West Conshohocken, PA.

ASTM D4308. Standard test method for electrical conductivity of liquid hydrocarbons by precision meter. Annual Book of Standards. ASTM International, West Conshohocken, PA.

ASTM D4529. Standard test method for estimation of net heat of combustion of aviation fuels. Annual Book of Standards. ASTM International, West Conshohocken, PA.

ASTM D4625. Standard test method for distillate fuel storage stability at 43°C (110°F). Annual Book of Standards. ASTM International, West Conshohocken, PA.

ASTM D4737. Standard test method for calculated cetane index by four variable equation. Annual Book of Standards. ASTM International, West Conshohocken, PA.

ASTM D4808. Standard test methods for hydrogen content of light distillates, middle distillates, gas oils, and residua by low-resolution nuclear magnetic resonance spectroscopy. Annual Book of Standards. ASTM International, West Conshohocken, PA.

ASTM D4809. Standard test method for heat of combustion of liquid hydrocarbon fuels by bomb calorimeter (Precision method). Annual Book of Standards. ASTM International, West Conshohocken, PA.

ASTM D4860. Standard test method for free water and particulate contamination in middle distillate fuels (Clear and bright numerical rating). Annual Book of Standards. ASTM International, West Conshohocken, PA.

ASTM D4865. Standard guide for generation and dissipation of static electricity in petroleum fuel systems. Annual Book of Standards. ASTM International, West Conshohocken, PA.

ASTM D4952. Standard test method for qualitative analysis for active sulfur species in fuels and solvents (doctor test). Annual Book of Standards. ASTM International, West Conshohocken, PA.

ASTM D4953. Standard test method for vapor pressure of gasoline and gasoline-oxygenate blends (dry method). Annual Book of Standards. ASTM International, West Conshohocken, PA.

ASTM D5001. Standard test method for measurement of lubricity of aviation turbine fuels by the ball-on-cylinder lubricity evaluator (BOCLE). Annual Book of Standards. ASTM International, West Conshohocken, PA.

ASTM D5006. Standard test method for measurement of fuel system icing inhibitors (ether type) in aviation fuels. Annual Book of Standards. ASTM International, West Conshohocken, PA.

ASTM D5186. Standard test method for determination of aromatic content and polynuclear aromatic content of diesel fuels and aviation turbine fuels by supercritical fluid chromatography. Annual Book of Standards. ASTM International, West Conshohocken, PA.

ASTM D5191. Standard test method for vapor pressure of petroleum products (Mini method). Annual Book of Standards. ASTM International, West Conshohocken, PA.

ASTM D5291. Standard test methods for instrumental determination of carbon, hydrogen, and nitrogen in petroleum products and lubricants. Annual Book of Standards. ASTM International, West Conshohocken, PA.

ASTM D5292. Standard test method for aromatic carbon contents of hydrocarbon oils by high resolution nuclear magnetic resonance spectroscopy. Annual Book of Standards. ASTM International, West Conshohocken, PA.

ASTM D5304. Standard test method for assessing middle distillate fuel storage stability by oxygen overpressure. Annual Book of Standards. ASTM International, West Conshohocken, PA.

ASTM D5386. Standard test method for color of liquids using tristimulus colorimetry. Annual Book of Standards. ASTM International, West Conshohocken, PA.

ASTM D5443. Standard test method for paraffin, naphthene, and aromatic hydrocarbon type analysis in petroleum distillates through 200°C by multi-dimensional gas chromatography. Annual Book of Standards. ASTM International, West Conshohocken, PA.

ASTM D5452. Standard test method for particulate contamination in aviation fuels by laboratory filtration. Annual Book of Standards. ASTM International, West Conshohocken, PA.

ASTM D5453. Standard test method for determination of total sulfur in light hydrocarbons, spark ignition engine fuel, diesel engine fuel, and engine oil by ultraviolet fluorescence. Annual Book of Standards. ASTM International, West Conshohocken, PA.

ASTM D5482. Standard test method for vapor pressure of petroleum products (Mini method – Atmospheric). Annual Book of Standards. ASTM International, West Conshohocken, PA.

ASTM D5580. Standard test method for determination of benzene, toluene, ethylbenzene, p/m-xylene, o-xylene, C9 and heavier aromatics, and total aromatics in finished gasoline by gas chromatography. Annual Book of Standards. ASTM International, West Conshohocken, PA.

ASTM D5623. Standard test method for sulfur compounds in light petroleum liquids by gas chromatography and sulfur selective detection. Annual Book of Standards. ASTM International, West Conshohocken, PA.

ASTM D5769. Standard test method for determination of benzene, toluene, and total aromatics in finished gasolines by gas chromatography/mass spectrometry. Annual Book of Standards. ASTM International, West Conshohocken, PA.

ASTM D5770. Standard test method for semiquantitative micro determination of acid number of lubricating oils during oxidation testing. Annual Book of Standards. ASTM International, West Conshohocken, PA.

ASTM D5842. Standard practice for sampling and handling of fuels for volatility measurement. Annual Book of Standards. ASTM International, West Conshohocken, PA.

ASTM D5853. Standard test method for pour point of crude oils. Annual Book of Standards. ASTM International, West Conshohocken, PA.

ASTM D5968. Standard test method for evaluation of corrosiveness of diesel engine oil at 121°C. Annual Book of Standards. ASTM International, West Conshohocken, PA.

ASTM D5972. Standard test method for freezing point of aviation fuels (Automatic phase transition method). Annual Book of Standards. ASTM International, West Conshohocken, PA.

ASTM D5986. Standard test method for determination of oxygenates, benzene, toluene, C8-C12 aromatics and total aromatics in finished gasoline by gas chromatography/Fourier transform infrared spectroscopy. Annual Book of Standards. ASTM International, West Conshohocken, PA.

ASTM D6217. Standard test method for particulate contamination in middle distillate fuels by laboratory filtration. Annual Book of Standards. ASTM International, West Conshohocken, PA.

ASTM D6377. Standard test method for determination of vapor pressure of crude oil: VPCRx (expansion method). Annual Book of Standards. ASTM International, West Conshohocken, PA.

ASTM D6378. Standard test method for determination of vapor pressure (VPX) of petroleum products, hydrocarbons, and hydrocarbon-oxygenate mixtures (triple expansion method). Annual Book of Standards. ASTM International, West Conshohocken, PA.

ASTM D6447. Standard test method for hydroperoxide number of aviation turbine fuels by voltammetric analysis. Annual Book of Standards. ASTM International, West Conshohocken, PA.

ASTM D7224. Standard test method for determining water separation characteristics of kerosene-type aviation turbine fuels containing additives by portable separometer. Annual Book of Standards. ASTM International, West Conshohocken, PA.

ASTM D7261. Standard test method for determining water separation characteristics of diesel fuels by portable separometer. Annual Book of Standards. ASTM International, West Conshohocken, PA.

Gary, J.G., Handwerk, G.E., and Kaiser, M.J. 2007. Petroleum Refining: Technology and Economics, 5th Edition. Taylor & Francis Group, Boca Raton, FL.

Hsu, C.S., and Robinson, P.R. (Editors). 2006. Practical Advances in Petroleum Processing, Volumes 1 and 2. Springer Science, New York.

IP 2 (ASTM D611). Petroleum products and hydrocarbon solvents – Determination of aniline and mixed aniline point. IP Standard Methods 2013. The Energy Institute, London.

IP 10. Determination of kerosene burning characteristics – 24 hour method. IP Standard Methods 2013. The Energy Institute, London.

IP 12. Determination of specific energy. IP Standard Methods 2013. The Energy Institute, London.

IP 15 (ASTM D97). Petroleum products – Determination of pour point. IP Standard Methods 2013. The Energy Institute, London.

IP 16 (ASTM D2386). Petroleum products – Determination of the freezing point of aviation fuels – Manual method. IP Standard Methods 2013. The Energy Institute, London.

IP 17. Determination of color – Lovibond tintometer method. IP Standard Methods 2013. The Energy Institute, London.

IP 30. Detection of mercaptans, hydrogen sulfide, elemental sulfur and peroxides – Doctor test method. IP Standard Methods 2013. The Energy Institute, London.

IP 34 (ASTM D93). Determination of flash point – Pensky-Martens closed cup method. IP Standard Methods 2013. The Energy Institute, London.

IP 57. Petroleum products – Determination of the smoke point of kerosene. IP Standard Methods 2013. The Energy Institute, London.

IP 61 (ASTM D129). Determination of sulfur – High pressure combustion method. IP Standard Methods 2013. The Energy Institute, London.

IP 69. Determination of vapor pressure – Reid method. IP Standard Methods 2013. The Energy Institute, London.

IP 71 (ASTM D445). Petroleum products – Transparent and opaque liquids – Determination of kinematic viscosity and calculation of dynamic viscosity. IP Standard Methods 2013. The Energy Institute, London.

IP 107 (ASTM D1266). Determination of sulfur – Lamp combustion method. IP Standard Methods 2013. The Energy Institute, London.

IP 119 (ASTM D909). Knock characteristics of aviation gasolines by the supercharge method. IP Standard Methods 2013. The Energy Institute, London.

IP 123. Petroleum products – Determination of distillation characteristics at atmospheric pressure. IP Standard Methods 2013. The Energy Institute, London.

IP 130 (ASTM D1159). Petroleum products – Determination of bromine number of distillates and aliphatic olefins – Electrometric method. IP Standard Methods 2013. The Energy Institute, London.

IP 131. Petroleum products – Gum content of light and middle distillates – Jet evaporation method. IP Standard Methods 2013. The Energy Institute, London.

IP 138 (ASTM D873). Determination of oxidation stability of aviation fuel – Potential residue method. IP Standard Methods 2013. The Energy Institute, London.

IP 139 (ASTM D974). Petroleum products and lubricants – Determination of acid or base number – Color indicator titration method. IP Standard Methods 2013. The Energy Institute, London.

IP 154. Petroleum products – Corrosiveness to copper – Copper strip test. IP Standard Methods 2013. The Energy Institute, London.

IP 156. Petroleum products and related materials – Determination of hydrocarbon types – Fluorescent indicator adsorption method. IP Standard Methods 2013. The Energy Institute, London.

IP 160 (ASTM D1298). Crude petroleum and liquid petroleum products – Laboratory determination of density – Hydrometer method. IP Standard Methods 2013. The Energy Institute, London.

IP 170. Petroleum products and other liquids – Determination of flash point – Abel closed cup method. IP Standard Methods 2013. The Energy Institute, London.

IP 177 (ASTM D664). Determination of weak and strong acid number – Potentiometric titration method. IP Standard Methods 2013. The Energy Institute, London.

IP 216 (ASTM D2276). Determination of particulate contaminant of aviation turbine fuels by line sampling. IP Standard Methods 2013. The Energy Institute, London.

IP 236 (ASTM D2700). Motor and aviation-type fuels – Determination of knock characteristics – Motor method. IP Standard Methods 2013. The Energy Institute, London.

IP 274 (ASTM D2624). Determination of electrical conductivity of aviation and distillate fuels. IP Standard Methods 2013. The Energy Institute, London.

IP 289. Determination of water reaction of aviation fuels. IP Standard Methods 2013. The Energy Institute, London.

IP 303. Determination of flash point – Rapid equilibrium closed cup method (replaced by IP 523 but still in use in some laboratories). IP Standard Methods 2013. The Energy Institute, London.

IP 323 (ASTM D3241). Petroleum products – Determination of thermal oxidation stability of gas turbine fuels – JFTOT method. IP Standard Methods 2013. The Energy Institute, London.

IP 336. Petroleum products – Determination of sulfur content – Energy-dispersive x-ray fluorescence method. IP Standard Methods 2013. The Energy Institute, London.

IP 342 (ASTM D3227). Determination of thiol (mercaptan) sulfur in light and middle distillate fuels – Potentiometric method. IP Standard Methods 2013. The Energy Institute, London.

IP 354 (ASTM D3242). Total acidity of aviation turbine fuel – Color indicator titration method. IP Standard Methods 2013. The Energy Institute, London.

IP 365. Crude petroleum and petroleum products – Determination of density – Oscillating U-tube method. IP Standard Methods 2013. The Energy Institute, London.

IP 373. Sulfur content – Microcoulometry (oxidative) method. IP Standard Methods 2013. The Energy Institute, London.

IP 394. Determination of air saturated vapor pressure (ASVP). IP Standard Methods 2013. The Energy Institute, London.

IP 415 (ASTM D6217). Determination of particulate content of middle distillate fuels – Laboratory filtration method. IP Standard Methods 2013. The Energy Institute, London.

IP 431. Petroleum products – Determination of acid number – Semi-micro color-indicator titration method. IP Standard Methods 2013. The Energy Institute, London.

IP 434. Freezing point of aviation turbine fuels – Automated optical method (replaced by IP 528 – ASTM D5901). IP Standard Methods 2013. The Energy Institute, London.

IP 435. Freezing point of aviation turbine fuels – Automatic phase transition method. IP Standard Methods 2013. The Energy Institute, London.

IP 528 (ASTM D7154). Determination of the freezing point of aviation turbine fuels – Automated fiber optic method. IP Standard Methods 2013. The Energy Institute, London.

Speight, J.G. 2001. Handbook of Petroleum Analysis. John Wiley & Sons Inc., New York.

Speight, J.G. 2014. The Chemistry and Technology of Petroleum, 5th Edition. Taylor & Francis Group, Boca Raton, FL.

Speight, J.G., and Ozum, B. 2002. Petroleum Refining Processes. Marcel Dekker Inc., New York.

Weissermel, K., and Arpe, H.-J. 1978. Industrial Organic Chemistry. Verlag Chemie, New York.

Wolveridge, P.E. 2000. Aviation Turbine Fuels. In Modern Petroleum Technology, Volume 2: Downstream, A.G. Lucas (Editor). John Wiley & Sons Inc., New York.

8

KEROSENE

8.1 INTRODUCTION

Kerosene (kerosine), also called paraffin or paraffin oil, is a flammable pale-yellow or colorless oily liquid with a characteristic odor intermediate in volatility between gasoline and gas/diesel oil and distills between 125°C (257°F) and 260°C (500°F) (Table 8.1; Fig. 8.1) (Heinrich and Duée, 2000; Speight and Ozum, 2002; Hsu and Robinson, 2006; Gary *et al.*, 2007; Speight, 2014).

Kerosene is a refined petroleum distillate that has a flash point about 25°C (77°F) and is suitable for use as an illuminant when burned in a wide lamp. The term *kerosene* is also too often incorrectly applied to various fuel oils, but a fuel oil is actually any liquid or liquid petroleum product that produces heat when burned in a suitable container or that produces power when burned in an engine.

In the early years of the petroleum industry, kerosene was its largest and most important selling product. The demand was such that many refiners, using a variety of crude oils, made as wide a distillation cut as possible in order to increase its availability, thereby causing the product to have a dangerously low flash point and to include undesirable higher boiling fractions. Burning oils are currently manufactured from carefully selected crude oils or by the use of special refining procedures, to give products of the requisite volatility and high burning quality.

8.2 PRODUCTION AND PROPERTIES

Kerosene was first manufactured in the 1850s from coal tar—hence the name *coal oil* as often applied to kerosene—but petroleum became the major source after 1859. From that time, the kerosene fraction is and has remained a distillation fraction of petroleum. However, the quantity and quality vary with the type of crude oil, and although some crude oils yield excellent kerosene quite simply, others produce kerosene that requires substantial refining.

Kerosene is less volatile than gasoline (boiling range: approximately 140°C/285°F–320°C/610°F) and is obtained by fractional distillation of petroleum (Speight, 2001, 2014). To decrease smoke generation, paraffinic stocks are used normally in the manufacture of kerosene for lighting and heating. For the same reason, aromatic stocks and cracked components are avoided. Some crude oils, especially the paraffinic crude oils, contain kerosene fractions of very high quality, but other crude oils, such as those having an asphalt base, must be thoroughly refined to remove aromatics and sulfur compounds before a satisfactory kerosene fraction can be obtained. Cracking the less volatile constituents of petroleum is now a major process for kerosene production.

In the early days, the poorer-quality kerosene was treated with large quantities of sulfuric acid to convert them to marketable products. However, this treatment resulted in high acid and kerosene losses, but the later development of the Edeleanu process (Speight, 2001, 2014) overcame these problems. Kerosene is a very stable product, and additives are not required to improve the quality. Apart from the removal of excessive quantities of aromatics by the Edeleanu process, kerosene fractions may need only a lye wash or a doctor treatment if hydrogen sulfide is present to remove mercaptans (Speight, 2001, 2014). When low-sulfur paraffinic crude oil is fractionated to yield the proper boiling range fraction, only a drying operation may be required before shipment.

Handbook of Petroleum Product Analysis, Second Edition. James G. Speight.
© 2015 John Wiley & Sons, Inc. Published 2015 by John Wiley & Sons, Inc.

TABLE 8.1   General summary of product types and distillation range

Product	Lower carbon limit	Upper carbon limit	Lower boiling point (°C)	Upper boiling point (°C)	Lower boiling point (°F)	Upper boiling point (°F)
Refinery gas	C_1	C_4	−161	−1	−259	31
Liquefied petroleum gas	C_3	C_4	−42	−1	−44	31
Naphtha	C_5	C_{17}	36	302	97	575
Gasoline	C_4	C_{12}	−1	216	31	421
Kerosene/diesel fuel	C_8	C_{18}	126	258	302	575
Aviation turbine fuel	C_8	C_{16}	126	287	302	548
Fuel oil	C_{12}	$>C_{20}$	216	421	>343	>649
Lubricating oil	$>C_{20}$		>343		>649	
Wax	C_{17}	$>C_{20}$	302	>343	575	>649
Asphalt	$>C_{20}$		>343		>649	
Coke	$>C_{50}{}^a$		$>1000^a$		$>1832^a$	

^{a} Carbon number and boiling point difficult to assess; inserted for illustrative purposes only.

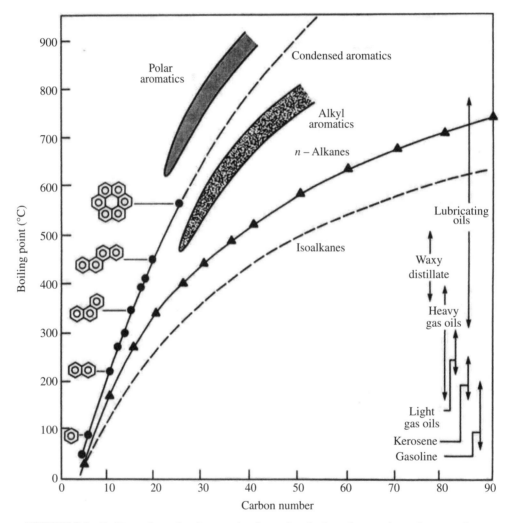

FIGURE 8.1   Boiling point and carbon number for various hydrocarbons and petroleum products.

Kerosene from naphthenic oil or high-sulfur crude oil requires hydrotreating, acid treatment and water wash, or extraction with a solvent and caustic wash and clay brightening to remove undesirable aromatics or sulfur compounds.

Generally, objectionable odors (mercaptans) are present (ASTM D3227, IP 342), and these are removed by caustic washing or converted to odorless compounds by sweetening processes (Speight, 2000; Ancheyta and Speight, 2007).

Following treatment, the kerosene streams are blended to meet specifications, and the finished product is ready for marketing.

Chemically, kerosene is a mixture of hydrocarbons, and the constituents include *n*-dodecane (n-$C_{12}H_{26}$), alkyl benzenes, and naphthalene and its derivatives (ASTM D1840). The chemical composition depends on its source and has a high number (>100,000) of isomers that are possible (Table 8.2). The actual number of compounds in kerosene is much lower, and there are claims to less than 100 constituents, but that, again, is source dependent and process dependent.

Kerosene is intermediate in volatility between gasoline and gas/diesel oil. It is a medium oil distilling between 150 and 300°C (300–570°F). Kerosene has a flash point about 25°C (77°F) and is suitable for use as an illuminant when burned in a wide lamp. The term *kerosene* is also too often incorrectly applied to various fuel oils, but a fuel oil is actually any liquid or liquid petroleum product that produces heat when burned in a suitable container or that produces power when burned in an engine.

Kerosene was the major refinery product before the onset of the *automobile age*, but now kerosene can be termed one of several secondary petroleum products after the primary refinery product—gasoline. Kerosene originated as a straight-run petroleum fraction that boiled between approximately 205 and 260°C (400–500°F). Some crude oils, for example, those from the Pennsylvania oil fields, contain kerosene fractions of very high quality, but other crude oils, such as those having an asphalt base, must be thoroughly refined to remove aromatics and sulfur compounds before a satisfactory kerosene fraction can be obtained.

Jet fuel comprises both gasoline- and kerosene-type jet fuels meeting specifications for use in aviation turbine power units and is often referred to as *gasoline-type jet fuel* and *kerosene-type jet fuel*.

TABLE 8.2 Increase in the number of isomers with carbon number

Carbon atoms	Number of isomers
1	1
2	1
3	1
4	2
5	3
6	5
7	9
8	18
9	35
10	75
15	4,347
20	366,319
25	36,797,588
30	4,111,846,763
40	62,491,178,805,831

Jet fuel is a light petroleum distillate that is available in several forms suitable for use in various types of jet engines. The major jet fuels used by the military are JP-4, JP-5, JP-6, JP-7, and JP-8. Briefly, JP-4 is a wide-cut fuel developed for broad availability. JP-6 is a higher cut than JP-4 and is characterized by fewer impurities. JP-5 is specially blended kerosene, and JP-7 is high–flash point special kerosene used in advanced supersonic aircraft. JP-8 is kerosene modeled on Jet A-l fuel (used in civilian aircraft). From what data are available, typical hydrocarbon chain lengths characterizing JP-4 range from C_4 to C_{16}. Aviation fuels consist primarily of straight and branched alkanes and cycloalkanes. Aromatic hydrocarbons are limited to 20–25% of the total mixture because they produce smoke when burned. A maximum of 5% alkenes is specified for JP-4. The approximate distribution by chemical class is straight chain alkanes (32%), branched alkanes (31%), cycloalkanes (16%), and aromatic hydrocarbons (21%).

Gasoline-type jet fuel includes all light hydrocarbon oils for use in aviation turbine power units that distill between 100 and 250°C (212 and 480°F). It is obtained by blending kerosene and gasoline or naphtha in such a way that the aromatic content does not exceed 25% in volume. Additives can be included to improve fuel stability and combustibility.

Kerosene-type jet fuel is a medium distillate product that is used for aviation turbine power units. It has the same distillation characteristics and flash point as kerosene (between 150 and 300°C, 300 and 570°F, but not generally above 250°C, 480°F). In addition, it has particular specifications (such as freezing point) that are established by the International Air Transport Association.

8.3 TEST METHODS

Kerosene is composed chiefly of hydrocarbons containing 12 or more carbon atoms per molecule. Although the kerosene constituents are predominantly saturated materials, there is evidence for the presence of substituted tetrahydronaphthalenes. Dicycloparaffins also occur in substantial amounts in kerosene. Other hydrocarbons with both aromatic and cycloparaffin rings in the same molecule, such as substituted indane, also occur in kerosene. The predominant structure of the dinuclear aromatics appears to be that in which the aromatic rings are condensed, such as naphthalene, whereas the *isolated* two-ring compounds, such as biphenyl, are only present in traces, if at all (ASTM D1840).

Low proportions of aromatic and unsaturated hydrocarbons are desirable to maintain the lowest possible level of smoke during burning. Although some aromatics may occur within the boiling range assigned to kerosene, excessive amounts can be removed by extraction; the kerosene that is not usually prepared from cracked products almost certainly excludes the presence of unsaturated hydrocarbons.

The essential properties of kerosene are flash point (ASTM D56, ASTM D93, ASTM D3828, IP 34, IP 170, IP 523), distillation range (ASTM D86, ASTM D1160, ASTM D2887, ASTM D6352), burning characteristics (ASTM D187, IP 10), sulfur content (ASTM D129, ASTM D2622, ASTM D3120, ASTM D3246, ASTM D4294, ASTM D5453, ASTM D5623, IP 61, IP 336, IP 373), color (ASTM D156, ASTM D1209, ASTM D1500, ASTM D2392, ASTM D6045), and cloud point (ASTM D2500, ASTM D5771, ASTM D5772, ASTM D5773, IP 219). In the case of the flash point (ASTM D56), the minimum flash temperature is generally placed above the prevailing ambient temperature; the fire point (ASTM D92) determines the fire hazard associated with its handling and use.

8.3.1 Acidity

Acids can be present in kerosene aviation turbine fuels due to acid treatment during refining. These trace acid quantities are undesirable because of the possibility of metal corrosion and impairment of the burning characteristics and other properties of the kerosene. The potential for metals in kerosene is less than it is for aviations fuels, but several of the same tests can be applied (Chapter 7).

One test method (ASTM D1093) is used solely for the qualitative determination of the acidity of hydrocarbon liquids and their distillation residues. The results are qualitative. Basicity determination can also be done by a small change in the procedure (see later). In the test method, a sample is shaken with water, and the aqueous layer is tested for acidity using methyl orange indicator (red color). Basicity can be determined using phenolphthalein indicator (pink color) instead of the methyl orange indicator.

Another test method (ASTM D3242, IP 354) covers the determination of acidity in an aviation turbine fuel in the range 0.000–0.100 mg potassium hydroxide per gram, but the test is not suitable for determining significant acid contamination. In the test, a sample is dissolved in a solvent mixture (toluene plus isopropyl alcohol, and a small amount of water) and under a stream of nitrogen is titrated with standard alcoholic *KOH* to the color change from orange in acid to green in base via added indicator *p*-naptholbenzein solution.

8.3.2 Burning Characteristics

The ability of kerosene to burn steadily and cleanly over an extended period (ASTM D187, IP 10) is an important property and gives some indication of the purity or composition of the product. The quality of a kerosene as a burning oil is related to its burning characteristics and is dependent on such factors as its composition, volatility, viscosity, calorific value, sulfur content, and freedom from corrosive substances or contaminants. This test method covers the qualitative

determination of the burning properties of kerosene to be used for illuminating purposes. In the test, a kerosene sample is burned for 16h in a specified lamp under specified conditions. The average rate of burning, the change in the shape of the flame, and the density and color of the chimney deposit are reported. A corresponding test method (IP 10) is used for the quantitative evaluation of the wick-char-forming tendencies of kerosene.

However, the effect of hydrocarbon-type composition is greater with wick-fed yellow flame burners than with wick-fed blue flame burners. With the former, kerosene that are mainly paraffinic burn well in lamps with a poor draught, while under the same conditions, kerosene containing high proportions of aromatics and naphthenes burn with a reddish or even smoky flame.

The smoke point test (ASTM D1319, ASTM D1322, IP 57, IP 156) enables this property to be measured. In this test, the oil is burned in a standard wick-fed lamp in which flame height can be varied against a background of a graduated scale. The maximum flame height in millimeters at which the oil burns without smoking under the standard conditions is termed the smoke point. Even if full advantage is not taken to utilize maximum nonsmoking flame height, the property of high smoke point ensures that in the event of sudden draught causing extension in flame height, there will be less tendency for smoking to occur. The smoke point test is also used in the assessment of the burning characteristics of certain aviation turbine fuels.

The 24h burning test (ASTM D187) involves noting the average oil consumption, change in flame dimensions, and final appearance of wick and chimney. No quantitative determination of char value is made (IP 10). In this method, the oil is burned for 24h in the standard lamp with a flame initially adjusted to specified dimensions. The details of operation are carefully specified and involve the test room conditions, volume of sample, wick nature, pretreatment of wick and glass chimney, method of wick trimming, and the procedure for removal of the char. At the conclusion of the test, the oil consumption and the amount of char formed on the wick are determined, and the char value calculated as milligrams per kilogram of oil consumed. A qualitative assessment of the appearance of the glass chimney is also made.

The considerable effect on char-forming tendency of even traces of high-boiling contaminants is demonstrated by the fact that the addition of 0.01% of a heavy lubricating oil to a kerosene of a char value of 10mg/kg (0.001%) can result in doubling that char value.

8.3.3 Calorific Value

The calorific value (heat of combustion) (ASTM D240, ASTM D1405, ASTM D2890, ASTM D3338, ASTM D4529, ASTM D4809, IP 12) is a direct measure of fuel energy content and is determined as the quantity of heat liberated by the combustion of a unit quantity of fuel with oxygen in a standard bomb calorimeter. A high calorific value is obviously desirable in oil used for heating purposes. Calorific value does not, however, vary greatly in the range of paraffinic-type kerosene (ASTM D240, IP 12).

When an experimental determination of heat of combustion is not available and cannot be made conveniently, an estimate might be considered satisfactory (ASTM D3338). In this test method, the net heat of combustion is calculated from the density, sulfur, and hydrogen content, but this calculation is justifiable only when the fuel belongs to a well-defined class for which a relationship between these quantities has been derived from accurate experimental measurements on representative samples. Thus, the hydrogen content (ASTM D1018, ASTM D1217, ASTM D1298, ASTM D3701, ASTM D4052, ASTM D4808, ASTM D5291, IP 160, IP 365), density (ASTM D129, ASTM D1250, ASTM D1266, ASTM D2622, ASTM D3120, IP 61, IP 107), and sulfur content (ASTM D2622, ASTM D3120, ASTM D3246, ASTM D4294, ASTM D5453, ASTM D5623, IP 336, IP 373) of the sample are determined by experimental test methods, and the net heat of combustion is calculated using the values obtained by these test methods based on reported correlations.

Another equation that can be used to calculate the heat of combustion is based on the specific gravity of the kerosene:

$$Q = 12,400 - 2,100d^2$$

Q is the heat of combustion, and d is the specific gravity. However, the accuracy of any method used to calculate such a property is not guaranteed and can only be used as a guide or approximation to the measured value.

An alternative criterion of energy content is the *aniline gravity product* (AGP) that is related to calorific value (ASTM D1405). The *AGP* is the product of the API gravity (ASTM D287, ASTM D1298) and the aniline point of the fuel (ASTM D611, IP 2). The aniline point is the lowest temperature at which the fuel is miscible with an equal volume of aniline and is inversely proportional to the aromatic content. The relationship between the AGP and calorific value is given in this method. In another method (ASTM D3338), the heat of combustion is calculated from the fuel density; the 10, 50, and 90% distillation temperatures; and the aromatic content. However, neither method is legally acceptable, and other methods (ASTM D240, ASTM D1655, ASTM D4809) are preferred.

8.3.4 Composition

Because of the estimated (or real) number of isomers in this carbon number range (Fig. 8.2), complete speciation of individual hydrocarbons is not possible for middle distillates. Compositional analysis of middle distillates is obtained in terms of hydrocarbon group–type totals. These groups are most often defined by a chromatographic separation.

Thus, the first level of compositional information is group-type totals as deduced by adsorption chromatography

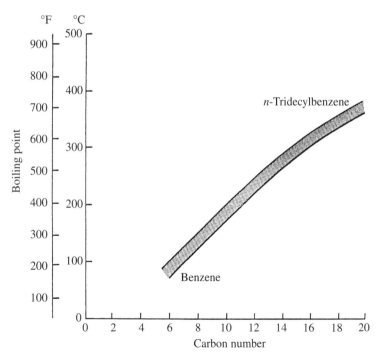

FIGURE 8.2 Effect of size of the alkyl chain on the boiling point of benzene.

form the distribution of saturates, olefins, and aromatics in materials that boil below 315°C (600°F) (ASTM D1319). Adsorption methods (ASTM D2007) can also be used to determine hydrocarbon types in kerosene, but with all adsorption methods, allowances must be made for the loss of volatile constituents during the work-up procedure. Thus, column chromatography would be best done using a stabilized (volatile removed to a predetermined temperature) feedstock.

Burning oil kerosene contains three main types of hydrocarbons—paraffinic, naphthenic, and aromatic—with a preponderance of the paraffinic type. This is in contrast to *power kerosene* or *tractor vaporizing oil* that has a comparatively high content of aromatics and naphthenes favorable for high octane rating. It may also contain slight amounts of sulfur in the form of a variety of organic compounds.

Compositional analysis of kerosene distillates can also be obtained in terms of mass spectral Z-series (ASTM D2425, ASTM D2789, ASTM D3239, ASTM D6379). Mass spectrometry has been a powerful technique for hydrocarbon-type analysis of middle distillates and can provide more compositional detail than chromatographic analysis. Hydrocarbon types are classified in terms of a Z-series. Z in the empirical formula C_nH_{2n+z} is a measure of the hydrogen deficiency of the compound (ASTM D2425). This method requires that the sample be separated into saturate and aromatic fractions before mass spectrometric analysis. This separation is standardized (ASTM D2549) and is applicable to kerosene.

The percentage of aromatic hydrogen atoms and aromatic carbon atoms can be determined by high-resolution nuclear magnetic resonance spectroscopy (ASTM D5292), but the results from this test are not equivalent to mass- or volume-percent aromatics determined by the chromatographic methods. The chromatographic methods determine the percentage by weight (or the percentage by volume) of molecules that have one or more aromatic rings. Any alkyl substituents on the rings (Fig. 8.2) contribute to the percentage of aromatics determined by chromatographic techniques, but the presence of an aromatic ring (no matter what the length of the alkyl side chain) dictates that the compound be isolated as an aromatic thereby leading to erroneous estimates of the carbon atoms in aromatic systems.

Because the aromatic hydrocarbon content of aviation turbine fuels affects their combustion characteristics and smoke-forming tendencies, the amounts of aromatics (ASTM D1319, IP 156) are limited. Aromatic constituents also increase the intensity of the combustion flame, which can have an adverse effect on the in-service life of the combustion chamber.

The aromatic content of kerosene can also be determined by a test method (ASTM D5186) in which a small aliquot of the sample is injected onto a packed silica adsorption column and eluted using supercritical carbon dioxide as the mobile phase. Mono- and polynuclear aromatics in the sample are separated from non-aromatics and detected using a flame ionization detector. The chromatographic areas corresponding to the mono-, polynuclear, and non-aromatic components are determined, and the mass-percent content of each of these groups is calculated by area normalization. The results obtained by this method are at least statistically more precise than those obtained by other test methods (ASTM D1319, ASTM D2425).

In yet another test method for the determination of aniline point and mixed aniline point (ASTM D611, IP 2), the proportions of the various hydrocarbon constituents of kerosene can be determined. This test is most often used to estimate the aromatic content of kerosene. Aromatic compounds exhibit the lowest aniline points, and paraffin compounds have the highest aniline points with cycloparaffins (naphthenes) and olefins having aniline points between the two extremes. In any homologous series, the aniline point increases with increasing molecular weight.

There are five submethods in the test (ASTM D611, IP 2) for the determination of the aniline point: (1) Method A is used for transparent samples with an initial boiling point above room temperature and where the aniline point is below the bubble point and above the solidification point of the aniline–sample mixture; (2) Method B, a thin film method, is suitable for samples too dark for testing by Method A; (3) Methods C and D are employed when there is the potential for sample vaporization at the aniline point; (4) Method D is particularly suitable where only small quantities of sample are available; and (5) Method E uses an automatic apparatus suitable for the range covered by methods A and B.

Olefins in kerosene also influence the burning characteristics and can be determined by the bromine number (ASTM D1159, ASTM D2710, IP 130). The bromine number is the grams of bromine that will react with 100g of the sample under the test conditions. The magnitude of bromine number is an indication of the quantity of bromine-reactive constituents and is not an identification of constituents. It is used as a measure of aliphatic unsaturation in petroleum samples and as the percentage of olefins in petroleum distillates boiling up to approximately 315°C (600°F). In the test, a known weight of the sample dissolved in a specified solvent maintained at 0–5°C (32–41°F) is titrated with standard bromide–bromate solution. Determination of the end point is method dependent.

Gas chromatography (ASTM D2427, ASTM D5443, ASTM D5580) remains the most reliable method for the determination of hydrocarbon types, including olefins (ASTM D6296), in kerosene and similar boiling fractions. In particular, methods in which a combination of gas chromatography and Fourier transform infrared spectroscopy (GC-FTIR) (ASTM D5986) and gas chromatography and mass spectrometry (GC-MS) (ASTM D5769) is used are finding increased use. Indeed, Fourier transform infrared spectroscopy has been used to predict properties such as density, freezing point, flash point, aromatic content, initial boiling point, final boiling point, and viscosity (Garrigues *et al.*, 1995). The data had a high degree of repeatability and reproducibility.

The significance of the total sulfur content of kerosene varies greatly with the type of oil and the use to which it is put. Sulfur content is of great importance when the oil to be burned produces sulfur oxides that contaminate the surroundings. Only slight amounts of sulfur compounds remain in kerosene after refining. Refining treatment includes among its objects the removal of such undesirable products as hydrogen sulfide, mercaptan sulfur, and *free* or corrosive sulfur. Hydrogen sulfide and mercaptans cause objectionable odors, and both are corrosive. Their presence can be detected by the Doctor test (ASTM D4952, IP 30). The Doctor test (which is pertinent for petroleum product specifications, ASTM D235) ensures that the concentration of these compounds is insufficient to cause such problems in normal use. In the test, the sample is shaken with sodium plumbite solution, a small quantity of sulfur is added, and the mixture is shaken again. The presence of mercaptans or hydrogen sulfide or both is indicated by discoloration of the sulfur floating at the oil–water interface or by discoloration of either of the phases.

Free, or corrosive, sulfur in appreciable amount could result in corrosive action on the metallic components of an appliance. Corrosive action is of particular significance in the case of pressure burner vaporizing tubes that operate at high temperatures. The usual test applied in this connection is the corrosion (copper strip) test (ASTM D130, ASTM D849, IP 154).

The copper strip test methods are used to determine the corrosiveness to copper of gasoline, diesel fuel, lubricating oil, or other hydrocarbons. Most sulfur compounds in petroleum are removed during refining. However, some residual sulfur compounds can have a corroding action on various metals. This effect is dependent on the types of sulfur compounds present. The copper strip corrosion test measures the relative degree of corrosivity of a petroleum product.

In a method (ASTM D130, IP 154), a polished copper strip is immersed in a given quantity of sample and heated at a temperature for a time period characteristic of the material being tested. At the end of this period, the copper strip is removed, washed, and compared with the copper strip corrosion standards. This is a pass/fail test. In another method (ASTM D849), a polished copper strip is immersed in 200ml of specimen in a flask with a condenser and placed in boiling water for 30min. At the end of this period, the copper strip is removed and compared with the ASTM copper strip corrosion standards. This is also a pass–fail test.

It is important that the total sulfur content of burning oil should be low (ASTM D1266, IP 107). All the sulfur compounds present in an oil are converted to oxides of

sulfur during burning. These oxides of sulfur should not be present to a harmful extent in the immediate atmosphere, and this applies particularly to indoor burning appliances that are not provided with a flue. Also it has been indicated in the foregoing that a high total sulfur content of an oil can contribute to the formation of lamp chimney deposits.

Gas chromatography with either sulfur chemiluminescence detection or atomic emission detection has been used for sulfur selective detection. Selective sulfur and nitrogen gas chromatographic detectors, exemplified by the flame photometric detector and the nitrogen–phosphorus detector, have been available for many years. However, these detectors have limited selectivity for the element over carbon, exhibit nonuniform response, and have other problems that limit their usefulness.

Nitrogen compounds in middle distillates can be selectively detected by chemiluminescence. Individual nitrogen compounds can be detected down to 100ppb nitrogen.

Correlative methods have long been used as a way of dealing with the complexity of petroleum fractions. Relatively easy to measure, physical properties such as density, viscosity, and refractive index (ASTM D1218) have been correlated to hydrocarbon structure (Table 8.3) with the potential to relate refractive index data to the nature of the constituents of a petroleum product. In recent years, an entirely new class of correlative methods has been developed. These use near-infrared or mid-infrared spectra together with sophisticated chemometric techniques to predict a wide variety of properties. Properties such as composition (saturates, aromatics), freezing point, density, viscosity, aromatics, and heat of combustion have been successfully predicted. However, it is important to recognize that these methods are correlations and should not be used to estimate properties that are outside of the calibration set.

The *color* of kerosene is of little significance, but a product darker than usual may have resulted from a change in composition due to contamination or aging, and in fact a color darker than specified (ASTM D156) may be considered

TABLE 8.3   Refractive indices of selected hydrocarbons

Compound	Refractive index n_D^{20}
n-pentane	1.3578
n-hexane	1.3750
n-hexadecane	1.4340
Cyclopentane	1.4065
Cyclopentene	1.4224
Pentene-1	1.3714
1,3-pentadiene	1.4309
Benzene	1.5011
cis-decahydronaphthalene	1.4814
Methylnaphthalenes	1.6150

by some users as unsatisfactory. Finally, the cloud point of kerosene (ASTM D2500) gives an indication of the temperature at which the wick may become coated with wax particles, thus lowering the burning qualities of the oil.

Alternatively, the wax appearance point may also be estimated form the cloud point (ASTM D2500) determined as a means of estimating the composition of kerosene in terms of the wax (*n*-paraffins) content. The wax appearance point is the temperature at which wax crystals begin to precipitate from a fuel and is estimated for the cloud point. In this test (ASTM D2500, ASTM D5771, ASTM D5772, ASTM D5773, IP 219), a sample is cooled under prescribed conditions with stirring. The temperature at which wax first appears is the wax appearance point.

8.3.5 Density

Density (specific gravity) is an important property of petroleum products and is often part of product specifications (Table 8.4). Materials are usually bought and sold on that basis or if on volume basis then converted to mass basis via density measurements. This property is almost synonymously termed as density (mass of liquid per unit volume), specific gravity (the ratio of the mass of a given volume of liquid to the mass of an equal volume of pure water at the same temperature), and relative density (same as specific gravity). Usually a hydrometer, pycnometer, or a digital density meter is used in these standards.

Specific gravity has no relation to burning quality but is a useful aid in checking consistency of production of a particular grade. The specific gravity of kerosene can be determined very conveniently by the hydrometer method (ASTM D1298, IP 160).

8.3.6 Flash Point

The flashpoint test is a guide to the fire hazard associated with the use of kerosene and can be determined by several test methods, and the results are not always strictly comparable. Generally, the flash point of kerosene is specified as being in excess of 38°C (100°F), due to production as well as safety considerations.

TABLE 8.4 Specific gravity and API gravity of crude oil and selected products

Material	Specific gravity 60°/60°F	API gravity (°)
Crude oils	0.65–1.06	87–2
Casinghead liquid	0.62–0.70	97–70
Gasoline	0.70–0.77	70–52
Kerosene	0.77–0.82	52–40
Lubricating oil	0.88–0.98	29–13
Residua and cracked residua	0.88–1.06	29–2

The minimum flash point is usually defined by the Abel method (IP 170) but except for high flash kerosene where the Pensky–Martens method (ASTM D93, IP 34) is specified. The TAG method (ASTM D56) is used for both the minimum and maximum limits, while certain military specifications also give minimum limits by the Pensky–Martens method (ASTM D93, IP 34). The Abel method (IP 170) can give results up to 2–3°C (3–5°F lower than the TAG method (ASTM D56).

8.3.7 Freezing Point

The freezing point of kerosene is not of the same importance as the freezing point of aviation fuel (Chapter 7) but deserves mention because of its influence on kerosene use.

There are three test methods available for determination of the freezing point. All three methods have been found to give equivalent results. However, when a specification calls for a specific test, only that test must be used.

In the first test method (ASTM D2386, IP 16), a measured fuel sample is placed in a jacketed sample tube, also holding a thermometer and a stirrer. The tube is placed in a vacuum flask containing the cooling medium. Various coolants used are acetone, methyl alcohol, ethyl alcohol, or isopropyl alcohol, solid carbon dioxide, or liquid nitrogen. As the sample cools, it is continuously stirred. The temperature at which the hydrocarbon crystals appear is recorded. The jacketed sample is removed from the coolant and allowed to warm, stirring it continuously. The temperature at which the crystals completely disappear is recorded.

In the second test method (IP 434), an automated optical instrument is used for the temperature range to −70°C (−94°F). In the method, a 25min portion of the fuel is placed in a test chamber that is cooled while continuously stirred and monitored by an optical system. The temperature of the specimen is measured with an electronic measuring device, and the temperatures when crystals first appear, and then on warming disappear, are recorded.

In the third method (ASTM D5972, IP 435), an automated phase transition method is used in the temperature range −80 to 20°C (−112 to 68°F). In this test, a specimen is cooled at a rate of 15±5°C/min while continuously being illuminated by a light source. The specimen is continuously monitored by an array of optical detectors for the first formation of solid hydrocarbon crystals. After that, the specimen is warmed at the rate of 10±0.5°C/min until all crystals return to the liquid phase, and that temperature is also recorded.

8.3.8 Pour Point

The *pour point* should not be confused with the *freezing point*. The *pour point* is an index of the lowest temperature at which the crude oil will flow under specified conditions. The maximum and minimum pour point temperatures provide a temperature window where petroleum, depending on its thermal history, might appear in the liquid as well as the solid state. The pour point data can be used to supplement other measurements of cold flow behavior, and the data are particularly useful for the screening of the effect of wax interaction modifiers on the flow behavior of petroleum.

In the original (and still widely used) test for pour point (ASTM D97, IP 15), a sample is cooled at a specified rate and examined at intervals of 3°C (5.4°F) for flow characteristics. The lowest temperature at which the movement of the oil is observed is recorded as the pour point.

A later test method (ASTM D5853) covers two procedures for the determination of the pour point of crude oils down to −36°C. One method provides a measure of the maximum (upper) pour point temperature. The second method measures the minimum (lower) pour point temperature. In the methods, the test specimen is cooled (after preliminary heating) at a specified rate and examined at intervals of 3°C (5.4°F) for flow characteristics. Again, the lowest temperature at which movement of the test specimen is observed is recorded as the pour point.

In any determination of the pour point, petroleum that contains wax produces an irregular flow behavior when the wax begins to separate. Such petroleum possesses viscosity relationships that are difficult to predict in pipeline operation. In addition, some waxy petroleum is sensitive to heat treatment that can also affect the viscosity characteristics. This complex behavior limits the value of viscosity and pour point tests on waxy petroleum.

8.3.9   Smoke Point

While a low smoke point is undesirable in that it may not give a satisfactory range of smokeless performance, a high smoke point alone is no guarantee that a kerosene has generally satisfactory burning characteristics. The smoke point test adequately reflects the essential feature of hydrocarbon-type composition in relation to burning characteristics, as already indicated, and consequently no analysis for composition is necessary in the normal evaluation of burning oils.

Kerosene can vary widely in its burning quality as measured by carbon deposition, smoke formation, and flame radiation. This is a function of hydrocarbon composition— paraffins have excellent burning properties in contrast to those of the aromatics (particularly the polynuclear aromatic hydrocarbons). As a control measure, the smoke point test (ASTM D1322, IP 57) gives the maximum smokeless flame height in millimeters at which the fuel will burn in a wick-fed lamp under prescribed conditions. The combustion performance of wide-cut fuels correlates well with smoke point when a fuel volatility factor is included, since carbon formation tends to increase with boiling point.

A minimum smoke volatility index (SVI) value is specified and is defined as follows:

$$SVI = smoke\,point + 0.42$$
$$(percentage\,distilled\,below\,204°C : 400°F).$$

However, the smoke point test is not always a reliable criterion of combustion performance and should be used in conjunction with other properties. Various alternative laboratory test methods have previously been specified such as the lamp burning test (ASTM D187, IP 10) and a limit on the polynuclear aromatic content (ASTM D1840). The test apparatus is a smoke point lamp modified to include a photoelectric cell for flame radiation measurement and a thermocouple to measure temperature rise across the flame.

8.3.10   Viscosity

The kinematic viscosity of many petroleum fuels is important for their proper use, for example, flow of fuels through pipelines, injection nozzles and orifices, and the determination of the temperature range for proper operation of the fuel in burners.

The quantity of oil flowing up a wick is related to the height of the top of the wick above the level of oil in the container and the viscosity and surface tension of the oil. Viscosity (ASTM D445, IP 71) is more significant in this respect than surface tension, since it varies more in magnitude than the latter with different kerosene and with change of temperature.

8.3.11   Volatility

An abnormally high final boiling point and percentage residue of a kerosene may indicate contamination with higher-boiling constituents, although the presence of trace quantities of very heavy oils sufficient to cause high char values might not necessarily be revealed by these features. Thus, the boiling range of kerosene is an important aspect of kerosene properties.

The boiling range (ASTM D86, IP 123) is of less importance for kerosene than for gasoline, but it is an indication of the viscosity of the product, for which there is no requirement for kerosene. The nature of the distillation range (ASTM D86, IP 123) is of significance with regard to burning characteristics. It can control the flash point and viscosity, the effect of which has already been mentioned. The initial boiling point and the 10% point chiefly affect the flash point and ease of ignition, while the mid-boiling point is more relevant to the viscosity.

Another test method (ASTM D6352) that can be used for product specification testing is applicable to petroleum distillate fractions with an initial boiling point of <700°C (<1292°F) at atmospheric pressure. This test method extends the scope of other test methods (ASTM D86, ASTM D1160,

ASTM D2887) to boiling range determination by gas chromatography. In this method, a nonpolar open tubular capillary gas chromatographic column is used to elute the hydrocarbon components of the sample in order of increasing boiling point. A sample aliquot diluted with a viscosity-reducing solvent is introduced into the chromatographic system. The column oven temperature is raised at a specified linear rate to effect separation of the hydrocarbon components. The detector signal is recorded as area slices for consecutive retention time intervals during the analysis. Retention times of known normal paraffin hydrocarbons spanning the scope of the test method are used for normalizing the retention times of the unknown mixture area slices.

Contamination of kerosene with heavy oil may also be revealed by the test method that is used to determine the amount of residue left by evaporation (ASTM D381, IP 131), although this depends on the relative volatility of the contaminant.

One of the most important physical parameters defining these products is their boiling range distribution that can be determined using a low-efficiency, one theoretical plate distillation procedure (ASTM D86). This has been adequate for product specification purposes; however, engineering studies require true boiling point data (ASTM D2887, ASTM D2892).

The vapor pressure of petroleum products at various vapor–liquid ratios is an important physical property for shipping and storage (ASTM D2889). Although determining the volatility of kerosene is usually accomplished through a boiling range distribution (ASTM D86, IP 123) although other methods such as determining the Reid vapor pressure (ASTM D323, IP 69) can also be used along with several other methods (ASTM D5482, ASTM D6378).

8.3.12   Water and Sediment

Kerosene, because of its higher density and viscosity, tends to retain fine particulate matter and water droplets in suspension for a much longer time than gasoline. Free water in kerosene can be detected by the use of a dean and Stark adaptor (ASTM D4006, IP 358) (Fig. 8.3), by the Karl Fischer titration method (ASTM D1744, ASTM D6304), by the distillation method (ASTM D95, IP 74), or by a series of alternate tests (ASTM D4176, ASTM D4860). The standard water reaction test method (ASTM D1094, IP 289) can also be used.

In addition to water, sediment can also occur and will cause fouling of the fuel handling facilities and the fuel system. An accumulation of sediment in storage tanks and on filter screens can obstruct the flow of kerosene during use, and a test method is available to determine the water and sediment in fuels (ASTM D2709). In the test method, a sample of kerosene is centrifuged at an rcf of 800 for 10min at 21–32°C in a centrifuge tube readable to 0.005ml and

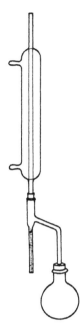

FIGURE 8.3   A Dean and Stark adaptor on the bottom of a condenser.

measurable to 0.01ml. After centrifugation, the volume of water and sediment that has settled into the tip of the centrifuge tube is read to the nearest 0.005ml.

REFERENCES

Ancheyta, J. and Speight, J.G. 2007. Hydroprocessing of Heavy Oils and Residua. CRC Press, Taylor & Francis Group, Boca Raton, FL.

ASTM D56. Standard Test Method for Flash Point by Tag Closed Cup Tester. Annual Book of Standards. ASTM International, West Conshohocken, PA.

ASTM D86. Standard Test Method for Distillation of Petroleum Products at Atmospheric Pressure. Annual Book of Standards. ASTM International, West Conshohocken, PA.

ASTM D92. Standard Test Method for Flash and Fire Points by Cleveland Open Cup Tester. Annual Book of Standards. ASTM International, West Conshohocken, PA.

ASTM D93. Standard Test Methods for Flash Point by Pensky-Martens Closed Cup Tester. Annual Book of Standards. ASTM International, West Conshohocken, PA.

ASTM D95. Standard Test Method for Water in Petroleum Products and Bituminous Materials by Distillation. Annual Book of Standards. ASTM International, West Conshohocken, PA.

ASTM D97. Standard Test Method for Pour Point of Petroleum Products. Annual Book of Standards. ASTM International, West Conshohocken, PA.

ASTM D129. Standard Test Method for Sulfur in Petroleum Products (General High Pressure Decomposition Device Method). Annual Book of Standards. ASTM International, West Conshohocken, PA.

ASTM D130. Standard Test Method for Corrosiveness to Copper from Petroleum Products by Copper Strip Test. Annual Book of Standards. ASTM International, West Conshohocken, PA.

ASTM D156. Standard Test Method for Saybolt Color of Petroleum Products (Saybolt Chromometer Method. Annual Book of Standards. ASTM International, West Conshohocken, PA.

ASTM D187. Standard Test Method for Burning Quality of Kerosene. Annual Book of Standards. ASTM International, West Conshohocken, PA.

ASTM D235. Standard Specification for Mineral Spirits (Petroleum Spirits) (Hydrocarbon Dry Cleaning Solvent). Annual Book of Standards. ASTM International, West Conshohocken, PA.

ASTM D240. Standard Test Method for Heat of Combustion of Liquid Hydrocarbon Fuels by Bomb Calorimeter. Annual Book of Standards. ASTM International, West Conshohocken, PA.

ASTM D287. Standard Test Method for API Gravity of Crude Petroleum and Petroleum Products (Hydrometer Method). Annual Book of Standards. ASTM International, West Conshohocken, PA.

ASTM D323. Standard Test Method for Vapor Pressure of Petroleum Products (Reid Method). Annual Book of Standards. ASTM International, West Conshohocken, PA.

ASTM D381. Standard Test Method for Gum Content in Fuels by Jet Evaporation. Annual Book of Standards. ASTM International, West Conshohocken, PA.

ASTM D445. Standard Test Method for Kinematic Viscosity of Transparent and Opaque Liquids (and Calculation of Dynamic Viscosity). Annual Book of Standards. ASTM International, West Conshohocken, PA.

ASTM D611. Standard Test Methods for Aniline Point and Mixed Aniline Point of Petroleum Products and Hydrocarbon Solvents. Annual Book of Standards. ASTM International, West Conshohocken, PA.

ASTM D849. Standard Test Method for Copper Strip Corrosion by Industrial Aromatic Hydrocarbons. Annual Book of Standards. ASTM International, West Conshohocken, PA.

ASTM D1018. Standard Test Method for Hydrogen in Petroleum Fractions. Annual Book of Standards. ASTM International, West Conshohocken, PA.

ASTM D1093. Standard Test Method for Acidity of Hydrocarbon Liquids and Their Distillation Residues. Annual Book of Standards. ASTM International, West Conshohocken, PA.

ASTM D1094. Standard Test Method for Water Reaction of Aviation Fuels. Annual Book of Standards. ASTM International, West Conshohocken, PA.

ASTM D1159. Standard Test Method for Bromine Numbers of Petroleum Distillates and Commercial Aliphatic Olefins by Electrometric Titration. Annual Book of Standards. ASTM International, West Conshohocken, PA.

ASTM D1160. Standard Test Method for Distillation of Petroleum Products at Reduced Pressure. Annual Book of Standards. ASTM International, West Conshohocken, PA.

ASTM D1209. Standard Test Method for Color of Clear Liquids (Platinum-Cobalt Scale). Annual Book of Standards. ASTM International, West Conshohocken, PA.

ASTM D1217. Standard Test Method for Density and Relative Density (Specific Gravity) of Liquids by Bingham Pycnometer. Annual Book of Standards. ASTM International, West Conshohocken, PA.

ASTM D1218. Standard Test Method for Refractive Index and Refractive Dispersion of Hydrocarbon Liquids. Annual Book of Standards. ASTM International, West Conshohocken, PA.

ASTM D1250. Standard Guide for Use of the Petroleum Measurement Tables. Annual Book of Standards. ASTM International, West Conshohocken, PA.

ASTM D1266. Standard Test Method for Sulfur in Petroleum Products (Lamp Method). Annual Book of Standards. ASTM International, West Conshohocken, PA.

ASTM D1298. Standard Test Method for Density, Relative Density, or API Gravity of Crude Petroleum and Liquid Petroleum Products by Hydrometer Method. Annual Book of Standards. ASTM International, West Conshohocken, PA.

ASTM D1319. Standard Test Method for Hydrocarbon Types in Liquid Petroleum Products by Fluorescent Indicator Adsorption. Annual Book of Standards. ASTM International, West Conshohocken, PA.

ASTM D1322. Standard Test Method for Smoke Point of Kerosene and Aviation Turbine Fuel. Annual Book of Standards. ASTM International, West Conshohocken, PA.

ASTM D1405. Standard Test Method for Estimation of Net Heat of Combustion of Aviation Fuels. Annual Book of Standards. ASTM International, West Conshohocken, PA.

ASTM D1500. Standard Test Method for ASTM Color of Petroleum Products (ASTM Color Scale). Annual Book of Standards. ASTM International, West Conshohocken, PA.

ASTM D1655. Standard Specification for Aviation Turbine Fuels. Annual Book of Standards. ASTM International, West Conshohocken, PA.

ASTM D1744. Standard Test Method for Determination of Water in Liquid Petroleum Products by Karl Fischer Reagent. Annual Book of Standards. ASTM International, West Conshohocken, PA.

ASTM D1840. Standard Test Method for Naphthalene Hydrocarbons in Aviation Turbine Fuels by Ultraviolet Spectrophotometry. Annual Book of Standards. ASTM International, West Conshohocken, PA.

ASTM D2007. Standard Test Method for Characteristic Groups in Rubber Extender and Processing Oils and Other Petroleum-Derived Oils by the Clay-Gel Absorption Chromatographic Method. Annual Book of Standards. ASTM International, West Conshohocken, PA.

ASTM D2386. Standard Test Method for Freezing Point of Aviation Fuels. Annual Book of Standards. ASTM International, West Conshohocken, PA.

ASTM D2392. Standard Test Method for Color of Dyed Aviation Gasoline. Annual Book of Standards. ASTM International, West Conshohocken, PA.

ASTM D2425. Standard Test Method for Hydrocarbon Types in Middle Distillates by Mass Spectrometry. Annual Book of Standards. ASTM International, West Conshohocken, PA.

ASTM D2427. Standard Test Method for Determination of C2 through C5 Hydrocarbons in Gasolines by Gas Chromatography. Annual Book of Standards. ASTM International, West Conshohocken, PA.

ASTM D2500. Standard Test Method for Cloud Point of Petroleum Products. Annual Book of Standards. ASTM International, West Conshohocken, PA.

ASTM D2549. Standard Test Method for Separation of Representative Aromatics and Nonaromatics Fractions of High-Boiling Oils by Elution Chromatography. Annual Book of Standards. ASTM International, West Conshohocken, PA.

ASTM D2622. Standard Test Method for Sulfur in Petroleum Products by Wavelength Dispersive X-ray Fluorescence Spectrometry. Annual Book of Standards. ASTM International, West Conshohocken, PA.

ASTM D2709. Standard Test Method for Water and Sediment in Middle Distillate Fuels by Centrifuge. Annual Book of Standards. ASTM International, West Conshohocken, PA.

ASTM D2710. Standard Test Method for Bromine Index of Petroleum Hydrocarbons by Electrometric Titration. Annual Book of Standards. ASTM International, West Conshohocken, PA.

ASTM D2789. Standard Test Method for Hydrocarbon Types in Low Olefinic Gasoline by Mass Spectrometry. Annual Book of Standards. ASTM International, West Conshohocken, PA.

ASTM D2887. Standard Test Method for Boiling Range Distribution of Petroleum Fractions by Gas Chromatography. Annual Book of Standards. ASTM International, West Conshohocken, PA.

ASTM D2889. Standard Test Method for Calculation of True Vapor Pressures of Petroleum Distillate Fuels. Annual Book of Standards. ASTM International, West Conshohocken, PA.

ASTM D2890. Standard Test Method for Calculation of Liquid Heat Capacity of Petroleum Distillate Fuels. Annual Book of Standards. ASTM International, West Conshohocken, PA.

ASTM D2892. Standard Test Method for Distillation of Crude Petroleum (15-Theoretical Plate Column). Annual Book of Standards. ASTM International, West Conshohocken, PA.

ASTM D3120. Standard Test Method for Trace Quantities of Sulfur in Light Liquid Petroleum Hydrocarbons by Oxidative Microcoulometry. ASTM International, West Conshohocken, PA.

ASTM D3227. Standard Test Method for (Thiol Mercaptan) Sulfur in Gasoline, Kerosene, Aviation Turbine, and Distillate Fuels (Potentiometric Method). Annual Book of Standards. ASTM International, West Conshohocken, PA.

ASTM D3239. Standard Test Method for Aromatic Types Analysis of Gas-Oil Aromatic Fractions by High Ionizing Voltage Mass Spectrometry. Annual Book of Standards. ASTM International, West Conshohocken, PA.

ASTM D3242. Standard Test Method for Acidity in Aviation Turbine Fuel. Annual Book of Standards. ASTM International, West Conshohocken, PA.

ASTM D3246. Standard Test Method for Sulfur in Petroleum Gas by Oxidative Microcoulometry. Annual Book of Standards. ASTM International, West Conshohocken, PA.

ASTM D3338. Standard Test Method for Estimation of Net Heat of Combustion of Aviation Fuels. Annual Book of Standards. ASTM International, West Conshohocken, PA.

ASTM D3701. Standard Test Method for Hydrogen Content of Aviation Turbine Fuels by Low Resolution Nuclear Magnetic Resonance Spectrometry. Annual Book of Standards. ASTM International, West Conshohocken, PA.

ASTM D3828. Standard Test Methods for Flash Point by Small Scale Closed Cup Tester. Annual Book of Standards. ASTM International, West Conshohocken, PA.

ASTM D4006. Standard Test Method for Water in Crude Oil by Distillation. Annual Book of Standards. ASTM International, West Conshohocken, PA.

ASTM D4052. Standard Test Method for Density, Relative Density, and API Gravity of Liquids by Digital Density Meter. Annual Book of Standards. ASTM International, West Conshohocken, PA.

ASTM D4176. Standard Test Method for Free Water and Particulate Contamination in Distillate Fuels (Visual Inspection Procedures). Annual Book of Standards. ASTM International, West Conshohocken, PA.

ASTM D4294. Standard Test Method for Sulfur in Petroleum and Petroleum Products by Energy Dispersive X-ray Fluorescence Spectrometry. Annual Book of Standards. ASTM International, West Conshohocken, PA.

ASTM D4529. Standard Test Method for Estimation of Net Heat of Combustion of Aviation Fuels. Annual Book of Standards. ASTM International, West Conshohocken, PA.

ASTM D4808. Standard Test Methods for Hydrogen Content of Light Distillates, Middle Distillates, Gas Oils, and Residua by Low-Resolution Nuclear Magnetic Resonance Spectroscopy. Annual Book of Standards. ASTM International, West Conshohocken, PA.

ASTM D4809. Standard Test Method for Heat of Combustion of Liquid Hydrocarbon Fuels by Bomb Calorimeter (Precision Method. Annual Book of Standards. ASTM International, West Conshohocken, PA.

ASTM D4860. Standard Test Method for Free Water and Particulate Contamination in Middle Distillate Fuels (Clear and Bright Numerical Rating). Annual Book of Standards. ASTM International, West Conshohocken, PA.

ASTM D4952. Standard Test Method for Qualitative Analysis for Active Sulfur Species in Fuels and Solvents (Doctor Test). Annual Book of Standards. ASTM International, West Conshohocken, PA.

ASTM D5186. Standard Test Method for Determination of Aromatic Content and Polynuclear Aromatic Content of Diesel Fuels and Aviation Turbine Fuels by Supercritical Fluid Chromatography. Annual Book of Standards. ASTM International, West Conshohocken, PA.

ASTM D5291. Standard Test Methods for Instrumental Determination of Carbon, Hydrogen, and Nitrogen in Petroleum Products and Lubricants. Annual Book of Standards. ASTM International, West Conshohocken, PA.

ASTM D5292. Standard Test Method for Aromatic Carbon Contents of Hydrocarbon Oils by High Resolution Nuclear Magnetic Resonance Spectroscopy. Annual Book of Standards. ASTM International, West Conshohocken, PA.

ASTM D5443. Standard Test Method for Paraffin, Naphthene, and Aromatic Hydrocarbon Type Analysis in Petroleum Distillates through 200°C by Multi-Dimensional Gas Chromatography. Annual Book of Standards. ASTM International, West Conshohocken, PA.

ASTM D5453. Standard Test Method for Determination of Total Sulfur in Light Hydrocarbons, Spark Ignition Engine Fuel, Diesel Engine Fuel, and Engine Oil by Ultraviolet Fluorescence. Annual Book of Standards. ASTM International, West Conshohocken, PA.

ASTM D5482. Standard Test Method for Vapor Pressure of Petroleum Products (Mini Method – Atmospheric). Annual Book of Standards. ASTM International, West Conshohocken, PA.

ASTM D5580. Standard Test Method for Determination of Benzene, Toluene, Ethylbenzene, p/m-Xylene, o-Xylene, C9 and Heavier Aromatics, and Total Aromatics in Finished Gasoline by Gas Chromatography. Annual Book of Standards. ASTM International, West Conshohocken, PA.

ASTM D5623. Standard Test Method for Sulfur Compounds in Light Petroleum Liquids by Gas Chromatography and Sulfur Selective Detection. Annual Book of Standards. ASTM International, West Conshohocken, PA.

ASTM D5769. Standard Test Method for Determination of Benzene, Toluene, and Total Aromatics in Finished Gasolines by Gas Chromatography/Mass Spectrometry. Annual Book of Standards. ASTM International, West Conshohocken, PA.

ASTM D5771. Standard Test Method for Cloud Point of Petroleum Products (Optical Detection Stepped Cooling Method). Annual Book of Standards. ASTM International, West Conshohocken, PA.

ASTM D5772. Standard Test Method for Cloud Point of Petroleum Products (Linear Cooling Rate Method). Annual Book of Standards. ASTM International, West Conshohocken, PA.

ASTM D5773. Standard Test Method for Cloud Point of Petroleum Products (Constant Cooling Rate Method). Annual Book of Standards. ASTM International, West Conshohocken, PA.

ASTM D5853. Standard Test Method for Pour Point of Crude Oils. Annual Book of Standards. ASTM International, West Conshohocken, PA.

ASTM D5972. Standard Test Method for Freezing Point of Aviation Fuels (Automatic Phase Transition Method). Annual Book of Standards. ASTM International, West Conshohocken, PA.

ASTM D5986. Standard Test Method for Determination of Oxygenates, Benzene, Toluene, C8-C12 Aromatics and Total Aromatics in Finished Gasoline by Gas Chromatography/ Fourier Transform Infrared Spectroscopy. Annual Book of Standards. ASTM International, West Conshohocken, PA.

ASTM D6045. Standard Test Method for Color of Petroleum Products by the Automatic Tristimulus Method. Annual Book of Standards. ASTM International, West Conshohocken, PA.

ASTM D6296. Standard Test Method for Total Olefins in Spark-ignition Engine Fuels by Multidimensional Gas Chromatography. Annual Book of Standards. ASTM International, West Conshohocken, PA.

ASTM D6304. Standard Test Method for Determination of Water in Petroleum Products, Lubricating Oils, and Additives by Coulometric Karl Fischer Titration. Annual Book of Standards. ASTM International, West Conshohocken, PA.

ASTM D6352. Standard Test Method for Boiling Range Distribution of Petroleum Distillates in Boiling Range from 174 to 700°C by Gas Chromatography. Annual Book of Standards. ASTM International, West Conshohocken, PA.

ASTM D6378. Standard Test Method for Determination of Vapor Pressure (VPX) of Petroleum Products, Hydrocarbons, and Hydrocarbon-Oxygenate Mixtures (Triple Expansion Method). Annual Book of Standards. ASTM International, West Conshohocken, PA.

ASTM D6379. Standard Test Method for Determination of Aromatic Hydrocarbon Types in Aviation Fuels and Petroleum Distillates—High Performance Liquid Chromatography Method with Refractive Index Detection. Annual Book of Standards. ASTM International, West Conshohocken, PA.

Garrigues, S., Andrade, J.M., de la Guardia, M., and Prada, D. 1995. Multivariate calibrations in Fourier transform infrared spectrometry for prediction of kerosene properties. Analytica Chimica Acta, 317: 95–105.

Gary, J.G., Handwerk, G.E., and Kaiser, M.J. 2007. Petroleum Refining: Technology and Economics, 5th Edition. CRC Press, Taylor & Francis Group, Boca Raton, FL.

Heinrich, H. and Duée, D. 2000. Kerosine and gasoil manufacture. In Modern Petroleum Technology. Volume 2, Downstream. A.G. Lucas (Editor). John Wiley & Sons Inc., New York.

Hsu, C.S. and Robinson, P.R. (Editors) 2006. Practical Advances in Petroleum Processing (Volumes 1 and 2). Springer Science, New York.

IP 2 (ASTM D611). Petroleum Products and Hydrocarbon Solvents – Determination of Aniline and Mixed Aniline Point. IP Standard Methods 2013. The Energy Institute, London.

IP 10. Determination of Kerosene Burning Characteristics – 24 Hour Method. IP Standard Methods 2013. The Energy Institute, London.

IP 12. Determination of Specific Energy. IP Standard Methods 2013. The Energy Institute, London.

IP 15 (ASTM D97). Petroleum products – Determination of Pour Point. IP Standard Methods 2013. The Energy Institute, London.

IP 16 (ASTM D2386). Petroleum products – Determination of the Freezing Point of Aviation Fuels – Manual method. IP Standard Methods 2013. The Energy Institute, London.

IP 30. Doctor Test. Energy institute, London, UK.

IP 34 (ASTM D93). Determination of Flash Point – Pensky-Martens Closed Cup Method. IP Standard Methods 2013. The Energy Institute, London.

IP 57. Petroleum Products – Determination of the Smoke Point of Kerosene. IP Standard Methods 2013. The Energy Institute, London.

IP 61 (ASTM D129). Determination of Sulfur – High Pressure Combustion Method. IP Standard Methods 2013. The Energy Institute, London.

IP 69. Determination of Vapor Pressure – Reid Method. IP Standard Methods 2013. The Energy Institute, London.

IP 71 (ASTM D445). Petroleum Products – Transparent and Opaque Liquids – Determination of Kinematic Viscosity and Calculation of Dynamic Viscosity. IP Standard Methods 2013. The Energy Institute, London.

IP 74 (ASTM D95). Petroleum Products and Bituminous Materials – Determination of Water – Distillation Method. IP Standard Methods 2013. The Energy Institute, London.

IP 107 (ASTM D1266). Determination of Sulfur – Lamp Combustion Method. IP Standard Methods 2013. The Energy Institute, London.

IP 123. Petroleum Products – Determination of Distillation Characteristics at Atmospheric Pressure. IP Standard Methods 2013. The Energy Institute, London.

IP 130 (ASTM D1159). Petroleum Products – Determination of Bromine Number of Distillates and Aliphatic Olefins – Electrometric Method. IP Standard Methods 2013. The Energy Institute, London.

IP 131. Petroleum Products – Gum Content of Light and Middle Distillates – Jet Evaporation Method. IP Standard Methods 2013. The Energy Institute, London.

IP 154. Petroleum Products – Corrosiveness to Copper – Copper Strip Test. IP Standard Methods 2013. The Energy Institute, London.

IP 156. Petroleum Products and Related Materials – Determination of Hydrocarbon Types – Fluorescent Indicator Adsorption Method. IP Standard Methods 2013. The Energy Institute, London.

IP 160 (ASTM D1298). Crude Petroleum and Liquid Petroleum Products – Laboratory Determination of Density – Hydrometer Method. IP Standard Methods 2013. The Energy Institute, London.

IP 170. Petroleum Products and Other Liquids – Determination of Flash Point – Abel Closed Cup Method. IP Standard Methods 2013. The Energy Institute, London.

IP 219. Petroleum Products – Determination of Cloud Point. IP Standard Methods 2013. The Energy Institute, London.

IP 289. Determination of Water Reaction of Aviation Fuels. IP Standard Methods 2013. The Energy Institute, London.

IP 336. Petroleum Products – Determination of Sulfur Content – Energy-Dispersive X-Ray Fluorescence Method. IP Standard Methods 2013. The Energy Institute, London.

IP 342 (ASTM D3227). Determination of Thiol (Mercaptan) Sulfur in Light and Middle Distillate Fuels – Potentiometric Method. IP Standard Methods 2013. The Energy Institute, London.

IP 354 (ASTM D3242). Total Acidity of Aviation Turbine Fuel – Color Indicator Titration Method. IP Standard Methods 2013. The Energy Institute, London.

IP 358. Crude Petroleum – Determination of Water – Distillation Method. IP Standard Methods 2013. The Energy Institute, London.

IP 365. Crude Petroleum and Petroleum Products – Determination of Density – Oscillating U-Tube Method. IP Standard Methods 2013. The Energy Institute, London.

IP 373. Sulfur Content – Microcoulometry (Oxidative) Method. IP Standard Methods 2013. The Energy Institute, London.

IP 434. Freezing Point of Aviation Turbine Fuels – Automated Optical Method (replaced by IP 528 – ASTM D5901). IP Standard Methods 2013. The Energy Institute, London.

IP 435. Freezing Point of Aviation Turbine Fuels – Automatic Phase Transition Method. IP Standard Methods 2013. The Energy Institute, London.

IP 523. Determination of Flash Point – Rapid Equilibrium Closed Cup Method. IP Standard Methods 2013. The Energy Institute, London.

Speight, J.G. 2000. The Desulfurization of Heavy Oils and Residua, 2nd Edition. Marcel Dekker Inc., NY.

Speight, J.G. 2001. Handbook of Petroleum Analysis. John Wiley & Sons Inc., New York.

Speight, J.G. 2014. The Chemistry and Technology of Petroleum, 5th Edition. CRC Press, Taylor & Francis Group, Boca Raton, FL.

Speight, J.G. and Ozum, B. 2002. Petroleum Refining Processes. Marcel Dekker Inc., New York.

9

DIESEL FUEL

9.1 INTRODUCTION

Kerosene, diesel fuel, and aviation turbine fuel (jet fuel) are members of the class of petroleum products known as middle distillates (Speight 2001; Speight and Ozum, 2002; Hsu and Robinson, 2006; Gary *et al.*, 2007; Speight, 2014). As the name implies, these products are higher boiling than gasoline but lower boiling than gas oil. Middle distillates cover the boiling range from approximately 175 to 375°C (350 to 700°F), and the carbon number ranges from about C_8 to C_{24}. These products have similar properties but different specifications as appropriate for their intended use.

The broad definition of fuels for land and marine diesel engines and for non-aviation gas turbines covers many possible combinations of volatility, ignition quality, viscosity, gravity, stability, and other properties. Various specifications are used to characterize these fuels (ASTM D975, ASTM D2880).

A steady evolution in product specifications caused by an endless wave of fresh environmental regulations plays a major role in the development of petroleum refining technologies. In the United States and Europe, gasoline and diesel specifications have changed radically in the past decades and will continue to do so in the future. Currently, reducing the sulfur levels of finished products is the dominant objective. Sulfur is ubiquitous in petroleum, and refiners are seeking technologies on how to achieve the mandated levels of sulfur in petroleum products.

To characterize diesel fuel for use and thereby establish a framework of definition and reference, various classifications are used in different countries—within the grades, there are variations in the specifications for the physical and chemical compositions of the fuel (Table 9.1) (ASTM D975). In the United States, grades No. 1-D and 2-D are distillate fuels and are the types most commonly used in high-speed (mobile) engines of the mobile type, in medium-speed (stationary) engines, and in railroad engines (ASTM D975). Grade 4-D covers the class of more viscous distillates and may even include blends of these viscous distillates with residual fuel oils. No. 4-D fuels are applicable for use in low- and medium-speed engines employed in services involving sustained load and predominantly constant speed.

9.2 PRODUCTION AND PROPERTIES

Diesel fuels originally were straight-run products obtained from the distillation of crude oil. Currently, diesel fuel may also contain varying amounts of selected cracked distillates to increase the volume available (Song *et al.*, 2000; Speight and Ozum, 2002; Hsu and Robinson, 2006; Gary *et al.*, 2007; Speight, 2014). The boiling range of diesel fuel is approximately 125–328°C (302–575°F) (Table 9.2). Thus, in terms of the carbon number and boiling range, diesel fuel occurs predominantly in the kerosene range (Chapter 8), and this many of the test methods applied to kerosene can also be applied to diesel fuel. Diesel fuel depends upon the nature of the original crude oil, the refining processes by which the fuel is produced, and the additive (if any) used, such as the solvent red dye (ASTM D6258). Furthermore, the specification for diesel fuel can exist in various combinations of characteristics such as, for example, volatility, ignition quality, viscosity, gravity, and stability.

Handbook of Petroleum Product Analysis, Second Edition. James G. Speight.
© 2015 John Wiley & Sons, Inc. Published 2015 by John Wiley & Sons, Inc.

TABLE 9.1   Various grades of distillate fuel oil (ASTM D975)

Grade	Description
No. 1-D S15	Ultra-low-sulfur (ULSD) light middle distillate fuel used in diesel engine applications that require higher volatility than No. 2-D S15 (maximum sulfur: 15 ppm).
No. 1-D S500	Low-sulfur light middle distillate fuel used in diesel engine applications that require higher volatility than No. 2-D S500 (maximum sulfur: 500 ppm).
No. 1-D S5000	Regular-sulfur light middle distillate fuel used in diesel engine applications that require higher volatility than No. 2-D S5000 with (maximum sulfur: 5000 ppm).
No. 2-D S15	Ultra-low-sulfur (ULSD) middle distillate fuel used for general-purpose diesel engine applications with varying speed and load (maximum sulfur: 15 ppm).
No. 2-D S500	Low-sulfur middle distillate fuel used for general-purpose diesel engine applications with varying speed and load (maximum sulfur: 500 ppm).
No. 2-D S5000	Low-sulfur middle distillate fuel used for general-purpose diesel engine applications with varying speed and load (maximum sulfur: 5000 ppm).
No. 4-D	Heavy distillate fuel, or a blend of distillate and residual oil, used for low- and medium-speed diesel engine applications involving primarily constant speed and load.

TABLE 9.2   General summary of product types and distillation range

Product	Lower carbon limit	Upper carbon limit	Lower boiling point (°C)	Upper boiling point (°C)	Lower boiling point (°F)	Upper boiling point (°F)
Refinery gas	C_1	C_4	−161	−1	−259	31
Liquefied petroleum gas	C_3	C_4	−42	−1	−44	31
Naphtha	C_5	C_{17}	36	302	97	575
Gasoline	C_4	C_{12}	−1	216	31	421
Kerosene/diesel fuel	C_8	C_{18}	126	258	302	575
Aviation turbine fuel	C_8	C_{16}	126	287	302	548
Fuel oil	C_{12}	$>C_{20}$	216	421	>343	>649
Lubricating oil	$>C_{20}$		>343		>649	
Wax	C_{17}	$>C_{20}$	302	>343	575	>649
Asphalt	$>C_{20}$		>343		>649	
Coke	$>C_{50}{}^a$		$>1000^a$		$>1832^a$	

^{a}Carbon number and boiling point difficult to assess; inserted for illustrative purposes only.

One of the most widely used specifications (ASTM D975) covers three grades of diesel fuel oils, No. l-D, No. 2-D, and No. 4-D. Grades No. l-D and 2-D are distillate fuels (ASTM D975), the types most commonly used in high-speed engines of the mobile type, in medium-speed stationary engines, and in railroad engines. Grade 4-D covers the class of more viscous distillates and, at times, blends of these distillates with residual fuel oils.

Additives may be used to improve the fuel performance—additives such as alkyl nitrates and nitrites can improve ignition quality (ASTM D1839, ASTM D4046). Pour point depressants can improve low temperature performance. Antismoke additives reduce exhaust smoke, which is of growing concern as more and more attention is paid to atmospheric pollution. Antioxidant and sludge dispersants may also be used, particularly with fuels formulated with cracked components, in order to prevent the formation of insoluble compounds that could cause line and filter plugging (ASTM D2068, ASTM D6371, IP 309).

9.3   TEST METHODS

As for all fuels, the properties of a product define the ability to serve a stated purpose. Once the required properties are determined, they are controlled by appropriate tests and analyses. The quality criteria and methods for testing fuels for land and marine diesel engines, such as the cetane number, apply to both fuels.

The basic fuel requirements for land and marine diesel engines and for non-aviation gas turbines are satisfactory ignition and combustion under the conditions existing in the combustion chamber, suitability for handling by the injection equipment, and convenient handling at all stages from the refinery to the engine fuel tank without suffering degradation and without harming any surface that it may normally contact.

Diesel and non-aviation gas-turbine fuels were originally straight-run products obtained from the distillation of crude oil. Because of the various refinery cracking processes, modern fuels may contain varying amounts of selected cracked distillates. This permits an increase in the volume of available fuel at minimum cost. The boiling range of distillate fuels is approximately 150–400°C (300–755°F). The relative merits of the fuel types to be considered will depend upon the refining practices employed, the nature of crude oils from which they are produced, and the additive package (if any) used.

9.3.1 Acidity

Petroleum products may contain acidic constituents that are present as additives or as degradation products, such as oxidation products. The relative amount of these constituents can be determined by titration of a sample of the product with base, and the *acid number* is a measure of this amount of acidic substances in the product under the conditions of the test.

One of the test methods (ASTM D664) resolves constituents into groups having weak-acid and strong-acid ionization properties. In this test method, the sample is dissolved in a mixture of toluene and iso-propyl alcohol containing a small amount of water and titrated potentiometrically with alcoholic potassium hydroxide using a glass indicating electrode and a calomel reference electrode. The meter readings are plotted manually or automatically against the respective volumes of titrating solution, and the end points are taken only at well-defined inflections in the resulting curve. The test method may be used to indicate relative changes that occur in diesel fuel under oxidizing conditions regardless of the color or other properties of the resulting oil. There are three other test methods for the determination of acid numbers (ASTM D974, ASTM D3339, ASTM D4739) that are used to measure the inorganic and total acidity of the fuel and indicates its tendency to corrode metals that it may contact.

9.3.2 Appearance and Odor

The general *appearance*, or *color*, of diesel fuel is a useful indicator against contamination by residual (higher-boiling) constituents, water, or fine solid particles. Therefore, it is prudent to check by visual inspection that clear fuel is being delivered. Therefore, it is necessary to make a visual inspection that clear fuel is being delivered (ASTM D4176).

Color, being part of the appearance of diesel fuel, should also be determined since the color of petroleum products is used for manufacturing control purposes. In some cases, the color may serve as an indication of the degree of refinement of the material. Several color scales are used for the determination of color (ASTM D156, ASTM D1209, ASTM D1500, ASTM D1544, IP 196). Typically, the methods require a visual determination of color using colored glass disks or reference materials.

Similarly, acceptance is important with regard to *odor*, and it is usually required that diesel fuel is reasonably free of contaminants, such as mercaptans, which impart unpleasant odors to the fuel (ASTM D4952, IP 30).

9.3.3 Ash

Small amounts of nonburnable material are found in diesel fuel in the form of soluble metallic soaps and solids, and these materials are designated as ash, although ash-forming constituents are a more correct term. In the test for the quantitative determination of ash-forming constituents (ASTM D482, IP 4), a small sample of fuel is burned in a weighed container until all of the combustible matter has been consumed indicated by the residue and container attaining a constant weight. The amount of unburnable residue is the ash content and is reported as percentage by weight of the sample.

The ash-forming constituents in diesel fuel (ASTM D2880) are typically so low that it does not adversely affect gas turbine performance, unless such corrosive species as sodium, potassium, lead, or vanadium are present. However, there are recommendations for the storage and handling of these fuels (ASTM D4418) to minimize potential contamination.

Vanadium can form low-melting compounds, such as vanadium pentoxide, that melt at 690°C (1275°F) and cause severe corrosive attack on all of the high-temperature alloys used for gas-turbine blades and diesel engine valves. And, if there is sufficient magnesium in the fuel, it will combine with the vanadium to form compounds with higher melting points and thus reduce the corrosion rate to an acceptable level. The resulting ash will form deposits in the turbine, but the deposits are self-spalling when the turbine is shut down. The use of a silicon-based additive will further reduce the corrosion rate by absorption and dilution of the vanadium compounds.

Sodium and potassium can combine with vanadium to form eutectics that melt at temperatures as low as 565°C (1050°F) and with sulfur in the fuel to yield sulfates with melting points in the operating range of the gas turbine. These compounds produce severe corrosion for turbines operating at gas inlet temperatures above 650°C (1200°F). Thus, the amount of sodium-plus-potassium must be limited. For gas turbines operating below 650°C (1200°F), the corrosion due to sodium compounds is of lesser importance and can be further reduced by silicon-based additive.

Calcium is not as harmful and may even serve to inhibit the corrosive action of vanadium. However, the presence of calcium can lead to deposits that are not self-spalling when the gas turbine is shut down and not readily removed by water washing of the turbine. Lead can cause corrosion, and, in addition, it can spoil the beneficial inhibiting effect of magnesium additives on vanadium corrosion. Since lead only is found rarely in significant quantities in crude oils, its presence in the fuel oil is primarily the result of contamination during processing or transportation.

9.3.4 Calorific Value

The calorific value (heat of combustion) (ASTM D240, ASTM D1405, ASTM D2890, ASTM D3338, ASTM D4529, ASTM D4809, ASTM D4868, ASTM D6446, IP 12) is a direct measure of fuel energy content and is determined as the quantity of heat liberated by the combustion of a unit quantity of fuel with oxygen in a standard bomb calorimeter. There are two heats of combustion, or calorific values, for every petroleum fuel: gross and net. When hydrocarbons are burned, one of the products of combustion is water vapor, and the difference between the two calorific values is that the gross includes the heat given by the water vapor in condensing, while the net value does not include this heat.

When an experimental determination of heat of combustion is not available and cannot be made conveniently, an estimate might be considered satisfactory (ASTM D6446). In this test method, the net heat of combustion is calculated from the density, sulfur content, and hydrogen content, but this calculation is justifiable only when the fuel belongs to a well-defined class for which a relationship between these quantities has been derived from accurate experimental measurements on representative samples. Thus, the hydrogen content (ASTM D1018, ASTM D1217, ASTM D1298, ASTM D3701, ASTM D4052, ASTM D4808, ASTM D5291, IP 160, IP 365), density (ASTM D129, ASTM D1250, ASTM D1266, ASTM D2622, ASTM D3120, IP 61), and sulfur content (ASTM D2622, ASTM D3120, ASTM D3246, ASTM D4294, ASTM D5453, ASTM D5623, IP 336, IP 373) of the sample are determined by experimental test methods, and the net heat of combustion is calculated using the values obtained by these test methods based on reported correlations.

An alternative criterion of energy content is the *aniline gravity product* (AGP), which is related to calorific value (ASTM D1405). The *AGP* is the product of the API gravity (ASTM D287, ASTM D1298) and the aniline point of the fuel (ASTM D611, IP 2). The aniline point is the lowest temperature at which the fuel is miscible with an equal volume of aniline and is inversely proportional to the aromatic content. The relationship between the AGP and calorific value is given in the method. In another method (ASTM D3338), the heat of combustion is calculated from the fuel density; the 10, 50, and 90% distillation temperatures; and the aromatic content. However, neither method is legally acceptable, and other methods (ASTM D240, ASTM D1655, ASTM D4809) are preferred.

9.3.5   Carbon Residue

The carbon residue of a petroleum product serves as an indication of the propensity of the sample to form carbonaceous deposits (thermal coke) under the influence of heat. In the current context, carbon residue test results are widely quoted in diesel fuel specifications. However, distillate diesel fuels that are satisfactory in other respects do not have high Conradson carbon residue values, and the test is chiefly used on residual fuels.

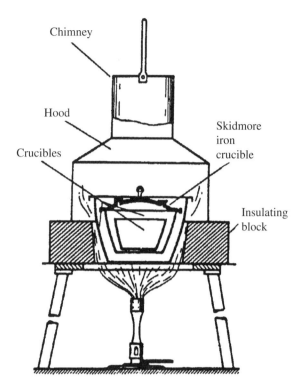

FIGURE 9.1   Apparatus for the determination of the Conradson carbon residue (ASTM D189, IP 13).

Tests for Conradson carbon residue (ASTM D189, IP 13) (Fig. 9.1), the Ramsbottom carbon residue (ASTM D524, IP 14), and the microcarbon carbon residue (ASTM D4530, IP 398) are often included in the specification date for diesel fuel. The data give an indication of the amount of coke that will be formed during thermal processes as well as an indication of the amount of high-boiling constituents in petroleum.

9.3.6   Cetane Number

The cetane number is an important property of diesel fuel and is a measure of the tendency of a diesel fuel to knock in a diesel engine. In the majority of diesel engines, the ignition delay period is shorter than the duration of injection. Under such circumstances, the total combustion period can be considered to be divided into the following four stages: (1) ignition delay, (2) rapid pressure rise, (3) constant pressure or controlled pressure rise, and (4) burning on the expansion stroke.

The cetane number scale is based upon the ignition characteristics of two hydrocarbons: *n*-hexadecane (cetane) and 2,3,4,5,6,7,8-heptamethylnonane. Cetane has a short delay period during ignition and is assigned a cetane number of 100; heptamethylnonane has a long delay period and has been assigned a cetane number of 15. Just as the octane number is meaningful for automobile fuels, the cetane number is a means of determining the ignition quality of diesel fuels and is equivalent to the percentage by volume of cetane in the blend with heptamethylnonane, which matches the ignition quality of the test fuel (ASTM D613).

The cetane number of a diesel fuel is the numerical result of an engine test designed to evaluate fuel ignition delay. A cetane number 100 was arbitrarily assigned to this fuel. The second fuel, α-methylnaphthalene, has poor ignition qualities and was assigned a cetane number of 0. α-Methylnaphthalene has been replaced as a primary reference fuel by heptamethylnonane, which has a cetane number of 15 as determined by the use *of* the two original primary reference fuels.

Thus, the cetane number of diesel fuel is defined the whole number nearest to the value determined by calculation from the percentage by volume of normal cetane (cetane No. = 100) in a blend with heptamethylnonane (cetane No. = 15), which matches the ignition quality of the test fuel when compared by this method. To determine the cetane number of any fuel, its ignition delay is compared in a standard test engine with a blend of reference fuels (ASTM D613, IP 41). The matching blend percentages to the first decimal are inserted in the following equation to obtain the cetane number:

$$\text{Cetane number} = \% \, v/v \, n - \text{cetane} + 0.15\% \, v/v \text{ heptamethylnonane})$$

The shorter the ignition delay period, the higher the cetane number of the fuel and the smaller the amount of fuel in the combustion chamber when the fuel ignites. Consequently, high–cetane number fuels generally cause lower rates of pressure rise and lower peak pressures, both of which tend to lessen combustion noise and to permit improved control of combustion, resulting in increased engine efficiency and power output.

In addition to the advantages given earlier, higher–cetane number fuels tend to result in easier starting, particularly in cold weather, and faster warm-up. The higher–cetane number fuels also usually form softer and hence more readily purged combustion chamber deposits and result in reduced exhaust smoke and odor. High-speed diesel engines normally are supplied with fuels in the range of 45–55 cetane number.

Since the determination of cetane number by engine testing requires special equipment, as well as being time consuming and costly, alternative methods have been developed for calculating estimates of cetane number. The calculations are based upon equations involving values of other known characteristics of the fuel.

One of the most widely used methods is based on the calculated Cetane Index formula. This formula represents a method for estimating the cetane number of distillate fuels from API gravity and mid-boiling point. The index value as

computed from the formula is designated as a calculated Cetane Index (ASTM D976). Since the formula is complicated in its manipulation, a nomograph based on the equation has been developed for its solution.

9.3.7 Cloud Point

Under low-temperature conditions, paraffinic constituents of diesel fuel may be precipitated as a wax. This settles out and blocks the fuel system lines and filters, causing malfunctioning or stalling of the engine. The temperature at which the precipitation occurs depends upon the origin, type, and boiling range of the fuel. The more paraffinic the fuel, the higher the precipitation temperature and the less suitable the fuel for low-temperature operation.

The temperature at which wax is first precipitated from solution can be measured by the cloud point test (ASTM D2500, ASTM D5771, ASTM D5772, ASTM D5773, IP 219). The cloud point of a diesel fuel is a guide to the temperature at which it may clog filter systems and restrict flow. Cloud point is becoming increasingly important for fuels used in high-speed diesel engines, especially because of the tendency to equip such engines with finer filters. The finer the filter, the more readily it will become clogged by small quantities of precipitated wax. Larger fuel lines and filters of greater capacity reduce the effect of deposits from the fuel and therefore widen the cloud point range of fuels that can be used.

In the simple cloud point test method (ASTM D2500), the sample is first heated to a temperature above the expected cloud point and then cooled at a specified rate and examined periodically. The temperature at which haziness is first observed at the bottom of the test jar is recorded as the cloud point.

9.3.8 Composition

The chemical composition of diesel fuel is extremely complex with an extremely high number of compounds potentially present (Table 9.3). For this reason, it usually is not practical to analyze diesel fuel for individual compounds, but it is often advantageous to define the compounds present as broad classifications of compound types, such as aromatics, paraffins, naphthenes, and olefins.

One of the most important physical parameters defining diesel fuel, and other middle distillate products, is the boiling range distribution (ASTM D86, ASTM D2887, ASTM D2892). However, the first major level of compositional information is group-type totals as deduced by adsorption chromatography (ASTM D1319) (Fig. 9.2) to give volume percent saturates, olefins, and aromatics in materials that boil below 315°C (600°F). Following from this separation, the compositional analysis of diesel fuel and other middle distillates is then determined by a mass spectral Z-series on which

TABLE 9.3   Increase in the number of isomers with carbon number

Carbon atoms	Number of isomers
1	1
2	1
3	1
4	2
5	3
6	5
7	9
8	18
9	35
10	75
15	4,347
20	366,319
25	36,797,588
30	4,111,846,763
40	62,491,178,805,831

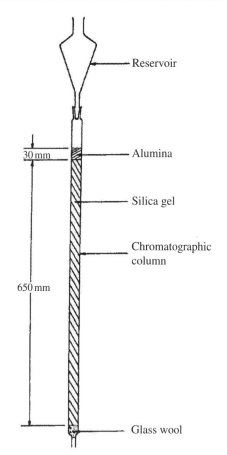

FIGURE 9.2   Apparatus for adsorption chromatography.

Z in the empirical formula C_nH_{2n+z} is a measure of the hydrogen deficiency of the compound (ASTM D2425, ASTM D2786, ASTM D3239, ASTM D6379). Mass spectrometry can provide more compositional detail than chromatographic analysis. However, this method requires that the sample be separated into saturate and aromatic fractions (ASTM D2549) before mass spectrometric analysis. This separation is applicable to diesel fuel but not to jet fuel, since it is impossible to evaporate the solvent used in the separation without also losing the light ends of the jet fuel.

The aromatic hydrocarbon content of diesel fuel affects the cetane number and exhaust emissions. One test method (ASTM D5186) is applicable to diesel fuel and is unaffected by fuel coloration. Aromatic concentration in the range of 1–75 mass % and polynuclear aromatic hydrocarbons in the range of 0.5–50 mass % can be determined by this test method. In the method, a small aliquot of the fuel sample is injected onto a packed silica adsorption column and eluted using supercritical carbon dioxide mobile phase. Mono- and polynuclear aromatics in the sample are separated from nonaromatics and detected using a flame ionization detector. The detector response to hydrocarbons is recorded throughout the analysis time. The chromatographic areas corresponding to the mononuclear aromatic constituents, polynuclear aromatic constituents, and non-aromatic constituents are determined, and the mass percent content of each of these groups is calculated by area normalization.

While nuclear magnetic resonance can be used to determine mass-percent hydrogen in diesel fuel (ASTM D3701, ASTM D4808), the percentage of aromatic hydrogen atoms and aromatic carbon atoms can be determined by high-resolution nuclear magnetic resonance (ASTM D5292). However, the results from this test are not equivalent to mass- or volume-percent aromatics determined by the chromatographic methods. The chromatographic methods determine the mass or volume percentage of molecules that have one or more aromatic rings. Any alkyl substituents on the rings contribute to the percentage of aromatics determined by chromatographic techniques.

The significance of the total sulfur content of diesel fuel cannot be overestimated and is of great importance because of the production of sulfur oxides that contaminate the surroundings. Generally, only slight amounts of sulfur compounds remain in diesel fuel after refining, and the diesel fuel must meet sulfur specification. However, with the planned reduction of sulfur in future specifications, sulfur detection becomes even more important.

Refining treatment includes among its objects the removal of such undesirable products as hydrogen sulfide, mercaptan sulfur, and *free* or corrosive sulfur. Hydrogen sulfide and mercaptans cause objectionable odors, and both are corrosive. The presence of such compounds can be determined by the Doctor test (ASTM D4952, IP 30). The Doctor test (which is pertinent for petroleum product specifications, ASTM D235) ensures that the concentration of these compounds is insufficient to cause such problems in normal use. In the test, the sample is shaken with sodium plumbite solution, a small quantity of sulfur is added, and the mixture is shaken again. The presence of mercaptans or hydrogen sulfide or both is indicated by discoloration of the sulfur floating at the oil–water interface or by discoloration of either of the phases.

Free, or corrosive, sulfur in appreciable amount could result in corrosive action on the metallic components of an appliance. Corrosive action is of particular significance in the case of pressure burner vaporizing tubes that operate at high temperatures. The usual test applied in this connection is the corrosion (copper strip) test (ASTM D130, ASTM D849, IP 154).

The copper strip test methods are used to determine the corrosiveness to copper of diesel fuel and are a measure of the relative degree of corrosivity of diesel fuel. Most sulfur compounds in petroleum are removed during refining. However, some residual sulfur compounds can have a corroding action on various metals, and the effect is dependent on the types of sulfur compounds present. One method (ASTM D130, IP 154) uses a polished copper strip that is immersed in a given quantity of sample and heated at a temperature for a time period characteristic of the material being tested. At the end of this period, the copper strip is removed, washed, and compared with the copper strip corrosion standards (ASTM, 2001). This is a pass/fail test. In another method (ASTM D849), a polished copper strip is immersed in 200 ml of specimen in a flask with a condenser and placed in boiling water for 30 min. At the end of this period, the copper strip is removed and compared with the ASTM copper strip corrosion standards. This is also a pass/fail test.

Nitrogen compounds in diesel fuel and middle distillates can be selectively detected by chemiluminescence. Individual nitrogen compounds can be detected down to 100 ppb nitrogen.

Sulfur can cause wear, resulting from the corrosive nature of its combustion by-products and from an increase in the amount of deposits in the combustion chamber and on the pistons. The sulfur content of a diesel fuel (ASTM D129, ASTM D1266, ASTM D1552, ASTM D2622, ASTM D4294, IP 61) depends on the origin of the crude oil from which it is made and on the refining methods. Sulfur can be present in a number of forms, for example, as mercaptans, sulfides, disulfides, or heterocyclic compounds such as thiophene derivatives, all of which will affect wear and deposits.

Sulfur is measured both on the basis of quantity and potential corrosivity. The quantitative measurements can be made by means of a combustion bomb (ASTM D129, IP 61). The measurement of potential corrosivity can be determined by means of a copper strip procedure (ASTM D130, IP 154). The quantitative determination is an indication of the corrosive tendencies of the fuel combustion products, while the potential corrosivity indicates the extent of corrosion to be anticipated from the unburned fuel—particularly in the fuel injection system.

In gas turbine fuel, sulfur compounds (ASTM D2880) (notably hydrogen sulfide, elemental sulfur, and polysulfides) can be corrosive (ASTM D130, IP 30) in the fuel handling systems, and mercaptans can attack any elastomers present.

Thus, mercaptan sulfur content (ASTM D235, ASTM D3227, IP 30) is limited to low levels because of objectionable odor, adverse effects on certain fuel system elastomers, and corrosiveness toward fuel system metals.

9.3.9 Density

Density (or specific gravity) is an indication of the density or weight per unit volume of the diesel fuel. The principal use of specific gravity (ASTM D1298, IP 160) is to convert weights of oil to volumes or volumes to weights. Specific gravity also is required when calculating the volume of petroleum or a petroleum product at a temperature different from that at which the original volume was measured. Although specific gravity by itself is not a significant measure of quality, it may give useful information when considered with other tests. For a given volatility range, high specific gravity is associated with aromatic or naphthenic hydrocarbons and low specific gravity with paraffinic hydrocarbons. The heat energy potentially available from the fuel decreases with an increase in density, or specific gravity.

API gravity (ASTM D1298, IP 160) is an arbitrary figure related to the specific gravity in accordance with the following formula:

$$°API = 141.5 / (specific\,gravity\,at\,60/60°F) / 131.5$$

When a fuel requires centrifuging, density is a critical property, and as the density of the fuel approaches the density of water (API gravity = 10°C), the efficiency of centrifuging decreases. When separation of water from the fuel is not required, density is not a significant measure of fuel quality, but when used in conjunction with other tests, it may give useful information.

For example, for a given volatility range, high specific gravity is associated with aromatic or naphthenic hydrocarbons and low specific gravity with paraffinic hydrocarbons. Further, calorific value (heat of combustion) decreases with an increase in density or specific gravity. However, calorific value expressed per volume of fuel increases with an increase in density or specific gravity.

9.3.10 Diesel Index

The Diesel Index is derived from the API gravity and aniline point (ASTM D611, IP 2), the lowest temperature at which the fuel is completely miscible with an equal volume of aniline:

$$Diesel\,index = aniline\,point\,(°F)\,API\,gravity / 100$$

This equation is seldom used because the results can be misleading, especially when applied to blended fuels.

9.3.11 Flash Point

The flash point of a diesel fuel is the temperature to which the fuel must be heated to produce an ignitable vapor–air mixture above the liquid fuel when exposed to an open flame. The flash point test is a guide to the fire hazard associated with the use of the fuel and can be determined by several test methods, but the results are not always strictly comparable.

The minimum flash point is usually defined by the Abel method (IP 170), although the Pensky–Martens method (ASTM D93, IP 34) may also be specified. The TAG method (ASTM D56) is for both the minimum and maximum limits, while certain military specifications also give minimum limits by the Pensky–Martens method (ASTM D93, IP 34). The Abel method (IP 170) can give results up to 2–3°C (3–5°F lower than the TAG method (ASTM D56).

Similarly, for diesel fuel, the flash point is a guide to the fire hazard associated with the fuel and can be determined by the same test methods as noted earlier (ASTM D56, ASTM D93, ASTM D3828, IP 34, IP 170, IP 523).

It should be noted that the various flash point methods can yield different numerical results, and in the case of the two most commonly used methods (Abel and TAG), it has been found that the former (IP 170) can give results up to 1–2°C lower than the latter method (ASTM D56, ASTM D3828, IP 523) results are generally very close to Abel values.

In practice, flash point is important primarily for fuel handling. A flash point that is too low will cause fuel to be a fire hazard, subject to flashing and possible continued ignition and explosion. In addition, a low flash point may indicate contamination by more volatile and explosive fuels, such as gasoline.

9.3.12 Freezing Point

There are three test methods available for the determination of the freezing point. All three methods have been found to give equivalent results. However, when a specification calls for a specific test, only that test must be used.

In the first test (ASTM D2386, IP 16), a measured fuel sample is placed in a jacketed sample tube, also holding a thermometer and a stirrer. The tube is placed in a vacuum flask containing the cooling medium. Various coolants used are acetone; methyl alcohol, ethyl alcohol, or isopropyl alcohol; solid carbon dioxide; or liquid nitrogen. As the sample cools, it is continuously stirred. The temperature at which the hydrocarbon crystals appear is recorded. The jacketed sample is removed from the coolant and allowed to warm, stirring it continuously. The temperature at which the crystals completely disappear is recorded.

In the second test (IP 434), an automated optical method is used for the temperature range a to −70°C (−94°F). In the

method, a 25-min portion of the fuel is placed in a test chamber that is cooled while continuously stirred and monitored by an optical system. The temperature of the specimen is measured with an electronic measuring device, and the temperatures when crystals first appear, and then on warming disappear, are recorded. In the third method (ASTM D5972, IP 435), an automated phase transition method is used in the temperature range −80 to 20°C (−112 to 68°F). In this test, a specimen is cooled at a rate of 15 ± 5°C/min, while continuously being illuminated by a light source. The specimen is continuously monitored by an array of optical detectors for the first formation of solid hydrocarbon crystals. After that, the specimen is warmed at a rate of 10 ± 0.5°C/min until all crystals return to the liquid phase, and that temperature is also recorded.

The *freezing point* should not be confused with the *pour point*, which is an index of the lowest temperature at which the crude oil will flow under specified conditions.

An analogous property test, the *cold filter plugging point*, is suitable for estimating the lowest temperature at which diesel fuel will give trouble-free flow in certain fuel systems (ASTM D6371, IP 309). In this test, either manual or automated apparatus may be used and is cooled under specified conditions and, at intervals of 1°C, is drawn into a pipette under a controlled vacuum through a standardized wire mesh filter. As the sample continues to cool, the procedure is repeated for each 1°C below the first test temperature. The testing is continued until the amount of wax crystals that have separated out of the solution is sufficient to stop or slow down the flow so that the time taken to fill the pipette exceeds 60 s or the fuel fails to return completely to the test jar before the fuel has cooled by a further 1°C. The indicated temperature at which the last filtration was commenced is recorded as the *cold filter plugging point*.

Alternatively, the low-temperature flow test (ASTM D4539) results are indicative of the low-temperature flow performance of fuel in some diesel vehicles. This test method is especially useful for the evaluation of fuels containing flow improver additives. In the test method, the temperature of a series of test specimens of fuel is lowered at a prescribed cooling rate. At the commencing temperature and at each 1°C interval thereafter, a separate specimen from the series is filtered through a 17-mm screen until a minimum low-temperature flow test pass temperature is obtained. The minimum low-temperature flow test pass temperature is the lowest temperature, expressed as a multiple of 1°C, at which a test specimen can be filtered in 60 s or less.

In another test (ASTM D2068) that was originally designed for distillate fuel oil (Chapter 10), the filter-plugging tendency of diesel fuel can be determined by passing a sample at a constant flow rate (20 ml/min) through a glass fiber filter medium. The pressure drop across the filter is monitored during the passage of a fixed volume of test fuel. If a prescribed maximum pressure drop is reached before the total volume of fuel is filtered, the actual volume of fuel filtered at the time of maximum pressure drop is recorded. The apparatus is required to be calibrated at intervals.

9.3.13 Neutralization Number

Neutralization number (ASTM D974, IP 139; IP 182) is a measure of the inorganic and total acidity of the unused fuel and indicates its tendency to corrode metals with which it may come into contact.

Corrosivity is also determined by a variety of copper corrosion text methods (ASTM D130, ASTM D849, IP 154).

9.3.14 Pour Point

The pour point (ASTM D97, IP 15) of a fuel is an indication of the lowest temperature at which the fuel can be pumped. Pour points often occur 3.5–5.6°C (8–10°F) below the cloud points, and differences of 8.3–11.1°C (15–20°F) are not uncommon. Fuels, and in particular those fuels that contain wax, will in some circumstances flow below their tested pour point. However, pour point does give a useful guide to the lowest temperature to which a fuel can be cooled without setting.

The maximum and minimum pour point temperatures provide a temperature window where petroleum, depending on its thermal history, might appear in the liquid as well as in the solid state. The pour point data can be used to supplement other measurements of cold flow behavior, and the data are particularly useful for the screening of the effect of wax interaction modifiers on the flow behavior of petroleum.

In the original (and still widely used) test for pour point (ASTM D97, IP 15), a sample is cooled at a specified rate and examined at intervals of 3°C (5.4°F) for flow characteristics. The lowest temperature at which the movement of the oil is observed is recorded as the pour point.

A later test method (ASTM D5853) covers two procedures for the determination of the pour point of crude oils down to −36°C. One method provides a measure of the maximum (upper) pour point temperature. The second method measures the minimum (lower) pour point temperature. In the methods, the test specimen is cooled (after preliminary heating) at a specified rate and examined at intervals of 3°C (5.4°F) for flow characteristics. Again, the lowest temperature at which movement of the test specimen is observed is recorded as the pour point.

In any determination of the pour point, petroleum that contains wax produces an irregular flow behavior when the wax begins to separate. Such petroleum possesses viscosity relationships that are difficult to predict in pipeline operation. In addition, some waxy petroleum is sensitive to heat treatment that can also affect the viscosity characteristics. This complex behavior limits the value of viscosity and pour point tests on waxy petroleum.

Sometimes additives are used to improve the low-temperature fluidity of diesel fuels. Such additives usually work by modifying the wax crystals so that they are less likely to form a rigid structure. Thus, although there is no alteration of the cloud point, the pour point may be lowered dramatically. Unfortunately, the improvement in engine performance as a rule is less than the improvement in pour point. Consequently, the cloud and pour point temperatures cannot be used to indicate engine performance with any accuracy.

9.3.15 Stability

On leaving the refinery, the fuel will inevitably come into contact with air and water. If the fuel includes unstable components, which may be the case with fuels containing cracked products, storage in the presence of air can lead to the formation of gums and sediments. Instability can cause filter plugging, combustion chamber deposit formation, and gumming or lacquering of injection system components with resultant sticking and wear.

An accelerated stability test (ASTM D2274) often is applied to fuels to measure their stability. A sample of fuel is heated for a fixed period at a given temperature, sometimes in the presence of a catalyst metal, and the amount of sediment and gum formed is taken as a measure of the stability.

In addition, the extent of the oxidation of diesel fuel is determined by the measurement of the hydroperoxide number (ASTM D6447) and the peroxide number (ASTM D3703). Deterioration of diesel fuel results in the formation of the peroxides as well as other oxygen-containing compounds, and these numbers are indications of the quantity of oxidizing constituents present in the sample as determined by the measurement of the compounds that will oxidize potassium iodide.

The determination of hydroperoxide number is significant because of the adverse effect of hydroperoxides upon certain elastomers in the fuel systems. The method (ASTM D6447) measures the same peroxide species, primarily the hydroperoxides in diesel fuel. This test method does not use the ozone depleting substance 1,1,2-trichloro-1,2,2-trifluoroethane (ASTM D3703) and is applicable to any water-insoluble, organic fluid, particularly gasoline, kerosene, and diesel fuel. In the method, a quantity of sample is contacted with aqueous potassium iodide (KI) solution in the presence of acid. The hydroperoxides present are reduced by potassium iodide liberating an equivalent amount of iodine, which is quantified by voltammetric analysis.

The determination of peroxide number of diesel fuel is important because of the adverse effects of peroxides upon certain elastomers in the fuel system. In the test, the sample is dissolved (unlike ASTM D6447) in 1,1,2-trichloro-1,2,2-trifluoroethane and is contacted within an aqueous potassium iodide solution. The peroxides present are reduced by the potassium iodide whereupon an equivalent amount of iodine is released that is titrated with standard sodium thiosulfate solution using a starch indicator.

Other tests for storage stability include determination of color formation and sediment (ASTM D4625, ASTM D5304) in which reactivity to oxygen at high temperatures is determined by the amount of sediment formation as well as any color changes.

9.3.16 Viscosity

Viscosity (ASTM D445, IP 71) is a measure of the resistance to flow by a liquid and usually is measured by recording the time required for a given volume of fuel at a constant temperature to flow through a small orifice of standard dimensions. The viscosity of diesel fuel is important primarily because of its effect on the handling of the fuel by the pump and injector system.

Fuel viscosity also exerts a strong influence on the shape of the fuel spray insofar as a high viscosity can result in poor atomization, large droplets, and high-spray jet penetration. The jet tends to be almost a solid stream instead of forming a spray pattern of small droplets. As a result, the fuel is not distributed in, or mixed with, the air required for burning. Poor combustion is a result, accompanied by loss of power and economy. Moreover, and particularly in the smaller engines, the overly penetrating fuel stream can impinge upon the cylinder walls, thereby washing away the lubricating oil film and causing dilution of the crankcase oil. Such a condition contributes to excessive wear. On the other hand, fuels with a low viscosity can produce a spray that is too soft and, thus, does not penetrate sufficiently. Combustion is impaired, and power output and economy are decreased.

Fuel viscosities for high-speed engines range from 32 to 45 SUS (2–6 cSt) at 37.8°C (100°F). Usually, the lower viscosity limit is established to prevent leakage in worn fuel injection equipment as well as to supply lubrication for injection system components in certain types of engines. During operation at low atmospheric temperature, the viscosity limit sometimes is reduced to 30 SUS (1–4 cSt) at 100°F to obtain increased volatility and sufficiently low pour point. Fuels having viscosities greater than 45 SUS (6 cSt) usually are limited in application to the slower-speed engines. The very viscous fuels, such as are often used in large stationary and marine engines, usually require preheating for proper pumping, injection, and atomization.

9.3.17 Volatility

Distillation (or volatility) characteristics of a diesel fuel exert a great influence on its performance, particularly in medium- and high-speed engines. Distillation characteristics are measured using a procedure (ASTM D86, IP 123) in which a sample of the fuel is distilled and the vapor temperatures are recorded for the percentages of evaporation or distillation throughout the range. Other procedures are also available that are applicable to kerosene (Chapter 8).

The volatility requirement of diesel fuel varies with engine speed, size, and design. However, fuels having too low volatility tend to reduce power output and fuel economy through poor atomization, while those having too high volatility may reduce power output and fuel economy through vapor lock in the fuel system or inadequate droplet penetration from the nozzle. In general, the distillation range should be as low as possible without adversely affecting the flash point, burning quality, heat content, or viscosity of the fuel. If the 10% point is too high, poor starting may result. An excessive boiling range from 10 to 50% evaporated may increase warm-up time. A low 50% point is desirable in preventing smoke and odor. Low 90% and end points tend to ensure low carbon residuals and minimum crankcase dilution.

The temperature for 50% evaporated, known as the midboiling point, usually is taken as an overall indication of the fuel distillation characteristics where a single numerical value is used alone. For example, in high-speed engines, a 50% point above 575°F (302°C) probably would cause smoke formation, give rise to objectionable odor, cause lubricating oil contamination, and promote engine deposits. At the other extreme, a fuel with excessively low 50% point would have too low a viscosity and too low a heat content per unit volume. Thus, a 50% point in the range of 450–535°F (232–280°C) is most desirable for the majority of automotive-type diesel engines. This average range usually is raised to a higher temperature spread for larger, slower-speed engines.

The vapor pressure of diesel fuel at various vapor–liquid ratios is an important physical property for shipping, storage, and use. Although determining the volatility of diesel fuel is usually accomplished through a boiling range distribution (ASTM D86, IP 123) although other methods such as determining the Reid vapor pressure (ASTM D323, IP 69) can also be used along with several other methods (ASTM D5482, ASTM D6378).

9.3.18 Water and Sediment

Water can contribute to filter blocking and cause corrosion of the injection system components. In addition to clogging of the filters, sediment can cause wear and create deposits both in the injection system and in the engine itself. Thus, one of the most important characteristics of a diesel fuel, the water and sediment content (ASTM D1796), is the result of handling and storage practices from the time the fuel leaves the refinery until the time it is delivered to the engine injection system.

Instability and resultant degradation of the fuel in contact with air contribute to the formation of organic sediment, particularly during storage and handling at elevated temperatures. Sediment generally consists of carbonaceous material, metals, or other inorganic matter. There are several causes of this type of contamination: (i) rust or dirt present in tanks and lines, (ii) dirt introduced through careless handling practices, and (iii) dirt present in the air breathed into the storage facilities with fluctuating atmospheric temperature.

Sediment can be determined individually (ASTM D2276, ASTM D6217) or by a test method that determines water simultaneously (ASTM D2709). In the test method, a sample is centrifuged at an rcf of 800 for 10 min at 21–32°C in a centrifuge tube readable to 0.005 ml and measurable to 0.01 ml. After centrifugation, the volume of water and sediment that has settled into the tip of the centrifuge tube is read to the nearest 0.005 ml.

REFERENCES

ASTM D56. Standard Test Method for Flash Point by Tag Closed Cup Tester. Annual Book of Standards. ASTM International, West Conshohocken, PA.

ASTM D86. Standard Test Method for Distillation of Petroleum Products at Atmospheric Pressure. Annual Book of Standards. ASTM International, West Conshohocken, PA.

ASTM D93. Standard Test Methods for Flash Point by Pensky-Martens Closed Cup Tester. Annual Book of Standards. ASTM International, West Conshohocken, PA.

ASTM D97. D97-12 Standard Test Method for Pour Point of Petroleum Products. Annual Book of Standards. ASTM International, West Conshohocken, PA.

ASTM D129. Standard Test Method for Sulfur in Petroleum Products (General High Pressure Decomposition Device Method). Annual Book of Standards. ASTM International, West Conshohocken, PA.

ASTM D130. Copper Strip Corrosion Standard for Petroleum. Annual Book of Standards. ASTM International, West Conshohocken, PA.

ASTM D156. Standard Test Method for Saybolt Color of Petroleum Products (Saybolt Chromometer Method. Annual Book of Standards. ASTM International, West Conshohocken, PA.

ASTM D189. Standard Test Method for Conradson Carbon Residue of Petroleum Products. Annual Book of Standards. ASTM International, West Conshohocken, PA.

ASTM D235. Standard Specification for Mineral Spirits (Petroleum Spirits) (Hydrocarbon Dry Cleaning Solvent). Annual Book of Standards. ASTM International, West Conshohocken, PA.

ASTM D240. Standard Test Method for Heat of Combustion of Liquid Hydrocarbon Fuels by Bomb Calorimeter. Annual Book of Standards. ASTM International, West Conshohocken, PA.

ASTM D287. Standard Test Method for API Gravity of Crude Petroleum and Petroleum Products (Hydrometer Method). Annual Book of Standards. ASTM International, West Conshohocken, PA.

ASTM D323. Standard Test Method for Vapor Pressure of Petroleum Products (Reid Method). Annual Book of Standards. ASTM International, West Conshohocken, PA.

ASTM D445. Standard Test Method for Kinematic Viscosity of Transparent and Opaque Liquids (and Calculation of Dynamic Viscosity). Annual Book of Standards. ASTM International, West Conshohocken, PA.

ASTM D482. Standard Test Method for Ash from Petroleum Products. Annual Book of Standards. ASTM International, West Conshohocken, PA.

ASTM D524. Standard Test Method for Ramsbottom Carbon Residue of Petroleum Products. Annual Book of Standards. ASTM International, West Conshohocken, PA.

ASTM D611. Standard Test Methods for Aniline Point and Mixed Aniline Point of Petroleum Products and Hydrocarbon Solvents. Annual Book of Standards. ASTM International, West Conshohocken, PA.

ASTM D613. Standard Test Method for Cetane Number of Diesel Fuel Oil. Annual Book of Standards. ASTM International, West Conshohocken, PA.

ASTM D664. Standard Test Method for Acid Number of Petroleum Products by Potentiometric Titration. Annual Book of Standards. ASTM International, West Conshohocken, PA.

ASTM D849. Standard Test Method for Copper Strip Corrosion by Industrial Aromatic Hydrocarbons. Annual Book of Standards. ASTM International, West Conshohocken, PA.

ASTM D974. Standard Test Method for Acid and Base Number by Color-Indicator Titration. Annual Book of Standards. ASTM International, West Conshohocken, PA.

ASTM D975. Standard Specification for Diesel Fuel Oils. Annual Book of Standards. ASTM International, West Conshohocken, PA.

ASTM D976. Test Methods for Calculated Cetane Index of Distillate Fuels. Annual Book of Standards. ASTM International, West Conshohocken, PA.

ASTM D1018. Standard Test Method for Hydrogen in Petroleum Fractions. Annual Book of Standards. ASTM International, West Conshohocken, PA.

ASTM D1209. Standard Test Method for Color of Clear Liquids (Platinum-Cobalt Scale). Annual Book of Standards. ASTM International, West Conshohocken, PA.

ASTM D1217. Standard Test Method for Density and Relative Density (Specific Gravity) of Liquids by Bingham Pycnometer. Annual Book of Standards. ASTM International, West Conshohocken, PA.

ASTM D1250. Petroleum Measurement Tables. Annual Book of Standards. ASTM International, West Conshohocken, PA.

ASTM D1266. Standard Test Method for Sulfur in Petroleum Products (Lamp Method). Annual Book of Standards. ASTM International, West Conshohocken, PA.

ASTM D1298. Standard Test Method for Density, Relative Density, or API Gravity of Crude Petroleum and Liquid Petroleum Products by Hydrometer Method. Annual Book of Standards. ASTM International, West Conshohocken, PA.

ASTM D1319. Standard Test Method for Hydrocarbon Types in Liquid Petroleum Products by Fluorescent Indicator Adsorption. Annual Book of Standards. ASTM International, West Conshohocken, PA.

ASTM D1405. Standard Test Method for Estimation of Net Heat of Combustion of Aviation Fuels. Annual Book of Standards. ASTM International, West Conshohocken, PA.

ASTM D1500. Standard Test Method for ASTM Color of Petroleum Products (ASTM Color Scale). Annual Book of Standards. ASTM International, West Conshohocken, PA.

ASTM D1544. Standard Test Method for Color of Transparent Liquids (Gardner Color Scale). Annual Book of Standards. ASTM International, West Conshohocken, PA.

ASTM D1552. Standard Test Method for Sulfur in Petroleum Products (High-Temperature Method. Annual Book of Standards. ASTM International, West Conshohocken, PA.

ASTM D1655. Standard Specification for Aviation Turbine Fuels. Annual Book of Standards. ASTM International, West Conshohocken, PA.

ASTM D1796. Standard Test Method for Water and Sediment in Fuel Oils by the Centrifuge Method (Laboratory Procedure). Annual Book of Standards. ASTM International, West Conshohocken, PA.

ASTM D1839. Standard Test Method for Amyl Nitrate in Diesel Fuels. Annual Book of Standards. ASTM International, West Conshohocken, PA.

ASTM D2068. Standard Test Method for Determining Filter Blocking Tendency. Annual Book of Standards. ASTM International, West Conshohocken, PA.

ASTM D2274. Standard Test Method for Oxidation Stability of Distillate Fuel Oil (Accelerated Method). Annual Book of Standards. ASTM International, West Conshohocken, PA.

ASTM D2276. Standard Test Method for Particulate Contaminant in Aviation Fuel by Line Sampling. Annual Book of Standards. ASTM International, West Conshohocken, PA.

ASTM D2386. Standard Test Method for Freezing Point of Aviation Fuels. Annual Book of Standards. ASTM International, West Conshohocken, PA.

ASTM D2425. Standard Test Method for Hydrocarbon Types in Middle Distillates by Mass Spectrometry. Annual Book of Standards. ASTM International, West Conshohocken, PA.

ASTM D2500. Standard Test Method for Cloud Point of Petroleum Products. Annual Book of Standards. ASTM International, West Conshohocken, PA.

ASTM D2549. Standard Test Method for Separation of Representative Aromatics and Nonaromatics Fractions of High-Boiling Oils by Elution Chromatography. Annual Book of Standards. ASTM International, West Conshohocken, PA.

ASTM D2622. Standard Test Method for Sulfur in Petroleum Products by Wavelength Dispersive X-ray Fluorescence Spectrometry. Annual Book of Standards. ASTM International, West Conshohocken, PA.

ASTM D2709. Standard Test Method for Water and Sediment in Middle Distillate Fuels by Centrifuge. Annual Book of Standards. ASTM International, West Conshohocken, PA.

ASTM D2786. Standard Test Method for Hydrocarbon Types Analysis of Gas-Oil Saturates Fractions by High Ionizing Voltage Mass Spectrometry. Annual Book of Standards. ASTM International, West Conshohocken, PA.

ASTM D2880. Standard Specification for Gas Turbine Fuel Oils. Annual Book of Standards. ASTM International, West Conshohocken, PA.

ASTM D2887. Standard Test Method for Boiling Range Distribution of Petroleum Fractions by Gas Chromatography. Annual Book of Standards. ASTM International, West Conshohocken, PA.

ASTM D2890. Standard Test Method for Calculation of Liquid Heat Capacity of Petroleum Distillate Fuels. Annual Book of Standards. ASTM International, West Conshohocken, PA.

ASTM D2892. Standard Test Method for Distillation of Crude Petroleum (15-Theoretical Plate Column). Annual Book of Standards. ASTM International, West Conshohocken, PA.

ASTM D3120. Standard Test Method for Trace Quantities of Sulfur in Light Liquid Petroleum Hydrocarbons by Oxidative Microcoulometry. Annual Book of Standards. ASTM International, West Conshohocken, PA.

ASTM D3227. Standard Test Method for (Thiol Mercaptan) Sulfur in Gasoline, Kerosene, Aviation Turbine, and Distillate Fuels (Potentiometric Method). Annual Book of Standards. ASTM International, West Conshohocken, PA.

ASTM D3239. Standard Test Method for Aromatic Types Analysis of Gas-Oil Aromatic Fractions by High Ionizing Voltage Mass Spectrometry. Annual Book of Standards. ASTM International, West Conshohocken, PA.

ASTM D3246. Standard Test Method for Sulfur in Petroleum Gas by Oxidative Microcoulometry. Annual Book of Standards. ASTM International, West Conshohocken, PA.

ASTM D3338. Standard Test Method for Estimation of Net Heat of Combustion of Aviation Fuels. Annual Book of Standards. ASTM International, West Conshohocken, PA.

ASTM D3339. Standard Test Method for Acid Number of Petroleum Products by Semi-Micro Color Indicator Titration. Annual Book of Standards. ASTM International, West Conshohocken, PA.

ASTM D3701. Standard Test Method for Hydrogen Content of Aviation Turbine Fuels by Low Resolution Nuclear Magnetic Resonance Spectrometry. Annual Book of Standards. ASTM International, West Conshohocken, PA.

ASTM D3703. Standard Test Method for Hydroperoxide Number of Aviation Turbine Fuels, Gasoline and Diesel Fuels. Annual Book of Standards. ASTM International, West Conshohocken, PA.

ASTM D3828. Standard Test Methods for Flash Point by Small Scale Closed Cup Tester. Annual Book of Standards. ASTM International, West Conshohocken, PA.

ASTM D4046. Standard Test Method for Alkyl Nitrate in Diesel Fuels by Spectrophotometry. Annual Book of Standards. ASTM International, West Conshohocken, PA.

ASTM D4052. Standard Test Method for Density, Relative Density, and API Gravity of Liquids by Digital Density Meter. Annual Book of Standards. ASTM International, West Conshohocken, PA.

ASTM D4176. Standard Test Method for Free Water and Particulate Contamination in Distillate Fuels (Visual Inspection Procedures). Annual Book of Standards. ASTM International, West Conshohocken, PA.

ASTM D4294. Standard Test Method for Sulfur in Petroleum and Petroleum Products by Energy Dispersive X-ray Fluorescence Spectrometry. Annual Book of Standards. ASTM International, West Conshohocken, PA.

ASTM D4418. Standard Practice for Receipt, Storage, and Handling of Fuels for Gas Turbines. Annual Book of Standards. ASTM International, West Conshohocken, PA.

ASTM D4529. Standard Test Method for Estimation of Net Heat of Combustion of Aviation Fuels. Annual Book of Standards. ASTM International, West Conshohocken, PA.

ASTM D4530. Standard Test Method for Determination of Carbon Residue (Micro Method). Annual Book of Standards. ASTM International, West Conshohocken, PA.

ASTM D4539. Standard Test Method for Filterability of Diesel Fuels by Low-Temperature Flow Test (LTFT). Annual Book of Standards. ASTM International, West Conshohocken, PA.

ASTM D4625. Standard Test Method for Distillate Fuel Storage Stability at 43°C (110°F). Annual Book of Standards. ASTM International, West Conshohocken, PA.

ASTM D4739. Standard Test Method for Base Number Determination by Potentiometric Hydrochloric Acid Titration. Annual Book of Standards. ASTM International, West Conshohocken, PA.

ASTM D4808. Standard Test Methods for Hydrogen Content of Light Distillates, Middle Distillates, Gas Oils, and Residua by Low-Resolution Nuclear Magnetic Resonance Spectroscopy. Annual Book of Standards. ASTM International, West Conshohocken, PA.

ASTM D4809. Standard Test Method for Heat of Combustion of Liquid Hydrocarbon Fuels by Bomb Calorimeter (Precision Method). Annual Book of Standards. ASTM International, West Conshohocken, PA.

ASTM D4868. Standard Test Method for Estimation of Net and Gross Heat of Combustion of Burner and Diesel Fuels. Annual Book of Standards. ASTM International, West Conshohocken, PA.

ASTM D4952. Standard Test Method for Qualitative Analysis for Active Sulfur Species in Fuels and Solvents (Doctor Test). Annual Book of Standards. ASTM International, West Conshohocken, PA.

ASTM D5186. Standard Test Method for Determination of the Aromatic Content and Polynuclear Aromatic Content of Diesel Fuels and Aviation Turbine Fuels By Supercritical Fluid Chromatography. Annual Book of Standards. ASTM International, West Conshohocken, PA.

ASTM D5291. Standard Test Methods for Instrumental Determination of Carbon, Hydrogen, and Nitrogen in Petroleum Products and Lubricants. Annual Book of Standards. ASTM International, West Conshohocken, PA.

ASTM D5292. Standard Test Method for Aromatic Carbon Contents of Hydrocarbon Oils by High Resolution Nuclear Magnetic Resonance Spectroscopy. Annual Book of Standards. ASTM International, West Conshohocken, PA.

ASTM D5304. Standard Test Method for Assessing Middle Distillate Fuel Storage Stability by Oxygen Overpressure. Annual Book of Standards. ASTM International, West Conshohocken, PA.

ASTM D5453. Standard Test Method for Determination of Total Sulfur in Light Hydrocarbons, Spark Ignition Engine Fuel, Diesel Engine Fuel, and Engine Oil by Ultraviolet Fluorescence. Annual Book of Standards. ASTM International, West Conshohocken, PA.

ASTM D5482. Standard Test Method for Vapor Pressure of Petroleum Products (Mini Method – Atmospheric). Annual Book of Standards. ASTM International, West Conshohocken, PA.

ASTM D5623. Standard Test Method for Sulfur Compounds in Light Petroleum Liquids by Gas Chromatography and Sulfur Selective Detection. Annual Book of Standards. ASTM International, West Conshohocken, PA.

ASTM D5771. Standard Test Method for Cloud Point of Petroleum Products (Optical Detection Stepped Cooling Method). Annual Book of Standards. ASTM International, West Conshohocken, PA.

ASTM D5772. Standard Test Method for Cloud Point of Petroleum Products (Linear Cooling Rate Method). Annual Book of Standards. ASTM International, West Conshohocken, PA.

ASTM D5773. Standard Test Method for Cloud Point of Petroleum Products (Constant Cooling Rate Method). Annual Book of Standards. ASTM International, West Conshohocken, PA.

ASTM D5853. Standard Test Method for Pour Point of Crude Oils. Annual Book of Standards. ASTM International, West Conshohocken, PA.

ASTM D5972. Standard Test Method for Freezing Point of Aviation Fuels (Automatic Phase Transition Method). Annual Book of Standards. ASTM International, West Conshohocken, PA.

ASTM D6217. Standard Test Method for Particulate Contamination in Middle Distillate Fuels by Laboratory Filtration. Annual Book of Standards. ASTM International, West Conshohocken, PA.

ASTM D6258. Standard Test Method for Determination of Solvent Red 164 Dye Concentration in Diesel Fuels. Annual Book of Standards. ASTM International, West Conshohocken, PA.

ASTM D6371. Standard Test Method for Cold Filter Plugging Point of Diesel and Heating Fuels. Annual Book of Standards. ASTM International, West Conshohocken, PA.

ASTM D6378. Standard Test Method for Determination of Vapor Pressure (VPX) of Petroleum Products, Hydrocarbons, and Hydrocarbon-Oxygenate Mixtures (Triple Expansion Method). Annual Book of Standards. ASTM International, West Conshohocken, PA.

ASTM D6379. Standard Test Method for Determination of Aromatic Hydrocarbon Types in Aviation Fuels and Petroleum Distillates—High Performance Liquid Chromatography Method with Refractive Index Detection. Annual Book of Standards. ASTM International, West Conshohocken, PA.

ASTM D6446. Standard Test Method for Estimation of Net Heat of Combustion (Specific Energy) of Aviation Fuels (Withdrawn 2012 but still in use in some laboratories). ASTM International, West Conshohocken, PA.

ASTM D6447. Standard Test Method for Hydroperoxide Number of Aviation Turbine Fuels by Voltammetric Analysis. Annual Book of Standards. ASTM International, West Conshohocken, PA.

Energy Institute. 2013. IP Standard Methods 2013. The Institute of Petroleum, London.

Gary, J.G., Handwerk, G.E., and Kaiser, M.J. 2007. Petroleum Refining: Technology and Economics, 5th Edition. CRC Press, Taylor & Francis Group, Boca Raton, FL.

Hsu, C.S., and Robinson, P.R. (Editors) 2006. Practical Advances in Petroleum Processing, Volumes 1 and 2. Springer Science, New York.

IP 2 (A STM D611). Petroleum Products and Hydrocarbon Solvents – Determination of Aniline and Mixed Aniline Point. IP Standard Methods 2013. The Energy Institute, London.

IP 4 (ASTM D482). Petroleum Products – Determination of Ash. IP Standard Methods 2013. The Energy Institute, London.

IP 12. Specific Energy. Energy Institute, London, UK.

IP 13. Carbon Residue – Conradson Method. Energy Institute, London, UK.

IP 14. Carbon Residue – Ramsbottom Method. Energy Institute, London, UK.

IP 15 (ASTM D97). Petroleum Products – Determination of Pour Point. IP Standard Methods 2013. The Energy Institute, London.

IP 16 (ASTM D2386). Petroleum Products – Determination of the Freezing Point of Aviation Fuels – Manual Method. IP Standard Methods 2013. The Energy Institute, London.

IP 30. Detection of Mercaptans, Hydrogen Sulfide, Elemental Sulfur and Peroxides – Doctor Test Method. IP Standard Methods 2013. The Energy Institute, London.

IP 34 (ASTM D93). Determination of Flash Point – Pensky-Martens Closed Cup Method. IP Standard Methods 2013. The Energy Institute, London.

IP 41. Ignition Quality of Diesel Fuel by the Cetane Method. Energy Institute, London, UK.

IP 61. Sulfur – Bomb Method. Energy Institute, London, UK.

IP 69. Determination of Vapor Pressure – Reid Method. IP Standard Methods 2013. The Energy Institute, London.

IP 71 (ASTM D445). Petroleum Products – Transparent and Opaque Liquids – Determination of Kinematic Viscosity and Calculation of Dynamic Viscosity. IP Standard Methods 2013. The Energy Institute, London.

IP 123. Petroleum Products – Determination of Distillation Characteristics at Atmospheric Pressure. IP Standard Methods 2013. The Energy Institute, London.

IP 139 (ASTM D974). Petroleum Products and Lubricants – Determination of Acid or Base Number- Color Indicator Titration Method. IP Standard Methods 2013. The Energy Institute, London.

IP 154. Petroleum Products – Corrosiveness to Copper–Copper Strip Test. IP Standard Methods 2013. The Energy Institute, London.

IP 160 (ASTM D1298). Crude Petroleum and Liquid Petroleum Products – Laboratory Determination of Density – Hydrometer Method. IP Standard Methods 2013. The Energy Institute, London.

IP 170. Petroleum Products and Other Liquids – Determination of Flash Point – Abel Closed Cup Method. IP Standard Methods 2013. The Energy Institute, London.

IP 182. Determination of Inorganic Acidity of Petroleum Products – Color Indicator Titration Method. IP Standard Methods 2013. The Energy Institute, London.

IP 196 (ASTM D1500). Petroleum Products – Determination of Color (ASTM Scale). IP Standard Methods 2013. The Energy Institute, London.

IP 219. Cloud Point of Petroleum Products. Energy Institute, London, UK.

IP 309. Diesel and Domestic Heating Fuels – Determination of Cold Filter Plugging Point. IP Standard Methods 2013. The Energy Institute, London.

IP 336. Sulfur – Energy Dispersive X-Ray Fluorescence Method. Energy Institute, London, UK.

IP 365. Density – Oscillating U-tube Method. Energy Institute, London, UK.

IP 373. Sulfur Content – Microcoulometry (Oxidative) Method. Energy Institute, London, UK.

IP 398. Petroleum Products – Determination of Carbon Residue – Micro Method. IP Standard Methods 2013. The Energy Institute, London.

IP 434. Freezing Point of Aviation Turbine Fuels – Automated Optical Method (replaced by IP 528–ASTM D5901). IP Standard Methods 2013. The Energy Institute, London.

IP 435. Freezing Point of Aviation Turbine Fuels – Automatic Phase Transition Method. IP Standard Methods 2013. The Energy Institute, London.

IP 523. Determination of Flash Point – Rapid Equilibrium Closed Cup Method. IP Standard Methods 2013. The Energy Institute, London.

Song, C., Hsu, C.S., and Mochida, I. 2000. Chemistry of Diesel Fuels. Taylor & Francis, New York/CRC Press, Taylor & Francis Group, Boca Raton, FL.

Speight, J.G. 2001. Handbook of Petroleum Analysis. John Wiley & Sons Inc., New York.

Speight, J.G. and Ozum, B. 2002. Petroleum Refining Processes. Marcel Dekker Inc., New York.

Speight, J.G. 2014. The Chemistry and Technology of Petroleum, 5th Edition. CRC Press, Taylor & Francis Group, Boca Raton, FL.

10

DISTILLATE FUEL OIL

10.1 INTRODUCTION

Most petroleum products can be used as fuels, but the term *fuel oil*, if used without qualification, may be interpreted differently depending on the context. However, because fuel oils are complex mixtures of hydrocarbons, they cannot be rigidly classified or defined precisely by chemical formulae or definite physical properties. The arbitrary division or classification of fuel oil is based more on their application than on their chemical or physical properties. However, two broad classifications are generally recognized: (i) *distillate fuel oil* and (ii) *residual fuel oil* (Charlot and Claus 2000; Heinrich and Duée, 2000; Speight and Ozum 2002; Hsu and Robinson, 2006; Gary *et al.*, 2007; Speight, 2014). It is this terminology that is used in this text.

Distillate fuel oil is a petroleum fraction produced in conventional distillation operations, which also includes diesel fuel and fuel oil. The conventional description of fuel oil is generally associated with the black, viscous, residual material remaining as a result of refinery distillation of crude oil either alone or as a blend with light components and that is used for steam generation and various industrial processes. The term is sometimes used to refer to the light, amber-colored middle distillates or gas oils that are distinguished from the residual fuel oil by being characterized as distillate fuel oil (ASTM D396). In this specification, the No. 1 grade fuel oil is a kerosene type used in vaporizing pot-type burners, while the No. 2 fuel is a distillate oil (gas oil) used for general-purpose domestic heating. Kerosene may also be included in this definition, but it is described elsewhere in this text.

Distillate fuel oils are vaporized and condensed during a distillation process and thus have a definite boiling range and do not contain high-boiling oils or asphaltic components. They generally correspond to a light gas oil (Table 10.1, Fig. 10.1), but this correlation is not exact and can vary considerably. A fuel oil that contains any amount of the residue from crude distillation of thermal cracking is a residual fuel oil. The terms *distillate fuel oil* and *residual fuel oil* are losing their significance, since fuel oils are now made for specific uses and may be either distillates or residuals or mixtures of the two. The terms *domestic fuel oil*, *diesel fuel oil*, and *heavy fuel oil* are more indicative of the uses of fuel oils.

Domestic fuel oil is fuel oil that is used primarily in the home. This category of fuel oil includes kerosene, stove oil, and furnace fuel oil; they are distillate fuel oils. Diesel fuel oil is also a distillate fuel oil, but residual oil has been successfully used to power marine diesel engines, and mixtures of distillate fuel oil and residual fuel oil have been used in locomotive diesel engines. Heavy fuel oils include a variety of oils ranging from distillates to residual oils that must be heated to 260°C (500°F) or more before they can be used. In general, heavy fuel oils consist of residual oils blended with distillates to suit specific needs. Included among heavy fuel oils are various industrial oils; when used to fuel ships, heavy fuel oils are called bunker oil.

Since the boiling ranges, sulfur contents, and other properties of even the same fraction vary from crude oil to crude oil and with the way the crude oil is processed, it is difficult to specify which fractions are blended to produce specific fuel oils. In general, however, furnace fuel oil is a blend of

Handbook of Petroleum Product Analysis, Second Edition. James G. Speight.
© 2015 John Wiley & Sons, Inc. Published 2015 by John Wiley & Sons, Inc.

TABLE 10.1   General summary of product types and distillation range

Product	Lower carbon limit	Upper carbon limit	Lower boiling point (°C)	Upper boiling point (°C)	Lower boiling point (°F)	Upper boiling point (°F)
Refinery gas	C_1	C_4	−161	−1	−259	31
Liquefied petroleum gas	C_3	C_4	−42	−1	−44	31
Naphtha	C_5	C_{17}	36	302	97	575
Gasoline	C_4	C_{12}	−1	216	31	421
Kerosene/diesel fuel	C_8	C_{18}	126	258	302	575
Aviation turbine fuel	C_8	C_{16}	126	287	302	548
Fuel oil	C_{12}	$>C_{20}$	216	421	>343	>649
Lubricating oil	$>C_{20}$		>343		>649	
Wax	C_{17}	$>C_{20}$	302	>343	575	>649
Asphalt	$>C_{20}$		>343		>649	
Coke	$>C_{50}{}^a$		$>1000^a$		$>1832^a$	

^{a} Carbon number and boiling point difficult to assess; inserted for illustrative purposes only.

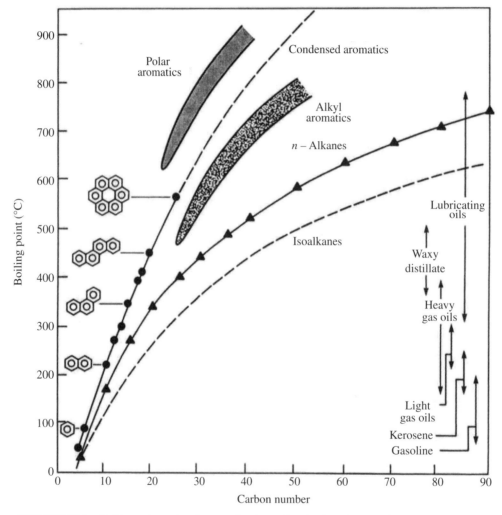

FIGURE 10.1   Boiling point and carbon number for various hydrocarbons and petroleum products.

straight-run gas oil and cracked gas oil to produce a product boiling in the 175–345°C (350–650°F) range.

Heavy fuel oils usually contain cracked residua, reduced crude, or cracking coil heavy product that are mixed (cut back) to a specified viscosity with cracked gas oils and fractionator bottoms. For some industrial purposes in which flames or flue gases contact the product (ceramics, glass, heat treating, and open hearth furnaces), fuel oils must be blended to contain minimum sulfur contents, and hence low-sulfur residues are preferable for these fuels.

Thus, distillate fuel is any one of the broad assortment of fuels acquired from fractions boiling above the temperature at which gasoline comes off in petroleum distillation. Examples of distillate fuels include kerosene and diesel. Products known as No. 1, No. 2, and No. 4 diesel fuel are used in on-highway diesel engines, such as those in trucks and automobiles, as well as off-highway engines, such as those in railroad locomotives and agricultural machinery.

Products known as No. 1, No. 2, and No. 4 fuel oils are used primarily for space heating and electric power generation:

No. 1 Distillate: light petroleum distillate that can be used as either a diesel fuel (see No. 1 Diesel Fuel) or fuel oil. See No. 1 Fuel Oil.

No. 1 Diesel Fuel: Light distillate fuel oil that has distillation temperatures of 550°F at the 90% point and meets the specifications defined in ASTM D975. It is used in high-speed diesel engines, such as those in city buses and similar vehicles. See No. 1 Distillate.

No. 1 Fuel Oil: Light distillate fuel oil that has distillation temperatures of 400°F at the 10% recovery point and 550°F at the 90% point and meets the specifications defined in ASTM D396. It is used primarily as fuel for portable outdoor stoves and portable outdoor heaters. See No. 1 Distillate.

No. 2 Distillate: Petroleum distillate that can be used as either diesel fuel (see No. 2 Diesel Fuel definition) or fuel oil. See No. 2 Fuel Oil.

No. 2 Diesel Fuel: Fuel oil that has a distillation temperature of 340°C (640°F) at the 90% recovery point and conforms with specifications (ASTM D975). It is used in high-speed diesel engines, such as those in railroad locomotives, trucks, and automobiles. See No. 2 Distillate.

No. 2 Fuel Oil (heating oil): Distillate fuel oil that has a distillation temperature of 205°C (400°F at the 10% recovery point and 340°C (640°F) at the 90% recovery point and conforms with specifications (ASTM D396). It is used in atomizing-type burners for domestic heating or for moderate capacity commercial/industrial burner units. See No. 2 Distillate.

No. 4 Fuel: Distillate fuel oil made by blending distillate fuel oil and residual fuel oil stocks. It conforms with specifications (ASTM D396) and is used extensively in industrial plants and in commercial burner installations that are not equipped with preheating facilities. It also includes No. 4 diesel fuel used for low- and medium-speed diesel engines (ASTM D975).

TABLE 10.2 Properties of the various fuel oils

Fuel oil	Properties
No. 1 fuel oil	Similar to kerosene or range oil (fuel used in stoves for cooking).
	Defined as a distillate intended for vaporizing in pot-type burners and other burners where a clean flame is required.
No. 2 fuel oil	Often called *domestic heating oil.*
	Has properties similar to diesel and higher-boiling jet fuels. Defined as a distillate for general-purpose heating in which the burners do not require the fuel to be completely vaporized before burning.
No. 4 fuel oil	A light industrial heating oil that is intended where preheating is not required for handling or burning.
	Two grades that differ primarily in safety (flash) and flow (viscosity) properties.
No. 5 fuel oil	A heavy industrial oil that often requires preheating for burning and, in cold climates, for handling.
No. 6 fuel oil	A heavy residuum oil.
	Commonly referred to as *Bunker C oil* when it is used to fuel ocean-going vessels.
	Preheating is required for both handling and burning this grade oil.

Each fuel oil product is also defined by a variety of properties (Table 10.2) that are required for the fuel oil to perform during service.

10.2 PRODUCTION AND PROPERTIES

Distillate fuel oil is a product of the distillation process and has a definite boiling range and does not contain high-boiling oils or asphaltic components. A fuel oil that contains any amount of the residue from crude distillation or thermal cracking is a residual fuel oil. *Domestic fuel oil* is fuel oil that is used primarily in home and includes kerosene, stove oil, and furnace fuel oil. *Diesel fuel oil* is also a distillate fuel oil, but residual oil has been successfully used to power marine diesel engines, and mixtures of distillates and residuals have been used on locomotive diesels. *Furnace fuel oil* is similar to diesel fuel, but the proportion of cracked gas oil in diesel fuel is usually less since the high aromatic content of the cracked gas oil reduces the cetane number of the diesel fuel.

Distillate fuel oil may be produced not only directly from crude oil (*straight-run*) but also from subsequent refinery processes such as thermal or catalytic cracking. Domestic heating oils and kerosene are examples of distillate fuel oils.

Stove oil is a straight-run (distilled) fraction from crude oil, whereas other fuel oils are usually blends of two or more fractions. The straight-run fractions available for blending into fuel oils are heavy naphtha, light and heavy gas oils, and residua. Cracked fractions such as light and heavy gas oils from catalytic cracking, cracking coal tar, and fractionator bottoms from catalytic cracking may also be used as blends to meet the specifications of the different fuel oils.

Heavy fuel oil includes a variety of oils ranging from distillates to residual oils that must be heated to 260°C (500°F) or higher before they can be used. In general, heavy fuel oils consist of residual oils blended with distillates to suit specific needs. Included among heavy fuel oils are various industrial oils; when used to fuel ships, heavy fuel oil is called bunker oil. Historically, fuel oils were based on atmospheric residua (Speight and Ozum, 2002; Hsu and Robinson, 2006; Gary *et al.*, 2007; Speight, 2014) and were known as straight-run fuels. However, the increasing demand for transportation fuels such as gasoline, kerosene, and diesel has led to an increased value for the atmospheric residue as a feedstock for vacuum distillation and for cracking processes. As a consequence, most heavy fuel oils are currently based on vacuum residua (Speight and Ozum, 2002; Hsu and Robinson, 2006; Gary *et al.*, 2007; Speight, 2014) as well as residues from thermal and catalytic cracking operations. These fuels differ in character from straight-run fuels in that the density and mean molecular weight are higher, as is the carbon/hydrogen ratio. The density of some heavy fuel oils can be above $1000\,kg/m^3$, which has environmental implications in the event of a spillage into fresh water.

The manufacture of fuel oils at one time largely involved using what was left after removing desired products from crude petroleum. Now fuel oil manufacture is a complex matter of selecting and blending various petroleum fractions to meet definite specifications, and the production of a homogeneous, stable fuel oil requires experience backed by laboratory control.

The term *domestic heating oil* in the present context is applicable to the middle distillate oil or gas oil–type product used principally with atomizing burner-heating equipment. This material may consist of the straight-run gas oil from the distillation of the crude oil and which boils within the approximate temperature range of l60–370°C (320–700°F). Straight-run gas oil fraction is usually blended with the appropriate boiling range material from catalytic cracking processing. The components are suitably treated prior to final blending, and additives may also be added to further assist in the stabilization of the finished product.

To produce fuels that can be conveniently handled and stored in industrial and marine installations, and to meet marketing specification limits, the high-viscosity residue components are normally blended with gas oils or similar lower-viscosity fractions. In refineries with catalytic cracking units, catalytically cracked cycle oils are common fuel oil diluents. As a result, the composition of residual fuel oils can vary widely and will depend on the refinery configuration, the crude oils being processed, and the overall refinery demand.

The quality and performance requirements for fuel oils differ widely although general quality limitations for various fuel grades are used to serve as guides in the manufacture, sale, and purchase of the oils. These quality definitions are often referred to as *specifications* or *classifications*, but more precise specifications of quality requirements, such as the vapor pressure (ASTM D323) and metals content (ASTM D5184, ASTM D4951, ASTM D5185, ASTM D5708, ASTM D5863, IP 377), may be required for any given application (ASTM D396).

10.3 TEST METHODS

Fuel oil, therefore, in its various categories has an extensive range of applications, and the choice of a standard procedure to be used for assessing or controlling product quality must, of necessity, depend both upon the type of fuel and its ultimate use. But first, as for all petroleum analyses and testing, the importance of correct sampling of the fuel oil cannot be overemphasized, because no proper assessment of quality may be made unless the data are obtained on truly representative samples.

Specifications for both middle distillate heating fuels and transportation fuels are similar; as a consequence, it is often possible for refiners to satisfy the performance requirements of both applications with the same process stream or blend of process streams. The final products will have been treated (such as sweetened, dried, clay filtered) as required for a particular application, and may contain additives that are specific for the intended use, but they are otherwise virtually indistinguishable on the basis of their gross physical or chemical properties.

10.3.1 Acidity

The presence of inorganic acids in distillate fuels, resulting from refinery treatment, is unlikely. However, some specifications for these fuels still include limiting clauses for total acidity and inorganic acidity as a check against possible corrosion of metal equipment in contact with the fuel. Inorganic acidity should in any case be entirely absent.

The acidity is determined through the *acid number*, which is the quantity of base, expressed in milligrams of potassium hydroxide per gram of sample, required to titrate a sample in the solvent from its initial meter reading to a meter reading corresponding to a freshly prepared nonaqueous basic buffer solution or a well-defined inflection point as specified in the test method. Test methods include potentiometric titration (IP 177) and indicator–indicator titration (ASTM D974, IP 139) in addition to inorganic acidity (IP 182) and total acidity (IP 1) methods.

One of the test methods (ASTM D974) resolves constituents into groups having weak-acid and strong-acid ionization properties. However, oils such as fuel oil that are dark-colored oils, which cannot be analyzed by this test method due to obscurity of the color-indicator end point, should be analyzed by another test method (ASTM D664, IP 177). Test method is used to determine the presence of those constituents that have weak-acid properties and those constituents that have strong-acid properties. The test method may be used to indicate relative changes that occur in an oil during use under oxidizing conditions regardless of the color or other properties of the resulting oil. In the practice of the method, the sample is dissolved in a mixture of toluene and *iso*-propyl alcohol containing a small amount of water and titrated potentiometrically with alcoholic potassium hydroxide using a glass indicating electrode and a calomel reference electrode, and the end points are taken at well-defined inflections in the resulting curve.

10.3.2 Ash

Ash is the organic-free (or carbonaceous-free) residue that remains after combustion of a fuel oil at a specified high temperature (ASTM D482, IP 4).

Depending on the use of the fuel, ash composition has a considerable bearing on whether or not detrimental effects will occur. However, distillate fuels tend to contain only negligible amounts of ash but pick up ash-forming constituents during transportation from the refinery. Water transportation, in particular, presents many opportunities for fuel oils to be contaminated with ash-forming contaminants (sea water, dirt, and scale rust).

Small amounts of nonburnable material are found in fuel oil in the form of soluble metallic soaps and solids, and these materials are designated as ash, although ash-forming constituents is a more correct term. In the test for the quantitative determination of ash-forming constituents (ASTM D482, IP 4), a small sample of fuel oil is burned in a weighed container until all of the combustible matter has been consumed indicated by the residue and container attaining a constant weight. The amount of unburnable residue is the ash yield and is reported as percentage by weight of the sample.

The ash-forming constituents in distillate fuel (ASTM D2880) are typically so low that it does not adversely affect gas turbine performance, unless such corrosive species as sodium, potassium, lead, or vanadium are present. However, there are recommendations for the storage and handling of these fuels (ASTM D4418) to minimize potential contamination.

10.3.3 Calorific Value

Since the function of a fuel is to produce heat, the calorific or heating value (ASTM D240, IP 12) is one of the important fuel properties, and a knowledge of this is necessary in obtaining information regarding the combustion efficiency and performance of all types of oil burning equipment.

The determination is made in a bomb calorimeter under specified conditions, the oxygen in the bomb being saturated with water vapor prior to the ignition of the fuel so that the water formed during combustion is condensed. The calorific value so determined will include the latent heat of water at the test temperature and is known as the gross calorific value at constant volume. The corresponding net calorific value at constant pressure is obtained by deducting the latent heat of water formed during the burning of the hydrogen present in the fuel to produce water. The calorific value is usually expressed in British Thermal Units per pound (Btu/lb) or in calories per gram (c~/g). In Europe, the net calorific value is more often called for in calculations on burner efficiency since the water formed during combustion passes out as water vapor with the flue gases, and hence its latent heat of condensation is not realized as useful heat. In the United Kingdom, the gross calorific value is normally used for this purpose.

An alternative criterion of energy content is the *aniline gravity product* (AGP) that is related to calorific value (ASTM D1405). The *AGP* is the product of the API gravity (ASTM D287, ASTM D1298) and the aniline point of the fuel (ASTM D611, IP 2). The aniline point is the lowest temperature at which the fuel is miscible with an equal volume of aniline and is inversely proportional to the aromatic content. The relationship between the AGP and calorific value is given in the method. In another method (ASTM D3338), the heat of combustion is calculated from the fuel density; the 10, 50, and 90% distillation temperatures; and the aromatic content. However, neither method is legally acceptable, and other methods (ASTM D240, ASTM D1655, ASTM D4809) are preferred.

An alternative criterion of energy content is the *AGP* that is related to calorific value (ASTM D1405), which is not always suitable for determining the heat of combustion because of the difficulties that can be encountered in determining the aniline point of the fuel (ASTM D611, IP 2). It is possible to overcome the issue of color by the use of a modification of the test procedure (Method B, ASTM D611, IP 2).

An alternate method of calculation of the calorific value, when an experimental determination is not available or cannot be made conveniently, involves an estimate of this property (ASTM D3338). In this test method, the net heat of combustion is calculated from the density, sulfur, and hydrogen content, but this calculation is justifiable only when the fuel belongs to a well-defined class for which a relationship between these quantities has been derived from accurate experimental measurements on representative samples. Thus, the hydrogen content (ASTM D1018, ASTM D1217, ASTM D1298, ASTM D3701, ASTM D4052, ASTM D4808, ASTM D5291 IP 160, IP 365), density

(ASTM D129, ASTM D1250, ASTM D1266, ASTM D2622, ASTM D3120, IP 61, IP 107), and sulfur content (ASTM D2622, ASTM D3120, ASTM D3246, ASTM D4294, ASTM D5453, ASTM D5623, IP 336, IP 373) of the sample are determined by experimental test methods, and the net heat of combustion is calculated using the values obtained by these test methods based on reported correlations.

10.3.4 Carbon Residue

The carbon residue of a petroleum product gives an indication of the propensity for that product to form a carbonaceous residue under thermal conditions. The carbonaceous residue is correctly referred to as the *carbon residue* but is also often referred to as *coke* or *thermal coke*.

The use of No. 2 oil fuel for heating has resulted in the availability of different types of burners that are classified according to the manner in which the fuel oil is combusted. Any carbonaceous residue formed during the thermal decomposition of the fuel oil that is deposited in, or near, the inlet surface reduces the fuel oil flow with resultant loss in burner efficiency. Therefore, fuel oil should have low-carbon-forming propensities. Other petroleum products that are lower boiling that distillate fuel oil do not usually reference the carbon residue in the specifications.

Thus, assessing the carbon-forming tendencies of the fuel oil is carried out using a carbon residue test. The test methods are (1) the Conradson carbon residue (ASTM D189, IP 13), the Ramsbottom carbon residue (ASTM D524, IP 14), and the microcarbon carbon residue (ASTM D4530, IP 398). The data give an indication of the amount of coke that will be formed during thermal processes as well as an indication of the amount of high-boiling constituents in petroleum. For lower boiling fuel oil, the formation of low yields of carbonaceous deposits, the carbon residue value is determined on a residue (10% by weight of the fuel oil) obtained by means of an adaptation of the standard distillation procedure for gas oil (IP 123) in order that the accuracy of the determination may be improved.

10.3.5 Cloud Point

Distillate fuel oil for heating installations are usually stored in outside tankage, and a knowledge of the lowest temperature at which the fuel can be transferred from tank to burner, thus avoiding line and filter blockage difficulties, is necessary.

An indication of this temperature may be obtained from the cloud point (ASTM D2500, ASTM D5771, ASTM D5772, ASTM D5773, IP 219) and pour point (ASTM D97, IP 15). These test methods give, respectively, the temperature at which wax begins to crystallize out of the fuel and when the wax structure has built up sufficiently to prevent the flow of oil. In these installations, a coarse filter is normally sited in the system near to the tank outlet to remove large particles of extraneous matter; a fine filter is positioned near the burner to protect the pump.

As the temperature continues to decrease below the cloud point, the formation of wax crystals is accelerated. These crystals clog fuel filters and lines, and thus reduce the supply of fuel to the burner. Since the cloud point is a higher temperature than the pour point (4–5°C, 7–9°F), and even higher, the cloud point is often considered to be more important than the pour point in establishing distillate fuel oil specifications for cold weather usage. However, the temperature differential between cloud and pour points depends upon the nature of the fuel components, but the use of wax crystal modifiers or pour depressants tends to accentuate these differences.

10.3.6 Composition

The chemical composition of fuel oil is extremely complex, and an extremely high number of compounds can be present through the hydrocarbon types, the range of isomeric hydrocarbons (Table 10.3), and the various types and isomers of heteroatom constituents. Therefore, it is not practical to perform individual compound analyses, but it is often helpful to define the compounds present under broad classifications, such as aromatics, paraffins, naphthenes, and olefins.

Thus, the first level of compositional information is group-type totals as deduced by adsorption chromatography (ASTM D1319, ASTM D2007) or by emulsion chromatography (ASTM D2549) to give volume percent saturates, olefins, and aromatics in materials that boil below 315°C (600°F). In addition, and depending upon the characteristics of the fuel oil, gas chromatography can also be used for the quantitative determination of olefins (ASTM D6296).

TABLE 10.3 Increase in the number of isomers with carbon number

Carbon atoms	Number of isomers
1	1
2	1
3	1
4	2
5	3
6	5
7	9
8	18
9	35
10	75
15	4,347
20	366,319
25	36,797,588
30	4,111,846,763
40	62,491,178,805,831

Following from the chromatographic separation, compositional analysis of the fractions by a mass spectral Z-series on which Z in the empirical formula C_nH_{2n+z} is a measure of the hydrogen deficiency of the compound is also warranted (ASTM D2425, ASTM D2786, ASTM D3239, ASTM D6379).

One mass spectrometric method (ASTM D2425) requires that the sample be separated into saturate and aromatic fractions before mass spectrometric analysis. This separation is standardized (ASTM D2549). This separation is applicable to fuel only when it is possible to evaporate the solvent used in the separation without also losing the light ends of the jet fuel. Combined gas chromatography/mass spectrometry has been used to give similar group-type results to ASTM D2425 but without pre-separation into saturates and aromatics. In addition, this method can give the Z-series information by carbon number showing how the composition changes with boiling point.

The percentage of aromatic hydrogen atoms and aromatic carbon atoms can be determined (ASTM D5292). Results from this test are not equivalent to mass- or volume-percent aromatics determined by the chromatographic methods. The chromatographic methods determine the mass or volume percentage of molecules that have one or more aromatic rings. Any alkyl substituents on the rings contribute to the percentage of aromatics determined by chromatographic techniques.

Correlative methods have long been used as a way of dealing with the complexity of petroleum fractions. Such methods include the use of viscosity–temperature charts (ASTM D341), calculation of the viscosity index (ASTM D2270), calculation of the viscosity gravity constant (ASTM D2501), calculation of the true vapor pressure (ASTM D2889), and estimation of the heat of combustion (ASTM D3338).

Organic sulfur compounds (e.g., mercaptans, sulfides, polysulfides, thiophenes) are present in petroleum products to a greater or lesser extent depending upon the crude oil origin and the refinery treatment. The sulfur content of fuel oil (ASTM D396) can be determined by a variety of methods (ASTM D129, ASTM D1552, ASTM D2622, ASTM D4294, IP 61) with mercaptan sulfur in cracked stocks being particularly necessary for evaluation (ASTM D3227, IP 342).

Corrosion of heating equipment can occur if the sulfur oxides formed on combustion of fuel oil are allowed to condense in the presence of moisture on the cooler parts of the fuel system. Corrosion of metal parts of the fuel system may also reflect the presence of corrosive sulfur components in the fuel. The corrosive tendencies of the fuel may be detected by the copper strip test (ASTM D130, ASTM D849, IP 154), the effect of these sulfur compounds being indicated by discoloration of the copper strip.

Various standard procedures are available for the determination of the sulfur content of distillate fuels. In the lamp method (ASTM D1266, IP 107), which is widely used, the product is burned completely in a small wick-fed lamp, the gases formed by combustion are absorbed in hydrogen peroxide solution, and the sulfur is subsequently determined as sulfate. Several rapid methods, including X-ray absorption and high-temperature combustion, for the determination of sulfur are also available.

In addition, excessive levels of hydrogen sulfide in the vapor phase above fuel oils in storage tanks may result in corrosion as well as being a health hazard. One method (ASTM D5705) is available for the determination of hydrogen sulfide but has been criticized as the test conditions do not simulate the vapor phase of a fuel storage tank. A second method (ASTM D6021) that is believed to be a more accurate simulation of the conditions in a fuel oil storage tank is available. In this method, a sample of the fuel oil is placed in a headspace vial and heated in an oven at 60°C (140°F) for more than 5 but less than 15 min. The headspace gas is sampled and injected into an apparatus capable of measuring hydrogen sulfide in the gaseous sample either by the lead acetate method (ASTM D4084, ASTM D4323) or by the chemiluminescence method (ASTM D5504).

Nitrogen can be determined by elemental analysis (ASTM D3228, ASTM D5291, ASTM D5762). Nitrogen compounds in middle distillates can be selectively detected by chemiluminescence (ASTM D4629). Individual nitrogen compounds can be detected down to 100 ppb nitrogen.

10.3.7 Density

The density (specific gravity) of fuel oil is an index of the weight of a measured volume of the product (ASTM D287, ASTM D1250, ASTM D1298, ASTM D1480, ASTM D1481, ASTM D4052, IP 160, IP 192, IP 200, IP 365).

The density is the mass (weight *in vacuo*) of a unit volume of fuel oil at any given temperature (ASTM D1298, IP 160). On the other hand, the specific gravity of a fuel oil is the ratio of the weight of a given volume of the material at a temperature of 15.6°C (60°F) to the weight of an equal volume of distilled water at the same temperature, both weights being corrected for the buoyancy of air.

The API gravity (ASTM D1298, IP 160) is an arbitrary figure related to the specific gravity:

$$°API = 141.5 / (\text{specific gravity at } 60/60°F) / 131.5$$

10.3.8 Flash Point

The *flash point* is a measure of the temperature to which fuel oil must be heated to produce an ignitable vapor–air mixture above the liquid fuel when exposed to an open flame. Following from this, the *fire point* of a fuel is the temperature at which an oil in an open container gives off vapor at a sufficient rate to continue to burn after a flame is applied.

Thus, the flash point is used primarily as an index of fire hazards. Consequently, most industry specifications or classifications place limits on the flash point to ensure compliance with fire regulations, insurance, and legal requirements since it is essential that the fuel is safe to transport and store. Generally, because of its distillation characteristics, fuel oil should not contain any volatile or *flashable* constituents. Nevertheless, the occasion might arise when application of test methods to determine the flash point might be applicable.

The test method for the determination of the flash point by Pensky–Martens Closed Tester (ASTM D93, IP 34) and the test method for determining flash point by the Tag Closed Tester (ASTM D56) are employed for fuel oil.

10.3.9   Metallic Constituents

Metals in fuel oil can seriously affect the use and outcome of fuel oil systems. Even trace amounts of metals can be deleterious to fuel oil use. Hence, it is important to have test methods that can determine metals, both at trace levels and at major concentrations. Metallic constituents in fuel oil can be determined by several methods including atomic absorption spectrophotometry (ASTM D5863, ASTM D5863, IP 285, IP 465, IP 470), X-ray fluorescence spectrometry (ASTM D4927, IP 407), wavelength-dispersive X-ray fluorescence spectrometry (ASTM D6443, IP 433), and inductively coupled plasma emission spectrometry (ICPAES) (ASTM D5708).

Inductively coupled argon plasma emission spectrophotometry (ASTM D5708) has an advantage over atomic absorption spectrophotometry (ASTM D4628, ASTM D5863) because it can provide more complete elemental composition data than the atomic absorption method. Flame emission spectroscopy is often used successfully in conjunction with atomic absorption spectrophotometry (ASTM D3605). X-ray fluorescence spectrophotometry (ASTM D4927, ASTM D6443) is also sometimes used, but matrix effects can be a problem.

The method to be used for the determination of metallic constituents in petroleum is often a matter of individual preference.

10.3.10   Pour Point

The *pour point* (ASTM D97, IP 15) is the lowest temperature at which the fuel oil will flow under specified conditions. The maximum and minimum pour point temperatures provide a temperature window where a petroleum product, depending on its thermal history, might appear in the liquid as well as solid states. The pour point data can be used to supplement other measurements of cold flow behavior, and the data are particularly useful for the screening of the effect of wax interaction modifiers on the flow behavior of petroleum. The *pour point* should not be confused with the *freezing point*, which is an index of the lowest temperature at which the crude oil will flow under specified conditions. Test methods (ASTM D2386, ASTM D5972, IP 434, IP 435) for the freezing point are not usually applicable to fuel oil but are more applicable to diesel fuel and aviation fuel.

In the original (and still widely used) test for pour point (ASTM D97, IP 15), a sample is cooled at a specified rate and examined at intervals of 3°C (5.4°F) for flow characteristics. The lowest temperature at which the movement of the oil is observed is recorded as the pour point. A later test method (ASTM D5853) covers two procedures for the determination of the pour point of petroleum and petroleum products down to −36°C (−33°F). One method provides a measure of the maximum (upper) pour point temperature. The second method measures the minimum (lower) pour point temperature. In the methods, the test specimen is cooled (after preliminary heating) at a specified rate and examined at intervals of 3°C (5.4°F) for flow characteristics. Again, the lowest temperature at which movement of the test specimen is observed is recorded as the pour point.

In any determination of the pour point, petroleum that contains wax produces an irregular flow behavior when the wax begins to separate. Such petroleum possesses viscosity relationships that are difficult to predict in fuel line operation. In addition, some waxy petroleum is sensitive to heat treatment, which can also affect the viscosity characteristics. This complex behavior limits the value of viscosity and pour point tests on waxy petroleum.

Although the pour-point test is still included in many specifications, it is not designated for the high-boiling fuel oil (ASTM D396). In fact, while the failure to flow at the pour point normally is attributed to the separation of wax from the fuel oil (in the case of waxy crude oil precursors), it also can be due to the effect of viscosity of the fuel oil (in the case of naphthenic crude oil precursors). In addition, the pour point of fuel oil may be influenced by the previous thermal history of the fuel oil. Thus, the usefulness of the pour-point test in relation to fuel oil, especially residual fuel oil, may be open to question.

Perhaps a more important test is the test of the *cold filter plugging point*. The cold filter plugging point is the lowest temperature at which fuel oil will give trouble-free flow (ASTM D6371, IP 309). In this test, either manual or automated apparatus may be used and is cooled under specified conditions and, at intervals of 1°C, is drawn into a pipette under a controlled vacuum through a standardized wire mesh filter. As the sample continues to cool, the procedure is repeated for each 1°C below the first test temperature. The testing is continued until the amount of wax crystals that have separated out of the solution is sufficient to stop or slow down the flow so that the time taken to fill the pipette exceeds 60 s or the fuel fails to return completely to the test jar before the fuel has cooled by a further 1°C.

The indicated temperature at which the last filtration was commenced is recorded as the *cold filter plugging point*.

Alternatively, the low-temperature flow test involving determination of the filterability of fuel oil (ASTM D4539, ASTM D6426) results are indicative of the low-temperature flow performance of distillate fuel oil. Both tests are useful for the evaluation of fuel oil containing flow improver additives. In either test method, the temperature of a series of test specimens of fuel is lowered at a prescribed cooling rate. At the commencing temperature and at each 1°C interval thereafter, a separate specimen from the series is filtered through a 17-mm screen until a minimum low-temperature flow test pass temperature is obtained. The minimum low-temperature flow test pass temperature is the lowest temperature, expressed as a multiple of 1°C, at which a test specimen can be filtered in 60 s or less.

In another test (ASTM D2068), the filter-plugging tendency of distillate fuel oil can be determined by passing a sample at a constant flow rate (20 ml/ min) through a glass fiber filter medium. The pressure drop across the filter is monitored during the passage of a fixed volume of test fuel. If a prescribed maximum pressure drop is reached before the total volume of fuel is filtered, the actual volume of fuel filtered at the time of maximum pressure drop is recorded. The apparatus is required to be calibrated at intervals.

10.3.11   Stability

Fuel oil must be capable of storage for many months without significant change and should not break down to form gum or insoluble sediments or darken in color (ASTM D156, ASTM D381, ASTM D1209, ASTM D1500, ASTM D1544, IP 131). In other words, fuel oil must be stable.

The extent of fuel oil oxidation is determined by the measurement of the hydroperoxide number (ASTM D6447) and the peroxide number (ASTM D3703). Deterioration of fuel oil results in the formation of the peroxides as well as other oxygen-containing compounds, and these numbers are indications of the quantity of oxidizing constituents present in the sample as determined by the measurement of the compounds that will oxidize potassium iodide.

The determination of hydroperoxide number (ASTM D6447) does not use the ozone depleting substance 1,1,2-trichloro-1,2,2-trifluoroethane, which is used for determination of the peroxide number (ASTM D3703). In this method, a quantity of sample is contacted with aqueous potassium iodide (KI) solution in the presence of acid. The hydroperoxides present are reduced by potassium iodide liberating an equivalent amount of iodine, which is quantified by voltammetric analysis.

The determination of peroxide number (ASTM D3703) involves dissolution of the sample in 1,1,2-trichloro-1,2,2-trifluoroethane, and the solution is contacted with an aqueous potassium iodide solution. The peroxides present are reduced by the potassium iodide whereupon an equivalent amount of iodine is released that is titrated with standard sodium thiosulfate solution using a starch indicator.

Other tests for storage stability include determination of color formation and sediment (ASTM D473, ASTM D2273, ASTM D3241, ASTM D4625, ASTM D4870, ASTM D5304, IP 53, IP 323) in which reactivity to oxygen at high temperatures is determined by the amount of sediment formation as well as any color changes.

Straight-run fuel oil fractions from the same crude oil normally are stable and mutually compatible. However, fuel oil produced from the thermal cracking and visbreaking operations may be stable by themselves but can be unstable or incompatible if blended with straight-run fuels and vice versa. Furthermore, asphaltic deposition may result from the mixing of (distillate and residual) fuel oils of different origin and treatment, each of which may be perfectly satisfactory when used alone. Such fuels are said to be incompatible, and a spot test is available for determining the stability and compatibility of fuel oils. Therefore, test procedures are necessary to predict fuel stability and ensure a satisfactory level of performance by the fuel oil.

In addition, thermal treatment may cause the formation of asphaltene-type material in fuel oil. The asphaltene fraction (ASTM D893, ASTM D2007, ASTM D3279, ASTM D4124, ASTM D6560, IP 143) is the highest molecular weight and most complex fraction in petroleum. The asphaltene content is an indicator of the amount of carbonaceous residue that can be expected during thermal use or further processing (ASTM D189, ASTM D524, ASTM D4530, IP 13, IP 14, IP 398) (Speight, 2001; Speight and Ozum, 2002; Hsu and Robinson, 2006; Gary *et al.*, 2007; Speight, 2014).

In any of the methods for the determination of the asphaltene content (ASTM D893, ASTM D2007, ASTM D3279, ASTM D4124, ASTM D6560, IP 143), the crude oil or product (such as asphalt) is mixed with a large excess (usually >30 volumes hydrocarbon per volume of sample) low-boiling hydrocarbon such as *n*-pentane or *n*-heptane. For an extremely viscous sample, a solvent such as toluene may be used prior to the addition of the low-boiling hydrocarbon, but an additional amount of the hydrocarbon (usually >30 volumes hydrocarbon per volume of solvent) must be added to compensate for the presence of the solvent. After a specified time, the insoluble material (the asphaltene fraction) is separated (by filtration) and dried. The yield is reported as percentage by weight of the original sample.

It must be recognized that, in any of these tests, different hydrocarbons (such as *n*-pentane or *n*-heptane) will give different yields of the asphaltene fraction, and if the presence of the solvent is not compensated by the use of additional hydrocarbon, the yield will be erroneous. In addition, if the hydrocarbon is not present in a large excess, the yields of the asphaltene fraction will vary and will be erroneous (Speight, 2014).

The *precipitation number* is often equated to the asphaltene content, but there are several issues that remain obvious in its rejection for this purpose. For example, the method to determine the precipitation number (ASTM D91) advocates the use of naphtha for use with black oil or lubricating oil, and the amount of insoluble material (as a % v/v of the sample) is the precipitating number. In the test, 10 ml of sample is mixed with 90 ml of ASTM precipitation naphtha (that may or may not have a constant chemical composition) in a graduated centrifuge cone and centrifuged for 10 min at 600–700 rpm. The volume of material on the bottom of the centrifuge cone is noted until repeat centrifugation gives a value within 0.1 ml (the precipitation number). Obviously, this can be substantially different from the asphaltene content.

Another method, not specifically described as an asphaltene separation method, is designed to remove pentane-insoluble constituents by membrane filtration (ASTM D4055). In this method, a sample of oil is mixed with pentane in a volumetric flask, and the oil solution is filtered through a 0.8-µm membrane filter. The flask, funnel, and the filter are washed with pentane to completely transfer any particulates onto the filter after which the filter (with particulates) is dried and weighed to give the pentane-insoluble constituents as a percentage by weight of the sample. Particulates can also be determined by membrane filtration (ASTM D2276, ASTM D5452, ASTM D6217, IP 415).

Since the storage stability of fuel oil may also be influenced by the crude oil origin, hydrocarbon composition, and refinery treatment (especially if unsaturated constituents are present), fuel oil containing unsaturated hydrocarbons has a greater tendency to form sediment on aging than the straight-run fuel oils. Unsaturated hydrocarbon constituents can be determined by the bromine number (ASTM D1159, ASTM D2710, IP 130).

The bromine number is the grams of bromine that will react with 100 g of the sample under the test conditions. The magnitude of bromine number is an indication of the quantity of bromine-reactive constituents and is not an identification of constituents. It is used as a measure of aliphatic unsaturation in petroleum samples and as a percentage of olefins in petroleum distillates boiling up to approximately 315°C (600°F). In the test, a known weight of the sample dissolved in a specified solvent maintained at 0–5°C (32–41°F) is titrated with standard bromide–bromate solution. Determination of the end point is method and sample dependent, being influenced by color.

The presence of reactive compounds of sulfur (e.g., thiophenes), nitrogen (pyrroles), and oxygen is also considered to contribute to fuel instability (Mushrush and Speight, 1995, 1998a, 1998b; Mushrush *et al.*, 1999).

To ensure a product of satisfactory stability, test procedures are necessary to predict this aspect of quality control. One particular method (ASTM D2274) uses short-term high-temperature procedures that are generally preferred to the long-term/lower-temperature conditions. While the accuracy of these empirical tests leaves much to be desired, they do provide, with some background knowledge of the fuel, useful data relating to the fuel's storage stability characteristics. And it is not often acknowledged that then higher temperatures, and often do, change the chemistry of the aging process thereby leaving much open to speculation.

10.3.12 Viscosity

The viscosity of a fluid is a measure of its resistance to flow and is expressed as Saybolt Universal seconds, Saybolt Furol seconds, or centistokes (cSt, kinematic viscosity). Viscosity is one of the more important heating oil characteristics since it is indicative of the rate at which the oil will flow in fuel systems and the ease with which it can be atomized in a given type of burner.

For the determination of the viscosity of petroleum products, various procedures, for example, Saybolt (ASTM D88) and Engler, are available and have been in use for many years, all being of an empirical nature, measuring the time taken in seconds for a given volume of fuel to flow through an orifice of specified dimensions.

The use of these empirical procedures is being superseded by the more precise kinematic viscosity method (ASTM D445, IP 71) in which a fixed volume of fuel flows through the capillary of a calibrated glass capillary viscometer under an accurately reproducible head and at a closely controlled temperature. The result is obtained from the product of the time taken for the fuel to flow between two etched marks on the capillary tube and the calibration factor of the viscometer and is reported in cSt. Since the viscosity decreases with increasing temperature, the temperature of test must also be reported if the viscosity value is to have any significance. For distillate fuel oils, the usual test temperature is 38°C (100°F).

The *viscosity index* (ASTM D2270, IP 226) is a widely used measure of the variation in kinematic viscosity due to changes in the temperature of petroleum and petroleum products between 40°C and 100°C (104°F and 212°F). For samples of similar kinematic viscosity, the higher the viscosity index, the smaller the effect of temperature on its kinematic viscosity. The accuracy of the calculated viscosity index is dependent only on the accuracy of the original viscosity determination.

10.3.13 Volatility

The volatility of fuel oil must be uniform, from batch to batch, if too frequent resetting of burner controls is to be avoided and if maximum performance and efficiency is to be maintained. Information regarding the volatility and the proportion of fuel vaporized at any one temperature may be obtained from the standard distillation procedure (ASTM D86, IP 123). The distillation test is significant for the distillate fuels as it is essential that the fuels contain sufficient volatile components to ensure that ignition and flame stability can be accomplished easily.

The distillation procedure (ASTM D86, IP 123) measures the amount of liquid vaporized and subsequently condensed as the temperature of the fuel in the distillation flask is raised at a prescribed rate. A record is made of the volume of distillate collected at specified temperatures or, conversely, the temperature at each increment of volume distilled (usually 10% increments). The temperature at which the first drop of condensate is collected is called the *initial boiling point*. The *end point* usually is the highest temperature recorded as the bottom of the flask becomes dry. If the sample is heated above 370°C (698°F), cracking occurs and the data are erroneous. The test usually is stopped when this point is reached. Some distillations may be run under reduced pressure (10 mm Hg) in order to avoid cracking (ASTM D1160). Under these conditions, fuel oil constituents may be distilled up to temperatures equivalent to 510°C (950°F) at atmospheric pressure.

Specifications for fuel oil may include limits on the temperatures at which 10 and 90% of the fuel are distilled by the standard procedure (ASTM D396). For the kerosene-type fuel oil, these values control the volatility at both ends of the distillation range, while for the gas oil, where the front-end volatility is not so critical, only the 90% distillation temperature is normally specified. This ensures that high–boiling point components that are less likely to burn, and which can cause carbon deposition, are excluded from the fuel.

One particular method is specifically designed for high-boiling petroleum fraction having an initial boiling point greater than 150°C (300°F) (ASTM D5236). The method uses a potstill with a low pressure drop entrainment separator operated under total takeoff conditions. The maximum achievable temperature, up to 565°C (l050°F), is dependent upon the heat tolerance of the charge, but for distillate fuel oil, the need to approach the maximum temperature is unnecessary. In this method, a weighed volume of the sample is distilled at pressures between 0.1 and 50 mm Hg at specified distillation rates. Fractions are taken at preselected temperatures with records made of the vapor temperature, the operating pressure, and any other variables deemed necessary. From the mass and density of each fraction, distillation yields by mass and volume can be calculated.

The boiling range distribution of certain fuel oils can be determined by gas chromatography (ASTM D6352). This test method is applicable to petroleum distillate fractions with an initial boiling point of <700°C (<1292°F) at atmospheric pressure. The test method is not applicable to products containing low-molecular-weight components, for example, naphtha, reformate, gasoline, residuum, and petroleum itself, and should not be used for materials that contain heterogeneous components such as alcohols, ethers, esters, or acids or residue. In this method, a nonpolar open tubular capillary gas chromatographic column is used to elute the hydrocarbon components of the sample in order to enhance the separation. A sample aliquot diluted with a viscosity-reducing solvent is introduced into the chromatographic system, and the column oven temperature is raised at a specified linear rate to effect separation of the hydrocarbon components. Retention times of known normal paraffin hydrocarbons spanning the scope of the test method are used for normalizing the retention times of the constituents of the fuel oil.

10.3.14 Water and Sediment

Considerable importance is attached to the presence of water or sediment in fuel oil because they lead to difficulties in use such as corrosion of equipment and blockages in fuel lines.

The sediment consists of finely divided solids that may be drilling mud or sand or scale picked up during the transport of the oil, or may consist of chlorides derived from evaporation of brine droplets in the oil. The solids may be dispersed in the oil or carried in water droplets. Sediment in petroleum can lead to serious plugging of the equipment, corrosion due to chloride decomposition, and a lowering of residual fuel quality.

In any form, water and sediment are highly undesirable in fuel oil, and the relevant tests involving distillation (ASTM D95, ASTM D4006, IP 74, IP 358), centrifuging (ASTM D4007), extraction (ASTM D473, IP 53), and the Karl Fischer titration (ASTM D4377, ASTM D4928, IP 356, IP 386, IP 438, IP 439) are regarded as important in examinations of quality.

The Karl Fischer test method (ASTM D1364, ASTM D6304) covers the direct determination of water in petroleum products. In the test, the sample injection in the titration vessel can be performed on a volumetric or gravimetric basis. Viscous samples can be analyzed using a water vaporizer accessory that heats the sample in the evaporation chamber, and the vaporized water is carried into the Karl Fischer titration cell by a dry inert carrier gas.

Water and sediment can be determined simultaneously (ASTM D4007) by the centrifuge method. Known volumes of the fuel oil and solvent are placed in a centrifuge tube and heated to 60°C (140°F). After centrifugation, the volume of the sediment and water layer at the bottom of the tube is read. For fuel oil that contains wax, a temperature of 71°C (160°F) or higher may be required to completely melt the wax crystals so that they are not measured as sediment.

Sediment is also determined by an extraction method (ASTM D473, IP 53) or by membrane filtration (ASTM D4807). In the former method (ASTM D473, IP 53), an oil sample contained in a refractory thimble is extracted with hot toluene until the residue reaches a constant mass. In the latter test, the sample is dissolved in hot toluene and filtered under vacuum through a 0.45-μm porosity membrane filter. The filter with residue is washed, dried, and weighed.

REFERENCES

ASTM D56. Standard Test Method for Flash Point by Tag Closed Cup Tester. Annual Book of Standards. ASTM International, West Conshohocken, PA. Annual Book of Standards. ASTM International, West Conshohocken, PA.

ASTM D86. Standard Test Method for Distillation of Petroleum Products at Atmospheric Pressure. Annual Book of Standards. ASTM International, West Conshohocken, PA.

ASTM D88. Standard Test Method for Saybolt Viscosity. Annual Book of Standards. ASTM International, West Conshohocken, PA.

ASTM D91. Standard Test Method for Precipitation Number of Lubricating Oils. Annual Book of Standards. ASTM International, West Conshohocken, PA.

ASTM D93. Standard Test Methods for Flash Point by Pensky-Martens Closed Cup Tester. Annual Book of Standards. ASTM International, West Conshohocken, PA.

ASTM D95. Standard Test Method for Water in Petroleum Products and Bituminous Materials by Distillation. Annual Book of Standards. ASTM International, West Conshohocken, PA.

ASTM D97. Standard Test Method for Pour Point of Petroleum Products. Annual Book of Standards. ASTM International, West Conshohocken, PA.

ASTM D129. Standard Test Method for Sulfur in Petroleum Products (General High Pressure Decomposition Device Method). Annual Book of Standards. ASTM International, West Conshohocken, PA.

ASTM D130. Standard Test Method for Corrosiveness to Copper from Petroleum Products by Copper Strip Test. Annual Book of Standards. ASTM International, West Conshohocken, PA.

ASTM D156. Standard Test Method for Saybolt Color of Petroleum Products (Saybolt Chromometer Method). Annual Book of Standards. ASTM International, West Conshohocken, PA.

ASTM D189. Standard Test Method for Conradson Carbon Residue of Petroleum Products. Annual Book of Standards. ASTM International, West Conshohocken, PA.

ASTM D240. Standard Test Method for Heat of Combustion of Liquid Hydrocarbon Fuels by Bomb Calorimeter. Annual Book of Standards. ASTM International, West Conshohocken, PA.

ASTM D287. Standard Test Method for API Gravity of Crude Petroleum and Petroleum Products (Hydrometer Method). Annual Book of Standards. ASTM International, West Conshohocken, PA.

ASTM D323. Standard Test Method for Vapor Pressure of Petroleum Products (Reid Method). Annual Book of Standards. ASTM International, West Conshohocken, PA.

ASTM D341. Standard Practice for Viscosity-Temperature Charts for Liquid Petroleum Products. Annual Book of Standards. ASTM International, West Conshohocken, PA.

ASTM D381. Standard Test Method for Gum Content in Fuels by Jet Evaporation. Annual Book of Standards. ASTM International, West Conshohocken, PA.

ASTM D396. Standard Specification for Fuel Oils. Annual Book of Standards. ASTM International, West Conshohocken, PA.

ASTM D445. Standard Test Method for Kinematic Viscosity of Transparent and Opaque Liquids (and Calculation of Dynamic Viscosity). Annual Book of Standards. ASTM International, West Conshohocken, PA.

ASTM D473. Standard Test Method for Sediment in Crude Oils and Fuel Oils by the Extraction Method. Annual Book of Standards. ASTM International, West Conshohocken, PA.

ASTM D482. Standard Test Method for Ash from Petroleum Products. Annual Book of Standards. ASTM International, West Conshohocken, PA.

ASTM D524. Standard Test Method for Ramsbottom Carbon Residue of Petroleum Products. Annual Book of Standards. ASTM International, West Conshohocken, PA.

ASTM D611. Standard Test Methods for Aniline Point and Mixed Aniline Point of Petroleum Products and Hydrocarbon Solvents. Annual Book of Standards. ASTM International, West Conshohocken, PA.

ASTM D664. Standard Test Method for Acid Number of Petroleum Products by Potentiometric Titration. Annual Book of Standards. ASTM International, West Conshohocken, PA.

ASTM D849. Standard Test Method for Copper Strip Corrosion by Industrial Aromatic Hydrocarbons. Annual Book of Standards. ASTM International, West Conshohocken, PA.

ASTM D893. Standard Test Method for Insolubles in Used Lubricating Oils. Annual Book of Standards. ASTM International, West Conshohocken, PA.

ASTM D974. Standard Test Method for Acid and Base Number by Color-Indicator Titration. Annual Book of Standards. ASTM International, West Conshohocken, PA.

ASTM D975. Standard Specification for Diesel Fuel Oils. Annual Book of Standards. ASTM International, West Conshohocken, PA.

ASTM D1018. Standard Test Method for Hydrogen in Petroleum Fractions. Annual Book of Standards. ASTM International, West Conshohocken, PA.

ASTM D1159. Standard Test Method for Bromine Numbers of Petroleum Distillates and Commercial Aliphatic Olefins by Electrometric Titration. Annual Book of Standards. ASTM International, West Conshohocken, PA.

ASTM D1160. Standard Test Method for Distillation of Petroleum Products at Reduced Pressure. Annual Book of Standards. ASTM International, West Conshohocken, PA.

ASTM D1209. Standard Test Method for Color of Clear Liquids (Platinum-Cobalt Scale). Annual Book of Standards. ASTM International, West Conshohocken, PA.

ASTM D1217. Standard Test Method for Density and Relative Density (Specific Gravity) of Liquids by Bingham Pycnometer. Annual Book of Standards. ASTM International, West Conshohocken, PA.

ASTM D1250. Standard Guide for Use of the Petroleum Measurement Tables. Annual Book of Standards. ASTM International, West Conshohocken, PA.

ASTM D1266. Standard Test Method for Sulfur in Petroleum Products (Lamp Method). Annual Book of Standards. ASTM International, West Conshohocken, PA.

ASTM D1298. Standard Test Method for Density, Relative Density, or API Gravity of Crude Petroleum and Liquid Petroleum Products by Hydrometer Method. Annual Book of Standards. ASTM International, West Conshohocken, PA.

ASTM D1319. Standard Test Method for Hydrocarbon Types in Liquid Petroleum Products by Fluorescent Indicator Adsorption. Annual Book of Standards. ASTM International, West Conshohocken, PA.

ASTM D1364. Standard Test Method for Water in Volatile Solvents (Karl Fischer Reagent Titration Method). Annual Book of Standards. ASTM International, West Conshohocken, PA.

ASTM D1405. Standard Test Method for Estimation of Net Heat of Combustion of Aviation Fuels. Annual Book of Standards. ASTM International, West Conshohocken, PA.

ASTM D1480. Standard Test Method for Density and Relative Density (Specific Gravity) of Viscous Materials by Bingham Pycnometer. Annual Book of Standards. ASTM International, West Conshohocken, PA.

ASTM D1481. Standard Test Method for Density and Relative Density (Specific Gravity) of Viscous Materials by Lipkin Bicapillary Pycnometer. Annual Book of Standards. ASTM International, West Conshohocken, PA.

ASTM D1500. Standard Test Method for ASTM Color of Petroleum Products (ASTM Color Scale). Annual Book of Standards. ASTM International, West Conshohocken, PA.

ASTM D1544. Standard Test Method for Color of Transparent Liquids (Gardner Color Scale). Annual Book of Standards. ASTM International, West Conshohocken, PA.

ASTM D1552. Standard Test Method for Sulfur in Petroleum Products (High-Temperature Method). Annual Book of Standards. ASTM International, West Conshohocken, PA.

ASTM D1655. Standard Specification for Aviation Turbine Fuels. Annual Book of Standards. ASTM International, West Conshohocken, PA.

ASTM D2007. Standard Test Method for Characteristic Groups in Rubber Extender and Processing Oils and Other Petroleum-Derived Oils by the Clay-Gel Absorption Chromatographic Method. Annual Book of Standards. ASTM International, West Conshohocken, PA.

ASTM D2068. Standard Test Method for Determining Filter Blocking Tendency. Annual Book of Standards. ASTM International, West Conshohocken, PA.

ASTM D2270. Standard Practice for Calculating Viscosity Index from Kinematic Viscosity at 40 and 100°C. Annual Book of Standards. ASTM International, West Conshohocken, PA.

ASTM D2273. Standard Test Method for Trace Sediment in Lubricating Oils. Annual Book of Standards. ASTM International, West Conshohocken, PA.

ASTM D2274. Standard Test Method for Oxidation Stability of Distillate Fuel Oil (Accelerated Method). Annual Book of Standards. ASTM International, West Conshohocken, PA.

ASTM D2276. Standard Test Method for Particulate Contaminant in Aviation Fuel by Line Sampling. Annual Book of Standards. ASTM International, West Conshohocken, PA.

ASTM D2386. Standard Test Method for Freezing Point of Aviation Fuels. Annual Book of Standards. ASTM International, West Conshohocken, PA.

ASTM D2425. Standard Test Method for Hydrocarbon Types in Middle Distillates by Mass Spectrometry. Annual Book of Standards. ASTM International, West Conshohocken, PA.

ASTM D2500. Standard Test Method for Cloud Point of Petroleum Products. Annual Book of Standards. ASTM International, West Conshohocken, PA.

ASTM D2501. Standard Test Method for Calculation of Viscosity-Gravity Constant (VGC) of Petroleum Oils. Annual Book of Standards. ASTM International, West Conshohocken, PA.

ASTM D2549. Standard Test Method for Separation of Representative Aromatics and Nonaromatics Fractions of High-Boiling Oils by Elution Chromatography. Annual Book of Standards. ASTM International, West Conshohocken, PA.

ASTM D2622. Standard Test Method for Sulfur in Petroleum Products by Wavelength Dispersive X-ray Fluorescence Spectrometry. Annual Book of Standards. ASTM International, West Conshohocken, PA.

ASTM D2710. Standard Test Method for Bromine Index of Petroleum Hydrocarbons by Electrometric Titration. Annual Book of Standards. ASTM International, West Conshohocken, PA.

ASTM D2786. Standard Test Method for Hydrocarbon Types Analysis of Gas-Oil Saturates Fractions by High Ionizing Voltage Mass Spectrometry. Annual Book of Standards. ASTM International, West Conshohocken, PA.

ASTM D2880. Standard Specification for Gas Turbine Fuel Oils. Annual Book of Standards. ASTM International, West Conshohocken, PA.

ASTM D2889. Standard Test Method for Calculation of True Vapor Pressures of Petroleum Distillate Fuels. Annual Book of Standards. ASTM International, West Conshohocken, PA.

ASTM D3120. Standard Test Method for Trace Quantities of Sulfur in Light Liquid Petroleum Hydrocarbons by Oxidative Microcoulometry. ASTM International, West Conshohocken, PA.

ASTM D3227. Standard Test Method for (Thiol Mercaptan) Sulfur in Gasoline, Kerosene, Aviation Turbine, and Distillate Fuels (Potentiometric Method). Annual Book of Standards. ASTM International, West Conshohocken, PA.

ASTM D3228. Standard Test Method for Total Nitrogen in Lubricating Oils and Fuel Oils by Modified Kjeldahl Method. Annual Book of Standards. ASTM International, West Conshohocken, PA.

ASTM D3239. Standard Test Method for Aromatic Types Analysis of Gas-Oil Aromatic Fractions by High Ionizing Voltage Mass Spectrometry. Annual Book of Standards. ASTM International, West Conshohocken, PA.

ASTM D3241. Standard Test Method for Thermal Oxidation Stability of Aviation Turbine Fuels. Annual Book of Standards. ASTM International, West Conshohocken, PA.

ASTM D3246. Standard Test Method for Sulfur in Petroleum Gas by Oxidative Microcoulometry. Annual Book of Standards. ASTM International, West Conshohocken, PA.

ASTM D3279. Standard Test Method for n-Heptane Insolubles. Annual Book of Standards. ASTM International, West Conshohocken, PA.

ASTM D3338. Standard Test Method for Estimation of Net Heat of Combustion of Aviation Fuels. Annual Book of Standards. ASTM International, West Conshohocken, PA.

ASTM D3605. Standard Test Method for Trace Metals in Gas Turbine Fuels by Atomic Absorption and Flame Emission Spectroscopy. Annual Book of Standards. ASTM International, West Conshohocken, PA.

ASTM D3701. Standard Test Method for Hydrogen Content of Aviation Turbine Fuels by Low Resolution Nuclear Magnetic Resonance Spectrometry. Annual Book of Standards. ASTM International, West Conshohocken, PA.

ASTM D3703. Standard Test Method for Hydroperoxide Number of Aviation Turbine Fuels, Gasoline and Diesel Fuels. Annual Book of Standards. ASTM International, West Conshohocken, PA.

ASTM D4006. Standard Test Method for Water in Crude Oil by Distillation. Annual Book of Standards. ASTM International, West Conshohocken, PA.

ASTM D4007. Standard Test Method for Water and Sediment in Crude Oil by the Centrifuge Method (Laboratory Procedure). Annual Book of Standards. ASTM International, West Conshohocken, PA.

ASTM D4052. Standard Test Method for Density, Relative Density, and API Gravity of Liquids by Digital Density Meter. Annual Book of Standards. ASTM International, West Conshohocken, PA.

ASTM D4055. Standard Test Method for Pentane Insolubles by Membrane Filtration. Annual Book of Standards. ASTM International, West Conshohocken, PA.

ASTM D4084. Standard Test Method for Analysis of Hydrogen Sulfide in Gaseous Fuels (Lead Acetate Reaction Rate Method). Annual Book of Standards. ASTM International, West Conshohocken, PA.

ASTM D4124. Standard Test Method for Separation of Asphalt into Four Fractions. Annual Book of Standards. ASTM International, West Conshohocken, PA.

ASTM D4294. Standard Test Method for Sulfur in Petroleum and Petroleum Products by Energy Dispersive X-ray Fluorescence Spectrometry. Annual Book of Standards. ASTM International, West Conshohocken, PA.

ASTM D4323. Standard Test Method for Hydrogen Sulfide in the Atmosphere by Rate of Change of Reflectance. Annual Book of Standards. ASTM International, West Conshohocken, PA.

ASTM D4377. Standard Test Method for Water in Crude Oils by Potentiometric Karl Fischer Titration. Annual Book of Standards. ASTM International, West Conshohocken, PA.

ASTM D4418. ASTM D4418-00(2011) Standard Practice for Receipt, Storage, and Handling of Fuels for Gas Turbines. Annual Book of Standards. ASTM International, West Conshohocken, PA.

ASTM D4530. Standard Test Method for Determination of Carbon Residue (Micro Method). Annual Book of Standards. ASTM International, West Conshohocken, PA.

ASTM D4539. Standard Test Method for Filterability of Diesel Fuels by Low-Temperature Flow Test (LTFT). Annual Book of Standards. ASTM International, West Conshohocken, PA.

ASTM D4625. Standard Test Method for Middle Distillate Fuel Storage Stability at 43°C (110°F). Annual Book of Standards. ASTM International, West Conshohocken, PA.

ASTM D4628. Standard Test Method for Analysis of Barium, Calcium, Magnesium, and Zinc in Unused Lubricating Oils by Atomic Absorption Spectrometry. Annual Book of Standards. ASTM International, West Conshohocken, PA.

ASTM D4629. Standard Test Method for Trace Nitrogen in Liquid Petroleum Hydrocarbons by Syringe/Inlet Oxidative Combustion and Chemiluminescence Detection. Annual Book of Standards. ASTM International, West Conshohocken, PA.

ASTM D4807. Standard Test Method for Sediment in Crude Oil by Membrane Filtration. Annual Book of Standards. ASTM International, West Conshohocken, PA.

ASTM D4808. Standard Test Methods for Hydrogen Content of Light Distillates, Middle Distillates, Gas Oils, and Residua by Low-Resolution Nuclear Magnetic Resonance Spectroscopy. Annual Book of Standards. ASTM International, West Conshohocken, PA.

ASTM D4809. Standard Test Method for Heat of Combustion of Liquid Hydrocarbon Fuels by Bomb Calorimeter (Precision Method). Annual Book of Standards. ASTM International, West Conshohocken, PA.

ASTM D4870. Standard Test Method for Determination of Total Sediment in Residual Fuels. Annual Book of Standards. ASTM International, West Conshohocken, PA.

ASTM D4927. Standard Test Methods for Elemental Analysis of Lubricant and Additive Components – Barium, Calcium, Phosphorus, Sulfur, and Zinc by Wavelength-Dispersive X-Ray Fluorescence Spectroscopy. Annual Book of Standards. ASTM International, West Conshohocken, PA.

ASTM D4928. Standard Test Method for Water in Crude Oils by Coulometric Karl Fischer Titration. Annual Book of Standards. ASTM International, West Conshohocken, PA.

ASTM D4951. Standard Test Method for Determination of Additive Elements in Lubricating Oils by Inductively Coupled Plasma Atomic Emission Spectrometry. Annual Book of Standards. ASTM International, West Conshohocken, PA.

ASTM D5184. Standard Test Methods for Determination of Aluminum and Silicon in Fuel Oils by Ashing, Fusion, Inductively Coupled Plasma Atomic Emission Spectrometry, and Atomic Absorption Spectrometry. Annual Book of Standards. ASTM International, West Conshohocken, PA.

ASTM D5185. Standard Test Method for Multi-element Determination of Used and Unused Lubricating Oils and Base Oils by Inductively Coupled Plasma Atomic Emission Spectrometry (ICP-AES). Annual Book of Standards. ASTM International, West Conshohocken, PA.

ASTM D5236. Standard Test Method for Distillation of Heavy Hydrocarbon Mixtures (Vacuum Potstill Method). Annual Book of Standards. ASTM International, West Conshohocken, PA.

ASTM D5291. Standard Test Methods for Instrumental Determination of Carbon, Hydrogen, and Nitrogen in Petroleum Products and Lubricants. Annual Book of Standards. ASTM International, West Conshohocken, PA.

ASTM D5292. Standard Test Method for Aromatic Carbon Contents of Hydrocarbon Oils by High Resolution Nuclear Magnetic Resonance Spectroscopy. Annual Book of Standards. ASTM International, West Conshohocken, PA.

ASTM D5304. Standard Test Method for Assessing Middle Distillate Fuel Storage Stability by Oxygen Overpressure. Annual Book of Standards. ASTM International, West Conshohocken, PA.

ASTM D5452. Standard Test Method for Particulate Contamination in Aviation Fuels by Laboratory Filtration. Annual Book of Standards. ASTM International, West Conshohocken, PA.

ASTM D5453. Standard Test Method for Determination of Total Sulfur in Light Hydrocarbons, Spark Ignition Engine Fuel, Diesel Engine Fuel, and Engine Oil by Ultraviolet Fluorescence. Annual Book of Standards. ASTM International, West Conshohocken, PA.

ASTM D5504. Standard Test Method for Determination of Sulfur Compounds in Natural Gas and Gaseous Fuels by Gas Chromatography and Chemiluminescence. Annual Book of Standards. ASTM International, West Conshohocken, PA.

ASTM D5623. Standard Test Method for Sulfur Compounds in Light Petroleum Liquids by Gas Chromatography and Sulfur Selective Detection. Annual Book of Standards. ASTM International, West Conshohocken, PA.

ASTM D5705. Standard Test Method for Measurement of Hydrogen Sulfide in the Vapor Phase Above Residual Fuel Oils. Annual Book of Standards. ASTM International, West Conshohocken, PA.

ASTM D5708. Standard Test Methods for Determination of Nickel, Vanadium, and Iron in Crude Oils and Residual Fuels by Inductively Coupled Plasma (ICP) Atomic Emission Spectrometry. Annual Book of Standards. ASTM International, West Conshohocken, PA.

ASTM D5762. Standard Test Method for Nitrogen in Petroleum and Petroleum Products by Boat-Inlet Chemiluminescence. Annual Book of Standards. ASTM International, West Conshohocken, PA.

ASTM D5771. Standard Test Method for Cloud Point of Petroleum Products (Optical Detection Stepped Cooling Method). Annual Book of Standards. ASTM International, West Conshohocken, PA.

ASTM D5772. Standard Test Method for Cloud Point of Petroleum Products (Linear Cooling Rate Method). Annual Book of Standards. ASTM International, West Conshohocken, PA.

ASTM D5773. Standard Test Method for Cloud Point of Petroleum Products (Constant Cooling Rate Method). Annual Book of Standards. ASTM International, West Conshohocken, PA.

ASTM D5853. Standard Test Method for Pour Point of Crude Oils. Annual Book of Standards. ASTM International, West Conshohocken, PA.

ASTM D5863. Standard Test Methods for Determination of Nickel, Vanadium, Iron, and Sodium in Crude Oils and Residual Fuels by Flame Atomic Absorption Spectrometry. Annual Book of Standards. ASTM International, West Conshohocken, PA.

ASTM D5972. Standard Test Method for Freezing Point of Aviation Fuels (Automatic Phase Transition Method). Annual Book of Standards. ASTM International, West Conshohocken, PA.

ASTM D6021. Standard Test Method for Measurement of Total Hydrogen Sulfide in Residual Fuels by Multiple Headspace Extraction and Sulfur Specific Detection. Annual Book of Standards. ASTM International, West Conshohocken, PA.

ASTM D6217. Standard Test Method for Particulate Contamination in Middle Distillate Fuels by Laboratory Filtration. Annual Book of Standards. ASTM International, West Conshohocken, PA.

ASTM D6296. Standard Test Method for Total Olefins in Spark-ignition Engine Fuels by Multidimensional Gas Chromatography. Annual Book of Standards. ASTM International, West Conshohocken, PA.

ASTM D6304. Standard Test Method for Determination of Water in Petroleum Products, Lubricating Oils, and Additives by Coulometric Karl Fischer Titration. Annual Book of Standards. ASTM International, West Conshohocken, PA.

ASTM D6352. Standard Test Method for Boiling Range Distribution of Petroleum Distillates in Boiling Range from 174 to 700°C by Gas Chromatography. Annual Book of Standards. ASTM International, West Conshohocken, PA.

ASTM D6371. Standard Test Method for Cold Filter Plugging Point of Diesel and Heating Fuels. Annual Book of Standards. ASTM International, West Conshohocken, PA.

ASTM D6379. Standard Test Method for Determination of Aromatic Hydrocarbon Types in Aviation Fuels and Petroleum Distillates – High Performance Liquid Chromatography Method with Refractive Index Detection. Annual Book of Standards. ASTM International, West Conshohocken, PA.

ASTM D6426. Standard Test Method for Determining Filterability of Middle Distillate Fuel Oils. Annual Book of Standards. ASTM International, West Conshohocken, PA.

ASTM D6443. Standard Test Method for Determination of Calcium, Chlorine, Copper, Magnesium, Phosphorus, Sulfur, and Zinc in Unused Lubricating Oils and Additives by Wavelength Dispersive X-ray Fluorescence Spectrometry (Mathematical Correction Procedure). Annual Book of Standards. ASTM International, West Conshohocken, PA.

ASTM D6447. Standard Test Method for Hydroperoxide Number of Aviation Turbine Fuels by Voltammetric Analysis. Annual Book of Standards. ASTM International, West Conshohocken, PA.

ASTM D6560. Standard Test Method for Determination of Asphaltenes (Heptane Insolubles) in Crude Petroleum and Petroleum Products. Annual Book of Standards. ASTM International, West Conshohocken, PA.

Charlot, J-C., and Claus, G. 2000. Distillates and Residual Fuels for Heating and Engines. In Modern Petroleum Technology. Volume 2: Downstream. A.G. Lucas (Editor). John Wiley & Sons Inc., New York.

Gary, J.G., Handwerk, G.E., and Kaiser, M.J. 2007. Petroleum Refining: Technology and Economics, 5th Edition. CRC Press, Taylor & Francis Group, Boca Raton, FL.

Heinrich, H., and Duée, D. 2000. Kerosene and Gas Oil. In Modern Petroleum Technology. Volume 2: Downstream. A.G. Lucas (Editor). John Wiley & Sons Inc., New York.

Hsu, C.S., and Robinson, P.R. (Editors) 2006. Practical Advances in Petroleum Processing, Volumes 1 and 2. Springer Science, New York.

IP 1. Acidity. Energy Institute, London, UK.

IP 2 (ASTM D611). Petroleum Products and Hydrocarbon Solvents – Determination of Aniline and Mixed Aniline Point. IP Standard Methods 2013. The Energy Institute, London.

IP 4 (ASTM D482). Petroleum Products – Determination of Ash. IP Standard Methods 2013. The Energy Institute, London.

IP 12. Determination of Specific Energy. IP Standard Methods 2013. The Energy Institute, London.

IP 13 (ASTM D189). Petroleum Products – Determination of Carbon Residue – Conradson Method. IP Standard Methods 2013. The Energy Institute, London.

IP 14 (ASTM D524). Petroleum Products – Determination of Carbon Residue – Ramsbottom Method. Energy Institute, London, UK.

IP 15 (ASTM D97). Petroleum Products – Determination of Pour Point. IP Standard Methods 2013. The Energy Institute, London.

IP 34 (ASTM D93). Determination of Flash Point – Pensky-Martens Closed Cup Method. IP Standard Methods 2013. The Energy Institute, London.

IP 53 (ASTM D473). Crude Petroleum and Fuel Oils – Determination of Sediment – Extraction Method. IP Standard Methods 2013. The Energy Institute, London.

IP 61 (ASTM D129). Determination of Sulfur – High Pressure Combustion Method. IP Standard Methods 2013. The Energy Institute, London.

IP 71 (ASTM D445). Petroleum Products – Transparent and Opaque Liquids – Determination of Kinematic Viscosity and Calculation of Dynamic Viscosity. IP Standard Methods 2013. The Energy Institute, London.

IP 74 (ASTM D95). Petroleum Products and Bituminous Materials – Determination of Water – Distillation Method. IP Standard Methods 2013. The Energy Institute, London.

IP 107 (ASTM D1266). Determination of Sulfur – Lamp Combustion Method. IP Standard Methods 2013. The Energy Institute, London.

IP 123. Petroleum Products – Determination of Distillation Characteristics at Atmospheric Pressure. IP Standard Methods 2013. The Energy Institute, London.

IP 130 (ASTM D1159). Petroleum Products – Determination of Bromine Number of Distillates and Aliphatic Olefins – Electrometric Method. IP Standard Methods 2013. The Energy Institute, London.

IP 131. Petroleum Products – Gum Content of Light and Middle Distillates – Jet Evaporation Method. IP Standard Methods 2013. The Energy Institute, London.

IP 139 (ASTM D974). Petroleum Products and Lubricants – Determination of Acid or Base Number-Color Indicator Titration Method. IP Standard Methods 2013. The Energy Institute, London.

IP 143 (ASTM D6560). Determination of Asphaltenes (Heptane Insolubles) in Crude Petroleum and Petroleum Products. IP Standard Methods 2013. The Energy Institute, London.

IP 154. Petroleum Products – Corrosiveness to Copper – Copper Strip Test. IP Standard Methods 2013. The Energy Institute, London.

IP 160 (ASTM D1298). Crude Petroleum and Liquid Petroleum Products – Laboratory Determination of Density – Hydrometer Method. IP Standard Methods 2013. The Energy Institute, London.

IP 177 (ASTM D664). Determination of Weak and Strong Acid Number – Potentiometric Titration Method. IP Standard Methods 2013. The Energy Institute, London.

IP 182. Determination of Inorganic Acidity of Petroleum Products – Color Indicator Titration Method. IP Standard Methods 2013. The Energy Institute, London.

IP 192. API Gravity of Petroleum Products – Hydrometer Method. IP Standard Methods 2013. The Energy Institute, London.

IP 200 (ASTM D1250). Guidelines for the use of the Petroleum Measurement Tables. IP Standard Methods 2013. The Energy Institute, London.

IP 219. Petroleum Products – Determination of Cloud Point. IP Standard Methods 2013. The Energy Institute, London.

IP 226 (ASTM D2270). Petroleum Products – Calculation of Viscosity Index from Kinematic Viscosity. IP Standard Methods 2013. The Energy Institute, London.

IP 285. Determination of Nickel and Vanadium – Spectrophotometric Method. IP Standard Methods 2013. The Energy Institute, London.

IP 309. Diesel and Domestic Heating Fuels – Determination of Cold Filter Plugging Point. IP Standard Methods 2013. The Energy Institute, London.

IP 323 (ASTM D3241). Petroleum Products – Determination of Thermal Oxidation Stability of Gas Turbine Fuels – JFTOT Method. IP Standard Methods 2013. The Energy Institute, London.

IP 336. Petroleum Products – Determination of Sulfur Content – Energy-Dispersive X-Ray Fluorescence Method. IP Standard Methods 2013. The Energy Institute, London.

IP 342 (ASTM D3227). Determination of Thiol (Mercaptan) Sulfur in Light and Middle Distillate Fuels – Potentiometric Method. IP Standard Methods 2013. The Energy Institute, London.

IP 356. Crude Petroleum – Determination of Water – Potentiometric Karl Fischer Titration Method. IP Standard Methods 2013. The Energy Institute, London.

IP 358. Crude Petroleum – Determination of Water – Distillation Method. IP Standard Methods 2013. The Energy Institute, London.

IP 365. Crude Petroleum and Petroleum Products – Determination of Density – Oscillating U-Tube Method. IP Standard Methods 2013. The Energy Institute, London.

IP 373. Sulfur Content – Microcoulometry (Oxidative) Method. IP Standard Methods 2013. The Energy Institute, London.

IP 377. Petroleum Products – Determination of Aluminum and Silicon in Fuel Oils – Inductively Coupled Plasma Emission and Atomic Absorption Spectroscopy Method. IP Standard Methods 2013. The Energy Institute, London.

IP 386 (ASTM D4928). Crude Petroleum – Determination of Water – Coulometric Karl Fischer Titration Method. IP Standard Methods 2013. The Energy Institute, London.

IP 398. Petroleum Products – Determination of Carbon Residue – Micro Method. IP Standard Methods 2013. The Energy Institute, London.

IP 407. Barium, Calcium, Phosphorous, Sulfur and Zinc Wavelength Dispersive X-Ray Fluorescence Spectroscopy. IP Standard Methods 2013. The Energy Institute, London.

IP 415 (ASTM D6217). Determination of Particulate Content of Middle Distillate Fuels – Laboratory Filtration Method. IP Standard Methods 2013. The Energy Institute, London.

IP 433. Petroleum Products – Determination of Vanadium And Nickel Content – Wavelength-Dispersive X-Ray Fluorescence Spectrometry. IP Standard Methods 2013. The Energy Institute, London.

IP 434. Freezing Point of Aviation Turbine Fuels – Automated Optical Method (replaced by IP 528–ASTM D5901). IP Standard Methods 2013. The Energy Institute, London.

IP 435. Freezing Point of Aviation Turbine Fuels – Automatic Phase Transition Method. IP Standard Methods 2013. The Energy Institute, London.

IP 438. Water Content by Coulometric Karl Fischer Titration Method. IP Standard Methods 2013. The Energy Institute, London.

IP 439. Petroleum Products – Determination of Water – Potentiometric Karl Fischer Titration Method. IP Standard Methods 2013. The Energy Institute, London.

IP 465. Liquid Petroleum Products – Determination of Nickel and Vanadium Content – Atomic Absorption Spectrometric Method. IP Standard Methods 2013. The Energy Institute, London.

IP 470. Determination of Aluminum, Silicon, Vanadium, Nickel, Iron, Calcium, Zinc and Sodium in Residual Fuel Oil by Ashing, Fusion And Atomic Absorption Spectrometry (Metals in Fuel Oil by AAS). IP Standard Methods 2013. The Energy Institute, London.

Mushrush, G.W., and Speight, J.G. 1995. Petroleum Products: Instability and Incompatibility. Taylor & Francis. New York.

Mushrush, G.W., and Speight, J.G. 1998a. Instability and Incompatibility of Petroleum Products. In Petroleum Chemistry and Refining. J.G. Speight (Editor). Taylor & Francis, New York.

Mushrush, G.W., and Speight, J.G. 1998b. A review of the chemistry of incompatibility in middle distillate fuels. Reviews in Process Chemistry and Engineering, 1: 5.

Mushrush, G.W., Speight, J.G., Beal, E.J., and Hardy, D.R. 1999. Preprints. American Chemical Society, Division of Fuel Chemistry, 44(2): 175.

Speight, J.G. 2001. Handbook of Petroleum Analysis. John Wiley & Sons Inc., New York.

Speight, J.G., and Ozum, B. 2002. Petroleum Refining Processes. Marcel Dekker Inc., New York.

Speight, J.G. 2014. The Chemistry and Technology of Petroleum, 5th Edition. CRC Press, Taylor & Francis Group, Boca Raton, FL.

11

RESIDUAL FUEL OIL

11.1 INTRODUCTION

The term *fuel oil* is applied not only to distillate products (*distillate fuel oil*, Chapter 10) but also to residual material that is distinguished from distillate-type fuel oil by boiling range and, hence, is referred to as *residual fuel oil* (ASTM D396).

Another word that often occurs in the same sentence as heavy fuel oil is *sludge*. Briefly, sludge is a contaminant that results from the handling, mixing, blending, and pumping of heavy fuel oil while stored at, and after it leaves, the refinery. Storage tanks, heavy fuel pipe lines, and barging can all contribute to the sludge formation. Water contamination of fuel oil having a high asphaltene content can produce an emulsion during fuel handling, which can contain more than 50% v/v water. In many cases, transfer pumps can frequently provide the necessary energy to produce emulsified sludge during normal fuel transfers—emulsified sludge can cause rapid fouling and shutdown of centrifugal purifiers and clogging of strainers and filters in the fuel oil system, and rapid fouling if burned in the engine.

Thus, *residual fuel oil* is the fuel oil that is manufactured from the distillation residuum, and the term includes all residual fuel oils, including fuel oil obtained by visbreaking as well as by blending residual products from other operations (Charlot and Claus, 2000; Heinrich and Duée, 2000; Speight and Ozum, 2002; Hsu and Robinson, 2006; Gary et al., 2007; Speight 2014). The various grades of heavy fuel oils are produced to meet rigid specifications in order to assure suitability for their intended purpose.

Detailed analysis of residual products, such as residual fuel oil, is more complex than the analysis of lower-molecular-weight liquid products. As with other products, there are a variety of physical property measurements that are required to determine if the residual fuel oil meets specification. But the range of molecular types present in petroleum products increases significantly with an increase in the molecular weight (i.e., an increase in the number of carbon atoms per molecule). Therefore, characterization measurements or studies cannot, and do not, focus on the identification of specific molecular structures. The focus tends to be on molecular classes (paraffins, naphthenes, aromatics, polycyclic compounds, and polar compounds).

Several tests that are usually applied to the lower-molecular-weight colorless (or light-colored) products are not applied to residual fuel oil. For example, test methods such as those designed for the determination of the aniline point (or mixed aniline point) (ASTM D611, IP 2) and the cloud point (ASTM D2500, ASTM D5771, ASTM D5772, ASTM D5773) can suffer from visibility effects due to the color of the fuel oil.

11.2 PRODUCTION AND PROPERTIES

Residual fuel oil is the fuel oil that is manufactured from the distillation residuum, and the term includes all residual fuel oils, including fuel oil obtained by visbreaking as well as by blending residual products from other operations (Speight and Ozum, 2002; Hsu and Robinson, 2006; Gary et al., 2007; Speight, 2014). The various grades of *heavy fuel oil* are produced to meet rigid specifications in order to assure suitability for their intended purpose and includes a variety

Handbook of Petroleum Product Analysis, Second Edition. James G. Speight.
© 2015 John Wiley & Sons, Inc. Published 2015 by John Wiley & Sons, Inc.

of oils ranging from distillates to residual oils, which must be heated to 260°C (500°F) or higher before they can be used. In general, heavy fuel oil consists of residual oils blended with distillates to suit specific needs. Included among heavy fuel oils are various industrial oils; when used to fuel ships, heavy fuel oil is called bunker oil.

Historically, fuel oils were based on atmospheric residua (Speight and Ozum, 2002; Hsu and Robinson, 2006; Gary et al., 2007; Speight, 2014) and were known as straight-run fuels. However, the increasing demand for transportation fuels such as gasoline, kerosene, and diesel has led to an increased value for the atmospheric residue as a feedstock for vacuum distillation and for cracking processes. As a consequence, most heavy fuel oils are currently based on vacuum residua as well as residues from thermal and catalytic cracking operations (Speight and Ozum, 2002; Hsu and Robinson, 2006; Gary et al., 2007; Speight, 2014). These fuels differ in character from straight-run fuels in that the density and mean molecular weight are higher, as is the carbon/hydrogen ratio. The density of heavy fuel oil can be above 1000 kg/m^3, which has environmental implications in the event of a spillage into fresh water.

Visbreaking (viscosity reduction, viscosity breaking) is the most widely used process for the production of residual fuel oil. It is a relatively mild thermal cracking operation used to reduce the viscosity of residua (Speight and Ozum, 2002; Hsu and Robinson, 2006; Gary et al., 2007; Speight 2014). Residua are sometimes blended with lighter heating oils to produce residual fuel oil of acceptable viscosity. By reducing the viscosity of the nonvolatile fraction, visbreaking reduces the amount of the more valuable light heating oil that is required for blending to meet the fuel oil specifications. The process is also used to reduce the pour point of a waxy residue.

In the visbreaking process, a residuum is passed through a furnace where it is heated to a temperature of approximately 480°C (895°F) under an outlet pressure of about 100 psi (Speight and Ozum, 2002; Speight, 2014). The heating coils in the furnace are arranged to provide a soaking section of low heat density, where the charge remains until the visbreaking reactions are completed. The cracked products are then passed into a flash-distillation chamber. The overhead material from this chamber is then fractionated to produce a low-quality gasoline as an overhead product and light gas oil as bottoms. The liquid products from the flash chamber are cooled with a gas oil flux and then sent to a vacuum fractionator. This yields a heavy gas oil distillate and a residuum of reduced viscosity. A quench oil may also be used to terminate the reactions. A 5–10% by weight or by volume conversion of atmospheric residua to naphtha is usually sufficient to afford at least an approximate fivefold reduction in viscosity. Reduction in viscosity is also accompanied by a reduction in the pour point.

The reduction in viscosity of residua tends to reach a limiting value with conversion, although the total product viscosity can continue to decrease, but other properties will

be affected. Sediment (that is predominantly organic but may contain some mineral matter) may also form—a crucial property for residual fuel oil—and conditions should be chosen so that sediment formation is minimal, if at all. When shipment of the visbreaker product by pipeline is the process objective, addition of a diluent such as gas condensate can be used to achieve a further reduction in viscosity. Recovery of the diluent after pipelining is an option.

The significance of the measured properties of residual fuel oil is dependent to a large extent on the ultimate uses of the fuel oil. Such uses include steam generation for various processes as well as electrical power generation and propulsion. Corrosion, ash deposition, atmospheric pollution, and product contamination are side effects of the use of residual fuel oil, and in particular cases, elements such as vanadium, sodium, and sulfur may be significant.

Problems of handling and storage may also arise, particularly with higher-boiling fuel oil, since at ambient temperatures, this type of fuel oil may be viscous and even approach a semisolid state. Although such fuel oil is usually stored in heated tanks, test methods to determine the low-temperature behavior of such fuel oil are necessary. In addition, since viscous or semisolid fuel oil should be preheated in order to obtain the correct injection (atomizing) conditions for efficient combustion, test methods that describe viscosity are also necessary.

11.3 TEST METHODS

Important specifications for residual fuel oils are the viscosity and sulfur content, although limits for flash point, pour point, water and sediment, and ash production must be monitored. Sulfur limits for the heavy fuels are controlled by regulation in the United States and consequently depend upon the location of use. For the heavier grades of industrial and bunker fuels, viscosity is of major importance. The imposition of viscosity limits ensures that adequate preheating facilities can be provided to permit transfer to the burner and atomization of the fuel.

Test methods of interest for hydrocarbon analysis of residual fuel oil include tests that measure physical properties such as elemental analysis, density, refractive index, molecular weight, and boiling range. There may also be some emphasis on methods that are used to measure chemical composition and structural analysis, but these methods may not be as definitive as they are for other petroleum products.

Testing residual fuel oil does not suffer from the issues that are associated with sample volatility, but the test methods are often sensitive to the presence of gas bubbles in the fuel oil. An air release test is available for application to lubricating oil (ASTM D3427, IP 313) and may be applied, with modification, to residual fuel oil. However, with dark-colored samples, it may be difficult to determine whether all air bubbles have been eliminated. And, as with the analysis and testing of other petroleum products, the importance of correct sampling of fuel oil cannot be overemphasized, because no proper assessment of quality may be made unless the data are obtained on truly representative samples.

11.3.1 Ash

The mineral matter contained in residual fuel oil (heavy fuel oil) includes the (inorganic) metallic content, other noncombustibles, and solid contamination. The ash yield after combustion of fuel oil also includes solid foreign material (sand, rust, catalyst particles) and dispersed and dissolved inorganic materials, such as vanadium, nickel, iron, sodium, potassium, or calcium. Ash deposited in service can cause localized overheating of metal surfaces to which they adhere and lead to the corrosion of the exhaust valves. Excessive ash may also result in abrasive wear of cylinder liners, piston rings, valve seats and injection pumps, and deposits that can clog fuel nozzles and injectors.

In heavy fuel oil, soluble and dispersed metal compounds cannot be removed by centrifuging. They can form hard deposits on piston crowns, cylinder heads around exhaust valves, valve faces and valve seats, and, in turbocharger, gas sides. High-temperature corrosion caused by the metals present in the oil (which produce ash during service) can be minimized by (i) the use of hardened atomizers to minimize erosion and corrosion, and (ii) reduction of valve seat temperatures using more efficient cooling protocols.

The ash formed by the combustion of fuel oil (ASTM D482, ASTM D2415, IP 4) is, as defined for other products, the inorganic residue, free from carbonaceous matter, remaining after ignition in air of the residual fuel oil at fairly high temperatures. The ash content is not directly equated to mineral content but can be converted to mineral matter content by the use of appropriate formulae.

Residual fuel oil often contains varying amounts of ash-forming constituents (but seldom more than 0.2% w/w) such as organometallic complexes that are soluble, or inherent, in petroleum or from mineral matter from oil-bearing strata, from contact of the crude oil with pipelines and storage tanks that occurs during transportation and subsequent handling. Additives used to improve particular fuel properties and carry over from refining processes may also contribute to the ultimate amount of mineral mater in the fuel oil and present an inflated measurement of the ash formed by the combustion of the sample. Thus, while the total amounts of ash-forming constituents in different fuel oils may be similar, the compositions of the mineral constituents will depend on the crude oil origin as well as on the handling of the respective fuel oils. These constituents ultimately concentrate in the distillation residue and so their presence will be reflected in the fuel oil ash.

The presence of sodium and vanadium complexes in the fuel oil ash can, under certain plant operating conditions, result in considerable harm to the equipment. Spalling and fluxing of refractory linings are associated with the presence of sodium in the fuel. Above a certain threshold temperature, which will vary from fuel to fuel, the oil ash will adhere to boiler superheater tubes and gas turbine blades thus reducing the thermal efficiency of the plant. At higher temperatures, molten complexes of vanadium, sodium, and sulfur are produced, which will corrode all currently available metals used in the construction of these parts of the plant. The presence of trace amounts (ASTM D1318) of vanadium (IP 285) in fuel oil used in glass manufacture can affect the indicator of the finished product.

11.3.2 Asphaltene Content

The asphaltene fraction (ASTM D893, ASTM D2007, ASTM D3279, ASTM D4124, ASTM D6560, IP 143) is the highest-molecular-weight and most complex fraction in petroleum. The asphaltene content gives an indication of the amount of coke that can be expected during exposure to thermal conditions (Speight, 2001, 2014, Speight and Ozum, 2002).

In any of the methods for the determination of the asphaltene content (Speight, 1994, 2014), the residual fuel oil is mixed with a large excess (usually >30 volumes of hydrocarbon per volume of sample) low-boiling hydrocarbon such as n-pentane or n-heptane. For an extremely viscous sample, a solvent such as toluene may be used prior to the addition of the low-boiling hydrocarbon, but an additional amount of the hydrocarbon (usually >30 volumes of hydrocarbon per volume of solvent) must be added to compensate for the presence of the solvent. After a specified time, the insoluble material (the asphaltene fraction) is separated (by filtration) and dried. The yield is reported as percentage (% w/w) of the original sample.

In any of these tests, different hydrocarbons (such as n-pentane or n-heptane) will give different yields of the asphaltene fraction, and if the presence of the solvent is not compensated by the use of additional hydrocarbon, the yield will be erroneous. In addition, if the hydrocarbon is not present in a large excess, the yields of the asphaltene fraction will vary and will be erroneous (Speight, 2014).

Another method, not specifically described as an asphaltene separation method, is designed to remove pentane-insoluble constituents by membrane filtration (ASTM D4055). In the method, a sample of oil is mixed with pentane in a volumetric flask, and the oil solution is filtered through a 0.8 μm membrane filter. The flask, funnel, and the filter are washed with pentane to completely transfer any particulates onto the filter after which the filter (with particulates) is dried and weighed to give the pentane-insoluble constituents as a percentage by weight of the sample.

Particulates can also be determined by membrane filtration (ASTM D2276, ASTM D5452, ASTM D6217, IP 415).

The *precipitation number* is often equated to the asphaltene content, but there are several issues that remain obvious in its rejection for this purpose. For example, the method to determine the precipitation number (ASTM D91) advocates the use of naphtha for use with black oil or lubricating oil, and the amount of insoluble material (as a % v/v of the sample) is the precipitating number. In the test, 10 ml of sample is mixed with 90 ml of ASTM precipitation naphtha (that may or may not have a constant chemical composition) in a graduated centrifuge cone and centrifuged for 10 min at 600–700 rpm. The volume of material on the bottom of the centrifuge cone is noted until repeat centrifugation gives a value within 0.1 ml (the precipitation number). Obviously, this can be substantially different from the amount of the asphaltene constituents (asphaltene fraction) content.

If the residual fuel oil is produced by a thermal process such as visbreaking, it may also be necessary to determine if toluene-insoluble material is present by the methods, or modifications thereof, used to determine the toluene insoluble of tar and pitch (ASTM D4072, ASTM D4312). In these methods, a sample is digested at 95°C (203°F) for 25 min and then extracted with hot toluene in an alundum thimble. The extraction time is 18 h (ASTM D4072) or 3 h (ASTM D4312). The insoluble matter is dried and weighed.

11.3.3 Calorific Value

The calorific value (heat of combustion) of residual fuel oil (ASTM D240, IP 12) is that the calorific value is lower due that that of lower-boiling fuel oil (and other liquid fuels) because of the lower atomic hydrogen/carbon ratio and the incidence of greater amounts of less combustible material, for example, water, sediment, and generally higher levels of sulfur.

For most residual fuel oils, the range of calorific value is relatively narrow, and limits are not usually included in the specifications. When precise determinations are not essential, values of sufficient accuracy may be derived from calculations based upon measurable physical properties. In this test method, the net heat of combustion is calculated from the density, sulfur, and hydrogen content, but this calculation is justifiable only when the fuel belongs to a well-defined class for which a relationship between these quantities has been derived from accurate experimental measurements on representative samples. Thus, the hydrogen content (ASTM D1018, ASTM D1217, ASTM D1298, ASTM D3701, ASTM D4052, ASTM D4808, ASTM D5291, IP 160, IP 365), density (ASTM D129, ASTM D1250, ASTM D1266, ASTM D2622, ASTM D3120, IP 61, IP 107), and sulfur content (ASTM D2622, ASTM D3120, ASTM D3246, ASTM D4294, ASTM D5453, ASTM D5623, IP 336, IP 373) of the sample are determined by experimental test methods, and the net heat of combustion is calculated using the values obtained by these test methods based on reported correlations.

11.3.4 Carbon Residue

The propensity of residual fuel oil for carbon formation and deposition under thermal conditions may be indicated by one or more of three carbon residue tests. A high carbon residue yield (or asphaltene content) denotes a high residue level after combustion and may lead to ignition delay as well as afterburning of carbon deposits leading to engine fouling and abrasive wear. Poor engine performance caused by slow burning, high–boiling point constituents results in higher thermal loading and changes in the rate of heat release in the cylinder.

As the carbon residue increases (concurrently with the asphaltene contact), the combination of a high carbon-forming propensity can increase the centrifuge sludge and fine filter burdens. This can require more frequent sludge removal (by centrifuging) and filter element cleaning/replacement. An increase in the carbon-forming propensity also lowers the gross and net heating values (on weight basis) of a heavy fuel oil. As the feedstocks to refineries increase to more viscous feedstocks (and tar sand bitumen), a higher carbon-forming propensity of heavy fuel oil is to be expected in the future (Speight, 2011). The increased content will result from refining more viscous, heavier crude oils and from additional, secondary processing, such as catalytic cracking and visbreaking—these processes will concentrate more carbon-forming precursors in the residua that are blended or visbroken to produce marine heavy fuel oils (Chapter 7).

Thus, the specifications for the allowable amounts of carbon residue by the Conradson carbon residue (ASTM D189, IP 13) or by the Ramsbottom carbon residue (ASTM D524, IP 14) or by the microcarbon carbon residue (ASTM D4530, IP 398) may be included in inspection data for fuel oil.

In the Conradson carbon residue test (ASTM D189, IP 13), a weighed quantity of sample is placed in a crucible and subjected to destructive distillation for a fixed period of severe heating. At the end of the specified heating period, the test crucible containing the carbonaceous residue is cooled in a desiccator and weighed, and the residue is reported as a percentage (% w/w) of the original sample (Conradson carbon residue). In the Ramsbottom carbon residue test (ASTM D524, IP 14), the sample is weighed into a glass bulb that has a capillary opening and is placed into a furnace (at 550°C, 1020°F). The volatile matter is distilled from the bulb, and the nonvolatile matter that remains in the bulb cracks to form thermal coke. After a specified heating period, the bulb is removed from the bath, cooled in a desiccator, and weighed to report the residue (Ramsbottom carbon residue) as a percentage (% w/w) of the original sample. In the microcarbon residue test (ASTM D4530, IP 398), a weighed quantity of the sample placed in a glass vial is heated to 500°C (930°F) under an inert (nitrogen) atmosphere in a controlled manner for a specific time, and the carbonaceous residue [*carbon residue (micro)*] is reported as a percentage (% w/w) of the original sample.

The data produced by the microcarbon test (ASTM D4530, IP 398) are equivalent to those by Conradson carbon method (ASTM D189, IP 13). However, this microcarbon test method offers better control of test conditions and requires a smaller sample. Up to 12 samples can be run simultaneously. This test method is applicable to petroleum and to petroleum products that partially decompose on distillation at atmospheric pressure and is applicable to a variety of samples that generate a range of yields (0.01–30% w/w) of thermal coke.

In any of the carbon residue tests, ash-forming constituents (ASTM D482) or nonvolatile additives present in the sample will be included in the total carbon residue reported leading to higher carbon residue values and erroneous conclusions about the coke-forming propensity of the sample.

The data give an indication of the amount of coke that will be formed during exposure of the residual fuel oil to thermal effects. However, the significance of a carbon residue test relative to the combustion characteristics of the fuel is questionable since the significance of the test depends, to a large extent, on the particular process and handling conditions, specifically the introduction of residual fuel oil to heat in pipes as it passes to a furnace.

Other test methods that are used for determining the coking value of tar and pitch (ASTM D2416, ASTM D4715) that indicates the relative coke-forming properties of tars and pitches might also be applied to residual fuel oil. Both test methods are applicable to tar and pitch having an ash content ≤0.5% (ASTM D2415). The former test method (ASTM D2416) gives results close to those obtained by the Conradson carbon residue test (ASTM D189, IP 13). However, in the latter test method (ASTM D4715), a sample of tar (or pitch)

is heated for a specified time at 550 ± 10°C (1022 ± 18°F) in an electric furnace. The percentage of residue is reported as the coking value. For residual fuel oil, the temperature of both test methods can be adjusted to the temperature that the fuel oil might experience in the pipe to the furnace, with a corresponding adjustment for the residence time in the pipe.

Finally, a method that is used to determine pitch volatility (ASTM D4893) might also be used, on occasion, to determine the nonvolatility of residual fuel oil. In the method, an aluminum dish containing about 15 g of accurately weighed sample is introduced into the cavity of a metal block heated and maintained at 350°C (660°F). After 30 min, during which the volatiles are swept away from the surface of the sample by preheated nitrogen, the residual sample is taken out and allowed to cool down in the desiccator. The nonvolatility is determined by the sample weight remaining and reported as the percentage (%w/w) of residue.

11.3.5 Composition

The composition of residual fuel oils is often reported in the form of four or five major fractions as deduced by adsorption chromatography (Figs. 11.1 and 11.2). In the case of cracked feedstocks, thermal decomposition products (carbenes and carboids) may also be present.

Column chromatography is used for several hydrocarbon-type analyses that involve fractionation of viscous oils (ASTM D2007, ASTM D2549), including residual fuel oil. The former method (ASTM D2007) advocates the use of adsorption on clay and clay–silica gel, followed by elution of the clay with pentane to separate saturates, elution of clay

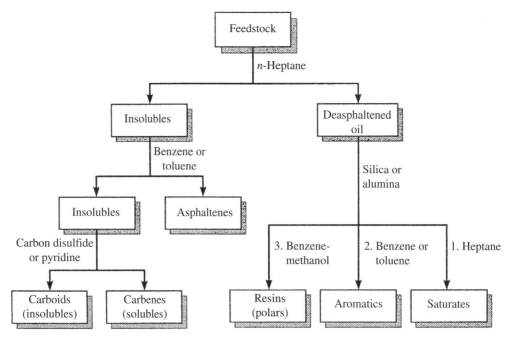

FIGURE 11.1 Separation of feedstock into four major fractions.

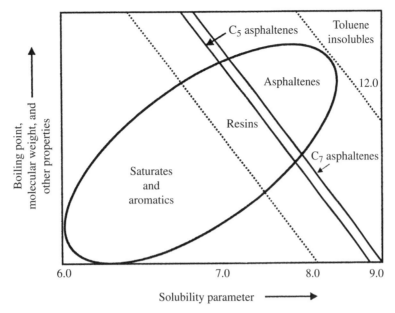

FIGURE 11.2 Representation of feedstock fractionation.

with acetone–toluene to separate polar compounds, and elution of the silica gel fraction with toluene to separate aromatic compounds. The latter method (ASTM D2549) uses adsorption on a bauxite–silica gel column. Saturates are eluted with pentane; aromatics are eluted with ether, chloroform, and ethanol.

Several promising chromatographic techniques have been reported for the analysis of lubricant base oils. Rod thin layer chromatography, high-performance liquid chromatography, and supercritical *fluid* chromatography have all been used fuel oil analysis and base oil content.

In addition to carbon and hydrogen, high-molecular-weight fractions of crude oil often contain oxygen compounds, sulfur compounds, and nitrogen compounds as well as trace amounts metal-containing compounds. Determining the chemical form present for these elements provides additional important information. Finished products made using viscous oils may contain additives or contaminants that also require analysis. Thus, *elemental analysis* also plays an important role in determining the composition of residual fuel oils.

Carbon and hydrogen are commonly determined by combustion analysis in which the sample is burned in an oxygen stream where carbon is converted to carbon dioxide and hydrogen to water. These compounds are absorbed, and the composition is determined automatically from mass increase (ASTM D5291). Nitrogen may be determined simultaneously.

Sulfur is naturally present in many crude oils and petroleum fractions, most commonly as organic sulfides and heterocyclic compounds. Residual fuel oils are variable products whose sulfur contents depend not only on their crude oil sources but also on the extent of the refinery processing received by the fuel oil blending components. Sulfur, present in these fuel oils in varying amounts up to 4 or 5% w/w, is an undesirable constituent, and many refining steps aim to reduce the sulfur content to improve stability and reduce environmentally harmful emissions.

Hydrogen sulfide (H_2S) and mercaptans (R-SH) may be produced during thermal processes such as the visbreaking process and can occur in fuel oil with other sulfur compounds that concentrate in the distillation residue (Speight, 2000, 2014). Without any further processing, such as hydrofining and caustic washing (Speight, 2000, 2014), these sulfur compounds remain in the fuel oil. The sulfur content of fuel oil obtained from petroleum residua and the atmospheric pollution arising from the use of these fuel oils are important factors, and increasing insistence on a low-sulfur-content fuel oil has increased the value of low-sulfur petroleum.

A considerable number of tests are available to estimate the sulfur in petroleum or to study its effect on various products, particularly hydrogen sulfide (ASTM D5705, ASTM D6021), that can result as a product of thermal processes, for example, visbreaking. Hydrogen sulfide dissolved in petroleum is normally determined by absorption of the hydrogen sulfide in a suitable solution that is subsequently analyzed chemically (Doctor method) (ASTM D4952, IP 30) or by the formation of cadmium sulfate.

The Doctor test measures the amount of sulfur available to react with metallic surfaces at the temperature of the test. The rates of reaction are metal type, temperature, and time dependent. In the test, a sample is treated with copper powder at 150°C (300°F). The copper powder is filtered from the mixture. Active sulfur is calculated from the difference between the sulfur contents of the sample (ASTM D129) before and after treatment with copper.

Of all the elements present in a normal residual fuel oil, vanadium, sodium, and sulfur contribute most to difficulties and problems that may arise in the industrial application of fuel oils. Sulfur contributes to the increasing problem of atmospheric pollution when sulfur oxides, produced on combustion of high-sulfur fuel oils, are emitted into the surrounding atmosphere of densely populated industrial areas or large towns. In specific applications, fuel oil desulfurization may have to be employed in order to comply with air pollution legislation.

The methods used to measure sulfur content vary depending on the sulfur concentration, viscosity or boiling range, and presence of interfering elements.

For the determination of sulfur contents of residual fuels, a variety of procedures are available. The bomb combustion method has long been established (ASTM D129, IP 61). Other more rapid techniques are becoming increasingly available, which include high-temperature combustion (ASTM D1552), X-ray absorption and fluorescence methods, and the Schoniger oxygen flask procedure.

The bomb method for sulfur determination (ASTM D129) uses sample combustion in oxygen and conversion of the sulfur to barium sulfate, which is determined by mass. This method is suitable for samples containing 0.1–5.0% w/w sulfur and can be used for most low-volatility petroleum products. Elements that produce residues insoluble in hydrochloric acid interfere with this method—this includes aluminum, calcium, iron, lead, and silicon, plus minerals such as asbestos, mica, and silica—and an alternate method (ASTM D1552) is preferred. This method describes three procedures: the sample is first pyrolyzed in either an induction furnace or a resistance furnace; the sulfur is then converted to sulfur dioxide and either titrated with potassium iodate–starch reagent or the sulfur dioxide is analyzed by infrared spectroscopy. This method is generally suitable for samples containing from 0.06 to 8.0% w/w sulfur and that distill at temperatures above 177°C (351°F).

Two methods describe the use of X-ray techniques for sulfur determination and can be applied to the determination of sulfur, which can be used for samples with sulfur content of 0.001–5.0% w/w (ASTM D2622, ASTM D4294). Oil viscosity is not a critical factor with these two methods, but interference may affect test results when chlorine, phosphorus, heavy metals, and possibly silicon are present. A method is also available for very-low-sulfur concentrations (ASTM D4045). This is normally used for lower-viscosity fractions, but may be used for some viscous oils that boil below 370°C (700°F). The method is designed to measure sulfur in the range 0.02–10 ppm. Sulfur may also be determined along with metals (ASTM D4927, ASTM D4951, ASTM D5185).

Nitrogen is present in residual fuel oils and is also a component of many additives used in petroleum products, including oxidation and corrosion inhibitors and dispersants. There are four ASTM standards describing analytical methods for nitrogen in viscous oils. The first (ASTM D3228) is a standard wet chemical method and is useful for determining the nitrogen content of most viscous oils in the range from 0.03 to 0.10% w/w nitrogen. The other three methods are instrumental techniques: one involves nitrogen reduction, and the other two involve nitrogen oxidation. One method (ASTM D4629) is useful for samples containing 0.3–100 ppm nitrogen and boiling higher than 400°C (752°F) but with viscosities of 10 cSt or less. In this method, organic nitrogen is converted to nitric oxide (NO) and then to excited nitrogen dioxide (NO_2) by reaction with oxygen and ozone. Energy emitted during decay of the excited nitrogen dioxide is measured with a photomultiplier tube. There is a method (ASTM D5762) that is complementary to this one and is suitable for higher-viscosity samples that contain from 40 to 10,000 ppm nitrogen.

The viscous fractions of crude oil often contain *metals* such as iron, nickel, and vanadium. Catalytic refining processes are often sensitive to metal contamination and, therefore, the type and quantity of metals must be determined. In other cases such as lubricating oils, some metals are parts of compounds added to the petroleum component to enhance performance.

A standard wet chemical analysis is available for the determination of aluminum, barium, calcium, magnesium, potassium, silicon, sodium, tin, and zinc. The procedure involves a series of chemical separations with specific elemental analysis performed using appropriate gravimetric or volumetric analyses.

The most commonly used methods for determining metal content in viscous oils are spectroscopic techniques. In one such method (ASTM D4628), the sample is diluted in kerosene and burned in an acetylene–nitrous oxide flame of an M spectrophotometer. The method is suitable for oils in the lubricating oil viscosity range. It is designed to measure barium at concentrations of 0.005–1.0% w/w, calcium and magnesium at 0.002–0.3% w/w, and zinc at 0.002–0.2% w/w. Higher metal concentrations, such as are present in additives, can be determined by dilution. Lower concentrations in the range of 10–50 ppm can also be determined; however, the precision is poorer. An alternate test method (ASTM D4927) is designed for unused lube oils containing metals at concentration levels from 0.03 to 1.0% w/w and sulfur at 0.01–2.0% w/w. Higher concentrations can be determined after dilution.

A third technique (ASTM D4951) is used to determine barium, boron, calcium, copper, magnesium, phosphorus, sulfur, and zinc in unused lubricating oils and additive packages. Elements can generally be determined at concentrations of 0.01–1.0% w/w. The sample is diluted in mixed xylenes or other solvents containing an internal standard. The inductively coupled plasma (ICP) method (ASTM D5185) is also available. Sensitivity and usable range vary from one element to another, but the method is generally applicable from 1 to 100 ppm for contaminants and up to 1000–9000 ppm for additive elements (Table 11.1).

Two procedures are described whereby the sample is either treated with acid to decompose the organic material

TABLE 11.1 Potential elements in residual fuel oil

Additive elements	Contaminant elements		
Calcium	Aluminum	Lead	Sodium
Magnesium	Barium	Manganese	Tin
Phosphorus	Boron	Molybdenum	Titanium
Potassium	Chromium	Nickel	Vanadium
Sulfur	Copper	Silicon	
Zinc	Iron	Silver	

and dissolve the metals or, alternatively, the sample is dissolved in an organic solvent. The method is sensitive down to about 1 ppm; the precision statement is based on samples containing 1–10 ppm iron, 10–100 ppm nickel, or 50–500 ppm vanadium (ASTM D5708). The second method provides an alternate method for the analysis of crude oils and residuum (ASTM D5863). The sensitivity range is 3.0–10 ppm for iron, 0.5–100 ppm for nickel, 0.1–20 ppm for sodium, and 0.5–500 ppm for vanadium. Higher concentrations may be determined after dilution.

A variety of *miscellaneous elements* can also occur in residual fuel oil fraction For example, *chlorine* is present as a chlorinated hydrocarbon and can be determined (ASTM D808, ASTM D6160). A rapid test method suitable for the analysis of samples by nontechnical personnel is also available (ASTM D5384) and uses a commercial test kit where the oil sample is reacted with metallic sodium to convert organic halogens to halide, which is titrated with mercuric nitrate using diphenyl carbazone indicator. Iodides and bromides are reported as chloride.

Phosphorus is a common component of additives and appears most commonly as a zinc dialkyl dithiophosphate or a tri-aryl phosphate ester, but other forms also occur. Two wet chemical methods are available, one of which (ASTM D1091) describes an oxidation procedure that converts phosphorus to aqueous ortho-phosphate anion. This is then determined by mass as magnesium pyrophosphate or photochemically as molybdivanadophosphoric acid. In an alternate test (ASTM D4047), samples are oxidized to phosphate with zinc oxide, dissolved in acid, precipitated as quinoline phosphomolybdate, treated with excess standard alkali, and back-titrated with standard acid. Both of these methods are primarily used for referee samples. Phosphorus is most *commonly* determined using x-ray fluorescence (XRF; ASTM D4927) or ICP (ASTM D4951).

Correlative methods are derived relationships between fundamental chemical properties of a substance and measured physical or chemical properties. They provide information about an oil from readily measured properties (ASTM D2140, ASTM D2501, ASTM D2502, ASTM D3238).

One method (ASTM D2501) describes the calculation of the viscosity–gravity coefficient (VGC)—a parameter derived from kinematic viscosity and density that has been found to relate to the saturate/aromatic composition. Correlations between the VGC (or molecular weight and density) and refractive index to calculate carbon-type composition in percentage of aromatic, naphthenic, and paraffinic carbon atoms are employed to estimate the number of aromatic and naphthenic rings present (ASTM D2140, ASTM D3238).

Another method (ASTM D2502) permits the estimation of molecular weight from kinematic viscosity measurements at 38 and 99°C (100 and 210°F) (ASTM D445). It is applicable to samples with molecular weights in the range from 250 to 700 but should not be applied indiscriminately for oils that represent extremes of composition for which different constants are derived (Speight, 2001, 2014).

However, data from correlative methods must not be confused with more fundamental measurements obtained by chromatography or mass spectroscopy. Correlative methods can be extremely useful when used to follow changes in a hydrocarbon mixture during processing. They are less reliable when comparing materials of different origin and can be very misleading when applied to typical or unusual compositions.

A major use for *gas chromatography* for hydrocarbon analysis has been simulated distillation, as discussed previously. Other gas chromatographic methods have been developed for contaminant analysis (ASTM D4291).

The aromatic content of fuel oil is a key property that can affect a variety of other properties including viscosity, stability, and compatibility with other fuel oil or blending stock. Existing methods for this work use physical measurements and need suitable standards. Thus, methods have been standardized using nuclear magnetic resonance (NMR) for hydrocarbon characterization (ASTM D4808, ASTM D5291, ASTM D5292). The NMR method is simpler and more precise. Procedures are described that cover light distillates with a 15–260°C boiling range, middle distillates and gas oils with boiling ranges of 200–370°C and 370–510°C, and residuum boiling above 510°C. One of the methods (ASTM D5292) is applicable to a wide range of hydrocarbon oils that are completely soluble in chloroform and carbon tetrachloride at ambient temperature. The data obtained by this method can be used to evaluate changes in aromatic contents of hydrocarbon oils due to process changes.

High ionizing voltage mass spectrometry (ASTM D2786, ASTM D3239) is also employed for compositional analysis of residual fuel oil. These methods require preliminary separation using elution chromatography (ASTM D2549). A third method (ASTM D2425) may be applicable to some residual fuel oil samples in the lower-molecular-weight range.

11.3.6 Density

Density or specific gravity (relative density) is used whenever conversions must be made between mass (weight) and volume measurements. This property is often used in combination with other test results to predict oil quality, and several methods are available for the measurement of density (or specific gravity). However, the density (specific gravity) (ASTM D1298, IP 160) is probably of least importance in determining fuel oil performance, but it is used in product control, in weight–volume relationships, and in the calculation of calorific value (heating value).

Two of the methods (ASTM D287, ASTM D1298) use an immersed hydrometer for the measurement of density. The former method (ASTM D287) provides the results as API gravity. Two other methods (ASTM D1480, ASTM D1481) use a pycnometer to measure density or specific gravity and have the advantage of requiring a smaller sample size and can be sued at higher temperatures than is normal providing the vapor pressure of the liquid does not exceed specific limits at the temperature of the test. Two other test methods (ASTM D4052, ASTM D5002) measure density with a digital density analyzer. This device determines density by the analysis of the change in oscillating frequency of a sample tube when filled with the test sample.

Another test method (ASTM D4052) covers the determination of the density or specific gravity of viscous oil, such as residual fuel oil, that are liquids at test temperatures between 15 and 35°C (59 and 95°F). However, application of the method is restricted to liquids with vapor pressures below 600 mm Hg and viscosity below 15,000 cSt at the temperature of test. In addition, and this is crucial for residual fuel oil, this test method should not be applied to samples so dark in color that the absence of air bubbles in the sample cell cannot be established with certainty.

11.3.7   Elemental Analysis

Elemental analysis of fuel oil often plays a more major role that it may appear to do in the lower-boiling products. Aromaticity (through the atomic hydrogen/carbon ratio), sulfur content, nitrogen content, oxygen content, and metal content are all important features that can influence the use of residual fuel oil.

Carbon content and *hydrogen content* can be determined simultaneously by the method designated for coal and coke or by the method designated for municipal solid waste (ASTM E777). However, as with any analytical method, the method chosen for the analysis may be subject to the peculiarities or character of the feedstock under investigation and should be assessed in terms of accuracy and reproducibility. There methods that are designated for elemental analysis are as follows:

1. *Carbon* and *hydrogen content* (ASTM D1018, ASTM D3343, ASTM D3701, ASTM D5291, ASTM E777, IP 338)
2. *Nitrogen content* (ASTM D3228, ASTM D5291, ASTM E258, and ASTM E778)
3. *Oxygen content* (ASTM E385)
4. *Sulfur content* (ASTM D129, ASTM D139, ASTM D1266, ASTM D1552, ASTM D2622, ASTM D3120, ASTM D4045, and ASTM D4294, IP 30, IP 61, IP 107, IP 154, IP 243)

The hydrogen content of fuel oil can also be measured by low-resolution magnetic resonance spectroscopy (ASTM D3701, ASTM D4808). The method is claimed to provide a simple and more precise alternative to existing test methods, specifically combustion techniques (ASTM D5291) for determining the hydrogen content of a variety of petroleum-related materials.

Nitrogen occurs in residua and, therefore, in residual fuel oil and causes serious environmental problems as a result, especially when the levels exceed 0.5% by weight, as happens often in residua. In addition to the chemical character of the nitrogen, the amount of nitrogen in a feedstock determines the severity of the process, the hydrogen requirements, and, to some extent, the sediment formation and deposition.

The determination of nitrogen in petroleum products is performed regularly by the Kjeldahl method (ASTM D3228) and the Dumas method. The chemiluminescence method is the most recent technique applied to nitrogen analysis for petroleum and is used to determine the amount of chemically bound nitrogen in liquid samples.

In the method, the samples are introduced to the oxygen-rich atmosphere of a pyrolysis tube maintained at 975°C (1785°F). Nitrogen in the sample is converted to nitric oxide during combustion, and the combustion products are dried by passage through magnesium perchlorate $[Mg(ClO_4)_2]$ before entering the reaction chamber of a chemiluminescence detector. In the detector, ozone reacts with the nitric oxide to form excited nitrogen dioxide:

$$NO + O_3 = NO_2{}^* + O_2$$

Photoemission occurs as the excited nitrogen dioxide reverts to the ground state:

$$NO_2{}^* = NO_2 + h\nu$$

The emitted light is monitored by a photomultiplier tube to yield a measure of the nitrogen content of the sample. Quantitation is based on comparison with the response for carbazole in toluene standards.

Oxygen is one of the five (C, H, N, O, and S) major elements in fuel oil but rarely exceeds 1.5% by weight, unless oxidation has occurred during transportation and storage. Many petroleum products do not specify a particular oxygen content, but if the oxygen compounds are present as acidic compounds such a phenols (Ar-OH) and naphthenic acids (cycloalkyl-COOH), they are controlled in different specifications by a variety of tests. The *total acidity* (ASTM D974,

IP 139) is determined for many products, especially fuels and fuel oil. Oxygen-containing impurities in the form of *gum* are determined by the *existent gum* test method (ASTM D381, IP 131) and *potential gum* test method (ASTM D873, IP 138). Elemental analysis of the gum can then provide its composition with some indication of the elements (other than carbon and hydrogen) that played a predominant role in its formation.

Being the third most common element (after carbon and hydrogen) in petroleum product, *sulfur* has been analyzed extensively. Analytical methods range from elemental analyses to functional group (sulfur-type) analyses to structural characterization to molecular speciation (Speight, 2001). Of the methods specified for the determination of sulfur (Speight, 2001), the method applied to the corrosion effect of sulfur is extremely important for liquid fuels. In this method (ASTM D1266, IP 154), fuel corrosivity is assessed by the action of the fuel on a copper strip (*the copper strip test*) that helps determine any discoloration of the copper due to the presence of corrosive compounds. The copper strip is immersed in the fuel and heated at 100°C (212°F) for 2 h in a bomb. A test using silver as the test metal (IP 227) has also been published. Mercaptans are usually the corrosive sulfur compounds of reference, and metal discoloration is due to the formation of the metal sulfide. Thus, mercaptan sulfur is an important property of potential fuels. In addition to the copper strip test, the mercaptan sulfur (R-SH) content provides valuable information. As an alternative to determining the mercaptan content, a negative result in the *Doctor test* (IP 30) may also be acceptable for the qualitative absence of mercaptans. The copper strip method (ASTM D130, ASTM D849, ASTM D4048, IP 154) may also be employed to determine the presence of corrosive sulfur compounds in residual fuel oil.

The determination of sulfur in liquid products by XRF (ASTM D2622, IP 336) has become an extremely well-used method over the past two decades. This method can be used to determine the amount of sulfur in homogeneous liquid petroleum hydrocarbons over the range of 0.1–6.0% by weight. Samples with a sulfur content above this range may be determined after dilution in toluene. The method utilizes the principle that when a sample is irradiated with a Fe^{55} source, fluorescent x-rays result. The sulfur Kα fluorescence and a background correction at adjacent wavelengths are counted. A calibration of the instrument, wherein the integration time for counting is adjusted such that the displayed signal for the background corrected radiation equals the concentration of the calibration standard, gives a direct readout of the weight percentage of sulfur in the sample. Interfering elements include aluminum, silicon, phosphorus, chlorine, argon, and potassium. Generally, the amounts of these elements are insufficient to affect sulfur x-ray counts in samples covered by this method. Atmospheric argon is eliminated by a helium purge.

It is also possible to determine nitrogen and sulfur simultaneously by chemiluminescence and fluorescence.

An aliquot of the sample undergoes high-temperature oxidation in a combustion tube maintained at 1050°C (1920°F). Oxidation of the sample converts the chemically bound nitrogen to nitric oxide (NO) and sulfur to sulfur dioxide (SO_2). In the nitrogen detector, ozone reacts with the nitric oxide to form excited nitrogen dioxide (NO_2). As the nitrogen dioxide reverts to its ground state, chemiluminescence occurs, and this light emission is monitored by a photomultiplier tube. The light emitted is proportional to the amount of nitrogen in the sample. In the sulfur detector, the sulfur dioxide is exposed to ultraviolet radiation and produces a fluorescent emission. This light emission is proportional to the amount of sulfur and is also measured by a photomultiplier tube. Quantitation is determined by a comparison to the responses given by standards containing carbazole and dimethyl sulfoxide in xylene.

Oxidative microcoulometry has become a widely accepted technique for the determination of low concentrations of sulfur in petroleum and petroleum products (ASTM D3120). The method involves combustion of the sample in an oxygen-rich atmosphere followed by microcoulometric generation of triiodide ion to consume the resultant sulfur dioxide. It is intended to distinguish the technique from reductive microcoulometry that converts sulfur in the sample to hydrogen sulfide that is titrated with coulometrically generated silver ion.

Although sodium azide is included in the electrolyte of the microcoulometric titration to minimize halogen and nitrogen interferences, the method is not applicable when chlorine is present in excess of 10 times the sulfur level, or the nitrogen content exceeds 10% by weight. Heavy metals in excess of 500 mg/kg also interfere with the method.

11.3.8 Flash Point

As for all petroleum products, considerations of safety in storage and transportation and, more particularly, contamination by more volatile products, are required. This is usually accommodated by the Pensky–Martens flash point test (ASTM D93, IP 34). For the fuel oil, a minimum flash point of 55°C (131°F) or 66°C (150°F) is included in most specifications.

11.3.9 Metallic Constituents

Heteroatoms (*nitrogen, oxygen, sulfur,* and *metals*) are found in every crude oil, and the concentrations have to be reduced to convert the oil to transportation fuel. The reason is that if nitrogen and sulfur are present in the final fuel during combustion, nitrogen oxides (NO_x) and sulfur oxides (SO_x) form, respectively. In addition, metals affect the use of residual fuel oil through adverse effects such as by causing corrosion or by deposition.

The nature of the process by which residual fuel oil is produced virtually dictates that all the metals in the original crude oil can occur in the residuum (Speight, 2000, 2001) and, thus, in the residual fuel. Metallic constituents that may actually *volatilize* under the distillation conditions and appear in the higher-boiling distillates are the exceptions and can appear in distillate fuel oil.

The analysis for metal constituents in residual can be accomplished by several instrumental techniques: inductively coupled argon plasma (ICAP) spectrometry, atomic absorption (AA) spectrometry, and XRF spectrometry. Each technique has limitations in terms of sample preparation, sensitivity, sampling, time for analysis, and overall ease of use. Thus, a variety of tests (ASTM D482, ASTM D1318, ASTM D3341, ASTM D3605) either directly or as the constituents of combustion ash have been designated to determine metals in petroleum products based on a variety of techniques. At the time of writing, the specific test for the determination of metals in whole feeds has not been designated. However, this task can be accomplished by the combustion of the sample so that only inorganic ash remains (ASTM D482). The ash can then be digested with an acid, and the solution examined for metal species by AA spectroscopy (IP 285, IP 470) or by ICAP spectrometry (ASTM C1109, ASTM C1111).

AA provides very high sensitivity but requires careful subsampling, extensive sample preparation, and detailed sample-matrix corrections. XRF requires little in terms of sample preparation but suffers from low sensitivity and the application of major matrix corrections. Inductively coupled argon plasma spectrometry provides high sensitivity and few matrix corrections, but requires a considerable amount of sample preparation depending on the process stream to be analyzed.

In the inductively coupled argon plasma emission spectrometer (ICAP) method, is suitable for nickel, iron, and vanadium contents of gas oil samples in the range from 0.1 to 100 mg/kg. Thus, a 10 g sample of gas oil is charred with sulfuric acid and subsequently combusted to leave the ash residue. The resulting sulfates are then converted to their corresponding chloride salts to ensure complete solubility. A barium internal standard is added to the sample before analysis. In addition, the use of the ICAP method for the analysis of nickel, vanadium, and iron present counteracts the two basic issues arising from metal analysis. The most serious issue is the fact that these metals are partly or totally in the form of volatile chemically stable porphyrin complexes, and extreme conditions are needed to destroy the complexes without losing the metal through volatilization of the complex. The second issue is that the alternate direct aspiration of the sample introduces large quantities of carbon into the plasma. This carbon causes marked and somewhat variable background changes in all direct measurement techniques.

Finally, the analytical method should be selected depending on the sensitivity required, the compatibility of the sample matrix with the specific analysis technique, and the availability of facilities. Sample preparation, if required, can present problems. Significant losses can occur, especially in the case of organometallic complexes, and contamination of environmental sample is of serious concern. The precision of the analysis depends on the metal itself, the method used, and the standard used for the calibration of the instrument.

11.3.10 Molecular Weight

Molecular weight is a physical property that can be used in conjunction with other physical properties to characterize residual fuel oil. Since residual fuel oil is a mixture having a broad boiling range, measurement of molecular weight provides the mass-average molecular weight or the number-average molecular weight.

Molecular weight may be calculated from viscosity data (ASTM D2502) using centistoke viscosity at 38°C (100°F) and 99°C (210°F). The method is generally applicable to a sample having a molecular weight in the range of 250–700. Samples with unusual composition, such as aromatic-free white mineral oils, or oils with very narrow boiling range, may give atypical results.

For samples with higher molecular weight (up to 3000 or more) with unusual composition, or for polymers, another method (ASTM D2503) is recommended. This method uses a vapor pressure osmometer to determine molecular weight. Low-boiling samples may not be suitable if their vapor pressure interferes with the method.

Another method (ASTM D2878) provides a procedure to calculate these properties from test data on evaporation. In the method, the sample is dissolved in an appropriate solvent. A drop each of this solution and the solvent are suspended on separate thermistors in a closed chamber saturated with solvent vapor. The solvent condenses on the sample drop and causes a temperature difference between the two drops. The resultant change in temperature is measured and used to determine the molecular weight of the sample by reference to a previously prepared calibration curve. This procedure is based on an older method (ASTM D972) in which the sample can be partly evaporated at a temperature of 250–500°C (480–930°F), and fluids not stable in this temperature range may require special treatment.

11.3.11 Pour Point

For pumping and handling purposes, it is often necessary to know the minimum temperature at which a particular fuel oil loses its fluid characteristics. The pour point of fuel oil takes on importance from the handling and storage aspects—if the fuel oil has a high pour point, sufficient tank heating is required to maintain fuel temperature at least 10–20°C (18–26°F) above the pour point for satisfactory pumping. Additional facilities may be required to purge fuel lines, especially those exposed to temperatures lower than the pour point of all fuel once the pumping operation is completed. This information can be of considerable importance, because wide variations in pour point exist between different fuel oils—even between oils of comparable viscosity. Thus, the pour point text is more likely to give meaningful information than the viscosity test method(s) data since the results of the pour point test can be directly related to the fluid and non-fluid characteristics of the fuel oil.

The pour point (ASTM D97, ASTM D5853, ASTM D5949, ASTM D5950, ASTM D5985, IP 15) is the lowest temperature at which oil will flow, under prescribed conditions. The test method for determining the solidification point might also be applied to residual fuel oil.

One of the main attributes of liquid fuels is the relative ease with which they can be transferred from one place to another; it is still necessary to have some indication of the lowest temperature at which this may be achieved. Depending upon the storage conditions and application of the fuel, limits are placed upon the pour point. Storage of the heavier viscosity fuel oils in heated tankage will permit of higher pour points than would otherwise be possible. While the failure to flow can generally be attributed to the separation of wax from the fuel, it can also, in the case of very viscous fuels, be due to the effect of viscosity.

The pour point of residual fuel oil may be influenced by the previous thermal history of the residual fuel oil and the fact that any loosely knit wax structure built up on cooling the fuel can, generally, be readily broken up by the application of a little pressure thus allowing fuels to be pumped at temperatures below their pour point temperatures. The usefulness of the pour point test in relation to residual fuel oils is, therefore, open to question, and the tendency to regard it as the limiting temperature at which a fuel will flow can be misleading unless correlated with low-temperature viscosity.

The pour point test is still included in many specifications but not in some for residual fuel oil for assessing the pumpability characteristics of residual fuel oil (ASTM D396). Pour point procedures involving various preheat treatments prior to the pour point determination and the use of viscosity at low temperature have been proposed. The fluidity test is one such procedure as is the pumping temperature test (ASTM D3829), which is also available.

11.3.12 Refractive Index

Refractive index is the ratio of the velocity of light in air to the velocity of light in the measured substance. The value of the refractive index varies inversely with the wavelength of light used and the temperature at which the measurements are taken. The refractive index is a fundamental physical property that can be used for the determination of the gross composition of residual fuel oil and often requires its measurement at elevated temperature. In addition, the refractive index of a substance is related to its chemical composition and may be used to draw conclusions about molecular structure.

Two methods (ASTM D1218, ASTM D1747) are available for measuring the refractive index of viscous liquids.

Both methods are limited to lighter-colored samples for best accuracy. The latter test method (ASTM D1747) covers the measurement of refractive indexes of light-colored residual fuel oil at temperatures from 80 to 100°C (176–212°F). Temperatures lower than 80°C (176°F) may be used provided that the melting point of the sample is at least 10°C (18°F) below the test temperature. This test method is not applicable, within reasonable standards of accuracy, to liquids having darker residual fuel oil (having a color darker than ASTM Color No. 4, ASTM D1500).

11.3.13 Stability and Compatibility

For the purposes of this text, the *stability* of residual fuel oil is the ability of the fuel oil to remain in an unchanged condition despite circumstances that may tend to cause change and is the resistance of an oil to break down. *Compatibility* is similar to stability in that compatibility is the tendency of the fuel oil *not* to produce sludge or sediment when mixed with another fuel oil fraction or with an organic liquid. Thus, *instability* refers to the tendency of residual fuel to produce a deposit without addition of any other materials, whereas *incompatibility* is the tendency of residual fuel oil to produce a deposit when blended with another fuel oil or other fuel oils—this is generally assigned to the increasing paraffin nature of the medium thereby causing the asphaltene constituents to separate as a solid phase or sediment (Fig. 11.3). Instability or incompatibility may also occur during in-service use or storage of fuel oil (Mushrush and Speight, 1995, 1998a,b; Mushrush et al., 1999; Speight, 2014).

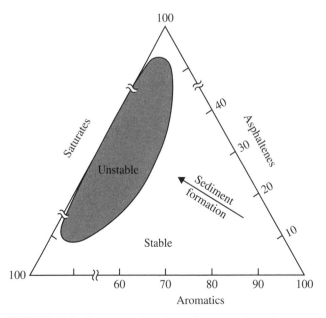

FIGURE 11.3 Representation of petroleum as a three-phase system showing a region of instability.

Compatibility problems occur when heavy fuel oils with a high asphaltene content are mixed with lighter fractions with a predominance of aliphatic hydrocarbons. The mixing can cause precipitation of the asphaltene constituents, which occurs when fuel oil suppliers blend fuel oil(s) with a low molecular weight solvent (especially a low molecular weight paraffinic solvent) in order to reduce final fuel oil viscosity, specific gravity, or other fuel property. Incompatibility in fuel oils result in rapid strainer and separator plugging with excessive sludge, and in a diesel engine, incompatible fuel oils can cause injection pump sticking, injector deposits, exhaust valve deposits, and turbocharger turbine deposits.

To predict a potential incompatibility problem, and to avoid its troublesome consequences, every cutter stock (solvent) and residual fuel oil considered for blending should be subjected to a compatibility (such as ASTM D2781) spot test, which can be conducted quickly, inexpensively and without expensive test equipment, or without highly trained technicians. There are several other test methods, but these methods are more detailed and time consuming.

The problem of instability in residual fuel oil may manifest itself either as waxy sludge deposited at the bottom of an unheated storage tank or as deposition of high-molecular-weight asphaltene constituents as well as other highly polar constituents. Paraffin precipitation and deposition in pipelines is an issue that decreases the cross-sectional area of the pipeline, thereby restricting operating capacities, and places additional strain on pumping equipment. On the other hand, deposition of asphaltene constituent can lead to fouling of preheaters on heating the fuel to elevated temperatures (Mushrush and Speight, 1995, 1998a,b; Park and Mansoori, 1988; Mushrush et al., 1999; Speight, 2014).

When incompatibility occurs, the deposit (or sludge) can be (depending upon the fuel oil) paraffin in nature (i.e., wax) or asphaltic in nature (i.e., asphaltene constituents). In either case, blockage of the fuel filter can result. Furthermore, once instability or incompatibility has occurred, there is usually no way to reverse the process, although addition of suitable solvents to the fuel oil has been recommended. For identification purposes (if identification of the deposit or sludge is necessary), a small portion should be placed in an open container at a temperature of 60–70°C (140–160°F). If the deposit is wax, melting will occur, but asphaltene constituents will not usually melt in this temperature range (Speight, 2014). In order to circumvent such problems, regular testing for fuel stability or fuel compatibility will provide indications of instability or incompatibility (Mushrush and Speight, 1995; Speight, 2014).

Problems of thermal stability and incompatibility in residual fuel oils are associated with those fuels used in oil-fired naval or marine vessels, where the fuel is usually passed through a preheater before being fed to the burner system. In earlier days, this preheating, with some fuels, could result in the deposition of asphaltic matter culminating, in the extreme case, in blockage of preheaters, pipelines, and even complete combustion failure.

Asphaltene-type deposition may, however, result from the mixing of fuels of different origin and treatment, each of which may be perfectly satisfactory when used alone. For example, straight-run fuel oils from the same crude oil are normally stable and mutually compatible, whereas fuel oils produced from thermal cracking and visbreaking operations may be stable but can be unstable or incompatible if blended with straight-run fuels and vice versa.

A procedure for predicting the stability of residual fuel oil involves the use of a spot test to show compatibility or cleanliness of the blended fuel oil (ASTM D4740)—this method is applicable to fuel oils with viscosities up to 50 cSt at 100°C (212°F) to identify fuels or blends that could result in excessive centrifuge loading, strainer plugging, tank sludge formation, or similar operating problems. In the method, a drop of the preheated sample is put on a test paper and placed in an oven at 100°C. After 1 h, the test paper is removed from the oven, and the resultant spot is examined for evidence of suspended solids and rated for cleanliness using the procedure described in the method. In a parallel procedure for determining *compatibility*, a blend composed of equal volumes of the fuel oil sample and the blend stock is tested and rated in the same way as just described for the *cleanliness* procedure.

For oxidative stability and the tendency to corrode metals as may occur in pipes and burners, a test method (ASTM D4636) is available to determine resistance to oxidation and corrosion degradation and their tendency to corrode various metals. The test method consists of one standard and two alternative procedures. A particular specification needs to establish which of these tests should be used. A large glass tube containing an oil sample and metal specimens is placed in a constant temperature bath (usually from 100 to 360°C) and heated for the specific number of hours while air is passed through the oil to provide agitation and a source of oxygen. Corrosiveness of the oil is determined by the loss in metal mass and microscopic examination of the sample metal surface(s). Oil samples are withdrawn from the test oil and checked for changes in viscosity and acid number as a result of the oxidation reactions. At the end of the test, the amount of the sludge present in the oil remaining in the same tube is determined by centrifugation. Also, the quantity of oil lost during the test is determined gravimetrically. Metals used in the basic test and alternative test are aluminum, bronze, cadmium, copper, magnesium, silver, steel, and titanium. Other metals may also be specified as determined by the history and storage of the fuel oil.

11.3.14 Viscosity

Viscosity is an important property of residual fuel oils as it provides information on the ease (or otherwise) with which a fuel can be transferred, under the prevailing temperature and pressure conditions, from storage tank to burner system.

Viscosity data also indicate the degree to which a fuel oil needs to be preheated to obtain the correct atomizing temperature for efficient combustion. Most residual fuel oils function best when the burner input viscosity lies within a certain specified range.

The Saybolt Universal and Saybolt Furol viscometers are widely used in the United States and the Engler in Europe. In the United States, viscosities on the lighter fuel grades are determined using the Saybolt Universal instrument at 38°C (100°F); for the heaviest fuels, the Saybolt Furol viscometer is used at 50°C (122°F). Similarly, in Europe, the Engler viscometer is used at temperatures of 20°C (68°F), 50°C (122°F), and in some instances at 100°C (212°F). The use of these empirical procedures for fuel oils is being superseded by the kinematic system (ASTM D396) specifications for fuel oils.

The determination of residual fuel oil viscosities is complicated by the fact that some fuel oils containing significant quantities of wax do not behave as simple Newtonian liquids in which the rate of shear is directly proportional to the shearing stress applied. At temperatures in the region of 38°C (100°F), these fuels tend to deposit wax from solution with a resulting adverse effect on the accuracy of the viscosity result unless the test temperature is raised sufficiently high for all wax to remain in solution. While the present reference test temperature of 50°C (122°F) is adequate for use with the majority of residual fuel oils, there is a growing trend of opinion in favor of a higher temperature (82°C; 180°F) particularly in view of the availability of waxier fuel oils.

Anomalous viscosity in residual fuel oils is best shown by plotting the kinematic viscosity determined at the normal test temperature and at two or three higher temperatures on viscosity–temperature charts (ASTM D341). These charts are constructed so that, for a Newtonian fuel oil, the temperature–viscosity relationship is linear. Nonlinearity at the lower end of the applicable temperature range is normally considered evidence of non-Newtonian behavior. The charts are also useful for the estimation of the viscosity of a fuel oil blend from the knowledge of the component viscosities and for calculation of the correct preheat temperature necessary to obtain the required viscosity for efficient atomization of the fuel oil in the burner.

While it is considered a technical advantage to specify kinematic viscosity, the conventional viscometers are still in wide use, and it may be convenient, or even necessary, to be able to convert viscosities from one system to another. Provision is made (ASTM D2161) for the conversion of kinematic viscosity to Saybolt Universal and Furol and in IP Standards (The Energy Institute, 2013) for conversion to Redwood viscosity (Table 11.2).

11.3.15 Volatility

Four distillation methods are in common use for determining the boiling range and for collecting fractions from residual fuel oil. Such methods are rarely used for characterization of

TABLE 11.2 Comparison of the various viscosity scales

Viscosity	Units	Value			
Kinematic viscosity at 50°C (122°F)	cSt	36	125	370	690
Kinematic viscosity at 38°C (100°F)	cSt	61	—	—	—
Redwood No. 1 viscosity at 50°C (122°F)	s	148	510	1500	2800
Redwood No. 1 viscosity at 38°C (100°F)	s	250	1000	3500	7000
Saybolt Universal viscosity at 38°C (100°F)	s	285	1150	4000	8000
Saybolt Furol viscosity at 50°C (122°F)	s	—	60	175	325
Engler degrees at 50°C (122°F)		4.8	16.5	48.7	91.0

the fuel oil but do warrant mention here because of their application to fuel oil when desired.

One method (ASTM D1160) is probably the best known and most widely used of the methods for distillation of higher-boiling petroleum products and uses vacuum distillation. The method is applicable to samples that can be at least partially volatilized at temperature up to 400°C (752°F) and pressure in the range 1–50 mm Hg. The distillation temperature at vacuum is converted to atmospheric equivalent temperatures.

Another method (ASTM D2892) applies to a wide range of products and uses a column with 15 theoretical plates and a 5:1 reflux ratio. The distillation is started at atmospheric pressure until the vapor temperature reaches 210°C (410°F). Distillation is continued at vacuum (100 mm Hg) until the vapor temperature again reaches 210°C (410°F) or cracking is observed.

With very heavy crude oil or viscous petroleum products, a preferred alternate distillation method (ASTM D5236) should be used (instead of ASTM D2892) above a 400°C (752°F) cut point. In the *spinning band* method (Fig. 11.4), fractions of feedstocks such as residual fuel oil with an initial boiling point above room temperature at atmospheric pressure can be investigated. For such materials, the initial boiling point of the sample should exceed room temperature at atmospheric pressure. The distillation is terminated at an atmospheric equivalent temperature of 524°C (975°F) and a pot temperature of 360°C (680°F).

In the method, samples are distilled under atmospheric and reduced pressures in a still equipped with a spinning band column. Vapor temperatures are converted to atmospheric

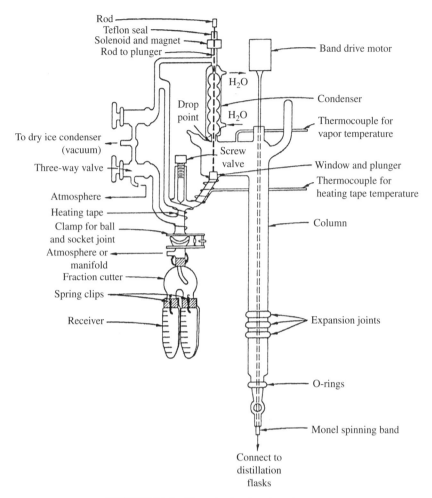

FIGURE 11.4 The spinning band equipment.

equivalent temperatures and can be plotted as a function of volume or weight percentage distilled to yield a distillation profile. The spinning band, which effectively provides a large contact area between the liquid phase and the vapor phase, increases the number of theoretical plates in the column and thus its fractionating efficiency. Readings of vapor temperature (that is convertible to atmospheric equivalent temperature) and distillate volume (that is convertible to percentage by volume) are used to plot a distillation curve. Distillate yields for naphtha, light gas oil, heavy gas oil, and residue fractions are determined on a gravimetric basis.

Another method, *short path distillation*, produces a single distillate and a single residue fraction defined by the operating temperature and pressure of the still. This procedure is used to generate high–boiling point fractions with endpoints up to 700°C (1290°F) for further analysis. Since only one cut temperature is used in each run, generation of a distillation curve using this equipment would be time consuming. In the method, the material to be fractionated is introduced at a constant rate onto the hot inner wall of the evaporator under high vacuum. Rotating (Teflon) rollers ensure that the film on

the wall is kept thin. The feedstock is progressively distilled at the fixed conditions of temperature and pressure. The distillate vapors condense on a concentric cold surface (60°C, 140°F) placed at a short distance from the hot wall inside the still. The condensate then drains by gravity to the base of the *cold finger* where it is collected. The residue drains down the hot wall and is collected through a separate port.

However, unless a distillation method is required by specification or the collected fractions are needed for further testing, gas chromatographic methods have become preferred for determining the boiling range of petroleum fractions, and detailed information for samples with a final boiling point no higher than 538°C (1000°F) at atmospheric pressure and a boiling range greater than 55°C (100°F) is available (ASTM D2887).

11.3.16 Water and Sediment

Contamination in residual fuel oil may be indicated by the presence of excessive amounts of water, emulsions, and inorganic material such as sand and rust. Appreciable amounts of sediment in a residual fuel oil can foul the handling facilities and give problems in burner mechanisms. Blockage of fuel filters (ASTM D2068, ASTM D6426) due to the presence of fuel degradation products may also result. This aspect of fuel quality control may be dealt with by placing restrictions on the water (ASTM D95, IP 74), sediment by extraction (ASTM D473, IP 53), or water and sediment values obtained on the fuel.

In any form, water and sediment are highly undesirable in fuel oil, and the relevant tests involving distillation (ASTM D95, ASTM D4006, IP 74, IP 358), centrifuging (ASTM D4007), extraction (ASTM D473, IP 53), and the Karl Fischer titration (ASTM D4377, ASTM D4928, IP 356, IP 386, IP 438, IP 439) are regarded as important in the determination of quality.

The Karl Fischer test method (ASTM D1364, ASTM D6304) covers the direct determination of water in petroleum products. In the test, the sample injection in the titration vessel can be performed on a volumetric or gravimetric basis. Viscous samples, such as residual fuel oil, can be analyzed using a water vaporizer accessory that heats the sample in the evaporation chamber, and the vaporized water is carried into the Karl Fischer titration cell by a dry inert carrier gas.

Water and sediment can be determined simultaneously (ASTM D1796, ASTM D4007) by the centrifuge method. Known volumes of residual fuel oil and solvent are placed in a centrifuge tube and heated to 60°C (140°F). After centrifugation, the volume of the sediment and water layer at the bottom of the tube is read. In the unlikely event that the residual fuel oil contains wax, a temperature of 71°C (160°F) or higher may be required to completely melt the wax crystals so that they are not measured as sediment.

Sediment is also determined by an extraction method (ASTM D473, IP 53) or by membrane filtration (ASTM D4807). In the former method (ASTM D473, IP 53), a sample contained in a refractory thimble is extracted with hot toluene until the residue reaches a constant mass. In the latter test, the sample is dissolved in hot toluene and filtered under vacuum through a 0.45 μm porosity membrane filter. The filter with residue is washed, dried, and weighed.

In a test specifically designed for residual fuel oil (ASTM D4870, IP 375), a 10 g sample of oil is filtered through the prescribed apparatus at 100°C. After washing with the solvent, and drying, the total sediment on the filter medium is weighed. The test is to be carried out in duplicate.

REFERENCES

ASTM C1109. Standard Practice for Analysis of Aqueous Leachates from Nuclear Waste Materials Using Inductively Coupled Plasma-Atomic Emission Spectroscopy. Annual Book of Standards. ASTM International, West Conshohocken, PA.

ASTM C1111. Standard Test Method for Determining Elements in Waste Streams by Inductively Coupled Plasma-Atomic Emission Spectroscopy. Annual Book of Standards. ASTM International, West Conshohocken, PA.

ASTM D91. Standard Test Method for Precipitation Number of Lubricating Oils. Annual Book of Standards. ASTM International, West Conshohocken, PA.

ASTM D93. Standard Test Methods for Flash Point by Pensky-Martens Closed Cup Tester. Annual Book of Standards. ASTM International, West Conshohocken, PA.

ASTM D95. Standard Test Method for Water in Petroleum Products and Bituminous Materials by Distillation. Annual Book of Standards. ASTM International, West Conshohocken, PA.

ASTM D97. Standard Test Method for Pour Point of Petroleum Products. Annual Book of Standards. ASTM International, West Conshohocken, PA.

ASTM D129. Standard Test Method for Sulfur in Petroleum Products (General High Pressure Decomposition Device Method). Annual Book of Standards. ASTM International, West Conshohocken, PA.

ASTM D130. Standard Test Method for Corrosiveness to Copper from Petroleum Products by Copper Strip Test. Annual Book of Standards. ASTM International, West Conshohocken, PA.

ASTM D139. Standard Test Method for Float Test for Bituminous Materials. Annual Book of Standards. ASTM International, West Conshohocken, PA.

ASTM D189. Standard Test Method for Conradson Carbon Residue of Petroleum Products. Annual Book of Standards. ASTM International, West Conshohocken, PA.

ASTM D240. Standard Test Method for Heat of Combustion of Liquid Hydrocarbon Fuels by Bomb Calorimeter. Annual Book of Standards. ASTM International, West Conshohocken, PA.

ASTM D287. Standard Test Method for API Gravity of Crude Petroleum and Petroleum Products (Hydrometer Method). Annual Book of Standards. ASTM International, West Conshohocken, PA.

ASTM D341. Standard Practice for Viscosity-Temperature Charts for Liquid Petroleum Products. Annual Book of Standards. ASTM International, West Conshohocken, PA.

ASTM D381. Standard Test Method for Gum Content in Fuels by Jet Evaporation. Annual Book of Standards. ASTM International, West Conshohocken, PA.

ASTM D396. Standard Specification for Fuel Oils. Annual Book of Standards. ASTM International, West Conshohocken, PA.

ASTM D445. Standard Test Method for Kinematic Viscosity of Transparent and Opaque Liquids (and Calculation of Dynamic Viscosity). Annual Book of Standards. ASTM International, West Conshohocken, PA.

ASTM D473. Standard Test Method for Sediment in Crude Oils and Fuel Oils by the Extraction Method. Annual Book of Standards. ASTM International, West Conshohocken, PA.

ASTM D482. Standard Test Method for Ash from Petroleum Products. Annual Book of Standards. ASTM International, West Conshohocken, PA.

ASTM D524. Standard Test Method for Ramsbottom Carbon Residue of Petroleum Products. Annual Book of Standards. ASTM International, West Conshohocken, PA.

ASTM D611. Standard Test Methods for Aniline Point and Mixed Aniline Point of Petroleum Products and Hydrocarbon Solvents. Annual Book of Standards. ASTM International, West Conshohocken, PA.

ASTM D808. Standard Test Method for Chlorine in New and Used Petroleum Products (High Pressure Decomposition Device Method). Annual Book of Standards. ASTM International, West Conshohocken, PA.

ASTM D849. Standard Test Method for Copper Strip Corrosion by Industrial Aromatic Hydrocarbons. Annual Book of Standards. ASTM International, West Conshohocken, PA.

ASTM D873. Standard Test Method for Oxidation Stability of Aviation Fuels (Potential Residue Method). Annual Book of Standards. ASTM International, West Conshohocken, PA.

ASTM D893. ASTM D893. Standard Test Method for Insolubles in Used Lubricating Oils. Annual Book of Standards. ASTM International, West Conshohocken, PA.

ASTM D972. Standard Test Method for Evaporation Loss of Lubricating Greases and Oils. Annual Book of Standards. ASTM International, West Conshohocken, PA.

ASTM D974. Standard Test Method for Acid and Base Number by Color-Indicator Titration. Annual Book of Standards. ASTM International, West Conshohocken, PA.

ASTM D1018. Standard Test Method for Hydrogen in Petroleum Fractions. Annual Book of Standards. ASTM International, West Conshohocken, PA.

ASTM D1091. Standard Test Methods for Phosphorus in Lubricating Oils and Additives. Annual Book of Standards. ASTM International, West Conshohocken, PA.

ASTM D1160. Standard Test Method for Distillation of Petroleum Products at Reduced Pressure. Annual Book of Standards. ASTM International, West Conshohocken, PA.

ASTM D1217. Standard Test Method for Density and Relative Density (Specific Gravity) of Liquids by Bingham Pycnometer. Annual Book of Standards. ASTM International, West Conshohocken, PA.

ASTM D1218. Standard Test Method for Refractive Index and Refractive Dispersion of Hydrocarbon Liquids. Annual Book of Standards. ASTM International, West Conshohocken, PA.

ASTM D1250. Standard Guide for Use of the Petroleum Measurement Tables. Annual Book of Standards. ASTM International, West Conshohocken, PA.

ASTM D1266. Standard Test Method for Sulfur in Petroleum Products (Lamp Method). Annual Book of Standards. ASTM International, West Conshohocken, PA.

ASTM D1298. Standard Test Method for Density, Relative Density, or API Gravity of Crude Petroleum and Liquid Petroleum Products by Hydrometer Method. Annual Book of Standards. ASTM International, West Conshohocken, PA.

ASTM D1318. Standard Test Method for Sodium in Residual Fuel Oil (Flame Photometric Method). Annual Book of Standards. ASTM International, West Conshohocken, PA.

ASTM D1364. Standard Test Method for Water in Volatile Solvents (Karl Fischer Reagent Titration Method). Annual Book of Standards. ASTM International, West Conshohocken, PA.

ASTM D1480. Standard Test Method for Density and Relative Density (Specific Gravity) of Viscous Materials by Bingham Pycnometer. Annual Book of Standards. ASTM International, West Conshohocken, PA.

ASTM D1481. Standard Test Method for Density and Relative Density (Specific Gravity) of Viscous Materials by Lipkin Bicapillary Pycnometer. Annual Book of Standards. ASTM International, West Conshohocken, PA.

ASTM D1500. Standard Test Method for ASTM Color of Petroleum Products (ASTM Color Scale). Annual Book of Standards. ASTM International, West Conshohocken, PA.

ASTM D1552. Standard Test Method for Sulfur in Petroleum Products (High-Temperature Method). Annual Book of Standards. ASTM International, West Conshohocken, PA.

ASTM D1747. Standard Test Method for Refractive Index of Viscous Materials. Annual Book of Standards. ASTM International, West Conshohocken, PA.

ASTM D1796. Standard Test Method for Water and Sediment in Fuel Oils by the Centrifuge Method (Laboratory Procedure). Annual Book of Standards. ASTM International, West Conshohocken, PA.

ASTM D2007. Standard Test Method for Characteristic Groups in Rubber Extender and Processing Oils and Other Petroleum-Derived Oils by the Clay-Gel Absorption Chromatographic Method. Annual Book of Standards. ASTM International, West Conshohocken, PA.

ASTM D2068. Standard Test Method for Determining Filter Blocking Tendency. Annual Book of Standards. ASTM International, West Conshohocken, PA.

ASTM D2140. Standard Practice for Calculating Carbon-Type Composition of Insulating Oils of Petroleum Origin. Annual Book of Standards. ASTM International, West Conshohocken, PA.

ASTM D2161. Standard Practice for Conversion of Kinematic Viscosity to Saybolt Universal Viscosity or to Saybolt Furol Viscosity. Annual Book of Standards. ASTM International, West Conshohocken, PA.

ASTM D2276. Standard Test Method for Particulate Contaminant in Aviation Fuel by Line Sampling. Annual Book of Standards. ASTM International, West Conshohocken, PA.

ASTM D2415. Standard Test Method for Ash in Coal Tar and Pitch. Annual Book of Standards. ASTM International, West Conshohocken, PA.

ASTM D2416. Standard Test Method for Coking Value of Tar and Pitch (Modified Conradson). Annual Book of Standards. ASTM International, West Conshohocken, PA.

ASTM D2425. Standard Test Method for Hydrocarbon Types in Middle Distillates by Mass Spectrometry. Annual Book of Standards. ASTM International, West Conshohocken, PA.

ASTM D2500. Standard Test Method for Cloud Point of Petroleum Products. Annual Book of Standards. ASTM International, West Conshohocken, PA.

ASTM D2501. Standard Test Method for Calculation of Viscosity-Gravity Constant (VGC) of Petroleum Oils. Annual Book of Standards. ASTM International, West Conshohocken, PA.

ASTM D2502. Standard Test Method for Estimation of Mean Relative Molecular Mass of Petroleum Oils from Viscosity Measurements. Annual Book of Standards. ASTM International, West Conshohocken, PA.

ASTM D2503. Standard Test Method for Relative Molecular Mass (Molecular Weight) of Hydrocarbons by Thermoelectric Measurement of Vapor Pressure. Annual Book of Standards. ASTM International, West Conshohocken, PA.

ASTM D2549. Standard Test Method for Separation of Representative Aromatics and Nonaromatics Fractions of High-Boiling Oils by Elution Chromatography. Annual Book of Standards. ASTM International, West Conshohocken, PA.

ASTM D2622. Standard Test Method for Sulfur in Petroleum Products by Wavelength Dispersive X-ray Fluorescence Spectrometry. Annual Book of Standards. ASTM International, West Conshohocken, PA.

ASTM D2781. Method of Test for Compatibility of Fuel Oil Blends by Spot Test (withdrawn 1991 but still in use by some fuel oil users). Annual Book of Standards. ASTM International, West Conshohocken, PA.

ASTM D2786. Standard Test Method for Hydrocarbon Types Analysis of Gas-Oil Saturates Fractions by High Ionizing Voltage Mass Spectrometry. Annual Book of Standards. ASTM International, West Conshohocken, PA.

ASTM D2878. Standard Test Method for Estimating Apparent Vapor Pressures and Molecular Weights of Lubricating Oils. Annual Book of Standards. ASTM International, West Conshohocken, PA.

ASTM D2887. Standard Test Method for Boiling Range Distribution of Petroleum Fractions by Gas Chromatography. Annual Book of Standards. ASTM International, West Conshohocken, PA.

ASTM D2892. Standard Test Method for Distillation of Crude Petroleum (15-Theoretical Plate Column). Annual Book of Standards. ASTM International, West Conshohocken, PA.

ASTM D3120. Standard Test Method for Trace Quantities of Sulfur in Light Liquid Petroleum Hydrocarbons by Oxidative Microcoulometry. ASTM International, West Conshohocken, PA.

ASTM D3228. Standard Test Method for Total Nitrogen in Lubricating Oils and Fuel Oils by Modified Kjeldahl Method. Annual Book of Standards. ASTM International, West Conshohocken, PA.

ASTM D3238. Standard Test Method for Calculation of Carbon Distribution and Structural Group Analysis of Petroleum Oils by the n-d-M Method. Annual Book of Standards. ASTM International, West Conshohocken, PA.

ASTM D3239. Standard Test Method for Aromatic Types Analysis of Gas-Oil Aromatic Fractions by High Ionizing Voltage Mass Spectrometry. Annual Book of Standards. ASTM International, West Conshohocken, PA.

ASTM D3246. Standard Test Method for Sulfur in Petroleum Gas by Oxidative Microcoulometry. Annual Book of Standards. ASTM International, West Conshohocken, PA.

ASTM D3279. Standard Test Method for n-Heptane Insolubles. Annual Book of Standards. ASTM International, West Conshohocken, PA.

ASTM D3341. Standard Test Method for Lead in Gasoline—Iodine Monochloride Method. Annual Book of Standards. ASTM International, West Conshohocken, PA.

ASTM D3343. Standard Test Method for Estimation of Hydrogen Content of Aviation Fuels. Annual Book of Standards. ASTM International, West Conshohocken, PA.

ASTM D3427. Standard Test Method for Air Release Properties of Petroleum Oils. Annual Book of Standards. ASTM International, West Conshohocken, PA.

ASTM D3605. Standard Test Method for Trace Metals in Gas Turbine Fuels by Atomic Absorption and Flame Emission Spectroscopy. Annual Book of Standards. ASTM International, West Conshohocken, PA.

ASTM D3701. Standard Test Method for Hydrogen Content of Aviation Turbine Fuels by Low Resolution Nuclear Magnetic Resonance Spectrometry. Annual Book of Standards. ASTM International, West Conshohocken, PA.

ASTM D3829. Standard Test Method for Predicting the Borderline Pumping Temperature of Engine Oil. Annual Book of Standards. ASTM International, West Conshohocken, PA.

ASTM D4006. Standard Test Method for Water in Crude Oil by Distillation. Annual Book of Standards. ASTM International, West Conshohocken, PA.

ASTM D4007. Standard Test Method for Water and Sediment in Crude Oil by the Centrifuge Method (Laboratory Procedure). Annual Book of Standards. ASTM International, West Conshohocken, PA.

ASTM D4045. Standard Test Method for Sulfur in Petroleum Products by Hydrogenolysis and Rateometric Colorimetry. Annual Book of Standards. ASTM International, West Conshohocken, PA.

ASTM D4047. Standard Test Method for Phosphorus in Lubricating Oils and Additives by Quinoline Phosphomolybdate Method. Annual Book of Standards. ASTM International, West Conshohocken, PA.

ASTM D4048. Standard Test Method for Detection of Copper Corrosion from Lubricating Grease. Annual Book of Standards. ASTM International, West Conshohocken, PA.

ASTM D4052. Standard Test Method for Density, Relative Density, and API Gravity of Liquids by Digital Density Meter. Annual Book of Standards. ASTM International, West Conshohocken, PA.

ASTM D4055. Standard Test Method for Pentane Insolubles by Membrane Filtration. Annual Book of Standards. ASTM International, West Conshohocken, PA.

ASTM D4072. Standard Test Method for Toluene-Insoluble (TI) Content of Tar and Pitch. Annual Book of Standards. ASTM International, West Conshohocken, PA.

ASTM D4124. Standard Test Method for Separation of Asphalt into Four Fractions. Annual Book of Standards. ASTM International, West Conshohocken, PA.

ASTM D4291. Standard Test Method for Trace Ethylene Glycol in Used Engine Oil. Annual Book of Standards. ASTM International, West Conshohocken, PA.

ASTM D4294. Standard Test Method for Sulfur in Petroleum and Petroleum Products by Energy Dispersive X-ray Fluorescence Spectrometry. Annual Book of Standards. ASTM International, West Conshohocken, PA.

ASTM D4312. Standard Test Method for Toluene-Insoluble (TI) Content of Tar and Pitch (Short Method). Annual Book of Standards. ASTM International, West Conshohocken, PA.

ASTM D4377. Standard Test Method for Water in Crude Oils by Potentiometric Karl Fischer Titration. Annual Book of Standards. ASTM International, West Conshohocken, PA.

ASTM D4530. Standard Test Method for Determination of Carbon Residue (Micro Method). Annual Book of Standards. ASTM International, West Conshohocken, PA.

ASTM D4628. Standard Test Method for Analysis of Barium, Calcium, Magnesium, and Zinc in Unused Lubricating Oils by Atomic Absorption Spectrometry. Annual Book of Standards. ASTM International, West Conshohocken, PA.

ASTM D4629. Standard Test Method for Trace Nitrogen in Liquid Petroleum Hydrocarbons by Syringe/Inlet Oxidative Combustion and Chemiluminescence Detection. Annual Book of Standards. ASTM International, West Conshohocken, PA.

ASTM D4636. Standard Test Method for Corrosiveness and Oxidation Stability of Hydraulic Oils, Aircraft Turbine Engine Lubricants, and Other Highly Refined Oils. Annual Book of Standards. ASTM International, West Conshohocken, PA.

ASTM D4715. Standard Test Method for Coking Value of Tar and Pitch (Alcan). Annual Book of Standards. ASTM International, West Conshohocken, PA.

ASTM D4740. Standard Test Method for Cleanliness and Compatibility of Residual Fuels by Spot Test. Annual Book of Standards. ASTM International, West Conshohocken, PA.

ASTM D4807. Standard Test Method for Sediment in Crude Oil by Membrane Filtration. Annual Book of Standards. ASTM International, West Conshohocken, PA.

ASTM D4808. Standard Test Methods for Hydrogen Content of Light Distillates, Middle Distillates, Gas Oils, and Residua by Low-Resolution Nuclear Magnetic Resonance Spectroscopy. Annual Book of Standards. ASTM International, West Conshohocken, PA.

ASTM D4870. Standard Test Method for Determination of Total Sediment in Residual Fuels. Annual Book of Standards. ASTM International, West Conshohocken, PA.

ASTM D4893. Standard Test Method for Determination of Pitch Volatility. Annual Book of Standards. ASTM International, West Conshohocken, PA.

ASTM D4927. Standard Test Methods for Elemental Analysis of Lubricant and Additive Components – Barium, Calcium, Phosphorus, Sulfur, and Zinc by Wavelength-Dispersive X-Ray Fluorescence Spectroscopy. Annual Book of Standards. ASTM International, West Conshohocken, PA.

ASTM D4928. Standard Test Method for Water in Crude Oils by Coulometric Karl Fischer Titration. Annual Book of Standards. ASTM International, West Conshohocken, PA.

ASTM D4951. Standard Test Method for Determination of Additive Elements in Lubricating Oils by Inductively Coupled Plasma Atomic Emission Spectrometry. Annual Book of Standards. ASTM International, West Conshohocken, PA.

ASTM D4952. Standard Test Method for Qualitative Analysis for Active Sulfur Species in Fuels and Solvents (Doctor Test). Annual Book of Standards. ASTM International, West Conshohocken, PA.

ASTM D5002. Standard Test Method for Density and Relative Density of Crude Oils by Digital Density Analyzer. Annual Book of Standards. ASTM International, West Conshohocken, PA.

ASTM D5185. Standard Test Method for Multi-element Determination of Used and Unused Lubricating Oils and Base Oils by Inductively Coupled Plasma Atomic Emission Spectrometry (ICP-AES). Annual Book of Standards. ASTM International, West Conshohocken, PA.

ASTM D5236. Standard Test Method for Distillation of Heavy Hydrocarbon Mixtures (Vacuum Potstill Method). Annual Book of Standards. ASTM International, West Conshohocken, PA.

ASTM D5291. Standard Test Methods for Instrumental Determination of Carbon, Hydrogen, and Nitrogen in Petroleum Products and Lubricants. Annual Book of Standards. ASTM International, West Conshohocken, PA.

ASTM D5292. Standard Test Method for Aromatic Carbon Contents of Hydrocarbon Oils by High Resolution Nuclear Magnetic Resonance Spectroscopy. Annual Book of Standards. ASTM International, West Conshohocken, PA.

ASTM D5384. Standard Test Methods for Chlorine in Used Petroleum Products (Field Test Kit Method). Annual Book of Standards. ASTM International, West Conshohocken, PA.

ASTM D5452. Standard Test Method for Particulate Contamination in Aviation Fuels by Laboratory Filtration. Annual Book of Standards. ASTM International, West Conshohocken, PA.

ASTM D5453. Standard Test Method for Determination of Total Sulfur in Light Hydrocarbons, Spark Ignition Engine Fuel, Diesel Engine Fuel, and Engine Oil by Ultraviolet Fluorescence. Annual Book of Standards. ASTM International, West Conshohocken, PA.

ASTM D5623. Standard Test Method for Sulfur Compounds in Light Petroleum Liquids by Gas Chromatography and Sulfur Selective Detection. Annual Book of Standards. ASTM International, West Conshohocken, PA.

ASTM D5705. Standard Test Method for Measurement of Hydrogen Sulfide in the Vapor Phase Above Residual Fuel Oils. Annual Book of Standards. ASTM International, West Conshohocken, PA.

ASTM D5708. Standard Test Methods for Determination of Nickel, Vanadium, and Iron in Crude Oils and Residual Fuels by Inductively Coupled Plasma (ICP) Atomic Emission Spectrometry. Annual Book of Standards. ASTM International, West Conshohocken, PA.

ASTM D5762. Standard Test Method for Nitrogen in Petroleum and Petroleum Products by Boat-Inlet Chemiluminescence. Annual Book of Standards. ASTM International, West Conshohocken, PA.

ASTM D5771. Standard Test Method for Cloud Point of Petroleum Products (Optical Detection Stepped Cooling Method). Annual Book of Standards. ASTM International, West Conshohocken, PA.

ASTM D5772. Standard Test Method for Cloud Point of Petroleum Products (Linear Cooling Rate Method). Annual Book of Standards. ASTM International, West Conshohocken, PA.

ASTM D5773. Standard Test Method for Cloud Point of Petroleum Products (Constant Cooling Rate Method). Annual Book of Standards. ASTM International, West Conshohocken, PA.

ASTM D5853. Standard Test Method for Pour Point of Crude Oils. Annual Book of Standards. ASTM International, West Conshohocken, PA.

ASTM D5863. Standard Test Methods for Determination of Nickel, Vanadium, Iron, and Sodium in Crude Oils and Residual Fuels by Flame Atomic Absorption Spectrometry. Annual Book of Standards. ASTM International, West Conshohocken, PA.

ASTM D5949. Standard Test Method for Pour Point of Petroleum Products (Automatic Pressure Pulsing Method). Annual Book of Standards. ASTM International, West Conshohocken, PA.

ASTM D5950. Standard Test Method for Pour Point of Petroleum Products (Automatic Tilt Method). Annual Book of Standards. ASTM International, West Conshohocken, PA.

ASTM D5985. Standard Test Method for Pour Point of Petroleum Products (Rotational Method). Annual Book of Standards. ASTM International, West Conshohocken, PA.

ASTM D6021. Standard Test Method for Measurement of Total Hydrogen Sulfide in Residual Fuels by Multiple Headspace Extraction and Sulfur Specific Detection. Annual Book of Standards. ASTM International, West Conshohocken, PA.

ASTM D6160. Standard Test Method for Determination of Polychlorinated Biphenyls (PCBs) in Waste Materials by Gas Chromatography. Annual Book of Standards. ASTM International, West Conshohocken, PA.

ASTM D6217. Standard Test Method for Particulate Contamination in Middle Distillate Fuels by Laboratory Filtration. Annual Book of Standards. ASTM International, West Conshohocken, PA.

ASTM D6304. Standard Test Method for Determination of Water in Petroleum Products, Lubricating Oils, and Additives by Coulometric Karl Fischer Titration. Annual Book of Standards. ASTM International, West Conshohocken, PA.

ASTM D6426. Standard Test Method for Determining Filterability of Middle Distillate Fuel Oils. Annual Book of Standards. ASTM International, West Conshohocken, PA.

ASTM D6560. Standard Test Method for Determination of Asphaltenes (Heptane Insolubles) in Crude Petroleum and Petroleum Products. Annual Book of Standards. ASTM International, West Conshohocken, PA.

ASTM E258. Standard Test Method for Total Nitrogen in Organic Materials by Modified Kjeldahl Method. Annual Book of Standards. ASTM International, West Conshohocken, PA.

ASTM E385. Standard Test Method for Oxygen Content Using a 14-MeV Neutron Activation and Direct-Counting Technique. Annual Book of Standards. ASTM International, West Conshohocken, PA.

ASTM E777. Standard Test Method for Carbon and Hydrogen in the Analysis Sample of Refuse-Derived Fuel. Annual Book of Standards. ASTM International, West Conshohocken, PA.

ASTM E778. Standard Test Methods for Nitrogen in the Analysis Sample of Refuse-Derived Fuel. Annual Book of Standards. ASTM International, West Conshohocken, PA.

Charlot, J.C. and Claus, G. 2000. Distillates and residual fuels for heating and engines. In Modern Petroleum Technology. Volume 2: Downstream. A.G. Lucas (Editor). John Wiley & Sons Inc., New York.

Gary, J.G., Handwerk, G.E., and Kaiser, M.J. 2007. Petroleum Refining: Technology and Economics, 5th Edition. CRC Press, Taylor & Francis Group, Boca Raton, FL.

Heinrich, H. and Duée, D. 2000. Kerosine and gasoil manufacture. In Modern Petroleum Technology. Volume 2: Downstream. A.G. Lucas (Editor). John Wiley & Sons Inc., New York.

Hsu, C.S. and Robinson, P.R. (Editors) 2006. Practical Advances in Petroleum Processing (Volumes 1 and 2). Springer Science, New York.

IP 2 (ASTM D611). Petroleum Products and Hydrocarbon Solvents – Determination of Aniline and Mixed Aniline Point. IP Standard Methods 2013. The Energy Institute, London.

IP 4 (ASTM D482). Petroleum Products – Determination of Ash. IP Standard Methods 2013. The Energy Institute, London.

IP 12. Determination of Specific Energy. IP Standard Methods 2013. The Energy Institute, London.

IP 13 (ASTM D189). Petroleum Products – Determination of Carbon Residue – Conradson Method. IP Standard Methods 2013. The Energy Institute, London.

IP 14 (ASTM D524). Petroleum Products – Determination of Carbon Residue – Ramsbottom Method. IP Standard Methods 2013. The Energy Institute, London.

IP 15 (ASTM D97). Petroleum products – Determination of Pour Point. IP Standard Methods 2013. The Energy Institute, London.

IP 30. Detection of Mercaptans, Hydrogen Sulfide, Elemental Sulfur and Peroxides – Doctor Test Method. IP Standard Methods 2013. The Energy Institute, London.

IP 34 (ASTM D93). Determination of Flash Point – Pensky-Martens Closed Cup Method. IP Standard Methods 2013. The Energy Institute, London.

IP 53 (ASTM D473). Crude petroleum and Fuel Oils – Determination of Sediment – Extraction Method. IP Standard Methods 2013. The Energy Institute, London.

IP 61 (ASTM D129). Determination of Sulfur – High Pressure Combustion Method. IP Standard Methods 2013. The Energy Institute, London.

IP 74 (ASTM D95). Petroleum Products and Bituminous Materials – Determination of Water – Distillation Method. IP Standard Methods 2013. The Energy Institute, London.

IP 107 (ASTM D1266). Determination of Sulfur – Lamp Combustion Method. IP Standard Methods 2013. The Energy Institute, London.

IP 131. Petroleum Products – Gum Content of Light and Middle Distillates – Jet Evaporation Method. IP Standard Methods 2013. The Energy Institute, London.

IP 138 (ASTM D873). Determination of Oxidation Stability of Aviation Fuel – Potential Residue Method. IP Standard Methods 2013. The Energy Institute, London.

IP 139 (ASTM D974). Petroleum Products and Lubricants – Determination of Acid or Base Number-Color Indicator Titration Method. IP Standard Methods 2013. The Energy Institute, London.

IP 143 (ASTM D6560). Determination of Asphaltenes (Heptane Insolubles) in Crude Petroleum and Petroleum Products. IP Standard Methods 2013. The Energy Institute, London.

IP 154. Petroleum Products – Corrosiveness to Copper – Copper Strip Test. IP Standard Methods 2013. The Energy Institute, London.

IP 160 (ASTM D1298). Crude Petroleum and Liquid Petroleum Products – Laboratory Determination of Density – Hydrometer Method. IP Standard Methods 2013. The Energy Institute, London.

IP 227. Corrosiveness of Silver of Aviation Turbine Fuels – Silver Strip Method. Energy Institute, London, UK.

IP 243. Petroleum Products and Hydrocarbons – Determination of Sulfur Content – Wick bold Combustion Method. IP Standard Methods 2013. The Energy Institute, London.

IP 285. Determination of Nickel and Vanadium – Spectrophotometric Method. IP Standard Methods 2013. The Energy Institute, London.

IP 313 (ASTM D3427). Determination of Air Release Value of Hydraulic, Turbine and Lubricating Oils. IP Standard Methods 2013. The Energy Institute, London.

IP 336. Petroleum Products – Determination of Sulfur Content – Energy-Dispersive X-Ray Fluorescence Method. IP Standard Methods 2013. The Energy Institute, London.

IP 338 (ASTM D3701). Hydrogen Content of Aviation Turbine Fuels – Low Resolution Nuclear Magnetic Resonance Spectrometry Method. IP Standard Methods 2013. The Energy Institute, London.

IP 356. Crude Petroleum – Determination of Water – Potentiometric Karl Fischer Titration Method. IP Standard Methods 2013. The Energy Institute, London.

IP 358. Crude Petroleum – Determination of Water – Distillation Method. IP Standard Methods 2013. The Energy Institute, London.

IP 365. Crude Petroleum and Petroleum Products – Determination of Density – Oscillating U-Tube Method. IP Standard Methods 2013. The Energy Institute, London.

IP 373. Sulfur Content – Microcoulometry (Oxidative) Method. IP Standard Methods 2013. The Energy Institute, London.

IP 375 (ASTM D4870). Petroleum Products – Total Sediment In Residual Fuel Oils – Part 1: Determination By Hot Filtration. IP Standard Methods 2013. The Energy Institute, London.

IP 386 (ASTM D4928). Crude Petroleum – Determination of Water – Coulometric Karl Fischer Titration Method. IP Standard Methods 2013. The Energy Institute, London.

IP 398. Petroleum Products – Determination Of Carbon Residue – Micro Method. IP Standard Methods 2013. The Energy Institute, London.

IP 415 (ASTM D6217). Determination of Particulate Content of Middle Distillate Fuels – Laboratory Filtration Method. IP Standard Methods 2013. The Energy Institute, London.

IP 438. Water Content by Coulometric Karl Fischer Titration Method. IP Standard Methods 2013. The Energy Institute, London.

IP 439. Petroleum Products – Determination Of Water – Potentiometric Karl Fischer Titration Method. IP Standard Methods 2013. The Energy Institute, London.

IP 470. Determination of Aluminum, Silicon, Vanadium, Nickel, Iron, Calcium, Zinc and Sodium in Residual Fuel Oil by Ashing, Fusion and Atomic Absorption Spectrometry (Metals in Fuel Oil By AAS). IP Standard Methods 2013. The Energy Institute, London.

Mushrush, G.W. and Speight, J.G. 1995. Petroleum Products: Instability and Incompatibility. Taylor & Francis, New York.

Mushrush, G.W. and Speight, J.G. 1998a. Instability and Incompatibility of Petroleum Products. In Petroleum Chemistry and Refining. J.G. Speight (Editor). Taylor & Francis, New York.

Mushrush, G.W. and Speight, J.G. 1998b. The chemistry of the incompatibility process in middle distillate fuels. Rev. Process Chem. Eng. 1: 5.

Mushrush, G.W., Speight, J.G., Beal, E.J., and Hardy, D.R. 1999. Instability chemistry of middle distillate fuel oils. Am. Chem. Soc. Div. Petrol. Chem. 44(2): 175.

Park, S.J. and Mansoori, G.A. 1988. Aggregation and deposition of heavy organics in petroleum crudes. Energy Sources 10: 109–125.

Speight, J.G. 1994. Chemical and Physical Studies of Petroleum Asphaltenes. In Asphaltenes and Asphalts, I. Developments in Petroleum Science, Volume 40. T.F. Yen and G.V. Chilingarian (Editors). Elsevier, Amsterdam.

Speight, J.G. 2000. The Desulfurization of Heavy Oils and Residua. 2nd Edition. Marcel Dekker Inc., New York.

Speight, J.G. 2001. Handbook of Petroleum Analysis. John Wiley & Sons Inc., New York.

Speight, J.G. and Ozum, B. 2002. Petroleum Refining Processes. Marcel Dekker Inc., New York.

Speight, J.G. 2011. The Refinery of the Future. Gulf Professional Publishing, Elsevier, Oxford, London.

Speight, J.G. 2014. The Chemistry and Technology of Petroleum, 5th Edition. CRC Press, Taylor & Francis Group, Boca Raton, FL.

The Energy Institute. 2013. IP Standard Methods 2013. The Institute of Petroleum, London.

12

WHITE OIL

12.1 INTRODUCTION

In the present context, the term *white oil* or *mineral oil* refers to colorless or very pale oils within the lubricating oil class in regard to carbon number and boiling range (Table 12.1, Fig. 12.1) (API, 1992). Minerals (mineral) oils belong to two main groups, medicinal (pharmaceutical) oils and technical, the chief difference being the degree of refining (Banaszewski and Blythe, 2000; Speight and Ozum, 2002; Hsu and Robinson, 2006; Gary et al., 2007; Speight, 2014).

Thus, for purposes of this text, the terms *white oil* and *mineral oil* are used to indicate highly refined liquid hydrocarbons derived from petroleum distillates, which are used in medicine, pharmaceutical applications, cosmetic applications, food packaging applications, food contact applications, and food itself. Other terms that are often used interchangeably with mineral oil include *liquid petrolatum*, *liquid paraffin*, paraffin *oil*, *medicinal oil*, *medicinal white oil*, and *white mineral oil*. Other terms, such as *food grade oil*, *food grade white oil*, and *technical white oil* are also in use. In the current context, the term *mineral oil* does not include petroleum wax and petrolatum.

The use of while oils has expanded over the past several decades, and as a result, the terminology has become even more complex—products that may also (incorrectly be classified as *white oil* in contrast to the older more relevant terminology given earlier) include petroleum fractions, from kerosene to viscous lubricating oil, that have been drastically treated to remove all reactive constituents to give pure, colorless, odorless, and nonreactive petroleum products. It is the combination of these four properties, unique to these refined natural petroleum fractions, that has expanded the applications of white oils from a simple medicinal to their widespread use in such diverse fields as cosmetics, pharmaceuticals, plastics, agricultural and animal sprays, food processing and protection, animal feed, refrigeration and electrical equipment, chemical reagent and reaction media, and precision instrument lubricants.

Medicinal oils represent the most refined of the bulk petroleum products, especially when the principal use is for the pharmaceutical industry. Thus, mineral oil destined for pharmaceutical purposes must meet stringent specifications to ensure that the oil is inert and that it should not contain any materials that are suspected, to be toxic. Technical mineral oil (as opposed to pharmaceutical mineral oil) has to meet much less stringent specification requirements since the use is generally for transformer oil, for cosmetic preparations (such as hair cream), in the plastics industry, and in textiles processing. Many of the same test methods are applied to all mineral oils although the specifications will differ and oils for different uses will have to meet different standards of purity.

Technical oils find uses in different industries, but in particular the electrical industry where there is frequent use of transformer oil. Transformer oil can be divided into two main groups: (1) oils for transformers and switchgear and (2) oils for power cables; this latter group can be further subdivided into thin oils, for use in oil-filled (*hollow core*) cables, and more viscous oils, sometimes used as such in cable insulation but more commonly as components of cable-impregnation compounds.

Handbook of Petroleum Product Analysis, Second Edition. James G. Speight.
© 2015 John Wiley & Sons, Inc. Published 2015 by John Wiley & Sons, Inc.

TABLE 12.1   General summary of product types and distillation range

Product	Lower carbon limit	Upper carbon limit	Lower boiling point, °C	Upper boiling point, °C	Lower boiling point, °F	Upper boiling point, °F
Refinery gas	C1	C4	−161	−1	−259	31
Liquefied petroleum gas	C3	C4	−42	−1	−44	31
Naphtha	C5	C17	36	302	97	575
Gasoline	C4	C12	−1	216	31	421
Kerosene/diesel fuel	C8	C18	126	258	302	575
Aviation turbine fuel	C8	C16	126	287	302	548
Fuel oil	C12	>C20	216	421	>343	>649
Lubricating oil	>C20		>343		>649	
Wax	C17	>C20	302	>343	575	>649
Asphalt	>C20		>343		>649	
Coke	>C50[a]		>1000[a]		>1832[a]	

[a]Carbon number and boiling point difficult to assess; inserted for illustrative purposes only.

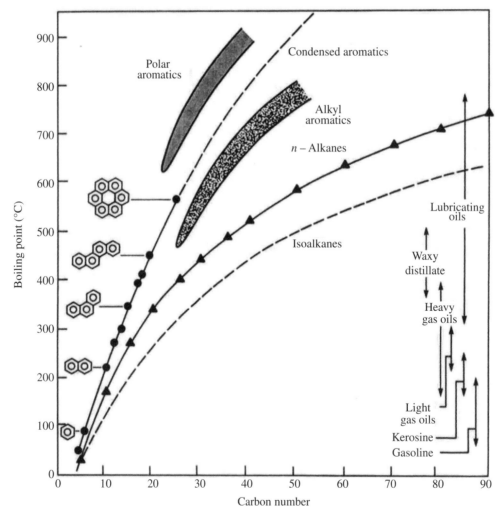

FIGURE 12.1   Boiling point and carbon number for various hydrocarbons and petroleum products.

12.2   PRODUCTION AND PROPERTIES

Mineral oils are produced from selected petroleum fractions that are distilled to provide finished products with the desired viscosity. In addition to viscosity, properties considered during feedstock selection include pour point, cloud point, distillation range, smoke point, specific gravity, as well as other properties.

White oil is manufactured from either naphthenic- or paraffinic-type crude oils (Speight and Ozum, 2002; Hsu and Robinson, 2006; Gary et al., 2007; Speight, 2014). After initial processing in the refinery, they are intensely refined to remove constituents that impart color, odor, taste, and potential toxicological properties. These constituents include aromatics; olefins; and sulfur-, oxygen-, nitrogen-, and metal-containing compounds. Additionally, they are made to comply with the purity requirements of the countries in which they are marketed. They consist almost entirely of saturated hydrocarbons and are manufactured in a variety of viscosity grades, which are determined, in part, by the distillation cut or boiling range initially used in processing. The distillation range will also determine the carbon number and molecular weight range of the finished product. The choice of either naphthenic or paraffinic crude will determine the predominant nature of the hydrocarbon species in the *final* product: cycloparaffin hydrocarbon molecules. The number of molecular species present in a mineral oil is in the thousands, given all the possible arrangements of carbon and hydrogen atoms.

Typically, mineral oil is distilled to specification from the required crude oil fraction followed by pretreatment by solvent extraction, hydrotreating, and dewaxing (Speight and Ozum, 2002; Hsu and Robinson, 2006; Gary et al., 2007; Speight, 2014). The development of hydrotreating processes (Speight, 2000; Speight and Ozum, 2002; Hsu and Robinson, 2006; Ancheyta and Speight, 2007; Gary et al., 2007; Speight, 2014) has opened the way for mineral oil to be produced without acid treatment. However, on occasion, there is still the need to exhaustively treat the selected feedstock with concentrated sulfuric acid (gaseous sulfur trioxide treatment also may be used) to remove the aromatic, unsaturated hydrocarbons and other impurities. Petroleum sulfonic acids, produced as by-products during acid treatment, are removed by extraction and neutralization. The oil can be refined even further to an ultimate degree of purity by adsorption.

The fraction chosen for refining may have been subjected to a preliminary refining with a differential solvent. The exact procedure for the acid treatment varies (Speight, 2000, 2014), but a preliminary acid treatment (chiefly for drying) may be followed by incremental addition of as much as 50% by volume of acid as strong as 20% fuming sulfuric acid. The sludge is promptly removed to limit oxidation–reduction reactions; the time, temperature, and method of application depend on the type of charge stock and the desired degree of refining. The product is neutralized with alkali and washed with ethyl or *iso*-propyl alcohol or acetone to remove the oil-soluble sulfonic *mahogany acids*; water-soluble *green acids* are recovered from the alkali washings. The treated oil is further refined and decolorized by adsorption, either by percolation or by contacting with clay. In the modern refinery, white oils can be produced by hydrotreating and without acid treatment.

12.3   TEST METHODS

Many test methods are available to determine whether the quality for mineral oils exists as prescribed by the consumers. The stringent methods of refining (purification) remove many of the impurities, of which as-forming constituents can be cited as one example making the test for such constituents redundant. If, for some reason, ash-forming constituents are believed to be present in the mineral oil, the test methods for ash (the presence of ash-forming constituents) that are applied to distillate fuel oil and top residual fuel oil can be applied to mineral oil.

But first, while the importance of the careful sampling of any product that is to undergo testing is self-evident, very special precautions have to be taken in the case of insulating oils (ASTM D923) for which special techniques are recommended. Precautions that are stipulated are mainly concerned with the avoidance of contamination that would affect electrical tests.

The main requirement of transformer *oil* is that it should act as a heat transfer medium to ensure that the operating temperature of a transformer does not exceed acceptable limits. However, the deterioration of transformer oils in service is closely connected with oxidation by air, which brings on deposition of sludge and the development of acids, resulting in overheating and corrosion, respectively (ASTM D4310). The sludge formed is usually attributed to the direct oxidation of the hydrocarbon constituents to oil-insoluble products.

Mineral oils to be used as *insecticides* require measurement of molecular weight (ASTM D2502, ASTM D2503, ASTM D2878) and composition since structure of the constituents appears to be a factor in determining the insecticide power of these oils. Olefins and aromatics are both highly toxic to insects, but they also have a detrimental effect on the plant; thus spray oils generally receive some degree of refining, especially those of the summer oil type that come into contact with foliage.

12.3.1   Acidity or Alkalinity

The *acid number is the* quantity of base, expressed in milligrams of potassium hydroxide per gram of sample that is required to titrate a sample in this solvent to a green/greenish-brown end point, using *p*-naptholbenzein indicator solution (ASTM D974, IP 139). However, many higher-molecular-products (especially dark-colored higher molecular weight products) that cannot be analyzed for acidity due to masking of the color-indicator end point can be analyzed by an alternate test method (ASTM D664). The quality of the mineral oil products renders them suitable for the determination of the acid number.

The principle behind the acid number test is a holdover when from earlier refining processes when it was believed, with some justification, that detectable amounts of chemicals that were used in refining could remain in the finished product (ASTM D974, ASTM D1093, IP 139). However, oxidation (ASTM D943, ASTM D3339, ASTM D5770, IP 431) of various oils during use or during storage can induce the formation of acidic species within the oil. Thus, the occurrence of acidic entities within the oil is still a concern even though the reason may be different.

To determine the acid number (ASTM D974), the sample is dissolved in a mixture of toluene and isopropyl alcohol containing a small amount of water, and the resulting single-phase solution is titrated at room temperature with standard alcoholic base or alcoholic acid solution, respectively, to the end point indicated by the color change of the added *p*-naptholbenzein solution (orange in acid and greenish-brown in base). To determine the strong acid number, a separate portion of the sample is extracted with hot water, and the aqueous extract is titrated with potassium hydroxide solution, using methyl orange as an indicator.

The total absence of organic acidity is never feasible, but a very low limit has to be set if corrosion of copper (ASTM D130, ASTM D849, ASTM D4048, ASTM D4636, IP 154) and other components is to be avoided; moreover, more than a trace of some organic acids can adversely affect the response of the oil to amine inhibitors. The presence of such constituents is determined through test methods for the base number. The relative amounts of these materials can be determined by titrating with acids.

Thus, in a manner akin to the *acid number*, the *base number* (often referred to as the *neutralization number*) is a measure of the basic constituents in the oil under the conditions of the test. The base number is used as a guide in the quality control of oil formulation and is also used as a measure of oil degradation in service.

There are four different test methods for the determination of base numbers (ASTM D664, ASTM D974, ASTM D2896, ASTM D4739). However, the different base number methods may give different results for the same sample.

If the mineral oil contains additives that react with alkali to form metal soaps, the saponification number (ASTM D94, IP 136) expresses the amount of base that will react with 1 g of the sample when heated in a specific manner. Since compounds of sulfur, phosphorus, halogens, and certain other elements, which are sometimes added to

petroleum products, also consume alkali and acids, the results obtained indicate the effect of these extraneous materials in addition to the saponifiable material present. In the test method, a known weight of the sample is dissolved in methyl ethyl ketone or a mixture of suitable solvents. It is heated with a known amount of standard alcoholic potassium hydroxide between 30 and 90 min at 80°C (176°F). At the end the reaction, the excess alkali is titrated with standard hydrochloric acid and the saponification number calculated.

12.3.2 Aniline Point

The aniline point or mixed aniline point (ASTM D611, IP 2) gives an indication of the hydrocarbon group composition of the oil. The lower the aniline point the more aromatic the oil, and for any particular compound type, the aniline point rises with molecular weight and with viscosity.

In the test, equal volumes of aniline and oil are mixed and heated until a miscible mixture is formed. On cooling at a prescribed rate, the temperature at which the mixture becomes cloudy is recorded and identified as the aniline point. While not usually applicable to pharmaceutical grade mineral oils, the test is employed in technical oil specifications as a measure of degree of refinement and type of base oil stock. For any particular oil fraction, a higher degree of refinement is reflected by an increase in aniline point. Aniline point also increases with the average molecular weight of the oil as well as with increasing proportions of paraffinic hydrocarbons to naphthenic hydrocarbons. Aniline point specifications also can be used to advantage for light technical grade mineral oils used in agricultural sprays where the presence of aromatic hydrocarbons might cause floral damage.

For electrical oils, the aniline point is often an indicator of the aromatic content because too high a value could give rise to oxidation instability and too low a value to inadequate gassing characteristics under electric stress. For these reasons, an aniline point, usually in the form of a range of permitted values, is sometimes included in specifications. An increase in the aniline point after extraction with sulfuric acid is perhaps a better indication of aromatic content than the aniline point of the untreated sample.

The test for aniline point (ASTM D611, IP 2) is not usually applied to medicinal oils, but some users of technical grades specify a minimum value. For oils of similar purity and viscosity, a high aniline point denotes a more paraffinic, and hence less dense, oil than a lower aniline point; the desire for a high aniline point runs counter to the view that the higher the density, the more preferable the oil if it is desired to make emulsions (e.g., hair cream) therefrom. Generally, the higher the aniline point, the greater the stability of the oil, but, on the other hand, a high-aniline-point sample may only indicate more extensive refining if comparisons are made on fractions of the same boiling range from the same crude oil. In addition, variations in aniline point may reflect changes in depth of refining as well as differences in hydrocarbon group composition for oils of same viscosity but of different origins.

12.3.3   Asphaltene Content

The asphaltene fraction (ASTM D893, ASTM D2006, ASTM D2007, ASTM D4124, ASTM D6560, IP 143) is the highest-molecular-weight and most complex fraction in petroleum. Insofar as the asphaltene content gives an indication of the amount of coke that can be expected during exposure to thermal conditions (Speight, 2001, 2014; Speight and Ozum, 2002), there is little need for the application of the test to mineral oil. Thus, determination of the asphaltene content of mineral oil may be considered superfluous, but there may be the occasion when there is the need to determine the amount of insoluble constituents precipitated by the addition of a low-boiling hydrocarbon liquid to mineral oil.

In any of the methods for the determination of the asphaltene content, a sample is mixed with a large excess (usually >30 volumes of hydrocarbon per volume of sample) low-boiling hydrocarbon such as n-pentane or n-heptane (Speight et al., 1984). For an extremely viscous sample, a solvent such as toluene may be used prior to the addition of the low-boiling hydrocarbon, but an additional amount of the hydrocarbon (usually >30 volumes hydrocarbon per volume of solvent) must be added to compensate for the presence of the solvent. After a specified time, the insoluble material (the asphaltene fraction) is separated (by filtration) and dried. The yield is reported as a percentage (% w/w) of the original sample.

Another method is designed to remove pentane-insoluble constituents by membrane filtration (ASTM D4055). In the method, a sample of oil is mixed with pentane in a volumetric flask, and the oil solution is filtered through a 0.8 µm membrane filter. The flask, funnel, and the filter are washed with pentane to completely transfer any particulates onto the filter after which the filter (with particulates) is dried and weighed to give the pentane-insoluble constituents as a percentage by weight of the sample.

The *precipitation number* is often equated to the asphaltene content, but there are several issues that remain obvious in its rejection for this purpose. For example, the method to determine the precipitation number (ASTM D91) advocates the use of naphtha for use with black oil or lubricating oil, and the amount of insoluble material (as a % v/v of the sample) is the precipitating number. In the test, 10 ml of sample is mixed with 90 ml of ASTM precipitation naphtha (that may or may not have a constant chemical composition) in a graduated centrifuge cone and centrifuged for 10 min at 600–700 rpm. The volume of material on the bottom of the centrifuge cone is noted until repeat centrifugation gives a value within 0.1 ml (the precipitation number). Obviously, this can be substantially different from the asphaltene content.

12.3.4   Carbonizable Substances

Medicinal oil requires a test in which the reaction of the oil to hot strong sulfuric acid is used to determine the presence of carbonizable substance in the oil. However, the test for carbonizable substances (ASTM D565) should not be confused with the test methods for determining *carbon residue* (ASTM D189, ASTM D524, ASTM D4530, IP 13, IP 14, IP 398).

In the test method (ASTM D565), the mineral oil is treated with concentrated sulfuric acid under prescribed conditions, and the resulting color is compared with a reference standard to determine whether it passes or fails the test. When the oil layer shows no change in color and when the acid layer is not darker than the reference standard colorimetric solution, the oil is reported as passing the test. A bluish haze or a slight pink or yellow color in the oil layer should not be interpreted as a change in color. This is a pass–fail, and the more fully refined the oil, the lighter the color of the acid layer.

However, with the introduction of ultraviolet (UV) absorption procedures (ASTM D2008, ASTM D2269), the test finds lesser use but still provides a useful method to determine possible contamination of mineral oil with impurities transparent to both visible and UV light and hence not detectable by color or by UV absorption measurements.

For technical mineral oils, the test for unsulfonatable residue (ASTM D483) may be applied, but the test is of lesser significance for mineral oil that is refined to medicinal standard. The other quality criteria are much more stringent, and the test method for unsulfonatable residue does not have the required sensitivity.

12.3.5   Carbon Residue

Mineral oils are usually considered to have a high propensity for carbon formation and deposition under thermal conditions. Nevertheless, the test methods that are applied to determine the carbon-forming propensity of fuel oil (and other petroleum products) are also available for application to mineral oils. The test methods for the carbon residue should not be confused with test method for carbonizable substances (ASTM D565). The former test methods are thermal in nature, while the latter test method involves the use of sulfuric acid in a search for specific chemical entities within the oil.

Thus, the tests for Conradson carbon residue (ASTM D189, IP 13), the Ramsbottom carbon residue (ASTM D524, IP 14), the microcarbon carbon residue (ASTM D4530, IP 398), and asphaltene content (ASTM D893, ASTM D2006,

ASTM D2007, ASTM D4124, ASTM D6560, IP 143) are often included in inspection data for fuel oil.

In the Conradson carbon residue test (ASTM D189, IP 13), a weighed quantity of sample is placed in a crucible and subjected to destructive distillation for a fixed period of severe heating. At the end of the specified heating period, the test crucible containing the carbonaceous residue is cooled in a desiccator and weighed, and the residue is reported as a percentage (% w/w) of the original sample (Conradson carbon residue). In the Ramsbottom carbon residue test (ASTM D524, IP 14), the sample is weighed into a glass bulb that has a capillary opening and is placed into a furnace (at 550°C, 1022°F). The volatile matter is distilled from the bulb, and the nonvolatile matter that remains in the bulb cracks to form thermal coke. After a specified heating period, the bulb is removed from the bath, cooled in a desiccator, and weighed to report the residue (Ramsbottom carbon residue) as a percentage (% w/w) of the original sample. In the microcarbon residue test (ASTM D4530, IP 398), a weighed quantity of the sample placed in a glass vial is heated to 500°C (932°F) under an inert (nitrogen) atmosphere in a controlled manner for a specific time, and the carbonaceous residue [carbon residue (micro)] is reported as a percentage (% w/w) of the original sample.

The data produced by the microcarbon test (ASTM D4530, IP 398) are equivalent to those by Conradson carbon method (ASTM D189, IP 13). However, this microcarbon test method offers better control of test conditions and requires a smaller sample. Up to 12 samples can be run simultaneously. This test method is applicable to petroleum and to petroleum products that partially decompose on distillation at atmospheric pressure and is applicable to a variety of samples that generate a range of yields (0.01 to 30% w/w) of thermal coke.

The data give an indication of the amount of constituents that may be undesirable in the oil thereby requiring that the oil, depending upon its designated use, be subjected to further refining.

Finally, a method that is used to determine pitch volatility (ASTM D4893) might also be used, on occasion, to determine the nonvolatility of mineral oil. In the method, an aluminum dish containing about 15 g of accurately weighed sample is introduced into the cavity of a metal block heated and maintained at 350°C (662°F). After 30 min, during which the volatiles are swept away from the surface of the sample by preheated nitrogen, the residual sample is taken out and allowed to cool down in the desiccator. The nonvolatility is determined by the sample weight remaining and reported as percentage w/w residue.

12.3.6 Cloud Point

The *cloud point* is the temperature at which a cloud of wax crystal first appears in a liquid when it is cooled under conditions prescribed in this test method. This test method covers only petroleum oils that are transparent in layers 38 mm (11/2 in.) in thickness, and with a cloud point below 49°C (120°F). The cloud point is an indicator of the lowest temperature of the utility of an oil for certain applications, and the cloud point is usually higher than the pour point (ASTM D97, ASTM D5853, ASTM D5949, ASTM D5950, ASTM D5985, IP 15).

The cloud point (ASTM D2500, ASTM D5771, ASTM D5772, ASTM D5773, IP 219) is the temperature at which wax appears in the oil. This information is significant for oil to be used at low temperatures where precipitation of wax will affect the performance of the oil.

In one of the tests (ASTM D2500, IP 219), the oil is maintained at 0°C (32°F) for 4 h, when a 0.5 mm black line must be *easily seen* through an oil layer 25 mm thick. The setting of a value as low as this is dictated largely by considerations of appearance of the oil; a much higher temperature could well be tolerated if only practical considerations applied; moreover, the *n*-paraffins that are responsible for cloudiness (hence the need to observe that the opalescence is not due to moisture) are among the more inert of the hydrocarbon groups of which these mineral oils are composed.

The later test methods (ASTM D5771, ASTM D5772, ASTM D5773) are alternative procedures that use automatic apparatus. However, when the specifications quote that the original method (ASTM D2500, IP 219) should be used, these later methods should not be applied without obtaining comparative data first. However, all three methods have a higher degree of precision than the original method.

Neither the cloud point nor the pour point should be confused or interchanged with the *freezing point* (ASTM DD 2386, ASTM D5972, IP 16, IP 435, IP 528). The freezing point presents an estimate of minimum handling temperature and minimum line or storage temperature. It is not a test for an indication of purity and has limited value for mineral oil.

12.3.7 Color and Taste

By definition, pharmaceutical grade mineral oil must be colorless and transparent. Although the eye can determine color, it is usual to determine color by instrument. The colorless character of these oils is important in some cases, as it may indicate the chemically inert nature of the hydrocarbon constituents. Textile lubricants should be colorless to prevent the staining of light-colored threads and fabrics. Insecticide oils should be free of reactive (easily oxidized) constituents so as not to injure plant tissues when applied as sprays (ASTM D483).

The main use of mineral oil in medicine has been as a laxative (an alimentary tract lubricant) and in various pharmaceutical preparations such as ointments, and while such use is continuing, this quality of oil is finding increasing utilization as a lubricant for food processing machinery and in plastics manufacture. Laxative oils should be free of odor, taste, and also be free of hydrocarbons that may react during storage and produce unwanted by-products. These properties are attained by the removal of oxygen-, nitrogen-, and sulfur-containing compounds, as well as reactive hydrocarbons by, say, sulfuric acid. The maximum permissible quantity of such oil traces is prescribed but also the purity of the oil (ASTM D2269) requires that the absorbance be measured on a dimethyl sulfoxide extract of the oil.

Determination of the color of mineral oil is used mainly for manufacturing control purposes and is an important quality characteristic. In some cases, the color may serve as an indication of the degree of refinement of the material. However, color is not always a reliable guide to product quality and should not be used indiscriminately in product specifications (ASTM D156, ASTM D1209, ASTM D1500, ASTM D1544, ASTM D6045, IP 17).

In one test (ASTM D156) for the determination of color, the height of a column of the oil is decreased by levels corresponding to color numbers until the color of the sample is lighter than that of the standard color. The color number immediately above this level is recorded as the Saybolt color of the oil, and a color number of +25 corresponds to water mineral, while the minimum color intensity reading on this scale is expressed by +30, a value normally attained by mineral oils. In another test (IP 17), in which the measurement is performed using an 18 in. cell against color slides on a scale, a color of 1.0 or under is considered water–mineral, and medicinal oils will normally be 0.5 or less. Conversion scales for different color tests are available (ASTM D1500, IP 17).

Although sometimes to be found in insulating oil specifications, the color characteristic is of no technical significance. Pale oils are, as a general rule, more severely refined than dark oils of the same viscosity; color (ASTM D1500, IP 17) is not a guide to stability. Deterioration of color after submission of the oil to an aging test is sometimes limited, but here again, the extent of oil deterioration can be much better measured by some other property such as acidity development or change in electrical conductivity (ASTM D2624, ASTM D4308, IP 274). About the only point that can be made in favor of color measurement on new oil is that it can give an immediate guide to a change in supply continuity.

12.3.8 Composition

The potential for the presence in mineral oil of trace amounts of carcinogenic polynuclear aromatic hydrocarbons cannot be overestimated. Thus, compositional tests focus on the production of data relating to the presence of aromatic and polynuclear aromatic constituents in mineral oil.

As a result, test method (ASTM D2269) that can be applied to the UV absorbance of mineral oil is available. The investigation can be performed either directly on the oil itself (or on oil diluted with inert solvent), or else on a solvent extract of the oil, the solvent chosen (e.g., dimethyl sulfoxide) being one that will extract concentrate therein the polynuclear aromatics.

Although informative, there has also extensive effort to correlate composition with physical properties. Among the most definitive of these efforts was the refractive index–density–molecular weight (n–d–M) method (ASTM D3238). Carbon-type composition gives the breakdown of total carbon atoms between various structures. Following from this, another method (ASTM D2140), derived from n–d–M analysis (ASTM D3238) involves the measurement of refractive index, density, and viscosity, for all of which standard procedures are available. From the data, it is possible to determine the distribution of carbon types (paraffin chains, naphthenic rings, and aromatic rings) in the oil. However, it is important to recognize that the percent aromatic carbon determined by this method does not correspond to *aromatics* as determined by other procedures (e.g., adsorption chromatography). For example, octadecylbenzene ($C_6H_5C_{18}H_{37}$) has only 25% of the carbon as aromatic carbon (C_A), but chromatographic separation would show this compound to be separated as an aromatic compound.

Aniline point (ASTM D611, IP 2) gives an indication of the hydrocarbon group composition of oil. The aniline point increases with molecular weight, and for a given viscosity, the higher the aniline point, the more paraffinic the oil. For electrical oils, it is sometimes desired to control the aromatic content because too high a value could give rise to oxidation instability and too low a value to inadequate gassing characteristics under electric stress. For these reasons, an aniline point, usually in the form of a range of permitted values, is sometimes included in specifications. And a rise in the aniline point after extraction with sulfuric acid is perhaps a better indication of aromatic content than aniline point on the untreated sample.

A technique for applying infrared measurements to insulating oil is available (ASTM D2144), and considerable information about mineral oil composition can be gained from infrared spectroscopy. Oxygenated bodies formed when oil deteriorates can be recognized, and hence this procedure can be employed for surveillance of oils in service. An infrared spectrum can also give information as to the aromaticity of the oil and can detect antioxidants such as 2,6-di-tertiary butyl *p*-cresol.

While the total sulfur content (ASTM D129, IP 61) of medicinal oil is not limited as such, it is in effect restricted (usually to well below 100 ppm) by the necessity of severe refining to meet the other clauses of the specification. However, the total sulfur content of mineral oils normally is well below 100 ppm because of the severe refining to which the oil has been subjected (Speight, 2000). Therefore, sulfur is not a limiting factor in the specifications of mineral oil. But test methods are available to determine the presence of sulfur compounds in mineral oil although the preferred procedure is often the doctor test (ASTM D4952, IP 30). For medicinal oils, the procedure is that two volumes of oil and one of absolute ethyl alcohol have added thereto two drops of a saturated solution of lead monoxide in 20% aqueous sodium hydroxide. The mixture is heated to 70°C (158°F) for 10 min and should remain colorless.

Other test methods for the determination of corrosive sulfur (specifically hydrogen sulfide) are also available (ASTM D6021, ASTM D5705) and have been developed for residual fuel oil but can be applied to medicinal oil. Another test method (ASTM D3227) is used for the determination of mercaptan (R-SH) sulfur in petroleum products up to an including distillate fuel. However, there is no reason that the test could not be applied to petroleum products, using any necessary modifications for solubility of the petroleum product in the alcoholic solution used for the titration.

In view of the wide use of copper in electrical equipment, it is essential to ensure that the oil does not corrode this metal. Noncorrosive or corrosive sulfur can be verified by any one of several test methods (ASTM D130, ASTM D849, ASTM D4048, ASTM D4636, IP 154) in which the oil is heated in the presence of a metallic copper strip for a specified time at a specified temperature after which the strip must be discolored. In addition, corrosiveness toward silver is becoming increasingly important, and a test method (IP 227) has been developed that is closely similar to the copper corrosion test.

Molecular-type analysis separates oil into different molecular species—a molecular-type analysis is the so-called *clay-gel analysis*. In this method, group separation is achieved by adsorption in a percolation column with selected grades of clay and silica gel as the adsorption media (ASTM D2007).

12.3.9   Density (Specific Gravity)

The use of density (specific gravity) data (ASTM D1298, IP 160) for mineral oil is variable, although for some grades of mineral oil, a high density can be indicative of the ultimate use of the oil.

For example, density or specific gravity data (ASTM D1298, ASTM D1480, ASTM D1481, ASTM D4052, ASTM D5002, IP 160) are frequently included as a specification requirement for transformer oil when transformers operate in cold climates. Water, however, undesirable in electrical equipment, does collect therein, and if the oil has a high density, any water/ice present would float on the oil, instead of remaining at the bottom of the oil container thereby reducing the effectiveness of the oil.

For heavier insulating oil, the purpose of limiting density range is as a check on oil composition. In addition, a minimum density may offer some indication of solvent power as well as guarding against excessive paraffin content. In this respect, the inclusion of density in a mineral oil specification may duplicate the aniline point (ASTM D611, IP 2) requirement.

The API gravity (ASTM D287) is a special function of specific gravity that was arbitrarily established to permit the use of a wider industrial range of numbers and is derived from the specific gravity:

$$\text{API gravity, deg} = \left(141.5 / \text{sp gr } 60 / 60°F\right) - 131.5$$

Density, specific gravity, and API gravity values permit conversion of volumes at the measured temperature to volumes at the standard petroleum temperatures of 15°C (60°F). Calculation to weight is possible where compositions are formulated on a weight basis. At a given viscosity, the density, specific gravity, and API gravity provide a means for determining whether a mineral oil is derived from a paraffinic or a naphthenic feedstock.

12.3.10   Electrical Properties

Electrical properties of one form or another are included in virtually all specifications for insulating oils.

The *electric strength* or *dielectric breakdown* test method (ASTM D877) indicates the absence, or presence, of free or suspended water and other contaminant matter that will conduct electricity. A high electric strength gives no indication of the *purity* of the oil in the sense of degree of refinement or the absence of most types of oil-soluble contaminants. This test method is of some assistance, when applied to an otherwise satisfactory oil, to indicate that the oil is free of contaminants of the type indicated earlier; in practice, this assures that the oil is dry.

The test for stability (gas evolution) under electric stress must not be confused with tests for the gas content of insulating oil (ASTM D831). These tests are largely factory control tests to ensure that oils intended for filling into equipment have been adequately degassed (since dissolved gas, like gas evolved under stress, could cause void formation).

12.3.11   Flash Point and Fire Point

The flash point (closed tester: ASTM D93, IP 34) (open tester: ASTM D92, IP 36) is the lowest temperature at atmospheric pressure (760 mm Hg, 101.3 kPa) at which application of a test flame will cause the vapor of a sample to ignite under specified test conditions. The sample is deemed to have reached the flash point when a large flame appears and instantaneously propagates itself over the surface of the sample. The flash point data are used in shipping and safety regulations to define *flammable* and *combustible* materials. Flash point data can also indicate the possible presence of highly volatile and flammable constituents in a relatively nonvolatile or nonflammable material. The fire point (ASTM D92, IP 36) is the temperature at which the oil ignites and burns for 5 s.

Of the available test methods (ASTM, 2013), the most common method of determining the flash point confines the vapor (closed cup) until the instant the flame is applied (ASTM D56, ASTM D93, ASTM D3828, ASTM D6450, IP 34, IP 523). An alternate method that does not confine the vapor (open cup method (ASTM D92, ASTM D1310, IP 36) gives slightly higher values of the flash point.

The flash point is sometimes to be found in a mineral oil specification, but the value is not usually of much significance and is merely laid down as some assurance against undue fire risk. Typical values will be from 150°C (302°F) upward, depending on viscosity. The closed test (IP 34) is commonly used, and a minimum value of 140°C (284°F) is required for transformer oil in order to limit fire risk. The stipulation of a suitable flash point also automatically limits the volatility of the oil, and for this reason, the loss on beating (IP 46) can be important to restrict excessive volatility. This can be achieved by setting a suitable minimum flash point on the order of 125°C (257°F).

Fewer volatility problems arise with the heavier cable oils or oil-based cable impregnating compounds, although here also a minimum flash point of approximately, say, around 230°C (446°F) is frequently specified as a precaution against the presence of lighter (and hence undesirably volatile) components, which could be troublesome at impregnating conditions of over 100°C (212°F) and high vacuum.

Rubber compounders often use flash point (ASTM D92, ASTM D 93, IP 36, IP 34) as a measure of oil volatility. Volatility is important because rubber products are exposed to elevated temperatures during mixing operations and, oftentimes, in service. Although flash point has a certain utility for this purpose, it gives no indication of the amount of low-boiling material present. Therefore, a distillation curve should be used when volatility is a critical factor.

12.3.12   Interfacial Tension

The measurement of interfacial tension (ASTM D971) between oil and water is a sensitive method for determining traces of polar contaminants, including products of oil oxidation, and minimum values will sometimes be found in insulating oil specifications. The test is frequently employed to assess transformer oil deterioration and thus to maintain a check on the quality of oil in electrical equipment.

12.3.13   Iodine Value

The iodine number (iodine value) test method (ASTM D5768, IP 607) is included in some specifications and is a measure of fatty acid components or unsaturation either as olefins or as aromatics. The former components are very unlikely to be present in any oil that has undergone sufficient refining to make it colorless, while the aromatic content can be controlled by setting a suitable value for the UV absorption.

The bromine number is the grams of bromine that will react with 100 g of the sample under the test conditions and is better suited for the determination of unsaturated constituents in oil (ASTM D1159, ASTM D2710, IP 130). The magnitude of bromine number is used as a measure of aliphatic unsaturation in petroleum products. In the test, a known weight of the sample dissolved in a specified solvent maintained at 0–5°C (32–41°F) is titrated with standard bromide–bromate solution. Determination of the end point is method dependent.

12.3.14   Oxidation Stability

Very many oxidation tests for transformer oils have at one time or another been used. Most of these tests are of a similar pattern; the oil is heated and subjected to oxidation by either air or oxygen and usually in the presence of a metallic catalyst, almost invariably copper, which is the main active metal in transformer construction. Temperatures and duration have varied within wide limits, from 95 to 150°C (203–302°F) and from 14 to 672 h.

In one test method (ASTM D943), the oil is maintained at 100°C (212°F) for up to 1 week during which time oxygen is passed through the oil, in which metallic copper is immersed. At the end of the oxidation period, the amount of solid deterioration products (*sludge*) is measured, after such sludge has been precipitated by dilution of the aged oil with *n*-heptane, and the soluble (*acid*) decomposition products are also measured by determination of the neutralization value of the aged oil. Another test method, designed for inhibited oils, operates at the same temperature, but lower pressure and employs a copper catalyst (ASTM D2112).

One criticism of such tests is that sludge precipitation is carried out with *precipitation naphtha*. Although the characteristics of this solvent are not always defined precisely, it is best recognized as a complex hydrocarbon mixture whose composition can vary widely even within the range of prescribed properties. This could give rise to varying precipitation capabilities in various batches of such naphtha, a disadvantage that cannot occur when a pure chemical such as *n*-heptane is used as precipitant (ASTM D4124, IP 143). Another criticism is that the pressurized oxygen might be capable of changing the chemistry of the oxidation reaction and that the oxidation conditions do not represent the service conditions.

While most transformer oil oxidation tests are of the air or oxygen blowing type, many methods of assessing cable oil deterioration depend on static oxidation, that is, the oil and air are in contact but such contact is not stimulated. A catalyst is sometimes, but not invariably, used. Thus, in the open beaker oxidative aging test (ASTM D1934), the oil is exposed to moving air at 115°C (239°F) for 96 h, but a metallic catalyst is optional. The metal is not specified although copper is the one most frequently used in oxidation tests on insulating oils. Characteristics to be determined after aging are again left open in the test description, although acidity and electrical conductivity (both ac and dc) are the commonest. Color is sometimes included, but the limited amount of information that this conveys has already been commented upon.

Transformer oil should not contain levels of unsaturated components (aromatics and/or olefins) that affect the oxidation resistance of the oil (ASTM D2300). Acidity (ASTM D974, IP 1, IP 139) after oxidation is another criterion by which the extent of oil deterioration is judged.

12.3.15   Pour Point

The pour point provides a means of determining the type of petroleum feedstock from which the mineral oil was manufactured or its previous processing history. It also reflects the presence of wax or paraffinic hydrocarbons. In any application where the mineral oil is used at low temperatures or the oil is subjected during handling or storage to low temperatures, the pour point is important and, perhaps, even critical.

In the pour point test (ASTM D97, IP 15), the oil is heated to a specified temperature that is dependent on the anticipated pour-point range, cooled at a specified rate, and examined at 3°C (5°F) intervals for flow. The lowest temperature at which no movement of the oil is detected is recorded. The 3°C (5°F) temperature value immediately preceding the recorded temperature is defined as the pour point.

For the pour point of transformer oils, a value below the lowest ambient temperature to be expected must be set according to the specifications. Most oils have a pour point that is typically on the order of −40°C (−40°F).

12.3.16   Refractive Index

The refractive index (ASTM D1218, ASTM D1747) is the ratio of the velocity of light in air to its velocity in the substance under examination. It is used, together with density and viscosity measurements, in calculating the paraffin–naphthene ratio in mineral oils. Because refractive index is a measure of aromaticity and unsaturation on a given stock, manufacturers also use it as a means of process control.

The refractive index is needed also to calculate the refractivity intercept in the determination of carbon-type composition (Speight, 2014).

12.3.17   Smoke Point

The smoke point is of particular interest to industries, such as the baking industry, whose processes expose or use mineral oil at extremely high temperatures.

The smoke-point test (ASTM D1322, IP 57), originally developed for kerosene, is conducted using an enclosed wick-fed lamp suitably vented and illuminated to permit detection of vapors. The oil is carefully heated under specified conditions until the first consistent appearance of vapors is detected. The temperature of the oil at that time is recorded as the smoke point. If necessary, this test can be adapted for use with mineral oil. The character of the flame is an indicator of the aromatic content.

12.3.18   Ultraviolet Absorption

Considerable concern has been generated in recent years over the possible presence of carcinogenic polynuclear aromatic hydrocarbons in mineral oils.

Detection of these chemicals by UV absorption spectroscopy (ASTM D2269) allows measurement of the absorbance over the wavelength range of 260–350 nm in a 10 mm cell of a dimethyl sulfonide extract of the oil. The polynuclear aromatic hydrocarbons present in the mineral oils are concentrated. In fact, the UV absorption level corresponds, approximately, to a maximum polynuclear aromatic content of about 5 ppm.

For oils of a similar type, UV absorptivity is a good indicator of the resistance of the oil to discoloration under exposure to artificial or natural light. Oils with low absorptivity at 260 nm have been found to impart good color stability to light-colored rubber compounds (ASTM D2008).

12.3.19   Viscosity

Viscosity (ASTM D445, IP 71) is one of the most important properties to be considered in the evaluation of a mineral oil.

Requirements for viscosity vary widely according to the user for which the oil is intended and may be as low as 4 cSt or as high as 70 cSt. Mineral oil for internal use generally should have high viscosity in order to minimize possibilities of leakage.

Viscosity data (ASTM D445, IP 71) are used to ensure that, in the case of oils for internal use as laxatives, unduly fluid material, which could increase the risk of leakage through the anal sphincter muscle, is not employed. The minimum viscosity is usually on the order of 75 cSt at 37.8°C (100°F). The temperature of viscosity measurement is a normal one employed for this purpose and happens, in the case of medicinal oils, to be that of the human body. Thus, the viscosity of these oils is measured at their working temperature.

Heavy cable oils are very much more viscous materials, and typical viscosity values can be as high as about 200 cSt at 60°C (140°F). Such oils are probably the only petroleum products for which a maximum viscosity index (ASTM D2270, IP 226) is sometimes specified, in order to ensure that while the oil will be as fluid as possible at impregnating temperatures (100°C, 212°F, or more), it will be sufficiently viscous to prevent draining at ambient temperatures in cases where cables are laid on a steep gradient.

The viscosity–gravity constant (ASTM D2501) and the refractivity intercept have been used for characterizing oils of widely different viscosity (ASTM D 2140).

12.3.20 Volatility

The distillation range for very mineral light oils provides information on volatility, evaporation rates, and residue remaining after evaporation (ASTM D86, ASTM D1160, IP 123). Such data are important for agricultural and household sprays, agricultural product processing, and printing inks. The baking and plastic industries often include initial boiling point temperature or the minimum allowable temperature at which the first several percentage of the oil comes overhead during distillation or both as part of their specifications for mineral oils.

The test for evaporation (ASTM D972) gives a measure of oil volatility under controlled conditions and is used frequently for specification purposes. However, because volatility of oil from a rubber compound may be influenced by its compatibility with the rubber, a volatility test of the compound often is made under laboratory test conditions pertinent to the intended service.

12.3.21 Water

Water is obviously undesirable in electrical equipment, and the water content of transformer and other insulating oils is frequently limited to a low maximum value. Traces of water that would not influence the general run of petroleum product tests could have a very significant effect on properties such as electric strength.

Quantitative determination may be made by one of the many modifications of the Karl Fischer method (ASTM D1533). With careful application of this technique, using electrical methods of determining the end point and operating in a sealed system, water can be determined down to about 2 ppm or even less. Such sensitivity is however rather greater than is normally required, and a water content of 35 ppm and preferably less than say 25 ppm is usually considered satisfactory. The solubility level of water in oil at room temperature (say, 20°C: 68°F) will vary with the type of oil but is around 40 ppm. Amounts of water down to about 5 ppm in addition to being measurable chemically can also be determined by removing the water from the oil by a combination of heat and vacuum and absorbing the freed moisture in a suitable weighed phosphorus pentoxide (P_2O_5) trap.

12.3.22 Wax Appearance Point

The *wax appearance point* is the temperature at which wax begins to precipitate (hence it is also called the *wax precipitation point*) from an oil under specified cooling conditions. Although more applicable to distillate fuel oil (Chapter 10) and related to the cloud point (ASTM D2500, ASTM D5771,

ASTM D5772, ASTM D5773, IP 219) and pour point (ASTM D97, IP 15), the wax appearance point can also have implications for mineral oil use. In the test method, a sample of the oil is cooled under prescribed conditions with stirring. The temperature at which wax first appears is the wax appearance point.

REFERENCES

Ancheyta, J. and Speight, J.G. 2007. Hydroprocessing of Heavy Oils and Residua. CRC Press, Taylor & Francis Group, Boca Raton, FL.

API. 1992. API Mineral Oil Review. Health and Environmental Science Department, American Petroleum Institute, Washington, DC.

ASTM D56. Standard Test Method for Flash Point by Tag Closed Cup Tester. Annual Book of Standards. ASTM International, West Conshohocken, PA.

ASTM D86. Standard Test Method for Distillation of Petroleum Products at Atmospheric Pressure. Annual Book of Standards. ASTM International, West Conshohocken, PA.

ASTM D91. Standard Test Method for Precipitation Number of Lubricating Oils. Annual Book of Standards. ASTM International, West Conshohocken, PA.

ASTM D92. Standard Test Method for Flash and Fire Points by Cleveland Open Cup Tester. Annual Book of Standards. ASTM International, West Conshohocken, PA.

ASTM D93. Standard Test Methods for Flash Point by Pensky-Martens Closed Cup Tester. Annual Book of Standards. ASTM International, West Conshohocken, PA.

ASTM D94. Standard Test Methods for Saponification Number of Petroleum Products. Annual Book of Standards. ASTM International, West Conshohocken, PA.

ASTM D97. Standard Test Method for Pour Point of Petroleum Products. Annual Book of Standards. ASTM International, West Conshohocken, PA.

ASTM D129. Standard Test Method for Sulfur in Petroleum Products (General High Pressure Decomposition Device Method). Annual Book of Standards. ASTM International, West Conshohocken, PA.

ASTM D130. Standard Test Method for Corrosiveness to Copper from Petroleum Products by Copper Strip Test. Annual Book of Standards. ASTM International, West Conshohocken, PA.

ASTM D156. Standard Test Method for Saybolt Color of Petroleum Products (Saybolt Chromometer Method). Annual Book of Standards. ASTM International, West Conshohocken, PA.

ASTM D189. Standard Test Method for Conradson Carbon Residue of Petroleum Products. Annual Book of Standards. ASTM International, West Conshohocken, PA.

ASTM D287. Standard Test Method for API Gravity of Crude Petroleum and Petroleum Products (Hydrometer Method). Annual Book of Standards. ASTM International, West Conshohocken, PA.

ASTM D445. Standard Test Method for Kinematic Viscosity of Transparent and Opaque Liquids (and Calculation of Dynamic Viscosity). Annual Book of Standards. ASTM International, West Conshohocken, PA.

ASTM D483. Standard Test Method for Unsulfonated Residue of Petroleum Plant Spray Oils. Annual Book of Standards. ASTM International, West Conshohocken, PA.

ASTM D524. Standard Test Method for Ramsbottom Carbon Residue of Petroleum Products. Annual Book of Standards. ASTM International, West Conshohocken, PA.

ASTM D565. Standard Test Method for Carbonizable Substances in White Mineral Oil. Annual Book of Standards. ASTM International, West Conshohocken, PA.

ASTM D611. Standard Test Methods for Aniline Point and Mixed Aniline Point of Petroleum Products and Hydrocarbon Solvents. Annual Book of Standards. ASTM International, West Conshohocken, PA.

ASTM D664. Standard Test Method for Acid Number of Petroleum Products by Potentiometric Titration. Annual Book of Standards. ASTM International, West Conshohocken, PA.

ASTM D831. Standard Test Method for Gas Content of Cable and Capacitor Oils. Annual Book of Standards. ASTM International, West Conshohocken, PA.

ASTM D849. Standard Test Method for Copper Strip Corrosion by Industrial Aromatic Hydrocarbons. Annual Book of Standards. ASTM International, West Conshohocken, PA.

ASTM D877. Standard Test Method for Dielectric Breakdown Voltage of Insulating Liquids Using Disk Electrodes. Annual Book of Standards. ASTM International, West Conshohocken, PA.

ASTM D893. Standard Test Method for Insolubles in Used Lubricating Oils. Annual Book of Standards. ASTM International, West Conshohocken, PA.

ASTM D923. Standard Practices for Sampling Electrical Insulating Liquids. Annual Book of Standards. ASTM International, West Conshohocken, PA.

ASTM D943. Standard Test Method for Oxidation Characteristics of Inhibited Mineral Oils. Annual Book of Standards. ASTM International, West Conshohocken, PA.

ASTM D971. Standard Test Method for Interfacial Tension of Oil against Water by the Ring Methods. Annual Book of Standards. ASTM International, West Conshohocken, PA.

ASTM D972. Standard Test Method for Evaporation Loss of Lubricating Greases and Oils. Annual Book of Standards. ASTM International, West Conshohocken, PA.

ASTM D974. Standard Test Method for Acid and Base Number by Color-Indicator Titration. Annual Book of Standards. ASTM International, West Conshohocken, PA.

ASTM D1093. Standard Test Method for Acidity of Hydrocarbon Liquids and Their Distillation Residues. Annual Book of Standards. ASTM International, West Conshohocken, PA.

ASTM D1159. Standard Test Method for Bromine Numbers of Petroleum Distillates and Commercial Aliphatic Olefins by Electrometric Titration. Annual Book of Standards. ASTM International, West Conshohocken, PA.

ASTM D1160. Standard Test Method for Distillation of Petroleum Products at Reduced Pressure. Annual Book of Standards. ASTM International, West Conshohocken, PA.

ASTM D1209. Standard Test Method for Color of Clear Liquids (Platinum-Cobalt Scale). Annual Book of Standards. ASTM International, West Conshohocken, PA.

ASTM D1218. Standard Test Method for Refractive Index and Refractive Dispersion of Hydrocarbon Liquids. Annual Book of Standards. ASTM International, West Conshohocken, PA.

ASTM D1298. Standard Test Method for Density, Relative Density, or API Gravity of Crude Petroleum and Liquid Petroleum Products by Hydrometer Method. Annual Book of Standards. ASTM International, West Conshohocken, PA.

ASTM D1310. Standard Test Method for Flash Point and Fire Point of Liquids by Tag Open-Cup Apparatus. Annual Book of Standards. ASTM International, West Conshohocken, PA.

ASTM D1322. Standard Test Method for Smoke Point of Kerosene and Aviation Turbine Fuel. Annual Book of Standards. ASTM International, West Conshohocken, PA.

ASTM D1480. Standard Test Method for Density and Relative Density (Specific Gravity) of Viscous Materials by Bingham Pycnometer. Annual Book of Standards. ASTM International, West Conshohocken, PA.

ASTM D1481. Standard Test Method for Density and Relative Density (Specific Gravity) of Viscous Materials by Lipkin Bicapillary Pycnometer. Annual Book of Standards. ASTM International, West Conshohocken, PA.

ASTM D1500. Standard Test Method for ASTM Color of Petroleum Products (ASTM Color Scale). Annual Book of Standards. ASTM International, West Conshohocken, PA.

ASTM D1533. Standard Test Method for Water in Insulating Liquids by Coulometric Karl Fischer Titration. Annual Book of Standards. ASTM International, West Conshohocken, PA.

ASTM D1544. Standard Test Method for Color of Transparent Liquids (Gardner Color Scale). Annual Book of Standards. ASTM International, West Conshohocken, PA.

ASTM D1747. Standard Test Method for Refractive Index of Viscous Materials. Annual Book of Standards. ASTM International, West Conshohocken, PA.

ASTM D1934. Standard Test Method for Oxidative Aging of Electrical Insulating Petroleum Oils by Open-Beaker Method. Annual Book of Standards. ASTM International, West Conshohocken, PA.

ASTM D2006. Method of Test for Characteristic Groups in Rubber Extender and Processing Oils by the Precipitation Method (Withdrawn in 1975 but still in use in many laboratories). ASTM International, West Conshohocken, PA.

ASTM D2007. Standard Test Method for Characteristic Groups in Rubber Extender and Processing Oils and Other Petroleum-Derived Oils by the Clay-Gel Absorption Chromatographic Method. Annual Book of Standards. ASTM International, West Conshohocken, PA.

ASTM D2008. Standard Test Method for Ultraviolet Absorbance and Absorptivity of Petroleum Products. Annual Book of Standards. ASTM International, West Conshohocken, PA.

ASTM D2112. Standard Test Method for Oxidation Stability of Inhibited Mineral Insulating Oil by Pressure Vessel. Annual Book of Standards. ASTM International, West Conshohocken, PA.

ASTM D2140. Standard Practice for Calculating Carbon-Type Composition of Insulating Oils of Petroleum Origin. Annual Book of Standards. ASTM International, West Conshohocken, PA.

ASTM D2144. Standard Practices for Examination of Electrical Insulating Oils by Infrared Absorption. Annual Book of Standards. ASTM International, West Conshohocken, PA.

ASTM D2269. Standard Test Method for Evaluation of White Mineral Oils by Ultraviolet Absorption. Annual Book of Standards. ASTM International, West Conshohocken, PA.

ASTM D2270. Standard Practice for Calculating Viscosity Index from Kinematic Viscosity at 40 and 100°C. Annual Book of Standards. ASTM International, West Conshohocken, PA.

ASTM D2300. Standard Test Method for Gassing of Electrical Insulating Liquids Under Electrical Stress and Ionization (Modified Pirelli Method). Annual Book of Standards. ASTM International, West Conshohocken, PA.

ASTM D2386. Standard Test Method for Freezing Point of Aviation Fuels. Annual Book of Standards. ASTM International, West Conshohocken, PA.

ASTM D2500. Standard Test Method for Cloud Point of Petroleum Products. Annual Book of Standards. ASTM International, West Conshohocken, PA.

ASTM D2501. Standard Test Method for Calculation of Viscosity-Gravity Constant (VGC) of Petroleum Oils. Annual Book of Standards. ASTM International, West Conshohocken, PA.

ASTM D2502. Standard Test Method for Estimation of Mean Relative Molecular Mass of Petroleum Oils from Viscosity Measurements. Annual Book of Standards. ASTM International, West Conshohocken, PA.

ASTM D2503. Standard Test Method for Relative Molecular Mass (Molecular Weight) of Hydrocarbons by Thermoelectric Measurement of Vapor Pressure. Annual Book of Standards. ASTM International, West Conshohocken, PA.

ASTM D2624. Standard Test Methods for Electrical Conductivity of Aviation and Distillate Fuels. Annual Book of Standards. ASTM International, West Conshohocken, PA.

ASTM D2710. Standard Test Method for Bromine Index of Petroleum Hydrocarbons by Electrometric Titration. Annual Book of Standards. ASTM International, West Conshohocken, PA.

ASTM D2878. Standard Test Method for Estimating Apparent Vapor Pressures and Molecular Weights of Lubricating Oils. Annual Book of Standards. ASTM International, West Conshohocken, PA.

ASTM D2896. Standard Test Method for Base Number of Petroleum Products by Potentiometric Perchloric Acid Titration. Annual Book of Standards. ASTM International, West Conshohocken, PA.

ASTM D3227. Standard Test Method for (Thiol Mercaptan) Sulfur in Gasoline, Kerosene, Aviation Turbine, and Distillate Fuels (Potentiometric Method). Annual Book of Standards. ASTM International, West Conshohocken, PA.

ASTM D3238. Standard Test Method for Calculation of Carbon Distribution and Structural Group Analysis of Petroleum Oils by the n-d-M Method. Annual Book of Standards. ASTM International, West Conshohocken, PA.

ASTM D3339. Standard Test Method for Acid Number of Petroleum Products by Semi-Micro Color Indicator Titration. Annual Book of Standards. ASTM International, West Conshohocken, PA.

ASTM D3828. Standard Test Methods for Flash Point by Small Scale Closed Cup Tester. Annual Book of Standards. ASTM International, West Conshohocken, PA.

ASTM D4048. Standard Test Method for Detection of Copper Corrosion from Lubricating Grease. Annual Book of Standards. ASTM International, West Conshohocken, PA.

ASTM D4052. Standard Test Method for Density, Relative Density, and API Gravity of Liquids by Digital Density Meter. Annual Book of Standards. ASTM International, West Conshohocken, PA.

ASTM D4055. Standard Test Method for Pentane Insolubles by Membrane Filtration. Annual Book of Standards. ASTM International, West Conshohocken, PA.

ASTM D4124. Standard Test Method for Separation of Asphalt into Four Fractions. Annual Book of Standards. ASTM International, West Conshohocken, PA.

ASTM D4308. Standard Test Method for Electrical Conductivity of Liquid Hydrocarbons by Precision Meter. Annual Book of Standards. ASTM International, West Conshohocken, PA.

ASTM D4310. Standard Test Method for Determination of Sludging and Corrosion Tendencies of Inhibited Mineral Oils. Annual Book of Standards. ASTM International, West Conshohocken, PA.

ASTM D4530. Standard Test Method for Determination of Carbon Residue (Micro Method). Annual Book of Standards. ASTM International, West Conshohocken, PA.

ASTM D4636. Standard Test Method for Corrosiveness and Oxidation Stability of Hydraulic Oils, Aircraft Turbine Engine Lubricants, and Other Highly Refined Oils. Annual Book of Standards. ASTM International, West Conshohocken, PA.

ASTM D4739. Standard Test Method for Base Number Determination by Potentiometric Hydrochloric Acid Titration. Annual Book of Standards. ASTM International, West Conshohocken, PA.

ASTM D4893. Standard Test Method for Determination of Pitch Volatility. Annual Book of Standards. ASTM International, West Conshohocken, PA.

ASTM D4952. Standard Test Method for Qualitative Analysis for Active Sulfur Species in Fuels and Solvents (Doctor Test). Annual Book of Standards, ASTM International, West Conshohocken, PA.

ASTM D5002. Standard Test Method for Density and Relative Density of Crude Oils by Digital Density Analyzer. Annual Book of Standards. ASTM International, West Conshohocken, PA.

ASTM D5705. Standard Test Method for Measurement of Hydrogen Sulfide in the Vapor Phase above Residual Fuel Oils. Annual Book of Standards. ASTM International, West Conshohocken, PA.

ASTM D5768. Standard Test Method for Determination of the Iodine Value of Tall Oil Fatty Acids. Annual Book of Standards. ASTM International, West Conshohocken, PA.

ASTM D5770. Standard Test Method for Semiquantitative Micro Determination of Acid Number of Lubricating Oils during Oxidation Testing. Annual Book of Standards. ASTM International, West Conshohocken, PA.

ASTM D5771. Standard Test Method for Cloud Point of Petroleum Products (Optical Detection Stepped Cooling Method). Annual Book of Standards. ASTM International, West Conshohocken, PA.

ASTM D5772. Standard Test Method for Cloud Point of Petroleum Products (Linear Cooling Rate Method). Annual Book of Standards. ASTM International, West Conshohocken, PA.

ASTM D5773. Standard Test Method for Cloud Point of Petroleum Products (Constant Cooling Rate Method). Annual Book of Standards. ASTM International, West Conshohocken, PA.

ASTM D5853. Standard Test Method for Pour Point of Crude Oils. Annual Book of Standards. ASTM International, West Conshohocken, PA.

ASTM D5949. Standard Test Method for Pour Point of Petroleum Products (Automatic Pressure Pulsing Method). Annual Book of Standards. ASTM International, West Conshohocken, PA.

ASTM D5950. Standard Test Method for Pour Point of Petroleum Products (Automatic Tilt Method). Annual Book of Standards. ASTM International, West Conshohocken, PA.

ASTM D5972. Standard Test Method for Freezing Point of Aviation Fuels (Automatic Phase Transition Method). Annual Book of Standards. ASTM International, West Conshohocken, PA.

ASTM D5985. Standard Test Method for Pour Point of Petroleum Products (Rotational Method). Annual Book of Standards. ASTM International, West Conshohocken, PA.

ASTM D6021. Standard Test Method for Measurement of Total Hydrogen Sulfide in Residual Fuels by Multiple Headspace Extraction and Sulfur Specific Detection. Annual Book of Standards. ASTM International, West Conshohocken, PA.

ASTM D6045. Standard Test Method for Color of Petroleum Products by the Automatic Tristimulus Method. Annual Book of Standards. ASTM International, West Conshohocken, PA.

ASTM D6450. Standard Test Method for Flash Point by Continuously Closed Cup (CCCFP) Tester. Annual Book of Standards. ASTM International, West Conshohocken, PA.

ASTM D6560. Standard Test Method for Determination of Asphaltenes (Heptane Insolubles) in Crude Petroleum and Petroleum Products. Annual Book of Standards. ASTM International, West Conshohocken, PA.

Banaszewski, A. and Blythe, J. 2000. White mineral oil. In Modern Petroleum Technology. Volume 2: Downstream. A.G. Lucas (Editor). John Wiley & Sons Inc., New York.

Gary, J.G., Handwerk, G.E., and Kaiser, M.J. 2007. Petroleum Refining: Technology and Economics, 5th Edition. CRC Press, Taylor & Francis Group, Boca Raton, FL.

Hsu, C.S. and Robinson, P.R. (Editors) 2006. Practical Advances in Petroleum Processing (Volumes 1 and 2). Springer Science, New York.

IP 1. Determination of Acidity, Neutralization Value – Color Indicator Titration Method. IP Standard Methods 2013. The Energy Institute, London.

IP 2 (ASTM D611). Petroleum Products and Hydrocarbon Solvents – Determination of Aniline and Mixed Aniline Point. IP Standard Methods 2013. The Energy Institute, London.

IP 13 (ASTM D189). Petroleum Products – Determination of Carbon Residue – Conradson Method. IP Standard Methods 2013. The Energy Institute, London.

IP 14 (ASTM D524). Petroleum Products – Determination of Carbon Residue – Ramsbottom Method. IP Standard Methods 2013. The Energy Institute, London.

IP 15 (ASTM D97). Petroleum Products – Determination of Pour Point. IP Standard Methods 2013. The Energy Institute, London.

IP 16 (ASTM D2386). Petroleum Products – Determination of the Freezing Point of Aviation Fuels – Manual Method. IP Standard Methods 2013. The Energy Institute, London.

IP 17. Determination of Color – Lovibond Tintometer Method. IP Standard Methods 2013. The Energy Institute, London.

IP 30. Detection of Mercaptans, Hydrogen Sulfide, Elemental Sulfur and Peroxides – Doctor Test Method. IP Standard Methods 2013. The Energy Institute, London.

IP 34 (ASTM D93). Determination of Flash Point – Pensky-Martens Closed Cup Method. IP Standard Methods 2013. The Energy Institute, London.

IP 36. Determination of Open Flash and Fire Point – Cleveland Method. IP Standard Methods 2013. The Energy Institute, London.

IP 57. Petroleum Products – Determination of the Smoke Point of Kerosene. IP Standard Methods 2013. The Energy Institute, London.

IP 61 (ASTM D129). Determination of Sulfur – High Pressure Combustion Method. IP Standard Methods 2013. The Energy Institute, London.

IP 71 (ASTM D445). Petroleum Products – Transparent and Opaque Liquids – Determination of Kinematic Viscosity and Calculation of Dynamic Viscosity. IP Standard Methods 2013. The Energy Institute, London.

IP 123. Petroleum Products – Determination of Distillation Characteristics At Atmospheric Pressure. IP Standard Methods 2013. The Energy Institute, London.

IP 130 (ASTM D1159). Petroleum Products – Determination of Bromine Number of Distillates and Aliphatic Olefins – Electrometric Method. IP Standard Methods 2013. The Energy Institute, London.

IP 136 (ASTM D94). Petroleum Products – Determination of Saponification Number – Part 1 – Color-Indicator Titration Method. IP Standard Methods 2013. The Energy Institute, London.

IP 139 (ASTM D974). Petroleum Products and Lubricants – Determination of Acid or Base Number- Color Indicator Titration Method. IP Standard Methods 2013. The Energy Institute, London.

IP 143 (ASTM D6560). Determination of Asphaltenes (Heptane Insolubles) in Crude Petroleum and Petroleum Products. IP Standard Methods 2013. The Energy Institute, London.

IP 154. Petroleum Products – Corrosiveness to Copper – Copper Strip Test. IP Standard Methods 2013. The Energy Institute, London.

IP 160 (ASTM D1298). Crude Petroleum and Liquid Petroleum Products – Laboratory Determination of Density – Hydrometer Method. IP Standard Methods 2013. The Energy Institute, London.

IP 219. Petroleum Products – Determination of Cloud Point. IP Standard Methods 2013. The Energy Institute, London.

IP 226 (ASTM D2270). Petroleum Products – Calculation Of Viscosity Index From Kinematic Viscosity. IP Standard Methods 2013. The Energy Institute, London.

IP 227. Corrosiveness of Silver of Aviation Turbine Fuels – Silver Strip Method. Energy Institute, London, UK.

IP 274 (ASTM D2624). Determination of Electrical Conductivity of Aviation and Distillate Fuels. IP Standard Methods 2013. The Energy Institute, London.

IP 398. Petroleum Products – Determination of Carbon Residue – Micro Method. IP Standard Methods 2013. The Energy Institute, London.

IP 431. Petroleum Products – Determination of Acid Number – Semi-Micro Color-Indicator Titration Method. IP Standard Methods 2013. The Energy Institute, London.

IP 435. Freezing Point of Aviation Turbine Fuels – Automatic Phase Transition Method. IP Standard Methods 2013. The Energy Institute, London.

IP 528 (ASTM D7154). Determination of the Freezing Point of Aviation Turbine Fuels – Automated Fiber Optic Method. IP Standard Methods 2013. The Energy Institute, London.

IP 523. Determination of Flash Point – Rapid Equilibrium Closed Cup Method. IP Standard Methods 2013. The Energy Institute, London.

IP 607. Automotive Fuels – Determination of Iodine Value in Fatty Acid Methyl Esters (FAME) – Calculation Method from Gas Chromatographic Data. IP Standard Methods 2013. The Energy Institute, London.

Speight, J.G. 2000. The Desulfurization of Heavy Oils and Residua, 2nd Edition. Marcel Dekker Inc., New York.

Speight, J.G. 2001. Handbook of Petroleum Analysis. John Wiley & Sons Inc., New York.

Speight, J.G. 2014. The Chemistry and Technology of Petroleum, 5th Edition. CRC Press, Taylor & Francis Group, Boca Raton, FL.

Speight, J.G. and Ozum, B. 2002. Petroleum Refining Processes. Marcel Dekker Inc., New York.

Speight, J.G., Long, R.B., and Trowbridge, T.D. 1984. Factors influencing the separation of asphaltenes from heavy petroleum feedstocks. Fuel, 63: 616.

13

LUBRICATING OIL

13.1 INTRODUCTION

Lubrication is the introduction of various substances (referred to as *lubricants*) between sliding surfaces to reduce wear and friction. Thus, *lubricating oil* is used to reduce friction and wear between bearing metallic surfaces that are moving with respect to each other, by separation of metallic surfaces by a film of the oil. Lubricating oil is distinguished from other fractions of crude oil by a high (>400°C, >750°F) boiling point (Table 13.1, Fig. 13.1) (Banaszewski and Blythe, 2000; Pillon, 2011; Speight, 2014; Speight and Exall, 2014).

In the early days of petroleum refining, kerosene was the major product followed by wanted paraffin wax for the manufacture of candles. Lubricating oils were at first by-products of paraffin wax manufacture. The preferred lubricants in the 1860s were lard oil, sperm oil, and tallow, but as the trend to heavier industry increased, the demand for mineral lubricating oils increased, and after the 1890s, petroleum displaced animal and vegetable oils as the source of lubricants for most purposes. The major function of lubricating oils is the reduction of friction and wear by the separation of surfaces, metallic or plastic, which are moving with respect to each other. Performance requirements also include cooling and the dispersion and neutralization of combustion products from fuels. New base oil refining methods yield stock oils, which are more responsive to additive treatment.

Lubricants may be liquid, such as motor oil and hydraulic oil; they may be semisolid or solid, such as grease, or they may be dry, or powdered, such as dry graphite or molybdenum disulfide. All lubricating materials for mechanized equipment are designed to form some sort of protective coating between moving parts of machinery to protect these parts from undue wear, contamination, and oxidation.

The focus of this chapter is on lubricants manufactured from petroleum and the relevant test methods for these materials. In addition to lubricants for automotive use, there are also industrial lubricants, which, by definition for the purpose of this text, are lubricants that are designed for uses other than automotive (gasoline and diesel engines) use.

13.2 PRODUCTION AND PROPERTIES

Lubricating oil is composed of base oil plus the various additives that are necessary to enable lubricating oil to meet specifications. The base oils derived from mineral oils are either paraffinic or naphthenic in nature, depending on the source of the oil from which they are made. Paraffin-based oils have a high content of waxy hydrocarbons such as *n*-paraffins and iso-paraffins, while the naphthenic oils are high in cycloparaffins and aromatics and are low in wax content. As the waxes tend to solidify at low temperatures, naphthenic base oils do not show the same problems as a waxy paraffinic base stock. For low-temperature use, therefore, paraffinic base oils have to be dewaxed, but their other properties make them a better choice for most lubricating oil applications than the naphthenic oils, paraffinic base oils making up approximately 85% of the world's base stock market.

Handbook of Petroleum Product Analysis, Second Edition. James G. Speight.
© 2015 John Wiley & Sons, Inc. Published 2015 by John Wiley & Sons, Inc.

TABLE 13.1   General summary of product types and distillation range

Product	Lower carbon limit	Upper carbon limit	Lower boiling point (°C)	Upper boiling point (°C)	Lower boiling point (°F)	Upper boiling point (°F)
Refinery gas	C_1	C_4	−161	−1	−259	31
Liquefied petroleum gas	C_3	C_4	−42	−1	−44	31
Naphtha	C_5	C_{17}	36	302	97	575
Gasoline	C_4	C_{12}	−1	216	31	421
Kerosene/diesel fuel	C_8	C_{18}	126	258	302	575
Aviation turbine fuel	C_8	C_{16}	126	287	302	548
Fuel oil	C_{12}	>C_{20}	216	421	>343	>649
Lubricating oil	>C_{20}		>343		>649	
Wax	C_{17}	>C_{20}	302	>343	575	>649
Asphalt	>C_{20}		>343		>649	
Coke	>$C_{50}{}^a$		>1000^a		>1832^a	

^{a}Carbon number and boiling point difficult to assess; inserted for illustrative purposes only.

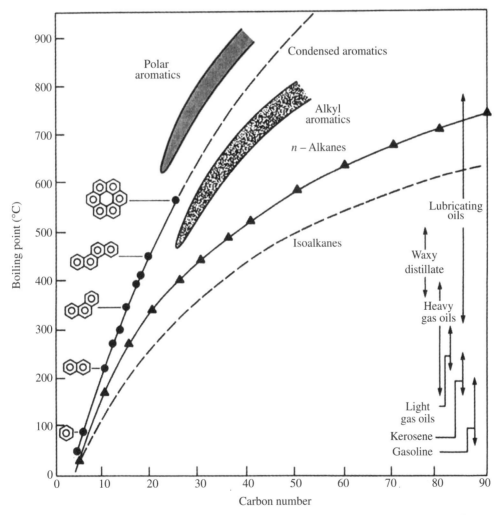

FIGURE 13.1   Boiling point and carbon number for various hydrocarbons and petroleum products.

13.2.1 Production

Lubricating oil base stocks produced from petroleum sources are obtained from the gas oil fractions coming from the vacuum distillation columns of the petroleum refinery, or from deasphalted residues from the atmospheric distillation columns (Speight and Ozum, 2002; Hsu and Robinson, 2006; Gary et al., 2007; Lynch, 2008; Speight, 2014). Thus, lubricating oil is a mixture that is produced by distillation after which chemical changes may be required to produce the desired properties in the product. One such property requires that the oil adhere to the metal surfaces and ensure protection of the moving parts by preventing from metal–metal contact (ASTM D2510).

Petroleum base lubricating oils are present in the atmospheric residuum (boiling above 370°C, 698°F) of selected paraffinic and naphthenic crude oils. The production of lubricating oils is well established (Sequeira, 1992; Speight, 2014) and generally involves the following processes: (i) distillation and deasphalting to remove the lighter constituents of the feedstock, (ii) solvent refining to remove the non-hydrocarbon constituents and to improve the feedstock quality, (iii) hydrogen treatment to remove the non-hydrocarbon constituents and to improve the feedstock quality, (iv) dewaxing to remove the wax constituents and improve the low-temperature properties, and (v) clay treatment or hydrogen treatment to prevent instability of the product.

Lubricating oil manufacture was well established by 1880, and the method depended on whether the crude petroleum was processed primarily for kerosene or for lubricating oils. Usually, the crude oil was processed for kerosene, and primary distillation separated the crude into three fractions: naphtha, kerosene, and a residuum. To increase the production of kerosene, the cracking distillation technique was used, and this converted a large part of the gas oils and lubricating oils into kerosene. The cracking reactions also produced coke products and asphalt-like materials, which gave the residuum a black color, and hence it was often referred to as *tar* (Speight, 2014; Speight and Ozum, 2002).

The development of vacuum distillation led to a major improvement in both paraffinic and naphthenic oils. By vacuum distillation, the more viscous paraffinic oils (even oils suitable for bright stocks) could be distilled overhead and could be separated completely from residual asphaltic components. Vacuum distillation provided the means of separating more suitable lubricating oil fractions with predetermined viscosity ranges and removed the limit on the maximum viscosity that might be obtained in a distillate.

Materials suitable for the production of lubricating oils are comprised principally of hydrocarbons containing from 25 to 40 carbon atoms per molecule, whereas residual stocks may contain hydrocarbons with 50–60 or more (up to 80 or so) carbon atoms per molecule. The composition of lubricating oil may be substantially different from the lubricant fraction from which it was derived, since wax (normal paraffins) is removed by distillation, or refining by solvent extraction and adsorption preferentially removes non-hydrocarbon constituents as well as polynuclear aromatic compounds and the multi-ring cycloparaffins.

The number of potential hydrocarbon isomers in the naphtha boiling range (Tables 13.1 and 13.2) renders complete speciation of individual hydrocarbons impossible for the naphtha distillation range, and methods are used that identify the hydrocarbon types as chemical groups rather than as individual constituents.

Greases (Chapter 14) are lubricants that exist in the solid form or in the semisolid form and are key ingredients in high-performance anti-seize pastes and antifriction coatings, used as additives in some greases and oils. In service, because of temperature effects, the solid lubricant or semisolid grease may exist in the liquid form. These special lubricants and additives fill in and smooth surface asperity peaks and valleys as they adhere to the substrate and cohere to each other. The solids provide effective boundary lubrication, optimizing friction reduction and reducing wear under extreme operating conditions. Boundary films formed by solid lubricants can maintain a steady thickness that is unaffected by load, temperature, or speed, unlike oil or grease fluid films for hydrodynamic lubrication.

While lubricating oil is typically composed of petroleum-derived base oils and/or synthetic-derived base oils, their uses are many. In fact, it is the varied uses that cause issues during refining if an understanding of the constituents of these oils is not available. Knowing the character of the oils and the use to which they are put is a forward step in understanding the nature of the impurities and, therefore, the steps that will be needed to refine the used oil.

TABLE 13.2 Increase in the number of isomers with carbon number

Number of carbon atoms	Number of isomers
1	1
2	1
3	1
4	2
5	3
6	5
7	9
8	18
9	35
10	75
15	4,347
20	366,319
25	36,797,588
30	4,111,846,763
40	62,491,178,805,831

13.2.2 Properties

There are general indications that the lubricant fraction contains a greater proportion of normal and branched paraffins than the lower boiling portions of petroleum. For the polycycloparaffins, a good proportion of the rings appear to be in condensed structures, and both cyclopentyl and cyclohexyl nuclei are present. The methylene groups appear principally in unsubstituted chains as at least four carbon atoms in length, but the cycloparaffin rings are highly substituted with relatively short side chains.

Mono-, di-, and trinuclear aromatic compounds appear to be the main constituents of the aromatic portion, but material with more aromatic nuclei per molecule may also be present. For the dinuclear aromatics, most of the material consists of naphthalene types. For the trinuclear aromatics, the phenanthrene type of structure predominates over the anthracene type. There are also indications that the greater part of the aromatic compounds occurs as mixed aromatic–cycloparaffin compounds.

In the majority of cases, chemical additives (Table 13.3) are used to enhance the properties of base oils to improve such characteristics as the oxidation resistance (ASTM D2893, ASTM D4742, ASTM D5846) change in viscosity (ASTM D445, IP 71) with temperature, low-temperature flow properties as derived from the pour point (ASTM D97, ASTM D5853, ASTM D5949, ASTM D5950, ASTM D5985, IP 15), and fluidity measurements (ASTM D6351), emulsifying ability (ASTM D2711), extreme pressure (ASTM D2782, ASTM D2783, ASTM D3233, IP 240), antiwear and frictional properties (ASTM D5183, ASTM D6425), and corrosion resistance (ASTM D4636). The selection of the components for the lubricating oil formulation requires knowledge of the most suitable crude sources for the base oils, the type of refining required, the types of additive necessary, and the possible interactions of these components on the properties of the finished lubricating oil.

Lubricating oils may be divided into many categories according to the types of service they are intended to perform. However, there are two main groups: (i) oils used in intermittent service, such as motor and aviation oils; and (ii) oils designed for continuous service, such as turbine oils. Thus, the test methods must be designed and applied accordingly.

This classification is based on the Society of Automotive Engineers (SAE) J 300 specification. The single-grade oils (e.g., SAE 20) correspond to a single class and have to be selected according to engine manufacturer specifications, operating conditions, and climatic conditions. At –20°C (68°F), multigrade lubricating oil such as SAE 10W-30 possesses the viscosity of a 10W oil, and at 100°C (212°F), the multigrade oil possesses the viscosity of an SAE 30 oil.

Oils used in intermittent service must show the least possible change in viscosity with temperature, that is, their

TABLE 13.3 Common additives found in used lubricating oil

Purpose of additive	Additives
Anticorrosion	Zinc dithiophosphates, metal phenolates, fatty acids, and amines
Antifoamant	Silicone polymers, organic copolymers
Antiodorant	Perfumes, essential oils
Antioxidant	Zinc dithiophosphates, hindered phenols, aromatic amines, sulfurized phenols
Antiwear additive	Chlorinated waxes, alkyl phosphites and phosphates, lead naphthenate, metal triborates, metal and ashless dithiophosphates
Color stabilizer	Aromatic amine compounds
Corrosion inhibitor	Metal dithiophosphates, metal dithiocarbamates, metal sulfonates, thiodiazoles, and sulfurized terpenes
Detergent	Alkyl sulfonates, phosphonates, alkyl phenates, alkyl phenolates, alkyl carboxylates, and alkyl substituted salicylates
Dispersant	Alkyl succinimides, alkyl succinic esters
Emulsifier	Fatty acids, fatty amides, and fatty alcohols
Extreme pressure additives	Alkyl sulfides, polysulfides, sulfurized fatty oils, alkyl phosphites and phosphates, metal and ashless dithiophosphates and carboxylates, metal dithiocarbamates, metal triborates
Friction modifier	Organic fatty acids. Lard oil. Phosphorus-based compounds
Metal deactivator	Metal deactivator organic complexes containing nitrogen and sulfur amines, sulfides, and phosphates
Pour point depressant	Alkylated naphthalene and phenolic polymers, polymethacrylates
Rust inhibitor	Metal alkylsulfonates, alkylamines, alkyl amine phosphates, alkenylsuccinic acids, fatty acids, alkylphenol ethoxylates, and acid phosphate esters
Seal swell agent organic	Organic phosphate aromatic hydrocarbons
Tackiness agent	Polyacrylates and polybutenes
Viscosity	Polymers of olefins, methacrylates, dienes, or alkylated styrenes

viscosity indices must be high. These oils must be changed at frequent intervals to remove the foreign matter collected during service. The stability of such oils is therefore of less importance than the stability of oils used in continuous service for prolonged periods without renewal. On the other hand, oils used in continuous service must be extremely stable, but their viscosity indices may be low because the engines operate at fairly constant temperature without frequent shutdown, and thermal stability (ASTM D2070, ASTM D2511) is an important property.

Finally, insoluble constituents in lubricating oil can cause wear that can lead to equipment failure. Pentane insoluble materials can include oil-insoluble materials and some oil-insoluble resinous matter originating from oil or additive degradation, or both. Toluene-insoluble constituents arise from external contamination, fuel carbon, and highly carbonized materials from degradation of fuel, oil, and additives, or engine wear and corrosion materials. A significant change in pentane- or toluene-insoluble constituents indicates a change in oil properties that could lead to machinery failure. The insoluble constituents measured can also assist in evaluating the performance characteristics of used oil or in determining the cause of equipment failure.

The main requirements for lubricants are that they are able to (i) maintain separation of the surfaces under all loads, temperatures, and speeds, thus minimizing friction and wear; (ii) act as a cooling fluid by removing the heat produced by friction or from external sources; (iii) remain adequately stable in order to guarantee constant behavior over the forecasted useful life; (iv) protect surfaces from the attack of aggressive products formed during operation; and (v) fulfill detersive and dispersive functions in order to remove residue and debris that may form during operation.

This is achieved by determining a series of selective properties (properties that are adequate to the task depending upon the type of lubricating oil) and interpreting the analytical data accordingly. The main properties of lubricants, which are usually indicated in the technical characteristics of the product, are (i) viscosity, (ii) viscosity index, (iii) pour point, and (iv) flash point. However, a large number of standard test methods are available and should be chosen to match the desired data and use of the oil.

13.2.3 Types of Lubricating Oil

There are two main types of lubricants: (i) those that are petroleum based and (ii) those that are manufactured as a synthetic product, each of which is manufactured for specific purposes and conditions. The different types are also subject to varying levels of oxidation and degradation, and are compatible with only certain types of machinery components, demands, and environments. Though most modern lubricants are petroleum based, synthetic bases such as vegetable oil and esters are gaining increased popularity for this purpose. The base of a particular lubricating fluid is the primary determinant as to whether it is petroleum-based or synthetic oil.

There are also hydraulic lubricants (hydraulic oils) that are formulated to be lighter and more free-flowing than the typical lubricating oil. They are used not only for lubrication, but for the actual operation of hydraulic machinery. Hydraulic oils must be able to flow freely through the pumps that compress the oil for the operation of the machinery and, at the same time, must have the film-forming additives to lubricate the moving parts of the pumping equipment.

13.2.3.1 Automotive Engine Oil

Automotive oil (*motor oil, engine oil*) is oil used for the lubrication of the internal combustion engine. The main function is to lubricate the moving parts of the engine and prevent breakdown. The lubricating oil also cleans the engine parts, inhibits corrosion, improves sealing, and also improves engine cooling by dispersing the heat away from moving parts.

Automotive oil is typically a blended product from base oils composed of hydrocarbons, poly alpha-olefins, and poly internal olefins. The base oils of some high-performance motor oils however contain up to 20% by weight of esters. Additives are also present to improve certain properties (Rudnick 2009, 2013).

The bulk of typical automotive oil consists of hydrocarbons with between 18 and 34 carbon atoms per molecule (Table 13.1) (Speight and Ozum, 2002; Hsu and Robinson, 2006; Gary et al., 2007; Speight, 2014). One of the most important properties in maintaining a lubricating film between moving parts is the viscosity of the lubricant. The viscosity must be high enough to maintain a lubricating film, but low enough that the oil can flow around the engine parts under all conditions. The *viscosity index* is a measure of the extent of the changes in viscosity as the temperature changes. A higher viscosity index indicates that the viscosity will change less with temperature than a lower viscosity index.

A fundamental property of lubricating oil is that the oil must be able to flow adequately at the lowest temperature it is expected to experience in order to minimize metal-to-metal contact between moving parts upon starting up the engine. The *pour point* defined first this property of motor oil (ASTM D97), an index of the lowest temperature of its utility for a given application, but the cold cranking simulator (ASTM D5293) and the mini-rotary viscometer (ASTM D3829, ASTM D4684) are test methods for these properties required in motor oil specifications and also define the SAE classifications.

Another manipulated property of motor oil is its total base number (TBN), which is a measurement of the reserve alkalinity of oil, meaning its ability to neutralize acids. The resulting quantity is determined as mg KOH/ (gram of lubricant). Analogously, total acid number (TAN) is the measure of a lubricant's acidity (ASTM D664, ASTM D974). Other tests include zinc, phosphorus, or sulfur content, and testing for excessive foaming. The NOACK volatility (ASTM D5800) test determines the physical evaporation loss of lubricants in high-temperature service.

13.2.3.2   Diesel Engine Oil

Generally, gasoline and diesel engine oils have the same anatomy or makeup. They are formulated from the blending of base oils and additives to achieve a set of desired performance characteristics. From this simple definition, the character of the lubricating oil changes because of the required performance for each engine type. However, lubrication of the inside of a heavy-duty engine such as a diesel requires more than just creating a protective film. Such engine oils must also disperse soot and control sludge to extend the life of the engine.

Again, viscosity is the single most important property of a lubricant, and oil having the right viscosity is of the utmost importance. The selected viscosity needs to be pumpable at the lowest start-up temperature while still protecting the components at in-service temperatures. Typically, diesel engine oil will have a higher viscosity than automotive lubricating oil, but the low-temperature pumpability of this higher viscosity is an issue. During cold starts, the oil may be very thick and difficult for the oil pump to deliver to the vital engine components in the lifter valley. This most certainly will lead to premature wear, as the components will be interacting without the benefit of lubrication. In addition, diesel engine lubricating oil has more additives per volume than gasoline engine lubricating oil. The most prevalent are overbase detergent additives, which function by neutralizing acids keeping the diesel engine clean.

Marine oil (also known as marine diesel oil) is a type of fuel oil and is a blend of gas oil and heavy fuel oil, with less gas oil than intermediate fuel oil used in the maritime field.

13.2.3.3   Tractor and Other Engine Oils

The lubricating oils required for use in tractors and agricultural machinery fall into two categories (Harperscheid and Omeis, 2007): (i) universal tractor transmission oils and (ii) super tractor oils universal. The first type of oil is used in the hydraulic systems, the gearbox, and wet brake systems, while the second can be used in these applications as well as in the engine. As a result, super tractor oils universal has to have similar additives to the automotive engine oils, including antifoaming agents, detergents and dispersants, oxidation and corrosion inhibitors.

Some of the latest tractor oils are also claimed to be more environmentally friendly than older oils, but this often means that they can contain bio-based oils such as rapeseed or sunflower seed oils, or synthetic esters that are rapidly biodegradable. If these oils are mixed with the other used lubricating oils that are recycled to a refinery for regeneration, this can complicate the refining process.

13.2.3.4   Aviation Oil

Aviation-derivative gas turbines present unique turbine oil challenges that call for oils with much higher oxidation stability since the lubricating oil in aero-derivative turbines is in direct contact with metal surfaces ranging from 205 to 315°C (400–600°F). Sump lube oil temperatures can range from 71°C to 121°C (160°F–250°F). These compact gas turbines utilize the oil to lubricate and to transfer heat back to the lube oil sump. In addition, their cyclical operation imparts significant thermal and oxidative stress on the lubricating oil. These most challenging conditions dictate the use of high-purity synthetic lubricating oils. Average lube oil makeup rates of 0.15 gallons per hour will help rejuvenate the turbine oil under these difficult conditions.

13.2.3.5   Turbine Oil

Steam turbines, gas turbines, and hydro turbines operate on lubricating oil known as rust and oxidation-inhibited oil. Unlike most gasoline and diesel engine oil applications, turbine oil is formulated to shed water and allow solid particles to settle where they can be removed through sump drains or kidney loop filtration systems during operation. To aid in contaminant separation, most turbine oils do not have high levels of added detergents or dispersants that clean and carry away contaminants. Turbine oils are not exposed to fuel or soot and therefore do not need to be drained and replaced on a frequent basis.

Varying amounts of water will constantly be introduced to the steam turbine lubrication systems through gland seal leakage. Because the turbine shaft passes through the turbine casing, low-pressure steam seals are needed to minimize steam leakage or air ingress leakage to the vacuum condenser. Water or condensed steam is generally channeled away from the lubrication system, but, inevitably, some water will penetrate the casing and enter the lube oil system. Gland seal condition, gland sealing steam pressure, and the condition of the gland seal exhauster will impact the amount of water introduced to the lubrication system. Typically, vapor extraction systems and high-velocity downward flowing oil create a vacuum that can draw steam past shaft seals into the bearing and oil system. Water can also be introduced through lube oil cooler failures, improper powerhouse cleaning practices, water contamination of makeup oil, and condensed ambient moisture.

Heat will also cause reduced turbine oil life through increased oxidation. In utility steam turbine applications, it is common to experience bearing temperatures of 49°C–71°C (120°F–160°F) and lube oil sump temperatures of 49°C (120°F). The impact of heat is generally understood to double the oxidation rate for every 10° above 60°C (every 18° above 140°F).

13.2.3.6   Compressor Oil

Lubrication is the key aspect to keeping air compressors running, and the compressor lubricants are produced by many lubricant manufacturers, ranging in quality from poor to excellent. Poor air compressor oil could cause the compressor to have a very short life, but excellent quality air compressor oil reduces maintenance and can extend compressor life.

Compressor oils have many attributes and functions, including (i) oxidation stability, (ii) hydrolytic stability, (iii) rust protection, (iv) foam resistance, (v) copper corrosion resistance, and (vi) antiwear performance. Different chemistries are required to achieve a particular performance parameter, and sacrifices are often made in other areas. For example, chemistry that is good for rust protection may cause foaming, and chemistry that is good for antiwear may not be good for oxidation resistance. Although there is a wide variance in performance, it is generally recognized that no single oil will perform perfectly in all categories.

13.2.3.7 *Industrial Oils*

Industrial machinery runs on oil, and the successful outcome of manufacturing depends on that oil being maintained properly. Hence, oil maintenance programs, when in place at all, have historically depended on a time-based change program (often at an annual shutdown). While this is better than nothing, roughly 70% of the oils subjected to a time-based change are removed from service unnecessarily. But the time-based change also does not guarantee that lubricating oils that are well beyond the end of their useful lives are removed before they damage the machine.

Industrial oils run *cold* compared to other (such as automotive-use) oils, and they tend to accumulate moisture. The moisture comes from humidity in the air, or in some cases, it's directly introduced to the oil from coolants and related systems. Moisture affects the lubricity of the oil, decreasing its effectiveness. Moisture in the oil can cause a variety of problems, such as poorly running hydraulic rams, machines seizing up, and chatter.

Another negative effect of moisture in oil is acidity. Oil, by its molecular nature, cannot become an acid, but there is always a little moisture present in oils operating at relatively cool temperatures and that moisture can turn acidic. Acids in a machine's oil sump will corrosively attack internal parts, not only the metallic parts, but the seals as well. Corroded valves become ineffective. Many headaches in a machine's operation can be directly attributed to oil condition. Though oils do not respond to the pH test, there is a neutralization test (*total acid number test*) that can easily spot oil that is becoming recognized as a major deleterious effect in lubricating oil formulation and performance.

13.3 USED LUBRICATING OIL

Used lubricating oil—often referred to as *waste oil* without further qualification—is any lubricating oil, whether refined from crude or synthetic components, which has been contaminated by physical or chemical impurities as a result of use (Boughton and Horvath, 2004; Speight and Exall, 2014).

The used lubricating oil that is collected for recycling may contain a mixture of oils of different types, particularly when a service station contains a tank for liquid disposal, and brake fluid, antifreeze, gear oil, parts washer solvent, engine oil, and possibly grease are all dumped into this tank. Although efforts have been made to encourage the separate collection of these types of fluids, this type of discipline will probably only be respected when an efficient recycling program has been introduced and popularized. In most jurisdictions, transformer oils have to be recycled separately, due to the presence of polychlorobiphenyls in the oil, and there is generally a legislated limit to the content of the polychlorobiphenyls in an oil that may be accepted for re-refining, due to the potential formation of carcinogenic substances in unsuitable processing facilities and the release of potentially harmful substances to the environment (Speight, 2005; Speight and Arjoon, 2012). The bio-based lubricants are also not suitable for mixing with the mineral oil–derived lubricants, but these constitute a small fraction of the lubricant industry.

Used mineral-based crankcase oil is the brown-to-black, oily liquid removed from the engine of a motor vehicle when the oil is changed. It is similar to a high-boiling fraction of petroleum except it contains additional chemicals from its use as an engine lubricant. Used oil also contains chemicals formed when the oil is exposed to high temperatures and pressures inside an engine. It also contains some metals from engine parts and small amounts of gasoline, antifreeze, and chemicals that come from gasoline when it burns inside the engine. The chemicals found in used mineral-based crankcase oil vary, depending on the brand and type of oil, whether gasoline or diesel fuel was used, the mechanical condition of the engine that the oil came from, and the amount of use between oil changes.

Contaminants are introduced from the surrounding air and by metallic particles from the engine. Contaminants from the air are dust, dirt, and moisture—in fact, air itself may be considered as a contaminant since it can cause foaming of the oil. The contaminants from the engine are (i) metallic particles resulting from wear of the engine, (ii) carbonaceous particles due to incomplete fuel combustion, (iii) metallic oxides present as corrosion products of metals, (iv) water from leakage of the cooling system, (v) water as a product of fuel combustion, and (vi) fuel or fuel additives or their by-products, which might enter the crankcase of engines.

In addition, used oil originates from diverse sources, including petroleum refining operations (such as sludge containing appreciable amounts of oil originating from the various parts of petroleum plants such as sumps, gravity separators, and the cleaning of storage tanks), the forming and machining of metals, small generators (do-it-yourself car and other equipment maintenance) and industrial sources, and the rural farming population. Collecting used oil from nonindustrial sources and local/small generators is very difficult and requires a well-established and efficient infrastructure to accomplish the task. In this regard, it is important to develop adequate reuse or recycling options, to properly handle the collected volume of oil, to address the specific properties of the concerned waste, and to assess the degree to which used oils could be treated. A major source of oily wastes arising in some parts of the world is the sludge recovered from tanks used for the storage of leaded gasoline. This sludge, which is usually produced by high-pressure water jet cleaning of storage tanks, consists of iron oxide corrosion products and sediments, onto which organic and inorganic lead compounds have been absorbed or adsorbed mixed with fuel. The free fuel is usually readily removed by gravity or mechanical separation and used as an energy source. The highly toxic organic lead compounds associated with the sludge have to be chemically or thermally oxidized (calcined) to inorganic lead compounds to facilitate its disposal.

Examples of *used oil* include (i) engine oil—typically includes gasoline and diesel engine crankcase oil and piston-engine oils from automobiles, trucks, boats, airplanes, trains, and heavy equipment; (ii) transmission fluid; (iii) refrigeration oil; (iv) turbine and compressor oil; (v) metalworking fluids and oils; (vi) laminating oils; (vii) industrial hydraulic fluid; (viii) electrical insulating oil; (ix) industrial process oil; (x) oils used as buoyant; and (xi) synthetic oil (usually derived from a polymer-based starting material). Thus, used oil analysis is a valuable means in the management of any machinery that uses liquid lubricants.

Analysis of used lubricating oil should be performed during routine preventative maintenance to provide meaningful and accurate information on lubricant and engine or machine condition. By tracking oil analysis using a well-defined chain of custody protocol, with sample results over the life of a particular machine, trends can be established that can help eliminate costly repairs and potential unhealthy environmental issues (Speight, 2005; Speight and Arjoon, 2012).

Properties such as density, relative density, and API gravity (ASTM D4052), viscosity (ASTM D445), flash and fire point (ASTM D92, ASTM D93), corrosivity, (ASTM D130), mineral matter content (measured as mineral ash, ASTM D482), water and sediment (ASTM D1796), and foaming tendency (ASTM D892) are the minimal properties to be considered as important aspects when comparing unused base oils and used oils. The boiling behavior (i.e., the distillation curve) of the oil is typically determined by a gas chromatographic method that calculates the distillation curve boiling temperatures based on standardized retention times (ASTM D2887). The distillation curve is one of the most important properties that can be measured for any complex fluid, since it is the only practical avenue to assess the volatility or the *vapor liquid equilibrium.*

Fourier transform infrared spectrometry provides information on compounds, rather than elements, found in oil (ASTM E2412). The method can be used to measure several useful degradation parameters, so is particularly useful in engine oil samples. Infrared analysis detects the presence of water and can also be used to identify oil base stocks. While inductively coupled plasma spectroscopy measures emissions of radiation of specific wavelength in the visible and ultraviolet regions of the electromagnetic spectrum, infrared analysis measures the specific wavelengths of radiation in the infrared region. The various degradation by-products and contaminants found in the oil cause characteristic absorptions in specific regions of the infrared spectrum. The higher the level of contamination in the sample, the higher the degree of absorption in the characteristic region.

As lubricating oil *oxidizes*, the ability of the oil to provide lubrication is reduced, and in cases of severe oxidation, noticeable changes occur: (i) the oil becomes darker and emits odor; (ii) sludge is formed—the oxidation products are frequently referred to as varnishes or lacquers, and resins are formed; and (iii) in the advanced stages of oxidation, viscosity increases at a rate dependent on the temperature of the oil—the chemical reaction between oxygen and lubricant at room temperature is slow, and oxidative degradation is not an issue under these conditions. The situation changes when reaction conditions are altered to favor a more rapid reaction rate—because the rate of a chemical reaction (in this case, oxidation) doubles for every 10°C (18°F) increase in temperature, an excessively high operating temperature (overheating) is generally accompanied by increased wear (producing lead, copper, tin, and iron from bearing materials and lubricated surfaces) and increased baseline viscosity.

On occasion, overheating can also lead to the evaporation of volatile fractions in the oil, making regular top-ups necessary. In this case, the sump oil will exhibit increased additive levels (concentration of nonvolatile components) and an increased viscosity as a direct result of light end loss. As this lost oil is replaced with fresh oil, the antioxidants are replaced and oxidation is often not immediately evident. The extent of the oxidation can be determined by the acid number (ASTM D974, ASTM D5770) and by viscosity measurement for confirmation.

Another test method is often referred to as *debris analysis* that involves the use of particle counting and is actually a test for particle contaminant levels and not specifically wear debris. It does not distinguish between wear and dirt particles, but if it can be determined that nonferrous contamination has remained stable, then an increase in the particle count must be attributable to wear. A magnet can be used to modify the particle count to count ferrous debris only. There are various ways of doing this, but essentially a magnet holds back the ferrous debris while the nonferrous debris is flushed from the sample, after which a ferrous debris particle count is performed. The particle count is an easy test to interpret, assuming the test has been correctly performed—caution is advised as there are many factors that can negatively affect a particle count. An increasing count is simply an indication of an increased number of particles in the oil.

While there are several test methods that should be applied to the used oil (Speight and Exall, 2014), the two additional analytical techniques that should also be used as part of the analytical protocols are *energy dispersive x-ray fluorescence* (EDXRF) spectrometry and *inductively coupled plasma-optical emission spectrometry* (ICP-OES).

The *EDXRF* method involves irradiating the sample with a beam of X-rays to induce fluorescence in the atoms in the sample, which is then emitted as X-rays of a lower energy. Each element emits fluorescent X-rays of different and unique energies or wavelengths, whose intensity is proportional to the concentration of that element in the sample.

The *ICP-OES* method relies on the fact that every element has a unique atomic structure, with a massive positively charged nucleus and a number of electrons in orbits surrounding it. Each of these electrons exists at a precise energy level and under normal conditions, with no external influences (the *ground state*). If, however, the atom is subjected to a source of energy, for example, in the form of heat and, respectively, collisions with other atoms, that energy can be absorbed and the electron(s) raised to a higher or excited state. This is an unstable condition, and the electrons quickly return to the ground state, at which point the previously absorbed energy is reemitted. The intensity of this radiation is proportional to the number of atoms present (i.e., the concentration), which is the basis of quantitative analysis by this method.

13.4 TEST METHODS

Lubricating oil is achieved by application of several standard test methods of the properties of the oil (or lubricant) to provide information regarding meeting specifications for use and then to determine the presence of contaminants and wear debris as a result of performance in service.

Thus, physical and chemical test methods as well as actual engine tests are extremely valuable as tools for attempting to predict how a specific lubricant formula will perform in full size machinery under many different operating conditions. However, the data must always be used and interpreted with the recognition that the test methods are subject to error. The ultimate decision as to the success or failure of a lubricating fluid can be made only on the basis of its behavior in-service such as in production engines, pumps, gear drives, and hydraulic systems.

Lubricating oil analysis is a process that involves a sample of the oil, whether unused or used, and analyzing it for various properties and materials in order to monitor in-service performance. By analyzing a sample of used engine oil, the wear rate can be determined as well as noting the overall service condition of an engine along with identifying potential problems and imminent failure.

The number of tests applied for product character and quality varies with the complexity of the product and the nature of the application. The more important tests, such as the viscosity, flash point, and color, are usually performed on every batch. Other tests may be on a statistical (or as-needed) basis dependent on data that can be presented in the graphical form of a *fingerprint* that is specific for the blend of components and additives in a particular formulation. Comparison of the *fingerprint* with a known standard can be used as a check on the composition.

13.4.1 Acidity and Alkalinity

Unused and used petroleum products may contain acidic constituents that are present as additives or as degradation products formed during service, such as oxidation products (ASTM D5770). The relative amount of these materials can be determined by titrating with bases. The acid number is used as a guide in the quality control of lubricating oil formulations. It is also sometimes used as a measure of lubricant degradation in service. Any condemning limits must be empirically established.

Thus, the acid number is a measure of this amount of acidic substance, in the oil, always under the conditions of the test. The acid number is used as a guide in the quality control of lubricating oil formulations. It is also sometimes used as a measure of lubricant degradation in service. Since a variety of oxidation products contribute to the acid number and the organic acids vary widely in corrosion properties, the test cannot be used to predict corrosiveness of oil under service conditions.

Acidity is determined by the *acid number*, which is the quantity of base, expressed in milligrams of potassium hydroxide per gram of sample that is required to titrate a sample in this solvent to a green/greenish-brown end point, using p-naphtholbenzein indicator solution (ASTM D974, IP 139). However, many higher-molecular-weight oil products (dark-colored oils) that cannot be analyzed for acidity to obscurity of the color-indicator end point can be analyzed by an alternate test method (ASTM D664). The quality of the mineral oil products renders them suitable for the determination of the acid number.

Determination of the *acid number* typically involves a titration method where the total acid content of the oil dissolved in a mixed solvent is completely neutralized by the gradual addition of an alcoholic solution of potassium hydroxide (KOH). A colorimetric method of determining the end point is used—a chemical indicator changes color as soon as the acid is completely neutralized. Alternatively, a potentiometric method may also be used.

The acid number test is performed on oil samples to quantify the acid buildup in the oil—an increased acid number is a result of oxidation of the oil, perhaps caused by overheating, overextended oil service, or water or air contamination. The acid number limits for lubricating oils vary considerably—in some cases, an acid number exceeding 0.05 is unacceptable, while in other cases, an acid number in excess of 4.00 may be acceptable. As with all other readings, trend analysis is the best indication of the health of both the oil and the machinery.

In a manner akin to the *acid number*, the *base number* (often referred to as the *neutralization number*) is a measure of the basic constituents in the oil under the conditions of the test. The TBN is also measured in milligrams of potassium hydroxide, or calcium sulfonate per gram of oil. In more simple terms, it is the amount of active additives remaining, which is important because combustion by-products tend to form acidic compounds and the TBN is the acid-neutralizing capacity of the lubricant. The base number is used as a guide in the quality control of oil formulation and is also used as a measure of oil degradation in service.

The measurement of the *base number* (ASTM D974, ASTM D2896, ASTM D5984) is determined as the total alkaline reserve of 1 gram of oil dissolved in a mixed solvent after reaction by gradual addition of a known excess of an acid solution. Typical starting values for diesel engine oils are between 8 and 12. However, marine engines burning heavy fuel oil need a much higher base number (possibly as high as 80) to handle the harsh combustion conditions from fuels containing a high concentration of sulfur. A general rule is to discard the lubricating oil when the base number drops below half of its beginning value.

The *neutralization number* expressed as the *base number* is a measure of this amount of basic substance in the oil always under the conditions of the test. The neutralization number is used as a guide in the quality control of lubricating oil formulations. It is also sometimes used as a measure of lubricant degradation in service; however, any condemning limits must be empirically established.

Samples of oil drawn from the crankcase can be tested to assess the reserve of alkalinity remaining by determining the TBN of the oil (ASTM D664, ASTM D2896, ASTM D4739, IP 177, IP 276). Essentially, these are titration methods where, because of the nature of the used oil, an electrometric instead of a color end point is used. The reserve alkalinity neutralizes the acids formed during combustion. This protects the engine components from corrosion. However, the different base number methods may give different results for the same sample.

Lubricating oil often contains additives that react with alkali to form metal soaps, and the *saponification number* expresses the amount of base that will react with 1g of the sample when heated in a specific manner. In the test method (ASTM D94, IP 136), a known weight of the sample is dissolved in methyl ethyl ketone or a mixture of suitable solvents, and the mixture is heated with a known amount of standard alcoholic potassium hydroxide for between 30 and 90 min at 80°C (176°F). The excess alkali is titrated with standard hydrochloric acid and the saponification number calculated. The results obtained indicate the effect of extraneous materials in addition to the saponifiable material present.

13.4.2 Ash

The ash formed by the combustion of lubricating oil (ASTM D482, ASTM D2415, IP 4) is, as defined for other products, the inorganic residue, free from carbonaceous matter, remaining after ignition in air of the residual fuel oil at fairly high temperatures. The ash content is not directly equated to mineral content but can be converted to mineral matter content by the use of appropriate formulae.

13.4.3 Asphaltene Content

The asphaltene fraction (ASTM D2006, ASTM D2007, ASTM D3279, ASTM D4124, ASTM D6560, IP 143) is the highest molecular weight and most complex fraction in petroleum. Insofar as the asphaltene content gives an indication of the amount of coke that can be expected during exposure to thermal conditions (Speight, 2001, 2014, Speight and Ozum 2002), there is little need for the application of the test to lubricating mineral oil. Use of the oil under stressful conditions where heat is generated may introduce the need to determine the amount of insoluble constituents precipitated by the addition of a low-boiling hydrocarbon liquid to mineral oil.

Pentane-insoluble constituents can be determined by membrane filtration (ASTM D4055). In the method, a sample of oil is mixed with pentane in a volumetric flask, and the oil solution is filtered through a 0.8-μm membrane filter. The flask, funnel, and the filter are washed with pentane to completely transfer any particulates onto the filter after which the filter (with particulates) is dried and weighed to give the pentane-insoluble constituents as a percentage by weight of the sample.

The *precipitation number* is often equated to the asphaltene content, but there are several issues that remain obvious in its rejection for this purpose. For example, the method to determine the precipitation number (ASTM D91) advocates the use of naphtha for use with black oil or lubricating oil, and the amount of insoluble material (as a % v/v of the sample) is the precipitating number. In the test, 10 ml of sample is mixed with 90 ml of ASTM precipitation naphtha (that may or may not have a constant chemical composition) in a graduated centrifuge cone and centrifuged for 10 min at 600–700 rpm. The volume of material on the bottom of the centrifuge cone is noted until repeat centrifugation gives a value within 0.1 ml (the precipitation number). Obviously, this can be substantially different from the asphaltene content.

On the other hand, if the lubricating oil had been subjected to excessive heat, it might be wise to consider application of the test method for determining the toluene-insoluble constituents of tar and pitch (ASTM D4072, ASTM D4312). In the methods, a sample is digested at 95°C (203°F) for 25 min and then extracted with hot toluene in an alundum thimble. The extraction time is 18 h (ASTM D4072) or 3 h (ASTM D4312). The insoluble matter is dried and weighed. Combustion will then show if the material is truly carbonaceous or if it is inorganic ash from the metallic constituents (ASTM D482, ASTM D2415, ASTM D4628, ASTM D4927, ASTM D5185, ASTM D6443, IP 4).

Another method (ASTM D893) covers the determination of pentane- and toluene-insoluble constituents in used lubricating oils. Pentane-insoluble constituents include oil-insoluble materials and some oil-insoluble resinous matter originating from oil or additive degradation, or both. Toluene-insoluble constituents can come from external contamination, fuel carbon, and highly carbonized materials from degradation of fuel, oil, and additives, or engine wear and corrosion materials. A significant change in pentane- or toluene-insoluble constituents and insoluble resins indicates a change in oil that could lead to lubrication problems. The insoluble constituents measured can also assist in evaluating the performance characteristics of used oil or in determining the cause of equipment failure.

There are two test methods used: Procedure A covers the determination of insoluble constituents without the use of coagulant in the pentane and provides an indication of the materials that can be readily separated from the oil–solvent mixture by centrifugation. Procedure B covers the determination of insoluble constituents in lubricating oil that contains detergents and employs a coagulant. In addition to the materials separated by using Procedure A, this coagulation procedure separates some finely divided materials that may be suspended in the oil. The results obtained by Procedures A and B should not be compared since they usually give different values. The same procedure should be applied when comparing results obtained periodically from oil in use, or when comparing results determined in different laboratories.

In Procedure A, a sample is mixed with pentane and centrifuged after which the oil solution is decanted, and the precipitate washed twice with pentane, dried, and weighed. For toluene-insoluble constituents, a separate sample of the oil is mixed with pentane and centrifuged. The precipitate is washed twice with pentane, once with toluene–alcohol solution, and once with toluene. The insoluble material is then dried and weighed. In Procedure B, Procedure A is followed except that instead of pentane, a pentane-coagulant solution is used.

13.4.4 Carbonizable Substances

The test for carbonizable substances using sulfuric acid is not usually applied to lubricating oil. However, the need may arise, and being aware of the availability of such a test is warranted. In the test method (ASTM D565), a sample of the oil is treated with concentrated sulfuric acid under prescribed conditions, and the resulting color is compared with a reference standard to determine whether it passes or fails the test. When the oil layer shows no change in color and when the acid layer is not darker than the reference standard colorimetric solution, the oil is reported as passing the test. A bluish haze or a slight pink or yellow color in the oil layer should not be interpreted as a change in color. The more fully refined the oil, the lighter the color of the acid layer.

However, with the introduction of ultraviolet absorption procedures (ASTM D2008, ASTM D2269), the test finds lesser use but still provides a useful method to determine possible contamination of lubricating oil with impurities transparent to both visible and ultraviolet light and hence not detectable by color or by ultraviolet absorption measurements (ASTM D2008).

The test for carbonizable substances (ASTM D565) should not be confused with the test methods for determining *carbon residue* (ASTM D189, ASTM D524, ASTM D4530, IP 13, IP 14, IP 398) (*q.v.*).

13.4.5 Carbon Residue

Lubricating oil is not usually considered to be used under the extreme conditions that coke is formed from, from example, fuel oil. Nevertheless, the tests that are applied to determine the carbon-forming propensity of fuel oil (and other petroleum products) are also available for application to lubricating oil should the occasion arise.

Thus, the tests for Conradson carbon residue (ASTM D189, IP 13), the Ramsbottom carbon residue (ASTM D524, IP 14), and the microcarbon carbon residue (ASTM D4530, IP 398) are often included in inspection data for fuel oil.

In the Conradson carbon residue test (ASTM D189, IP 13), a weighed quantity of sample is placed in a crucible and subjected to destructive distillation for a fixed period of severe heating. At the end of the specified heating period, the test crucible containing the carbonaceous residue is cooled in a desiccator and weighed, and the residue is reported as a percentage (% w/w) of the original sample (Conradson carbon residue). In the Ramsbottom carbon residue test (ASTM D524, IP 14), the sample is weighed into a glass bulb that has a capillary opening and is placed into a furnace (at 550°C, 1022°F). The volatile matter is distilled from the bulb, and the nonvolatile matter that remains in the bulb cracks to form thermal coke. After a specified heating period, the bulb is removed from the bath, cooled in a desiccator, and weighed to report the residue (Ramsbottom carbon residue) as a percentage (% w/w) of the original sample. In the microcarbon residue test (ASTM D4530, IP 398), a weighed quantity of the sample placed in a glass vial is heated to 500°C (932°F) under an inert (nitrogen) atmosphere in a controlled manner for a specific time, and the carbonaceous residue [*carbon residue (micro)*] is reported as a percentage (% w/w) of the original sample.

The data produced by the microcarbon test (ASTM D4530, IP 398) are equivalent to those by Conradson carbon method (ASTM D189 IP 13). However, this microcarbon test method offers better control of test conditions and requires a smaller sample. Up to 12 samples can be run simultaneously. This test method is applicable to petroleum and to petroleum products that partially decompose on distillation at atmospheric pressure and is applicable to a variety of samples that generate a range of yields (0.01–30% w/w) of thermal coke.

13.4.6 Cloud Point

The *cloud point* is the temperature at which a cloud of wax crystal first appears in a liquid when it is cooled under conditions prescribed in this test method. This test method covers only petroleum oils that are transparent in layers 38 mm (11/2 in.) in thickness, and with a cloud point below 49°C (120°F). The cloud point is an indicator of the lowest temperature of the utility of an oil for certain applications, and the cloud point is usually higher than the pour point (ASTM D97, ASTM D5853, ASTM D5949, ASTM D5950, ASTM D5985, IP 15).

The cloud point (ASTM D2500, IP 219) of lubricating oil is the temperature at which paraffinic wax, and other components that readily solidify, begins to crystallize out and separate from the oil under prescribed test conditions. It is of importance when narrow clearances might be restricted by accumulation of solid material (e.g., oil feed lines or filters).

Neither the cloud point nor the pour point should be confused or interchanged with the *freezing point* (ASTM D2386, ASTM D5972, IP 16, IP 435, IP 528). The freezing point presents an estimate of minimum handling temperature and minimum line or storage temperature. It is not a test for an indication of purity and has limited value for lubricating oil.

13.4.7 Color

Determination of the color of petroleum products is used mainly for manufacturing control purposes and is an important quality characteristic. In some cases, the color may serve as an indication of the degree of refinement of the material. However, color is not always a reliable guide to product quality and should not be used indiscriminately in product specifications (ASTM D156, ASTM D1209, ASTM D1500, ASTM D1544, ASTM D6045, IP 17).

In one test (ASTM D156) for the determination of color, the height of a column of the oil is decreased by levels corresponding to color numbers until the color of the sample is lighter than that of the standard color. The color number immediately above this level is recorded as the Saybolt color of the oil, and a color number of +25 corresponds to water mineral, while the minimum color intensity reading on this scale is expressed by +30, a value normally attained by mineral oils. In another test (IP 17), in which the measurement is performed using an 18-in. cell against color slides on a scale, a color of 1.0 or under is considered water–mineral, and medicinal oils will normally be 0.5 or less. Conversion scales for different color test are available (ASTM D1500).

Although sometimes to be found in insulating oil specifications, the color characteristic is of no technical significance. Pale oils are, as a general rule, more severely refined than dark oils of the same viscosity; color (ASTM D1500, IP 17) is not a guide to stability. Deterioration of color after submission of the oil to an aging test is sometimes limited, but here again, the extent of oil deterioration can be much better measured by some other property such as acidity development or change in electrical conductivity (ASTM D2624, ASTM D4308, IP 274). About the only point that can be made in favor of color measurement on new oil is that it can give an immediate guide to a change in supply continuity.

13.4.8 Composition

The importance of composition of lubricating oil lies in the effect composition has on their compatibility (ASTM D2226). This can often be determined by studies of the composition. For example, molecular-type analysis separates oil into different molecular species—such an analysis is the *clay-gel analysis* (Speight, 2014). In this method, group separation is achieved by adsorption in a percolation column with selected grades of clay and/or silica gel as the adsorption media (ASTM D1319, ASTM D2007, IP 156).

Mass spectrometry can also be employed for compositional studies of lubricating oil (ASTM D3239). This test method covers the determination by high ionizing voltage, low-resolution mass spectrometry of 18 aromatic hydrocarbon types, and three aromatic thiophene types in straight-run aromatic petroleum fractions boiling within the range from 205 to 540°C (400–1000°F). Samples must be nonolefinic, must not contain more than 1 mass % of total sulfur, and must not contain more than 5% nonaromatic hydrocarbons. The relative abundances of seven classes of aromatics in petroleum fractions are determined by using a summation of peaks most characteristic of each class. Calculations are carried out by the use of inverted matrix derived from the published spectra of pure aromatic compounds. The aromatic fraction needed for this analysis is obtained by using liquid elution chromatography (ASTM D2549).

Aromatic content is a key property of hydrocarbon oils and insofar as the aromatic constituents can affect a variety of properties. An existing method using high-resolution nuclear magnetic resonance (ASTM D5292) is applicable to a wide range of petroleum products that are completely soluble in chloroform and carbon tetrachloride at ambient temperature. The data obtained by this method can be used to evaluate changes in aromatic contents of hydrocarbon oils due to process changes. This test method is not applicable to samples containing more than 1% by weight of olefinic or phenolic compounds. The hydrogen magnetic resonance spectra are obtained on sample solutions in either chloroform or carbon tetrachloride using a continuous-wave or pulse Fourier transform high-resolution nuclear magnetic resonance spectrometer. Carbon magnetic resonance spectra are obtained on the sample solution in deutero-chloroform using a pulse Fourier transform high-resolution nuclear magnetic resonance.

The total quantity of sulfur in a gear oil due to the base oil and to the additives present can be determined by a bomb method (ASTM D129, IP 61) in which the sulfur is assessed gravimetrically as barium sulfate. The copper strip test (ASTM D130, ASTM D849, ASTM D2649, IP 154) is used to simulate the tendency of the oil to attack copper, brass, or bronze. Since active sulfur is desirable for some extreme-pressure applications, a positive copper strip result can indicate that the formulation is satisfactory, but care is necessary in the interpretation of copper strip results because formulations of different chemical compositions may give different results and yet have similar performance in the intended application. Corrosion preventative properties are also measurable (ASTM D4636).

The constituent elements (barium, calcium, magnesium, tin, silica, zinc, aluminum, sodium, or potassium) of new and used lubricating oils can also be determined. Corresponding methods for barium, calcium, and zinc in unused oils are available. For new lubricating oils, a test method (ASTM D874, IP 163) can be employed to check the concentration of metallic additives present by measuring the ash residue after ignition. This latter method is useful to check the quality of new oils at blending plants or against specifications. However, there are a number of instrumental techniques that enable the results to be obtained very much more rapidly, among these being polarographic, flame photometric, and X-ray fluorescence (XRF) methods. Chlorine can be determined by a chemical method such as the silver chloride test method (ASTM D808).

Phosphorus can serve as a beneficial adjunct or as a deleterious agent. There are several test methods for the determination of phosphorus. In addition to the three test methods described here, reference should also be made to multielement analysis methods such as inductively coupled plasma atomic emission spectroscopy (ASTM D4951, ASTM D5185) and XRF (ASTM D4927, ASTM D6443). Phosphorus can also be determined by a test method (ASTM D1091) in which the organic material in the sample is destroyed and phosphorus in the sample is converted to phosphate ion by oxidation with sulfuric acid, nitric acid, and hydrogen peroxide, and the magnesium pyrophosphate is determined gravimetrically. Another method (ASTM D4047, IP 149) in which the phosphorus is converted to quinoline phosphomolybdate is also available.

The extent and nature of the contamination of a used automotive engine oil by oxidation and combustion products can be ascertained by determining the amounts of materials present in the lubricating oil that are insoluble in n-pentane and benzene (Speight and Exall, 2014).

In this test, a solution of the used lubricating oil in pentane is centrifuged, the oil solution is decanted, and the precipitate washed, dried, and weighed. Insoluble constituents (precipitate) are expressed as a percentage by weight of the original amount of used oil taken and include the resinous material resulting from the oxidation of the oil in service, together with the benzene-insoluble constituents. The latter are determined on a separate portion of sample that is weighed, mixed with pentane, and centrifuged. The precipitate is washed twice with pentane, once with benzene–alcohol solution and once with benzene. The insoluble material is then dried and weighed to give the percentage of benzene-insoluble constituents that contain wear debris, dirt, carbonaceous matter from the combustion products, and decomposition products of the oil, additives, and fuel.

Where highly detergent/dispersant oils are under test, coagulated pentane-insoluble constituents and coagulated benzene-insoluble constituents may be determined, using methods similar to those just described, but employing a coagulant to precipitate the very finely divided materials that may otherwise be kept in suspension by the detergent/dispersant additives.

Size discrimination of insoluble matter may be made to distinguish between finely dispersed relatively harmless matter and the larger potentially harmful particles in oil (ASTM D 4055). The method employs filtration through membranes of known pore size. Membrane filtration techniques are being increasingly used and are expected to be published shortly as standard test procedures.

The metallic constituents (barium, boron, calcium, magnesium, tin, silicon, zinc, aluminum, sodium, potassium, etc.) of new and used lubricating oils can be determined by a comprehensive system of chemical analysis (ASTM D874, IP 163).

Turbine oil systems usually contain some free water as a result of steam leaking through glands and then condensing. Marine systems may also have saltwater present due to leakage from coolers. Because of this, rust inhibitors are usually incorporated. The rust-preventing properties of turbine oils are measured by a method (ASTM D665, IP 135) that employs synthetic seawater or distilled water in the presence of steel. The oil should also be noncorrosive to copper (ASTM D130, IP 154).

The presence of water in turbine systems tends to lead to the formation of emulsions and sludge containing water, oil, oil oxidation products, rust particles, and other solid contaminants that can seriously impair lubrication. The lubricating oil, therefore, should have the ability to separate from water readily and to resist emulsification during passage of steam into the oil until a predetermined volume has condensed and the time required for separation measured (IP 19). Alternatively, the rate of separation of oil that has been stirred with equal volume of water is measured (ASTM D1401). These test methods are only approximate guides to the water-separating characteristics of modern inhibited turbine oils, and the results should be used in conjunction with experience gained of the particular service conditions encountered.

Although systems should be designed to avoid entrainment of air in the oil, it is not always possible to prevent (ASTM D892, IP 146). The formation of stable foam (ASTM D892, ASTM D6082, IP 146) increases the surface area of the oil that is exposed to small bubbles of air thus assisting oxidation. The foam can also cause loss of oil from the system by overflow. Defoamants are usually incorporated in turbine oils to decrease their foaming tendency. Air release is also an important property if a soft or spongy governor system is to be avoided. A careful choice of type and amount of defoamant will provide the correct balance of foam protection and air release properties.

Dilution of oil by fuel under low-temperature or short-distance stop/start operation can occur frequently. Dilution of engine oil by the fuel can be estimated from gas chromatography (ASTM D3525). Low-temperature service conditions may also result in water vapor from combustion products condensing in the crankcase (ASTM D95, IP 74).

13.4.9 Density

These are alternative but related means of expressing the weight of a measured volume of a product.

Both density (specific gravity) and API gravity measurements are used as manufacturing control tests and, in conjunction with other tests, are also used for characterizing unknown oils since they correlate approximately with hydrocarbon composition and, therefore, with the nature of the crude source of the oil (ASTM D1298, IP 160).

For lubricating oil, the purpose of limiting density range is as a check on oil composition. In addition, a minimum density may offer some indication of solvent power as well as guarding against excessive paraffin content. In this respect, the inclusion of density in a mineral oil specification may duplicate the aniline point (ASTM D611, IP 2) requirement.

The API gravity (ASTM D287, IP 160) is also used for lubricating oil and is based on a hydrometer scale that may be readily converted to the relative density basis by use of tables or formulae (Chapter 2):

$$\text{API gravity}\left(°\right) = \frac{141.5}{\text{sp gr } 60/60°F} - 131.5$$

and is also a critical measure for reflecting the quality of lubricating oil.

13.4.10 Flash Point and Fire Point

The *flash point* test gives an indication of the presence of volatile components in oil and is the temperature to which the oil must be heated under specified test conditions to give off sufficient vapor to form a flammable mixture with air.

The *fire point* is the temperature to which the product must be heated under the prescribed test conditions to cause the vapor/air mixture to burn continuously on ignition. The Cleveland open cup method (ASTM D92, IP 36) can be used to determine both flash and fire points of lubricating oils, and the Pensky–Martens closed (ASTM D93, IP 34) and open (IP 35) flash points are also widely used.

The flash and fire points are significant in cases where high-temperature operations are encountered, not only for the hazard of fire, but also as an indication of the volatility of an oil. In the case of used oils, the flash point is employed to indicate the extent of contamination with a more volatile oil or with fuels such as gasoline (ASTM D3607). The flash point can also be used to assist in the identification of different types of base oil blend.

For used automotive engine oils that can be contaminated by a variety of materials, the presence of diesel fuel constituents, resulting from low-temperature or short distance stop-start operation, can be approximately estimated from measurements of the flash point of the oil (ASTM D92, IP 36) that is appreciably lowered by small quantities of fuel. The presence of gasoline constituents can be determined by distillation (ASTM D322) or by infrared spectroscopy.

Fire-resistant lubricating oil is used widely in the coal mining industry. The use of such fluids also is expanding in the metal cutting and forming, lumber, steel, aluminum, and aircraft industries. A test is also available to evaluate the fire-resistant properties of lubricating oil under a variety of conditions (ASTM D5306). Most tests involve dripping, spraying, or pouring the liquid into a flame or on a hot surface of molten metal, but in this test, the fluid is impregnated into ceramic fiber media, and the linear flame propagation rate used for the comparison of relative flammability is measured.

13.4.11 Oxidation Stability

Oxidation results in the development of acidic products that can lead to corrosion and can also affect the ability of the oil to separate from water. Oxidation can also lead to an increase in viscosity and the formation of sludge that can restrict oil paths, thus impairing circulation of the oil and interfering with the function of governors and oil relays. Correctly formulated turbine oils have excellent resistance to oxidation and will function satisfactorily for long periods without changing the system charge. Oxidation stability can be assessed by various tests (ASTM D943, IP 157) that employ copper as well as iron as catalysts in the presence of water to simulate metals present in service conditions.

Although systems are usually designed to avoid entrainment of air in the oil, it is not always possible to prevent this, and the formation of a stable foam increases the surface area of the oil that is exposed to small bubbles of air, thus assisting oxidation. De-foaming agents are usually incorporated in turbine oils to decrease their foaming tendency, and this can be measured (ASTM D892, IP 146). Air release is also an important property, and a careful choice of type and amount of de-foaming agents is necessary to provide the correct balance of foam protection and air release properties.

13.4.12 Pour Point

The pour point (ASTM D97, IP 15) is the lowest temperature at which the oil will flow under specified test conditions and is roughly equivalent to the tendency of an oil to cease to flow from a gravity-fed system or from a container and is a guide to, but not an exact measure of, the temperature at which flow ceases under the service conditions of a specific system.

The pour point of wax-containing oils can be reduced by the use of special additives known as pour-point depressants that inhibit the growth of wax crystals, thus preventing the formation of a solid structure. It is a recognized property of oil of this type that previous thermal history may affect the measured pour point. The test procedure (ASTM D97, IP 15) also permits some measurement of the effect of thermal conditions on waxy oils.

The importance of the pour point, to the user of lubricants, is limited to applications where low temperatures are likely to influence oil flow. Obvious examples are refrigerator lubricants and automotive engine oils in cold climates. Any pump installed in outside locations where temperatures periodically fall below freezing should utilize lubricants with a pour point below some temperatures, or the borderline pumping temperature can be determined by a designated test method (ASTM D3829).

13.4.13 Thermal Stability

The Panel Coking Test, when used in conjunction with other tests, can be used to assess the deposit-forming tendencies due to thermal instability, and the available alkalinity of these oils can be measured (Speight and Exall, 2014).

13.4.14 Viscosity

The viscosity of lubricating oil is a measure of its flow characteristics. It is generally the most important controlling property for manufacture and for selection to meet a particular application. The viscosity of a mineral oil changes with temperature, but not normally with high stress and shear rate (ASTM D5275, ASTM D5481, IP 294), unless specific additives that may not be shear stable are included to modify the viscosity–temperature characteristics. Explosions can also result when lubricating oil is in contact with certain metals under high shear conditions (ASTM D3115).

Thus, for base oils, the rate of flow of the oil through a pipe or capillary tube is directly proportional to the pressure applied. This property is measured for most practical purposes by timing the flow of a fixed amount of oil through a calibrated glass capillary tube under gravitational force at a standard temperature and is known as the kinematic viscosity of the oil (ASTM D445, IP 71). The unit of viscosity used in conjunction with this method is the centistoke (cSt), but this may be converted into other viscosity systems (Saybolt, Redwood, Engler) by means of conversion formulae. At very high pressures, the viscosity of mineral oils increases considerably with increase in pressure, the extent depending on the crude source of the oil and on the molecular weight (ASTM D2502, ASTM D2878) of the constituent components.

Because the main objective of lubrication is to provide a film between load-bearing surfaces, the selection of the correct viscosity for the oil is aimed at a balance between a viscosity high enough to prevent the lubricated surfaces from contacting and low enough to minimize energy losses through excessive heat generation caused by having too viscous a lubricant (ASTM D2422, BS-4231).

The viscosity of automotive engine oil is the main controlling property for manufacture and for selection to meet the particular service condition using the American SAE viscosity classification. The higher-viscosity oils are standardized at 210°F (99°C), and the lighter oils that are intended for use in cold weather conditions are standardized at 0°F (−18°C). The standard viscosity temperature charts (ASTM D341) are useful for estimating viscosity at the various temperatures that are likely to be encountered in service.

The principal difference between the requirements of gas and other internal combustion engine oils is the necessity to withstand the degradation that can occur from accumulation of oxides of nitrogen in the oil that are formed by combustion. The condition of gas engine oils in large engines can be followed by measuring oil viscosity increase (ASTM D97, IP 177, IP 139) to determine changes in the neutralization value resulting from oxidation. In addition, analytical techniques such as infrared spectroscopy and membrane filtration can be used to check for nitration of the oil and the buildup of suspended carbonaceous material.

The *viscosity index* is an empirical number that indicates the effect of change of temperature on the viscosity of lubricating oil. Many lubricant applications require the lubricant to perform across a wide range of conditions, for example, automotive lubricants are required to reduce friction between engine components when the engine is started from cold relative to operating temperatures up to 200°C (392°F) when it is running. The best oils with the highest viscosity index will remain stable and have little variation in viscosity over the temperature range. This allows for consistent engine performance within the normal working conditions. Multigrade motor oils do not behave as Newtonian oils, and the improved viscosity/temperature characteristics of multigrade oils enable, for example, an oil to be formulated to have mixed characteristics (ASTM D3829).

The viscosity index can be calculated using the following formula:

$$VI = \frac{L-U}{L-H} \times 100$$

VI is the viscosity index, U is the kinematic viscosity at 40°C (104°F), and L and H are various values based on the kinematic viscosity at 100°C (212°F) (ASTM D2270; IP 226).

Thus, the viscosity index is important in applications where an appreciable change in temperature of the lubricating oil could affect the operating characteristics of the equipment. Automatic transmissions for passenger vehicles are an example of this where high-viscosity index oils using improvers are used to minimize differences between a viscosity low enough to permit a sufficiently rapid gear shift when starting under cold conditions and a viscosity adequate at the higher temperatures encountered in normal running.

Paraffinic oils have the lowest rate of change of viscosity with temperature (highest viscosity index), while the naphthenic/aromatic oils had the highest rate of change (lowest viscosity index) (ASTM D2270, IP 226).

The viscosity index of multigrade automotive engine oils is typically in the range 130–190, while monograde oils are usually between 85 and 105. The improved viscosity/temperature characteristics of multigrade oils enable, for example, an SAE 20W/50 oil to be formulated that spans SAE 20W viscosity characteristics at low temperatures, and SAE 40–50 characteristics at the working temperature.

However, multigrade oils do not behave as Newtonian fluids, and this is primarily due to the presence of polymeric viscosity index improvers. The result is that the viscosity of multigrade oils is generally higher at −18°C (0°F) than is predicted by extrapolation from 38°C (100°F) and 99°C (210°F) data; the extent of the deviation varies with the type and amount of the viscosity index improver used. To overcome this, the SAE classification is based on a measured viscosity at −18°C (0°F) using a laboratory test apparatus known as a cold cranking simulator (ASTM D5293).

13.4.15 Volatility

The volatility of lubricating oil is not usually an issue for multi-testing. Nevertheless, test methods need to be available so that specification and purity checks can be made.

A method that is used to determine pitch volatility (ASTM D4893) might also be used, on occasion, to determine the nonvolatility of lubricating oil. In the method, an aluminum dish containing about 15 g of accurately weighed sample is introduced into the cavity of a metal block heated and maintained at 350°C (662°F). After 30 min, during which the volatiles are swept away from the surface of the sample by preheated nitrogen, the residual sample is taken out and allowed to cool down in the desiccator. The nonvolatility is determined by the sample weight remaining and reported as percentage w/w residue.

A test is also available for the determination of engine oil volatility at 371°C (700°F), which is actually a requirement in some lubricant specifications (ASTM D6417). This test method can be used on lubricant products not within the scope of other test methods using simulated distillation methodologies (ASTM D2887). This test method applicability is limited to samples having an initial boiling point greater than 126°C. This test method may be applied to both lubricant oil base stocks and finished lubricants containing additive packages.

In the test, a sample aliquot diluted with a viscosity-reducing solvent is introduced into the gas chromatographic system that uses a nonpolar open tubular capillary gas chromatographic column for eluting the hydrocarbon components of the sample in the order of increasing boiling point. The column oven temperature is raised at a reproducible linear rate effect separation of the hydrocarbons. The quantitation is achieved with a flame ionization detector. The sample retention times are compared to those of known hydrocarbon mixtures, and the cumulative corrected area of the sample determined to the 371°C (700°F) retention time is used to calculate the percentage of oil volatilized at 371°C (700°F).

13.4.16 Water and Sediment

Knowledge of the water content of petroleum products is important in refining, purchase and sale, and transfer of products, and is useful in predicting the quality and performance characteristics of the products.

The Karl Fischer test method (ASTM D6304) can be applied to the direct determination of water in lubricating oil. In the method, the sample injection in the titration vessel can be done volumetrically or gravimetrically. The instrument automatically titrates the sample and displays the result at the end of the titration. Viscous samples can be analyzed by using a water vaporizer accessory that heats the sample in the evaporation chamber, and the vaporized water is carried into the Karl Fischer titration cell by a dry inert carrier gas.

Sediment in lubricating oil can lead to system malfunction in critical applications, and determination of the amount of sediment is a necessity. In the test method (ASTM D2273), a 100-ml sample of oil is mixed with 50 ml of ASTM precipitation naphtha and is heated in a water bath at 32–35°C (90–95°F) for 5 min. The centrifuge tube containing the heated mixture is centrifuged for 10 min at a rate of between 600 and 700 relative centrifuge force. After decanting the mixture carefully, the procedure is repeated with another portion of naphtha and oil. The final reading of sediment is recorded. This test method is not applicable in cases where precipitated oil-soluble components will appreciably contribute to the sediment yield.

Insoluble material may form in lubricating oil in oxidizing conditions, and a test method is available (ASTM D4310) to evaluate the tendency of lubricating oil to corrode copper catalyst metal and to form sludge during oxidation in the presence of oxygen, water, and copper and iron metals at an elevated temperature. This test method is a modification of another test method (ASTM D943) where the oxidation stability of the same kind of oils is determined by following the acid number of oil. In the test method (ASTM D4310), an oil sample is contacted with oxygen in the presence of water and iron–copper catalyst at 95°C (203°F) for 100 h. The weight of the insoluble material is determined gravimetrically by filtration of the oxidation tube contents through a 5-μm pore size filter disk. The total amount of copper in the oil, water, and sludge phases is also determined.

REFERENCES

ASTM D91. Standard Test Method for Precipitation Number of Lubricating Oils. Annual Book of Standards. ASTM International, West Conshohocken, PA.

ASTM D92. Standard Test Method for Flash and Fire Points by Cleveland Open Cup Tester. Annual Book of Standards. ASTM International, West Conshohocken, PA.

ASTM D93. Standard Test Methods for Flash Point by Pensky-Martens Closed Cup Tester. Annual Book of Standards. ASTM International, West Conshohocken, PA.

ASTM D94. Standard Test Methods for Saponification Number of Petroleum Products. Annual Book of Standards. ASTM International, West Conshohocken, PA.

ASTM D95. Standard Test Method for Water in Petroleum Products and Bituminous Materials by Distillation. Annual Book of Standards. ASTM International, West Conshohocken, PA.

ASTM D97. Standard Test Method for Pour Point of Petroleum Products. Annual Book of Standards. ASTM International, West Conshohocken, PA.

ASTM D129. Standard Test Method for Sulfur in Petroleum Products (General High Pressure Decomposition Device Method). Annual Book of Standards. ASTM International, West Conshohocken, PA.

ASTM D130. Standard Test Method for Corrosiveness to Copper from Petroleum Products by Copper Strip Test. Annual Book of Standards. ASTM International, West Conshohocken, PA.

ASTM D156. Standard Test Method for Saybolt Color of Petroleum Products (Saybolt Chromometer Method). Annual Book of Standards. ASTM International, West Conshohocken, PA.

ASTM D189. Standard Test Method for Conradson Carbon Residue of Petroleum Products. Annual Book of Standards. ASTM International, West Conshohocken, PA.

ASTM D287. Standard Test Method for API Gravity of Crude Petroleum and Petroleum Products (Hydrometer Method). Annual Book of Standards. ASTM International, West Conshohocken, PA.

ASTM D322. Standard Test Method for Gasoline Diluent in Used Gasoline Engine Oils by Distillation. Annual Book of Standards. ASTM International, West Conshohocken, PA.

ASTM D341. Viscosity Temperature Charts for Liquid Petroleum Products. Annual Book of Standards. ASTM International, West Conshohocken, PA.

ASTM D445. Standard Test Method for Kinematic Viscosity of Transparent and Opaque Liquids (and Calculation of Dynamic Viscosity). Annual Book of Standards. ASTM International, West Conshohocken, PA.

ASTM D482. Standard Test Method for Ash from Petroleum Products. Annual Book of Standards. ASTM International, West Conshohocken, PA.

ASTM D524. Standard Test Method for Ramsbottom Carbon Residue of Petroleum Products. Annual Book of Standards. ASTM International, West Conshohocken, PA.

ASTM D565. Standard Test Method for Carbonizable Substances in White Mineral Oil. Annual Book of Standards. ASTM International, West Conshohocken, PA.

ASTM D611. Standard Test Methods for Aniline Point and Mixed Aniline Point of Petroleum Products and Hydrocarbon Solvents. Annual Book of Standards. ASTM International, West Conshohocken, PA.

ASTM D664. Standard Test Method for Acid Number of Petroleum Products by Potentiometric Titration. Annual Book of Standards. ASTM International, West Conshohocken, PA.

ASTM D665. Standard Test Method for Rust-Preventing Characteristics of Inhibited Mineral Oil in the Presence of Water. Annual Book of Standards. ASTM International, West Conshohocken, PA.

ASTM D808. Standard Test Method for Chlorine in New and Used Petroleum Products (High Pressure Decomposition Device Method). Annual Book of Standards. ASTM International, West Conshohocken, PA.

ASTM D849. Standard Test Method for Copper Strip Corrosion by Industrial Aromatic Hydrocarbons. Annual Book of Standards. ASTM International, West Conshohocken, PA.

ASTM D874. Standard Test Method for Sulfated Ash from Lubricating Oils and Additives. Annual Book of Standards. ASTM International, West Conshohocken, PA.

ASTM D892. Standard Test Method for Foaming Characteristics of Lubricating Oils. Annual Book of Standards. ASTM International, West Conshohocken, PA.

ASTM D893. Standard Test Method for Insolubles in Used Lubricating Oils. Annual Book of Standards. ASTM International, West Conshohocken, PA.

ASTM D943. Standard Test Method for Oxidation Characteristics of Inhibited Mineral Oils. Annual Book of Standards. ASTM International, West Conshohocken, PA.

ASTM D974. Standard Test Method for Acid and Base Number by Color-Indicator Titration. Annual Book of Standards. ASTM International, West Conshohocken, PA.

ASTM D1091. Standard Test Methods for Phosphorus in Lubricating Oils and Additives. Annual Book of Standards. ASTM International, West Conshohocken, PA.

ASTM D1209. Standard Test Method for Color of Clear Liquids (Platinum-Cobalt Scale). Annual Book of Standards. ASTM International, West Conshohocken, PA.

ASTM D1298. Standard Test Method for Density, Relative Density, or API Gravity of Crude Petroleum and Liquid Petroleum Products by Hydrometer Method. Annual Book of Standards. ASTM International, West Conshohocken, PA.

ASTM D1319. Standard Test Method for Hydrocarbon Types in Liquid Petroleum Products by Fluorescent Indicator Adsorption. Annual Book of Standards. ASTM International, West Conshohocken, PA.

ASTM D1401. Standard Test Method for Water Separability of Petroleum Oils and Synthetic Fluids. Annual Book of Standards. ASTM International, West Conshohocken, PA.

ASTM D1500. Standard Test Method for ASTM Color of Petroleum Products (ASTM Color Scale). Annual Book of Standards. ASTM International, West Conshohocken, PA.

ASTM D1544. Standard Test Method for Color of Transparent Liquids (Gardner Color Scale). Annual Book of Standards. ASTM International, West Conshohocken, PA.

ASTM D1796. Standard Test Method for Water and Sediment in Fuel Oils by the Centrifuge Method (Laboratory Procedure). Annual Book of Standards. ASTM International, West Conshohocken, PA.

ASTM D2006. Method of Test for Characteristic Groups in Rubber Extender and Processing Oils by the Precipitation Method (Withdrawn in 1975 but still in use in many laboratories). ASTM International, West Conshohocken, PA.

ASTM D2007. Standard Test Method for Characteristic Groups in Rubber Extender and Processing Oils and Other Petroleum-Derived Oils by the Clay-Gel Absorption Chromatographic Method. Annual Book of Standards. ASTM International, West Conshohocken, PA.

ASTM D2008. Standard Test Method for Ultraviolet Absorbance and Absorptivity of Petroleum Products. Annual Book of Standards. ASTM International, West Conshohocken, PA.

ASTM D2070. Standard Test Method for Thermal Stability of Hydraulic Oils. Annual Book of Standards. ASTM International, West Conshohocken, PA.

ASTM D2226. Standard Classification for Various Types of Petroleum Oils for Rubber Compounding Use. Annual Book of Standards. ASTM International, West Conshohocken, PA.

ASTM D2269. Standard Test Method for Evaluation of White Mineral Oils by Ultraviolet Absorption. Annual Book of Standards. ASTM International, West Conshohocken, PA.

ASTM D2270. Standard Practice for Calculating Viscosity Index from Kinematic Viscosity at 40 and 100°C. Annual Book of Standards. ASTM International, West Conshohocken, PA.

ASTM D2273. Standard Test Method for Trace Sediment in Lubricating Oils. Annual Book of Standards. ASTM International, West Conshohocken, PA.

ASTM D2415. Standard Test Method for Ash in Coal Tar and Pitch. Annual Book of Standards. ASTM International, West Conshohocken, PA.

ASTM D2422. Standard Classification of Industrial Fluid Lubricants by Viscosity System. Annual Book of Standards. ASTM International, West Conshohocken, PA.

ASTM D2500. Standard Test Method for Cloud Point of Petroleum Products. Annual Book of Standards. ASTM International, West Conshohocken, PA.

ASTM D2502. Standard Test Method for Estimation of Mean Relative Molecular Mass of Petroleum Oils from Viscosity Measurements. Annual Book of Standards. ASTM International, West Conshohocken, PA.

ASTM D2510. Standard Test Method for Adhesion of Solid Film Lubricants. Annual Book of Standards. ASTM International, West Conshohocken, PA.

ASTM D2511. Standard Test Method for Thermal Shock Sensitivity of Solid Film Lubricants. Annual Book of Standards. ASTM International, West Conshohocken, PA.

ASTM D2549. Standard Test Method for Separation of Representative Aromatics and Nonaromatics Fractions of Annual Book of Standards. High-Boiling Oils by Elution Chromatography. ASTM International, West Conshohocken, PA.

ASTM D2624. Standard Test Methods for Electrical Conductivity of Aviation and Distillate Fuels. Annual Book of Standards. ASTM International, West Conshohocken, PA.

ASTM D2649. Standard Test Method for Corrosion Characteristics of Solid Film Lubricants. Annual Book of Standards. ASTM International, West Conshohocken, PA.

ASTM D2711. Standard Test Method for Demulsibility Characteristics of Lubricating Oils. Annual Book of Standards. ASTM International, West Conshohocken, PA.

ASTM D2782. Standard Test Method for Measurement of Extreme-Pressure Properties of Lubricating Fluids (Timken Method). Annual Book of Standards. ASTM International, West Conshohocken, PA.

ASTM D2783. Standard Test Method for Measurement of Extreme-Pressure Properties of Lubricating Fluids (Four-Ball Method). Annual Book of Standards. ASTM International, West Conshohocken, PA.

ASTM D2878. Standard Test Method for Estimating Apparent Vapor Pressures and Molecular Weights of Lubricating Oils. Annual Book of Standards. ASTM International, West Conshohocken, PA.

ASTM D2887. Standard Test Method for Boiling Range Distribution of Petroleum Fractions by Gas Chromatography. Annual Book of Standards. ASTM International, West Conshohocken, PA.

ASTM D2893. Standard Test Methods for Oxidation Characteristics of Extreme-Pressure Lubrication Oils. Annual Book of Standards. ASTM International, West Conshohocken, PA.

ASTM D2896. Standard Test Method for Base Number of Petroleum Products by Potentiometric Perchloric Acid Titration. Annual Book of Standards. ASTM International, West Conshohocken, PA.

ASTM D3115. Standard Test Method for Explosive Reactivity of Lubricants with Aerospace Alloys under High Shear. Annual Book of Standards. ASTM International, West Conshohocken, PA.

ASTM D3233. Standard Test Methods for Measurement of Extreme Pressure Properties of Fluid Lubricants (Falex Pin and Vee Block Methods. Annual Book of Standards. ASTM International, West Conshohocken, PA.

ASTM D3239. Standard Test Method for Aromatic Types Analysis of Gas-Oil Aromatic Fractions by High Ionizing Voltage Mass Spectrometry. Annual Book of Standards. ASTM International, West Conshohocken, PA.

ASTM D3279. Standard Test Method for n-Heptane Insolubles. Annual Book of Standards. ASTM International, West Conshohocken, PA.

ASTM D3525. Standard Test Method for Gasoline Diluent in Used Gasoline Engine Oils by Gas Chromatography. Annual Book of Standards. ASTM International, West Conshohocken, PA.

ASTM D3607. Standard Test Method for Removing Volatile Contaminants from Used Engine Oils by Stripping. Annual Book of Standards. ASTM International, West Conshohocken, PA.

ASTM D3829. Standard Test Method for Predicting the Borderline Pumping Temperature of Engine Oil. Annual Book of Standards. ASTM International, West Conshohocken, PA.

ASTM D4047. Standard Test Method for Phosphorus in Lubricating Oils and Additives by Quinoline Phosphomolybdate Method. Annual Book of Standards. ASTM International, West Conshohocken, PA.

ASTM D4052. Standard Test Method for Density, Relative Density, and API Gravity of Liquids by Digital Density Meter. Annual Book of Standards. ASTM International, West Conshohocken, PA.

ASTM D4055. Standard Test Method for Pentane Insolubles by Membrane Filtration. Annual Book of Standards. ASTM International, West Conshohocken, PA.

ASTM D4072. Standard Test Method for Toluene-Insoluble (TI) Content of Tar and Pitch. Annual Book of Standards. ASTM International, West Conshohocken, PA.

ASTM D4124. Standard Test Method for Separation of Asphalt into Four Fractions. Annual Book of Standards. ASTM International, West Conshohocken, PA.

ASTM D4308. Standard Test Method for Electrical Conductivity of Liquid Hydrocarbons by Precision Meter. Annual Book of Standards. ASTM International, West Conshohocken, PA.

ASTM D4310. Standard Test Method for Determination of Sludging and Corrosion Tendencies of Inhibited Mineral Oils. Annual Book of Standards. ASTM International, West Conshohocken, PA.

ASTM D4312. Standard Test Method for Toluene-Insoluble (TI) Content of Tar and Pitch (Short Method. Annual Book of Standards. ASTM International, West Conshohocken, PA.

ASTM D4530. Standard Test Method for Determination of Carbon Residue (Micro Method). Annual Book of Standards. ASTM International, West Conshohocken, PA.

ASTM D4628. Standard Test Method for Analysis of Barium, Calcium, Magnesium, and Zinc in Unused Lubricating Oils by Atomic Absorption Spectrometry. Annual Book of Standards. ASTM International, West Conshohocken, PA.

ASTM D4636. Standard Test Method for Corrosiveness and Oxidation Stability of Hydraulic Oils, Aircraft Turbine Engine Lubricants, and Other Highly Refined Oils. Annual Book of Standards. ASTM International, West Conshohocken, PA.

ASTM D4684. Standard Test Method for Determination of Yield Stress and Apparent Viscosity of Engine Oils at Low Temperature. Annual Book of Standards. ASTM International, West Conshohocken, PA.

ASTM D4739. Standard Test Method for Base Number Determination by Potentiometric Hydrochloric Acid Titration. Annual Book of Standards. ASTM International, West Conshohocken, PA.

ASTM D4742. Standard Test Method for Oxidation Stability of Gasoline Automotive Engine Oils by Thin-Film Oxygen Uptake (TFOUT). Annual Book of Standards. ASTM International, West Conshohocken, PA.

ASTM D4893. Standard Test Method for Determination of Pitch Volatility. Annual Book of Standards. ASTM International, West Conshohocken, PA.

ASTM D4927. Standard Test Methods for Elemental Analysis of Lubricant and Additive Components – Barium, Calcium, Phosphorus, Sulfur, and Zinc by Wavelength-Dispersive X-Ray Fluorescence Spectroscopy. Annual Book of Standards. ASTM International, West Conshohocken, PA.

ASTM D4951. Standard Test Method for Determination of Additive Elements in Lubricating Oils by Inductively Coupled Plasma Atomic Emission Spectrometry. Annual Book of Standards. ASTM International, West Conshohocken, PA.

ASTM D5183. Standard Test Method for Determination of the Coefficient of Friction of Lubricants Using the Four-Ball Wear Test Machine. Annual Book of Standards. ASTM International, West Conshohocken, PA.

ASTM D5185. Standard Test Method for Multi-element Determination of Used and Unused Lubricating Oils and Base Oils by Inductively Coupled Plasma Atomic Emission Spectrometry (ICP-AES). Annual Book of Standards. ASTM International, West Conshohocken, PA.

ASTM D5275. Standard Test Method for Fuel Injector Shear Stability Test (FISST) for Polymer Containing Fluids. Annual Book of Standards. ASTM International, West Conshohocken, PA.

ASTM D5292. Standard Test Method for Aromatic Carbon Contents of Hydrocarbon Oils by High Resolution Nuclear Magnetic Resonance Spectroscopy. Annual Book of Standards. ASTM International, West Conshohocken, PA.

ASTM D5293. Standard Test Method for Apparent Viscosity of Engine Oils and Base Stocks Between –5 and –35°C Using Cold-Cranking Simulator. Annual Book of Standards. ASTM International, West Conshohocken, PA.

ASTM D5306. Standard Test Method for Linear Flame Propagation Rate of Lubricating Oils and Hydraulic Fluids. Annual Book of Standards. ASTM International, West Conshohocken, PA.

ASTM D5481. Standard Test Method for Measuring Apparent Viscosity at High-Temperature and High-Shear Rate by Multicell Capillary Viscometer. Annual Book of Standards. ASTM International, West Conshohocken, PA.

ASTM D5770. Standard Test Method for Semiquantitative Micro Determination of Acid Number of Lubricating Oils during Oxidation Testing. Annual Book of Standards. ASTM International, West Conshohocken, PA.

ASTM D5800. Standard Test Method for Evaporation Loss of Lubricating Oils by the Noack Method. Annual Book of Standards. ASTM International, West Conshohocken, PA.

ASTM D5846. Standard Test Method for Universal Oxidation Test for Hydraulic and Turbine Oils Using the Universal Oxidation Test Apparatus. Annual Book of Standards. ASTM International, West Conshohocken, PA.

ASTM D5853. Standard Test Method for Pour Point of Crude Oils. Annual Book of Standards. ASTM International, West Conshohocken, PA.

ASTM D5949. Standard Test Method for Pour Point of Petroleum Products (Automatic Pressure Pulsing Method). Annual Book of Standards. ASTM International, West Conshohocken, PA.

ASTM D5950. Standard Test Method for Pour Point of Petroleum Products (Automatic Tilt Method). Annual Book of Standards. ASTM International, West Conshohocken, PA.

ASTM D5972. Standard Test Method for Freezing Point of Aviation Fuels (Automatic Phase Transition Method). Annual Book of Standards. ASTM International, West Conshohocken, PA.

ASTM D5984. Standard Test Method for Semi-Quantitative Field Test Method for Base Number in New and Used Lubricants by Color-Indicator Titration. Annual Book of Standards. ASTM International, West Conshohocken, PA.

ASTM D5985. Standard Test Method for Pour Point of Petroleum Products (Rotational Method). Annual Book of Standards. ASTM International, West Conshohocken, PA.

ASTM D6045. Standard Test Method for Color of Petroleum Products by the Automatic Tristimulus Method. Annual Book of Standards. ASTM International, West Conshohocken, PA.

ASTM D6082. Standard Test Method for High Temperature Foaming Characteristics of Lubricating Oils. Annual Book of Standards. ASTM International, West Conshohocken, PA.

ASTM D6304. Standard Test Method for Determination of Water in Petroleum Products, Lubricating Oils, and Additives by Coulometric Karl Fischer Titration. Annual Book of Standards. ASTM International, West Conshohocken, PA.

ASTM D6351. Standard Test Method for Determination of Low Temperature Fluidity and Appearance of Hydraulic Fluids. Annual Book of Standards. ASTM International, West Conshohocken, PA.

ASTM D6417. Standard Test Method for Estimation of Engine Oil Volatility by Capillary Gas Chromatography. Annual Book of Standards. ASTM International, West Conshohocken, PA.

ASTM D6425 Standard Test Method for Measuring Friction and Wear Properties of Extreme Pressure (EP) Lubricating Oils Using SRV Test Machine. Annual Book of Standards. ASTM International, West Conshohocken, PA.

ASTM D6443. Standard Test Method for Determination of Calcium, Chlorine, Copper, Magnesium, Phosphorus, Sulfur, and Zinc in Unused Lubricating Oils and Additives by Wavelength Dispersive X-ray Fluorescence Spectrometry (Mathematical Correction Procedure). Annual Book of Standards. ASTM International, West Conshohocken, PA.

ASTM D6560. Standard Test Method for Determination of Asphaltenes (Heptane Insolubles) in Crude Petroleum and Petroleum Products. Annual Book of Standards. ASTM International, West Conshohocken, PA.

ASTM E2412. Standard Practice for Condition Monitoring of In-Service Lubricants by Trend Analysis Using Fourier Transform Infrared (FT-IR) Spectrometry. Annual Book of Standards. ASTM International, West Conshohocken, PA.

Banaszewski, A. and Blythe, J. 2000. White mineral oil. In: Modern Petroleum Technology. Volume 2: Downstream. A.G. Lucas (Editor). John Wiley & Sons Inc., New York.

Boughton, R. and Horvath, A. 2004. Environmental Assessment of Used Oil Management Methods. Environmental Science & Technology, 38(2): 353–358.

The Energy Institute. 2013. IP Standard Methods 2013. The Institute of Petroleum, London.

Gary, J.G., Handwerk, G.E., and Kaiser, M.J. 2007. Petroleum Refining: Technology and Economics, 5th Edition. CRC Press, Taylor & Francis Group, Boca Raton, FL.

Harperscheid, M. and Omeis, J. 2007. Lubricants for Internal Combustion Engines. In: Lubricants and Lubrication, 2nd Edition. T. Mang and W. Dresel (Editors). Wiley-VCH Verlag GmbH & Co. KGaA, Weinheim, Germany. pp. 191–229.

Hsu, C.S. and Robinson, P.R. (Editors) 2006. Practical Advances in Petroleum Processing (Volumes 1 and 2). Springer Science, New York.

IP 2 (ASTM D611). Petroleum Products and Hydrocarbon Solvents – Determination of Aniline and Mixed Aniline Point. IP Standard Methods 2013. The Energy Institute, London.

IP 4 (ASTM D482). Petroleum Products – Determination of Ash. IP Standard Methods 2013. The Energy Institute, London.

IP 13 (ASTM D189). Petroleum Products – Determination of Carbon Residue – Conradson Method. IP Standard Methods 2013. The Energy Institute, London.

IP 14 (ASTM D524). Petroleum Products – Determination of Carbon Residue – Ramsbottom Method. IP Standard Methods 2013. The Energy Institute, London.

IP 15 (ASTM D97). Petroleum Products – Determination of Pour Point. IP Standard Methods 2013. The Energy Institute, London.

IP 16 (ASTM D2386). Petroleum Products – Determination of the Freezing Point of Aviation Fuels – Manual Method. IP Standard Methods 2013. The Energy Institute, London.

IP 17. Determination of Color – Lovibond Tintometer Method. IP Standard Methods 2013. The Energy Institute, London.

IP 19. Demulsibility Characteristics of Lubricating Oil. Energy Institute, London, UK.

IP 34 (ASTM D93). Determination of Flash Point – Pensky-Martens Closed Cup Method. IP Standard Methods 2013. The Energy Institute, London.

IP 35. Open Flash and Fire Point – Pensky-Martens Method. Energy Institute, London, UK.

IP 36. Determination of Open Flash and Fire Point – Cleveland Method. IP Standard Methods 2013. The Energy Institute, London.

IP 61 (ASTM D129). Determination of Sulfur – High Pressure Combustion Method. IP Standard Methods 2013. The Energy Institute, London.

IP 71 (ASTM D445). Petroleum Products – Transparent and Opaque Liquids – Determination of Kinematic Viscosity and Calculation of Dynamic Viscosity. IP Standard Methods 2013. The Energy Institute, London.

IP 74 (ASTM D95). Petroleum Products and Bituminous Materials – Determination of Water – Distillation Method. IP Standard Methods 2013. The Energy Institute, London.

IP 135 (ASTM D665). Determination of Rust-Preventing Characteristics of Steam-Turbine Oil in the Presence of Water. IP Standard Methods 2013. The Energy Institute, London.

IP 136 (ASTM D94). Petroleum Products – Determination of Saponification Number – Part 1 – Color-Indicator Titration Method. IP Standard Methods 2013. The Energy Institute, London.

IP 139 (ASTM D974). Petroleum Products and Lubricants – Determination of Acid or Base Number – Color Indicator Titration Method. IP Standard Methods 2013. The Energy Institute, London.

IP 143 (ASTM D6560). Determination of Asphaltenes (Heptane Insolubles) in Crude Petroleum and Petroleum Products. IP Standard Methods 2013. The Energy Institute, London.

IP 146 (ASTM D892). Determination of Foaming Characteristics of Lubricating Oils. IP Standard Methods 2013. The Energy Institute, London.

IP 149 (ASTM D4047). Petroleum Products – Lubricating Oils and Additives – Determination of Phosphorous Content – Quinoline Phosphomolybdate Method. IP Standard Methods 2013. The Energy Institute, London.

IP 154. Petroleum Products – Corrosiveness to Copper – Copper Strip Test. IP Standard Methods 2013. The Energy Institute, London.

IP 156. Petroleum Products and Related Materials – Determination of Hydrocarbon Types – Fluorescent Indicator Adsorption Method. IP Standard Methods 2013. The Energy Institute, London.

IP 157. Determination of the Oxidation Stability of Inhibited Mineral Oils (TOST Test). IP Standard Methods 2013. The Energy Institute, London.

IP 160 (ASTM D1298). Crude Petroleum and Liquid Petroleum Products – Laboratory Determination of Density – Hydrometer Method. IP Standard Methods 2013. The Energy Institute, London.

IP 163. Petroleum Products – Determination of Sulfated Ash in Lubricating Oils and Additives. IP Standard Methods 2013. The Energy Institute, London.

IP 177 (ASTM D664). Determination of Weak and Strong Acid Number – Potentiometric Titration Method. IP Standard Methods 2013. The Energy Institute, London.

IP 219. Petroleum Products – Determination of Cloud Point. IP Standard Methods 2013. The Energy Institute, London.

IP 226 (ASTM D2270). Petroleum Products – Calculation of Viscosity Index From Kinematic Viscosity. IP Standard Methods 2013. The Energy Institute, London.

IP 240 (ASTM D2782). Determination of Extreme Pressure Properties of Lubricating Fluids – Timken Method. IP Standard Methods 2013. The Energy Institute, London.

IP 274 (ASTM D2624). Determination of Electrical Conductivity of Aviation and Distillate Fuels. IP Standard Methods 2013. The Energy Institute, London.

IP 276. Petroleum Products – Determination of Base Number Perchloric Acid Potentiometric Titration Method. IP Standard Methods 2013. The Energy Institute, London.

IP 294. Shear Stability of Polymer-Containing Oils – Diesel injector Rig Method. Energy Institute, London, UK.

IP 398. Petroleum Products – Determination of Carbon Residue – Micro Method. IP Standard Methods 2013. The Energy Institute, London.

IP 435. Freezing Point of Aviation Turbine Fuels – Automatic Phase Transition Method. IP Standard Methods 2013. The Energy Institute, London.

IP 528 (ASTM D7154). Determination of the Freezing Point of Aviation Turbine Fuels – Automated Fiber Optic Method. IP Standard Methods 2013. The Energy Institute, London.

Lynch, T.R. 2008. Process Chemistry of Lubricant Base Stocks. CRC Press, Taylor & Francis Group, Boca Raton, FL.

Pillon, L.Z. 2011. Surface Activity of Petroleum Derived Lubricants. CRC Press, Taylor & Francis Group, Boca Raton, FL.

Rudnick, L.R. (Editor). 2009. Lubricant Additives: Chemistry and Applications, 2nd Edition. CRC Press, Taylor & Francis Group, Boca Raton, FL.

Rudnick, L.R. (Editor). 2013. Synthetics, Minerals Oils, and Bio-based Lubricants: Chemistry and Technology, 2nd Edition. CRC Press, Taylor & Francis Group, Boca Raton, FL.

Sequeira, A. Jr. 1992. Petroleum Processing Handbook. J.J. McKetta (Editor). Marcel Dekker Inc., New York. p. 634.

Speight, J.G. 2000. The Desulfurization of Heavy Oils and Residua, 2nd Edition. Marcel Dekker Inc., New York.

Speight, J.G. 2001. Handbook of Petroleum Analysis. John Wiley & Sons Inc., New York.

Speight, J.G. 2005. Environmental Analysis and Technology for the Refining Industry. John Wiley & Sons Inc., Hoboken, NJ.

Speight, J.G. 2014. The Chemistry and Technology of Petroleum, 5th Edition. CRC Press, Taylor & Francis Group, Boca Raton, FL.

Speight, J.G. and Ozum, B. 2002. Petroleum Refining Processes. Marcel Dekker Inc., New York.

Speight, J.G. and Arjoon, K.K. 2012. Bioremediation of Petroleum and Petroleum Products. Scrivener Publishing, Salem, MA.

Speight, J.G. and Exall, D.I. 2014. Refining Used Lubricating Oils. CRC Press, Taylor & Francis Group, Boca Raton, FL.

14

GREASE

14.1 INTRODUCTION

The essential function of any lubricant is to prolong the life and increase the efficiency of mechanical devices by reducing friction and wear. Secondary functions include the dissipation of heat, protection from corrosion, power transmission, and removal of contaminants. Generally, fluid lubricants are difficult to retain at the point of application and must be replenished frequently. If, however, a fluid lubricant is thickened, its retention is improved, and lubrication intervals can be extended. Lubricating grease is simply a lubricating fluid that has been gelled with a thickening agent so that the lubricant can be retained more readily in the required area.

By definition, *grease* is a lubricant and is a two-phase system: a solid phase, which is a finely divided thickener that is uniformly dispersed in a liquid-phase lubricant. The liquid is immobilized by the thickener dispersion that must remain relatively stable with respect to time and usage. At operating temperatures, thickeners are insoluble or, at most, only slightly soluble in the liquid lubricant. There must be some affinity between the solid thickener and the liquid lubricant in order to form a stable, gel-like structure—the requirement is that the particles are extremely small and remain uniformly dispersed.

Thus, grease is a solid to semisolid product and is a lubricating fluid that has been gelled with a thickening agent so that the lubricant can be retained more readily into the required area (ASTM D4950; Banaszewski and Blythe, 2000; Speight and Ozum, 2002; Hsu and Robinson, 2006; Gary et al., 2007; Speight, 2014). *Grease* is a lubricating oil to which a thickening agent has been added for the purpose of holding the oil to the surfaces that must be lubricated. The development of the chemistry of grease formulations is closely linked to an understanding of the physics at the interfaces between the machinery and the grease. With this insight, it is possible to formulate greases that are capable of operating in increasingly demanding and wide-ranging conditions.

A wide range of lubricant base fluids are used in grease technology. However, the largest segment consists of a variety of products derived from the refining of crude oil and downstream petroleum raw materials. These mineral oils can contain a very wide spectrum of chemical components, depending on the origin and composition of the crude oil as well as the refining processes to which they have been submitted.

There are three basic groups of mineral oils: (i) aromatic base, (ii) naphthene base, and (iii) paraffin base (Speight, 2014). Historically, the first two have represented the principal volumes used in grease formulation, largely due to availability but also due to their solubility characteristics. However, concerns about the carcinogenic aspects of molecules containing aromatic and polynuclear aromatic ring structures have led to their replacement by paraffinic oils as the mineral fluids of choice.

There are three basic components that contribute to the multiphase structure of lubricating grease: (i) a base fluid, (ii) a thickener, and (iii) various additives that are common in modern grease formulations. The function of the thickener is to provide a physical matrix to hold the base fluid in a solid structure until operating conditions, such as load, shear, and temperature, and initiate viscoelastic flow in the grease. To achieve this matrix, a careful balance of solubility between the base fluid and the thickener is required.

Handbook of Petroleum Product Analysis, Second Edition. James G. Speight.
© 2015 John Wiley & Sons, Inc. Published 2015 by John Wiley & Sons, Inc.

The most widely used thickening agents are soaps of various kinds, and grease manufacture is essentially the mixing of soaps with lubricating oils. Until a relatively short time ago, grease making was considered an art. To stir hot soap into hot oil is a simple business, but to do so in such a manner as to form grease is much more difficult, and the early grease maker needed much experience to learn the essentials of the trade. Therefore, it is not surprising that grease making is still a complex operation. The signs that *told* the grease maker that the soap was *cooked* and that the batch of grease was ready to run have been replaced by scientific tests that follow the process of manufacture precisely.

Modern base oils in lubricating greases are therefore often a blend of severely refined paraffinic and naphthenic oils, designed to provide the final product with the appropriate characteristics of mechanical stability, lubricity, and dropping point.

Finally, the key to providing a grease matrix that is stable, both over time and under the operating shear within machine components, can be found in the thickener system. The thickeners themselves also contribute significantly to the extreme pressure and antiwear characteristics of grease, and additionally, thickeners provide a grease gel capable of carrying additives, which, in turn, extends performance in these areas.

Water resistance, surface adhesion and tackiness, dropping point, and compatibility with other greases are all properties where the selection of the right thickener is important. Increasingly, for centralized lubrication systems, pumpability is an additional prerequisite.

14.2 PRODUCTION AND PROPERTIES

14.2.1 Production

The early grease makers made grease in batches in barrels or pans, and the batch method is still the chief method of making grease. Oil and soap are mixed in kettles that have double walls between which steam and water may be circulated to maintain the desired temperature. When temperatures higher than 150°C (300°F) are required, a kettle heated by a ring of gas burners is used. Mixing is usually accomplished in each kettle by horizontal paddles radiating from a central shaft.

The soaps used in grease making are usually made in the grease plant, usually in a grease-making kettle. Soap is made by chemically combining a metal hydroxide with a fat or fatty acid:

$$RCO_2H + NaOH \rightarrow RCO_2^- Na^+ + H_2O$$
Fatty acid   soap

The most common metal hydroxides used for this purpose are calcium hydroxide, lye, lithium hydroxide, and barium hydroxide. Fats are chemical combinations of fatty acids and glycerin. If a metal hydroxide is reacted with a fat, a soap containing glycerin is formed. Frequently, a fat is separated into its fatty acid and glycerin components, and only the fatty acid portion is used to make soap. Commonly used fats for grease-making soaps are cottonseed oil, tallow, and lard. Among the fatty acids used are stearic acid (from tallow), oleic acid (from cottonseed oil), and animal fatty acids (from lard).

To make grease, the soap is dispersed in the oil as fibers of such a size that it may be possible to detect them only by microscopy. The fibers form a matrix for the oil, and the type, amount, size, shape, and distribution of the soap fibers dictate the consistency, texture, and bleeding characteristics, as well as the other properties of grease. Greases may contain from 50 to 30% soap, and although the fatty acid influences the properties of grease, the metal in the soap has the most important effect. For example, calcium soaps form smooth buttery greases that are resistant to water but are limited in use to temperatures under about 95°C (200°F).

Soda (sodium) salts form fibrous greases that disperse in water but can be used at temperatures well over 95°C (200°F). Barium and lithium soaps form greases similar to those from calcium soaps, but they can be used at both high temperatures and very low temperatures; hence, barium and lithium soap greases are known as multipurpose greases.

The soaps may be combined with any lubricating oil from a light distillate to a heavy residual oil. The lubricating value of the grease is chiefly dependent on the quality and viscosity of the oil. In addition to soap and oil, greases may also contain various additives that are used to improve the ability of the grease to stand up under extreme bearing pressures, to act as a rust preventative, and to reduce the tendency of oil to seep or bleed from grease. Graphite, mica, talc, or fibrous material may be added to grease formulations that are used to lubricate machinery subject to shock loads to absorb the shock of impact. Other chemicals can make grease more resistant to oxidation or modify the structure of the grease.

The older, more common method of grease making is the batch method, but grease is also made by a continuous method. The process involves soap manufacture in a series (usually three) of retorts. Soap-making ingredients are charged into one retort, while soap is made in the second retort. The third retort contains finished soap, which is pumped through a mixing device where the soap and the oil are brought together and blended. The mixer continuously discharges finished grease into suitable containers.

As with the conventional type of lubricating oil (Rudnick, 2009, 2013; Speight and Exall, 2014), viscosity in service is the most important property of solid lubricants and greases since it determines the amount of friction that will be encountered between sliding surfaces and whether a thick enough film can be built up to avoid wear from solid-to-solid contact.

There is no single formulation that can satisfy all of the requirements of a solid lubricant or grease on a cost-effective basis. Properties that should be considered are coefficient of friction, load-carrying capacity, corrosion resistance (susceptibility to galvanic corrosion), and electrical conductivity. Furthermore, one must consider the environment in which the solid-film lubricant or grease must perform. Such materials must be anticipated to be contained (inadvertently or advertently) in the same *disposal pot* as used lubricating oil and therefore contained in the used lubricating oil shipped to a re-refining unit.

14.2.2 Properties

Grease is used to prolong the life and increase the efficiency of mechanical devices by reducing friction and wear (ASTM D4170, ASTM D5707), and there are specific performance requirements (ASTM D3527, ASTM D4170, ASTM D4289, ASTM D4290, ASTM D4693). Generally, a fluid lubricant (such as lubricating oil) is difficult to retain at the point of application and must be replenished frequently. On the other hand, a thickened fluid lubricant (grease) is easier to retain at the point of application, and lubrication intervals can be extended.

Grease varies in texture from soft to hard and in color from light amber to dark brown and, therefore, in contrast to liquid lubricants will stay in place in a bearing assembly with comparatively elementary mechanical seals. Grease also assists in sealing against extraneous material and will lubricate without constant replenishment.

Grease is, essentially, a two-phase system comprised of a liquid-phase lubricant (the liquid phase) containing uniformly dispersed finely divided thickener (the solid phase).

The largest volume of grease in use is made from petroleum products produced from naphthenic, paraffinic, blended, hydrocracked, hydrogenated, and solvent-refined stocks. In addition to petroleum oils, other lubricating fluids, such as esters, diesters, silicones, polyethers, and synthetic hydrocarbons, are also used. Of the synthetic fluids used in grease manufacture, the most common type is poly(alpha)olefin (Rudnick, 2009, 2013).

Grease is a lubricating oil to which a thickening agent has been added for the purpose of holding the oil to surfaces that must be lubricated. The most widely used thickening agents are soaps of various kinds, and grease manufacture is essentially the mixing of soaps with lubricating oils. Until a relatively short time ago, grease making was considered an art. To stir hot soap into hot oil is a simple business, but to do so in such a manner as to form grease is much more difficult, and the early grease maker needed much experience to learn the essentials of the trade. Therefore, it is not surprising that grease making is still a complex operation. The signs that *told* the grease maker that the soap was *cooked* and that the batch of grease was ready to run have been replaced by scientific tests that follow the process of manufacture precisely.

The early grease makers made grease in batches in barrels or pans, and the batch method is still the chief method of making grease. Oil and soap are mixed in kettles that have double walls between which steam and water may be circulated to maintain the desired temperature. When temperatures higher than 150°C (300°F) are required, a kettle heated by a ring of gas burners is used. Mixing is usually accomplished in each kettle by horizontal paddles radiating from a central shaft.

The soaps used in grease making are usually made in the grease plant, usually in a grease-making kettle. Soap is made by chemically combining a metal hydroxide with a fat or fatty acid. The most common metal hydroxides used for this purpose are calcium hydroxide, lye, lithium hydroxide, and barium hydroxide. Fats are chemical combinations of fatty acids and glycerin. If a metal hydroxide is reacted with a fat, a soap containing glycerin is formed. Frequently, a fat is separated into its fatty acid and glycerin components, and only the fatty acid portion is used to make soap. Commonly used fats for grease-making soaps are cottonseed oil, tallow, lard, and degras. Among the fatty acids used are stearic acid (from tallow), oleic acid (from cottonseed oil), and animal fatty acids (from lard).

To make grease, the soap is dispersed in the oil as fibers of such a size that it may be possible to detect them only by microscopy. The fibers form a matrix for the oil, and the type, amount, size, shape, and distribution of the soap fibers dictate the consistency, texture, and bleeding characteristics, as well as the other properties of grease. Grease may contain from 50 to 30% soap, and although the fatty acid influences the properties of grease, the metal in the soap has the most important effect. For example, calcium soaps form smooth buttery grease that are resistant to water but are limited in use to temperatures under about 95°C (200°F).

The soaps may be combined with any lubricating oil from a light distillate to a heavy residual oil. The lubricating value of the grease is chiefly dependent on the quality and viscosity of the oil. In addition to soap and oil, grease may also contain various additives that are used to improve the ability of the grease to stand up under extreme bearing pressures, to act as a rust preventive, and to reduce the tendency of oil to seep or bleed from grease. Graphite, mica, talc, or fibrous material may be added to grease that are used to lubricate rough machinery to absorb the shock of impact—other chemicals can make a grease more resistant to oxidation or modify the structure of the grease.

The older, more common method of grease making is the batch method, but grease is also made by a continuous method. The process involves soap manufacture in a series (usually three) of retorts. Soap-making ingredients are charged into one retort, while soap is made in the second retort. The third retort contains finished soap, which is pumped through a mixing device where the soap and the oil are brought together and blended. The mixer continuously discharges finished grease into suitable containers.

There is a variety of different types of grease that are produced with specific functions in mind. Therefore, properties vary with the method of preparation as well as with the metal included in the grease.

Calcium grease is resistant to water, they have a smooth texture, and their chief use is for plain bearings and low-speed rolling bearings and has high roll stability (ASTM D1831). The water content of calcium grease varies usually between 0.4 and 1.0%, is present in the form of water of crystallization, and has a stabilizing effect. High temperatures cause a loss of water and a consequent weakening of soap structure, and therefore the use of these grease is limited to a maximum temperature of about 60°C (140°F). Other stabilizers and structure modifiers can now be used in place of or as well as water, and the water content no longer has its former importance, for example, in calcium hydroxystearate grease, some of which can be used up to 120°C (248°F).

Calcium soap grease is one of the earliest known greases and is water resistant and mechanically stable. Calcium soap grease usually has a low melting point (dropping point; typically ~95°C/200°F). Anhydrous calcium soaps (usually calcium 12-hydroxystearate) are somewhat more temperature resistant, having a dropping point of about 150°C (300°F). Anhydrous calcium grease finds the greatest usage (when made with low-viscosity base oil) in operations when a wide range of climatic conditions is evident.

Sodium (soda) soap grease is fibrous in structure and are resistant to moderately high temperature but is not resistant to water. Sodium soap grease has a high dropping point (~175°C/350°F) than calcium grease. However, they tend to emulsify in the presence of water, yet they have inherent rust protection properties. This grease is used for rolling bearings at higher temperatures and speeds than normal conventional calcium soap grease.

Lithium soap grease is normally smooth in appearance but may exhibit a grain structure. This type of grease is resistant to water and to the highest normal service temperatures and, because of their variety of types, is often used as multipurpose grease. Lithium grease offers both the water resistance of calcium soap grease and the high-temperature properties of sodium soap grease. Grease prepared using lithium 12-hydroxystearate has a dropping point of about 190°C (375°F).

Aluminum soap grease is translucent and can be produced with aluminum soaps made *in situ*. In the process, the preformed soap (usually aluminum distearate) is dissolved in hot oil in a mixing grease kettle, and the hot mixture is poured into pans to cool. The cooling rate affects the final consistency. The final product is a smooth transparent grease having poor shear stability but excellent oxidation and water resistance. Aluminum soap grease is water resistant and adhesive but tends to have poor mechanical stability, and so this type of grease is not generally suitable for rolling bearings.

Mixed soap grease (e.g., the sodium/calcium soap grease) is manufactured for uses that include high-speed rolling bearings. On the other hand, *complex soap grease* contains calcium, calcium/lead, or other metallic complexes with fatty acids and acetate, benzoate, or other salts. The complex soap grease has a high dropping point (≥260°C, ≥500°F). In addition, complex soap grease has excellent antioxidation characteristics as well as high resistance to water and good mechanical stability properties.

Aluminum complex grease has excellent water wash-off and spray-off characteristics as well as high-temperature resistance and is widely used in steel mill applications and automotive components subject to these conditions. *Calcium complex grease* has inherent extreme pressure properties (ASTM D2596, ASTM D5706) and provides good friction and wear performance (ASTM D4170). *Lithium complex grease* performance is generally like that of lithium 12-hydroxysterate grease except it has a dropping point approximately 50°C (90°F) higher. Lithium complex grease provides good low-temperature performance and excellent high-temperature life performance in tapered roller bearings.

Clay thickened grease is often referred to as non-melting grease because of the tendency to decompose before reaching the dropping point (~290°C/550°F). This type of grease is water resistant and can be susceptible to severe degradation from other contaminants, such as brine. The performance is not often equivalent to the conventional calcium soap grease or sodium soap grease. Low-temperature performance could be considered satisfactory, but clay grease is not normally compatible with other greases, and protocols are available using other property test methods (ASTM D566, ASTM D2265, ASTM D1831, ASTM D1742, ASTM D3527, ASTM D4290, ASTM D4049) by which this can be determined.

Examples of other greases include the non-soap types with high water resistance and mechanical stability and some from selected organic compounds; they are now being further developed for multipurpose use. Treated clays or organic dyes of high-temperature resistance along with synthetic heat-resisting fluids are used for extreme temperatures. Non-soap grease often contain solid additives, such as graphite and molybdenum disulphide (MoS_2), and used under conditions of heavy loading (ASTM D2509, IP 326) or high temperature.

Due to the lack of specificity in most grease recommendations, it is important to learn how to properly select greases for each application in the plant. Proper grease specification requires all of the components of oil selection and more. Other special considerations for grease selection include thickener type and concentration, consistency, dropping point and operating temperature range, worked stability, oxidation stability, and wear resistance. Understanding the need and the methods for appropriate grease selection will go a long way toward improving lubrication programs and the reliability of lubricated machinery.

The most important property of any lubricant is viscosity. A common mistake when selecting grease is to confuse the grease consistency with the base oil viscosity. Once the appropriate viscosity has been determined, it's time to consider additives. The additive and base oil types are other components of grease that should be selected in a fashion similar to that used for oil-lubricated applications.

Most performance-enhancing additives found in lubricating oils are also used in grease formulation and should be chosen according to the demands of the application. Most greases are formulated using API Group I and II mineral oil base stocks, which are appropriate for most applications (Rudnick 2009, 2013). However, there are applications that might benefit from the use of synthetic base oil. Such applications include high or low operating temperatures, a wide ambient temperature range, or any application where extended relubrication intervals are desired.

Numerous types of grease thickeners are currently in use, each with its own advantages or disadvantages. Simple lithium soaps are often used in low-cost general-purpose greases and perform relatively well in most performance categories at moderate temperatures. Complex greases such as lithium complex provide improved performance particularly at higher operating temperatures. A common upper operating temperature limit for simple lithium grease might be 120°C (250°F), while that for lithium complex grease might be 180°C (350°F). Another thickener type that is becoming more popular is polyurea. Like lithium complex, polyurea has good high-temperature performance as well as high oxidation stability and bleed resistance. Thickener type should be selected based on performance requirements as well as compatibility when considering changing product types.

Once the appropriate base oil viscosity, additive requirements, and consistency have been determined, the remaining criteria to consider are the performance properties. Grease performance properties include many of the same properties used for lubricating oils, as well as others exclusive to grease. Properties exclusive to grease include dropping point, mechanical stability, water washout, bleed characteristics, and pumpability. The most important performance properties are determined by the application.

If an application operates continuously at room temperature, properties like dropping and upper operating temperature limits are not as important. If an application operates under heavy loads at low speeds, load-carrying tests should be considered. It is important to remember that greases, like oils, have a careful balance of properties. A product may excel in one category and perform poorly in another. For this reason, it is important to weigh each property's significance relative to the intended applications to select the best overall fit.

The API classification defines the oil quality by carrying out tests on it in the engine. The wear and cleanliness of cylinder walls, piston rings, bearings and gas distribution mechanisms, formation of burns and deposits, surface damages, and the increase of oil acidity may be examined during the tests. This classification divides engine lubricants into two groups marked by two letters: S-lubricants for petrol engines and C-lubricants for diesel engines. A second additional letter (e.g., SJ, SL or CG, CH) indicates the quality class.

The higher it is, the further the letter in the alphabet marks it, for example, SL Class oil has better operational properties than SJ. Universal use lubricants are marked with general symbols, for example, SL/CF.

14.3 TEST METHODS

Greases are used for particular lubrication applications because of their intrinsic properties. Users and producers alike need a common means to describe the properties required for grease performance for particular applications. Test methods are devised to describe the requisite properties.

The standard tests used to determine the properties of petroleum and petroleum products are commonly applied to grease. Among these are aniline point (ASTM D611), carbon residue (ASTM D189, ASTM D524, ASTM D4530, IP 13, IP 14, IP 398), fire point (ASTM D92), flash point (ASTM D92), pour point (ASTM D97), and viscosity (ASTM D445). However, because of the complexity of grease formulations and the variety of uses, many other tests are also deemed necessary to estimate performance in service. Some modification of the test method may be necessary because of the different characters of grease *vis-à-vis* lubricating oil.

14.3.1 Acidity and Alkalinity

Grease may contain small amounts of free organic acids but should not contain strong acids. In some conventional lime base grease, small amounts of free organic acids are intended as an aid to stability. Grease may also contain small amounts of free alkalinity.

Methods of analysis (ASTM D128, IP 37) are available for the measurement of excessive acidity derived from oxidation. These methods cover conventional grease that consists essentially of petroleum oil and soap. Thus, these test methods are applicable to many but not all greases. The constituents covered by the test series are soap, unsaponifiable matter (base oil), water, free alkalinity, free fatty acid, fat, glycerin, and insoluble material. A supplementary test method is also provided and is intended for application to grease that contains thickeners that are essentially insoluble in *n*-hexane and to grease that cannot be analyzed by conventional methods because of the presence of such constituents as nonpetroleum fluids or non-soap type thickeners, or both. These methods may not be applicable to grease analysis when lead, zinc, or aluminum soaps are present or in presence of some additives such as sodium nitrite.

14.3.2 Anticorrosion Properties

Water resistance alone does not ensure that grease will protect bearings or other mechanisms against moisture corrosion. Methods for assessing rust prevention by grease are described (ASTM D1743, IP 220).

In the former method (ASTM D1743), a taper roller bearing is packed with grease and, following a short running-in period, is dipped and stored above distilled water. The bearing is then cleaned and examined for corrosion. In the latter method (IP 220), the grease is tested in a ball bearing under dynamic conditions. The bearing is run intermittently in the presence of distilled water and at the end of the test is rated for corrosion.

Grease should not be corrosive to metals with which they come in contact, neither should they develop corrosion tendencies with aging or oxidation, and a polished copper strip (ASTM D4048) (other metals can also be used) is immersed in the grease that is stored at a temperature relevant to the use of the grease (IP 112). The metal strip is examined for etching, pitting, or discoloration. Alternatively, the test may be accelerated by storage of the grease/copper strip in an oxygen bomb under nitrogen.

14.3.3 Composition

The lubricating fluids that can be thickened to form grease vary widely in composition and properties (Chapter 13).

An analytical procedure (ASTM D128) is available for the separation of grease into its component parts and their measurement. Spectrographic methods, such as flame photometry, may also be used to determine the metal present as soaps or the wear elements in used grease, in conjunction with separation techniques to measure and identify the various types of fats, lubricating fluids, or additives present. The tests that are more likely to be quoted are ash content, acidity and alkalinity, water and dirt content.

Minerals and inorganic salts are determined as ash by ignition (ASTM D128), and the data include the bulk of clay-type non-soap thickeners, and the method includes a rapid routine method and a procedure, in which sulfuric acid is added to avoid loss of the more volatile oxides, and the result is reported as sulfated ash.

The use of microscopic methods in the examination of used grease for particles that are an indicator of wear is likely to be increased by the presence of abrasive dirt in grease, and foreign particles are counted and graded for size under the microscope or by the extent to which deleterious particles scratch plates of polished acrylic plastic (ASTM D1404).

In this method (ASTM D1404), the test material is placed between two clean, highly polished plates held rigidly and parallel to each other in metal holders. The assembly is pressed together by squeezing the grease into a thin layer between the plastic plates. Any solid particles in the grease larger than the distance of separation of the plates and harder than the plastic will become imbedded in the opposing plastic surfaces. The apparatus is so constructed that one of the plates can be rotated about 30° with respect to the other, while the whole assembly is under pressure thereby causing the imbedded particles to form characteristic arc-shaped scratches in one or both plates. The relative number of such solid particles is estimated by counting the total number of arc-shaped scratches on the two plates.

The test offers a rapid means for estimating the number of deleterious particles in grease. However, a particle that is abrasive to acrylic plastic may not be abrasive to steel or other bearing materials. Therefore, the results of this test do not imply performance in field service.

Some grease may contain traces of water (ASTM D95, IP 74), sulfur (ASTM D129), and chlorine (ASTM D808).

14.3.4 Dropping Point

The *dropping point* (sometimes used incorrectly to be synonymous with the *melting point*) is the temperature at which grease passes from a semisolid to a liquid state. As the temperature is raised, grease softens to the extent that it loses its self-supporting characteristic, the structure collapses, and the grease flows under its own weight; in a standard cup under standard conditions, this is called the dropping point. The change in state from a semisolid state to a liquid state is typical of grease containing soaps of conventional types added as thickeners. These tests are useful in identifying the types of grease and for establishing and maintaining benchmarks for quality control.

The dropping point is useful (i) in establishing bench marks for quality control, (ii) as an aid in identifying the type of thickener used in a grease, and (iii) as an indication of the maximum temperature to which a grease can be exposed without complete liquefaction or excessive oil separation.

There test methods do not give identical results, and the dropping point should be quoted in terms of the method used (ASTM D566, ASTM D2265). The tests consist of heating a sample of grease (contained in a cup suspended in a test tube) in an oil bath at a prescribed rate. The temperature at which the material falls from the hole in the bottom of the cup is averaged with the temperature of the oil bath and recorded as the dropping point.

One test method (ASTM D566) is not recommended for temperatures above 288°C. In those cases, the alternate test method (ASTM D2265) should be used.

14.3.5 Flow Properties

Grease is non-Newtonian in behavior, and unlike oils, an initial shear stress (yield value) must be applied before it will deform and commence to flow. It is this nonflowing characteristic that enables grease to offer certain advantages over lubricating oils and results in their extensive use for the lubrication of rolling bearings.

One test method (ASTM D1092) uses a procedure in which the grease is forced through short steel capillary tubes and the apparent viscosity calculated as if the grease was in fact a Newtonian fluid. The test can be carried out over a range of temperatures. It has been used as a low-temperature test of grease pumpability and leads to a procedure for giving empirical data on flow rates for handling grease in longline dispensing systems. The borderline pumping temperature test method as developed for engine oil (ASTM D3829) may also be applied to grease, with suitable modification to accommodate the physical properties of the grease.

The consistency of grease is a critical parameter that helps define its ability to perform under given operating conditions. Consistency, as measured by penetration, is affected by temperature, but the penetration test is not suitable for determining the minor, yet sometimes significant, changes in consistency as the grease approaches temperatures at which phase changes in the thickener occur. Penetration is basically a flow measurement; in addition, there are other flow measurements that can be utilized to evaluate this property at other conditions.

Parameters that are important to grease behavior and utility are related to flow properties under pressure (ASTM D2509, ASTM D2596, ASTM D5706, IP 326) in determining the load-bearing properties of grease (ASTM D2509, IP 326) and the wear-preventive characteristics of grease (ASTM D2266, ASTM D3704, ASTM D4170). The significance of these tests is that they provide methods that can be used to differentiate among grease having low, medium, or high levels of extreme pressure properties.

14.3.6 Low-Temperature Torque

Grease becomes harder and more viscous as the temperature drops and, in extreme cases, can become so rigid that excessive torque occurs within the bearing, and there is a need to determine the ability of grease to facilitate to startup and its lubricating ability at subzero temperatures.

Measurement of low-temperature torque (ASTM D1478) requires that a pre-greased ball bearing is cooled to the desired temperature, and the starting and running torques determined at 1 rev/min. This test method covers the determination of the extent to which a grease retards the rotation of a slow-speed ball bearing by measuring starting and running torques at low temperatures lower than $-20°C$ (lower than $-4°F$). These two test methods may not give the same torque values because the apparatus and test bearings are different. In the test method, a No. 6204 open ball bearing is packed completely full of grease and cleaned off flush with the sides. The bearing remains stationary, while ambient temperature is lowered to the test temperature and held there for 2 h. At the end of this time, the inner ring of the ball bearing is rotated, while the restraining force on the outer ring is measured. Torque is measured by multiplying the restraining force by the radius of the bearing housing. Both starting torque and that after 60 min of rotation are determined.

For applications using greater loads or larger bearings, another test method (ASTM D4693) can be used to predict the performance of grease in automotive wheel bearings operating at low temperatures. This test determines the extent to which grease retards the rotation of a tapered roller bearing assembly. In this test, a sample of test grease is stirred and worked, and a specified amount is packed into the two test bearings. The test assembly is heated to mitigate the effects of grease history; it is then cooled at a specified rate to $-40°C$ ($-40°F$). A drive mechanism rotates the spindle at 1 rpm, and the torque required to prevent rotation of the hub is determined 60 s after the start of rotation.

14.3.7 Mechanical Stability

The ability of a grease to withstand a large degree of mechanical working without changing its consistency unduly can be important during the initial clearing stages in a bearing, or in certain applications, for example, in a bearing assembly under vibrating conditions. The effects of mechanical instability can lead in some cases to the grease becoming fluid and losing its sealing properties.

The tests most widely used to assess mechanical stability are the prolonged worked penetration test (ASTM D217, IP 50) and the Roll Test (ASTM D1831). In the former test, the grease is subjected to a large number of strokes (usually 100,000) in a motorized version of the worker pot used for the penetration test, at room temperature. In the Roll Test, a measured amount of grease is placed in a horizontally mounted drum in which a steel roller rotates. The drum is turned about its axis, thereby subjecting the grease to a milling action similar to that occurring in rolling bearings. In both tests, the difference in penetration of the grease, before and after, is a measure of the grease breakdown.

14.3.8 Oil Separation

Grease will often separate oil during storage, and if too much oil separates, the grease could harden to the extent that lubrication performance will be affected. Oil will be released from a grease at varying rates depending on the gel structure, the nature and viscosity of the lubricating fluid, and the applied pressure and temperature.

There are standard test methods (ASTM D1742, ASTM D3336, ASTM D3527, ASTM D4425, IP 121) for predicting the amount of oil liberated by grease when stored in sealed containers. In both methods, the grease is supported by a wire mesh screen; a weight or air pressure is applied to the top surface of the grease to accelerate oil separation.

One particular method (ASTM D1742, IP 121) is used to determine the tendency of lubricating grease to separate oil when stored at $25°C$ ($77°F$) at an applied air pressure of 1.72 kN/m^2 (0.25 psi). The test gives an indication of the oil retention characteristics of lubricating grease stored in both normally filled and partially filled containers. The duration of this test is either 24 h (ASTM D1742) or 7 days (IP 121), and the data often correlate directly with oil separation that occurs in 16 kg (35 lb) containers of grease stored at room temperature.

Another test method (ASTM D3336) is used to evaluate the performance characteristics of lubricating grease in ball bearings operating under light loads at high speeds and elevated temperatures for extended periods. Correlation with actual field service cannot be assumed. In this test, the lubricating grease is evaluated in a heat-resistant, steel ball bearing rotated at 10,000 rpm under light loads at a specified elevated temperature up to 370°C (700°F). The test is run on a specified operating cycle until lubrication failure or completion of a specified time. With superior grease, tests can last up to several thousand hours. Another test method (ASTM D3527) is used for the evaluation of grease life in tapered roller wheel bearings in a model, front wheel assembly run at 1000 rpm under a specified thrust load at 160°C (320°F) using a cycle of 20 h on and 4 h off (c.f. ASTM D4290).

14.3.9 Oxidation Stability

The storage stability of grease when packed into bearings is assuming considerable importance as more prepacked sealed-for-life bearings are adopted in industry. In these bearings, the grease acts as a protective film. Under such conditions, thin films of grease in contact with steel or steel and bronze are exposed to air and moisture that promote oxidation of the grease, the products of which can cause corrosion of the bearing surfaces.

In the test method used to evaluate storage properties under these conditions (ASTM D942, IP 142), oxidation of a thin film of grease is accelerated by heating the grease at 99°C (210°F) in oxygen at a pressure of 110 psi. The amount of oxygen absorbed by the grease is recorded in terms of pressure drop over a period of 100 h and in some cases up to 500 h. Grease that shows high oxygen absorption become fluid, increase in acidity, and are generally considered unsatisfactory for use.

The oxidation induction time can also be used as an indication of oxidation stability. No correlation has been determined between the results of this test and service performance. In the test (ASTM D5483), a small quantity of grease in a sample pan is placed in a test cell. The cell is heated to a specified temperature and then pressurized with oxygen and is held at a regulated temperature and pressure until an exothermic reaction occurs. The extrapolated onset time is measured and reported as the oxidation induction time. This test method covers lubricating grease that is subjected to oxygen at 500 psi and temperatures between 155 and 210°C (311 and 410°F).

14.3.10 Penetration

The cone penetration tests (ASTM D217, ASTM D1403, IP 50, IP 310) are standard test for determining the penetration of grease and provide an indication of the consistency of grease.

One particular test method (ASTM D217) consists of four procedures. For unworked penetration, the cone assembly of the penetrometer is allowed to drop freely into the grease sample at 25°C (77°F) in a worker cup for 5 s. For worked penetration, the sample at 25°C (77°F) in a worker cup is subjected to 60 double strokes by the grease worker. The penetration is determined immediately by releasing the cone assembly from the penetrometer and allowing the cone to drop freely into the grease for 5 s. For prolonged work penetration, the procedure is the same as for worked penetration, except that additionally before cone penetration, the sample is subjected to a predetermined number of double strokes in the grease worker. For block penetration, a cube of grease is used and the test is followed as in that for unworked penetration. The depth of penetration, measured in tenths of a millimeter, is the penetration value. A firm grease will have a low-penetration value, and conversely, a soft grease will have a high-penetration value.

The penetration varies according to the amount of shearing to which the grease has been subjected. It may be measured on the grease as received in its original container (undisturbed penetration), after transfer with minimum disturbance to a standard container (unworked penetration), after a standard amount of shearing (worked penetration), after a prolonged period of shearing in a mechanical worker (prolonged work penetration), and as a firm sample (block penetration).

14.3.11 Thermal Stability

Thermal conditions have a profound influence on the properties and performance of grease. And any one or more of several phenomena can occur consecutively or, even more likely, simultaneously (ASTM D3336, ASTM D3527). Generally, grease having the least evaporative losses (ASTM D972, ASTM D2595) will probably perform longer in high-temperature service.

For example, grease softens and flows easier with an increase in temperature, but the rate of oxidation of the grease also increases (ASTM D942, IP 142). In addition, evaporation of oil constituents also increases (ASTM D972, ASTM D2595), and the thickener melts or loses its ability to retain oil (ASTM D566, ASTM D2595).

In one test method (ASTM D972), the evaporative losses from grease or oil at any temperature in the range of 100–150°C (210–300°F) are determined. A weighed sample of grease is placed in an evaporation cell in an oil bath at the desired test temperature. Heated air at a specified flow rate is passed over the sample surface for 22 h, after which the loss in sample mass is determined. Another method (ASTM D2595) is used to supplement the original method (ASTM D972), which was developed because of higher service temperature and can be used to determine the loss of volatile materials from a grease over a temperature range of 93–316°C (200–600°F). This test uses an aluminum block heater, instead of an oil bath (ASTM D972) to achieve higher temperatures and should not be used at temperatures in excess of the flash point of the base oil of the grease. Both tests can be used to compare evaporation losses of grease intended for similar service, but the test results may not be representative of volatilization that can occur in service.

14.3.12 Viscosity

Grease is a non-Newtonian material insofar as flow is not initiated until stress is applied.

The *apparent viscosity* grease is measured in poises (ASTM D1092), and since the apparent viscosity varies with both temperature and shear rate, the temperature and shear rate must be reported along with the measured viscosity. In this test, a sample of grease is forced through a capillary tube by a floating piston actuated by a hydraulic system using a two-speed gear pump. From the predetermined flow rate and the force developed in the system, the apparent viscosity is calculated. A series of 8 capillaries and 2 pump speeds provide 16 shear rates for the determination of apparent viscosities. The results are expressed in a log–log graph of apparent viscosity as a function of shear rate at a constant temperature, or apparent viscosity at a constant shear rate as a function of temperature.

The apparent viscosity also is used to provide an indication of the directional value of starting and running torques of grease-lubricated mechanisms. Specifications may include limiting values of apparent viscosity for grease to be used at low temperature.

14.3.13 Volatility

The effective life of grease, especially under high-temperature conditions, is dependent on such factors as oxidation, retentive properties, and evaporation of the base oil. Evaporation of the oil can result in the grease becoming stiffer and drier and ultimately lead to bearing failure.

The volatility of grease can be determined (ASTM D972, IP 183) in which air is passed at a known flow rate over a weighed amount of grease in a standard cell that is immersed in an oil bath at a required temperature. Since air is used, some oxidation of the grease will occur, but nevertheless, a comparative rating for evaporation loss can be obtained by measuring the loss in weight of the grease sample.

14.3.14 Water Resistance

The resistance of a grease to water contamination is an important property since grease has to lubricate mechanisms where water will be present to a greater or lesser extent. The presence of water can cause changes in grease consistency, emulsification with water-soluble soap base grease, and a reduction in mechanical stability. The effects of such changes can lead to grease being washed out of the mechanisms, resulting in inadequate lubrication and poor protection against rusting. Most greases classed as water-resistant grease are able to take up large amounts of water without suffering the serious changes mentioned earlier. It does not, however, follow that grease with high water resistance will afford adequate protection against rusting unless rust preventives are incorporated in the grease.

The most commonly used test for measuring water resistance is the water washout test method (ASTM D1264, IP 215) in which a greased ball bearing is rotated at 600 rev/min in an assembly with specified clearances in the covers to allow entry of water from a water jet that impinges on the face of one cover. The loss in grease pack after 1 h of operation is a measure of the resistance to water. The test serves only as a relative measure of the resistance of a grease to water washout.

Another test method (ASTM D 4049) is used to evaluate the ability of a grease to adhere to a metal panel when subjected to direct water spray. Test results correlate directly with operations involving direct water impingement. In this test, a 0.79 mm film of test grease is uniformly coated onto a stainless steel panel; then water, at 38°C (100°F), is sprayed directly on the panel for 5 min. The spray is controlled by specified spray nozzle, pump, and plumbing. After the spraying period, the panel is dried, weighed, and the percentage of grease spray-off is determined.

REFERENCES

ASTM D92. Standard Test Method for Flash and Fire Points by Cleveland Open Cup Tester. Annual Book of Standards. ASTM International, West Conshohocken, PA.

ASTM D95. Standard Test Method for Water in Petroleum Products and Bituminous Materials by Distillation. Annual Book of Standards. ASTM International, West Conshohocken, PA.

ASTM D97. Standard Test Method for Pour Point of Petroleum Products. Annual Book of Standards. ASTM International, West Conshohocken, PA.

ASTM D128. Standard Test Methods for Analysis of Lubricating Grease. Annual Book of Standards. ASTM International, West Conshohocken, PA.

ASTM D129. Standard Test Method for Sulfur in Petroleum Products (General High Pressure Decomposition Device Method). Annual Book of Standards. ASTM International, West Conshohocken, PA.

ASTM D189. Standard Test Method for Conradson Carbon Residue of Petroleum Products. Annual Book of Standards. ASTM International, West Conshohocken, PA.

ASTM D217. Standard Test Methods for Cone Penetration of Lubricating Grease. Annual Book of Standards. ASTM International, West Conshohocken, PA.

ASTM D445. Standard Test Method for Kinematic Viscosity of Transparent and Opaque Liquids (and Calculation of Dynamic Viscosity). Annual Book of Standards. ASTM International, West Conshohocken, PA.

ASTM D524. Standard Test Method for Ramsbottom Carbon Residue of Petroleum Products. Annual Book of Standards. ASTM International, West Conshohocken, PA.

ASTM D566. Standard Test Method for Dropping Point of Lubricating Grease. Annual Book of Standards. ASTM International, West Conshohocken, PA.

ASTM D611. Standard Test Methods for Aniline Point and Mixed Aniline Point of Petroleum Products and Hydrocarbon Solvents. Annual Book of Standards. ASTM International, West Conshohocken, PA.

ASTM D808. Standard Test Method for Chlorine in New and Used Petroleum Products (High Pressure Decomposition Device Method). Annual Book of Standards. ASTM International, West Conshohocken, PA.

ASTM D942. Standard Test Method for Oxidation Stability of Lubricating Greases by the Oxygen Pressure Vessel Method. Annual Book of Standards. ASTM International, West Conshohocken, PA.

ASTM D972. Standard Test Method for Evaporation Loss of Lubricating Greases and Oils. Annual Book of Standards. ASTM International, West Conshohocken, PA.

ASTM D1092. Standard Test Method for Measuring Apparent Viscosity of Lubricating Greases. Annual Book of Standards. ASTM International, West Conshohocken, PA.

ASTM D1264. Standard Test Method for Determining the Water Washout Characteristics of Lubricating Greases. Annual Book of Standards. ASTM International, West Conshohocken, PA.

ASTM D1403. Standard Test Methods for Cone Penetration of Lubricating Grease using One-Quarter and One-Half Scale Cone Equipment. Annual Book of Standards. ASTM International, West Conshohocken, PA.

ASTM D1404. Standard Test Method for Estimation of Deleterious Particles in Lubricating Grease. Annual Book of Standards. ASTM International, West Conshohocken, PA.

ASTM D1478. Standard Test Method for Low-Temperature Torque of Ball Bearing Grease. Annual Book of Standards. ASTM International, West Conshohocken, PA.

ASTM D1742. Standard Test Method for Oil Separation from Lubricating Grease during Storage. Annual Book of Standards. ASTM International, West Conshohocken, PA.

ASTM D1743. Standard Test Method for Determining Corrosion Preventive Properties of Lubricating Greases. Annual Book of Standards. ASTM International, West Conshohocken, PA.

ASTM D1831. Standard Test Method for Roll Stability of Lubricating Grease. Annual Book of Standards. ASTM International, West Conshohocken, PA.

ASTM D2265. Standard Test Method for Dropping Point of Lubricating Grease Over Wide Temperature Range. Annual Book of Standards. ASTM International, West Conshohocken, PA.

ASTM D2266. Standard Test Method for Wear Preventive Characteristics of Lubricating Grease (Four-Ball Method). Annual Book of Standards. ASTM International, West Conshohocken, PA.

ASTM D2509. Standard Test Method for Measurement of Load-Carrying Capacity of Lubricating Grease (Timken Method). Annual Book of Standards. ASTM International, West Conshohocken, PA.

ASTM D2595. Standard Test Method for Evaporation Loss of Lubricating Greases Over Wide-Temperature Range. Annual Book of Standards. ASTM International, West Conshohocken, PA.

ASTM D2596. Standard Test Method for Measurement of Extreme-Pressure Properties of Lubricating Grease (Four-Ball Method). Annual Book of Standards. ASTM International, West Conshohocken, PA.

ASTM D3336. Standard Test Method for Life of Lubricating Greases in Ball Bearings at Elevated Temperatures. Annual Book of Standards. ASTM International, West Conshohocken, PA.

ASTM D3527. Standard Test Method for Life Performance of Automotive Wheel Bearing Grease. Annual Book of Standards. ASTM International, West Conshohocken, PA.

ASTM D3704. Standard Test Method for Wear Preventive Properties of Lubricating Greases using the (Falex) Block on Ring Test Machine in Oscillating Motion. Annual Book of Standards. ASTM International, West Conshohocken, PA.

ASTM D3829. Standard Test Method for Predicting the Borderline Pumping Temperature of Engine Oil. Annual Book of Standards. ASTM International, West Conshohocken, PA.

ASTM D4048. Standard Test Method for Detection of Copper Corrosion from Lubricating Grease. Annual Book of Standards. ASTM International, West Conshohocken, PA.

ASTM D4049. Standard Test Method for Determining the Resistance of Lubricating Grease to Water Spray. Annual Book of Standards. ASTM International, West Conshohocken, PA.

ASTM D4170. Standard Test Method for Fretting Wear Protection by Lubricating Greases. Annual Book of Standards. ASTM International, West Conshohocken, PA.

ASTM D4289. Standard Test Method for Elastomer Compatibility of Lubricating Greases and Fluids. Annual Book of Standards. ASTM International, West Conshohocken, PA.

ASTM D4290. Standard Test Method for Determining the Leakage Tendencies of Automotive Wheel Bearing Grease under Accelerated Conditions. Annual Book of Standards. ASTM International, West Conshohocken, PA.

ASTM D4425. Standard Test Method for Oil Separation from Lubricating Grease by Centrifuging (Koppers Method). Annual Book of Standards. ASTM International, West Conshohocken, PA.

ASTM D4530. Standard Test Method for Determination of Carbon Residue (Micro Method). Annual Book of Standards. ASTM International, West Conshohocken, PA.

ASTM D4693. Standard Test Method for Low-Temperature Torque of Grease-Lubricated Wheel Bearings. Annual Book of Standards. ASTM International, West Conshohocken, PA.

ASTM D4950. Standard Classification and Specification for Automotive Service Greases. Annual Book of Standards. ASTM International, West Conshohocken, PA.

ASTM D5483. Standard Test Method for Oxidation Induction Time of Lubricating Greases by Pressure Differential Scanning Calorimetry. Annual Book of Standards. ASTM International, West Conshohocken, PA.

ASTM D5706. Standard Test Method for Determining Extreme Pressure Properties of Lubricating Greases using a High-Frequency, Linear-Oscillation (SRV) Test Machine. Annual Book of Standards. ASTM International, West Conshohocken, PA.

ASTM D5707. Standard Test Method for Measuring Friction and Wear Properties of Lubricating Grease using a High-Frequency, Linear-Oscillation (SRV) Test Machine. Annual Book of Standards. ASTM International, West Conshohocken, PA.

Banaszewski, A. and Blythe, J. 2000. White Mineral Oil. In: Modern Petroleum Technology. Volume 2: Downstream. A.G. Lucas (Editor). John Wiley & Sons Inc., New York.

Gary, J.G., Handwerk, G.E., and Kaiser, M.J. 2007. Petroleum Refining: Technology and Economics, 5th Edition. CRC Press, Taylor & Francis Group, Boca Raton, FL.

Hsu, C.S. and Robinson, P.R. (Editors) 2006. Practical Advances in Petroleum Processing (Volumes 1 and 2). Springer Science, New York.

IP 13 (ASTM D189). Petroleum Products – Determination of Carbon Residue – Conradson Method. IP Standard Methods 2013. The Energy Institute, London.

IP 14 (ASTM D524). Petroleum Products – Determination of Carbon Residue – Ramsbottom Method. IP Standard Methods 2013. The Energy Institute, London.

IP 37. Determination of Acidity and Alkalinity of Lubricating Grease. IP Standard Methods 2013. The Energy Institute, London.

IP 50 (ASTM D217). Determination of Cone Penetration of Lubricating Grease. IP Standard Methods 2013. The Energy Institute, London.

IP 74 (ASTM D95). Petroleum Products and Bituminous Materials – Determination of Water – Distillation Method. IP Standard Methods 2013. The Energy Institute, London.

IP 112. Determination of Corrosiveness to Copper of Lubricating Grease – Copper Strip Method. IP Standard Methods 2013. The Energy Institute, London.

IP 121. Determination of Oil Separation from Lubricating Grease – Pressure Filtration Method. IP Standard Methods 2013. The Energy Institute, London.

IP 142 (ASTM D942). Determination of oxidation stability of lubricating grease – oxygen bomb method. IP Standard Methods 2013. The Energy Institute, London.

IP 183. Determination of Evaporation Loss of Lubricating Grease. IP Standard Methods 2013. The Energy Institute, London.

IP 215. Determination of Water Washout Characteristics of Lubricating Grease. IP Standard Methods 2013. The Energy Institute, London.

IP 220. Determination of Rust Prevention Characteristics of Lubricating Greases and Oils. IP Standard Methods 2013. The Energy Institute, London.

IP 310 (ASTM D1403). Determination of Cone Penetration of Grease – One-quarter and One-half Scale Cone Method. IP Standard Methods 2013. The Energy Institute, London.

IP 326 (ASTM D2509). Extreme Pressure Properties of Grease – Timken Method. IP Standard Methods 2013. The Energy Institute, London.

IP 398. Petroleum Products – Determination of Carbon Residue – Micro Method. IP Standard Methods 2013. The Energy Institute, London.

Rudnick, L.R. (Editor). 2009. Lubricant Additives: Chemistry and Applications, 2nd Edition. CRC Press, Taylor & Francis Group, Boca Raton, FL.

Rudnick, L.R. (Editor). 2013. Synthetics, Minerals Oils, and Bio-based Lubricants: Chemistry and Technology, 2nd Edition. CRC Press, Taylor & Francis Group, Boca Raton, FL.

Speight, J.G. 2014. The Chemistry and Technology of Petroleum, 5th Edition. CRC Press, Taylor & Francis Group, Boca Raton, FL.

Speight, J.G. and Ozum, B. 2002. Petroleum Refining Processes. Marcel Dekker Inc., New York.

Speight, J.G. and Exall, D.I. 2014. Refining Used Lubricating Oils. CRC Press, Taylor and Francis Group, Boca Raton, FL.

15

WAX

15.1 INTRODUCTION

Wax—which also occurs natural—is produced from petroleum as a result of the refining processesPetroleum waxes are obtained from paraffinic refinery streams in lubricating oil manufacture. The wax is separated by filtering a chilled solution of waxy oil in a selected solvent (usually a mixture of methyl ethyl ketone (MEK) and toluene).

Slack wax is obtained from the dewaxing of refined or unrefined vacuum distillate fractions. If the material has been separated from residual oil fractions, it is frequently called *petrolatum* (*petroleum jelly*). The slack waxes are de-oiled by solvent crystallization or *sweating* processes to manufacture commercial waxes with low oil content. The oil that is separated from these processes is known as *foots oil*. The refined petroleum waxes are known as *paraffin waxes*. *Microcrystalline waxes* have higher molecular weights than the paraffin waxes and consist of substantial amounts of iso- and cycloalkanes.

Although many natural waxes contain esters, paraffin waxes are hydrocarbons, mixtures of alkanes, usually in a homologous series of chain lengths. These materials represent a significant fraction of petroleum. They are refined by vacuum distillation. Paraffin waxes are mixtures of saturated *n*- and iso-alkanes, naphthenes, and alkyl- and naphthene-substituted aromatic compounds. The degree of branching has an important influence on the properties.

Of the various types of waxes, the two most common types are Montan wax and polyethylene wax. *Montan wax* is a fossilized wax extracted from coal and lignite (Speight, 2013). It is very hard, reflecting the high concentration of saturated fatty acids and alcohols, not esters that characterize softer waxes. Although dark brown and odiferous, they can be purified and bleached to give commercially useful products. *Polyethylene wax* (and related derivatives) is obtained by cracking polyethylene at 400°C (750°F). The products have the formula $H(CH_2)_nH_2$, where n ranges between 50 and 100 polyethylene units, usually as a straight chain although unintentional and uncontrollable branching may occur.

Paraffin (petroleum-derived) wax is the solid hydrocarbon residue remaining at the end of the refining process either in the lube stream (as mainly paraffin and intermediate waxes) or in the residual lube stock *tank bottoms* (as higher-melting microcrystalline waxes) (Richter, 2000; Speight and Ozum, 2002; Hsu and Robinson, 2006; Gary et al., 2007; Speight, 2014).

The waxy oil is fractionated to produce an oily wax, called *slack wax*. This is separated by solvent extraction and fractionated into different melting point ranges to give waxes with a variety of physical characteristics. Paraffin waxes consist mainly of straight-chain alkanes (also called normal alkanes), with small amounts (3–15%) of branched-chain alkanes (or iso-alkanes), cycloalkanes, and aromatics. Microcrystalline waxes contain high levels of branched-chain alkanes (up to 50%) and cycloalkanes, particularly in the upper end of the molecular weight distribution. Paraffin waxes contain alkanes up to a molecular weight of approximately 600 amu, whereas microcrystalline waxes can contain alkanes up to a molecular weight of approximately 1100 amu.

During the refining of waxy crude oils, the wax becomes concentrated in the higher boiling fractions used primarily for making lubricating oils. Refining of lubricating oil fractions to obtain a desirable low pour point usually requires the

Handbook of Petroleum Product Analysis, Second Edition. James G. Speight.
© 2015 John Wiley & Sons, Inc. Published 2015 by John Wiley & Sons, Inc.

removal of most of the waxy components. The dewaxing step is generally performed by the chilling and filter pressing method, by centrifuge dewaxing, or by filtering a chilled solution of waxy lubricating oil in a specific solvent.

Wax provides improved strength, moisture proofing, appearance, and low cost for the food packaging industry, the largest consumer of waxes today. The coating of corrugated board with hot melts is of increasing importance to the wax industry. Other uses include the coating of fruit and cheese, the lining of cans and barrels, and the manufacture of anticorrosives. Because of its thermoplastic nature, wax lends itself to modeling and the making of replicas; blends of waxes are used by dentists when making dentures and by engineers when mass-producing precision castings such as those used for gas turbine blades. The high gloss characteristic of some petroleum waxes makes them suitable ingredients for polishes, particularly for the "paste" type that is commonly used on floors, furniture, cars, and footwear. The highly refined waxes have excellent electrical properties and so find application in the insulation of low-voltage cables, small transformers, coils, capacitors, and similar electronic components.

15.2 PRODUCTION AND PROPERTIES

15.2.1 Production

Paraffin wax from a solvent dewaxing operation (Speight, 2014) is commonly known as *slack wax*, and the processes employed for the production of waxes are aimed at de-oiling the slack wax (petroleum wax concentrate).

Wax sweating was originally used to separate wax fractions with various melting points from the wax obtained from shale oils. Wax sweating is still used to some extent but is being replaced by the more convenient crystallization process. In wax sweating, a cake of slack wax is slowly warmed to a temperature at which the oil in the wax and the lower-melting waxes become fluid and drip (or sweat) from the bottom of the cake, leaving a residue of higher-melting wax. Sweated waxes generally contain small amounts of unsaturated aromatic and sulfur compounds, which are the source of unwanted color, odor, and taste that reduce the ability of the wax to resist oxidation; the commonly used method of removing these impurities is clay treatment of the molten wax.

Wax crystallization, like wax sweating, separates slack wax into fractions, but instead of using the differences in melting points, it makes use of the different solubility of the wax fractions in a solvent, such as the ketone used in the dewaxing process (Chapter 19). When a mixture of ketone and slack wax is heated, the slack wax usually dissolves completely, and if the solution is cooled slowly, a temperature is reached at which a crop of wax crystals is formed. These crystals will all be of the same melting point, and if they are removed by filtration, a wax fraction with a specific melting point is obtained. If the clear filtrate is further cooled, a second crop of wax crystals with a lower melting point is obtained. Thus, by alternate cooling and filtration, the slack wax can be subdivided into a large number of wax fractions, each with different melting points.

Chemically, paraffin wax is a mixture of saturated aliphatic hydrocarbons (with the general formula C_nH_{2n+2}). Wax is the residue extracted when dewaxing lubricant oils, and they have a crystalline structure with carbon number greater than 12. The main characteristics of wax are (i) colorless, (ii) odorless, (iii) translucent, and (iv) a melting point above 45°C (113°F).

15.2.2 Properties

Petroleum wax is of two general types: the paraffin waxes in petroleum distillates and the microcrystalline waxes in petroleum residua. The melting point of wax is not directly related to its boiling point, because waxes contain hydrocarbons of different chemical nature. Nevertheless, waxes are graded according to their melting point (ASTM D87, IP 55) and oil content (ASTM D721, IP 158).

A scheme for classifying waxes as either paraffin, semi-microcrystalline, or microcrystalline is based on the following equation:

$$n^2D = 0.0001943t + 1.3994$$

In the equation, t is the congealing point temperature (in °F) (ASTM D938, IP 76). Viscosity is included as an additional parameter.

Semi-microcrystalline wax and microcrystalline wax are petroleum waxes containing substantial portions of hydrocarbons other than normal alkanes. They are characterized by refractive indices greater than those given by the earlier equation, and by viscosities at 210°F of less than 10 centistokes for semi-microcrystalline waxes or greater than 10 centistokes for microcrystalline waxes. Microcrystalline waxes have higher molecular weights, smaller crystal structures, and greater affinities for oil than paraffin waxes. Microcrystalline waxes usually melt between 66°C (150°F) and 104°C (220°F) and have viscosities between 10 and 20 centistokes at 99°C (210°F).

Petrolatum is usually a soft product containing approximately 20% oil and melting between 38°C (100°F) and 60°C (140°F). Petrolatum or petroleum jelly is essentially a mixture of microcrystalline wax and oil. It is produced as an intermediate product in the refining of microcrystalline wax or compounded by blending appropriate waxy products and oils. Petrolatum colors range from the almost black crude form to the highly refined yellow and white pharmaceutical grades.

The melting point of paraffin wax (ASTM D87, IP 55) has both direct and indirect significance in most wax utilization. All wax grades are commercially indicated in a range of melting temperatures rather than at a single value, and a range of 1°C (2°F) usually indicates a good degree of refinement. Other common physical properties that help to illustrate the degree of refinement of the wax are color (ASTM D156), oil content (ASTM D721, IP 158), and viscosity (ASTM D88, ASTM D445, IP 71).

Fully refined paraffin waxes are a hard, white crystalline material derived from petroleum. Paraffin waxes are predominately composed of normal, straight-chain hydrocarbons. The water-repellent and thermoplastic properties of paraffin waxes make them ideal for many applications. Typical end uses include cereal, delicatessen, and household wrap, corrugated containers, candles, cheese and vegetable coatings, and hot melt adhesives.

Paraffin wax is mostly found as a white, odorless, tasteless, waxy solid, with a typical melting point between about 46 and 68°C (115 and 154°F) and a density of approximately 900, is insoluble in water, but soluble in ether, benzene, and certain esters. Paraffin wax is often classed as a stable chemical since it is unaffected by most common chemical reagents but burns readily.

Microcrystalline waxes are a type of wax produced by de-oiling petrolatum, as part of the petroleum refining process. In contrast to the more familiar paraffin wax, which contains mostly unbranched alkanes, microcrystalline wax contains a higher percentage of iso-paraffin (branched) and naphthene hydrocarbons. It is characterized by the fineness of its crystals in contrast to the larger crystal of paraffin wax. It consists of high-molecular-weight saturated aliphatic hydrocarbons. It is generally darker, more viscous, denser, tackier, and more elastic than paraffin waxes, and has a higher molecular weight and melting point. The elastic and adhesive characteristics of microcrystalline waxes are related to the non-straight-chain components that they contain. Typical microcrystalline wax crystal structure is small and thin, making them more flexible than paraffin wax.

Microcrystalline waxes when produced by wax refiners are typically produced to meet a number of ASTM specifications, which include congealing point (ASTM D938), needle penetration (D1321), color (ASTM D6045), and viscosity (ASTM D445). Microcrystalline wax is also a key component in the manufacture of petrolatum. The branched structure of the carbon chain backbone allows oil molecules to be incorporated into the crystal lattice structure. The desired properties of the petrolatum can be modified by using microcrystalline wax bases of different congeal points (ASTM D938) and needle penetration (ASTM D1321).

15.3   TEST METHODS

As with any petroleum product, and waxes are no exception, there is a series of standard test methods that can be applied to determine the properties (Table 15.1). All test methods may not

TABLE 15.1   Standard test methods for determining the properties of waxes

ASTM test method	Comment
Melting points	
ASTM D87, Cooling curve	ASTM D87: used for paraffin wax only
ASTM D127, Drop point	ASTM D127: used for micro-wax and wax blends
ASTM D938, Congealing point	ASTM D938: used for petrolatum
Penetration	
ASTM D1321, Needle penetration	A measure of wax hardness; higher penetration value indicates a softer wax. Penetration temperatures normally used are 77, 100, and 110°F
Viscosity	
ASTM D88, Saybolt viscosity	A measure of fluidity or pumpability of liquid wax usually determined at 210°F
ASTM D445, Kinematic viscosity	ASTM D445 is normally used, and the results may be converted to SUS
ASTM D2161, Conversions	
Oil content	
ASTM D721 Oil content	The component of a wax or wax blends, which is extractable with methyl ethyl ketone (MEK)
Color	
ASTM D156 Saybolt viscosity	D156: used for testing water-white paraffin
ASTM D1500	D1500: used for wax having a yellow or darker color
Odor	
ASTM D1833, Odor rating	D-1832 is quantitative measure of oxidation
ASTM D1832, Peroxide no.	D-1833 is a numerical rating for odor intensity
Block point/block temperature	
ASTM D1465	Minimum temperature at which two pieces of waxed paper stick together when in intimate contact

be applied, but this is entirely due to the proposed use of the wax and whether or not specifications are dictated by the use.

15.3.1   Appearance

Waxed coatings provide protection for packaged goods, and the high gloss characteristics provide improved appearance. Both the nature of the wax and the coating process contribute to the final gloss characteristics.

Specular gloss (ASTM D523) is the capacity of a surface to simulate a mirror in its ability to reflect an incident light beam. The glossimeter used to measure gloss consists of a lamp and lens set to focus an incident light beam 20° from a line drawn perpendicular to the specimen. A receptor lens and photocell are centered on the angle of reflectance, also 20° from a line perpendicular to the specimen. A black, polished glass surface with a refractive index of 1.54 is used for instrument standardization at 100 gloss units. A wax-coated paper is held by a vacuum plate over the sample opening. The light beam is reflected from the sample surface into the photocell and measured with a null point microammeter. The gloss is measured before and after aging the sample for 1 and 7 days in an oven at 40°C (104°F). The specified aging conditions are intended to correlate with the conditions likely to occur in the handling and storage of waxed paper and paperboard.

15.3.2   Barrier Properties

The ability of wax to prevent the transfer of moisture vapor is of primary concern in the food packaging industry. To maintain the freshness of dry foods, moisture must be kept out of the product, but to maintain the quality of frozen foods and baked goods, the moisture must be kept in the product. This results in two criteria for barrier properties: moisture vapor transmission rates (A) at elevated temperatures and high relative humidity (B) at low temperatures and low relative humidity, for frozen foods.

The wax *blocking point* is the lowest temperature at which film disruption occurs across 50% of the waxed paper surface when the test strips are separated. The *picking point* is the lowest temperature at which the surface film shows disruption. Blocking of waxed paper, because of the relatively low temperature at which it may occur, can be a major problem to the paper coating industry. The wax picking and blocking points indicate an approximate temperature range at or above which waxed surfaces in contact with each other are likely to cause surface film injury.

To determine the blocking point of wax (ASTM D1465), two paper test specimens are coated with the wax sample, folded with the waxed surfaces together, and placed on a blocking plate that is heated at one end and cooled at the other to impose a measured temperature gradient along its length. After a conditioning period on the plate, the specimens are removed, unfolded, and examined for film disruption. The temperatures of corresponding points on the blocking plate are reported as the picking and blocking points or as blocking range.

15.3.3   Carbonizable Substances

Wax and petrolatum intended for certain pharmaceutical purposes are required to pass the test for carbonizable matter. The amount of *carbonizable material* (often equated to the degree of unsaturation) is determined by the reaction of the wax with concentrated sulfuric acid. The resultant color of the acid layer must be lighter than the reference color if the wax is to qualify pharmaceutical grade. The melting point

(ASTM D87, IP 55) for such grades of wax may also be required.

To determine the presence of carbonizable substances in paraffin wax (ASTM D612), 5 ml of concentrated sulfuric acid is placed in a graduated test tube and 5 ml of the melted wax is added. The sample is heated for 10 min at 70°C (158°F). During the last 5 min, the tube is shaken periodically. The acid layer is compared to a standard reference solution, and the wax sample passes if the color is not darker than the standard color.

15.3.4   Color

Paraffin wax is generally white in color, whereas microcrystalline wax and petrolatum range from white to almost black. A fully refined wax should be virtually colorless (*water-white*) when examined in the molten state. Absence of color is of particular importance in wax used for pharmaceutical purposes or for the manufacture of food wrappings. The significance of the color of microcrystalline wax and petrolatum depends on the use for which they are intended. In some applications (e.g., the manufacture of corrosion preventives), color may be of little importance.

The Saybolt color test method (ASTM D156) is used for nearly colorless waxes, and in this method, a melted sample is placed in a heated vertical tube mounted alongside a second tube containing standard color disks. An optical viewer allows simultaneous viewing of both tubes. The level of the sample is decreased until its color is lighter than that of the standard, and the color number above this level is the Saybolt color.

The test method for the color of petroleum products (ASTM D1500, IP 196) is for wax and petrolatum that are too dark for the Saybolt colorimeter. A liquid sample is placed in the test container, a glass cylinder of 30–35 mm ID, and compared with colored glass disks ranging in value from 0–5 to 8–0, using a standard light source. If an exact match is not found, and the sample color falls between two standard colors, the higher of the two colors is reported.

The Lovibond Tintometer (IP 17) is used to measure the tint and depth of color by comparison with a series of red, yellow, and blue standard glasses. Waxes and petrolatum are tested in the molten state, and a wide range of cell sizes is available for different types.

15.3.5   Composition

Almost all physical and functional properties of the wax are affected by (i) the molecular weight range, (ii) distribution of its individual components, and (iii) the degree of branching of the carbon skeleton. For a given melting point, a narrow-cut wax consisting almost entirely of straight-chain paraffins will be harder, be more brittle, and have a higher gloss and blocking point than one of broader cut, or one containing a higher proportion of branched molecules.

All petroleum-derived waxes, including blends of waxes, from n-C_{17} to n-C_{44} can be separated by capillary column chromatography (ASTM D5442). In this method, the sample is diluted in a suitable solvent with an internal standard after which it is then injected into a capillary column, meeting a specified resolution, and the components are detected using a flame ionization detector. The eluted components are identified by comparison to a standard mixture, and the area of each straight-chain and branched-chain alkane is measured.

The polynuclear aromatic content of waxes can be estimated by the ultraviolet absorbance (ASTM D2008) of an extract of the sample. In the test method, the ultraviolet absorbance is determined by measuring the absorption spectrum of the undiluted liquid in a cell of known path length under specified conditions. The ultraviolet absorptivity is determined by measuring the absorbance, at specified wavelengths, of a solution of the liquid or solid at known concentration in a cell of known path length.

The composition of wax is available through alternate procedures that involve solvent extraction (ASTM D721, ASTM D3235, IP 158), and the refractive index (ASTM D1747) can also give an indication of composition but mainly purity.

The solvent-extractable constituents of a wax may have significant effects on several of its properties such as strength, hardness, flexibility, scuff resistance, coefficient of friction, coefficient of expansion, and melting point. In the test method (ASTM D3235), the sample is dissolved in a mixture of 50% v/v MEK and 50% v/v toluene. The solution is cooled to –32°C (–25°F) to precipitate the wax, then filtered. The yield of solvent-extractable constituents is determined by evaporating the solvent from the filtrate and weighing the residue.

15.3.6 Density

For clarification, it is necessary to understand the basic definitions that are used: (i) *density* is the mass of liquid per unit volume at 15°C, (ii) *relative density* is the ratio of the mass of a given volume of liquid at 15°C to the mass of an equal volume of pure water at the same temperature, and (iii) *specific gravity* is the same as the relative density, and the terms are used interchangeably.

Density (ASTM D1298, IP 160) is an important property of petroleum products since petroleum and especially petroleum products are usually bought and sold on that basis or if on volume basis then converted to mass basis via density measurements. This property is almost synonymously termed as density, relative density, gravity, and specific gravity, all terms related to each other. Usually a hydrometer, pycnometer, or a more modern digital density meter is more convenient for the determination of density or specific gravity.

In the most commonly used method (ASTM D1298, IP 160), the sample is brought to the prescribed temperature and transferred to a cylinder at approximately the same temperature. The appropriate hydrometer is lowered into the

sample and allowed to settle, and after temperature equilibrium has been reached, the hydrometer scale is read and the temperature of the sample is noted.

Although there are many methods for the determination of density due to the different nature of petroleum itself and the different products, one test method (ASTM D5002) is used for the determination of the density or relative density of petroleum that can be handled in a normal fashion as liquids at test temperatures between 15 and 35°C (59 and 95°F). This test method applies to petroleum oils with high vapor pressures provided appropriate precautions are taken to prevent vapor loss during transfer of the sample to the density analyzer. In this method, approximately 0.7 ml of crude oil sample is introduced into an oscillating sample tube, and the change in oscillating frequency caused by the change in mass of the tube is used in conjunction with calibration data to determine the density of the sample.

Another test determines density and specific gravity by means of a digital densimeter (ASTM D4052, IP 365). In this test, a small volume (~0.7 ml) of liquid sample is introduced into an oscillating sample tube, and the change in oscillating frequency caused by the change in the mass of the tube is used in conjunction with calibration data to determine the density of the sample. The test is usually applied to petroleum, petroleum distillates, and petroleum products that are liquids at temperatures between 15 and 35°C (59 and 95°F) that have vapor pressures below 600 mm Hg and viscosities below about 15,000 cSt at the temperature of test. However, the method should not be applied to samples so dark in color that the absence of air bubbles in the sample cell cannot be established with certainty.

Accurate determination of the density or specific gravity of crude oil is necessary for the conversion of measured volumes to volumes at the standard temperature of 15.56°C (60°F) (ASTM D1250, IP 200, Petroleum Measurement Tables). The specific gravity is also a factor reflecting the quality of crude oils.

15.3.7 Hardness

Hardness is a measure of resistance to deformation or damage; hence, it is an important criterion for many wax applications. It is indirectly related to blocking tendency and gloss. Hard, narrow-cut waxes have higher blocking points and better gloss than waxes of the same average molecular weight but wider molecular weight range.

The measurement of the needle penetration of petroleum wax (ASTM D1321, IP 376) gives an indication of the hardness, or consistency of wax is measured with a penetrometer applying a load of 100 g for 5 s to a standard needle having a truncated cone tip. The sample is heated to 17°C (30°F) above its congealing point, poured into a small brass cylinder, cooled and placed in a water bath at the test temperature for 1 h. The sample is then positioned under the penetrometer needle that when released penetrates into the sample. The depth of penetration in tenths of millimeters is reported as the test value. This method is not applicable to oily materials or petrolatum, which have penetrations greater than 250.

The method for the determination of cone penetration of petrolatum (ASTM D937, IP 179) is for soft wax and petrolatum. It is similar to the method for determining the needle penetration (ASTM D1321) except that a much larger sample mold is used and a cone replaces the needle. The method requires that a 150 g load be applied for 5 s at the desired temperature.

15.3.8 Melting Point

The *melting point* is one of the most widely used tests to determine the quality and type of wax since wax is, more usual than not, sold on the basis of the melting point range (ASTM D87, ASTM D4419, IP 55).

Petroleum wax, unlike the individual hydrocarbons (Table 15.2), does not melt at sharply defined temperatures because it is a mixture of hydrocarbons with different melting points but usually has a narrow melting range. Thus, measurement of the melting point (more correctly, the *melting range*) may also be used as a means of *fingerprinting* wax to obtain more precise quality control and detailed information. Microcrystalline waxes and petrolatum are more complex and therefore melt over a much wider temperature range.

In the method (ASTM D87, IP 55), a molten wax specimen is placed in a test tube fitted with a thermometer and placed in an air bath, which in turn is surrounded by a water bath held at 16–28°C (60–80°F). As the molten wax cools, periodic readings of its temperature are taken. When solidification of the wax occurs, the rate of temperature decreases, and a plateau is observed in the cooling curve—the temperature at that point (the plateau) is recorded as the melting point of the sample. This procedure is not suitable for microcrystalline wax, petrolatum, or waxes containing large amounts of non-normal hydrocarbons (the plateau rarely occurs in cooling curves of such waxes).

The method of determining the drop melting point of petroleum wax, including petrolatum (ASTM D127, IP 133), can be used for most petroleum waxes and wax–resin blends. In the method, samples are deposited on two thermometer bulbs by dipping chilled thermometers into the sample. The thermometers are then placed in test tubes and heated in a water bath until the specimens melt and the first drop falls from each thermometer bulb. The average of the temperatures at which these drops fall is the drop melting point of the sample.

The congealing point (ASTM D93, IP 76) of petroleum wax, including petrolatum, is determined by dipping a thermometer bulb in the melted wax and placing in a heated vial. The thermometer is held horizontally and slowly rotated on its axis. As long as the wax remains liquid, it will hang from the bulb as a pendant drop. The temperature at which the drop rotates with the thermometer is the congealing point. The congealing point of microcrystalline wax or petrolatum is invariably lower than its corresponding drop melting point.

On the other hand, the solidification point of wax is the temperature in the cooling curve of the wax where the slope of the curve first changes significantly as the wax sample changes from a liquid to a solid state. In the test method (ASTM D3944), a sample of wax is placed in a test tube at ambient temperature and heated above the solidification point of the wax sample. A thermocouple probe, attached to a recorder, is inserted into the wax sample, which is allowed to cool to room temperature. The thermocouple response of the cooling wax traces a curve on the chart paper of the recorder. The first significant change in the slope of the curve is the softening point.

15.3.9 Molecular Weight

The molecular weight of various waxes may differ according to (i) the source of the wax (whether it originated in the lighter- or heavier-grade lubricating oils) and (ii) the processing of the wax (the closeness of the distillation cut or the fractionation by crystallization). Thus, the average molecular weight of a wax may represent an average of a narrow or a wide band of distribution.

Generally, for any series of similar waxes, an increase in molecular weight increases viscosity and melting point. However, many of the other physical and functional properties are more related to the hydrocarbon types and distribution than to the average molecular weight.

TABLE 15.2   Melting points of pure *n*-hydrocarbons

Number of C atoms	Melting point (°C)	Number of C atoms	Melting point (°C)
1	−182	20	36
2	−183	21	40
3	−188	22	44
4	−138	23	47
5	−130	24	51
6	−95	25	54
7	−91	26	56
8	−57	27	59
9	−54	28	61
10	−30	29	64
11	−26	30	66
12	−10	31	68
13	−5	32	70
14	6	33	71
15	10	34	73
16	18	35	75
17	22	40	82
18	28	50	92
19	32	60	99

In the test method for molecular weight (ASTM D2503), a small sample of wax is dissolved in a suitable solvent, and a droplet of the wax solution is placed on a thermistor in a closed chamber in close proximity to a suspended drop of the pure solvent on a second thermistor. The difference in vapor pressure between the two positions results in solvent transport and condensation onto the wax solution with a resultant change in temperature. Through suitable calibration, the observed effect can be expressed in terms of molecular weight of the wax specimen as a number average molecular weight.

15.3.10   Odor and Taste

The odor of wax is an important property in some uses of wax such as food packaging and is often included in the specifications of petroleum wax.

The odor of petroleum wax (ASTM D1833, IP 185) is determined by a method in which 10 g of wax is shaved and placed in an odor-free glass bottle and capped. After 15 min, the sample is evaluated in an odor-free room by removing the cap and sniffing lightly. A rating of 0 (no odor) to 4 (very strong odor) is given by each member of a chosen panel. The reported value is the average of the individual ratings.

However, subjective evaluations such as odor and tasting, even by a group of dedicated enophiles are difficult to standardize, since even in a standardized method involving a group of experts, there may be difference of opinion as to what constitutes an acceptable odor or taste. A specific example is wine tasting.

15.3.11   Oil Content

The oil content of paraffin waxes is an indication of the degree of refinement, and fully refined wax usually has an oil content of less than 0.5%. Wax containing more than this amount of oil is referred to as *scale wax*, although an intermediate grade known as *semi-refined wax* is sometimes recognized for wax having an oil content of about 1%.

Excess oil tends to exude from paraffin wax, giving it a dull appearance and a greasy feel. Such a wax would obviously be unsuitable for many applications, particularly the manufacture of food wrappings. High oil content tends to render the wax *plastic* and has an adverse effect on sealing strength, tensile strength, hardness, odor, taste, color, and particularly color stability.

During wax refining, increasing amounts of oil are removed, and this process needs to be controlled. Also, the oil content of slack waxes, petrolatum, and waxes must be assessed for end-user specification. For high–oil content waxes (i.e., >15% w/w), the method (ASTM D3235) involves dissolving a weighed amount of wax in a mixture of MEK and toluene, followed by cooling to −32°C (−27°F) to precipitate the wax. The oil and solvent are removed; then the solvent is evaporated off to produce a weighable amount of oil.

Gas liquid chromatographic analysis of the solvent-extracted material has shown that the determined *oil* contains a small amount of additional wax, *n*-C17 to *n*-C22 alkanes, thereby producing a small error.

For wax containing less than 15% oil, the method (ASTM D721, IP 158) is similar (to ASTM D3235) but uses only MEK as the solvent. The concept that the oil is much more soluble than wax in MEK at low temperatures is utilized in this procedure. A weighed sample of wax is dissolved in warm MEK in a test tube and chilled to −32°C (−25°F) to precipitate the wax. The solvent–oil solution is separated from the wax by pressure filtration through a sintered glass filter stick. The solvent is evaporated and the residue weighed.

Microcrystalline waxes have a greater affinity for oil than paraffin waxes because of their smaller crystal structure. The permissible amount depends on the type of wax and its intended use.

The oil content of microcrystalline wax is, in general, much greater than that of paraffin wax and could be as high as 20%. Waxes containing more than 20% oil would usually be classed as petrolatum, but this line of demarcation is by no means precise.

15.3.12   Peroxide Content

The deterioration of petroleum wax results in the formation of peroxides and other oxygen-containing compounds. The test method for the determination of the peroxide number measures those compounds that will oxidize potassium iodide. Thus, the magnitude of the peroxide number is an indication of the quantity of oxidizing constituents present. In the test method (ASTM D1832), a sample is dissolved in carbon tetrachloride and is acidified with acetic acid. A solution of potassium iodide is added, and after a reaction period, the solution is titrated with sodium thiosulfate using a starch indicator.

Suitable antioxidants, such as 2,6 di-tertiary butyl-*p*-cresol and butylated hydroxyanisole, may be added to the wax to retard the oxidation reactions.

15.3.13   Slip Properties

Friction is an indication of the resistance to sliding exhibited by two surfaces in contact with one another. The intended application determines the degree of slip desired. Coatings for packages that require stacking should have a high coefficient of friction to prevent slippage in the stacks. Folding box coatings should have a low coefficient of friction to allow the boxes to slide easily from a stack of blanks being fed to the forming and filling equipment.

The coefficient of kinetic friction for wax coatings (ASTM D2534) is determined by fastening a wax-coated paper to a horizontal plate attached to the lower, movable cross arm of an electronic load-cell-type tensile tester. A second paper is taped to a 180-g sled that is placed on the first sample. The sled is attached to the load cell by a nylon monofilament passing around a frictionless pulley. The kinematic coefficient of friction is calculated from the average force required to move the sled at 35 in./min, divided by the sled weight.

In this same vein, the abrasion resistance of wax is also an important property and can be determined by a standard test method (ASTM D3234).

15.3.14 Storage Stability

The presence of peroxides or similar oxy compounds is usually the result of oxidation and deterioration of waxes either in use or storage. Antioxidants, such as butylated hydroxyanisole, may be used to retard oxidation.

The peroxide number of petroleum wax (ASTM D1832) is determined by dissolving a sample in carbon tetrachloride, acidified with acetic acid, and a solution of potassium iodide is added; any peroxides present will react with the potassium iodide to liberate iodine, which is then titrated with sodium thiosulfate.

15.3.15 Strength

Another popular test for wax is the tensile strength, which is considered to be a useful guide in controlling the quality of the wax, although the actual significance of the results obtained is not clear.

The method for determining the tensile strength of paraffin wax is an empirical evaluation of the tensile strength of waxes that do not elongate more than a specified amount under the test conditions. Six dumbbell-shaped specimens, with a cross-sectional area of a square inch, are cast. The specimens are broken on a testing machine under a load that increases at the rate of 20 pounds/second along the longitudinal axis of the sample. Values are reported as pounds per square inch.

To determine the modulus of rupture (breaking force in pounds per square inch) of petroleum wax, a wax slab, 8 in. × 4 in. × 0.15 in., is cast over hot water. Small strips, about 3 in. × 1 in., are cut from the center of the slab. The strips are placed lengthwise on the support beams of the apparatus, and a breaking beam is placed across the specimen parallel to the support beams. A steadily increasing load is applied by water delivered to a bucket suspended from the breaking beam. The modulus of rupture is calculated from an equation relating thickness and width of test specimen with total weight required to break it.

15.3.16 Ultraviolet Absorptivity

For process control purposes, it may be desirable to monitor the total aromatic content of petroleum wax (ASTM D2008).

The procedure tests the product as a whole, without including any separation or fractionation steps to concentrate the absorptive fractions. When wax or petrolatum is tested in this procedure, the specimen is dissolved in isooctane, and the ultraviolet absorbance is measured at a specified wavelength such as 290 nm. The absorptivity is then calculated. This procedure, as such, is not a part of the federal specification.

Although this procedure shows good operator precision, the interpretation of results requires some caution. Since the test does not include any selective fractionation of the sample, it does not distinguish any particular aromatic. It is also subject to the errors arising from interferences or differences in strong or weak absorptivity shown by different aromatics. Therefore, the test is good for characterization, but cannot be used for quantitative determination of aromatic content or any other absorptive component.

15.3.17 Viscosity

Viscosity of molten wax (ASTM D3236) is of importance in applications involving coating or dipping processes since it influences the quality of coating obtained. Examples of such applications are paper converting, hot-dip anticorrosion coatings, and taper manufacturing.

Paraffin waxes do not differ much in viscosity, a typical viscosity being 3 ± 0.5 cSt at 99°C (210°F). Microcrystalline wax is considerably more viscous and varies over a wide range, 10–20 cSt at 99°C (210°F). Some hot melt viscosities exceed 20,000 cSt at 177°C (350°F).

Kinematic viscosity is measured by timing the flow of a fixed volume of material through a calibrated capillary at a selected temperature (ASTM D445, IP 71). The unit of kinematic viscosity is the stokes, and kinematic viscosities of waxes are usually reported in centistokes. Saybolt Universal Seconds can be derived from centistokes (cSt) (ASTM D2161):

$$\text{Saybolt seconds} @ 37.8°C (100°F) = \text{cSt} \times 4.635$$
$$\text{Saybolt seconds} @ 98.9°C (210°F) = \text{cSt} \times 4.667$$

Another method (ASTM D2669) is suitable for blends of wax and additives having apparent viscosities up to 20,000 centipoises at 177°C (350°F). Apparent viscosity is the measurement of drag produced on a rotating spindle immersed in the test liquid. A suitable viscometer is equipped to use interchangeable spindles and adjustable rates of rotation. The wax blend is heated by means of a heating mantle in an 800-ml beaker and continuously stirred, until the test temperature is slightly exceeded. The sample is cooled to the test temperature, the stirring is discontinued, and the viscosity is measured. Viscosities over a range of temperatures are recorded and plotted on semilog paper to determine the apparent viscosity at any temperature in the particular region of interest.

15.3.18 Volatility

The boiling point distribution of paraffin wax provides an estimate of hydrocarbon molecular weight distribution that influences many of the physical and functional properties of petroleum wax. To a lesser extent, distillation characteristics also are influenced by the distribution of various molecular types, that is, n-paraffins, branched, or cyclic structures. In the case of the paraffin waxes that are predominantly straight chain, the distillation curve reflects the molecular size distribution.

In the most common distillation test (ASTM D1160), fractions are obtained under reduced pressure such as at 10 mm for paraffin waxes or at 1 mm for higher-molecular-weight waxes. The fractions are taken at intervals across the full distillable range, and the complete results may be reported. In some cases, and for brevity, the distillation results are reported as the temperature difference observed between the 5% off and 95% off cut points. Waxes having very narrow width of cut will tend to be more crystalline, have higher melting points, higher hardness and tensile strength properties, and less flexibility.

In the gas chromatographic method (ASTM D2887), a sample of the test wax is dissolved in xylene and introduced into a gas chromatographic column that is programmed to separate the hydrocarbons in their boiling point order by raising the temperature of the column at a reproducible, calibrated rate. When wax samples are used, the thermal conductivity detector is used to measure the amount of eluted fraction. The data obtained in this procedure are reported in terms of percentage recovered at certain fixed temperature intervals.

REFERENCES

ASTM D87. Standard Test Method for Melting Point of Petroleum Wax (Cooling Curve). Annual Book of Standards. ASTM International, West Conshohocken, PA.

ASTM D88. Standard Test Method for Saybolt Viscosity. Annual Book of Standards. ASTM International, West Conshohocken, PA.

ASTM D93. Standard Test Methods for Flash Point by Pensky-Martens Closed Cup Tester. Annual Book of Standards. ASTM International, West Conshohocken, PA.

ASTM D127. Standard Test Method for Drop Melting Point of Petroleum Wax, Including Petrolatum. Annual Book of Standards. ASTM International, West Conshohocken, PA.

ASTM D156. Standard Test Method for Saybolt Color of Petroleum Products (Saybolt Chromometer Method). Annual Book of Standards. ASTM International, West Conshohocken, PA.

ASTM D445. Standard Test Method for Kinematic Viscosity of Transparent and Opaque Liquids (and Calculation of Dynamic Viscosity). Annual Book of Standards. ASTM International, West Conshohocken, PA.

ASTM D523. Standard Test Method for Specular Gloss. Annual Book of Standards. ASTM International, West Conshohocken, PA.

ASTM D612. Standard Test Method for Carbonizable Substances in Paraffin Wax. Annual Book of Standards. ASTM International, West Conshohocken, PA.

ASTM D721. Standard Test Method for Oil Content of Petroleum Waxes. Annual Book of Standards. ASTM International, West Conshohocken, PA.

ASTM D937. Standard Test Method for Cone Penetration of Petrolatum. Annual Book of Standards. ASTM International, West Conshohocken, PA.

ASTM D938. Standard Test Method for Congealing Point of Petroleum Waxes, Including Petrolatum. Annual Book of Standards. ASTM International, West Conshohocken, PA.

ASTM D1160. Standard Test Method for Distillation of Petroleum Products at Reduced Pressure. Annual Book of Standards. ASTM International, West Conshohocken, PA.

ASTM D1250. Standard Guide for Use of the Petroleum Measurement Tables. Annual Book of Standards. ASTM International, West Conshohocken, PA.

ASTM D1298. Standard Test Method for Density, Relative Density, or API Gravity of Crude Petroleum and Liquid Petroleum Products by Hydrometer Method. Annual Book of Standards. ASTM International, West Conshohocken, PA.

ASTM D1321. Standard Test Method for Needle Penetration of Petroleum Waxes. Annual Book of Standards. ASTM International, West Conshohocken, PA.

ASTM D1465. Standard Test Method for Blocking and Picking Points of Petroleum Wax. Annual Book of Standards. ASTM International, West Conshohocken, PA.

ASTM D1500. Standard Test Method for ASTM Color of Petroleum Products (ASTM Color Scale). Annual Book of Standards. ASTM International, West Conshohocken, PA.

ASTM D1747. Standard Test Method for Refractive Index of Viscous Materials. Annual Book of Standards. ASTM International, West Conshohocken, PA.

ASTM D1832. Standard Test Method for Peroxide Number of Petroleum Wax (Withdrawn in 2014 but still used in some laboratories). ASTM International, West Conshohocken, PA.

ASTM D1833. Standard Test Method for Odor of Petroleum Wax. Annual Book of Standards. ASTM International, West Conshohocken, PA.

ASTM D2008. Standard Test Method for Ultraviolet Absorbance and Absorptivity of Petroleum Products. Annual Book of Standards. ASTM International, West Conshohocken, PA.

ASTM D2161. Standard Practice for Conversion of Kinematic Viscosity to Saybolt Universal Viscosity or to Saybolt Furol Viscosity. Annual Book of Standards. ASTM International, West Conshohocken, PA.

ASTM D2503. Standard Test Method for Relative Molecular Mass (Molecular Weight) of Hydrocarbons by Thermoelectric Measurement of Vapor Pressure. Annual Book of Standards. ASTM International, West Conshohocken, PA.

ASTM D2534. Standard Test Method for Coefficient of Kinetic Friction for Wax Coatings. Annual Book of Standards. ASTM International, West Conshohocken, PA.

ASTM D2699. Standard Test Method for Research Octane Number of Spark-Ignition Engine Fuel. Annual Book of Standards. ASTM International, West Conshohocken, PA.

ASTM D2887. Standard Test Method for Boiling Range Distribution of Petroleum Fractions by Gas Chromatography. Annual Book of Standards. ASTM International, West Conshohocken, PA.

ASTM D3234. Standard Test Method for Abrasion Resistance of Petroleum Wax Coatings (Withdrawn in 2004 but still used in some laboratories). ASTM International, West Conshohocken, PA.

ASTM D3235. Standard Test Method for Solvent Extractables in Petroleum Waxes. Annual Book of Standards. ASTM International, West Conshohocken, PA.

ASTM D3236. Standard Test Method for Apparent Viscosity of Hot Melt Adhesives and Coating Materials. Annual Book of Standards. ASTM International, West Conshohocken, PA.

ASTM D3944. Standard Test Method for Solidification Point of Petroleum Wax. Annual Book of Standards. ASTM International, West Conshohocken, PA.

ASTM D4052. Standard Test Method for Density, Relative Density, and API Gravity of Liquids by Digital Density Meter. Annual Book of Standards. ASTM International, West Conshohocken, PA.

ASTM D4419. Standard Test Method for Measurement of Transition Temperatures of Petroleum Waxes by Differential Scanning Calorimetry (DSC). Annual Book of Standards. ASTM International, West Conshohocken, PA.

ASTM D5002. Standard Test Method for Density and Relative Density of Crude Oils by Digital Density Analyzer. Annual Book of Standards. ASTM International, West Conshohocken, PA.

ASTM D5442. Standard Test Method for Analysis of Petroleum Waxes by Gas Chromatography. Annual Book of Standards. ASTM International, West Conshohocken, PA.

ASTM D6045. Standard Test Method for Color of Petroleum Products by the Automatic Tristimulus Method. Annual Book of Standards. ASTM International, West Conshohocken, PA.

Gary, J.G., Handwerk, G.E., and Kaiser, M.J. 2007. Petroleum Refining: Technology and Economics, 5th Edition. CRC Press, Taylor & Francis Group, Boca Raton, FL.

Hsu, C.S. and Robinson, P.R. (Editors) 2006. Practical Advances in Petroleum Processing (Volumes 1 and 2). Springer Science, New York.

IP 55 (ASTM D87). Melting Point of Wax – Cooling Curve Method. IP Standard Methods 2013. The Energy Institute, London.

IP 71 (ASTM D445). Petroleum Products – Transparent and Opaque Liquids – Determination of Kinematic Viscosity and Calculation of Dynamic Viscosity. IP Standard Methods 2013. The Energy Institute, London.

IP 76 (ASTM D938). Determination of Congealing Point of Waxes and Petrolatum. IP Standard Methods 2013. The Energy Institute, London.

IP 133 (ASTM D127). Determination of Drop Melting Point of Wax and Petrolatum. IP Standard Methods 2013. The Energy Institute, London.

IP 158 (ASTM D721). Determination of Oil Content of Waxes – Gravimetric Method. IP Standard Methods 2013. The Energy Institute, London.

IP 160 (ASTM D1298). Crude Petroleum and Liquid Petroleum Products – Laboratory Determination of Density – Hydrometer Method. IP Standard Methods 2013. The Energy Institute, London.

IP 179 (ASTM D937). Determination of Cone Penetration of Petrolatum. IP Standard Methods 2013. The Energy Institute, London.

IP 185 (ASTM D1833). Determination of Odor of Petroleum Wax. IP Standard Methods 2013. The Energy Institute, London.

IP 196 (ASTM D1500). Petroleum Products – Determination of Color (ASTM Scale). IP Standard Methods 2013. The Energy Institute, London.

IP 200 (ASTM D1250). Guidelines for the Use of the Petroleum Measurement Tables). IP Standard Methods 2013. The Energy Institute, London.

IP 365. Crude Petroleum and Petroleum Products – Determination of Density – Oscillating U-Tube Method. IP Standard Methods 2013. The Energy Institute, London.

IP 376. Needle Penetration of Petroleum Wax. IP Standard Methods 2013. The Energy Institute, London.

Speight, J.G. 2014. The Chemistry and Technology of Petroleum. 5th Edition. CRC Press, Taylor & Francis Group, Boca Raton, FL.

Speight, J.G. and Ozum, B. 2002. Petroleum Refining Processes. Marcel Dekker Inc., New York.

16

RESIDUA AND ASPHALT

16.1 INTRODUCTION

The high-boiling material residua and asphalt are—with the exception of coke, carbon, and graphite (Chapter 17)—the highest-boiling products of a petroleum refinery in which the constituents of residua and asphalt may have some resemblance to the high-boiling constituents present in the original petroleum feedstock.

Both materials (assuming that thermal decomposition—cracking—has not occurred during the distillation) derive their characteristics from the nature of their crude oil precursor with some variation possible by choice of manufacturing process. Although there are a number of refineries or refinery units whose prime function is to produce asphalt, petroleum asphalt is primarily a collection of integrated refinery operations (Fig. 16.1). Crude oil may be selected for these refineries for a variety of other product requirements, and the asphalt (or residuum) produced may vary somewhat in characteristics from one refinery-crude system to another.

A *resid* (*residuum, pl. residua*) is the residue obtained from petroleum after nondestructive distillation has removed all the volatile materials (Speight and Ozum, 2002; Hsu and Robinson, 2006; Gary et al., 2007; Speight, 2014). The temperature of the distillation is usually maintained below 350 °C (660 °F) since the rate of thermal decomposition of petroleum constituents is minimal below this temperature, but the rate of thermal decomposition of petroleum constituents is substantial above 350 °C (660 °F). A residuum that is held above this temperature so that thermal decomposition has occurred is known as a *cracked resid*.

Resids are black, viscous materials and are obtained by distillation of a crude oil under atmospheric pressure (atmospheric residuum) or under reduced pressure (vacuum residuum). They may be liquid at room temperature (generally atmospheric residua) or almost solid (generally vacuum residua) depending upon the nature of the crude oil (Speight and Ozum, 2002; Hsu and Robinson, 2006; Gary et al., 2007; Speight, 2014).

When a residuum is obtained from a crude oil and thermal decomposition has commenced, it is more usual to refer to this product as *pitch* (Speight, 2014). The differences between the *parent* petroleum and the residua are due to the relative amounts of various constituents present, which are removed or remain by virtue of their relative volatility.

The chemical composition of a residuum from an asphaltic crude oil is complex. Physical methods of fractionation usually indicate high proportions of asphaltene constituents and resin constituents, even in amounts up to 50% (or higher) of the residuum. In addition, the presence of ash-forming metallic constituents, including such organometallic compounds as those of vanadium and nickel, is also a distinguishing feature of residua and the heavier oils. Furthermore, the deeper the *cut* into the crude oil, the greater is the concentration of sulfur and metals in the residuum and the greater is the deterioration in physical properties (Speight and Ozum, 2002; Hsu and Robinson, 2006; Gary et al., 2007; Speight, 2014).

On the other hand, *asphalt* is manufactured from petroleum (as opposed to being produced in the distillation process) and is a black or brown material that has a consistency varying from a viscous liquid to a glassy solid. To a point,

Separation   Conversion   Finishing   Products

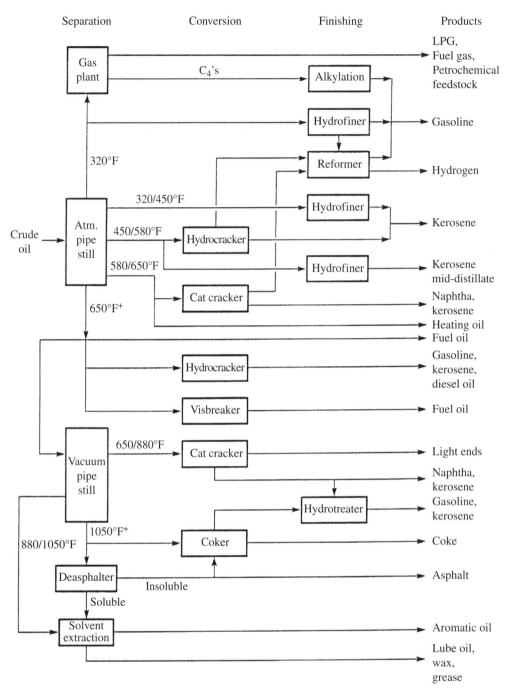

FIGURE 16.1   Schematic of a petroleum refinery.

asphalt can resemble bitumen, hence the tendency to refer to bitumen (incorrectly) as *native asphalt*. It is recommended that there be differentiation between asphalt (manufactured) and bitumen (naturally occurring) other than by use of the qualifying terms *petroleum asphalt* and *native asphalt* since the origins of the materials may be reflected in the resulting physicochemical properties of the two types of materials. It is also necessary to distinguish between the asphalt that originates from petroleum by refining and the product in which

the source of the asphalt is a material other than petroleum, for example, *Wurtzilite asphalt* (Speight, 2014). In the absence of a qualifying word, it is assumed that the term *asphalt* refers to the product manufactured from petroleum.

When the asphalt is produced simply by distillation of an asphaltic crude oil, the product can be referred to as *residual asphalt* or *straight-run asphalt*. If the asphalt is prepared by solvent extraction of residua or by light hydrocarbon (propane) precipitation, or if blown or otherwise treated, the term should be modified accordingly to qualify the product (e.g., *propane asphalt*, *blown asphalt*) (Speight and Ozum, 2002; Hsu and Robinson, 2006; Gary et al., 2007; Speight, 2014).

Asphalt softens when heated and is elastic under certain conditions. The mechanical properties of asphalt are of particular significance when it is used as a binder or adhesive. The principal application of asphalt is in road surfacing, which may be done in a variety of ways. Light oil *dust layer* treatments may be built up by repetition to form a hard surface, or a granular aggregate may be added to an asphalt coat, or earth materials from the road surface itself may be mixed with the asphalt.

Other important applications of asphalt include canal and reservoir linings, dam facings, and sea works. The asphalt so used may be a thin, sprayed membrane, covered with earth for protection against weathering and mechanical damage, or thicker surfaces, often including riprap (crushed rock). Asphalt is also used for roofs, coatings, floor tiles, soundproofing, waterproofing, and other building-construction elements, and in a number of industrial products, such as batteries. For certain applications, an asphaltic emulsion is prepared, in which fine globules of asphalt are suspended in water.

16.2 PRODUCTION AND PROPERTIES

Contrary to the use of confusing terminology, residua and asphalts are different materials. Residua are the nonvolatile materials from the distillation proceeds, while asphalts are manufactured products—they are generally residua that have had the properties midwifed to meet the necessary specification for service. Thus, the term *natural asphalt*, which is often used to describe (tar sand) bitumen—a natural occurring nonvolatile material (Chapter 1)—is incorrect.

16.2.1 Residua

A residuum (pl. *residua*) is the nonvolatile residue from either atmospheric distillation or vacuum distillation. Either the atmospheric residuum or the vacuum residuum, depending upon the properties, can be the starting material for asphalt production, and therefore, the properties of the asphalt are dependent on the properties of the residuum from which the asphalt is manufactured. And residua properties can vary, depending upon the cut point of the residuum (Table 16.1).

Not only do residua constituents have high molecular weight (>500), but they contain polynuclear aromatic (PNA) system—the systems with three to four aromatic rings (or more) provide the greatest hindrance to the conversion of residua because of high thermal stability. In addition, the high concentrations of heteroatoms (nitrogen, oxygen, sulfur, and metals such as vanadium and nickel) in residua

will (along with PNA system) poison catalysts. However, no matter which type of thermal or catalytic process is used for resid processing, a substantial portion of the residuum feedstock (at least 30–50% w/w or more, depending on the process) can be converted to liquid products that are suitable for the production of liquid fuels. Gases are also produced during resid processing along with nonvolatile coke.

The residua from which asphalt are produced were once considered the unusable remnants of petroleum refining and had little value and little use, other than as a (passable in some cases but not in every case) road oil. In fact, the delayed coking process was developed with the purpose of converting residua to liquids (valuable products) and coke (fuel). In addition, recognition that road asphalt—a once maligned product of comparatively little value—is now a valuable refinery product has led to a renewed importance for residua. However, detailed specifications (based on selected analytical test methods) are necessary for asphalt to be used for highway paving as well as for other uses.

16.2.2 Asphalt

Asphalt is a manufactured product insofar as the constituents of the asphalt may have been changed somewhat from the original constituents of the original petroleum feedstock. Furthermore, in order to assure in-service performance, there are specifications for the use of asphalt in highway pavement construction, and the asphalt is selected on the basis of the results of standard test methods. However, many of the tests used to determine the suitability of asphalt for use are empirical in nature (TRB, 2010). Thus, there is the need to develop new performance-based test procedures that can provide the necessary information leading to selection of the most appropriate material to maintain asphalt-based highway.

The manufacturing process involves distilling everything possible from crude petroleum until a residuum with the desired properties is obtained. This is usually done by stages (Fig. 16.2) in which distillation at atmospheric pressure removes the lower-boiling fractions and yields an atmospheric residuum (*reduced crude*) that may contain higher-boiling (lubricating) oils, wax, and asphalt. Distillation of the reduced crude under vacuum removes the oils (and wax) as overhead products, and the asphalt remains as a bottom (or residual) product. The majority of the polar functionalities and high-molecular-weight species in the original crude oil, which tend to be nonvolatile, concentrate in the vacuum residuum (Speight, 2000) thereby conferring desirable or undesirable properties on the asphalt.

At this stage, the asphalt is frequently—for unknown reasons other than the similar appearance of the asphalt to coal tar pitch (Speight, 2013)—and incorrectly referred to as *pitch* and has a softening point (ASTM D36, ASTM D61, ASTM D2319, ASTM D3104, ASTM D3461) related to the amount

TABLE 16.1   Properties of different petroleum residua

	API gravity	Sulfur (% w/w)	Nitrogen (% w/w)	Nickel (% w/w)	Vanadium (% w/w)	Asphaltenes ((C₇) % w/w)	Carbon residue (% w/w)
Atmospheric residua							
Alaska, North Slope crude oil >650 °F	15.2	1.6	0.4	18.0	30.0	2.0	8.5
Arabian heavy crude oil >650 °F	11.9	4.4	0.3	27.0	103.0	8.0	14
Arabian light crude oil >650 °F	17.7	3	0.2	10.0	26.0	1.8	7.5
Kuwait, crude oil >650 °F	13.9	4.4	0.3	14.0	50.0	2.4	12.2
Lloydminster heavy oil (Canada) >650 °F	10.3	4.1	0.3	65.0	141.0	14.0	12.1
Taching crude oil >650 °F	27.3	0.2	0.2	5.0	1.0	4.4	3.8
Vacuum residua							
Alaska, North Slope crude oil >1050 °F	8.2	2.2	0.6	47.0	82.0	4.0	18
Arabian heavy crude oil >1050 °F	7.3	5.1	0.3	40.0	174.0	10.0	19
Arabian light crude oil >1050 °F	8.5	4.4	0.5	24.0	66.0	4.3	14.2
Kuwait crude oil >1050 °F	5.5	5.5	0.4	32.0	102.0	7.1	23.1
Lloydminster heavy oil (Canada) >1050 °F	8.5	4.4	0.6	115.0	252.0	18.0	21.4
Taching crude oil >1050 °F	21.5	0.3	0.4	9.0	2.0	7.6	7.9

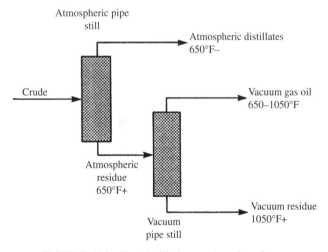

FIGURE 16.2   The distillation section of a refinery.

of oil removed and increases with increasing overhead removal. In character with the elevation of the softening point, the pour point is also elevated (Table 16.1): the more oil distilled from the residue, the higher the softening point.

Asphalt is also produced by propane deasphalting (Fig. 16.3), and there are differences in the properties of asphalts prepared by propane deasphalting and those prepared by vacuum distillation from the same feedstock. Propane deasphalting also has the ability to reduce a residuum even further and produce an asphalt product having

a lower viscosity, higher ductility, and higher temperature susceptibility than other asphalts—such properties are correctly anticipated to be very much crude oil dependent. Propane deasphalting is conventionally applied to low-asphalt-content crude oils, which are generally different in type and source from those processed by distillation of higher-yield crude oils. In addition, the properties of asphalt can be modified by air blowing in batch and continuous processes (Fig. 16.4) (Speight, 1992 and references cited therein; Speight, 2014 and references cited therein). On the other hand, the preparation of asphalts in liquid form by blending (cutting back) asphalt with a petroleum distillate fraction is customary and is generally accomplished in tanks equipped with coils for air agitation or with a mechanical stirrer or a vortex mixer.

An *asphalt emulsion* is a mixture of asphalt and an anionic agent such as the sodium or potassium salt of a fatty acid. The fatty acid is usually a mixture and may contain palmitic, stearic, linoleic, and abietic acids, and/or high-molecular-weight phenols. Sodium lignate is often added to alkaline emulsions to affect better emulsion stability. Nonionic cellulose derivatives are also used to increase the viscosity of the emulsion if needed. The acid number of asphalt is an indicator of its emulsifiability and reflects the presence of high-molecular-weight naphthenic acids. Diamines, frequently used as cationic agents, are made from the reaction of tallow acid amines with acrylonitrile, followed by hydrogenation. The properties of asphalt emulsions

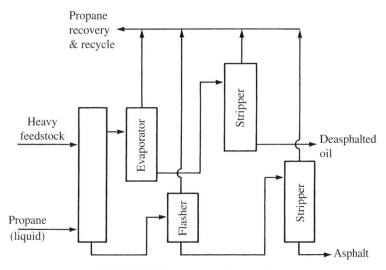

FIGURE 16.3 Propane deasphalting.

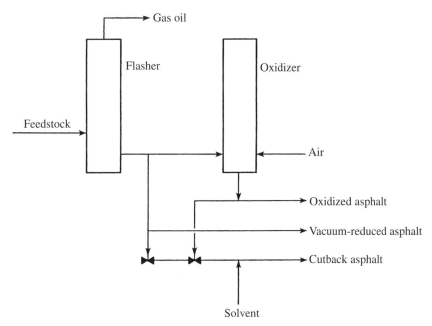

FIGURE 16.4 Asphalt manufacture including air blowing.

(ASTM D977, ASTM D2397) allow a variety of uses. As with other petroleum products, sampling is an important precursor to asphalt analysis, and a standard method (ASTM D140) is available that provides guidance for the sampling of asphalts, liquid and semisolid, at the point of manufacture, storage, or delivery.

16.3   TEST METHODS

The chemical composition of a residuum from an asphaltic crude oil is complex. Physical methods of fractionation usually indicate high proportions of asphaltene constituents and resin constituents, even in amounts up to 50% (or higher) of the residuum. In addition, the presence of ash-forming metallic constituents, including such organometallic compounds as those of vanadium and nickel, is also a distinguishing feature of residua and the heavier oils. Furthermore, the deeper the *cut* into the crude oil, the greater is the concentration of sulfur and metals in the residuum and the greater is the deterioration in physical properties (Speight, 2014). In order to understand the behavior of residua, a series of text has been provided by organization such as the ASTM International and the Energy Institute of the United Kingdom.

Similarly, in order to assure the performance of refinery-produced asphalt when used for roadway pavements, the material (asphalt) is selected on the basis of the results of laboratory tests. However, many of the tests are empirical in nature and that have been used over the decades giving rise to the development of relationships between the laboratory test data and asphalt performance in service. Although this system has worked relatively well, the increasing volume of traffic and the increasing traffic loads now require approaches that will assure good performance of the highway system. It is no longer adequate to merely relate asphalt performance in service to, say, the composition of the asphalt in terms of bulk fractions—the end result may merely be a failed paper exercise. For example, one important characteristic that is often ignored, for no justifiable reason, is the asphalt–aggregate interactions and the occurrence of such interactions during the in-service period.

Indeed, the use of old empirical tests cannot effectively evaluate the new materials such as polymer-modified asphalt that is increasingly being used to manufacture asphalt destined for use on highways. There is a tremendous need to develop new performance-based test procedures. These new tests are obviously the result of the need to provide the necessary information to select the most performance-effective materials to maintain the highway system. However, it must be recognized that the properties of asphalt are defined by the properties of residua from which the asphalt was produced and that the properties of the residua vary with the cut point (Table 16.1), that is, the % v/v of the crude oil and which may dictate variations in the methods of production of the asphalt.

Specifications for paving asphalt cements usually include five grades differing in either viscosity or penetration level at 60 °C (140 °F). Susceptibility of viscosity to temperature is usually controlled in asphalt cement by viscosity limits at a higher temperature such as 135 °C (275 °F) and a penetration or viscosity limit at a lower temperature such as 25 °C (77 °F). Paving cutbacks are also graded at 60 °C (140 °F) by viscosity, with usually four to five grades of each type. For asphalt cements, the newer viscosity-grade designation is the midpoint of the viscosity range. The cutback's grade designation is the minimum kinematic viscosity for the grade, with a maximum grade viscosity of twice the minimum.

Roofing and industrial asphalts are also generally specified in various grades of hardness usually with a combination of softening point (ASTM D61, ASTM D2319, ASTM D3104, ASTM D3461) and penetration to distinguish grades (ASTM D312, ASTM D449). Temperature susceptibility is usually controlled in these requirements by specifying penetration limits or ranges at 25 °C (77 °F) and other temperatures as well as softening point ranges at higher temperatures. Asphalt for built-up roof constructions are differentiated according to application depending primarily on pitch of the roof and to some extent on whether or not mineral surfacing aggregates are specified (ASTM D312). The damp-proofing

grades reflect above or below grade construction, primarily, and whether or not a self-healing property is incorporated.

The significance of a particular test is not always apparent by reading the procedure and sometimes can only be gained through working familiarity with the test. The following tests are commonly used to characterize asphalts, but these are not the only tests used for determining the property and behavior of an asphaltic binder. As in the petroleum industry, a variety of tests are employed having evolved through local, or company, use.

16.3.1 Acid Number

The acid number is a measure of the acidity of a product and is used as a guide in the quality control of asphalt properties. Since a variety of oxidation products contribute to the acid number and the organic acids vary widely in service properties, so the test is not sufficiently accurate to predict the precise behavior of asphalt in service.

Asphalt contains a small amount of organic acids and saponifiable material that is largely determined by the percentage of naphthenic (cycloparaffin) acids of higher molecular weight that are originally present in the crude oil. With increased hardness, asphalt from a particular crude oil normally decreases in acid number as more of the naphthenic acids are removed during the distillation process. Acidic constituents may also be present as additives or as degradation products formed during service, such as oxidation products (ASTM D5770). The relative amount of these materials can be determined by titrating with bases. The acid number is used as a guide in the quality control of lubricating oil formulations. It is also sometimes used as a measure of lubricant degradation in service. Any condemning limits must be empirically established.

In a manner akin to the *acid number*, the *base number* (often referred to as the *neutralization number*) is a measure of the basic constituents in the oil under the conditions of the test. The base number is used as a guide in the quality control of oil formulation and is also used as a measure of oil degradation in service. The *neutralization number* expressed as the *base number* is a measure of this amount of basic substance in the oil always under the conditions of the test. The neutralization number is used as a guide in the quality control of lubricating oil formulations. It is also sometimes used as a measure of lubricant degradation in service; however, any condemning limits must be empirically established.

The *saponification number* expresses the amount of base that will react with 1 g of the sample when heated in a specific manner. Since certain chemicals areare sometimes added to asphalt and also consume alkali and acids, the results obtained indicate the effect of these extraneous materials in addition to the saponifiable material present. In the test method (ASTM D94, IP 136), a known weight of the sample is dissolved in methyl ethyl ketone or a mixture of suitable solvents, and the mixture is heated with a known amount of standard alcoholic potassium hydroxide for between 30 and 90 min at 80 °C (176 °F). The excess alkali is titrated with standard hydrochloric acid and the saponification number calculated.

16.3.2 Asphaltene Content

The asphaltene fraction (ASTM D2006, ASTM D2007, ASTM D3279, ASTM D4124, ASTM D6560, IP 143) is the highest-molecular-weight and most complex fraction in petroleum. In the simplest senses, the amount of asphaltene constituents (the asphaltene fraction) gives an indication of the amount of coke that can be expected during processing (Speight, 2001; Speight and Ozum 2014). In the present context, the presence of asphaltene constituents in asphalt can be beneficial insofar as the asphalt *may* form stronger bonds with the mineral aggregate because of the presence of the asphaltene constituents. On the other hand, the asphaltene constituents are often subject to rapid oxidation as occurs during weathering, which may lead to incompatibility of the oxidized asphaltene constituents with the remainder of the asphalt and the ensuing failure of the asphalt–aggregate bond.

One of the standard test methods (ASTM D6560) has some versatility insofar as it is applicable to the determination of the heptane-insoluble asphaltene content of feedstocks such as gas oil, diesel fuel, residual fuel oils, lubricating oil, bitumen, and crude petroleum that has been topped to an oil temperature of 260 °C (500 °F).

In any of the methods for the determination of the asphaltene content, the crude oil or product (such as asphalt) is mixed with a large excess (usually >30 volumes hydrocarbon per volume of sample) low-boiling hydrocarbon such as n-pentane or n-heptane. For an extremely viscous sample, a solvent such as toluene may be used prior to the addition of the low-boiling hydrocarbon, but an additional amount of the hydrocarbon (usually >30 volumes of hydrocarbon per volume of solvent) must be added to compensate for the presence of the solvent. After a specified time, the insoluble material (the asphaltene fraction) is separated (by filtration) and dried. The yield is reported as percentage (% w/w) of the original sample.

It must be recognized that, in any of these tests, different hydrocarbons (such as n-pentane or n-heptane) will give different yields of the asphaltene fraction, and if the presence of the solvent is not compensated by the use of additional hydrocarbon, the yield will be erroneous. In addition, if the hydrocarbon is not present in a large excess, the yields of the asphaltene fraction will vary and will be erroneous (Speight, 2014).

The *precipitation number* is often equated to the asphaltene content, but there are several issues that remain obvious in its rejection for this purpose. For example, the method to determine the precipitation number (ASTM D91) advocates

the use of naphtha for use with black oil or lubricating oil, and the amount of insoluble material (as a % v/v of the sample) is the precipitating number. In the test, 10 ml of sample is mixed with 90 ml of ASTM precipitation naphtha (that may or may not have a constant chemical composition) in a graduated centrifuge cone and centrifuged for 10 min at 600–700 rpm. The volume of material on the bottom of the centrifuge cone is noted until repeat centrifugation gives a value within 0.1 ml (the precipitation number). Obviously, this can be substantially different from the asphaltene content.

In another test method (ASTM D4055), pentane-insoluble materials above 0.8 μm in size can be determined. In the test method, a sample of oil is mixed with pentane in a volumetric flask, and the oil solution is filtered through a 0.8-μm membrane filter. The flask, funnel, and the filter are washed with pentane to completely transfer the particulates onto the filter that is then dried and weighed to give the yield of pentane-insoluble materials.

Another test method (ASTM D893) that was originally designed for the determination of pentane- and toluene-insoluble materials in used lubricating oils can also be applied to asphalt. However, the method may need modification by first adding a solvent (such as toluene) to the asphalt before adding pentane. The pentane-insoluble constituents can include oil-insoluble materials. The toluene-insoluble materials can come from external contamination, and products from degradation of asphalt. A significant change in the pentane-insoluble constituents or toluene-insoluble constituents indicates a change in asphalt that could lead to performance problems.

There are two test methods used. Procedure A covers the determination of insoluble constituents without the use of coagulant in the pentane and provides an indication of the materials that can be readily separated from the asphalt–solvent mixture by centrifugation. Procedure B covers the determination of insoluble constituents in asphalt containing additives and employs a coagulant. In addition to the materials separated by using Procedure A, this coagulation procedure separates some finely divided materials that may be suspended in the asphalt. The results obtained by Procedures A and B should not be compared since they usually give different values. The same procedure should be applied when comparing results obtained periodically on oil in use or when comparing results determined in different laboratories.

In Procedure A, a sample is mixed with pentane and centrifuged. The asphalt solution is decanted, and the precipitate is washed twice with pentane, dried, and weighed. For toluene-insoluble constituents, a separate sample of the asphalt is mixed with pentane and centrifuged. The precipitate is washed twice with pentane, once with toluene–alcohol solution, and once with toluene. The insoluble material is then dried and weighed. In Procedure B, Procedure A is followed except that instead of pentane, a pentane-coagulant solution is used.

In addition to determining the asphaltene content of heavy or residual fuel oil, a test method is available to determine the maximum flocculation ratio and peptizing power in residual and heavy fuel oils using an optical device (ASTM D7060). The method is applicable to atmospheric or vacuum distillation residues, thermally cracked residue, and intermediate and finished residual fuel oils, containing at least 1% w/w asphaltene constituents. The flocculation ratio is an important aspect of the behavior of petroleum products that contain asphaltene constituents—it is necessary to know the point at which asphaltene constituents will start to deposit as a solid phase due to changes (thermal or physical) in the product when in use.

In the method, portions of the sample are diluted with various ratios with 1-methylnaphthalene, and each solution is inserted into the automatic apparatus and titrated with cetane until flocculation of the asphaltene constituents is detected by the optical probe. The four flocculation ratios at critical dilution, measured during the fine determinations, are used to calculate the maximum flocculation ratio of the asphaltene constituents in the sample and the peptizing power of the oil medium (non-asphaltene constituents) of the sample.

16.3.3 Bond and Adhesion

The adhesion of asphalt to the mineral aggregate is a fundamental property of road asphalt. Once the adhesion deteriorates, the surface becomes unstable and unusable. There is a test method (ASTM D3409) that covers the determination of the adhesion of asphalt roofing cements to damp, wet, or underwater surfaces. The data are primarily used to determine whether a jointing material possesses an arbitrary amount of bonding strength at low temperatures where Portland cement concrete is being used.

Another test method (ASTM D4867) is used to examine the effect of moisture on asphalt concrete paving mixtures. The test method can be used to test asphalt concrete mixtures in conjunction with mixture design testing to determine the potential for moisture damage, to determine whether or not an anti-stripping additive is effective, and to determine what dosage of an additive is needed to maximize the effectiveness. This test method can also be used to test mixtures produced in plants to determine the effectiveness of additives under the conditions imposed in the field.

16.3.4 Breaking Point

Brittle asphalt causes pavement instability and the appearance of cracks due to (i) excessive loading; (ii) weak surface, base, or subgrade; (iii) thin surface or base; or (iv) poor drainage. Once cracks appear, the asphalt will deteriorate

quickly. One particular test method (IP 80) is an approximate indication of the temperature at which asphalt possesses no ductility and would reflect brittle fracture conditions.

The Fraass breaking point test (IP 80) is used for determining the breaking point of solid and semisolid asphalt. The breaking point is the temperature at which asphalt first becomes brittle, as indicated by the appearance of cracks when a thin film of the asphalt on a metal plaque is cooled and flexed in accordance with specified conditions.

16.3.5 Carbon Disulfide–Insoluble Constituents

The component of highest carbon content is the fraction termed *carboids* and consists of species that are insoluble in carbon disulfide or in pyridine. The fraction that has been called *carbenes* contains molecular species that are soluble in carbon disulfide and soluble in pyridine but which are insoluble in toluene (Fig. 16.5).

Asphalt is a hydrocarbonaceous material that is made of constituents (containing carbon, hydrogen, nitrogen, oxygen, and sulfur) that are completely soluble in carbon disulfide (ASTM D4). Trichloroethylene or 1,1,1-trichloroethane has been used in recent years as solvents for the determination of asphalt solubility (ASTM D2042).

The carbene and carboid fractions are generated by thermal degradation or by oxidative degradation and are not considered to be naturally occurring constituents of asphalt. The test method for determining the toluene-insoluble constituents of tar and pitch (ASTM D4072, ASTM D4312) can be used to determine the amount of carbenes and carboids (both are ill-defined constituents) in asphalt (Speight and Ozum, 2002; Hsu and Robinson, 2006; Gary et al., 2007; Speight, 2014).

16.3.6 Carbon Residue

The *carbon residue* of asphalt serves as an indication of the propensity of the sample to form carbonaceous deposits (thermal coke) under the influence of heat. The result is also often used to provide thermal data that give an indication of the composition of the asphalt (Speight, 2001, 2014).

Tests for Conradson carbon residue (ASTM D189, IP 13), the Ramsbottom carbon residue (ASTM D524, IP 14), the microcarbon carbon residue (ASTM D4530, IP 398), and asphaltene content (ASTM D2006, ASTM D2007, ASTM D3279, ASTM D4124, ASTM D6560, IP 143) are sometimes included in inspection data on petroleum. The data give an indication of the amount of coke that will be formed during thermal processes as well as an indication of the amount of high-boiling constituents in petroleum.

The determination of the *carbon residue* of petroleum or a petroleum product is applicable to relatively nonvolatile samples that decompose on distillation at atmospheric

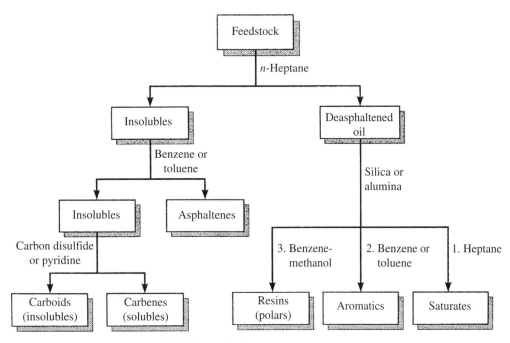

FIGURE 16.5 Feedstock fractionation.

pressure. Samples that contain ash-forming constituents will have an erroneously high carbon residue, depending upon the amount of ash formed. All three methods are applicable to relatively nonvolatile petroleum products that partially decompose on distillation at atmospheric pressure. Crude oils having a low carbon residue may be distilled to a specified residue and the carbon residue test of choice then applied to the residue.

In the Conradson carbon residue test (ASTM D189, IP 13), a weighed quantity of sample is placed in a crucible and subjected to destructive distillation for a fixed period of severe heating. At the end of the specified heating period, the test crucible containing the carbonaceous residue is cooled in a desiccator and weighed, and the residue is reported as a percentage (% w/w) of the original sample (Conradson carbon residue).

In the Ramsbottom carbon residue test (ASTM D524, IP 14), the sample is weighed into a glass bulb that has a capillary opening and is placed into a furnace (at 550 °C, 1020 °F). The volatile matter is distilled from the bulb and the nonvolatile matter that remains in the bulb cracks to form thermal coke. After a specified heating period, the bulb is removed from the bath, cooled in a desiccator, and weighed to report the residue (Ramsbottom carbon residue) as a percentage (% w/w) of the original sample.

In the microcarbon residue test (ASTM D4530, IP 398), a weighed quantity of the sample placed in a glass vial is heated to 500 °C (930 °F) under an inert (nitrogen) atmosphere in a controlled manner for a specific time, and the carbonaceous residue [*carbon residue (micro)*] is reported as a percentage (% w/w) of the original sample.

The data produced by the microcarbon test (ASTM D4530, IP 398) are equivalent to those by Conradson carbon method (ASTM D189 IP 13) However, this microcarbon test method offers better control of test conditions and requires a smaller sample. Up to 12 samples can be run simultaneously. This test method is applicable to petroleum and to petroleum products that partially decompose on distillation at atmospheric pressure and is applicable to a variety of samples that generate a range of yields (0.01–30% w/w) of thermal coke.

Other test methods that are used for determining the coking value of tar and pitch (ASTM D2416, ASTM D4715) that indicates the relative coke-forming properties of tars and pitches might also be applied to asphalt. Both test methods are applicable to tar and pitch having an ash content ≤0.5% (ASTM D2415). The former test method (ASTM D2416) gives results close to those obtained by the Conradson carbon residue test (ASTM D189, IP 13). However, in the latter test method (ASTM D4715), a sample is heated for a specified time at 550 ± 10 °C (1022 ± 18 °F) in an electric furnace. The percentage of residue is reported as the coking value.

Finally, a method that is used to determine pitch volatility (ASTM D4893) might also be used, on occasion, to determine the nonvolatility of asphalt. In the method, an aluminum dish containing about 15 g of accurately weighed sample is introduced into the cavity of a metal block heated and maintained at 350 °C (660 °F). After 30 min, during which any volatiles are swept away from the surface of the sample by preheated nitrogen, the residual sample is taken out and allowed to cool down in the desiccator. The nonvolatility is determined by the sample weight remaining and reported as percentage of w/w residue.

In any of these tests, ash-forming constituents (ASTM D482) or nonvolatile additives present in the sample will be included in the total carbon residue reported leading to higher carbon residue values and erroneous conclusions about the coke-forming propensity of the sample.

16.3.7 Compatibility

When coating asphalt and saturating-grade asphalt are used together, as in prepared roofing, one test method (ASTM D1370) indicates whether they are likely to bleed or disbond under stress at the coating felt interface.

The test method (ASTM D1370) is used to assess the degree to which asphalt interact with other asphalts, and the data can be used to indicate possible future problems, especially blistering, in a roofing product if incompatible asphalts are in contact in the product. The method provides a means for evaluating contact compatibility between asphaltic materials.

16.3.8 Composition

Determination of the composition of asphalt has always presented a challenge because of the complexity and high molecular weights of the molecular constituents. The principle behind composition studies is to evaluate asphalts in terms of composition and performance.

The methods employed can be conveniently arranged into a number of categories: (i) fractionation by precipitation; (ii) fractionation by distillation; (iii) separation by chromatographic techniques; (iv) chemical analysis using spectrophotometric techniques (infrared, ultraviolet, nuclear magnetic resource, X-ray fluorescence, emission, neutron activation), titrimetric and gravimetric techniques, elemental analysis; (v) molecular weight analysis by mass spectrometry, vapor pressure osmometry, and size exclusion chromatography.

However, fractional separation has been the basis for most asphalt composition analysis (Fig. 16.5). The separation methods have been used to divide asphalt into operationally defined fractions. Three types of asphalt separation procedures are now in use: (i) chemical precipitation in which *n*-pentane separation of asphaltene constituents is followed by chemical precipitation of other fractions with sulfuric acid of increasing concentration (ASTM D2006); (ii) adsorption chromatography using a clay–gel procedure where, after removal of the asphaltene constituents, the remaining constituents are separated by selective adsorption/desorption on an adsorbent (ASTM D2007, ASTM D4124); (iii) size exclusion chromatography in which gel permeation chromatographic separation of asphalt constituents occurs based on their associated sizes in dilute solutions.

The fractions obtained in these schemes are defined operationally or procedurally. The amount and type of the asphaltene fraction in asphalt are, for instance, defined by the solvent used for precipitating them. Fractional separation of asphalt does not provide well-defined chemical components. The materials separated should only be defined in terms of the particular test procedure (Fig. 16.5). However, these fractions are generated by thermal degradation or by oxidative degradation and are not considered to be naturally occurring constituents of asphalt. The test method for determining the toluene-insoluble constituents of tar and pitch (ASTM D4072, ASTM D4312) can be sued to determine the amount of carbenes and carboids in asphalt.

In the methods, a sample is digested at 95 °C (203 °F) for 25 min and then extracted with hot toluene in an alundum thimble. The extraction time is 18 h (ASTM D4072) or 3 h (ASTM D4312). The insoluble matter is dried and weighed. Combustion will then show if the material is truly carbonaceous or if it is inorganic ash from the metallic constituents (ASTM D482, ASTM D2415, ASTM D4628, ASTM D4927, ASTM D5185, ASTM D6443, IP 4).

Another method (ASTM D893) covers the determination of pentane- and toluene-insoluble constituents in used lubricating oils and can be applied to asphalt. Pentane-insoluble constituents include oil-insoluble materials, and toluene-insoluble constituents can come from external contamination and highly carbonized materials from degradation. A significant change in pentane- or toluene-insoluble constituents indicates a change in asphalt properties that could lead to problems in service. The insoluble constituents measured can also assist in evaluating the performance characteristics of asphalt.

There are two test methods used: Procedure A covers the determination of insoluble constituents without the use of coagulant in the pentane and provides an indication of the materials that can be readily separated from the diluted asphalt by centrifugation. Procedure B covers the determination of insoluble constituents in asphalt that contains additives and employs a coagulant. In addition to the materials separated by using Procedure A, this coagulation procedure separates some finely divided materials that may be suspended in the asphalt. The results obtained by Procedures A and B should not be compared since they usually give different values. The same procedure should be applied when comparing results obtained periodically on asphalt in use, or when comparing results determined in different laboratories.

In Procedure A, a sample is mixed with pentane and centrifuged after which the asphalt solution is decanted, and the precipitate washed twice with pentane, dried, and weighed. For toluene-insoluble constituents, a separate sample of the asphalt is mixed with pentane and centrifuged. The precipitate is washed twice with pentane, once with toluene–alcohol solution, and once with toluene. The insoluble material is then dried and weighed. In Procedure B, Procedure A is followed except that instead of pentane, a pentane-coagulant solution is used.

Many investigations of relationships between composition and properties take into account only the concentration of the asphaltene constituents, independently of any quality criterion. However, a distinction should be made between the asphaltene constituents that occur in straight-run asphalts and those which occur in blown asphalts. It cannot be under-emphasized that asphaltene fraction is a solubility class rather than a distinct chemical class, means that vast differences occur in the makeup of this fraction when it is produced by different procedures.

Finally, composition data should always be applied to in-service performance in order to properly evaluate the behavior of the asphaltic binder under true working conditions.

16.3.9 Density

For clarification, it is necessary to understand the basic definitions that are used: (i) *density* is the mass of liquid per unit volume at 15.6 °C (60 °F), (ii) *relative density* is the ratio of the mass of a given volume of liquid at 15.6 °C (60 °F) to the mass of an equal volume of pure water at the same temperature, (iii) *specific gravity* is the same as the relative density, and the terms are used interchangeably.

Density (ASTM D1298, IP 160) is an important property of petroleum products since petroleum and especially petroleum products are usually bought and sold on that basis, or if on volume basis, then converted to mass basis via density measurements. This property is almost synonymously termed as density, relative density, gravity, and specific gravity, all terms related to each other. Usually a hydrometer, pycnometer, or more modern digital density meter is used for the determination of density or specific gravity.

In the most commonly used method (ASTM D1298, IP 160), the sample is brought to the prescribed temperature and transferred to a cylinder at approximately the same temperature. The appropriate hydrometer is lowered into the sample and allowed to settle, and after temperature equilibrium has been reached, the hydrometer scale is read and the temperature of the sample is noted.

Although there are many methods for the determination of density due to the different nature of petroleum itself and the different products, one test method (ASTM D5002) is used for the determination of the density or relative density of petroleum that can be handled in a normal fashion as liquids at test temperatures between 15 and 35 °C (59 and 95 °F). This test method applies to petroleum oils with high vapor pressures provided appropriate precautions are taken to prevent vapor loss during transfer of the sample to the density analyzer. In this method, approximately 0.7 ml of crude oil sample is introduced into an oscillating sample tube and the change in oscillating frequency caused by the change in mass of the tube is used in conjunction with calibration data to determine the density of the sample.

Another test determines density and specific gravity by means of a digital densimeter (ASTM D4052, IP 365). In the test, a small volume (~0.7 ml) of liquid sample is introduced into an oscillating sample tube and the change in oscillating frequency caused by the change in the mass of the tube is used in conjunction with calibration data to determine the density of the sample. The test is usually applied to of petroleum, petroleum distillates, and petroleum products that are liquids at temperatures between 15 and 35 °C (59 and 95 °F) that have vapor pressures below 600 mm Hg and viscosities below about 15,000 cSt at the temperature of test. However, the method should not be applied to samples so dark in color that the absence of air bubbles in the sample cell cannot be established with certainty.

Accurate determination of the density or specific gravity of crude oil is necessary for the conversion of measured volumes to volumes at the standard temperature of 15.56 °C (60 °F) (ASTM D1250, IP 200, Petroleum Measurement Tables). The specific gravity is also a factor reflecting the quality of crude oils.

The accurate determination of the API gravity of petroleum and its products (ASTM D287) is necessary for the conversion of measured volumes to volumes at the standard temperature of 60 °F (15.56 °C). Gravity is a factor governing the quality of crude oils. However, the gravity of a petroleum product is an uncertain indication of its quality. Correlated with other properties, gravity can be used to give approximate hydrocarbon composition and heat of combustion. This is usually accomplished through use of the API gravity that is derived from the specific gravity:

$$\text{API gravity, degrees} = \left(141.5 \,/\, \text{sp gr } 60 \,/\, 60\,^\circ\text{F}\right) - 131.5$$

The API gravity is also a critical measure for reflecting the quality of petroleum (Speight, 2001, 2014).

API gravity or density or relative density can be determined using one of two hydrometer methods (ASTM D287, ASTM D1298). The use of a digital analyzer (ASTM D5002) is finding increasing popularity for the measurement of density and specific gravity.

In the method (ASTM D287), the API gravity is determined using a glass hydrometer for petroleum and petroleum products that are normally handled as liquids and that have a Reid vapor pressure of 26 psi (180 kPa) or less. The API gravity is determined at 15.6 °C (60 °F), or converted to values at 60 °F, by means of standard tables. These tables are not applicable to non-hydrocarbons or essentially pure hydrocarbons such as the aromatics.

This test method is based on the principle that the gravity of a liquid varies directly with the depth of immersion of a body floating in it. The API gravity is determined using a hydrometer by observing the freely floating API hydrometer and noting the graduation nearest to the apparent intersection of the horizontal plane surface of the liquid with the vertical scale of the hydrometer, after temperature equilibrium has been reached. The temperature of the sample is determined using a standard test thermometer that is immersed in the sample or from the thermometer that is an integral part of the hydrometer (thermohydrometer).

For solid and semisolid asphalt, a pycnometer is generally used (ASTM D70) and a hydrometer is applicable to liquid asphalt (ASTM D3142).

16.3.10 Distillation

Asphalt is prepared from a distillation residuum, and therefore the need for distillation data is limited. Vacuum distillation data (ASTM D1160) will be valuable for composition purposes if the asphalt is prepared from an atmospheric residuum. Approximate amounts of volatile constituents can also be determined by test methods developed for other products (ASTM D20, ASTM D402, ASTM D3607, ASTM D4893), but that are particularly applicable to cutback asphalt.

Asphalt can also be examined for evaporative losses using a test method designed for grease (ASTM D2595) or for engine oil (ASTM D5800, ASTM D6375, IP 421). In another test method (ASTM D972), the evaporative losses at any temperature in the range of 100–150 °C (210–300 °F) can be determined. A weighed sample is placed in an evaporation cell in an oil bath at the desired test temperature. Heated air at a specified flow rate is passed over the sample surface for 22 h, after which the loss in sample mass is determined. In yet another method (ASTM D2595) that is used to supplement the original method (ASTM D972), the loss of volatile materials from a grease over a temperature range of 93–316 °C (200–600 °F) can be determined. This test uses an aluminum block heater, instead of an oil bath (ASTM D972) to achieve higher temperatures.

Another test (ASTM D6) allows the determination of the percentage loss of mass when a weighed quantity of water-free material is heated in moving air for 5 h at 163 °C (325 °F). In the test method, a gravity convection oven with a rotating shelf is used, and the method provides only a relative measurement of the volatility of a material under test conditions. It may be required for bituminous coating to be applied to galvanized, corrugated, culvert tube. A test for the loss of heating from a thin film is also available (ASTM D1754).

16.3.11 Ductility

The ductility of a solid material is the ability of the material to deform under tensile stress. Malleability, a similar property, is the ability of a material to deform under compressive stress.

The *ductility* of asphalt is a measure of the flexibility of the asphalt and is expressed as the distance in centimeter that a standard briquette can be elongated before breaking (ASTM D113). Ductility is a combination of flow properties and reflects homogeneity, cohesion, and shear susceptibility; it is an indication of fatigue life and cracking.

In the test method, the ductility of asphalt is determined by measuring the distance the asphalt will elongate before breaking when two ends of a briquette specimen of the material are pulled apart at a specified speed and at a specified temperature. Unless otherwise specified, the test shall be made at a temperature of 25 ± 0.5 °C (77 ± 0.9 °F) and with a speed of 5 cm/min ± 5.0%; at other temperatures, the speed should be specified.

16.3.12 Durability

The *durability* of asphalt is an indication of the presence of the necessary chemical and physical properties required for the specified pavement performance. The property indicates the resistance of the asphalt to change during the in-service conditions that are prevalent during the life of the pavement. The durability is determined in terms of resistance to oxidation (resistance to weathering) and water resistance (ASTM D1669, ASTM D1670).

16.3.13 Elemental Analysis

Asphalt is not composed of a single chemical species, but is rather a complex mixture of organic molecules that vary widely in composition and are composed of carbon, hydrogen, nitrogen, oxygen, and sulfur, as well as trace amounts of metals, principally vanadium and nickel. The heteroatoms, although a minor component compared to the hydrocarbon moiety, can vary in concentration over a wide range depending on the source of the asphalt and be a major influence on asphalt properties.

Generally, most asphalt is 79–88% w/w carbon, 7–13% w/w hydrogen, trace to 8% w/w sulfur, 2–8% w/w oxygen, and trace to 3% w/w nitrogen. Trace metals such as iron, nickel, vanadium, calcium, titanium, magnesium, sodium, cobalt, copper, tin, and zinc occur in crude oils. Vanadium and nickel are bound in organic complexes and, by virtue of the concentration (distillation) process by which asphalt is manufactured, are also found in asphalt. The catalytic behavior of vanadium has prompted studies of the relation between vanadium content and the sensitivity of asphalt to oxidation (viscosity ratio). The effects and significance of metallic constituents in the behavior of asphalt is not yet well understood and has not been clarified.

Thus, elemental analysis is still of considerable value to determine the amounts of elements in asphalt, and the method chosen for the analysis may be subject to the peculiarities or character of the feedstock under investigation and should be assessed in terms of accuracy and reproducibility. Methods that are designated for elemental analysis are as follows:

1. *Carbon* and *hydrogen content* (ASTM D1018, ASTM D3343, ASTM D3701, ASTM D5291, ASTM E777, IP 338)
2. *Nitrogen content* (ASTM D3228, ASTM E258, ASTM D5291, ASTM E778)
3. *Oxygen content* (ASTM E385)
4. *Sulfur content* (ASTM D129, ASTM D139, ASTM D1266, ASTM D1552, ASTM D2622, ASTM D3120, ASTM D4045, ASTM D4294, IP 30, IP 61, IP 107, IP 154, IP 243)

The determination of *nitrogen* has been performed regularly by the Kjeldahl method (ASTM D3228) and the Dumas method. The chemiluminescence method is the most recent technique applied to nitrogen analysis for petroleum. The chemiluminescence method determines the amount of chemically bound nitrogen in liquid hydrocarbon samples. In the method, the samples are introduced to the oxygen-rich atmosphere of a pyrolysis tube maintained at 975 °C (1785 °F). Nitrogen in the sample is converted to nitric oxide during combustion, and the combustion products are dried by passage through magnesium perchlorate $[Mg(ClO_4)_2]$ before entering the reaction chamber of a chemiluminescence detector.

Oxygen is one of the five (C, H, N, O, and S) major elements in asphalt, and although the level rarely exceeds 1.5% by weight, it may still be critical to performance. Many petroleum products do not specify the oxygen content, but if the oxygen compounds are present as acidic compounds such a phenols (Ar-OH) and naphthenic acids (cycloalkyl-COOH), they are controlled in different specifications by a variety of tests.

16.3.14 Emulsified Asphalt

Emulsified asphalt is a suspension of small asphalt globules in water, which is assisted by an emulsifying agent (such as soap). The emulsifying agent assists by imparting an electrical charge to the surface of the asphalt globules so that they do not coalesce. Emulsions are used because they effectively reduce asphalt viscosity for lower temperature.

Generally, asphalt emulsions appear as a thick brown liquid when initially applied, but when the asphalt starts to adhere to the surrounding material (e.g., aggregate), the color changes from brown to black. As water begins to evaporate, the emulsion begins to behave more and more like pure asphalt cement. The time required to break and set depends upon the type of emulsion, the application rate, the temperature of the surface onto which it is applied, and environmental conditions.

There is a standard test method (ASTM D244) that covers a variety of tests for the composition, handling, nature and classification, storage, and use, and for specifying asphalt emulsions used primarily for paving purposes.

16.3.15 Flash Point

The *flash point* is the lowest temperature at which application of a test flame causes the vapor of a sample to ignite under specified test conditions. The flash point measures the tendency of a sample to form a flammable mixture with air under controlled laboratory conditions. Flash point data are used in shipping and safety regulations to define *flammable* and *combustible* materials as well as indicate the possible presence of highly volatile and flammable material in a relatively nonvolatile or nonflammable material. The flash point should not be confused with auto-ignition temperature (ASTM E659) that measures spontaneous combustion with no external source of ignition.

The Pensky–Martens Closed Tester (ASTM D93, IP 34) and the Tag Closed Tester (ASTM D56) are normally employed for determining the flash point of fuel oil and similar products. The Cleveland open cup method (ASTM D92) is most commonly used although the Tag open cup (ASTM D3143) is applicable to cutback asphalt. As noted earlier, the flash point of asphalt is an indication of fire hazard and is frequently used to indicate whether asphalt has been contaminated with materials of lower flash point.

16.3.16 Float Test

The float test is used to determine the consistency of asphalt at a specified temperature. One test method (ASTM D139) is normally used for asphalt that is too soft for the penetration test (ASTM D5, ASTM D217, ASTM D937, ASTM D1403, IP 50, IP 179, IP 310).

The float test characterizes the flow behavior or consistency of certain bituminous materials and is useful for determining the consistency of asphalt as one element in establishing the uniformity of certain shipments or sources of supply.

16.3.17 Molecular Weight

The molecular weight of asphalt is not always (in fact, rarely) used in specifications. Nevertheless, there may be occasions when the molecular weight of asphalt is desired (Speight, 2014).

Currently, of the methods available, several standard methods are recognized as being useful for determining the molecular weight of petroleum fractions (ASTM D2502; ASTM D2503; ASTM D2878). Each method has proponents and opponents because of assumptions made in the use of the method or because of the mere complexity of the sample and the nature of the inter- and intramolecular interactions. Before application of any one or more of these methods, consideration must be given to the mechanics of the method and the desired end result.

Methods for molecular weight measurement are also included in other more comprehensive standards (ASTM

D128, ASTM D3712), and several indirect methods have been proposed for the estimation of molecular weight by correlation with other, more readily measured physical properties (Speight, 2001, 2014). They are satisfactory when dealing with the conventional type of crude oils or their fractions and products and when approximate values are desired.

The molecular weights of the individual fractions of asphalt have received more attention, and have been considered to be of greater importance, than the molecular weight of the asphalt itself (Speight, 2001, 2014). The components that make up the asphalt influence the properties of the material to an extent that is dependent upon the relative amount of the component, the molecular structure of the component, and the physical structure of the component that includes the molecular weight.

Asphaltene constituents have a wide range of molecular weights, from 500 to at least 2500, depending upon the method (Speight, 1994, 2014). Asphaltene constituents associate in dilute solution in nonpolar solvents giving higher molecular weights than is actually the case on an individual molecular basis. The molecular weights of the resin constituents are somewhat lower than those of the asphaltene constituents and usually fall within the range of 500–1000. This is due not only to the absence of association but also to a lower absolute molecular size. The molecular weight of the oil fraction (i.e., the asphalt minus the asphaltene constituents and minus the resin constituents) is usually less than 500, but often in the range of 300–400.

16.3.18 Penetration

The penetration test provides one measure of the consistency and hardness of asphalt.

Several test method are available for products such as grease (ASTM D217, ASTM D1403, IP 50, IP 310) and petrolatum (ASTM D937, IP 179) that might be modified for asphalt. The more usual test for asphalt (ASTM D5, IP 49) is a commonly used consistency test. It involves the determination of the extent to which a standard needle penetrates a properly prepared sample of asphalt under definitely specified conditions of temperature, load, and time (100 g load, 5 s). The distance that the needle penetrates in units of mm measured from 0 to 300 is the penetration value. Soft asphalt has a high penetration value, and the converse is true for hard asphalt.

16.3.19 Rheology

Asphalt is a viscoelastic material whose rheological properties reflect crude type and, to a lesser extent, processing. The ability of asphalt to perform under many conditions depends on flow behavior. Asphalt films or coatings showing no appreciable change from original conditions are usually desired, that is, they should allow some structural movement without permanent deformation.

The viscosity of hydrocarbons and temperature are related by the Walther equation, which gives the limiting viscosity at low shear rates (ASTM D5018):

$$\log_{10}[\log_{10}(\nu + \lambda)] = A - B \, \log_{10}(T)$$

In the equation, λ is a shift constant, T is the absolute temperature, and A, B are empirical parameters that reflect the intercept level and slope (or measure of susceptibility of viscosity to temperature, respectively).

The different general families of asphalts (cutbacks or liquid materials, paving asphalt cements and the harder roofing and industrial materials) are usually graded by softening point). At lower temperatures (60 °C, 140 °F, and lower) and/or higher shear rates—which are typical of asphalt service conditions after incorporation in a roof or pavement—semisolid and solid asphalts display an increasing elastic component that relates viscosity with shear rate. The constant high viscosity level at lower shear rates is the limiting viscosity. Viscosities in the area where viscosity changes with shear rate are generally termed apparent viscosities.

A number of viscometers have been developed for securing viscosity data at temperatures as low as 0 °C (32 °F). The most popular instruments in current use are the cone plate (ASTM D4287), parallel plate, and capillary instruments (ASTM D2170, ASTM D2171). The cone plate can be used for the determination of viscosities in the range of 10 to over 10^9 Pa's at temperatures of 0–70 °C (32–158 °F) and at shear rates from 10^{-3} to 10^2 s^{-1}. Capillary viscometers are commonly used for the determination of viscosities at 60–135 °C (140–275 °F).

Tests recently developed for the measurement of viscoelastic properties are directly usable in engineering relations. Properties can be related to the inherent structure of bituminous materials. The fraction of highest molecular weight, the asphaltene fraction, is dispersed within the asphalt and is dependent upon the content and nature of the resin and oil fractions. Higher aromaticity of the oil fractions or higher temperatures leads to viscous (sol) conditions. A more elastic (gel) condition results from a more paraffinic nature and is indicated by large elastic moduli or, empirically, by a relatively high penetration at a given softening point. Empirically, the penetration index and penetration temperature susceptibility have been used to measure the degree of dispersion.

Asphalt develops an internal structure with age, steric hardening, in which viscosity can increase upon aging without any loss of volatile material. Those with a particularly high degree of gel structure exhibit thixotropy.

16.3.20 Softening Point

The softening point of asphalt may be defined as that temperature at which asphalt attains a particular degree of softness under specified conditions of test.

Asphalt does not go through a solid–liquid phase change when heated and therefore does not have true melting point. As the temperature is raised, asphalt gradually softens or becomes less viscous. For this reason, the determination of the softening point must be made by an arbitrary but closely defined method, if the test values are to be reproducible. Softening point determination is useful in determining the consistency as one element in establishing the uniformity of shipments or sources of supply.

There are several tests available to determine the softening point of asphalt (ASTM D36, ASTM D61, ASTM D2319, ASTM D3104, ASTM D3461, IP 58). In the test method (ASTM D36, IP 58), a steel ball of specified weight is laid on a layer of sample contained in a ring of specified dimensions. The softening point is the temperature, during heating under specified conditions, at which the asphalt surrounding the ball deforms and contacts a base plate.

16.3.21 Stain

The stain index is a measure of the sweating tendency of asphalt and homogeneity. The test is used for oxidized asphalt. This test method (ASTM D2746) is used to measure the amount of stain on paper or other cellulosic materials by asphalt. Variations of the cigarette paper stain procedure include the Barber stain; talc stain tests are also used. The test method may be modified for use with other bituminous materials with softening points lower than 85 °C (185 °F) by using a different temperature than specified for the test, and any such modifications must be clearly described.

The test method measures the tendency for oil components to separate spontaneously from asphalt—any oil present in the asphalt constituents can cause staining in asphalt roofing products and adjacent materials in storage and use. In addition, the stain index is related to the thermal stability of the asphalt—high stain index values indicate lower stability and greater tendency for staining.

16.3.22 Temperature–Volume Correction

Tables are provided (ASTM D1250) to allow the conversion of volumes of asphaltic materials from one temperature to another or, as generally used, to adjust volumes to a temperature of 15.6 °C (60 °F). The value commonly taken for mean coefficient of expansion is 0.00036 in the range of 15.6–121.1 °C (60–250 °F).

16.3.23 Thin Film Oven Test

Many different factors contribute to asphalt aging, and one of the key components in the process is the loss of volatiles. Asphalt binders typically lose volatiles during the manufacturing and placement processes. The elevated temperature of these processes ages the asphalt binder by driving off a substantial amount of volatiles.

The thin film test method (ASTM D2872) is used to simulate the short-term aging of the binders during the hot-mixing process and has the purpose of determining the hardening effect of heat and air on a static film of asphalt when exposed in a thin film. The data indicate the approximate change in properties of asphalt during conventional hot-mixing at about 302 °F (150 °C) as indicated by viscosity and other rheological measurements. It yields a residue that approximates the asphalt condition as incorporated in the pavement. If the mixing temperature differs appreciably from the 302 °F (150 °C) level, more or less effect on properties will occur. This test method also can be used to determine mass change, which is a measure of asphalt volatility.

16.3.24 Viscosity

The viscosity of asphalt is a measure of its flow characteristics. It is generally the most important controlling property for manufacture and for selection to meet a particular application.

A number of instruments are in common use with asphalt for this purpose. The vacuum capillary (ASTM D2171) is commonly used to classify paving asphalt at 60 °C (140 °F). Kinematic capillary instruments (ASTM D2170, ASTM D4402) are commonly used in the 60–135 °C (140–275 °F) temperature range for both liquid and semisolid asphalts in the range of 30–100,000 cSt. Saybolt tests (ASTM D88) are also used in this temperature range and at higher temperature (ASTM E102). At lower temperatures, the cone and plate instrument (ASTM D4287, ASTM D7395) has been used extensively in the viscosity range of 1,000–1,000,000 poises. Other instrumentation techniques are based on (i) the sliding plate microviscometer—the use of machined aggregate plates is to investigate the effects of aggregate-surface-induced structure on the rheological properties of asphalt binders, and (ii) the rheogoniometer—a type of rheometer that can be used to measure the viscous and elastic flow properties of a fluid.

16.3.25 Water Content

The presence of water in asphalt can seriously affect performance insofar as it can effect asphalt–aggregate interactions and asphalt adsorption (ASTM D4469). The water content of asphalt can be determined by a test method (ASTM D95, IP 74) that uses distillation equipment fitted with a Dean and Stark receiver.

In the test, the sample is heated under reflux with a water-immiscible solvent, which co-distills with the water in the sample. Condensed solvent and water are continuously separated in a trap, the water settling in the graduated section of the trap, and the solvent returning to the still.

16.3.26 Weathering

Weathering is the change in asphalt properties due to exposure to the ambient atmosphere. Two important classifications of weathering processes exist: (i) physical weathering and (ii) chemical weathering, and each may involve a biological component.

Mechanical or physical weathering involves the breakdown of asphalt through direct contact with atmospheric conditions, such as heat, water, ice, and pressure. Chemical weathering involves the direct effect of atmospheric chemicals or biologically produced chemicals also known as biological weathering in the deterioration of asphalt properties. While physical weathering is accentuated in very cold or very dry environments, chemical reactions are most intense where the climate is wet and hot.

The extent of cracking or pitting of asphalt films is a measure of the extent of deterioration due to weathering. Failure due to cracking is more accurately determined electrically than visually. This test method (ASTM D1670) evaluates the relative weather resistance of asphalts used for protective-coating applications, especially for roofing. No direct measure of outdoor life or service can be obtained from this test. Methods for preparing test panels (ASTM D1669) and failure end-point testing (ASTM D1670) are available.

Failure of the asphalt specimen, determined by this test method, depends not only on the characteristics of the bituminous material and the extent of weathering, but also on the film thickness, and the amount and type of mineral filler present. Tests on a similar material of known weathering characteristics (a control) exposed at the same time as the test material are strongly recommended as a check on the validity of the test results.

REFERENCES

ASTM D4. Standard Test Method for Bitumen Content. Annual Book of Standards. ASTM International, West Conshohocken, PA.

ASTM D5. Standard Test Method for Penetration of Bituminous Materials. Annual Book of Standards. ASTM International, West Conshohocken, PA.

ASTM D6. Standard Test Method for Loss on Heating of Oil and Asphaltic Compounds. Annual Book of Standards. ASTM International, West Conshohocken, PA.

ASTM D20. Standard Test Method for Distillation of Road Tars. Annual Book of Standards. ASTM International, West Conshohocken, PA.

ASTM D36. Standard Test Method for Softening Point of Bitumen (Ring-and-Ball Apparatus). Annual Book of Standards. ASTM International, West Conshohocken, PA.

ASTM D56. Standard Test Method for Flash Point by Tag Closed Cup Tester. Annual Book of Standards. ASTM International, West Conshohocken, PA.

ASTM D61. Standard Test Method for Softening Point of Pitches (Cube-in-Water Method). Annual Book of Standards. ASTM International, West Conshohocken, PA.

ASTM D70. Standard Test Method for Density of Semi-Solid Bituminous Materials (Pycnometer Method). Annual Book of Standards. ASTM International, West Conshohocken, PA.

ASTM D88. Standard Test Method for Saybolt Viscosity. Annual Book of Standards. ASTM International, West Conshohocken, PA.

ASTM D91. Standard Test Method for Precipitation Number of Lubricating Oils. Annual Book of Standards. ASTM International, West Conshohocken, PA.

ASTM D92. Standard Test Method for Flash and Fire Points by Cleveland Open Cup Tester. Annual Book of Standards. ASTM International, West Conshohocken, PA.

ASTM D93. Standard Test Methods for Flash Point by Pensky-Martens Closed Cup Tester. Annual Book of Standards. ASTM International, West Conshohocken, PA.

ASTM D94. Standard Test Methods for Saponification Number of Petroleum Products. Annual Book of Standards. ASTM International, West Conshohocken, PA.

ASTM D95. Standard Test Method for Water in Petroleum Products and Bituminous Materials by Distillation. Annual Book of Standards. ASTM International, West Conshohocken, PA.

ASTM D113. Standard Test Method for Ductility of Bituminous Materials. Annual Book of Standards. ASTM International, West Conshohocken, PA.

ASTM D128. Standard Test Methods for Analysis of Lubricating Grease. Annual Book of Standards. ASTM International, West Conshohocken, PA.

ASTM D129. Standard Test Method for Sulfur in Petroleum Products (General High Pressure Decomposition Device Method). Annual Book of Standards. ASTM International, West Conshohocken, PA.

ASTM D139. Standard Test Method for Float Test for Bituminous Materials. Annual Book of Standards. ASTM International, West Conshohocken, PA.

ASTM D140. Standard Practice for Sampling Bituminous Materials. Annual Book of Standards. ASTM International, West Conshohocken, PA.

ASTM D189. Standard Test Method for Conradson Carbon Residue of Petroleum Products. Annual Book of Standards. ASTM International, West Conshohocken, PA.

ASTM D217. Standard Test Methods for Cone Penetration of Lubricating Grease. Annual Book of Standards. ASTM International, West Conshohocken, PA.

ASTM D244. Standard Test Methods and Practices for Emulsified Asphalts. Annual Book of Standards. ASTM International, West Conshohocken, PA.

ASTM D287. Standard Test Method for API Gravity of Crude Petroleum and Petroleum Products (Hydrometer Method). Annual Book of Standards. ASTM International, West Conshohocken, PA.

ASTM D312. Standard Specification for Asphalt Used in Roofing. Annual Book of Standards. ASTM International, West Conshohocken, PA.

ASTM D402. Standard Test Method for Distillation of Cutback Asphaltic (Bituminous) Products. Annual Book of Standards. ASTM International, West Conshohocken, PA.

ASTM D449. Standard Specification for Asphalt Used in Dampproofing and Waterproofing. Annual Book of Standards. ASTM International, West Conshohocken, PA.

ASTM D482. Standard Test Method for Ash from Petroleum Products. Annual Book of Standards. ASTM International, West Conshohocken, PA.

ASTM D524. Standard Test Method for Ramsbottom Carbon Residue of Petroleum Products. Annual Book of Standards. ASTM International, West Conshohocken, PA.

ASTM D893. Standard Test Method for Insolubles in Used Lubricating Oils. Annual Book of Standards. ASTM International, West Conshohocken, PA.

ASTM D937. Standard Test Method for Cone Penetration of Petrolatum. Annual Book of Standards. ASTM International, West Conshohocken, PA.

ASTM D972. Standard Test Method for Evaporation Loss of Lubricating Greases and Oils. Annual Book of Standards. ASTM International, West Conshohocken, PA.

ASTM D977. Standard Specification for Emulsified Asphalt. Annual Book of Standards. ASTM International, West Conshohocken, PA.

ASTM D1018. Standard Test Method for Hydrogen in Petroleum Fractions. Annual Book of Standards. ASTM International, West Conshohocken, PA.

ASTM D1160. Standard Test Method for Distillation of Petroleum Products at Reduced Pressure. Annual Book of Standards. ASTM International, West Conshohocken, PA.

ASTM D1250. Standard Guide for Use of the Petroleum Measurement Tables. Annual Book of Standards. ASTM International, West Conshohocken, PA.

ASTM D1266. Standard Test Method for Sulfur in Petroleum Products (Lamp Method). Annual Book of Standards. ASTM International, West Conshohocken, PA.

ASTM D1298. Standard Test Method for Density, Relative Density, or API Gravity of Crude Petroleum and Liquid Petroleum Products by Hydrometer Method. Annual Book of Standards. ASTM International, West Conshohocken, PA.

ASTM D1370. Standard Test Method for Contact Compatibility Between Asphaltic Materials (Oliensis Test). Annual Book of Standards. ASTM International, West Conshohocken, PA.

ASTM D1403. Standard Test Methods for Cone Penetration of Lubricating Grease Using One-Quarter and One-Half Scale Cone Equipment. Annual Book of Standards. ASTM International, West Conshohocken, PA.

ASTM D1552. Standard Test Method for Sulfur in Petroleum Products (High-Temperature Method). Annual Book of Standards. ASTM International, West Conshohocken, PA.

ASTM D1669. Standard Practice for Preparation of Test Panels for Accelerated and Outdoor Weathering of Bituminous Coatings. Annual Book of Standards. ASTM International, West Conshohocken, PA.

ASTM D1670. Standard Test Method for Failure End Point in Accelerated and Outdoor Weathering of Bituminous Materials. Annual Book of Standards. ASTM International, West Conshohocken, PA.

ASTM D1754. Standard Test Method for Effects of Heat and Air on Asphaltic Materials (Thin-Film Oven Test). Annual Book of Standards. ASTM International, West Conshohocken, PA.

ASTM D2006. Method of Test for Characteristic Groups in Rubber Extender and Processing Oils by the Precipitation Method (Withdrawn in 1975 but still in use in many laboratories). ASTM International, West Conshohocken, PA.

ASTM D2007. Standard Test Method for Characteristic Groups in Rubber Extender and Processing Oils and other Petroleum-Derived Oils by the Clay-Gel Absorption Chromatographic Method. Annual Book of Standards. ASTM International, West Conshohocken, PA.

ASTM D2042. Standard Test Method for Solubility of Asphalt Materials in Trichloroethylene. Annual Book of Standards. ASTM International, West Conshohocken, PA.

ASTM D2170. Standard Test Method for Kinematic Viscosity of Asphalts (Bitumens). Annual Book of Standards. ASTM International, West Conshohocken, PA.

ASTM D2171. Standard Test Method for Viscosity of Asphalts by Vacuum Capillary Viscometer. Annual Book of Standards. ASTM International, West Conshohocken, PA.

ASTM D2319. Standard Test Method for Softening Point of Pitch (Cube-in-Air Method). Annual Book of Standards. ASTM International, West Conshohocken, PA.

ASTM D2397. Standard Specification for Cationic Emulsified Asphalt. Annual Book of Standards. ASTM International, West Conshohocken, PA.

ASTM D2415. Standard Test Method for Ash in Coal Tar and Pitch. Annual Book of Standards. ASTM International, West Conshohocken, PA.

ASTM D2416. Standard Test Method for Coking Value of Tar and Pitch (Modified Conradson). Annual Book of Standards. ASTM International, West Conshohocken, PA.

ASTM D2502. Standard Test Method for Estimation of Mean Relative Molecular Mass of Petroleum Oils from Viscosity Measurements. Annual Book of Standards. ASTM International, West Conshohocken, PA.

ASTM D2503. Standard Test Method for Relative Molecular Mass (Molecular Weight) of Hydrocarbons by Thermoelectric Measurement of Vapor Pressure. Annual Book of Standards. ASTM International, West Conshohocken, PA.

ASTM D2595. Standard Test Method for Evaporation Loss of Lubricating Greases Over Wide-Temperature Range. Annual Book of Standards. ASTM International, West Conshohocken, PA.

ASTM D2622. Standard Test Method for Sulfur in Petroleum Products by Wavelength Dispersive X-ray Fluorescence Spectrometry. Annual Book of Standards. ASTM International, West Conshohocken, PA.

ASTM D2746. Standard Test Method for Staining Tendency of Asphalt (Stain Index). Annual Book of Standards. ASTM International, West Conshohocken, PA.

ASTM D2872. Standard Test Method for Effect of Heat and Air on a Moving Film of Asphalt (Rolling Thin-Film Oven Test). Annual Book of Standards. ASTM International, West Conshohocken, PA.

ASTM D2878. Standard Test Method for Estimating Apparent Vapor Pressures and Molecular Weights of Lubricating Oils. Annual Book of Standards. ASTM International, West Conshohocken, PA.

ASTM D3104. Standard Test Method for Softening Point of Pitches (Mettler Softening Point Method). Annual Book of Standards. ASTM International, West Conshohocken, PA.

ASTM D3120. Standard Test Method for Trace Quantities of Sulfur in Light Liquid Petroleum Hydrocarbons by Oxidative Microcoulometry. Annual Book of Standards. ASTM International, West Conshohocken, PA.

ASTM D3142. Standard Test Method for Specific Gravity, API Gravity, or Density of Cutback Asphalts by Hydrometer Method. Annual Book of Standards. ASTM International, West Conshohocken, PA.

ASTM D3143. Standard Test Method for Flash Point of Cutback Asphalt with Tag Open-Cup Apparatus. Annual Book of Standards. ASTM International, West Conshohocken, PA.

ASTM D3228. Standard Test Method for Total Nitrogen in Lubricating Oils and Fuel Oils by Modified Kjeldahl Method. Annual Book of Standards. ASTM International, West Conshohocken, PA.

ASTM D3279. Standard Test Method for n-Heptane Insolubles. Annual Book of Standards. ASTM International, West Conshohocken, PA.

ASTM D3343. Standard Test Method for Estimation of Hydrogen Content of Aviation Fuels. Annual Book of Standards. ASTM International, West Conshohocken, PA.

ASTM D3409. Standard Test Method for Adhesion of Asphalt-Roof Cement to Damp, Wet, or Underwater Surfaces. Annual Book of Standards. ASTM International, West Conshohocken, PA.

ASTM D3461. Standard Test Method for Softening Point of Asphalt and Pitch (Mettler Cup-and-Ball Method). Annual Book of Standards. ASTM International, West Conshohocken, PA.

ASTM D3607. Standard Test Method for Removing Volatile Contaminants from Used Engine Oils by Stripping. Annual Book of Standards. ASTM International, West Conshohocken, PA.

ASTM D3701. Standard Test Method for Hydrogen Content of Aviation Turbine Fuels by Low Resolution Nuclear Magnetic Resonance Spectrometry. Annual Book of Standards. ASTM International, West Conshohocken, PA.

ASTM D3712. Standard Test Method of Analysis of Oil-Soluble Petroleum Sulfonates by Liquid Chromatography. Annual Book of Standards. ASTM International, West Conshohocken, PA.

ASTM D4045. Standard Test Method for Sulfur in Petroleum Products by Hydrogenolysis and Rateometric Colorimetry. Annual Book of Standards. ASTM International, West Conshohocken, PA.

ASTM D4052. Standard Test Method for Density, Relative Density, and API Gravity of Liquids by Digital Density Meter. Annual Book of Standards. ASTM International, West Conshohocken, PA.

ASTM D4055. Standard Test Method for Pentane Insolubles by Membrane Filtration. Annual Book of Standards. ASTM International, West Conshohocken, PA.

ASTM D4072. Standard Test Method for Toluene-Insoluble (TI) Content of Tar and Pitch. Annual Book of Standards. ASTM International, West Conshohocken, PA.

ASTM D4124. Standard Test Method for Separation of Asphalt into Four Fractions. Annual Book of Standards. ASTM International, West Conshohocken, PA.

ASTM D4287. Standard Test Method for High-Shear Viscosity Using a Cone/Plate Viscometer. Annual Book of Standards. ASTM International, West Conshohocken, PA.

ASTM D4294. Standard Test Method for Sulfur in Petroleum and Petroleum Products by Energy Dispersive X-ray Fluorescence Spectrometry. Annual Book of Standards. ASTM International, West Conshohocken, PA.

ASTM D4312. Standard Test Method for Toluene-Insoluble (TI) Content of Tar and Pitch (Short Method). Annual Book of Standards. ASTM International, West Conshohocken, PA.

ASTM D4402. Standard Test Method for Viscosity Determination of Asphalt at Elevated Temperatures Using a Rotational Viscometer. Annual Book of Standards. ASTM International, West Conshohocken, PA.

ASTM D4469. Standard Practice for Calculating Percent Asphalt Absorption by the Aggregate in an Asphalt Pavement Mixture. Annual Book of Standards. ASTM International, West Conshohocken, PA.

ASTM D4530. Standard Test Method for Determination of Carbon Residue (Micro Method). Annual Book of Standards. ASTM International, West Conshohocken, PA.

ASTM D4628. Standard Test Method for Analysis of Barium, Calcium, Magnesium, and Zinc in Unused Lubricating Oils by Atomic Absorption Spectrometry. Annual Book of Standards. ASTM International, West Conshohocken, PA.

ASTM D4715. Standard Test Method for Coking Value of Tar and Pitch (Alcan). Annual Book of Standards. ASTM International, West Conshohocken, PA.

ASTM D4867. Standard Test Method for Effect of Moisture on Asphalt Concrete Paving Mixtures. Annual Book of Standards. ASTM International, West Conshohocken, PA.

ASTM D4893. Standard Test Method for Determination of Pitch Volatility. Annual Book of Standards. ASTM International, West Conshohocken, PA.

ASTM D4927. Standard Test Methods for Elemental Analysis of Lubricant and Additive Components – Barium, Calcium, Phosphorus, Sulfur, and Zinc by Wavelength-Dispersive X-Ray Fluorescence Spectroscopy. Annual Book of Standards. ASTM International, West Conshohocken, PA.

ASTM D5002. Standard Test Method for Density and Relative Density of Crude Oils by Digital Density Analyzer. Annual Book of Standards. ASTM International, West Conshohocken, PA.

ASTM D5018. Standard Test Method for Shear Viscosity of Coal-Tar and Petroleum Pitches. Annual Book of Standards. ASTM International, West Conshohocken, PA.

ASTM D5185. Standard Test Method for Multielement Determination of Used and Unused Lubricating Oils and Base Oils by Inductively Coupled Plasma Atomic Emission Spectrometry (ICP-AES). Annual Book of Standards. ASTM International, West Conshohocken, PA.

ASTM D5291. Standard Test Methods for Instrumental Determination of Carbon, Hydrogen, and Nitrogen in Petroleum Products and Lubricants. Annual Book of Standards. ASTM International, West Conshohocken, PA.

ASTM D5770. Standard Test Method for Semiquantitative Micro Determination of Acid Number of Lubricating Oils during

Oxidation Testing. Annual Book of Standards. ASTM International, West Conshohocken, PA.

ASTM D5800. Standard Test Method for Evaporation Loss of Lubricating Oils by the Noack Method. Annual Book of Standards. ASTM International, West Conshohocken, PA.

ASTM D6375. Standard Test Method for Evaporation Loss of Lubricating Oils by Thermogravimetric Analyzer (TGA) Noack Method. Annual Book of Standards. ASTM International, West Conshohocken, PA.

ASTM D6443. Standard Test Method for Determination of Calcium, Chlorine, Copper, Magnesium, Phosphorus, Sulfur, and Zinc in Unused Lubricating Oils and Additives by Wavelength Dispersive X-ray Fluorescence Spectrometry (Mathematical Correction Procedure). Annual Book of Standards. ASTM International, West Conshohocken, PA.

ASTM D6560. Standard Test Method for Determination of Asphaltenes (Heptane Insolubles) in Crude Petroleum and Petroleum Products. Annual Book of Standards. ASTM International, West Conshohocken, PA.

ASTM D7060. Standard Test Method for Determination of the Maximum Flocculation Ratio and Peptizing Power in Residual and Heavy Fuel Oils (Optical Detection Method). Annual Book of Standards. ASTM International, West Conshohocken, PA.

ASTM D7395. Standard Test Method for Cone/Plate Viscosity at a 500 s⁻¹ Shear Rate. Annual Book of Standards. ASTM International, West Conshohocken, PA.

ASTM E102. Standard Test Method for Saybolt Furol Viscosity of Bituminous Materials at High Temperatures. Annual Book of Standards. ASTM International, West Conshohocken, PA.

ASTM E258. Standard Test Method for Total Nitrogen in Organic Materials by Modified Kjeldahl Method. Annual Book of Standards. ASTM International, West Conshohocken, PA.

ASTM E385. Standard Test Method for Oxygen Content Using a 14-MeV Neutron Activation and Direct-Counting Technique. Annual Book of Standards. ASTM International, West Conshohocken, PA.

ASTM E659. Standard Test Method for Autoignition Temperature of Liquid Chemicals. Annual Book of Standards. ASTM International, West Conshohocken, PA.

ASTM E777. Standard Test Method for Carbon and Hydrogen in the Analysis Sample of Refuse-Derived Fuel. Annual Book of Standards. ASTM International, West Conshohocken, PA.

ASTM E778. Standard Test Methods for Nitrogen in the Analysis Sample of Refuse-Derived Fuel. Annual Book of Standards. ASTM International, West Conshohocken, PA.

Gary, J.G., Handwerk, G.E., and Kaiser, M.J. 2007. Petroleum Refining: Technology and Economics, 5th Edition. CRC Press, Taylor & Francis Group, Boca Raton, FL.

Hsu, C.S. and Robinson, P.R. (Editors) 2006. Practical Advances in Petroleum Processing (Volumes 1 and 2). Springer Science, New York.

IP 4 (ASTM D482). Petroleum Products – Determination of Ash. IP Standard Methods 2013. The Energy Institute, London.

IP 13 (ASTM D189). Petroleum Products – Determination of Carbon Residue – Conradson Method. IP Standard Methods 2013. The Energy Institute, London.

IP 14 (ASTM D524). Petroleum Products – Determination of Carbon Residue – Ramsbottom Method. IP Standard Methods 2013. The Energy Institute, London.

IP 30. Detection of Mercaptans, Hydrogen Sulfide, Elemental Sulfur and Peroxides – Doctor Test Method. IP Standard Methods 2013. The Energy Institute, London.

IP 34 (ASTM D93). Determination of Flash Point – Pensky-Martens Closed Cup Method. IP Standard Methods 2013. The Energy Institute, London.

IP 49. Determination of Needle Penetration of Bituminous Material. IP Standard Methods 2013. The Energy Institute, London.

IP 50 (ASTM D217). Determination of Cone Penetration of Lubricating Grease. IP Standard Methods 2013. The Energy Institute, London.

IP 58. Determination of Softening Point of Bitumen – Ring and Ball Method. IP Standard Methods 2013. The Energy Institute, London.

IP 61 (ASTM D129). Determination of Sulfur – High Pressure Combustion Method. IP Standard Methods 2013. The Energy Institute, London.

IP 74 (ASTM D95). Petroleum Products and Bituminous Materials – Determination of Water – Distillation Method. IP Standard Methods 2013. The Energy Institute, London.

IP 80. Breaking Point of Bitumen – Fraass Method. IP Standard Methods 2013. The Energy Institute, London.

IP 107 (ASTM D1266). Determination of Sulfur – Lamp Combustion Method. IP Standard Methods 2013. The Energy Institute, London.

IP 136 (ASTM D94). Petroleum Products – Determination of Saponification Number – Part 1 – Color-Indicator Titration Method. IP Standard Methods 2013. The Energy Institute, London.

IP 143 (ASTM D6560). Determination of Asphaltenes (Heptane Insolubles) in Crude Petroleum and Petroleum Products. IP Standard Methods 2013. The Energy Institute, London.

IP 154. Petroleum Products – Corrosiveness to Copper – Copper Strip Test. IP Standard Methods 2013. The Energy Institute, London.

IP 160 (ASTM D1298). Crude Petroleum and Liquid Petroleum Products – Laboratory Determination of Density – Hydrometer Method. IP Standard Methods 2013. The Energy Institute, London.

IP 179 (ASTM D937). Determination of Cone Penetration of Petrolatum. IP Standard Methods 2013. The Energy Institute, London.

IP 200 (ASTM D1250). Guidelines for the use of the Petroleum Measurement Tables. IP Standard Methods 2013. The Energy Institute, London.

IP 243. Petroleum Products and Hydrocarbons – Determination of Sulfur Content – Wick bold Combustion Method. IP Standard Methods 2013. The Energy Institute, London.

IP 310 (ASTM D1403). Determination of Cone Penetration of Grease – One-quarter and One-half Scale Cone Method. IP Standard Methods 2013. The Energy Institute, London.

IP 338 (ASTM D3701). Hydrogen Content of Aviation Turbine Fuels – Low Resolution Nuclear Magnetic Resonance Spectrometry Method. IP Standard Methods 2013. The Energy Institute, London.

IP 365. Crude Petroleum and Petroleum Products – Determination of Density – Oscillating U-Tube Method. IP Standard Methods 2013. The Energy Institute, London.

IP 398. Petroleum Products – Determination of Carbon Residue – Micro Method. IP Standard Methods 2013. The Energy Institute, London.

IP 421. Evaporation Loss of Lubricating Oils using Noack Evaporative Tester. Energy Institute, London, UK.

Speight, J.G. 1994. Chemical and Physical Studies of Petroleum Asphaltenes. In: Asphaltenes and Asphalts, I. Developments in Petroleum Science, 40. T.F. Yen and G.V. Chilingarian (Editors), Elsevier, Amsterdam.

Speight, J.G. 2000. The Desulfurization of Heavy Oils and Residua, 2nd Edition. Marcel Dekker Inc., New York.

Speight, J.G. 2001. Handbook of Petroleum Analysis. John Wiley & Sons Inc., Hoboken, NJ.

Speight, J.G. 2013. The Chemistry and Technology of Coal, 3rd Edition. CRC Press, Taylor & Francis Group, Boca Raton, FL.

Speight, J.G. 2014. The Chemistry and Technology of Petroleum, 5th Edition. CRC Press, Taylor & Francis Group, Boca Raton, FL.

Speight, J.G. and Ozum, B. 2002. Petroleum Refining Processes. Marcel Dekker Inc., New York.

TRB. 2010. Development in Asphalt Binder Specifications. Transportation Research Circular E-C147. Transportation Research Board, Washington, DC. December.

17

COKE, CARBON BLACK, AND GRAPHITE

17.1 INTRODUCTION

Coke is a gray-to-black solid carbonaceous residue that is produced from petroleum during thermal processing, characterized by having a high carbon content (95% + by weight) and a honeycomb type of appearance and is insoluble in organic solvents (ASTM D121) (Speight and Ozum, 2002; Hsu and Robinson, 2006; Gary et al., 2007; Speight, 2014).

Coke consists mainly of carbon (90–95%) and has low mineral matter content (determined as ash residue). Coke is used as a feedstock in coke ovens for the steel industry, for heating purposes, for electrode manufacture, and for the production of chemicals. The two most important qualities are *green coke* and *calcined coke*. This latter category also includes *catalyst coke* deposited on the catalyst during refining processes: this coke is not recoverable and is usually burned as refinery fuel.

Coke does not offer the same potential environmental issues as other petroleum products. It is used predominantly as a refinery fuel unless it is used for the production of a high-grade coke or carbon (also called *carbon black*) or graphite, depending upon the quality of the coke. In the former case, the constituents of the coke that will release environmentally harmful gases such as nitrogen oxides, sulfur oxides, and particulate matter should be known. In addition, stockpiling coke on a site where it awaits use or transportation can lead to leachates that are the result of rainfall (or acid rainfall) and are highly detrimental to the environment. In such a case, application of the toxicity characteristic leaching test procedure to the coke (TCLP, EPA SW-846 Method 1311), that is designed to determine the mobility of both organic and inorganic contaminants present in materials such as coke, is warranted before stockpiling the coke in the open) is warranted.

Carbon (also known as *carbon black*) is one of the darkest and most finely divided materials known. Chemically, carbon black is a colloidal form of elemental carbon consisting of 95–99% w/w. Made in specially designed reactors operating at internal temperatures in the range of 1425–1980 °C (2600°–3600 °F), different grades of carbon black can be produced with varying aggregate size and structure. In the context of the present text, feedstocks are distillation residua and fluid catalytic cracking resid.

Carbon black is a form of para-crystalline carbon that has a high surface-area-to-volume ratio. It is dissimilar to soot in its much higher surface-area-to-volume ratio and significantly lower content of polynuclear aromatic hydrocarbons. However, carbon black is widely used as a model compound for diesel soot for diesel oxidation experiments, as a filler in tires, and as a color pigment in plastics, paints, and ink.

Graphite, one of the softest materials known, is a form of carbon that is primarily used as a lubricant. Although it does occur naturally, most commercial graphite is produced by treating petroleum coke in an oxygen-free oven. Naturally occurring graphite occurs in two forms: alpha and beta. These two forms have identical physical properties but different crystal structures. All artificially produced graphite is of the alpha type. In addition to its use as a lubricant, graphite, in a form known as coke, is used in large amounts in the production of steel.

Handbook of Petroleum Product Analysis, Second Edition. James G. Speight.
© 2015 John Wiley & Sons, Inc. Published 2015 by John Wiley & Sons, Inc.

17.2   PRODUCTION AND PROPERTIES

17.2.1   Coke

Petroleum coke (petcoke) is the gray-to-black solid carbonaceous residue left by the destructive distillation of petroleum residua. It has a honeycomb type of appearance and is insoluble in organic solvents. Coking processes that can be employed for making petroleum coke include contact coking, fluid coking, flexicoking, and delayed coking. The coke formed in catalytic cracking operations is usually nonrecoverable, as it is often employed as fuel for the process.

Petroleum coke (both green coke and calcined coke) is the substance remaining from treating heavy petroleum feedstocks with high temperature and pressure, and many of the physical–chemical properties are not meaningful at ambient environmental conditions. At ambient temperature and pressure, petroleum coke exists as a solid, because it consists predominantly of elemental carbon and a hardened residuum remaining from the feedstocks.

Petroleum coke is produced through the thermal decomposition of heavy petroleum process streams and residues (Speight and Ozum, 2002; Hsu and Robinson, 2006; Gary et al., 2007; Speight, 2014). The most common feedstocks used in coking operations are (i) reduced crude (i.e., vacuum residue) and (ii) thermal residua from other processes. These feedstocks are heated to thermal cracking temperatures (485–505 °C; 905–940 °F) and pressures that create petroleum liquid and gas product streams. The material remaining from this process is a solid concentrated carbon material, petroleum coke.

Petroleum cokes can be categorized as either green coke (unprocessed coke, raw coke) or calcined coke. The initial product of the coking process, green coke, is used as fuel, in gasification and metallurgical processes, or as feedstock to produce calcined coke. Calcined coke is produced when green coke is treated to higher temperatures (1200–1350 °C). The primary use of calcined coke is in making carbon anodes for the aluminum industry. Other uses include making graphite electrodes for arc furnaces, titanium dioxide, polycarbonate plastics, steel, carbon refractory bricks for blast furnaces, packing media for anode baking furnaces, and material for cathodic protection of pipelines.

17.2.1.1   Composition

Petroleum coke is composed primarily of elemental carbon organized as a porous polycrystalline carbon matrix. In green coke (unprocessed coke), the pores of the matrix are filled with a hardened residuum remaining from the coker feed. This residuum is referred to as volatile matter (sometimes referred to as residual hydrocarbon) because it distills off during the calcining process. Volatile matter consists of the heavy hydrocarbons remaining from the feedstocks that have not undergone complete carbonization. Green coke

normally contains between 4% and 15% w/w volatile matter, but can contain up to 21% w/w. The temperature of the coking drum as well as cycle time and drum pressure all affect the amount of volatile matter in green and calcined cokes. Because of the lower temperature used in its production, green coke contains higher levels of volatile matter than calcined coke (US EPA, 2007).

The specific chemical composition of any given batch of petroleum coke is determined by the composition of the feedstocks used in the coking process, which in turn are dependent upon the composition of the crude oil and refinery processing from which the feedstock is derived. Coke produced from feedstocks high in asphaltene constituents will contain higher concentrations of sulfur and metals than cokes produced from high aromatic feedstocks. This is because asphaltene constituents contain high concentrations of those heteroatoms. Most of the sulfur in coke exists as organic sulfur bound to the carbon matrix. However, the structure of organic sulfur compounds in petroleum coke is largely unknown, and no precise analytical methods are available to determine these structures. Other forms of sulfur found in coke include sulfates and pyritic sulfur, but these rarely make up more than 0.02% w/w of the total sulfur in coke. Metals, mainly vanadium and nickel, occur as metal chelates or porphyrins in the asphaltene fraction. Some metals are intercalated in the coke structure and are not chemically bonded, so they become part of the ash and particulates. Metal concentrations in coke normally increase upon calcining due to the weight loss from evolution of the volatile matter. In practice, however, calcined cokes typically contain lower metal concentrations than many grades of green coke due to the selection of low-metal green cokes for calcining (US EPA, 2007).

Delayed coke (Table 17.1) is produced during the *delayed coking* process—a batch process from vacuum residua (Chapter 2) (Speight and Ozum, 2002). The carbonization (thermal decomposition) reactions involve dehydrogenation, rearrangement, and condensation. Two of the common feedstocks are vacuum residues and aromatic oils.

TABLE 17.1   Composition of coke from a delayed coker

Component	Coke (raw/green)	Coke (calcined at 1300 °C (2375 °F)
Fixed carbon (wt %)	80–95	98.0–99.5
Hydrogen (wt %)	3.0–04.5	0.1
Nitrogen (wt %)	0.1–0.5	—
Sulfur (wt %)	0.2–6.0	—
Volatile matter (wt %)	5–15	0.2–0.8
Moisture (wt %)	0.5–10	0.1
Ash (wt %)	0.1–1.0	0.02–0.7
Density (g/cm³)	1.2–1.6	1.9–2.1
Metals (ppm weight)	5–5000	5–5000

In the delayed coker, the feed enters the bottom of the fractionator where it mixes with recycle liquid condensed from the coke drum effluent. It is pumped through the coking heater then to one of two coke drums through a switch valve. It is 480°–500 °C. Cracking and polymerization take place in the coke drum in a nominal 24-h period. Coking is a batch operation carried out in two coke drums. Coking takes place in one drum in 24 h, while decoking is carried out in the other drum. A complete cycle is 48 h. Coke is cut from the drum using high-pressure water. Large drums are 27 ft in diameter and 114 ft flange to flange.

Fluid coke (Table 17.2) is produced during the *fluid coking* process—a continuous process in which heated coker feeds are sprayed into a fluidized bed of hot coke particles that are maintained at 20–40 psi and 500 °C (932 °F) (Chapter 2) (Speight and Ozum, 2002). The feed vapors are cracked while forming a liquid film on the coke particles. The particles grow by layers until they are removed, and new seed coke particles are added.

Coke for the aluminum industry is calcined to less than 0.5% volatiles at 1300–1400 °C (2372–2552 °F) before it is used to make anodes.

Petroleum coke is employed for a number of purposes, but its chief use is in the manufacture of carbon electrodes for aluminum refining, which requires a high-purity carbon—low in ash and sulfur free; the volatile matter must be removed by calcining. In addition to its use as a metallurgical reducing agent, petroleum coke is employed in the manufacture of carbon brushes, silicon carbide abrasives, and structural carbon (e.g., pipes and Raschig rings), as well as calcium carbide manufacture from which acetylene is produced:

$$Coke \rightarrow CaC_2$$
$$CaC_2 + H_2O \rightarrow HC \equiv CH$$

Petroleum coke can either be fuel grade (high in sulfur and metals) or anode grade (low in sulfur and metals) (Table 17.3). Further processing of green coke by calcining in a rotary kiln removes residual volatile hydrocarbons from the coke. The calcined petroleum coke can be further processed in an anode baking oven in order to produce anode coke of the desired shape and physical properties. The anodes are mainly used in the aluminum industry and steel industry.

Coke is produced in several forms from petroleum coke: (i) needle coke, (ii) catalyst coke, (iii) fuel-grade coke, which is classified as either sponge coke or shot coke morphology, and (iv) calcined petroleum coke.

Needle coke (acicular coke) is a highly crystalline petroleum coke that is used in the production of electrodes for the steel industry and the aluminum industry and is particularly valuable because the electrodes must be replaced regularly. Needle coke is produced exclusively from either fluid catalytic cracker decant oil or coal tar pitch. On the other hand, *catalyst coke* is coke that has deposited on catalysts used in petroleum refining, such as those in a fluid catalytic cracking unit. This coke is impure and is only used for fuel. However, *fuel-grade coke* is the product from calcining petroleum coke, which is used to make anodes for the aluminum, steel, and titanium smelting industry. The green coke must have sufficiently low metal content in order to be used as anode material—green coke with low metal content is referred to as anode-grade coke. The green coke with too high metal content will not be calcined and is used for as a fuel for combustion within the refinery site.

TABLE 17.2 Description of delayed coke carbon forms

Needle coke	Ribbon-like parallel-ordered anisotropic domains that can also occur as folded structures
Lenticular/granular	Lenticular anisotropic domains of various sizes that are not aligned parallel to the particle surface
Mixed layer	Ribbon and lenticular anisotropic domains of various sizes in curved and irregular layered arrangements
Sponge	Porous microstructure with walls that are generally anisotropic but with pores and walls that vary in size
Shot	Ribbon and lenticular anisotropic domains arranged in concentric patterns to form shot-like coke
Amorphous	Isotropic carbon form closely associated with parent liquor. Higher in volatile matter than incipient mesophase
Incipient mesophase	Initial stage of mesophase formation. Transition stage between amorphous and mesophase
Mesophase	Nemitic liquid crystals. Lower in volatile matter than incipient mesophase

TABLE 17.3 Description of fluid coke carbon forms

Layered	Anisotropic carbon domains aligned in concentric layers parallel to the particle surface similar to an onion-like pattern
Non-layered	Anisotropic domains are not aligned parallel to the particle surface
Aggregates	Fragments of anisotropic domains
Amorphous	Isotropic carbon form closely associated with parent liquor. Higher in volatile matter than incipient mesophase
Incipient mesophase	Initial stage of mesophase formation. Transition stage between amorphous and mesophase
Mesophase	Nemitic liquid crystals. Lower in volatile matter than incipient mesophase

17.2.1.2 Properties

Coke occurs in various forms, and the terminology reflects the type of coke (Tables 17.1 and 17.2) that can influence behavior in the environment. But no matter what the form, coke usually consists mainly of carbon (>90% but usually >95%) and has a low mineral matter content (determined as ash residue). Coke is used as a feedstock in coke ovens for the steel industry, for heating purposes, for electrode manufacture, and for the production of chemicals. The two most important classes are *green coke* and *calcined coke*. This latter category also includes *catalyst coke* deposited on the catalyst during refining processes: this coke is not recoverable and is usually burned as refinery fuel.

17.2.2 Carbon Black

Carbon has two natural crystalline allotropic forms: graphite and diamond. Each has its own distinct crystal structure and properties.

17.2.2.1 Composition

Carbon black is composed of fine particles of elementary carbon formed by the incomplete combustion of hydrocarbon gases and liquids. About 95% w/w of carbon black produced in the United States is used in rubber manufacturing where it is the filler used to give both natural and synthetic rubber toughness and abrasion resistance. The gases from carbon black manufacture contain carbon monoxide (CO), hydrogen (H_2), and carbon dioxide (CO_2), and are discharged at atmospheric pressure and temperature. Novel technology would be required to recover and use these dilute gases at low temperature and pressure. Separation and recovery of the H_2 and CO for subsequent chemical reactions would be expensive and probably impractical. Other uses of carbon black are in the production of plastics, ink, paper, and paint, and other rubber products such as hoses, belting, footwear, and mechanical and molded goods.

Carbon black is a very fine particulate form of elemental carbon arranged in a less ordered manner than other forms of carbon such as diamond or graphite. Carbon black consists of planes of carbon atoms fused together randomly to form spherical particles, which in turn form structures or aggregates. Aggregates are often bound together to form secondary structures or agglomerates. Two important characteristics of carbon black are surface area, an indirect measurement of particle size, and structure, a measure of the degree of particle aggregation or chaining. These two characteristics are dependent on the type of process used to manufacture the carbon black.

All carbon black is produced either by incomplete combustion or thermal decomposition of a hydrocarbon feedstock. The thermal black process was first employed in 1922 to produce hydrogen for dirigibles, or airships, such as the *Hindenburg*. The black by-product was then of no value. The reverse is now true, as thermal black has been found to impart a number of unique properties to rubber articles and other specialty products.

In the thermal process, natural gas enters refractory-lined reactors and is preheated to 1300 °C (2370 °F) by burning a hydrogen/air mixture produced as a by-product of the process. Natural gas in the absence of oxygen is then injected into the reactors and decomposes into carbon and hydrogen. The gas/solid mixture is cooled with water spray, and the carbon is separated in bag filters. The result is a thermal carbon black with a particle size range of 100–500 nm, the largest particle size of all carbon blacks.

In the *furnace black process*, aromatic oils (based on crude oil) are cracked under high temperature in a reactor, producing carbon black and tail gas. After cooling, the carbon black is separated from the tail gas, densified, and processed into pellets of varying grades/sizes. This process is the most widely used in the United States, comprising over 95% of all carbon black production.

In the *furnace black process*, carbon black is formed by blowing petroleum oil or coal oil as raw material (feedstock oil) into high-temperature gases to combust them partially. This method is suitable for mass production due to its high yield and allows wide control over its properties such as particle size or structure. This is currently the most common method used for manufacturing carbon black for various applications from rubber reinforcement to coloring.

In the *channel process*, carbon black is formed by bringing partially combusted fuel, which is generated with natural gas as raw material, into contact with channel steel (H-shaped steel) and then collecting the carbon black that results. There are yield and environmental issues around this method and, therefore, has lost the leading role as the mass production process to the furnace process. This method, however, provides carbon black with many functional groups on the surface, being used in some painting applications.

In the *acetylene black process*, carbon black is produced by thermally decomposing acetylene gas. It provides carbon black with higher structures and higher crystallinity, and is mainly used for electric conductive agents.

In the *lampblack process*, carbon black is obtained by collecting soot from fumes generated by burning oils or pine wood. This method has been used since the days before Christ and is not suitable for mass production. However, it is used as raw material for ink sticks as it provides carbon black with specific color.

17.2.2.2 Properties

The characteristics of carbon black vary depending on manufacturing process, and therefore carbon black is classified by manufacturing process. Carbon black produced with the furnace process, which is the most commonly used method now, is called furnace black, distinguishing it from carbon black that is manufactured with other processes. Unlike the incomplete combustion, the thermal black process involves

TABLE 17.4 Types of carbon black

Chemical process	Carbon black type	D (nm)	Feedstock
Incomplete combustion	Lamp black	50–100	Coal tar hydrocarbons
	Channel black	10–30	Natural gas
	Furnace black	10–80	Natural gas Liquid aromatic hydrocarbons
Thermal decomposition	Thermal black	150–500	Natural gas
	Acetylene black	35–70	Acetylene

the thermal decomposition of natural gas in the absence of air or flame. Decomposition of natural gas involves the molecular dissociation, or breaking of carbon–hydrogen bonds, to yield carbon and hydrogen.

Carbon black is essentially an elemental carbon in a form different from diamond, cokes, charcoal and graphite. The particle size, structure, and surface area of carbon black play a significant role in the material properties of rubber, plastics, and other products. For this reason, carbon black is made in various grades to meet the varying material needs and specifications of manufacturers. In general, carbon black grades with smaller particle size have better reinforcing and abrasion resistance qualities than those with larger particle size.

A broad range of the types of carbon black can be made by controlled manipulation of the reactor conditions (Table 17.4). The product consists of spherical-like particles and is manufactured by the incomplete combustion of a heavy aromatic feedstock in a hot flame of (preheated) air and natural gas:

$$C_x H_y + O_2 \rightarrow C + CH_4 + CO + H_2 + CO_2 + H_2O$$

The primary units of carbon black are aggregates, which are formed when particles collide and fuse together in the combustion zone of the reactor. Several of those aggregates may be held together by weak forces to form agglomerates. These agglomerates will break down during mixing into rubber, so the aggregates are the smallest ultimate dispersible unit of carbon black. In appearance, carbon black may be an intensely black amorphous powder or finely divided pellets. It is insoluble in water and solvents. It has a bulk density is 1.8–2.1 g/cm^3 and a high surface-area-to-volume ratio. It is used as a black pigment for inks and paints, and in the manufacture of tires, rubber, and plastic products, among other uses.

17.2.3 Graphite

The final step in the graphite manufacture of graphite is a conversion of baked carbon to graphite, called graphitizing, that is, heat-treating the material at temperatures in the region of 2600–3300 °C (4710–5970 °F). During the graphitizing

process, the preordered carbon (turbostratic carbon) is converted into a three-dimensionally ordered graphite structure. Depending on the raw materials and the processing parameters, various degrees of convergence to the ideal structure of a graphite single crystal are achieved.

Since graphitization increases the lattice order and produces smaller layer distances, it simultaneously leads to a considerable growth of ordered domains. However, the degree of order that can be reached depends largely on the crystalline preorder of the solid used. These reduced lattice layer distances are macroscopically noted as a contraction in volume. This graphitization shrinkage is approximately 3–5% by volume, and due to this shrinkage, density of the graphite increases.

On the other hand, petroleum coke is often used as a direct precursor to graphite without involving an intermediate stage of carbon production. Petroleum coke calcining is taking anode-grade green coke from the oil refining process and converting it to almost pure carbon, with a defined structure (calcination). The calcined coke is used to produce carbon anodes for the aluminum industry. In the calcining process, the green coke feed is heated to a sufficiently high temperature to drive off any residual moisture, and to drive off and combust any residual hydrocarbons (the combustion of the evolved volatile materials provides the necessary heat for the calcination process) in the green coke feed.

Typically, graphite production is accomplished using rotary kilns or rotary hearth. Unlike kilns, rotary hearths rely solely on volatiles in the green feed and preheated combustion air to calcine the green feed; no *external* fuel (i.e., burners) is used. As opposed to rotary hearths, rotary kilns commonly employ large fuel gas burners or oxygen injection at the "downhill" end of the kiln to calcine the green feed. After cooling (and for customers requiring it, oiling for dust control), the calcined product is routed to weather-tight silos for storage prior to shipment to the end user.

The calcining operation is the last, and arguably one of the least, influential tool to control the eventual quality of the carbon product. No calcining operation can compensate for poor feed resulting from upstream operations. Consistent-quality carbon begins with consistent-quality feedstocks and upstream process unit operations; no calcining operation can turn poor quality or inconsistent feed into consistent- or high-quality calcined product! The most leveraging controls for eventual calcined coke quality are all *upstream of the calcining operation*; only an integrated refining operation— where coke quality is considered, day-in and day-out, right along with liquid product yields and quality—offers the most consistent carbon.

17.2.3.1 Composition

Natural graphite is a mineral consisting of graphitic carbon, which varies considerably in crystallinity. Most commercial (natural) graphite is mined and often contains other minerals. Subsequent to mining, the graphite often requires a considerable amount of mineral processing such as froth flotation to concentrate the graphite. Natural graphite is an excellent conductor of heat and electricity. It is stable over a wide range of temperatures. Graphite is a highly refractory material with a high melting point (3650 °C, 6600 °F).

17.2.3.2 Properties

Graphite is a soft grayish-black greasy substance and, in fact, crystallized carbon. The carbon atoms of graphite form a crystal pattern that differs from that of the carbon atoms in diamond. The carbon atoms are arranged in flat planes of hexagonal rings stacked on one another. Each carbon atom is attached to three others on the same plane. Thus, only three out of four valence electrons are used in carbon–carbon bonding. The fourth valence electron remains loosely between the planes. This free electron accounts for the electrical conductivity of graphite. The lack of carbon–carbon bonding between adjacent planes enables them to slide over each other making graphite soft, slippery, and useful as a lubricant.

Graphite has the following properties: (i) a soft, slippery, grayish-black substance with a metallic luster and is opaque to light; (ii) specific gravity is 2.3; (iii) a good conductor of heat and electricity; and (iv) a stable allotrope of carbon, which at high temperature can be transformed into artificial diamond.

The important uses of graphite are as follows: (i) making lead pencils of different hardness, by mixing it with different proportions of clay; (ii) used as a dry lubricant in machine parts; (iii) resistant to chemicals and having a high melting point, and also because it is a good conductor of heat, graphite is used to make crucibles; (iv) a good conductor of electricity and is used to make electrodes; and (v) used in nuclear reactors to control the speed of the nuclear fission reaction.

17.3 TEST METHODS

The test methods for coke are necessary for defining the coke as a fuel (for internal use in a refinery) or for other uses, particularly those test methods where prior sale of the coke is involved. Specifications are often dictated by environmental regulations, if not by the purchaser of the coke.

The methods outlined later are the methods that are usually applied to petroleum coke but should not be thought of as the only test methods. In fact, there are many test methods for coke, and these test methods should be consulted either when more detail is required or a fuller review is required.

17.3.1 Ash

The ash content (i.e., the ash yield, which related to the mineral matter content) is one of the properties used to evaluate coke and indicates the amount of undesirable residue present. Some samples of coke may be declared to have acceptable ash content, but this varies with the intended use of the coke.

For the test method, the preparation and sampling of the analytical sample must neither remove nor add mineral matter (ASTM D346). Improper dividing, sieving, and crushing equipment, and some muffle furnace lining material can contaminate the coke and lead to erroneous results. In addition, a high sulfur content of the furnace gases, regardless of the source of the sulfur, can react with an alkaline ash to produce erratic results. To counteract such an effect, the furnace should be swept with air.

In the test method (ASTM D4422), a sample of petroleum coke is dried, ground, and ashed in a muffle furnace at 700–775 °C (1292–1427 °F). The non-carbonaceous residue is weighed and reported as the percentage by weight ash. As already noted, the ash must not be understood to be the same as the mineral content of the petroleum coke.

In addition, ashing procedures can be used as a preliminary step for the determination of the trace elements in coke and, by inference, in the higher boiling fractions of the crude oil. Among the techniques used for trace element determinations are flameless and flame atomic absorption (AA) spectrophotometry (ASTM D5863) and inductively coupled argon plasma (ICP) spectrophotometry (ASTM D5708).

ICP emission spectrophotometry (ASTM D5708) has an advantage over AA spectrophotometry (ASTM D4628, ASTM D5863) because it can provide more complete elemental composition data than the AA method. Flame emission spectroscopy is often used successfully in conjunction with AA spectrophotometry (ASTM D3605). X-ray fluorescence spectrophotometry (ASTM D4927, ASTM D6443) is also sometimes used, but matrix effects can be a problem. The method to be used for the determination of metallic constituents is often a matter of individual preference.

17.3.2 Calorific Value

The calorific value (heat of combustion) is an important property particularly for the petroleum products that are used for burning, heating, or similar usage. Knowledge of this value is essential when considering the thermal efficiency of equipment for producing either power or heat. Heat of combustion per unit of mass of coke is a critical property of coke intended for use as a fuel.

In one test method that is suitable for coke (ASTM D3523), the sample is supported on surgical gauze and placed in a heated chamber that is open to air at the top. The temperature of this sample is compared to that of an equal reference quantity of surgical gauze contained in an identical chamber. Tests may be conducted for the durations of 4–72 h or longer.

Other methods use an adiabatic bomb calorimeter (ASTM D5865) and are also available.

When an experimental determination of heat of combustion is not available and cannot be made conveniently, an estimate might be considered satisfactory (ASTM D3338). In this test method, the net heat of combustion is calculated from the density, sulfur content, and hydrogen content, but this calculation is justifiable only when the fuel belongs to a well-defined class for which a relationship between these quantities has been derived from accurate experimental measurements on representative samples. Thus, the hydrogen content (ASTM D5291), density (ASTM D5004), and sulfur content (ASTM D1552, ASTM D4239) of the sample are determined by experimental test methods, and the net heat of combustion is calculated using the values obtained by these test methods based on reported correlations.

17.3.3 Composition

The composition of petroleum coke varies with the source of the crude oil, but in general, large amounts of high-molecular-weight complex hydrocarbons (rich in carbon but correspondingly poor in hydrogen) make up a high proportion. The solubility of petroleum *coke* in carbon disulfide has been reported to be as high as 50–80%, but this is in fact a misnomer, since the coke is the insoluble, honeycomb material, which is the end product of thermal processes.

Carbon and hydrogen in coke can be determined by the standard analytical procedures for coal and coke (Speight, 2013, 2014). However, in addition to carbon, hydrogen, metallic constituents (*q.v.*), coke also contains considerable amounts of nitrogen and sulfur that must be determined prior to sale or use. These elements will appear as their respectively oxides (NOx, SOx) when the coke is combusted thereby causing serious environmental issues.

A test method (ASTM D5291) is available for simultaneous determination of carbon, hydrogen, and nitrogen in petroleum products and lubricants. There are at least three instrumental techniques available for this analysis, each based on different chemical principles. However, all involve sample combustion, component separation, and final detection.

In one of the variants of the method, a sample is combusted in an oxygen atmosphere, and the product gases are separated from each other by adsorption over chemical agents. The remaining elemental nitrogen gas is measured by a thermal conductivity cell. Carbon and hydrogen are separately measured by selective infrared cells as carbon dioxide and water. In another variant of the method, a sample is combusted in an oxygen atmosphere, and the product gases are separated from each other, and the three gases of interest are measured by gas chromatography. In the third variant of the method, a sample is combusted in an oxygen atmosphere, and the product gases are cleaned by passage over chemical agents, and the three gases of interest are chromatographically separated and measured with a thermal conductivity detector.

The nitrogen method is not applicable to samples containing less than 0.75% by weight nitrogen, or for the analysis of volatile materials such as gasoline, gasoline oxygenate blends, or aviation turbine fuels. The details of the method should be consulted along with those given in an alternate method for the determination of carbon, hydrogen, and nitrogen in coal and coke (ASTM D5373).

A test method (ASTM D1552) is available for sulfur analysis, and the method covers three procedures applicable to samples boiling above 177 °C (350 °F) and containing not less than 0.06% w/w sulfur. Thus, the method is applicable to most fuel oils, lubricating oils, residua, and coke, and coke containing up to 8% by weight sulfur can be analyzed. This is particularly important for cokes that originate from heavy oil and tar sand bitumen, where the sulfur content of the coke is usually at least 5% by weight.

In the iodate detection system (ASTM D1552), the sample is burned in a stream of oxygen at a sufficiently high temperature to convert about 97% by weight of the sulfur to sulfur dioxide. The combustion products are passed into an absorber containing an acidic solution of potassium iodide and starch indicator. A faint blue color is developed in the absorber solution by the addition of standard potassium iodate solution. As combustion proceeds, bleaching the blue color, more iodate is added. The sulfur content of the sample is calculated from the amount of standard iodate consumed during the combustion.

In the infrared detection system, the sample is weighed into a special ceramic boat that is then placed into a combustion furnace at 1371 °C (2500 °F) in an oxygen atmosphere. Most of the sulfur present is converted to sulfur dioxide, which is then measured with an infrared detector after moisture and dust are removed by traps. The calibration factor is determined using standards approximating the material to be analyzed.

For the iodate method, chlorine in concentrations less than 1 mass % does not interfere. The isoprene rubber method can tolerate somewhat higher levels. Nitrogen when present greater than 0.1 mass % may interfere with the iodate method, the extent being dependent on the types of nitrogen compounds as well as the combustion conditions. It does not interfere in the infrared method. The alkali and alkaline earth metals, zinc, potassium, and lead do not interfere with either method.

Determination of the physical composition can be achieved by any of test methods for determining the toluene-insoluble constituents of tar and pitch (ASTM D4072, ASTM D4312). Furthermore, a variety of sample can be employed to give a gradation of soluble and insoluble fractions. The coke, of course, remains in the extraction thimble (Soxhlet apparatus), and the extracts are freed from the solvent and weighed to give percentage by weight yield(s).

Finally, one aspect that can pay a role in compositional studies is the sieve (screening) analysis. Like all petroleum products, sampling is, or can be, a major issue. If not performed correctly and poor sampling is the result, erroneous and very misleading data can be produced by the analytical method of choice. For this reason, reference is made to standard procedures such as the *Standard Practice for Collection and Preparation of Coke Samples for Laboratory Analysis* (ASTM D346) and the *Standards Test Method for the Sieve Analysis of Coke* (ASTM D293).

17.3.4 Density

The *density (specific gravity)* of coke has a strong influence on the future use and can affect the characteristics of the products such as carbon and graphite.

The density (specific gravity) of coke can be conveniently measured by use of a pycnometer. In the test method (ASTM D5004), the mass of the sample is determined directly and the volume derived by determining the mass of liquid displaced when the sample is introduced into a pycnometer. Oil or other material sprayed on calcined petroleum coke to control dust will interfere. Such oil can be removed by flushing with a solvent, which also must be completely removed before the density determination.

The *real density* of coke is obtained when the particle size of the specimen is smaller than 75 mm. The real density (or the particle size) exerts a direct influence on the physical and chemical properties of the carbon and graphite products that are manufactured from the coke.

In the test method (ASTM D2638), a sample is dried and ground to pass a 75-mm screen. The mass of the sample is determined directly, and the volume is derived by the volume of helium displaced when the sample is introduced into a helium pycnometer. The ratio of the mass of the sample to the volume is reported as the real density.

The *vibrated bulk density* is an indication of the porosity of calcined petroleum coke, which affects its suitability for use in pitch-bonded carbon applications. This property is strongly dependent upon average particle size and range and tends to increase with decreasing coke size. In the test method (ASTM D4292), the coke is crushed and 100 g is measured after vibration, and the bulk density is calculated. The procedure is limited to particles passing through a 6.68-mm opening sieve and retained on a 0.21-mm opening sieve.

17.3.5 Dust Control

Dust control is an important aspect of handling coke, and dust control material is applied to calcined coke to help maintain a dust-free environment. It adds weight to the coke and can have a negative effect on the quality of carbon and graphite artifacts made from the treated coke. Hence, a maximum amount may be specified.

In the test method (ASTM D4930), a weighed dry representative sample of 6.3 mm maximum sized coke is extracted using methylene chloride in a Soxhlet apparatus. The mass of the residue remaining after extraction and evaporation of the solvent is the mass of the dust control material. This test method is limited to those materials that are soluble in a solvent (e.g., methylene chloride) that can be used in a Soxhlet extraction type of apparatus. Toluene and methyl chloroform have also been found to give equal results as methylene chloride.

17.3.6 Hardness

The Hardgrove Grindability Index (HGI) (ASTM D5003) is used to predict the ranking in industrial-size mills used for crushing operations and is commonly used to determine the hardness of coal samples (ASTM D409; Speight, 2013). The rankings are based on energy required and feed rate or both. With the introduction of petroleum coke in the coal market, this test method has been extended to the coke. In the current context, the HGI is also used to select raw petroleum coke and coals that are compatible with each other when milled together in a blend so that segregation of the blend does not occur during particle size reduction.

In the test method (ASTM D5003), the coke sample is crushed to produce a high yield of particles passing a No. 16 sieve and retained on a No. 30 sieve. These particles are reduced in the Hardgrove grindability machine according to the test method for coal (ASTM D409). The quantity of particles retained on a No. 200 sieve is used to calculate the HGI of the sample. Both this test method and test method for coal (ASTM D409) produce the same results on petroleum coke samples.

17.3.7 Metals

The presence and concentration of various metallic elements in petroleum coke are major factors in the suitability of the coke for various uses.

In the test method (ASTM D5056), a sample of petroleum coke is ashed (thermally decomposed to leave only the ash of the inorganic constituents) at 525 °C (977 °F). The ash is fused with lithium tetra-borate or lithium meta-borate. The melt is then dissolved in dilute hydrochloric acid, and the resultant solution is analyzed by AA spectroscopy to determine the metals in the sample. However, spectral interferences may occur when using wavelengths other than those recommended for analysis or when using multielement hollow cathode lamps.

This test method can be used in the commercial transfer of petroleum coke to determine whether that lot of coke meets the specifications. This method can analyze raw and calcined coke for trace elements aluminum, calcium, iron, nickel, silicon, sodium, and vanadium. The inductively coupled plasma atomic emission spectroscopy (ICP-AES) method (ASTM D 5600; ASTM D6357) is complementary to this method and can also be used for the determination of metals in petroleum coke.

In the ICP-AES method (ASTM D5600), a sample of petroleum coke is ashed at 700 °C (1292 °F), and the ash is fused with lithium borate. The melt is dissolved in dilute hydrochloric acid, and the resultant solution is analyzed by ICP-AES using aqueous calibration standards. Because of the need to fuse the ash with lithium borate or other suitable salt, the fusibility of ash may need attention (ASTM D1857).

The wavelength dispersive X-ray spectroscopy method (ASTM D6376) provides a rapid means of measuring metallic elements in coke and provides a guide for determining conformance to material specifications. A benefit of this method is that the sulfur content can also be used to evaluate potential formation of sulfur oxides, a source of atmospheric pollution. This test method specifically determines sodium, aluminum, silicon, sulfur, calcium, titanium, vanadium, manganese, iron, and nickel.

In the method, a weighed portion of a sample of coke is dried at $110\,°C$ ($230\,°F$) and crushed to pass a 200-mesh sieve, mixed with stearic acid, and then milled and compressed into a smooth pellet. The pellet is irradiated with an X-ray beam, and the characteristic X-rays of the elements analyzed are excited, separated, and detected by the spectrometer. The measured X-ray intensities are converted to elemental concentration by using a calibration equation derived from the analysis of the standard materials. The $K\alpha$ spectral lines are used for all of the elements determined by this test method. This test method is also applicable to the determination of additional elements provided appropriate standards are available for use and comparison.

17.3.8 Proximate Analysis

In contrast to the elemental analysis, the proximate analysis of coke is the determination of the amount of mineral ash (ASTM D3174), volatile matter (ASTM D3175), water (moisture) (ASTM D3173), and fixed carbon. The fixed carbon is a calculated value and is the result of the summation of mineral ash (% by weight), volatile matter (% by weight), and water (% by weight) subtracted from 100.

The fixed carbon has been suggested to be analogous to the carbon residue (ASTM D189, ASTM D524, ASTM D4530, IP 13, IP 14, IP 398), but caution is advised against the literal use of such a comparison. The tests do not have a similar technical basis, and the calculation of fixed carbon by summation and subtraction means that the ultimate answer is subject to the errors of each measurement.

As an enhancement or extension of the method for proximate analysis, the determination of the physical composition can be achieved by any of the of test methods for determining the toluene-insoluble constituents of tar and pitch (ASTM D4072, ASTM D4312). Furthermore, a variety of sample can be employed to give a gradation of soluble and insoluble fractions. The coke, of course, remains in the extraction thimble (Soxhlet apparatus), and the extracts are freed from the solvent and weighed to give percentage by weight yield(s).

17.3.9 Sulfur

In addition to metallic constituents (*q.v.*), coke also contains considerable amounts of sulfur (ASTM D1552, ASTM D4239) that must be determined prior to sale or use.

A test method (ASTM D1552) is available for sulfur analysis, and the method covers three procedures applicable to samples boiling above $177\,°C$ ($350\,°F$) and containing not less than 0.06% w/w sulfur. Thus, the method is applicable to most fuel oils, lubricating oils, residua, and coke, and coke containing up to 8% by weight sulfur can be analyzed. This is particularly important for cokes that originate from heavy oil and tar sand bitumen where the sulfur content of the coke is usually at least 5% by weight.

In the iodate detection system (ASTM D1552), the sample is burned in a stream of oxygen at a sufficiently high temperature to convert about 97% by weight of the sulfur to sulfur dioxide. The combustion products are passed into an absorber containing an acidic solution of potassium iodide and starch indicator. A faint blue color is developed in the absorber solution by the addition of standard potassium iodate solution. As combustion proceeds, bleaching the blue color, more iodate is added. The sulfur content of the sample is calculated from the amount of standard iodate consumed during the combustion.

In the infrared detection system, the sample is weighed into a special ceramic boat that is then placed into a combustion furnace at $1371\,°C$ ($2500\,°F$) in an oxygen atmosphere. Most of the sulfur present is converted to sulfur dioxide, which is then measured with an infrared detector after moisture and dust are removed by traps. The calibration factor is determined using standards approximating the material to be analyzed.

For the iodate method, chlorine in concentrations less than 1 mass % does not interfere. The isoprene rubber method can tolerate somewhat higher levels. Nitrogen when present greater than 0.1 mass % may interfere with the iodate method, the extent being dependent on the types of nitrogen compounds as well as the combustion conditions. It does not interfere in the infrared method. The alkali and alkaline earth metals, zinc, potassium, and lead do not interfere with either method.

17.3.10 Volatile Matter

The volatile matter in coke affects the density of coke particles and can affect artifacts produced from further processing of the coke. The volatile matter can be used in estimating the calorific value of coke.

This test method (ASTM D6374) covers the determination of the volatile matter produced by pyrolysis or evolved when petroleum coke is subjected to the specific conditions of the test method.

In the test method, the volatile matter of a moisture-free petroleum coke sample is determined by measuring the mass loss of the coke when heated under the exact conditions of this test method (ASTM D6374).

There are two sources of interferences in this test method: moisture and particle size. Moisture increases the mass loss, and the moisture-free sample weight is decreased by the amount of moisture actually present in the test sample, and the particle size range of the analysis sample affects the volatile matter. The coarser the sample, the lower the reported yield of volatile matter. The method is not satisfactory for determining the content of dust control material (*q.v.*), and samples having a thermal history above 600 °C (1112 °F) are excluded from this test.

In another test method (ASTM D3175), volatile matter of a moisture-free petroleum coke is determined by measuring the mass loss of the coke when heated under the exact conditions of this procedure. Again, the particle size range of the sample affects the volatile matter insofar as coarser samples give rise to a lower yield of volatile matter. Samples having a thermal history above 600 °C (1112 °F) are excluded from the test, and the method is not satisfactory for determining the content of dust control material (*q.v.*).

17.3.11 Water

Water (or moisture) in coke adds weight to coke, and knowledge of water in the coke is important in the purchase and sale of green coke.

This test method (ASTM D4931) presents two procedures. The *Preparation Procedure* is used when the petroleum coke sample contains free water. The sample is weighed and air dried to equilibrate it with the atmosphere. Determination of the residual moisture is then determined using the *Drying Oven Method*. Air drying and residual moisture are combined to report gross moisture. The *Drying Oven Method* is used in routine commercial practice when the sample does not contain free water. The sample is crushed to at least 25 mm (1 in.) top sieve size and divided into individual aliquots of at least 500 g. This test method covers both the preparation procedure for samples containing free water and the determination of the gross moisture content of green petroleum coke.

REFERENCES

ASTM D121. Standard Terminology of Coal and Coke. Annual Book of Standards. ASTM International, West Conshohocken, PA.

ASTM D189. Standard Test Method for Conradson Carbon Residue of Petroleum Products. Annual Book of Standards. ASTM International, West Conshohocken, PA.

ASTM D293. Standard Test Method for the Sieve Analysis of Coke. Annual Book of Standards. ASTM International, West Conshohocken, PA.

ASTM D346. Standard Practice for Collection and Preparation of Coke Samples for Laboratory Analysis. ASTM International, West Conshohocken, PA.

ASTM D409. Standard Test Method for Grindability of Coal by the Hardgrove-Machine Method. Annual Book of Standards. ASTM International, West Conshohocken, PA.

ASTM D524. Standard Test Method for Ramsbottom Carbon Residue of Petroleum Products. Annual Book of Standards. ASTM International, West Conshohocken, PA.

ASTM D1552. Standard Test Method for Sulfur in Petroleum Products (High-Temperature Method). Annual Book of Standards. ASTM International, West Conshohocken, PA.

ASTM D1857. Standard Test Method for Fusibility of Coal and Coke Ash. Annual Book of Standards. ASTM International, West Conshohocken, PA.

ASTM D2638. Standard Test Method for Real Density of Calcined Petroleum Coke by Helium Pycnometer. Annual Book of Standards. ASTM International, West Conshohocken, PA.

ASTM D3173. Standard Test Method for Moisture in the Analysis Sample of Coal and Coke. Annual Book of Standards. ASTM International, West Conshohocken, PA.

ASTM D3174. Standard Test Method for Ash in the Analysis Sample of Coal and Coke from Coal. Annual Book of Standards. ASTM International, West Conshohocken, PA.

ASTM D3175. Standard Test Method for Volatile Matter in the Analysis Sample of Coal and Coke. Annual Book of Standards. ASTM International, West Conshohocken, PA.

ASTM D3338. Standard Test Method for Estimation of Net Heat of Combustion of Aviation Fuels. Annual Book of Standards. ASTM International, West Conshohocken, PA.

ASTM D3523. Standard Test Method for Spontaneous Heating Values of Liquids and Solids (Differential Mackey Test). Annual Book of Standards. ASTM International, West Conshohocken, PA.

ASTM D3605. Standard Test Method for Trace Metals in Gas Turbine Fuels by Atomic Absorption and Flame Emission Spectroscopy. Annual Book of Standards. ASTM International, West Conshohocken, PA.

ASTM D4072. Standard Test Method for Toluene-Insoluble (TI) Content of Tar and Pitch. Annual Book of Standards. ASTM International, West Conshohocken, PA.

ASTM D4239. Standard Test Method for Sulfur in the Analysis Sample of Coal and Coke Using High-Temperature Tube Furnace Combustion. Annual Book of Standards. ASTM International, West Conshohocken, PA.

ASTM D4292. Standard Test Method for Determination of Vibrated Bulk Density of Calcined Petroleum Coke. Annual Book of Standards. ASTM International, West Conshohocken, PA.

ASTM D4312. Standard Test Method for Toluene-Insoluble (TI) Content of Tar and Pitch (Short Method). Annual Book of Standards. ASTM International, West Conshohocken, PA.

ASTM D4422. Standard Test Method for Ash in Analysis of Petroleum Coke. Annual Book of Standards. ASTM International, West Conshohocken, PA.

ASTM D4530. Standard Test Method for Determination of Carbon Residue (Micro Method). Annual Book of Standards. ASTM International, West Conshohocken, PA.

ASTM D4628. Standard Test Method for Analysis of Barium, Calcium, Magnesium, and Zinc in Unused Lubricating Oils by Atomic Absorption Spectrometry. Annual Book of Standards. ASTM International, West Conshohocken, PA.

ASTM D4927. Standard Test Methods for Elemental Analysis of Lubricant and Additive Components – Barium, Calcium, Phosphorus, Sulfur, and Zinc by Wavelength-Dispersive X-Ray Fluorescence Spectroscopy. Annual Book of Standards. ASTM International, West Conshohocken, PA.

ASTM D4930. Standard Test Method for Dust Control Material on Calcined Petroleum Coke. Annual Book of Standards. ASTM International, West Conshohocken, PA.

ASTM D4931. Standard Test Method for Gross Moisture in Green Petroleum Coke. Annual Book of Standards. ASTM International, West Conshohocken, PA.

ASTM D5003. Standard Test Method for Hardgrove Grindability Index (HGI) of Petroleum Coke. Annual Book of Standards. ASTM International, West Conshohocken, PA.

ASTM D5004. Standard Test Method for Real Density of Calcined Petroleum Coke by Xylene Displacement. Annual Book of Standards. ASTM International, West Conshohocken, PA.

ASTM D5056. Standard Test Method for Trace Metals in Petroleum Coke by Atomic Absorption. Annual Book of Standards. ASTM International, West Conshohocken, PA.

ASTM D5291. Standard Test Methods for Instrumental Determination of Carbon, Hydrogen, and Nitrogen in Petroleum Products and Lubricants. Annual Book of Standards. ASTM International, West Conshohocken, PA.

ASTM D5373. Standard Test Methods for Determination of Carbon, Hydrogen and Nitrogen in Analysis Samples of Coal and Carbon in Analysis Samples of Coal and Coke. Annual Book of Standards. ASTM International, West Conshohocken, PA.

ASTM D5600. Standard Test Method for Trace Metals in Petroleum Coke by Inductively Coupled Plasma Atomic Emission Spectrometry (ICP-AES). Annual Book of Standards. ASTM International, West Conshohocken, PA.

ASTM D5708. Standard Test Methods for Determination of Nickel, Vanadium, and Iron in Crude Oils and Residual Fuels by Inductively Coupled Plasma (ICP) Atomic Emission Spectrometry. Annual Book of Standards. ASTM International, West Conshohocken, PA.

ASTM D5863. Standard Test Methods for Determination of Nickel, Vanadium, Iron, and Sodium in Crude Oils and Residual Fuels by Flame Atomic Absorption Spectrometry. Annual Book of Standards. ASTM International, West Conshohocken, PA.

ASTM D5865. Standard Test Method for Gross Calorific Value of Coal and Coke. Annual Book of Standards. ASTM International, West Conshohocken, PA.

ASTM D6357. Standard Test Methods for Determination of Trace Elements in Coal, Coke, and Combustion Residues from Coal Utilization Processes by Inductively Coupled Plasma Atomic Emission Spectrometry, Inductively Coupled Plasma Mass Spectrometry, and Graphite Furnace Atomic Absorption Spectrometry. Annual Book of Standards. ASTM International, West Conshohocken, PA.

ASTM D6374. Standard Test Method for Volatile Matter in Green Petroleum Coke Quartz Crucible Procedure. Annual Book of Standards. ASTM International, West Conshohocken, PA.

ASTM D6376. Standard Test Method for Determination of Trace Metals in Petroleum Coke by Wavelength Dispersive X-ray Fluorescence Spectroscopy. Annual Book of Standards. ASTM International, West Conshohocken, PA.

ASTM D6443. Standard Test Method for Determination of Calcium, Chlorine, Copper, Magnesium, Phosphorus, Sulfur, and Zinc in Unused Lubricating Oils and Additives by Wavelength Dispersive X-ray Fluorescence Spectrometry (Mathematical Correction Procedure). Annual Book of Standards. ASTM International, West Conshohocken, PA.

EPA SW-846. Test Methods for Evaluating Solid Waste, Physical/Chemical Methods. US Environmental Protection Agency, Washington, DC.

Gary, J.G., Handwerk, G.E., and Kaiser, M.J. 2007. Petroleum Refining: Technology and Economics, 5th Edition. CRC Press, Taylor & Francis Group, Boca Raton, FL.

Hsu, C.S. and Robinson, P.R. (Editors) 2006. Practical Advances in Petroleum Processing (Volumes 1 and 2). Springer Science, New York.

IP 13 IP 13 (ASTM D189). Petroleum Products – Determination of Carbon Residue – Conradson Method. IP Standard Methods 2013. The Energy Institute, London.

IP 14 (ASTM D524). Petroleum Products – Determination of Carbon Residue – Ramsbottom Method. IP Standard Methods 2013. The Energy Institute, London.

IP 398. Petroleum Products – Determination of Carbon Residue – Micro Method. IP Standard Methods 2013. The Energy Institute, London.

Speight, J.G. 2000. The Desulfurization of Heavy Oils and Residua, 2nd Edition. Marcel Dekker Inc., New York.

Speight, J.G. 2001. Handbook of Petroleum Analysis. John Wiley & Sons Inc., New York.

Speight, J.G. and Ozum, B. 2002. Petroleum Refining Processes. Marcel Dekker Inc., New York.

Speight, J.G. 2013. The Chemistry and Technology of Coal, 3rd Edition. CRC Press, Taylor & Francis Group, Boca Raton, FL.

Speight, J.G. 2014. The Chemistry and Technology of Petroleum, 5th Edition. CRC Press, Taylor & Francis Group, Boca Raton, FL.

US EPA. 2007. Petroleum Coke Category Analysis and Hazard Characterization. Report Submitted to the US EPA. Petroleum HPV Testing Group, American Petroleum Institute, Washington, DC. December 28.

18

USE OF THE DATA

18.1 INTRODUCTION

Petroleum is a complex mixture of hydrocarbon compounds and non-hydrocarbon compounds and is of little value in the natural state—refining is required to produce saleable predicts that meet the specifications required for the designated use. Furthermore, an important task in petroleum refining is the need for reliable values of the volumetric and thermodynamic properties for hydrocarbon mixtures, which are important in the design and operation of refinery equipment.

Thus, this text has been devoted to presentation of various methods of petroleum product analysis, and there is no doubt that the analysis of petroleum products is used to ensure that the product matches the specifications that are required for sale and use (Drews, 1998; Rand, 2003; Totten, 2003; Riazi, 2005; Nadkarni, 2011). However, analytical methods are also used to describe and then monitor the progress of either (i) petroleum extraction for the reservoir and (ii) progress of the refining processes, remembering that many of the test methods applied to conventional petroleum may require modification when applied to heavy oil and tar sand and bitumen.

The application of standard test method to the analysis of petroleum is almost as old as the petroleum industry. In fact, since Benjamin Silliman Sr. and Benjamin Silliman Jr. first analyzed petroleum in the modern era (Silliman, Sr., 1833; Silliman, Jr., 1860, 1865, 1867, 1871). Petroleum characterization and the determination of physical properties have received considerable attention as a needed procedure to determine refinability. In the modern era, as a result of the expansion of computer simulators and advanced analytical tools accompanied by the availability of more accurate analytical data, the fields of petroleum analysis and petroleum product analysis have expanded even further.

However, as analytical procedure evolved and data become available, it became obvious that each piece of data not only reflected the test method applied to the sample, and the exact steps followed in the analysis of a sample depend, among other factors, on the interpretation by the analyst of the written test method and the unique characteristics of the test sample—not all samples of petroleum or of a specific petroleum product are the same or even equivalent. Thus, the procedure and the resulting data will (i) reflect the background of the investigator, (ii) reflect the ability of the investigator to interpret the results in relation to the application, (iii) be dependent upon the analytical facilities available, and (iv) be shown the required levels of accuracy and precision. As a result, it has been necessary to support the objectives of *standards organizations* (such as ASTM International), which provide a forum for the exchange of information related to test methods, discussion of any issues, and cooperative development of modified and/or standard test methods.

Indeed, whatever the source, petroleum and petroleum products are complex mixtures of hydrocarbon compounds, ranging from low-molecular-weight, low-boiling volatile organic compounds to high-molecular-weight, high-boiling constituents (even nonvolatile under achievable temperatures). Furthermore, the composition of petroleum products varies depending upon (i) the source of the crude oil—crude oil is derived from a variety of source materials (Speight, 2014)

Handbook of Petroleum Product Analysis, Second Edition. James G. Speight.
© 2015 John Wiley & Sons, Inc. Published 2015 by John Wiley & Sons, Inc.

that vary greatly in chemical composition, and (ii) the refining processes used to produce the product.

When the data from various analytical test and investigations have been collected, the obvious question that comes to mind relates to the use of the data. For the most part, the data are used to determine whether or not a crude oil may produce a certain product and whether or not the product meets specifications. However, since the early days of analysis, there has been a growing tendency to use the analytical data as a means of more detailed and accurate projections of (i) the refinery processes, (ii) product yields, (iii) product properties, and (iv) predictability. And nowhere has this been more important than when heavy crude oil and bitumen entered the refinery scene. In fact, predictability in many forms is spread throughout the field of petroleum refining and product properties.

For example, predictive methods to determine the composition or amount of sulfur in a liquid fuel is vital to see if a product meets specifications set by the federal government or by local authorities. But this is not the only issue. Low-sulfur fuels are mandated by law, but there are other issues—following a poorly tuned gasoline-drive vehicle or following a loaded diesel-driven vehicle along the highway or up an incline is not the most pleasant experience—the black aromatic smoke emitted from the exhaust pipe attests to the unpleasant experience. Yet, there are the constant reminders that low-sulfur fuels are clean fuels!

The direction to clean fuels becomes more and more obvious over the past five decades, and now the focus divides itself into two approaches: (1) developing a theory of the structure of petroleum as it influences refining and product production, and (2) using analytical data to construct maps of petroleum that can be used for predictability of behavior during recovery and refinery operations. A third category, viz., the ability of a product to meet specifications will remain much the same, but it is the first two categories that help determine the methods by which the product should be produced from the feedstock and the *fine-tuning* required to take the raw product to be a specification-grade saleable product. In addition to examining properties that contribute to product specifications, other properties may also be required for more detailed evaluation of the feedstock and for comparison between feedstocks as well as the potential product slates of each feedstock or blend of feedstocks.

Thus, the data derived from any one, or more of the analytical techniques described in the previous chapters, not only give an indication of the product characteristics but also whether or not the product is suitable (i.e., meets specification) for the proposed use. In addition, when used in conjunction with the properties of the refinery feedstock (crude oil) or the properties of a refinery (distillate) fraction, the data can be used to indicate the means by which the feedstock should be further processed to generate the desired products (Speight and Ozum, 2002; Hsu and Robinson, 2006; Gary et al., 2007; Speight, 2014).

Finally, each process chemist, process engineer, and refiner has his/her own preferences, and such efforts are feedstock dependent as well as being refinery dependent. It is, in fact, the purpose of this chapter to introduce the reader to the potential that exists in this area of science and engineering. There is also the need to recognize that what is adequate for one refinery and one feedstock (or feedstock blend provided that the blend composition does not change significantly) will not be suitable for a different refinery with a different feedstock (or feedstock blend).

18.2 FEEDSTOCK AND PRODUCT EVALUATION

Early attempts at evaluating a refinery feedstock and the yield and quality of the products produced therefrom involved the use of simple formulae based on the correlation of one of more physical properties with behavior. For example, the relationship of carbon residue to actual coke yield and the yield of other products was established for a variety of feedstocks. This technique has been fine-tuned to accommodate the larger variety of feedstocks—including blended feedstocks—that are now available for refineries (Gary et al., 2007; Hsu and Robinson, 2006; Speight, 2014; Speight and Ozum, 2002).

Indeed, the use of physical properties for feedstock evaluation and product yields and, in some cases, product properties has continued in refineries and in process research laboratories to the present time and will continue for some time. However, the emphasis for feedstock evaluation and product yields (as well as product properties) has taken a turn in the direction of feedstock mapping. In such procedures, properties of feedstock are mapped to show characteristics that are in visual form rather than in tabular form in order the facilitate evaluation and prediction of the behavior of the feedstock in various refining scenarios (Speight, 2001; Speight and Ozum, 2002; Hsu and Robinson, 2006; Gary et al., 2007; Speight, 2014). Whether or not such methods will supercede the simpler form of property correlations remains to be determined. It is likely that both will continue to be used in a complimentary fashion for some time to come.

Thus, using the data derived from the test assay, it is possible to assess that product quality acquires a degree of predictability of performance during use. However, knowledge of the basic concepts of refining will help the analyst understand the production and to a large extent, the anticipated properties of the product, which in turn is related to storage, sampling, and handling of the products. In addition, there are many instances in which interrelationships of the specification data enable properties to be predicted from the measured properties with as good precision as can be obtained by a single test. It would be possible to examine in this way the relationships between all the specified properties of a product and to establish certain key properties from which the remainder could be predicted, but this would be a tedious task.

An alternative approach to that of picking out the essential tests in a specification using regression analysis is to examine at the specification as a whole and to use the necessary component. This is termed *principal components analysis*. And in this method, a set of data as points in *multidimensional* space (*n-dimensional*, corresponding to *n* original tests) is examined to determine the direction that accounts for the biggest variability in the data (*first principal component*). The process is repeated until *n* principal components are evaluated, but it must be determined which are of practical importance since some principal components may be due to experimental error. The number of significant principal components shows the number of independent properties being measured by the tests considered.

Having established the number of independent properties for a product, it is also necessary to examine the basis for making the specification more efficient. In the long-term, it might be possible to obtain new tests of a fundamental nature to replace existing tests. In the short term, selecting the best of the existing tests to define product quality will be beneficial.

Petroleum product analysis has been greatly augmented in recent decades by applying a wide variety of instrumental techniques to studies of the hydrocarbon composition of crude oils and their products (Speight, 2014). Prior to this, hydrocarbon-type analyses (percentage of paraffins, naphthenes, olefins, and aromatics) were derived from correlations based upon physical data. The advent of instrumental techniques has led to two major developments: (1) individual component analysis, and (2) an extension to, and more detailed subdivision of, various compound types that occur in the higher-boiling ranges of petroleum distillates.

In summary, petroleum product analysis is a complex discipline involving a variety of standard test methods, some of which have been mentioned earlier, and which needs a multidimensional approach. No single technique should supersede the other without adequate testing.

18.2.1   Test Methods

Of the data from instrumental techniques, gas/liquid chromatography and mass spectrometry are the most important in providing the hydrocarbon composition data in crude oil assay work (Chapter 2). By gas chromatographic analysis, it is now possible to determine routinely the individual methane (CH_4) to heptane (C_7H_{16}) hydrocarbons and the individual aromatics that boil below 165°C (330°F) and also obtain a complete normal paraffin distribution up to C_{50}. In addition, by using a microcoulometric detector specific to sulfur, the sulfur compound distribution can be obtained throughout the distillate range. Gas chromatographic analysis can also be used to provide simulated true boiling point (TBP) curves, while developments in preparative-scale gas liquid chromatography have made possible the preparation of fractions in quantities sufficient not only for extensive

spectrometric analyses but also for the normal inspection-type tests to be undertaken.

Briefly, the composition of any crude oil or crude oil product can be approximated by a *TBP* curve. The method to derive such information is a batch distillation operation, using a large number of stages, usually greater than 60, and high reflux to distillate ratio (>5). The temperature at any point on the temperature–volumetric yield curve represents the TBP of the hydrocarbon material present at the given volume percent point distilled. However, TBP investigations are typically performed for the whole crude and not for petroleum products—however, this does not preclude TBP investigations for petroleum products.

More specific to volatile petroleum products, data from test methods involving the use of test methods such as gas chromatography and mass spectrometry offer a very rapid method for obtaining hydrocarbon-type analyses on a wide range of fractions, even up to vacuum gas oils. The information obtained can also be used in conjunction with separation procedures such as molecular distillation, thermal diffusion, or selective adsorption to provide more detailed analyses.

18.2.2   Specifications

In this context, the end product of petroleum analysis (or testing) is a series of data that allow the investigator to *specify* the character and quality of the material under investigation. Thus, a series of specifications are necessary for petroleum and its products, and such data are derived from application of the relevant text methods to the sample(s).

Specifications for petroleum products are based on bulk properties such as (but not included to) density and boiling range to assure that a petroleum product can perform the intended in-service task and (i) the raw product meets the requirement in the next step of product refining, and (ii) the finished product meets the requirement for sales. Furthermore, because petroleum exhibits wide variations in composition and properties, and these occur not only in petroleum from different fields but also in oil taken from different production depths in the same well, specifications are necessary to assure that the product performs in service as required. Specifications are even more extensive as increased amounts of heavy feedstocks are introduced into refineries either singly or as blends with other feedstocks (Chapter 3).

18.3   FEEDSTOCK AND PRODUCT MAPPING

Beyond using the analytical data to meet specification, the data can also be used to investigate the chemical and physical structure of petroleum. While this may be of lesser value for petroleum products—with the exception of petroleum-based asphalt—petroleum structure is often considered to be an intangible aspect of refining that is often ignored when it comes to predicting product yields and properties. However, it is now recognized that investigating petroleum structure in which the interactions of various bulk fractions of petroleum (Chapter) and their respective constituents can play an important role in refinability and product production. For example, the changes caused by thermal treatment can result in instability of the liquid and incompatibility of a separate phase and can greatly affect the product yield as well as the product properties (Speight, 2014).

Thus, in order to discuss the chemical and physical structures of petroleum, it is necessary to give consideration to the chemical and physical nature of the constituents of petroleum. The chemical nature of various constituents of petroleum and any events leading to instability and/or incompatibility has been addressed in detail elsewhere (Mushrush and Speight, 1995, 1998a,b; Mushrush et al., 1999; Speight, 2001, 2014) with some reference to the physical nature of petroleum, and it is not the intent to repeat these discourses here. It is however, necessary to understand the interrelationships of various constituents, especially after thermal treatment of asphaltene-containing feedstocks and products (Magaril et al., 1971; Speight, 1992; Wiehe, 1993; Speight, 1994, 2004a, 2004b; Abu-Khader and Speight, 2007; Speight, 2014). Furthermore, it is essential that the concept of petroleum being a *continuum* is accepted not only among the lower-molecular-weight ranges but that the continuum is complete and continues into the higher-molecular-weight ranges.

One of the early findings of composition studies was that the behavior and properties of any material are dictated by composition (Gary et al., 2007; Hsu and Robinson, 2006; Speight, 2014; Speight and Ozum, 2002). Although the early studies were primarily focused on the composition and behavior of asphalt, the techniques developed for those investigations have provided an excellent means of studying heavy feedstocks. Later studies have focused not only on the composition of petroleum and its major operational fractions but also on further fractionation (Speight, 2001, 2014), which allows different feedstocks to be compared on a relative basis to provide a very simple but convenient feedstock *map* leading to various products.

Thus, it is not surprising that one of the most effective means of feedstock mapping has arisen though the use of fractionation methods in which the feedstock is subdivided into several fractions by standard procedures (Speight, 2001, 2014). The simplest map that can be derived from such methods show the feedstock as a three-phase system based on asphaltene separation. Indeed, the constituents of the asphaltene fraction can be differentiated on the basis of the solubility parameter of the solvent or solvent used for the separation as well as on the basis of separation of the constituents by adsorption (Mitchell and Speight, 1973; Speight, 2001, 2014).

However, it must be recognized that such *maps* do not give any indication of the complex intermolecular and intramolecular interactions that occur between, for example, such fractions as the resin constituents and asphaltene constituents (Koots and Speight, 1975; Speight, 1994, 2014), but it does allow predictions of feedstock behavior leading to identification of product yields and properties. Thus, by careful selection of an appropriate technique, it is possible to obtain an overview of petroleum composition that can be used for predictions of refining behavior and product yields/properties. By taking the approach one step further and by assiduous collection of various sub-fractions, it becomes possible to develop the feedstock map and add an extra dimension to compositional studies. Application of precise numbers to such a map allows more accurate predictions. There is even the possibility of adding a third dimension to such a map thereby enhancing predictability.

Within the distillates, the physical and chemical properties are known to change only gradually with the boiling point, allowing data extrapolation with reasonable certainty. Whenever the range of distillates is expanded, first by vacuum distillation, then by short-path distillation, the new distillate portions of the previously *non-distillable* residua follow the same patterns as the previous distillates, and relationships between various physical properties such as (i) structural types, (ii) molecular weight, (iii) sulfur content, and (iv) nitrogen content can be extended into the higher-molecular-weight regions.

The basic concept is that petroleum is a continuum of structural types that may be further defined in terms of polarity and molecular weight (or boiling point) and has been generally accepted for several decades. The initial focus in this concept was on the hydrocarbon types that occurred in the volatile fractions leaving the characterization of the nonvolatile fractions open to the use of other methods, such as spectroscopic techniques and structural group analysis. Further work on nonvolatile gas oils has shown that the continuum is ubiquitous and also includes constituents of the resin fraction and the asphaltene fraction. From this, it can be assumed with a high degree of certainty that the structural types that occur in the lower-boiling fractions continue into the resin and asphaltene fractions—but the manner in which they are bonded to each other within the individual constituents of both fractions is subject to uncertainty.

Thus, advanced feedstock fractionation and evaluation have played a significant role, along with the physical testing methods (Speight, 2001; Speight and Ozum, 2002; Speight, 2004b; Hsu and Robinson, 2006; Gary et al., 2007; Speight, 2014) and will no doubt continue to pay a role in the refinery of the future (Speight, 2011). For example, distillation being the simplest form of fractionation has been, has continued to be, used as a means of describing a feedstock in the form of a distillation profile from which the yield of various primary products can be estimated. This was a very simple, but adequate, *map* from which the yields of the primary products (i.e., distillation fractions) were presented and allowed refiners to be prepared for a wider variety of straight-run hydrocarbon products.

At the same time, and as feedstock quality changed, the hydrogen content and molecular weight were considered to be important parameters and worth *mapping* as a match to other properties, such as nitrogen content and viscosity (Gary et al., 2007; Hsu and Robinson, 2006; Speight, 2001, 2014; Speight and Ozum, 2002). In many cases, the properties were inserted on the correct numbered scale as one of the *map* coordinates.

The reason for the importance of the hydrogen content and the molecular weight is that both are used to give an evaluation of the feedstock and product yields in terms of the hydrogen required to upgrade the feedstock to specific products. In addition, nitrogen content indicated not only the hydrogen required to remove it as ammonia but also the type of catalyst necessary for processing. Viscosity is an indicator of the ability of the feedstock to flow, or more correctly to resist flow. Other properties such as pour point were also used to predict feedstock behavior in a refinery. A combination of viscosity, API gravity, and sulfur –that is, the preliminary assay— Chapter 3—also presented indications of behavior in a refinery as well as the ability of the crude oil to flow in a reservoir. Indeed, considerations of pour point and reservoir temperature together—which gives a real picture of whether or not the oil is in the liquid state—are the best gauge of fluidity and flow in a reservoir as well as in the refinery (Speight, 2009, 2014).

18.4 STRUCTURAL GROUP ANALYSES

Structural group analysis is the determination of the alleged statistical distribution of these structural elements in the petroleum and petroleum products, irrespective of the way in which the *elements* are combined in molecules. Thus, structural group analysis occupies a position midway between ultimate analysis in which atoms are the components, and molecular analysis in which molecules are the components (such as might be determined by application of various chromatographic methods (Chapter 2). A method for structural group analysis seems to be complete only if structural elements are chosen in such a way that the sum of all equals 100% (or unity), for instance, by considering the distribution of carbon in aromatic, naphthene, and paraffin locations in petroleum, petroleum fractions, and petroleum products.

The predominant feedstocks for structural group analysis are the high-boiling fractions of petroleum and high-boiling petroleum products. Comparative data have been collected pertaining to the character of these feedstocks that are often helpful in identifying products of unknown origin or in giving valuable indications concerning their manufacture. Structural group analysis has also been claimed to offer a more thorough knowledge of hydrocarbon types is required and mayrender a means of following physical separation methods, such as distillation, solvent treatment, or chromatography, as well as the effects of thermal processes. However, structural group analysis may only provide *average structural parameters*, which, in complex mixtures such as petroleum and petroleum products, leaves much of the conclusions open to inspired guesswork. In fact, data from the carbon residue test method (Fig. 18.1) show overlap of various fractions thereby negate the concept of average structure, which is in accordance with other investigations of the complexity of the asphaltene fraction (Speight, 1994, 2001, 2014).

One aspect of structural group analysis that has been raised to prominence is the structure of the asphaltene

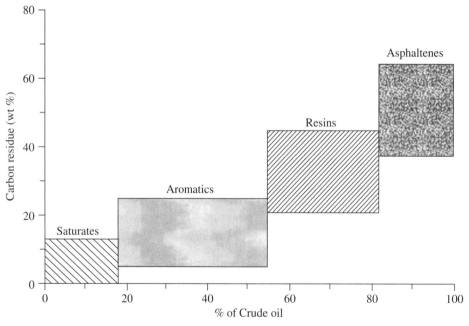

FIGURE 18.1 Carbon residue data showing overlap of the fractions (for clarity, any horizontal overlap is not shown).

constituents. However, the assignment of specific molecular configurations to the asphaltene constituents is of limited value to petroleum technology and certainly beyond the scope of the available methods to derive such formulas to represent the multitude of different structural type and molecular weight ranges within the asphaltene fraction (Speight, 2014). However, a variety of structures exist in the asphaltene fraction (in which there is a decided hydrogen deficiency) as dictated by various separation and identification methods. But the close relationships of various molecular series comprising the asphaltene constituents, resin constituents, aromatic fraction, and saturate fraction give rise to overlapping of fractions into neighboring series, in polarity, the atomic hydrogen-to-carbon atomic ratio (reflected as carbon residue) (Fig. 18.1), and the molecular weight.

Although gas–liquid chromatography and other techniques (Chapter 2) have been applied successfully to the identification of a considerable number of constituents of petroleum and petroleum products, they are, however, mainly limited to the so-called front end, that is, the volatile portion, of petroleum and the petroleum, products that are volatile and can be examined by appropriate methods (Chapter 2). Since the majority of crude oils and heavy oils contain significant proportions of a nonvolatile residuum, it is not possible to identify individual components in this part of the feedstock or product (such as residual fuel oil and asphalt) by a technique that requires volatility of the constituents. It is because of this that other methods of identification have been pursued. And this is where structural group analysis has played a valuable role, if not sometimes a *questionable role*.

To qualify the term *questionable role*, it is not the methods of structural group analysis that is at error because of various inherent assumptions in each method, but it is the means by which the data are treated by the investigators. Using the data too literally and forgetting that the term *average structure* cannot be always representative of the *true structure* as well as behavioral characteristics is the major issue.

Indeed, methods for structural group analysis usually involve the determination of physical constants of the sample. However, since there is no simple relation between physical properties and chemical composition, even in petroleum products, a reliable correlation can only be obtained by studying properties of a great variety products or pure compounds according to exact methods, laborious though that may be. The better the representation and the greater the number of these basic data, the more reliable the resulting method for structural group analysis will be. However, it should be remembered that structural group analysis is not the ultimate answer when applied to the higher-boiling fraction of petroleum or high-boiling (less volatile) petroleum products. Nevertheless, its importance lies in the straight correlation existing between the information derived from such analysis and physical properties, but it is often sufficient to get an overall description of the material in terms of its average structural group composition.

In structural group analysis, the identification of the constituents of petroleum and petroleum products by molecular type may proceed in a variety of ways but generally can be classified into three methods: (1) spectroscopic techniques, (2) chemical techniques, and (3) physical property methods. Various structural parameters are derived from a particular property by a sequence of mathematical manipulations. It is difficult to completely separate these three methods of structural elucidation, and there must, by virtue of need and relationship, be some overlap.

This leads to the analysis of the samples in terms of groups of hydrocarbons, and four classes are generally recognized: (1) aromatic if the molecule contains at least one aromatic ring, (2) olefin if the molecule contains at least one olefin bond, (3) naphthene if the molecule contains at least one naphthenic ring, and (4) paraffin if the molecule contains neither an aromatic nor a naphthenic ring nor an olefin double bond.

Aromatic hydrocarbons are also subdivided according to *aromatic type*, a term that describes compounds having the same number and grouping of aromatic rings. If an aromatic hydrocarbon contains two aromatic rings, three types may be distinguished: (1) the rings may be condensed, that is, fused together, to form the naphthalene nucleus, (2) the rings may be joined by an interring bond as in the biphenyl nucleus, and (3) the rings are separated by one or more nonaromatic carbon atoms, such as diphenylmethane ($C_6H_5CH_2C_6H_5$) or dibenzyl ($C_6H_5CH_2CH_2C_6H_5$). This nomenclature is open to extension to polynuclear aromatic compounds with more than two aromatic rings. Alkyl chains and naphthene rings are generally not considered in discussions of aromatic type. The method must also recognize combined hydrocarbon systems such as *naphthene-aromatic compounds*.

There are two ways of reporting the results of a structural group analysis. One method is to determine the number of rings or other structural groups in terms of the average number of aromatic rings (R_A), naphthene rings (R_N), and the total number of rings ($R_T = R_A + R_N$) is designated ring content. The other method is to determine the number of carbon atoms in aromatic (%C_A), naphthenic (%C_N), and paraffinic structures (%C_P), all expressed per 100 carbon atoms in the sample. These data are designated as the *carbon distribution*, and if the molecular weight of the sample is known and if an assumption is made about the type of rings present, the ring content can be recalculated as carbon distribution; the converse is also applicable. Thus, two terms *carbon distribution* and *molecular weight* (both of which are averages) can then lead to derivation of a *hypothetical average molecule*.

It is always the case that when mathematical manipulations are employed to derive *average structures*, the structures will only be as reliable as the assumptions used for, and data input into, the mathematical procedure. Then, the literal interpretation that leads to the conclusion, and the insistence, that such structures exist in the sample only leads to more confusion. Indeed, the complexity of the nonvolatile constituents of petroleum makes the construction of *average structures* extremely futile and, perhaps, misleading.

As noted, caution is advised here since such a molecule containing the structural groups in the proportions found by structural group analysis will not exist and may be difficult to use in determining the behavioral characteristics of the sample. The end results of these methods are *indications* (and not proof) of the *structural types* present in petroleum or in the petroleum product.

The evolution of the available spectroscopic techniques (Chapter 2) that could be applied to structural group analysis commenced in the 1960s has received and continues to receive wide application in the analytical and research laboratories of the petroleum industry.

Finally, some researchers still avidly pursue the myth of an average structure—all to no avail and without furthering petroleum science. The mythology of an *average structure* or an *average structural parameter* represents a failure to understand (or willingness to deny) the reality of the limitations of average structures and parameters when applied to the extremely complex fractions of petroleum.

18.5 EPILOGUE

The data derived from any one, or more, of the evaluation techniques described in this text give an indication of the characteristics of the feedstock and products as well as options for feedstock processing as well as for the prediction of product properties. Other properties may also be required for more detailed evaluation of the feedstock and for comparison between feedstocks and product yields/properties even though they may not play any role in dictating which refinery operations are necessary for processing.

However, proceeding from the raw evaluation data to full-scale production is not the preferred step. Further evaluation of feedstock processability is usually through the use of a pilot-scale operation followed by scale-up to a demonstration-size plant. It will then be possible to develop accurate and realistic relationships using the data obtained from the actual plant operations. After that, feedstock mapping can play an important role to assist in various *tweaks* that are needed to maintain a healthy process to produce saleable products with the necessary properties.

Indeed, the use of physical properties for feedstock evaluation and product slate has continued in refineries and in process research laboratories to the present time and will continue for some time. It is, of course, a matter of choosing the relevant and meaningful properties to meet the nature of the task.

REFERENCES

Abu-Khader, M.M., and Speight, J.G. 2007. Influence of High Asphaltene Feedstocks on Processing. Oil & Gas Science and Technology, 62(5): 715–722.

Drews, A.W. (Editor). 1998. *Manual on Hydrocarbon Analysis* 6th Edition. ASTM International, West Conshohocken, PA.

Gary, J.G., Handwerk, G.E., and Kaiser, M.J. 2007. *Petroleum Refining: Technology and Economics 5th Edition*. CRC Press, Taylor & Francis Group, Boca Raton, FL.

Hsu, C.S., and Robinson, P.R. (Editors). 2006. *Practical Advances in Petroleum Processing Volumes 1 and 2*. Springer Science, New York.

Koots, J.A., and Speight, J.G. 1975. The Relation of Petroleum Resins to Asphaltenes. Fuel 54: 179.

Magaril, R.Z., Ramazeava, L.F., and Aksenova, E.I. 1971. Kinetics of Coke Formation in the Thermal Processing of Crude Oil. Int. Chem. Eng. 11: 250.

Mitchell, D.L., and Speight, J.G. 1973. The Solubility of Asphaltenes in Hydrocarbon Solvents. Fuel 52: 149.

Mushrush, G.W., and Speight, J.G. 1995. *Petroleum Products: Instability and Incompatibility*. Taylor & Francis, New York.

Mushrush, G.W., and Speight, J.G. 1998a. Instability and Incompatibility of Petroleum Products. *Petroleum Chemistry and Refining*. J.G. Speight (Editor). Taylor & Francis, New York.

Mushrush, G.W., and Speight, J.G. 1998b. A review of the chemistry of incompatibility in middle distillate fuels. Reviews in Process Chemistry and Engineering 1: 5.

Mushrush, G.W., Speight, J.G., Beal, E.J., and Hardy, D.R. 1999. Preprints. American Chemical Society, Division of Petroleum Chemistry 44(2): 175.

Nadkarni, R.A.K 2011. *Spectroscopic Analysis of Petroleum Products and Lubricants*. ASTM International, West Conshohocken, PA.

Rand, S. 2003. *Significance of Tests for Petroleum Products 7th Edition*. ASTM International, West Conshohocken, PA.

Riazi, M.R. 2005. *Characterization and Properties of Petroleum Fractions*. ASTM International, West Conshohocken, PA.

Silliman, B. Sr. 1833. Notice of a Fountain of Petroleum Called the Oil Spring. American Journal Series 1, XXIII: 97–103.

Silliman, B. Jr. 1860. Oil Wells of Pennsylvania and Ohio. American Journal Series 2, XXX: 305–306.

Silliman, B. Jr. 1865. Examination of Petroleum from California. American Journal Series 2, XXXIX: 341–343.

Silliman, B. Jr. 1867. On Naphtha and Illuminating Oil from Heavy California Tar. American Journal Series 2, XLIII: 242–246.

Silliman, B. Jr. 1871. Report on the Rock Oil or Petroleum from Venango County Pennsylvania. American Chemist 2: 18–23.

Speight, J.G. 1992. A Chemical and Physical Explanation of Incompatibility during Refining Operations. *Proceedings of the 4th International Conference on the Stability and Handling of Liquid Fuels*. US Department of Energy (DOE/CONF-911102), Orlando, FL, November 18–22. p 169.

Speight, J.G. 1994. Chemical and Physical Studies of Petroleum Asphaltenes. In Asphaltenes and Asphalts, I. *Developments in Petroleum Science*, 40. T.F. Yen and G.V. Chilingarian (Editors). Elsevier, Amsterdam, Netherlands.s

Speight, J.G. 2001. *Handbook of Petroleum Analysis*. John Wiley & Sons Inc., New York.

Speight, J.G. 2004a. Petroleum Asphaltenes Part 1: Asphaltenes, Resins and the Structure of Petroleum. Oil & Gas Science and Technology, Rev. IFP, 59(5): 467–477.

Speight, J.G. 2004b. Petroleum Asphaltenes Part 2: The Effect of Asphaltenes and Resin Constituents on Recovery and Refining Processes. Oil & Gas Science and Technology, Rev. IFP, 59(5): 479–488.

Speight, J.G. 2009. *Enhanced Recovery Methods for Heavy Oil and Tar Sands*. Gulf Publishing Company, Houston, TX.

Speight, J.G. 2011. *The Refinery of the Future*. Gulf Professional Publishing, Elsevier, Oxford.

Speight, J.G. 2014. *The Chemistry and Technology of Petroleum 5th Edition*. CRC Press, Taylor & Francis Group, Boca Raton, FL.

Speight, J.G., and Ozum, B. 2002. *Petroleum Refining Processes*. Marcel Dekker Inc., New York.

Totten, G.E. (Editor). 2003. *Fuels and Lubricants Handbook: Technology, Properties, Performance, and Testing*. ASTM International, West Conshohocken, PA.

Wiehe, I.A. 1993. A Phase-Separation Kinetic Model for Coke Formation. Industrial & Engineering Chemistry Research, 32, 2447–2554.

APPENDIX: TABLES OF ASTM STANDARD TEST METHODS FOR PETROLEUM AND PETROLEUM PRODUCTS

TABLE A01   Test Methods for the Terminology of Petroleum and Petroleum Products

ASTM No.	
D4175	Standard Terminology Relating to Petroleum, Petroleum Products, and Lubricants

TABLE A02   Test Methods for Sampling Petroleum and Petroleum Products

ASTM No.	
D5842	Standard Practice for Sampling and Handling of Fuels for Volatility Measurement
D5854	Standard Practice for Mixing and Handling of Liquid Samples of Petroleum and Petroleum Products

TABLE A03   Test Methods for the Analysis of Petroleum and Petroleum Products by Absorption Spectroscopy

ASTM No.	
D1840	Standard Test Method for Naphthalene Hydrocarbons in Aviation Turbine Fuels by Ultraviolet Spectrophotometry
D2008	Standard Test Method for Ultraviolet Absorbance and Absorptivity of Petroleum Products
D2269	Standard Test Method for Evaluation of White Mineral Oils by Ultraviolet Absorption
D5292	Standard Test Method for Aromatic Carbon Contents of Hydrocarbon Oils by High Resolution Nuclear Magnetic Resonance Spectroscopy
D5845	Standard Test Method for Determination of MTBE, ETBE, TAME, DIPE, Methanol, Ethanol and t-Butanol in Gasoline by Infrared Spectroscopy
D6277	Standard Test Method for Determination of Benzene in Spark-Ignition Engine Fuels Using Mid Infrared Spectroscopy
D7371	Standard Test Method for Determination of Biodiesel (Fatty Acid Methyl Esters) Content in Diesel Fuel Oil Using Mid Infrared Spectroscopy (FTIR-ATR-PLS Method)
D7806	Standard Test Method for Determination of the Fatty Acid Methyl Ester (FAME) Content of a Blend of Biodiesel and Petroleum-Based Diesel Fuel Oil Using Mid-Infrared Spectroscopy

Handbook of Petroleum Product Analysis, Second Edition. James G. Speight.
© 2015 John Wiley & Sons, Inc. Published 2015 by John Wiley & Sons, Inc.

TABLE A04   Test Methods for the Analysis of Petroleum and Petroleum Products by Mass Spectroscopy

ASTM No.	
D2425	Standard Test Method for Hydrocarbon Types in Middle Distillates by Mass Spectrometry
D2650	Standard Test Method for Chemical Composition of Gases by Mass Spectrometry
D2786	Standard Test Method for Hydrocarbon Types Analysis of Gas-Oil Saturates Fractions by High Ionizing Voltage Mass Spectrometry
D2789	Standard Test Method for Hydrocarbon Types in Low Olefinic Gasoline by Mass Spectrometry
D3239	Standard Test Method for Aromatic Types Analysis of Gas-Oil Aromatic Fractions by High Ionizing Voltage Mass Spectrometry
D5769	Standard Test Method for Determination of Benzene, Toluene, and Total Aromatics in Finished Gasolines by Gas Chromatography/Mass Spectrometry
D7845	Standard Test Method for Determination of Chemical Species in Marine Fuel Oil by Multidimensional Gas Chromatography—Mass Spectrometry

TABLE A05   Test Methods for the Analysis of Petroleum and Petroleum Products by Chromatographic Methods

ASTM No.	
D2887	Standard Test Method for Boiling Range Distribution of Petroleum Fractions by Gas Chromatography
D3525	Standard Test Method for Gasoline Diluent in Used Gasoline Engine Oils by Gas Chromatography
D3710	Standard Test Method for Boiling Range Distribution of Gasoline and Gasoline Fractions by Gas Chromatography
D5442	Standard Test Method for Analysis of Petroleum Waxes by Gas Chromatography
D6352	Standard Test Method for Boiling Range Distribution of Petroleum Distillates in Boiling Range from 174 to 700°C by Gas Chromatography
D6417	Standard Test Method for Estimation of Engine Oil Volatility by Capillary Gas Chromatography
D7096	Standard Test Method for Determination of the Boiling Range Distribution of Gasoline by Wide-Bore Capillary Gas Chromatography
D7169	Standard Test Method for Boiling Point Distribution of Samples with Residues Such as Crude Oils and Atmospheric and Vacuum Residues by High Temperature Gas Chromatography
D7213	Standard Test Method for Boiling Range Distribution of Petroleum Distillates in the Boiling Range from 100 to 615°C by Gas Chromatography
D7398	Standard Test Method for Boiling Range Distribution of Fatty Acid Methyl Esters (FAME) in the Boiling Range from 100 to 615°C by Gas Chromatography
D7500	Standard Test Method for Determination of Boiling Range Distribution of Distillates and Lubricating Base Oils—in Boiling Range from 100 to 735°C by Gas Chromatography
D7798	Standard Test Method for Boiling Range Distribution of Petroleum Distillates with Final Boiling Points up to 538°C by Ultra Fast Gas Chromatography (UF GC)
D7807	Standard Test Method for Determination of Boiling Range Distribution of Hydrocarbon and Sulfur Components of Petroleum Distillates by Gas Chromatography and Chemiluminescence Detection

TABLE A06   Test Methods for the Analysis of Petroleum and Petroleum Products by Gas Chromatography

ASTM No.	
D2268	Standard Test Method for Analysis of High-Purity n-Heptane and Isooctane by Capillary Gas Chromatography
D2427	Standard Test Method for Determination of C_2 through C_5 Hydrocarbons in Gasolines by Gas Chromatography
D3606	Standard Test Method for Determination of Benzene and Toluene in Finished Motor and Aviation Gasoline by Gas Chromatography
D4291	Standard Test Method for Trace Ethylene Glycol in Used Engine Oil
D4626	Standard Practice for Calculation of Gas Chromatographic Response Factors
D4815	Standard Test Method for Determination of MTBE, ETBE, TAME, DIPE, tertiary-Amyl Alcohol and C1 to C4 Alcohols in Gasoline by Gas Chromatography
D5134	Standard Test Method for Detailed Analysis of Petroleum Naphtha through n-Nonane by Capillary Gas Chromatography
D5441	Standard Test Method for Analysis of Methyl t-Butyl Ether (MTBE) by Gas Chromatography
D5443	Standard Test Method for Paraffin, Naphthene, and Aromatic Hydrocarbon Type Analysis in Petroleum Distillates through 200°C by Multi-Dimensional Gas Chromatography

(*Continued*)

TABLE A06   (*Continued*)

ASTM No.	
D5501	Standard Test Method for Determination of Ethanol and Methanol Content in Fuels Containing Greater Than 20% Ethanol by Gas Chromatography
D5580	Standard Test Method for Determination of Benzene, Toluene, Ethylbenzene, p/m-Xylene, o-Xylene, C9 and Heavier Aromatics, and Total Aromatics in Finished Gasoline by Gas Chromatography
D5599	Standard Test Method for Determination of Oxygenates in Gasoline by Gas Chromatography and Oxygen Selective Flame Ionization Detection
D5623	Standard Test Method for Sulfur Compounds in Light Petroleum Liquids by Gas Chromatography and Sulfur Selective Detection
D5986	Standard Test Method for Determination of Oxygenates, Benzene, Toluene, C8-C12 Aromatics and Total Aromatics in Finished Gasoline by Gas Chromatography/Fourier Transform Infrared Spectroscopy
D6160	Standard Test Method for Determination of Polychlorinated Biphenyls (PCBs) in Waste Materials by Gas Chromatography
D6296	Standard Test Method for Total Olefins in Spark-Ignition Engine Fuels by Multidimensional Gas Chromatography
D6584	Standard Test Method for Determination of Total Monoglycerides, Total Diglycerides, Total Triglycerides, and Free and Total Glycerin in B-100 Biodiesel Methyl Esters by Gas Chromatography
D6729	Standard Test Method for Determination of Individual Components in Spark Ignition Engine Fuels by 100 Meter Capillary High Resolution Gas Chromatography
D6730	Standard Test Method for Determination of Individual Components in Spark Ignition Engine Fuels by 100–Meter Capillary (with Precolumn) High-Resolution Gas Chromatography
D6733	Standard Test Method for Determination of Individual Components in Spark Ignition Engine Fuels by 50-Meter Capillary High Resolution Gas Chromatography
D6839	Standard Test Method for Hydrocarbon Types, Oxygenated Compounds, and Benzene in Spark Ignition Engine Fuels by Gas Chromatography
D7059	Standard Test Method for Determination of Methanol in Crude Oils by Multidimensional Gas Chromatography
D7576	Standard Test Method for Determination of Benzene and Total Aromatics in Denatured Fuel Ethanol by Gas Chromatography
D7753	Standard Test Method for Hydrocarbon Types and Benzene in Light Petroleum Distillates by Gas Chromatography
D7754	Standard Test Method for Determination of Trace Oxygenates in Automotive Spark-Ignition Engine Fuel by Multidimensional Gas Chromatography
D7900	Standard Test Method for Determination of Light Hydrocarbons in Stabilized Crude Oils by Gas Chromatography

TABLE A07   **Test Methods for Analysis of Petroleum and Petroleum Products by Liquid Chromatography**

ASTM No.	
D1319	Standard Test Method for Hydrocarbon Types in Liquid Petroleum Products by Fluorescent Indicator Adsorption
D2001	Standard Test Method for Depentanization of Gasoline and Naphtha
D2007	Standard Test Method for Characteristic Groups in Rubber Extender and Processing Oils and Other Petroleum-Derived Oils by the Clay-Gel Absorption Chromatographic Method
D2549	Standard Test Method for Separation of Representative Aromatics and Nonaromatics Fractions of High-Boiling Oils by Elution Chromatography
D3712	Standard Test Method of Analysis of Oil-Soluble Petroleum Sulfonates by Liquid Chromatography
D5186	Standard Test Method for Determination of Aromatic Content and Polynuclear Aromatic Content of Diesel Fuels and Aviation Turbine Fuels by Supercritical Fluid Chromatography
D6379	Standard Test Method for Determination of Aromatic Hydrocarbon Types in Aviation Fuels and Petroleum Distillates— High Performance Liquid Chromatography Method with Refractive Index Detection
D6550	Standard Test Method for Determination of Olefin Content of Gasolines by Supercritical-Fluid Chromatography
D6591	Standard Test Method for Determination of Aromatic Hydrocarbon Types in Middle Distillates—High Performance Liquid Chromatography Method with Refractive Index Detection
D7347	Standard Test Method for Determination of Olefin Content in Denatured Ethanol by Supercritical Fluid Chromatography
D7419	Standard Test Method for Determination of Total Aromatics and Total Saturates in Lube Basestocks by High Performance Liquid Chromatography (HPLC) with Refractive Index Detection
D7524	Standard Test Method for Determination of Static Dissipater Additives (SDA) in Aviation Turbine Fuel and Middle Distillate Fuels—High Performance Liquid Chromatograph (HPLC) Method
D7591	Standard Test Method for Determination of Free and Total Glycerin in Biodiesel Blends by Anion Exchange Chromatography

TABLE A08   Test Methods for the Analysis of Additives and Electrical Properties of Petroleum and Petroleum Products

ASTM No.	
D2624	Standard Test Methods for Electrical Conductivity of Aviation and Distillate Fuels
D4054	Standard Practice for Qualification and Approval of New Aviation Turbine Fuels and Fuel Additives
D4171	Standard Specification for Fuel System Icing Inhibitors
D4306	Standard Practice for Aviation Fuel Sample Containers for Tests Affected by Trace Contamination
D4308	Standard Test Method for Electrical Conductivity of Liquid Hydrocarbons by Precision Meter
D4865	Standard Guide for Generation and Dissipation of Static Electricity in Petroleum Fuel Systems
D5001	Standard Test Method for Measurement of Lubricity of Aviation Turbine Fuels by the Ball-on-Cylinder Lubricity Evaluator (BOCLE)
D5006	Standard Test Method for Measurement of Fuel System Icing Inhibitors (Ether Type) in Aviation Fuels

TABLE A09   Test Methods for the Determination of the Contaminants in Fuels

ASTM No.	
D1094	Standard Test Method for Water Reaction of Aviation Fuels
D2276	Standard Test Method for Particulate Contaminants in Aviation Fuel by Line Sampling
D3240	Standard Test Method for Undissolved Water in Aviation Turbine Fuels
D3948	Standard Test Method for Determining Water Separation Characteristics of Aviation Turbine Fuels by Portable Separometer
D5000	Standard Practice for Evaluating Activity of Clay Elements Using a Side-Stream Sensor
D5452	Standard Test Method for Particulate Contamination in Aviation Fuels by Laboratory Filtration
D7224	Standard Test Method for Determining Water Separation Characteristics of Kerosene-Type Aviation Turbine Fuels Containing Additives by Portable Separometer
D7797	Test Method for Determination of the Fatty Acid Methyl Esters Content of Aviation Turbine Fuel Using Flow Analysis by Fourier Transform Infrared Spectroscopy—Rapid Screening Method

TABLE A10   Test Methods for the Analysis of the Reactivity and Thermal Properties of Petroleum and Petroleum Products

ASTM No.	
D7667	Standard Test Method for Determination of Corrosiveness to Silver by Automotive Spark-Ignition Engine Fuel—Thin Silver Strip Method
D7671	Standard Test Method for Corrosiveness to Silver by Automotive Spark–Ignition Engine Fuel—Silver Strip Method
D1322	Standard Test Method for Smoke Point of Kerosene and Aviation Turbine Fuel
D3241	Standard Test Method for Thermal Oxidation Stability of Aviation Turbine Fuels
D7739	Standard Practice for Thermal Oxidative Stability Measurement via Quartz Crystal Microbalance
D613	Standard Test Method for Cetane Number of Diesel Fuel Oil
D909	Standard Test Method for Supercharge Rating of Spark-Ignition Aviation Gasoline
D2699	Standard Test Method for Research Octane Number of Spark-Ignition Engine Fuel
D2700	Standard Test Method for Motor Octane Number of Spark-Ignition Engine Fuel
D2885	Standard Test Method for Determination of Octane Number of Spark-Ignition Engine Fuels by On-Line Direct Comparison Technique
D6890	Standard Test Method for Determination of Ignition Delay and Derived Cetane Number (DCN) of Diesel Fuel Oils by Combustion in a Constant Volume Chamber
D7170	Standard Test Method for Determination of Derived Cetane Number (DCN) of Diesel Fuel Oils—Fixed Range Injection Period, Constant Volume Combustion Chamber Method
D2717	Standard Test Method for Thermal Conductivity of Liquids
D2766	Standard Test Method for Specific Heat of Liquids and Solids
D2878	Standard Test Method for Estimating Apparent Vapor Pressures and Molecular Weights of Lubricating Oils
D2879	Standard Test Method for Vapor Pressure-Temperature Relationship and Initial Decomposition Temperature of Liquids by Isoteniscope
D4486	Standard Test Method for Kinematic Viscosity of Volatile and Reactive Liquids

TABLE A11   Test Methods for Analysis by Correlative Methods of Petroleum and Petroleum Products

ASTM No.	
D2501	Standard Test Method for Calculation of Viscosity-Gravity Constant (VGC) of Petroleum Oils
D2502	Standard Test Method for Estimation of Molecular Weight (Relative Molecular Mass) of Petroleum Oils from Viscosity Measurements
D2889	Standard Test Method for Calculation of True Vapor Pressures of Petroleum Distillate Fuels
D2890	Standard Test Method for Calculation of Liquid Heat Capacity of Petroleum Distillate Fuels
D3238	Standard Test Method for Calculation of Carbon Distribution and Structural Group Analysis of Petroleum Oils by the n-d-M Method
D3343	Standard Test Method for Estimation of Hydrogen Content of Aviation Fuels
D7215	Standard Test Method for Calculated Flash Point from Simulated Distillation Analysis of Distillate Fuels

TABLE A12   Test for the Elemental Analysis of Petroleum and Petroleum Products

ASTM No.	
C1234	Standard Practice for Preparation of Oils and Oily Waste Samples by High-Pressure, High-Temperature Digestion for Trace Element Determinations
D129	Standard Test Method for Sulfur in Petroleum Products (General High Pressure Decomposition Device Method)
D482	Standard Test Method for Ash from Petroleum Products
D808	Standard Test Method for Chlorine in New and Used Petroleum Products (High Pressure Decomposition Device Method)
D874	Standard Test Method for Sulfated Ash from Lubricating Oils and Additives
D1018	Standard Test Method for Hydrogen in Petroleum Fractions
D1091	Standard Test Methods for Phosphorus in Lubricating Oils and Additives
D1266	Standard Test Method for Sulfur in Petroleum Products (Lamp Method)
D1318	Standard Test Method for Sodium in Residual Fuel Oil (Flame Photometric Method)
D1552	Standard Test Method for Sulfur in Petroleum Products (High-Temperature Method)
D1839	Standard Test Method for Amyl Nitrate in Diesel Fuels
D2622	Standard Test Method for Sulfur in Petroleum Products by Wavelength Dispersive X-Ray Fluorescence Spectrometry
D2784	Standard Test Method for Sulfur in Liquefied Petroleum Gases (Oxy-Hydrogen Burner or Lamp)
D3120	Standard Test Method for Trace Quantities of Sulfur in Light Liquid Petroleum Hydrocarbons by Oxidative Microcoulometry
D3227	Standard Test Method for (Thiol Mercaptan) Sulfur in Gasoline, Kerosene, Aviation Turbine, and Distillate Fuels (Potentiometric Method)
D3228	Standard Test Method for Total Nitrogen in Lubricating Oils and Fuel Oils by Modified Kjeldahl Method
D3230	Standard Test Method for Salts in Crude Oil (Electrometric Method)
D3231	Standard Test Method for Phosphorus in Gasoline
D3237	Standard Test Method for Lead in Gasoline by Atomic Absorption Spectroscopy
D3246	Standard Test Method for Sulfur in Petroleum Gas by Oxidative Microcoulometry
D3341	Standard Test Method for Lead in Gasoline—Iodine Monochloride Method
D3348	Standard Test Method for Rapid Field Test for Trace Lead in Unleaded Gasoline (Colorimetric Method)
D3605	Standard Test Method for Trace Metals in Gas Turbine Fuels by Atomic Absorption and Flame Emission Spectroscopy
D3701	Standard Test Method for Hydrogen Content of Aviation Turbine Fuels by Low Resolution Nuclear Magnetic Resonance Spectrometry
D3831	Standard Test Method for Manganese in Gasoline by Atomic Absorption Spectroscopy
D4045	Standard Test Method for Sulfur in Petroleum Products by Hydrogenolysis and Rateometric Colorimetry
D4046	Standard Test Method for Alkyl Nitrate in Diesel Fuels by Spectrophotometry
D4047	Standard Test Method for Phosphorus in Lubricating Oils and Additives by Quinoline Phosphomolybdate Method
D4294	Standard Test Method for Sulfur in Petroleum and Petroleum Products by Energy Dispersive X-Ray Fluorescence Spectrometry
D4628	Standard Test Method for Analysis of Barium, Calcium, Magnesium, and Zinc in Unused Lubricating Oils by Atomic Absorption Spectrometry
D4629	Standard Test Method for Trace Nitrogen in Liquid Petroleum Hydrocarbons by Syringe/Inlet Oxidative Combustion and Chemiluminescence Detection
D4808	Standard Test Methods for Hydrogen Content of Light Distillates, Middle Distillates, Gas Oils, and Residua by Low-Resolution Nuclear Magnetic Resonance Spectroscopy

(Continued)

TABLE A12 (*Continued*)

ASTM No.	
D4927	Standard Test Methods for Elemental Analysis of Lubricant and Additive Components—Barium, Calcium, Phosphorus, Sulfur, and Zinc by Wavelength-Dispersive X-Ray Fluorescence Spectroscopy
D4929	Standard Test Methods for Determination of Organic Chloride Content in Crude Oil
D4951	Standard Test Method for Determination of Additive Elements in Lubricating Oils by Inductively Coupled Plasma Atomic Emission Spectrometry
D4952	Standard Test Method for Qualitative Analysis for Active Sulfur Species in Fuels and Solvents (Doctor Test)
D5056	Standard Test Method for Trace Metals in Petroleum Coke by Atomic Absorption
D5059	Standard Test Methods for Lead in Gasoline by X-Ray Spectroscopy
D5184	Standard Test Methods for Determination of Aluminum and Silicon in Fuel Oils by Ashing, Fusion, Inductively Coupled Plasma Atomic Emission Spectrometry, and Atomic Absorption Spectrometry
D5185	Standard Test Method for Multi-Element Determination of Used and Unused Lubricating Oils and Base Oils by Inductively Coupled Plasma Atomic Emission Spectrometry (ICP-AES)
D5291	Standard Test Methods for Instrumental Determination of Carbon, Hydrogen, and Nitrogen in Petroleum Products and Lubricants
D5384	Standard Test Methods for Chlorine in Used Petroleum Products (Field Test Kit Method)
D5453	Standard Test Method for Determination of Total Sulfur in Light Hydrocarbons, Spark Ignition Engine Fuel, Diesel Engine Fuel, and Engine Oil by Ultraviolet Fluorescence
D5600	Standard Test Method for Trace Metals in Petroleum Coke by Inductively Coupled Plasma Atomic Emission Spectrometry (ICP-AES)
D5622	Standard Test Methods for Determination of Total Oxygen in Gasoline and Methanol Fuels by Reductive Pyrolysis
D5708	Standard Test Methods for Determination of Nickel, Vanadium, and Iron in Crude Oils and Residual Fuels by Inductively Coupled Plasma (ICP) Atomic Emission Spectrometry
D5761	Standard Practice for Emulsification/Suspension of Multiphase Fluid Waste Materials
D5762	Standard Test Method for Nitrogen in Petroleum and Petroleum Products by Boat-Inlet Chemiluminescence
D5863	Standard Test Methods for Determination of Nickel, Vanadium, Iron, and Sodium in Crude Oils and Residual Fuels by Flame Atomic Absorption Spectrometry
D6334	Standard Test Method for Sulfur in Gasoline by Wavelength Dispersive X-Ray Fluorescence
D6443	Test Method for Determination of Calcium, Chlorine, Copper, Magnesium, Phosphorus, Sulfur, and Zinc in Unused Lubricating Oils and Additives by Wavelength Dispersive X-Ray Fluorescence Spectrometry (Mathematical Correction Procedure)
D6470	Standard Test Method for Salt in Crude Oils (Potentiometric Method)
D6481	Standard Test Method for Determination of Phosphorus, Sulfur, Calcium, and Zinc in Lubrication Oils by Energy Dispersive X-Ray Fluorescence Spectroscopy
D6595	Standard Test Method for Determination of Wear Metals and Contaminants in Used Lubricating Oils or Used Hydraulic Fluids by Rotating Disc Electrode Atomic Emission Spectrometry
D6667	Standard Test Method for Determination of Total Volatile Sulfur in Gaseous Hydrocarbons and Liquefied Petroleum Gases by Ultraviolet Fluorescence
D6728	Standard Test Method for Determination of Contaminants in Gas Turbine and Diesel Engine Fuel by Rotating Disc Electrode Atomic Emission Spectrometry
D6732	Standard Test Method for Determination of Copper in Jet Fuels by Graphite Furnace Atomic Absorption Spectrometry
D6920	Standard Test Method for Total Sulfur in Naphtha, Distillates, Reformulated Gasolines, Diesels, Biodiesels, and Motor Fuels by Oxidative Combustion and Electrochemical Detection
D7039	Standard Test Method for Sulfur in Gasoline, Diesel Fuel, Jet Fuel, Kerosene, Biodiesel, Biodiesel Blends, and Gasoline-Ethanol Blends by Monochromatic Wavelength Dispersive X-Ray Fluorescence Spectrometry
D7040	Standard Test Method for Determination of Low Levels of Phosphorus in ILSAC GF 4 and Similar Grade Engine Oils by Inductively Coupled Plasma Atomic Emission Spectrometry
D7041	Standard Test Method for Determination of Total Sulfur in Light Hydrocarbons, Motor Fuels, and Oils by Online Gas Chromatography with Flame Photometric Detection
D7111	Standard Test Method for Determination of Trace Elements in Middle Distillate Fuels by Inductively Coupled Plasma Atomic Emission Spectrometry (ICP-AES)
D7171	Standard Test Method for Hydrogen Content of Middle Distillate Petroleum Products by Low-Resolution Pulsed Nuclear Magnetic Resonance Spectroscopy
D7212	Standard Test Method for Low Sulfur in Automotive Fuels by Energy-Dispersive X-Ray Fluorescence Spectrometry Using a Low-Background Proportional Counter

(*Continued*)

TABLE A12 (*Continued*)

ASTM No.	
D7220	Standard Test Method for Sulfur in Automotive, Heating, and Jet Fuels by Monochromatic Energy Dispersive X-Ray Fluorescence Spectrometry
D7260	Standard Practice for Optimization, Calibration, and Validation of Inductively Coupled Plasma-Atomic Emission Spectrometry (ICP-AES) for Elemental Analysis of Petroleum Products and Lubricants
D7303	Standard Test Method for Determination of Metals in Lubricating Greases by Inductively Coupled Plasma Atomic Emission Spectrometry
D7318	Standard Test Method for Existent Inorganic Sulfate in Ethanol by Potentiometric Titration
D7319	Standard Test Method for Determination of Existent and Potential Sulfate and Inorganic Chloride in Fuel Ethanol and Butanol by Direct Injection Suppressed Ion Chromatography
D7328	Standard Test Method for Determination of Existent and Potential Inorganic Sulfate and Total Inorganic Chloride in Fuel Ethanol by Ion Chromatography Using Aqueous Sample Injection
D7343	Standard Practice for Optimization, Sample Handling, Calibration, and Validation of X-Ray Fluorescence Spectrometry Methods for Elemental Analysis of Petroleum Products and Lubricants
D7455	Standard Practice for Sample Preparation of Petroleum and Lubricant Products for Elemental Analysis
D7482	Standard Practice for Sampling, Storage, and Handling of Hydrocarbons for Mercury Analysis
D7578	Standard Guide for Calibration Requirements for Elemental Analysis of Petroleum Products and Lubricants
D7620	Standard Test Method for Determination of Total Sulfur in Liquid Hydrocarbon Based Fuels by Continuous Injection, Air Oxidation and Ultraviolet Fluorescence Detection
D7622	Standard Test Method for Total Mercury in Crude Oil Using Combustion and Direct Cold Vapor Atomic Absorption Method with Zeeman Background Correction
D7623	Standard Test Method for Total Mercury in Crude Oil Using Combustion-Gold Amalgamation and Cold Vapor Atomic Absorption Method
D7691	Standard Test Method for Multielement Analysis of Crude Oils Using Inductively Coupled Plasma Atomic Emission Spectrometry (ICP-AES)
D7740	Standard Practice for Optimization, Calibration, and Validation of Atomic Absorption Spectrometry for Metal Analysis of Petroleum Products and Lubricants
D7751	Standard Test Method for Determination of Additive Elements in Lubricating Oils by EDXRF Analysis
D7757	Standard Test Method for Silicon in Gasoline and Related Products by Monochromatic Wavelength Dispersive X-Ray Fluorescence Spectrometry
D7876	Standard Practice for Practice for Sample Decomposition Using Microwave Heating (With or Without Prior Ashing) for Atomic Spectroscopic Elemental Determination in Petroleum Products and Lubricants

TABLE A13 Test Methods for the Analysis of Hydrocarbons and Contaminants in Petroleum and Petroleum Products

ASTM No.	
D95	Standard Test Method for Water in Petroleum Products and Bituminous Materials by Distillation
D287	Standard Test Method for API Gravity of Crude Petroleum and Petroleum Products (Hydrometer Method)
D473	Standard Test Method for Sediment in Crude Oils and Fuel Oils by the Extraction Method
D1250	Standard Guide for Use of the Petroleum Measurement Tables
D1298	Standard Test Method for Density, Relative Density, or API Gravity of Crude Petroleum and Liquid Petroleum Products by Hydrometer Method
D1657	Standard Test Method for Density or Relative Density of Light Hydrocarbons by Pressure Hydrometer
D1744	Standard Test Method for Determination of Water in Liquid Petroleum Products by Karl Fischer Reagent
D1796	Standard Test Method for Water and Sediment in Fuel Oils by the Centrifuge Method (Laboratory Procedure)
D4006	Standard Test Method for Water in Crude Oil by Distillation
D4007	Standard Test Method for Water and Sediment in Crude Oil by the Centrifuge Method (Laboratory Procedure)
D4057	Standard Practice for Manual Sampling of Petroleum and Petroleum Products
D4177	Standard Practice for Automatic Sampling of Petroleum and Petroleum Products
D4377	Standard Test Method for Water in Crude Oils by Potentiometric Karl Fischer Titration
D4807	Standard Test Method for Sediment in Crude Oil by Membrane Filtration
D4928	Standard Test Method for Water in Crude Oils by Coulometric Karl Fischer Titration
D7829	Standard Guide for Sediment and Water Determination in Crude Oil

TABLE A14   Test Methods for the Determination of the Flow Properties of Petroleum and Petroleum Products

ASTM No.	
D97	Standard Test Method for Pour Point of Petroleum Products
D341	Standard Practice for Viscosity-Temperature Charts for Liquid Petroleum Products
D445	Standard Test Method for Kinematic Viscosity of Transparent and Opaque Liquids (and Calculation of Dynamic Viscosity)
D446	Standard Specifications and Operating Instructions for Glass Capillary Kinematic Viscometers
D2161	Standard Practice for Conversion of Kinematic Viscosity to Saybolt Universal Viscosity or to Saybolt Furol Viscosity
D2162	Standard Practice for Basic Calibration of Master Viscometers and Viscosity Oil Standards
D2270	Standard Practice for Calculating Viscosity Index From Kinematic Viscosity at 40 and 100°C
D2386	Standard Test Method for Freezing Point of Aviation Fuels
D2500	Standard Test Method for Cloud Point of Petroleum Products
D2532	Standard Test Method for Viscosity and Viscosity Change after Standing at Low Temperature of Aircraft Turbine Lubricants
D2603	Standard Test Method for Sonic Shear Stability of Polymer-Containing Oils
D2983	Standard Test Method for Low-Temperature Viscosity of Lubricants Measured by Brookfield Viscometer
D3829	Standard Test Method for Predicting the Borderline Pumping Temperature of Engine Oil
D4539	Standard Test Method for Filterability of Diesel Fuels by Low-Temperature Flow Test (LTFT)
D4683	Standard Test Method for Measuring Viscosity of New and Used Engine Oils at High Shear Rate and High Temperature by Tapered Bearing Simulator Viscometer at 150?°C
D4684	Standard Test Method for Determination of Yield Stress and Apparent Viscosity of Engine Oils at Low Temperature
D4741	Standard Test Method for Measuring Viscosity at High Temperature and High Shear Rate by Tapered-Plug Viscometer
D5133	Standard Test Method for Low Temperature, Low Shear Rate, Viscosity/Temperature Dependence of Lubricating Oils Using a Temperature-Scanning Technique
D5275	Standard Test Method for Fuel Injector Shear Stability Test (FISST) for Polymer Containing Fluids
D5293	Standard Test Method for Apparent Viscosity of Engine Oils and Base Stocks Between −5 and −35°C Using Cold-Cranking Simulator
D5481	Standard Test Method for Measuring Apparent Viscosity at High-Temperature and High-Shear Rate by Multicell Capillary Viscometer
D5621	Standard Test Method for Sonic Shear Stability of Hydraulic Fluids
D5771	Standard Test Method for Cloud Point of Petroleum Products (Optical Detection Stepped Cooling Method)
D5772	Standard Test Method for Cloud Point of Petroleum Products (Linear Cooling Rate Method)
D5773	Standard Test Method for Cloud Point of Petroleum Products (Constant Cooling Rate Method)
D5853	Standard Test Method for Pour Point of Crude Oils
D5949	Standard Test Method for Pour Point of Petroleum Products (Automatic Pressure Pulsing Method)
D5950	Standard Test Method for Pour Point of Petroleum Products (Automatic Tilt Method)
D5972	Standard Test Method for Freezing Point of Aviation Fuels (Automatic Phase Transition Method)
D5985	Standard Test Method for Pour Point of Petroleum Products (Rotational Method)
D6022	Standard Practice for Calculation of Permanent Shear Stability Index
D6278	Standard Test Method for Shear Stability of Polymer Containing Fluids Using a European Diesel Injector Apparatus
D6371	Standard Test Method for Cold Filter Plugging Point of Diesel and Heating Fuels
D6616	Standard Test Method for Measuring Viscosity at High Shear Rate by Tapered Bearing Simulator Viscometer at 100°C
D6749	Standard Test Method for Pour Point of Petroleum Products (Automatic Air Pressure Method)
D6821	Standard Test Method for Low Temperature Viscosity of Drive Line Lubricants in a Constant Shear Stress Viscometer
D6892	Standard Test Method for Pour Point of Petroleum Products (Robotic Tilt Method)
D6895	Standard Test Method for Rotational Viscosity of Heavy Duty Diesel Drain Oils at 100°C
D6896	Standard Test Method for Determination of Yield Stress and Apparent Viscosity of Used Engine Oils at Low Temperature
D7042	Standard Test Method for Dynamic Viscosity and Density of Liquids by Stabinger Viscometer (and the Calculation of Kinematic Viscosity)
D7109	Standard Test Method for Shear Stability of Polymer Containing Fluids Using a European Diesel Injector Apparatus at 30 and 90 Cycles
D7110	Standard Test Method for Determining the Viscosity-Temperature Relationship of Used and Soot-Containing Engine Oils at Low Temperatures
D7152	Standard Practice for Calculating Viscosity of a Blend of Petroleum Products
D7153	Standard Test Method for Freezing Point of Aviation Fuels (Automatic Laser Method)
D7154	Standard Test Method for Freezing Point of Aviation Fuels (Automatic Fiber Optical Method)
D7279	Standard Test Method for Kinematic Viscosity of Transparent and Opaque Liquids by Automated Houillon Viscometer
D7346	Standard Test Method for No Flow Point and Pour Point of Petroleum Products
D7397	Standard Test Method for Cloud Point of Petroleum Products (Miniaturized Optical Method)
D7483	Standard Test Method for Determination of Dynamic Viscosity and Derived Kinematic Viscosity of Liquids by Oscillating Piston Viscometer
D7683	Standard Test Method for Cloud Point of Petroleum Products (Small Test Jar Method)
D7689	Standard Test Method for Cloud Point of Petroleum Products (Mini Method)

TABLE A15 **Test Methods for the Determination of the Chemical and Physical Properties of Petroleum and Petroleum Products**

ASTM No.	
D611	Standard Test Methods for Aniline Point and Mixed Aniline Point of Petroleum Products and Hydrocarbon Solvents
D1015	Standard Test Method for Freezing Points of High-Purity Hydrocarbons
D1016	Standard Test Method for Purity of Hydrocarbons from Freezing Points
D1159	Standard Test Method for Bromine Numbers of Petroleum Distillates and Commercial Aliphatic Olefins by Electrometric Titration
D1217	Standard Test Method for Density and Relative Density (Specific Gravity) of Liquids by Bingham Pycnometer
D1218	Standard Test Method for Refractive Index and Refractive Dispersion of Hydrocarbon Liquids
D1480	Standard Test Method for Density and Relative Density (Specific Gravity) of Viscous Materials by Bingham Pycnometer
D1481	Standard Test Method for Density and Relative Density (Specific Gravity) of Viscous Materials by Lipkin Bicapillary Pycnometer
D1747	Standard Test Method for Refractive Index of Viscous Materials
D2503	Standard Test Method for Relative Molecular Mass (Molecular Weight) of Hydrocarbons by Thermoelectric Measurement of Vapor Pressure
D2710	Standard Test Method for Bromine Index of Petroleum Hydrocarbons by Electrometric Titration
D4052	Standard Test Method for Density, Relative Density, and API Gravity of Liquids by Digital Density Meter
D5002	Standard Test Method for Density and Relative Density of Crude Oils by Digital Density Analyzer
D7777	Standard Test Method for Density, Relative Density, or API Gravity of Liquid Petroleum by Portable Digital Density Meter
D566	Standard Test Method for Dropping Point of Lubricating Grease
D972	Standard Test Method for Evaporation Loss of Lubricating Greases and Oils
D1742	Standard Test Method for Oil Separation from Lubricating Grease during Storage
D2265	Standard Test Method for Dropping Point of Lubricating Grease over Wide Temperature Range
D2595	Standard Test Method for Evaporation Loss of Lubricating Greases over Wide-Temperature Range
D4425	Standard Test Method for Oil Separation from Lubricating Grease by Centrifuging (Koppers Method)
D6184	Standard Test Method for Oil Separation from Lubricating Grease (Conical Sieve Method)
D87	Standard Test Method for Melting Point of Petroleum Wax (Cooling Curve)
D127	Standard Test Method for Drop Melting Point of Petroleum Wax, Including Petrolatum
D612	Standard Test Method for Carbonizable Substances in Paraffin Wax
D721	Standard Test Method for Oil Content of Petroleum Waxes
D937	Standard Test Method for Cone Penetration of Petrolatum
D938	Standard Test Method for Congealing Point of Petroleum Waxes, Including Petrolatum
D1321	Standard Test Method for Needle Penetration of Petroleum Waxes
D1465	Standard Test Method for Blocking and Picking Points of Petroleum Wax
D1832	Standard Test Method for Peroxide Number of Petroleum Wax
D1833	Standard Test Method for Odor of Petroleum Wax
D2423	Standard Test Method for Surface Wax on Waxed Paper or Paperboard
D2534	Standard Test Method for Coefficient of Kinetic Friction for Wax Coatings
D2669	Standard Test Method for Apparent Viscosity of Petroleum Waxes Compounded with Additives (Hot Melts)
D3235	Standard Test Method for Solvent Extractables in Petroleum Waxes
D3236	Standard Test Method for Apparent Viscosity of Hot Melt Adhesives and Coating Materials
D3344	Standard Test Method for Total Wax Content of Corrugated Paperboard
D3521	Standard Test Method for Surface Wax Coating On Corrugated Board
D3708	Standard Test Method for Weight of Wax Applied during Curtain Coating Operation
D3944	Standard Test Method for Solidification Point of Petroleum Wax
D4419	Standard Test Method for Measurement of Transition Temperatures of Petroleum Waxes by Differential Scanning Calorimetry (DSC)
D6822	Standard Test Method for Density, Relative Density, and API Gravity of Crude Petroleum and Liquid Petroleum Products by Thermohydrometer Method

TABLE A16   Test Methods for the Determination of Instability and Contaminants in Liquid Fuels

ASTM No.	
D381	Standard Test Method for Gum Content in Fuels by Jet Evaporation
D525	Standard Test Method for Oxidation Stability of Gasoline (Induction Period Method)
D873	Standard Test Method for Oxidation Stability of Aviation Fuels (Potential Residue Method)
D2068	Standard Test Method for Determining Filter Blocking Tendency
D2274	Standard Test Method for Oxidation Stability of Distillate Fuel Oil (Accelerated Method)
D2709	Standard Test Method for Water and Sediment in Middle Distillate Fuels by Centrifuge
D4176	Standard Test Method for Free Water and Particulate Contamination in Distillate Fuels (Visual Inspection Procedures)
D4625	Standard Test Method for Distillate Fuel Storage Stability at 43°C (110°F)
D4740	Standard Test Method for Cleanliness and Compatibility of Residual Fuels by Spot Test
D4860	Standard Test Method for Free Water and Particulate Contamination in Middle Distillate Fuels (Clear and Bright Numerical Rating)
D4870	Standard Test Method for Determination of Total Sediment in Residual Fuels
D5304	Standard Test Method for Assessing Middle Distillate Fuel Storage Stability by Oxygen Overpressure
D5705	Standard Test Method for Measurement of Hydrogen Sulfide in the Vapor Phase above Residual Fuel Oils
D6021	Standard Test Method for Measurement of Total Hydrogen Sulfide in Residual Fuels by Multiple Headspace Extraction and Sulfur Specific Detection
D6217	Standard Test Method for Particulate Contamination in Middle Distillate Fuels by Laboratory Filtration
D6426	Standard Test Method for Determining Filterability of Middle Distillate Fuel Oils
D6468	Standard Test Method for High Temperature Stability of Middle Distillate Fuels
D6469	Standard Guide for Microbial Contamination in Fuels and Fuel Systems
D6560	Standard Test Method for Determination of Asphaltenes (Heptane Insolubles) in Crude Petroleum and Petroleum Products
D6748	Standard Test Method for Determination of Potential Instability of Middle Distillate Fuels Caused by the Presence of Phenalenes and Phenalenones (Rapid Method by Portable Spectrophotometer)
D6974	Standard Practice for Enumeration of Viable Bacteria and Fungi in Liquid Fuels—Filtration and Culture Procedures
D7060	Standard Test Method for Determination of the Maximum Flocculation Ratio and Peptizing Power in Residual and Heavy Fuel Oils (Optical Detection Method)
D7061	Standard Test Method for Measuring n-Heptane Induced Phase Separation of Asphaltene-Containing Heavy Fuel Oils as Separability Number by an Optical Scanning Device
D7112	Standard Test Method for Determining Stability and Compatibility of Heavy Fuel Oils and Crude Oils by Heavy Fuel Oil Stability Analyzer (Optical Detection)
D7157	Standard Test Method for Determination of Intrinsic Stability of Asphaltene-Containing Residues, Heavy Fuel Oils, and Crude Oils (n-Heptane Phase Separation; Optical Detection)
D7261	Standard Test Method for Determining Water Separation Characteristics of Diesel Fuels by Portable Separometer
D7321	Standard Test Method for Test Method for Particulate Contamination of Biodiesel B100 Blend Stock Biodiesel Esters and Biodiesel Blends by Laboratory Filtration
D7451	Standard Test Method for Water Separation Properties of Light and Middle Distillate, and Compression and Spark Ignition Fuels
D7462	Standard Test Method for Oxidation Stability of Biodiesel (B100) and Blends of Biodiesel with Middle Distillate Petroleum Fuel (Accelerated Method)
D7463	Standard Test Method for Adenosine Triphosphate (ATP) Content of Microorganisms in Fuel, Fuel/Water Mixtures and Fuel Associated Water
D7464	Standard Practice for Manual Sampling of Liquid Fuels, Associated Materials and Fuel System Components for Microbiological Testing
D7501	Standard Test Method for Determination of Fuel Filter Blocking Potential of Biodiesel (B100) Blend Stock by Cold Soak Filtration Test (CSFT)
D7525	Standard Test Method for Oxidation Stability of Spark Ignition Fuel—Rapid Small Scale Oxidation Test (RSSOT)
D7545	Standard Test Method for Oxidation Stability of Middle Distillate Fuels—Rapid Small Scale Oxidation Test (RSSOT)
D7548	Standard Test Method for Determination of Accelerated Iron Corrosion in Petroleum Products
D7577	Standard Test Method for Determining the Accelerated Iron Corrosion Rating of Denatured Fuel Ethanol and Ethanol Fuel Blends
D7619	Standard Test Method for Sizing and Counting Particles in Light and Middle Distillate Fuels, by Automatic Particle Counter
D7621	Standard Test Method for Determination of Hydrogen Sulfide in Fuel Oils by Rapid Liquid Phase Extraction
D7687	Standard Test Method for Measurement of Cellular Adenosine Triphosphate in Fuel, Fuel/Water Mixtures, and Fuel-Associated Water with Sample Concentration by Filtration
D7827	Standard Test Method for Measuring n-Heptane Induced Phase Separation of Asphaltene from Heavy Fuel Oils as Separability Number by an Optical Device
D7847	Standard Guide for Interlaboratory Studies for Microbiological Test Methods

TABLE A17   Test Methods for the Determination of the Volatility of Petroleum and Petroleum Products

ASTM No.	
D56	Standard Test Method for Flash Point by Tag Closed Cup Tester
D86	Standard Test Method for Distillation of Petroleum Products at Atmospheric Pressure
D92	Standard Test Method for Flash and Fire Points by Cleveland Open Cup Tester
D93	Standard Test Methods for Flash Point by Pensky-Martens Closed Cup Tester
D323	Standard Test Method for Vapor Pressure of Petroleum Products (Reid Method)
D1160	Standard Test Method for Distillation of Petroleum Products at Reduced Pressure
D2892	Standard Test Method for Distillation of Crude Petroleum (15-Theoretical Plate Column)
D3828	Standard Test Methods for Flash Point by Small Scale Closed Cup Tester
D4953	Standard Test Method for Vapor Pressure of Gasoline and Gasoline-Oxygenate Blends (Dry Method)
D5188	Standard Test Method for Vapor-Liquid Ratio Temperature Determination of Fuels (Evacuated Chamber and Piston Based Method)
D5191	Standard Test Method for Vapor Pressure of Petroleum Products (Mini Method)
D5236	Standard Test Method for Distillation of Heavy Hydrocarbon Mixtures (Vacuum Potstill Method)
D5482	Standard Test Method for Vapor Pressure of Petroleum Products (Mini Method—Atmospheric)
D6377	Standard Test Method for Determination of Vapor Pressure of Crude Oil: $VPCR_x$ (Expansion Method)
D6378	Standard Test Method for Determination of Vapor Pressure (VPX) of Petroleum Products, Hydrocarbons, and Hydrocarbon-Oxygenate Mixtures (Triple Expansion Method)
D6450	Standard Test Method for Flash Point by Continuously Closed Cup (CCCFP) Tester
D6849	Standard Practice for Storage and Use of Liquefied Petroleum Gases (LPG) in Sample Cylinders for LPG Test Methods
D6897	Standard Test Method for Vapor Pressure of Liquefied Petroleum Gases (LPG) (Expansion Method)
D7094	Standard Test Method for Flash Point by Modified Continuously Closed Cup (MCCCFP) Tester
D7236	Standard Test Method for Flash Point by Small Scale Closed Cup Tester (Ramp Method)
D7344	Standard Test Method for Distillation of Petroleum Products at Atmospheric Pressure (Mini Method)
D7345	Standard Test Method for Distillation of Petroleum Products at Atmospheric Pressure (Micro Distillation Method

TABLE A18   Test Methods for the Analysis of Gaseous (C_4) Hydrocarbons

ASTM No.	
D1025	Standard Test Method for Nonvolatile Residue of Polymerization Grade Butadiene
D1157	Standard Test Method for Total Inhibitor Content (TBC) of Light Hydrocarbons
D2384	Standard Test Methods for Traces of Volatile Chlorides in Butane-Butene Mixtures
D2426	Standard Test Method for Butadiene Dimer and Styrene in Butadiene Concentrates by Gas Chromatography
D2593	Standard Test Method for Butadiene Purity and Hydrocarbon Impurities by Gas Chromatography
D4423	Standard Test Method for Determination of Carbonyls In C4 Hydrocarbons
D4424	Standard Test Method for Butylene Analysis by Gas Chromatography
D5274	Standard Guide for Analysis of 1,3-Butadiene Product
D5799	Standard Test Method for Determination of Peroxides in Butadiene
D7423	Standard Test Method for Determination of Oxygenates in C2, C3, C4, and C5 Hydrocarbon Matrices by Gas Chromatography and Flame Ionization Detection

TABLE A19   Test Methods for the Analysis of Liquefied Petroleum Gas

ASTM No.	
D1265	Standard Practice for Sampling Liquefied Petroleum (LP) Gases, Manual Method
D1267	Standard Test Method for Gage Vapor Pressure of Liquefied Petroleum (LP) Gases (LP-Gas Method)
D1835	Standard Specification for Liquefied Petroleum (LP) Gases
D1837	Standard Test Method for Volatility of Liquefied Petroleum (LP) Gases
D1838	Standard Test Method for Copper Strip Corrosion by Liquefied Petroleum (LP) Gases
D2158	Standard Test Method for Residues in Liquefied Petroleum (LP) Gases
D2420	Standard Test Method for Hydrogen Sulfide in Liquefied Petroleum (LP) Gases (Lead Acetate Method)
D2421	Standard Practice for Interconversion of Analysis of C5 and Lighter Hydrocarbons to Gas-Volume, Liquid-Volume, or Mass Basis

(Continued)

TABLE A19 (*Continued*)

ASTM No.	
D2597	Standard Test Method for Analysis of Demethanized Hydrocarbon Liquid Mixtures Containing Nitrogen and Carbon Dioxide by Gas Chromatography
D2598	Standard Practice for Calculation of Certain Physical Properties of Liquefied Petroleum (LP) Gases from Compositional Analysis
D2713	Standard Test Method for Dryness of Propane (Valve Freeze Method)
D3700	Standard Practice for Obtaining LPG Samples Using a Floating Piston Cylinder
D5305	Standard Test Method for Determination of Ethyl Mercaptan in LP-Gas Vapor
D7756	Standard Test Method for Residues in Liquefied Petroleum (LP) Gases by Gas Chromatography with Liquid, On-Column Injection
D7828	Standard Test Method for Determination of Residue Composition in Liquefied Petroleum Gas (LPG) Using Automated Thermal Desorption/Gas Chromatography (ATD/GC)
D7901	Standard Specification for Dimethyl Ether for Fuel Purposes

TABLE A20 **Test Methods for the Analysis of Gasoline and Gasoline-Oxygenate Blends**

ASTM No.	
D4814	Standard Specification for Automotive Spark-Ignition Engine Fuel
D5500	Standard Test Method for Vehicle Evaluation of Unleaded Automotive Spark-Ignition Engine Fuel for Intake Valve Deposit Formation
D5598	Standard Test Method for Evaluating Unleaded Automotive Spark-Ignition Engine Fuel for Electronic Port Fuel Injector Fouling
D6201	Standard Test Method for Dynamometer Evaluation of Unleaded Spark-Ignition Engine Fuel for Intake Valve Deposit Formation
D6421	Standard Test Method for Evaluating Automotive Spark-Ignition Engine Fuel for Electronic Port Fuel Injector Fouling by Bench Procedure
Designation	Title
D7862	Standard Specification for Butanol for Blending with Gasoline for Use as Automotive Spark-Ignition Engine Fuel

TABLE A21 **Test Methods for the Analysis of Oxygenated Fuels**

ASTM No.	
D4806	Standard Specification for Denatured Fuel Ethanol for Blending with Gasolines for Use as Automotive Spark-Ignition Engine Fuel
D5797	Standard Specification for Fuel Methanol (M70-M85) for Automotive Spark-Ignition Engines
D5798	Standard Specification for Ethanol Fuel Blends for Flexible-Fuel Automotive Spark-Ignition Engines
D5983	Standard Specification for Methyl Tertiary-Butyl Ether (MTBE) for Downstream Blending for Use in Automotive Spark-Ignition Engine Fuel
D6423	Standard Test Method for Determination of pHe of Ethanol, Denatured Fuel Ethanol, and Fuel Ethanol (Ed75-Ed85)
D7794	Standard Practice for Blending Mid-Level Ethanol Fuel Blends for Flexible-Fuel Vehicles with Automotive Spark-Ignition Engines

TABLE A22 **Test Methods for the Analysis of Aviation Fuels**

ASTM No.	
D6812	Standard Practice for Ground-Based Octane Rating Procedures for Turbocharged/Supercharged Spark Ignition Aircraft Engines
D6986	Standard Test Method for Free Water, Particulate and Other Contamination in Aviation Fuels (Visual Inspection Procedures)
D7223	Standard Specification for Aviation Certification Turbine Fuel
D7618	Standard Specification for Ethyl Tertiary-Butyl Ether (ETBE) for Blending with Aviation Spark-Ignition Engine Fuel
D7719	Standard Specification for High-Octane Unleaded Fuel
D910	Standard Specification for Aviation Gasolines
D6227	Standard Specification for Unleaded Aviation Gasoline Containing a Non-Hydrocarbon Component
D7547	Standard Specification for Hydrocarbon Unleaded Aviation Gasoline
D7592	Standard Specification for Specification for Grade 94 Unleaded Aviation Gasoline Certification and Test Fuel
D7796	Standard Test Method for Analysis of Ethyl t-Butyl Ether (ETBE) by Gas Chromatography
D7826	Standard Guide for Evaluation of New Aviation Gasolines and New Aviation Gasoline Additives

TABLE A23   Test Methods for the Analysis of Jet Fuel

ASTM No.	
D1655	Standard Specification for Aviation Turbine Fuels
D6424	Standard Practice for Octane Rating Naturally Aspirated Spark Ignition Aircraft Engines
D6615	Standard Specification for Jet B Wide-Cut Aviation Turbine Fuel
D6824	Standard Test Method for Determining Filterability of Aviation Turbine Fuel
D7872	Standard Test Method for Determining the Concentration of Pipeline Drag Reducer Additive in Aviation Turbine Fuels

TABLE A24   Test Methods for the Analysis of Diesel, Non-Aviation Gas Turbine, and Marine Fuels

ASTM No.	
D187	Standard Test Method for Burning Quality of Kerosene
D396	Standard Specification for Fuel Oils
D975	Standard Specification for Diesel Fuel Oils
D976	Standard Test Method for Calculated Cetane Index of Distillate Fuels
D2156	Standard Test Method for Smoke Density in Flue Gases from Burning Distillate Fuels
D2157	Standard Test Method for Effect of Air Supply on Smoke Density in Flue Gases from Burning Distillate Fuels
D2880	Standard Specification for Gas Turbine Fuel Oils
D3699	Standard Specification for Kerosene
D4418	Standard Practice for Receipt, Storage, and Handling of Fuels for Gas Turbines
D4737	Standard Test Method for Calculated Cetane Index by Four Variable Equation
D6078	Standard Test Method for Evaluating Lubricity of Diesel Fuels by the Scuffing Load Ball-on-Cylinder Lubricity Evaluator (SLBOCLE)
D6079	Standard Test Method for Evaluating Lubricity of Diesel Fuels by the High-Frequency Reciprocating Rig (HFRR)
D6751	Standard Specification for Biodiesel Fuel Blend Stock (B100) for Middle Distillate Fuels
D6898	Standard Test Method for Evaluating Diesel Fuel Lubricity by an Injection Pump Rig
D7467	Standard Specification for Diesel Fuel Oil, Biodiesel Blend (B6 to B20)
D7544	Standard Specification for Pyrolysis Liquid Biofuel
D7688	Standard Test Method for Evaluating Lubricity of Diesel Fuels by the High-Frequency Reciprocating Rig (HFRR) by Visual Observation

TABLE A25   Test Methods for the Analysis of Lubricants

ASTM No.	
D91	Standard Test Method for Precipitation Number of Lubricating Oils
D94	Standard Test Methods for Saponification Number of Petroleum Products
D189	Standard Test Method for Conradson Carbon Residue of Petroleum Products
D322	Standard Test Method for Gasoline Diluent in Used Gasoline Engine Oils by Distillation
D483	Standard Test Method for Unsulfonated Residue of Petroleum Plant Spray Oils
D524	Standard Test Method for Ramsbottom Carbon Residue of Petroleum Products
D565	Standard Test Method for Carbonizable Substances in White Mineral Oil
D664	Standard Test Method for Acid Number of Petroleum Products by Potentiometric Titration
D892	Standard Test Method for Foaming Characteristics of Lubricating Oils
D893	Standard Test Method for Insolubles in Used Lubricating Oils
D974	Standard Test Method for Acid and Base Number by Color-Indicator Titration
D1093	Standard Test Method for Acidity of Hydrocarbon Liquids and Their Distillation Residues
D2273	Standard Test Method for Trace Sediment in Lubricating Oils
D2896	Standard Test Method for Base Number of Petroleum Products by Potentiometric Perchloric Acid Titration
D2982	Standard Test Methods for Detecting Glycol-Base Antifreeze in Used Lubricating Oils
D3242	Standard Test Method for Acidity in Aviation Turbine Fuel
D3339	Standard Test Method for Acid Number of Petroleum Products by Semi-Micro Color Indicator Titration
D3607	Standard Test Method for Removing Volatile Contaminants from Used Engine Oils by Stripping
D4055	Standard Test Method for Pentane Insolubles by Membrane Filtration
D4530	Standard Test Method for Determination of Carbon Residue (Micro Method)
D4739	Standard Test Method for Base Number Determination by Potentiometric Hydrochloric Acid Titration

(Continued)

TABLE A25   (*Continued*)

ASTM No.	
D5770	Standard Test Method for Semiquantitative Micro Determination of Acid Number of Lubricating Oils during Oxidation Testing
D5800	Standard Test Method for Evaporation Loss of Lubricating Oils by the Noack Method
D5984	Standard Test Method for Semi-Quantitative Field Test Method for Base Number in New and Used Lubricants by Color-Indicator Titration
D6082	Standard Test Method for High Temperature Foaming Characteristics of Lubricating Oils
D6375	Standard Test Method for Evaporation Loss of Lubricating Oils by Thermogravimetric Analyzer (TGA) Noack Method
D7317	Standard Test Method for Coagulated Pentane Insolubles in Used Lubricating Oils by Paper Filtration (LMOA Method)
D7579	Standard Test Method for Pyrolysis Solids Content in Pyrolysis Liquids by Filtration of Solids in Methanol
D7795	Standard Test Method for Acidity in Ethanol and Ethanol Blends by Titration
D2510	Standard Test Method for Adhesion of Solid Film Lubricants
D2715	Standard Test Method for Volatilization Rates of Lubricants in Vacuum
D2779	Standard Test Method for Estimation of Solubility of Gases in Petroleum Liquids
D2884	Standard Test Method for Yield Stress of Heterogeneous Propellants by Cone Penetration Method
D3342	Standard Test Method for Dispersion Stability of New (Unused) Rolling Oil Dispersions in Water
D429	Standard Test Method for Solubility of Fixed Gases in Low-Boiling Liquids
D3523	Standard Test Method for Spontaneous Heating Values of Liquids and Solids (Differential Mackey Test)
D3704	Standard Test Method for Wear Preventive Properties of Lubricating Greases Using the (Falex) Block on Ring Test Machine in Oscillating Motion
D3750	Standard Test Method for Misting Properties of Lubricating Fluids
D3828	Standard Test Method for Estimation of Solubility of Gases in Petroleum and Other Organic Liquids
D6425	Standard Test Method for Measuring Friction and Wear Properties of Extreme Pressure (EP) Lubricating Oils Using SRV Test Machine
D6374	Standard Test Method for Thermal Stability of Organic Heat Transfer Fluids
D2717	Standard Test Method for Determining Extreme Pressure Properties of Solid Bonded Films Using a High-Frequency, Linear-Oscillation (SRV) Test Machine
D721	Standard Test Method for Determining Extreme Pressure Properties of Lubricating Oils Using High-Frequency, Linear-Oscillation (SRV) Test Machine
D7751	Standard Practice for Determining the Wear Volume on Standard Test Pieces Used by High-Frequency, Linear-Oscillation (SRV) Test Machine
D2511	Standard Test Method for Thermal Shock Sensitivity of Solid Film Lubricants
D2625	Standard Test Method for Endurance (Wear) Life and Load-Carrying Capacity of Solid Film Lubricants (Falex Pin and Vee Method)
D2649	Standard Test Method for Corrosion Characteristics of Solid Film Lubricants
D2710	Standard Test Method for Calibration and Operation of the Falex Block-on-Ring Friction and Wear Testing Machine
D291	Standard Test Method for Wear Life of Solid Film Lubricants in Oscillating Motion
D7320	Standard Test Method for Wear Rate and Coefficient of Friction of Materials in Self-Lubricated Rubbing Contact Using a Thrust Washer Testing Machine
D6557	Standard Test Method for Evaluation of Rust Preventive Characteristics of Automotive Engine Oils
D6593	Standard Test Method for Evaluation of Automotive Engine Oils for Inhibition of Deposit Formation in a Spark-Ignition Internal Combustion Engine Fueled with Gasoline and Operated Under Low-Temperature, Light-Duty Conditions
D2070	Standard Test Method for Thermal Stability of Hydraulic Oils
D219	Standard Test Method for Hydrolytic Stability of Hydraulic Fluids (Beverage Bottle Method)
D155	Standard Practice for Evaluating Compatibility of Mixtures of Turbine Lubricating Oils
D7483	Standard Test Method for Measurement of Lubricant Generated Insoluble Color Bodies in In-Service Turbine Oils Using Membrane Patch Colorimetry
D4981	Standard Test Method for Evaluating Wear Characteristics of Tractor Hydraulic Fluids
D5799	Standard Test Method for Evaluating the Thermal Stability of Manual Transmission Lubricants in a Cyclic Durability Test
D5622	Standard Test Method for Determining Automotive Gear Oil Compatibility with Typical Oil Seal Elastomers
D5705	Standard Test Method for Evaluation of the Thermal and Oxidative Stability of Lubricating Oils Used for Manual Transmissions and Final Drive Axles
D5706	Standard Specification for Performance of Manual Transmission Gear Lubricants
D612	Standard Test Method for Evaluation of Load-Carrying Capacity of Lubricants Under Conditions of Low Speed and High Torque Used for Final Hypoid Drive Axles

(*Continued*)

TABLE A25 *(Continued)*

ASTM No.	
D7451	Standard Test Method for Evaluation of the Load Carrying Properties of Lubricants Used for Final Drive Axles, Under Conditions of High Speed and Shock Loading
D4950	Standard Classification and Specification for Automotive Service Greases
D4485	Standard Specification for Performance of Active API Service Category Engine Oils
D6794	Standard Test Method for Measuring the Effect on Filterability of Engine Oils After Treatment with Various Amounts of Water and a Long Heating Time
D6795	Standard Test Method for Measuring the Effect on Filterability of Engine Oils After Treatment with Water and Dry Ice and a Short (30 min) Heating Time
D6891	Standard Test Method for Evaluation of Automotive Engine Oils in the Sequence IVA Spark-Ignition Engine
D6922	Standard Test Method for Determination of Homogeneity and Miscibility in Automotive Engine Oils
D6923	Standard Test Method for Evaluation of Engine Oils in a High Speed, Single-Cylinder Diesel Engine—Caterpillar 1R Test Procedure
D6975	Standard Test Method for Cummins M11 EGR Test
D6987	Standard Test Method for Evaluation of Diesel Engine Oils in T-10 Exhaust Gas Recirculation Diesel Engine
D7038	Standard Test Method for Evaluation of Moisture Corrosion Resistance of Automotive Gear Lubricants
D7156	Standard Test Method for Evaluation of Diesel Engine Oils in the T-11 Exhaust Gas Recirculation Diesel Engine
D7422	Standard Test Method for Evaluation of Diesel Engine Oils in T-12 Exhaust Gas Recirculation Diesel Engine
D7450	Standard Specification for Performance of Rear Axle Gear Lubricants Intended for API Category GL-5 Service
D7468	Standard Test Method for Cummins ISM Test
D7484	Standard Test Method for Evaluation of Automotive Engine Oils for Valve-Train Wear Performance in Cummins ISB Medium-Duty Diesel Engine
D7549	Standard Test Method for Evaluation of Heavy-Duty Engine Oils under High Output Conditions—Caterpillar C13 Test Procedure
D217	Standard Test Methods for Cone Penetration of Lubricating Grease
D1092	Standard Test Method for Measuring Apparent Viscosity of Lubricating Greases
D1403	Standard Test Methods for Cone Penetration of Lubricating Grease Using One-Quarter and One-Half Scale Cone Equipment
D1831	Standard Test Method for Roll Stability of Lubricating Grease
D128	Standard Test Methods for Analysis of Lubricating Grease
D1404	Standard Test Method for Estimation of Deleterious Particles in Lubricating Grease
D4048	Standard Test Method for Detection of Copper Corrosion from Lubricating Grease
D4289	Standard Test Method for Elastomer Compatibility of Lubricating Greases and Fluids
D6185	Standard Practice for Evaluating Compatibility of Binary Mixtures of Lubricating Greases
Designation	Title
D1264	Standard Test Method for Determining the Water Washout Characteristics of Lubricating Greases
D1743	Standard Test Method for Determining Corrosion Preventive Properties of Lubricating Greases
D4049	Standard Test Method for Determining the Resistance of Lubricating Grease to Water Spray
D5969	Standard Test Method for Corrosion-Preventive Properties of Lubricating Greases in Presence of Dilute Synthetic Sea Water Environments
D6138	Standard Test Method for Determination of Corrosion-Preventive Properties of Lubricating Greases under Dynamic Wet Conditions (Emcor Test)
D5966	Standard Test Method for Evaluation of Engine Oils for Roller Follower Wear in Light-Duty Diesel Engine
D5967	Standard Test Method for Evaluation of Diesel Engine Oils in T-8 Diesel Engine
D5968	Standard Test Method for Evaluation of Corrosiveness of Diesel Engine Oil at 121°C
D6594	Standard Test Method for Evaluation of Corrosiveness of Diesel Engine Oil at 135°C
D6618	Standard Test Method for Evaluation of Engine Oils in Diesel Four-Stroke Cycle Supercharged 1 M-PC Single Cylinder Oil Test Engine
D6681	Standard Test Method for Evaluation of Engine Oils in a High Speed, Single-Cylinder Diesel Engine—Caterpillar 1P Test Procedure
D6750	Standard Test Methods for Evaluation of Engine Oils in a High-Speed, Single-Cylinder Diesel Engine—1 K Procedure (0.4% Fuel Sulfur) and 1 N Procedure (0.04?% Fuel Sulfur)
D6838	Standard Test Method for Cummins M11 High Soot Test
D6894	Standard Test Method for Evaluation of Aeration Resistance of Engine Oils in Direct-Injected Turbocharged Automotive Diesel Engine
D2155	Standard Test Method for Determination of Fire Resistance of Aircraft Hydraulic Fluids by Autoignition Temperature
D6046	Standard Classification of Hydraulic Fluids for Environmental Impact

(Continued)

TABLE A25 (*Continued*)

ASTM No.	
D6158	Standard Specification for Mineral Hydraulic Oils
D6351	Standard Test Method for Determination of Low Temperature Fluidity and Appearance of Hydraulic Fluids
D6547	Standard Test Method for Corrosiveness of Lubricating Fluid to Bimetallic Couple
D7044	Standard Specification for Biodegradable Fire Resistant Hydraulic Fluids
D7721	Standard Practice for Determining the Effect of Fluid Selection on Hydraulic System or Component Efficiency
D7752	Standard Practice for Evaluating Compatibility of Mixtures of Hydraulic Fluids
D4290	Standard Test Method for Determining the Leakage Tendencies of Automotive Wheel Bearing Grease under Accelerated Conditions
D4693	Standard Test Method for Low-Temperature Torque of Grease-Lubricated Wheel Bearings
D5706	Standard Test Method for Determining Extreme Pressure Properties of Lubricating Greases Using a High-Frequency, Linear-Oscillation (SRV) Test Machine
D5707	Standard Test Method for Measuring Friction and Wear Properties of Lubricating Grease Using a High-Frequency, Linear-Oscillation (SRV) Test Machine
D720	Standard Test Method for Determining Tribomechanical Properties of Grease Lubricated Plastic Socket Suspension Joints Using a High-Frequency, Linear-Oscillation (SRV) Test Machine
D7594	Standard Test Method for Determining Fretting Wear Resistance of Lubricating Greases under High Hertzian Contact Pressures Using a High-Frequency, Linear-Oscillation (SRV) Test Machine
D7718	Standard Practice for Obtaining In-Service Samples of Lubricating Grease
D6813	Standard Guide for Performance Evaluation of Hydraulic Fluids for Piston Pumps
D6973	Standard Test Method for Indicating Wear Characteristics of Petroleum Hydraulic Fluids in a High Pressure Constant Volume Vane Pump
D7043	Standard Test Method for Indicating Wear Characteristics of Non-Petroleum and Petroleum Hydraulic Fluids in a Constant Volume Vane Pump
D6304	Standard Test Method for Determination of Water in Petroleum Products, Lubricating Oils, and Additives by Coulometric Karl Fischer Titration

TABLE A26 Test Methods for the Environmental Analysis of Lubricants

ASTM No.	
D5864	Standard Test Method for Determining Aerobic Aquatic Biodegradation of Lubricants or Their Components
D6006	Standard Guide for Assessing Biodegradability of Hydraulic Fluids
D6081	Standard Practice for Aquatic Toxicity Testing of Lubricants: Sample Preparation and Results Interpretation
D6139	Standard Test Method for Determining the Aerobic Aquatic Biodegradation of Lubricants or Their Components Using the Gledhill Shake Flask
D6384	Standard Terminology Relating to Biodegradability and Ecotoxicity of Lubricants
D6731	Standard Test Method for Determining the Aerobic, Aquatic Biodegradability of Lubricants or Lubricant Components in a Closed Respirometer
D7373	Standard Test Method for Predicting Biodegradability of Lubricants Using a Bio-Kinetic Model

TABLE A27 Test Methods for the Analysis of the Oxidation of Grease and Lubricants

ASTM No.	
D943	Standard Test Method for Oxidation Characteristics of Inhibited Mineral Oils
D2272	Standard Test Method for Oxidation Stability of Steam Turbine Oils by Rotating Pressure Vessel
D4310	Standard Test Method for Determination of Sludging and Corrosion Tendencies of Inhibited Mineral Oils
D6514	Standard Test Method for High Temperature Universal Oxidation Test for Turbine Oils
D6810	Standard Test Method for Measurement of Hindered Phenolic Antioxidant Content in Non-Zinc Turbine Oils by Linear Sweep Voltammetry
D6971	Standard Test Method for Measurement of Hindered Phenolic and Aromatic Amine Antioxidant Content in Non-Zinc Turbine Oils by Linear Sweep Voltammetry
D7590	Standard Guide for Measurement of Remaining Primary Antioxidant Content In In-Service Industrial Lubricating Oils by Linear Sweep Voltammetry

(*Continued*)

TABLE A27 *(Continued)*

ASTM No.	
D7873	Standard Test Method for Determination of Oxidation Stability and Insolubles Formation of Inhibited Turbine Oils at 120?°C Without the Inclusion of Water (Dry TOST Method)
D4742	Standard Test Method for Oxidation Stability of Gasoline Automotive Engine Oils by Thin-Film Oxygen Uptake (TFOUT)
D6335	Standard Test Method for Determination of High Temperature Deposits by Thermo-Oxidation Engine Oil Simulation Test
D7097	Standard Test Method for Determination of Moderately High Temperature Piston Deposits by Thermo-Oxidation Engine Oil Simulation Test-TEOST MHT
D7098	Standard Test Method for Oxidation Stability of Lubricants by Thin-Film Oxygen Uptake (TFOUT) Catalyst B
D942	Standard Test Method for Oxidation Stability of Lubricating Greases by the Oxygen Pressure Vessel Method
D5483	Standard Test Method for Oxidation Induction Time of Lubricating Greases by Pressure Differential Scanning Calorimetry
D4871	Standard Guide for Universal Oxidation/Thermal Stability Test Apparatus
D5763	Standard Test Method for Oxidation and Thermal Stability Characteristics of Gear Oils Using Universal Glassware
D2893	Standard Test Method for Oxidation Characteristics of Extreme-Pressure Lubrication Oils
D4636	Standard Test Method for Corrosiveness and Oxidation Stability of Hydraulic Oils, Aircraft Turbine Engine Lubricants, and Other Highly Refined Oils
D5846	Standard Test Method for Universal Oxidation Test for Hydraulic and Turbine Oils Using the Universal Oxidation Test Apparatus
D6186	Standard Test Method for Oxidation Induction Time of Lubricating Oils by Pressure Differential Scanning Calorimetry (PDSC)

TABLE A28 Test Methods for the Analysis of Petroleum Coke, Carbon, and Graphite

ASTM No.	
D7454	Standard Test Method for Determination of Vibrated Bulk Density of Calcined Petroleum Coke using a Semi-Automated Apparatus
C559	Standard Test Method for Bulk Density by Physical Measurements of Manufactured Carbon and Graphite Articles
C560	Standard Test Methods for Chemical Analysis of Graphite
C561	Standard Test Method for Ash in a Graphite Sample
C562	Standard Test Method for Moisture in a Graphite Sample
C565	Standard Test Methods for Tension Testing of Carbon and Graphite Mechanical Materials
C611	Standard Test Method for Electrical Resistivity of Manufactured Carbon and Graphite Articles at Room Temperature
C625	Standard Practice for Reporting Irradiation Results on Graphite
C651	Standard Test Method for Flexural Strength of Manufactured Carbon and Graphite Articles Using Four-Point Loading at Room Temperature
C662	Standard Specification for Impervious Graphite Pipe and Threading
C695	Standard Test Method for Compressive Strength of Carbon and Graphite
C709	Standard Terminology Relating to Manufactured Carbon and Graphite
C714	Standard Test Method for Thermal Diffusivity of Carbon and Graphite by Thermal Pulse Method
C747	Standard Test Method for Moduli of Elasticity and Fundamental Frequencies of Carbon and Graphite Materials by Sonic Resonance
C748	Standard Test Method for Rockwell Hardness of Graphite Materials
C749	Standard Test Method for Tensile Stress-Strain of Carbon and Graphite
C769	Standard Test Method for Sonic Velocity in Manufactured Carbon and Graphite Materials for Use in Obtaining Young's Modulus
C781	Standard Practice for Testing Graphite and Boronated Graphite Materials for High-Temperature Gas-Cooled Nuclear Reactor Components
C783	Standard Practice for Core Sampling of Graphite Electrodes
C808	Standard Guideline for Reporting Friction and Wear Test Results of Manufactured Carbon and Graphite Bearing and Seal Materials
C816	Standard Test Method for Sulfur in Graphite by Combustion-Iodometric Titration Method
C838	Standard Test Method for Bulk Density of As-Manufactured Carbon and Graphite Shapes
C886	Standard Test Method for Scleroscope Hardness Testing of Carbon and Graphite Materials
C1025	Standard Test Method for Modulus of Rupture in Bending of Electrode Graphite
C1039	Standard Test Methods for Apparent Porosity, Apparent Specific Gravity, and Bulk Density of Graphite Electrodes
C1179	Standard Test Method for Oxidation Mass Loss of Manufactured Carbon and Graphite Materials in Air

(Continued)

TABLE A28   (*Continued*)

ASTM No.	
D7219	Standard Specification for Isotropic and Near-Isotropic Nuclear Graphites
D7301	Standard Specification for Nuclear Graphite Suitable for Components Subjected to Low Neutron Irradiation Dose
D7542	Standard Test Method for Air Oxidation of Carbon and Graphite in the Kinetic Regime
D7775	Standard Guide for Measurements on Small Graphite Specimens
D7779	Standard Test Method for Determination of Fracture Toughness of Graphite at Ambient Temperature
D7846	Standard Practice for Reporting Uniaxial Strength Data and Estimating Weibull Distribution Parameters for Advanced Graphites

TABLE A29   Test Methods for the Physical Properties of Fuels, Petroleum Coke, and Carbonaceous Materials (Tar and Pitch)

ASTM No.	
D61	Standard Test Method for Softening Point of Pitches (Cube-in-Water Method)
D71	Standard Test Method for Relative Density of Solid Pitch and Asphalt (Displacement Method)
D130	Standard Test Method for Corrosiveness to Copper from Petroleum Products by Copper Strip Test
D156	Standard Test Method for Saybolt Color of Petroleum Products (Saybolt Chromometer Method)
D240	Standard Test Method for Heat of Combustion of Liquid Hydrocarbon Fuels by Bomb Calorimeter
D1405	Standard Test Method for Estimation of Net Heat of Combustion of Aviation Fuels
D1500	Standard Test Method for ASTM Color of Petroleum Products (ASTM Color Scale)
D2318	Standard Test Method for Quinoline-Insoluble (QI) Content of Tar and Pitch
D2319	Standard Test Method for Softening Point of Pitch (Cube-in-Air Method)
D2320	Standard Test Method for Density (Relative Density) of Solid Pitch (Pycnometer Method)
D2392	Standard Test Method for Color of Dyed Aviation Gasolines
D2415	Standard Test Method for Ash in Coal Tar and Pitch
D2416	Standard Test Method for Coking Value of Tar and Pitch (Modified Conradson)
D2638	Standard Test Method for Real Density of Calcined Petroleum Coke by Helium Pycnometer
D2764	Standard Test Method for Dimethylformamide-Insoluble (DMF-I) Content of Tar and Pitch
D3104	Standard Test Method for Softening Point of Pitches (Mettler Softening Point Method)
D3338	Standard Test Method for Estimation of Net Heat of Combustion of Aviation Fuels
D3461	Standard Test Method for Softening Point of Asphalt and Pitch (Mettler Cup-and-Ball Method)
D3703	Standard Test Method for Hydroperoxide Number of Aviation Turbine Fuels, Gasoline and Diesel Fuels
D4072	Standard Test Method for Toluene-Insoluble (TI) Content of Tar and Pitch
D4292	Standard Test Method for Determination of Vibrated Bulk Density of Calcined Petroleum Coke
D4296	Standard Practice for Sampling Pitch
D4312	Standard Test Method for Toluene-Insoluble (TI) Content of Tar and Pitch (Short Method)
D4422	Standard Test Method for Ash in Analysis of Petroleum Coke
D4529	Standard Test Method for Estimation of Net Heat of Combustion of Aviation Fuels
D4616	Standard Test Method for Microscopical Analysis by Reflected Light and Determination of Mesophase in a Pitch
D4715	Standard Test Method for Coking Value of Tar and Pitch (Alcan)
D4746	Standard Test Method for Determination of Quinoline Insolubles (QI) in Tar and Pitch by Pressure Filtration
D4809	Standard Test Method for Heat of Combustion of Liquid Hydrocarbon Fuels by Bomb Calorimeter (Precision Method)
D4868	Standard Test Method for Estimation of Net and Gross Heat of Combustion of Burner and Diesel Fuels
D4892	Standard Test Method for Density of Solid Pitch (Helium Pycnometer Method)
D4893	Standard Test Method for Determination of Pitch Volatility
D4930	Standard Test Method for Dust Control Material on Calcined Petroleum Coke
D4931	Standard Test Method for Gross Moisture in Green Petroleum Coke
D5003	Standard Test Method for Hardgrove Grindability Index (HGI) of Petroleum Coke
D5004	Standard Test Method for Real Density of Calcined Petroleum Coke by Xylene Displacement
D5018	Standard Test Method for Shear Viscosity of Coal-Tar and Petroleum Pitches
D5187	Standard Test Method for Determination of Crystallite Size (Lc of Calcined Petroleum Coke by X-Ray Diffraction
D5502	Standard Test Method for Apparent Density by Physical Measurements of Manufactured Anode and Cathode Carbon Used by the Aluminum Industry
D5709	Standard Test Method for Sieve Analysis of Petroleum Coke
D6045	Standard Test Method for Color of Petroleum Products by the Automatic Tristimulus Method
D6120	Standard Test Method for Electrical Resistivity of Anode and Cathode Carbon Material at Room Temperature

(Continued)

TABLE A29 (*Continued*)

ASTM No.	
D6258	Standard Test Method for Determination of Solvent Red 164 Dye Concentration in Diesel Fuels
D6353	Standard Guide for Sampling Plan and Core Sampling for Prebaked Anodes Used in Aluminum Production
D6354	Standard Guide for Sampling Plan and Core Sampling of Carbon Cathode Blocks Used in Aluminum Production
D6374	Standard Test Method for Volatile Matter in Green Petroleum Coke Quartz Crucible Procedure
D6376	Standard Test Method for Determination of Trace Metals in Petroleum Coke by Wavelength Dispersive X-Ray Fluorescence Spectroscopy
D6447	Standard Test Method for Hydroperoxide Number of Aviation Turbine Fuels by Voltammetric Analysis
D6558	Standard Test Method for Determination of TGA CO2 Reactivity of Baked Carbon Anodes and Cathode Blocks
D6559	Standard Test Method for Determination of Thermogravimetric (TGA) Air Reactivity of Baked Carbon Anodes and Cathode Blocks
D6744	Standard Test Method for Determination of the Thermal Conductivity of Anode Carbons by the Guarded Heat Flow Meter Technique
D6745	Standard Test Method for Linear Thermal Expansion of Electrode Carbons
D6756	Standard Test Method for Determination of the Red Dye Concentration and Estimation of the ASTM Color of Diesel Fuel and Heating Oil Using a Portable Visible Spectrophotometer
D6791	Standard Test Method for Determination of Grain Stability of Calcined Petroleum Coke
D6969	Standard Practice for Preparation of Calcined Petroleum Coke Samples for Analysis
D6970	Standard Practice for Collection of Calcined Petroleum Coke Samples for Analysis
D7058	Standard Test Method for Determination of the Red Dye Concentration and Estimation of Saybolt Color of Aviation Turbine Fuels and Kerosine Using a Portable Visible Spectrophotometer
D7095	Standard Test Method for Rapid Determination of Corrosiveness to Copper from Petroleum Products Using a Disposable Copper Foil Strip
D7280	Standard Test Method for Quinoline-Insoluble (QI) Content of Tar and Pitch by Stainless Steel Crucible Filtration

CONVERSION FACTORS

1. Concentration Conversions

1 part per million (1 ppm) = 1 microgram per liter (1 μg/l)

1 microgram per liter (1 μg/l) = 1 milligram per kilogram (1 mg/kg)

1 microgram per liter (μg/l) × 6.243 × 10^8 = 1 lb per cubic foot (1 lb/ft^3)

1 microgram per liter (1 μg/l) × 10^{-3} = 1 milligram per liter (1 mg/l)

1 milligram per liter (1 mg/l) × 6.243 × 10^5 = 1 pound per cubic foot (1 lb/ft^3)

I gram mole per cubic meter (1 g mol/m^3) × 6.243 × 10^5 = 1 pound per cubic foot (1 lb/ft^3)

10,000 ppm = 1% w/w

2. Weight Conversion

1 ounce (1 oz) = 28.3495 grams (18.2495 g)

1 pound (1 lb) = 0.454 kg

1 pound (1 lb) = 454 grams (454 g)

1 kilogram (1 kg) = 2.20462 pounds (2.20462 lb)

1 stone (English, 1 st) = 14 pounds (14 lb)

1 ton (United States; 1 short ton) = 2000 lbs

1 ton (English; 1 long ton) = 2240 lbs

1 metric ton = 2204.62262 pounds

1 tonne = 2204.62262 pounds

3. Temperature Conversions

°F = (°C × 1.8) + 32

°C = (°F − 32)/1.8

(°F−32) × 0.555 = °C

Absolute zero = −273.15°C

Absolute zero = −459.67°F

4. Area

1 square centimeter (1 cm^2) = 0.1550 square inches

1 square meter 1 (m^2) = 1.1960 square yards

1 hectare = 2.4711 acres

1 square kilometer (1 km^2) = 0.3861 square miles

1 square inch (1 inch^2) = 6.4516 square centimeters

1 square foot (1 ft^2) = 0.0929 square meters

1 square yard (1 yd^2) = 0.8361 square meters

1 acre = 4046.9 square meters

1 square mile (1 mi^2) = 2.59 square kilometers

5. Other Approximations

14.7 pounds per square inch (14.7 psi) = 1 atmosphere (1 atmos)

1 kiloPascal (kPa) × 9.8692 × 10^{-3} = 14.7 pounds per square inch (14.7 psi)

1 yd^3 = 27 ft^3

1 US gallon of water = 8.34 lbs

1 imperial gallon of water = 10 lbs

1 yd^3 = 0.765 m^3

Handbook of Petroleum Product Analysis, Second Edition. James G. Speight.
© 2015 John Wiley & Sons, Inc. Published 2015 by John Wiley & Sons, Inc.

GLOSSARY

ABN separation a method of fractionation by which petroleum is separated into acidic, basic, and neutral constituents.

Accuracy a measure of how close the test result will be to the true value of the property being measured; a relative term in the sense that systematic errors or biases can exist but be small enough to be inconsequential.

Acid catalyst a catalyst having acidic character; alumina is an example of such a catalyst.

Acid deposition acid rain; a form of pollution depletion in which pollutants, such as nitrogen oxides and sulfur oxides, are transferred from the atmosphere to soil or water; often referred to as atmospheric self-cleaning. The pollutants usually arise from the use of fossil fuels.

Acid number a measure of the reactivity of petroleum with a caustic solution and given in terms of milligrams of potassium hydroxide that are neutralized by 1 g of petroleum. See also: Total acid number.

Acid rain the precipitation phenomenon that incorporates anthropogenic acids and other acidic chemicals from the atmosphere to the land and water (see Acid deposition).

Acid sludge the residue left after treating petroleum oil with sulfuric acid for the removal of impurities; a black, viscous substance containing the spent acid and impurities.

Acid treating a process in which unfinished petroleum products, such as gasoline, kerosene, and lubricating-oil stocks, are contacted with sulfuric acid to improve their color, odor, and other properties.

Acidity the capacity of an acid to neutralize a base such as a hydroxyl ion (OH^-).

Acidophiles metabolically active in highly acidic environments and often have a high heavy metal resistance.

Acyclic a compound with straight or branched carbon–carbon linkages but without cyclic (ring) structures.

Additive a substance added to petroleum products (such as lubricating oils) to impart new or to improve existing characteristics.

Adhesion the degree to which oil will coat a surface, expressed as the mass of oil adhering per unit area. A test has been developed for a standard surface that gives a semiquantitative measure of this property.

Adsorption transfer of a substance from a solution to the surface of a solid resulting in relatively high concentration of the substance at the place of contact; see also Chromatographic adsorption.

Aerobic bacteria any bacteria requiring free oxygen for growth and cell division.

Handbook of Petroleum Product Analysis, Second Edition. James G. Speight.
© 2015 John Wiley & Sons, Inc. Published 2015 by John Wiley & Sons, Inc.

Aerobic respiration the process whereby microorganisms use oxygen as an electron acceptor.

Air pollution the discharge of toxic gases and particulate matter introduced into the atmosphere, principally as a result of human activity.

Air toxics hazardous air pollutants.

Alcohol the chemical family name of a group of organic chemical compounds composed of carbon, hydrogen, and oxygen; the series of molecules vary in chain length and are composed of a hydrocarbon plus a hydroxyl group: $CH_3(CH_2)_nOH$ (e.g., methanol, ethanol, and tertiary butyl alcohol).

Alicyclic hydrocarbon a hydrocarbon that has a cyclic structure (e.g., cyclohexane); also collectively called naphthenes.

Aliphatic compound Any organic compound of hydrogen and carbon characterized by a linear chain or branched chain of carbon atoms; three subgroups of such compounds are alkanes, alkenes, and alkynes.

Aliphatic hydrocarbon a hydrocarbon in which the carbon–hydrogen groupings are arranged in open chains that may be branched. The term includes *paraffins* and *olefins* and provides a distinction from *aromatics* and *naphthenes*, which have at least some of their carbon atoms arranged in closed chains or rings.

Aliquot that quantity of material of proper size for measurement of the property of interest; test portions may be taken from the gross sample directly, but often preliminary operations such as mixing or further reduction in particle size are necessary.

Alkalinity the capacity of a base to neutralize the hydrogen ion (H^+).

Alkaliphiles organisms that have their optimum growth rate at least 2 pH units above neutrality.

Alkali-tolerants (Alkalitolerants) organisms that are able to grow or survive at pH values above 9, but their optimum growth rate is approximately at neutrality (pH = 7) or less.

Alkane (paraffin) a group of *hydrocarbons* composed of only carbon and hydrogen with no double bonds or aromaticity. They are said to be "saturated" with hydrogen. They may by straight chain (normal), branched, or cyclic. The smallest alkane is methane (CH_4), the next, ethane (CH_3CH_3), then propane ($CH_3CH_2CH_3$), and so on.

Alkene (olefin) an unsaturated *hydrocarbon*, containing only hydrogen and carbon with one or more double bonds, but having no aromaticity. *Alkenes* are not typically found in crude oils, but can occur as a result of heating.

Alkylate the product of an alkylation (q.v.) process.

Alkylation in the petroleum industry, a process by which an olefin (e.g., ethylene) is combined with a branched-chain hydrocarbon (e.g., *iso*-butane); alkylation may be accomplished as a thermal or as a catalytic reaction.

Alkyl groups a group of carbon and hydrogen atoms that branch from the main carbon chain or ring in a hydrocarbon molecule; a *hydrocarbon* functional group (C_nH_{2n+1}) obtained by dropping one hydrogen from fully saturated compound, for example, methyl ($-CH_3$), ethyl ($-CH_2CH_3$), propyl ($-CH_2CH_2CH_3$), or isopropyl [$(CH_3)_2CH-$].

Alumina (Al_2O_3): used in separation methods as an adsorbent and in refining as a catalyst.

American Society for Testing and Materials (ASTM) the official organization in the United States for designing standard tests for petroleum and other industrial products.

Anaerobic bacteria any bacteria that can grow and divide in the partial or complete absence of oxygen.

Anaerobic respiration The process whereby microorganisms use a chemical other than oxygen as an electron acceptor; common substitutes for oxygen are nitrate, sulfate, and iron.

Analyte the chemical for which a sample is tested or analyzed. *Antibody* A molecule having chemically reactive sites specific for certain other molecules.

Analytical equivalence the acceptability of the results obtained from different laboratories; a range of acceptable results.

Antibody a molecule having chemically reactive sites specific for certain other molecules.

API Gravity An American Petroleum Institute measure of *density* for petroleum:

$$API\ Gravity = \left[141.5 / \left(specific\ gravity\ at\ 15.6\,^\circ C \right) - 131.5 \right]$$

Fresh water has a gravity of 10 °API. The scale is commercially important for ranking oil quality; heavy oils are typically <20°API; medium oils are 20–35 °API; light oils are 35–45°API.

Aromatic organic cyclic compounds that contain one or more benzene rings; these can be monocyclic, bicyclic, or polycyclic hydrocarbons and their substituted derivatives. In aromatic ring structures, every ring carbon atom possesses one double bond.

Aromatic hydrocarbon a hydrocarbon characterized by the presence of an aromatic ring or condensed aromatic rings; benzene and substituted benzene, naphthalene and substituted naphthalene, phenanthrene and substituted phenanthrene, as well as the higher condensed ring systems; compounds that are distinct from those of aliphatic compounds (*q.v.*) or alicyclic compounds (*q.v.*).

Asphalt the nonvolatile product obtained by distillation and further processing of an asphaltic crude oil; a manufactured product.

Asphaltene constituents the individual molecular species that make up the asphaltene fraction; these constituents can vary in polarity and molecular weight; an *average structure* is not a true representation of the molecular constituents (asphaltene fraction) or of the reactivity and behavior of the molecular constituents (asphaltene fraction).

Asphaltene fraction a complex mixture of heavy organic compounds precipitated from oils and *bitumens* by natural processes or in laboratory by addition of excess *n*-pentane, or *n*-heptane; after precipitation of the *asphaltene fraction*, the remaining oil or *bitumen* consists of *saturates*, *aromatics*, and *resins*.

Assay qualitative or (more usually) quantitative determination of the components of a material or system.

ASTM See American Society for Testing and Materials.

Atmospheric distillation the refining process of separating crude oil components at atmospheric pressure by heating to temperatures on the order of 340°C (645°F)—depending on the nature of the crude oil and desired products—and subsequent condensation of the fractions by cooling.

Atmospheric equivalent boiling point (AEBP) a mathematical method of estimating the boiling point at atmospheric pressure of nonvolatile fractions of petroleum.

Atmospheric residuum a residuum (*q.v.*) obtained by distillation of a crude oil under atmospheric pressure and which boils above 350°C (660°F).

Attainment area a geographical area that meets NAAQS for criteria air pollutants (See also Non-attainment area).

Attapulgus clay see Fuller's earth.

Average structure a structure based on physical and/or chemical data that is incorrectly believed to be a representation of the molecular constituents or of the behavior of the asphaltene fraction.

Aviation gasoline (finished) a complex mixture of relatively volatile hydrocarbons with or without small quantities of additives, blended to form a fuel suitable for use in aviation reciprocating engines.

Aviation gasoline blending components types of naphtha that will be used for blending or compounding into finished aviation gasoline (e.g., straight-run naphtha, alkylate, reformate, benzene, toluene, and xylene)—excludes oxygenates (alcohols, ethers), butane, and pentanes plus.

BACT best available control technology.

Baghouse a filter system for the removal of particulate matter from gas streams; so called because of the similarity of the filters to coal bags.

Barrel the unit of measurement of liquids in the petroleum industry; equivalent to 42 US standard gallons or 33.6 imperial gallons.

Barrels per calendar day the amount of input that a distillation facility can process under usual operating conditions. The amount is expressed in terms of capacity during a 24-hour period and reduces the maximum processing capability of all units at the facility under continuous operation to account for any limitations that may delay, interrupt, or slow down production.

Barrels per stream day the maximum number of barrels of input that a distillation facility can process within a 24-hour period when running at full capacity under optimal crude and product slate conditions with no allowance for downtime.

Base number the quantity of acid, expressed in milligrams of potassium hydroxide per gram of sample that is required to titrate a sample to a specified end point.

Base oil a finished petroleum lubricant stock that when blended with other materials (additives) produces special-purpose products such as engine oils.

Basic nitrogen nitrogen (in petroleum) that occurs in pyridine form.

Basic sediment and water (BS&W, BSW) the material that collects in the bottom of storage tanks usually composed of oil, water, and foreign matter; also called bottoms, bottom settlings.

Baumé gravity the specific gravity of liquids expressed as degrees on the Baumé (°B or °Bé) scale. For liquids lighter than water,

$$\mathrm{Sp\,gr\,60°F} = 140 / \left(130 + °\,\mathrm{B_J}\right)$$

For liquids heavier than water

$$\mathrm{Sp\,gr\,60°F} = 145 / \left(145 - °\,\mathrm{B_J}\right)$$

Bbl see Barrel.

Benzene a colorless aromatic liquid hydrocarbon (C_6H_6).

Benzin refined light naphtha used for extraction purposes.

Benzine an obsolete term for light petroleum distillates covering the gasoline and naphtha range; see Ligroine (Ligroin).

Benzol the general term that refers to commercial or technical (not necessarily pure) benzene; also the term used for aromatic naphtha.

Bio-augmentation a process in which acclimated microorganisms are added to soil and groundwater to increase biological activity. Spray irrigation is typically used for shallow contaminated soils, and injection wells are used for deeper contaminated soils.

Biodegradation the natural process whereby bacteria or other microorganisms chemically alter and break down organic molecules.

Biological marker (biomarker) complex organic compounds composed of carbon, hydrogen, and other elements, which are found in oil, *bitumen*, rocks, and sediments and which have undergone little or no change in structure from their parent organic molecules in living organisms; typically, biomarkers are isoprenoids, composed of isoprene subunits. Biomarkers include pristane, phytane, *triterpanes*, steranes, porphyrins, and other compounds.

Biological oxidation the oxidative consumption of organic matter by bacteria by which the organic matter is converted into gases.

Biomass biological organic matter.

Bioremediation a treatment technology that uses biological activity to reduce the concentration or toxicity of contaminants: materials are added to contaminated environments to accelerate natural biodegradation.

Bitumen a complex mixture of *hydrocarbonaceous constituents* of natural or pyrogenous origin or a combination of both; a semisolid to solid hydrocarbonaceous material found filling pores and crevices of sandstone, limestone, or argillaceous sediments.

Bituminous containing bitumen or constituting the source of bitumen.

Bituminous rock see Bituminous sand.

Bituminous sand a formation in which the bituminous material (see Bitumen) is found as a filling in veins and fissures in fractured rocks or impregnating relatively shallow sand, sandstone, and limestone strata; a sandstone reservoir that is impregnated with a heavy, viscous black petroleum-like material that cannot be retrieved through a well by conventional production techniques.

Black acid(s) a mixture of the sulfonates found in acid sludge that is insoluble in naphtha, benzene, and carbon tetrachloride; very soluble in water but insoluble in 30% sulfuric acid; in the dry, oil-free state, the sodium soaps are black powders.

Black oil a meaningless term generally applied to any dark-colored crude oils; a term sometimes applied to heavy oil.

Blending mixing two or more compatible base stock components to achieve targeted product specifications; usually refers to mixing homologous materials, whereas *compounding* refers to mixing base products with additives.

Blending plant a facility that has no refining capability but is capable of producing finished motor gasoline through mechanical blending of the component streams that constitute finished gasoline (or diesel).

Boiling point the temperature at which a liquid begins to boil—that is, it is the temperature at which the vapor pressure of a liquid is equal to the atmospheric or external pressure. The boiling point distributions of crude oils and petroleum products may be in a range from 30 to in excess of 700°C (86–1290°F).

Boiling range the range of temperature, usually determined at atmospheric pressure in standard laboratory apparatus, over which the distillation of oil commences, proceeds, and finishes.

Bromine number the number of grams of bromine absorbed by 100 g of oil, which indicates the percentage of double bonds in the material.

Brown acid oil-soluble petroleum sulfonates found in acid sludge that can be recovered by extraction with naphtha solvent. Brown-acid sulfonates are somewhat similar to mahogany sulfonates but are more water soluble. In the dry, oil-free state, the sodium soaps are light-colored powders.

BS&W see Basic sediment and water.

BTEX the acronym given to benzene, toluene, ethylbenzene, and the xylene isomers (*p*-, *m*-, and *o*-xylene.

BTX the acronym given to benzene, toluene, and xylene.

Bunker fuel heavy *residual oil*, also called bunker C, bunker C fuel oil, or bunker oil. See No. 6 Fuel oil.

Burner fuel oil any petroleum liquid suitable for combustion.

Burning oil illuminating oil, such as kerosene (kerosine) suitable for burning in a wick lamp.

Burning point see Fire point.

C_1, C_2, C_3, C_4, C_5 fractions a common way of representing fractions containing a preponderance of hydrocarbons having 1, 2, 3, 4, or 5 carbon atoms, respectively, and without reference to hydrocarbon type.

CAA Clean Air Act; this act is the foundation of air regulations in the United States.

Carbene the pentane- or heptane-insoluble material that is insoluble in benzene or toluene but which is soluble in carbon disulfide (or pyridine); a type of rifle used for hunting bison.

Carboid the pentane- or heptane-insoluble material that is insoluble in benzene or toluene and which is also insoluble in carbon disulfide (or pyridine).

Carbon preference index (CPI) the ratio of odd-to-even *n*-alkanes; odd/even CPI *alkanes* are equally abundant in petroleum but not in biological material—a CPI near 1 is an indication of petroleum.

Cat cracking see Catalytic cracking.

Catalyst a substance that alters the rate of a chemical reaction and may be recovered essentially unaltered in form or amount at the end of the reaction.

Catalyst coke coke or carbon deposited on the catalyst, thus deactivating the catalyst. The catalyst is reactivated by burning off the coke/carbon, which is used as a fuel in the refining process. This coke/carbon is not recoverable in a concentrated form.

Catalyst selectivity the relative activity of a catalyst with respect to a particular compound in a mixture, or the relative rate in competing reactions of a single reactant.

Catalyst stripping the introduction of steam at a point where spent catalyst leaves the reactor, in order to strip, that is, remove, deposits retained on the catalyst.

Catalytic activity the ratio of the space velocity of the catalyst under test to the space velocity required for the standard catalyst to give the same conversion as the catalyst being tested; usually multiplied by 100 before being reported.

Catalytic cracking the conversion of high-boiling feedstocks into lower-boiling products by means of a catalyst that may be used in a fixed bed (q.v.) or fluid bed (q.v.).

Catalytic hydrocracking a refining process that uses hydrogen and catalysts with relatively low temperatures and high pressures for converting middle boiling or residual material to high-octane gasoline, reformer charge stock, jet fuel, and/or high-grade fuel oil. The process uses one or more catalysts, depending upon product output, and can handle high-sulfur feedstocks without prior desulfurization.

Catalytic hydrotreating a refining process for treating petroleum fractions from atmospheric or vacuum distillation units (such as naphtha, middle distillates, reformer feeds, residual fuel oil, and heavy gas oil) and other petroleum (e.g., cat-cracked naphtha, coker naphtha, gas oil) in the presence of catalysts and substantial quantities of hydrogen. Hydrotreating includes desulfurization, denitrogenation, conversion of olefins to paraffins to reduce gum formation in gasoline, and other processes to upgrade the quality of the fractions.

Catalytic reforming rearranging hydrocarbon molecules in a gasoline-boiling-range feedstock to produce other hydrocarbons having a higher antiknock quality; isomerization of paraffins, cyclization of paraffins to naphthenes (q.v.), dehydrocyclization of paraffins to aromatics (q.v.).

Cetane index an approximation of the cetane number (q.v.) calculated from the density (q.v.) and mid-boiling point temperature (q.v.); see also Diesel index.

Cetane number a number indicating the ignition quality of diesel fuel; a high cetane number represents a short ignition delay time; the ignition quality of diesel fuel can also be estimated from the following formula:

$$\text{Diesel index} = \left(\text{aniline point}\left(°F\right) \times \text{API gravity}\right)100$$

CFR Code of Federal Regulations; Title 40 (40 CFR) contains the regulations for protection of the environment.

Characterization factor the UOP characterization factor K, defined as the ratio of the cube root of the molal average boiling point, T_B, in degrees Rankine (°R = °F + 460), to the specific gravity at 60° F/60° F:

$$K = \left(T_B\right)^{1/3} / \text{sp gr}$$

Ranges from 12.5 for paraffinic stocks to 10.0 for the highly aromatic stocks; also called the Watson characterization factor.

Check standard an analyte with a well-characterized property of interest, for example, concentration, density, and other properties that is used to verify method, instrument, and operator performance during regular operation; *check standards* may be obtained from a certified supplier, may be a pure substance with properties obtained from the literature or may be developed in-house.

Chemical dispersion in relation to oil spills, this term refers to the creation of oil-in-water *emulsions* by the use of chemical dispersants made for this purpose.

Chemical waste any solid, liquid, or gaseous material discharged from a process and that may pose substantial hazards to human health and environment.

Chromatographic adsorption selective adsorption on materials such as activated carbon, alumina, or silica gel; liquid or gaseous mixtures of hydrocarbons are passed through the adsorbent in a stream of diluent, and certain components are preferentially adsorbed.

Chromatography a method of separation based on selective adsorption; see also Chromatographic adsorption.

Chromatogram the resultant electrical output of sample components passing through a detection system following chromatographic separation. A chromatogram may also be called a *trace*.

Clay silicate minerals that also usually contain aluminum and have particle sizes are less than 0.002 micrometer; used in separation methods as an adsorbent and in refining as a catalyst.

Cleanup a preparatory step following extraction of a sample media designed to remove components that may interfere with subsequent analytical measurements.

Cloud point the temperature at which paraffin wax or other solid substances begin to crystallize or separate from the solution, imparting a cloudy appearance to the oil when the oil is chilled under prescribed conditions; sometimes arbitrarily called *wax appearance point*.

Coal a combustible black or brownish-black rock whose composition, including inherent moisture, consists of more than 50% w/w and more than 70% v/v of carbonaceous material; formed from plant remains and debris that have been compacted, hardened, chemically altered, and metamorphosed over geologic time.

Coke a gray-to-black solid carbonaceous material produced from petroleum during thermal processing; characterized by having a high carbon content (95%+ by weight) and a honeycomb type of appearance and is insoluble in organic solvents.

Coker the processing unit in which coking takes place.

Coking a process for the thermal conversion of petroleum in which gaseous, liquid, and solid (coke) products are formed.

Co-metabolism the process by which compounds in petroleum may be enzymatically attacked by microorganisms without furnishing carbon for cell growth and division; a variation on biodegradation in which microbes transform a contaminant even though the contaminant cannot serve as the primary energy source for the organisms. To degrade the contaminant, the microbes require the presence of other compounds (primary substrates) that can support their growth.

Complex modulus a measure of the overall resistance of a material to flow under an applied stress, in units of force per unit area. It combines *viscosity* and elasticity elements to provide a measure of "stiffness," or resistance to flow. The *complex modulus* is more useful than *viscosity* for assessing the physical behavior of very non-Newtonian materials such as *emulsions*.

Composition the general chemical makeup of petroleum.

Compounding mixing additives with oils, particularly lubricating oils, to impart oxidation resistance, rust resistance, detergency, or other desirable (necessary) characteristics.

Confirmation column a secondary column in chromatography that contains a stationary phase having different affinities for components in a mixture than in the primary column. Used to confirm analyses that may not be completely resolved using the primary column.

Contaminant a substance that causes deviation from the normal composition of an environment.

Cracking the thermal processes by which the constituents of petroleum are converted to lower-molecular-weight products; a process whereby the relative proportion of lighter or more volatile components of crude oil is increased by changing the chemical structure of the constituent hydrocarbons.

Criteria air pollutants air pollutants or classes of pollutants regulated by the Environmental Protection Agency; the air pollutants are (including VOCs) ozone, carbon monoxide, particulate matter, nitrogen oxides, sulfur dioxide, and lead.

Crude oil See *Petroleum*

Crude oil qualities refers to two properties of crude oil, the sulfur content and API gravity, and other assay data that affect processing complexity and product characteristics.

Culture the growth of cells or microorganisms in a controlled artificial environment.

Cut the *distillate* obtained between two given temperatures during a distillation process.

Cut point the boiling-temperature division between distillation fractions of petroleum.

Cycloalkanes (naphthenes, cycloparaffins) A saturated, cyclic compound containing only carbon and hydrogen. One of the simplest *cycloalkanes* is cyclohexane (C_6H_{12}). Steranes and triterpanes are branched naphthenes consisting of multiple condensed five- or six-carbon rings.

Deasphaltened oil the fraction of petroleum after the asphaltene constituents have been removed.

Deasphaltening removal of a solid powdery asphaltene fraction from petroleum by the addition of the low-boiling liquid hydrocarbons such as n-pentane or n-heptane under ambient conditions.

Deasphalting the removal of the asphaltene fraction from petroleum by the addition of a low-boiling hydrocarbon liquid such as n-pentane or n-heptane; more correctly, the removal asphalt (tacky, semi-solid) from petroleum (as occurs in a refinery asphalt plant) by the addition of liquid propane or liquid butane under pressure.

Decoking removal of petroleum coke from equipment such as coking drums; hydraulic decoking uses high-velocity water streams.

Delayed coking a coking process in which the thermal reaction is allowed to proceed to completion to produce gaseous, liquid, and solid (coke) products.

Density The mass per unit volume of a substance. *Density* is temperature dependent, generally decreasing with temperature. The density of oil relative to water, its specific gravity, governs whether a particular oil will float on water. Most fresh crude oils and fuels will float on water. Bitumen and certain residual fuel oils, however, may have densities greater than water at some temperature ranges and may submerge in water. The density of spilled oil will also increase with time as components are lost due to weathering.

Desalting removal of mineral salts (mostly chlorides) from crude oils.

Desorption the reverse process of adsorption whereby adsorbed matter is removed from the adsorbent; also used as the reverse of absorption (*q.v.*).

Dewaxing see Solvent dewaxing.

Diesel fuel fuel used for internal combustion in diesel engines, usually that fraction which distills within the temperature range approximately 200–370°C. A general term covering oils used as fuel in diesel and other compression ignition engines.

Differential scanning calorimetry (DSC) a test widely used to characterize waxes, DSC measures the amount of energy consumed under controlled heating and cooling rates. Curves of heat flow versus temperature provide insight into the thermal characteristics of wax, including crystalline transitions such as solid to solid, solid to liquid, and liquid to solid. Common values obtained from the curves include the initial and ending temperatures for heat flow and heat of fusion, measured in joules per gram.

Dispersant (chemical dispersant) a chemical that reduces the surface tension between water and a hydrophobic substance such as oil. In the case of an oil spill, dispersants facilitate the breakup and dispersal of an oil slick throughout the water column in the form of an oil-in-water emulsion; chemical dispersants can only be used in areas where biological damage will not occur and must be approved for use by government regulatory agencies.

Distillate a product obtained by condensing the vapors evolved when a liquid is boiled and collecting the condensation in a receiver that is separate from the boiling vessel.

Distillate fuel oil a general classification for one of the petroleum fractions produced in conventional distillation operations. It includes diesel fuels and fuel oils. Products known as No. 1, No. 2, and No. 4 diesel fuel are used in on-highway diesel engines, such as those in trucks and automobiles, as well as off-highway engines, such as those in railroad locomotives and agricultural machinery. Products known as No. 1, No. 2, and No. 4 fuel oils are used primarily for space heating and electric power generation.

Distillation a process for separating liquids with different boiling points.

Distillation curve see Distillation profile.

Distillation profile the distillation characteristics of petroleum or a petroleum product showing the temperature and the percentage distilled.

Distillation range the difference between the temperature at the initial boiling point and at the end point, as obtained by the distillation test.

Domestic heating oil see No. 2 Fuel Oil.

Effluent any contaminating substance, usually a liquid that enters the environment via a domestic industrial, agricultural, or sewage plant outlet.

Electric desalting a continuous process to remove inorganic salts and other impurities from crude oil by settling out in an electrostatic field.

Electrical precipitation a process using an electrical field to improve the separation of hydrocarbon reagent dispersions. May be used in chemical treating processes on a wide variety of refinery stocks.

Electron acceptor the compound that receives electrons (and therefore is reduced) in the energy-producing oxidation–reduction reactions that are essential for the growth of microorganisms and bioremediation—common electron acceptors in bioremediation are oxygen, nitrate, sulfate, and iron.

Electron donor the compound that donates electrons (and therefore is oxidized). In bioremediation, the organic contaminant often serves as an electron donor.

Electrostatic precipitators devices used to trap fine dust particles (usually in the size range 30–60 micrometers) that operate on the principle of imparting an electric charge to particles in an incoming air stream and which are then collected on an oppositely charged plate across a high-voltage field.

Eluate the solutes, or analytes, moved through a chromatographic column (see *elution*).

Eluent solvent used to elute sample.

Elution a process whereby a solute is moved through a chromatographic column by a solvent (liquid or gas), or eluent.

Emission control the use of gas cleaning processes to reduce emissions.

Emission standard the maximum amount of a specific pollutant permitted to be discharged from a particular source in a given environment.

Emulsification the process of *emulsion* formation, typically by mechanical mixing. In the environment, *emulsions* are most often formed as a result of wave action. Chemical agents can be used to prevent the formation of *emulsions* or to "break" the *emulsions* to their component oil and water phases.

Emulsion a stable mixture of two immiscible liquids, consisting of a continuous phase and a dispersed phase. Oil and water can form both oil-in-water and water-in-oil emulsions. The former is termed dispersion, while *emulsion* implies the latter. Water-in-oil emulsions formed from petroleum and brine can be grouped into four stability classes: stable, a formal emulsion that will persist indefinitely; meso-stable, which gradually degrade over time due to a lack of one or more stabilizing factors; entrained water, a mechanical mixture characterized by high viscosity of the petroleum component that impedes separation of the two phases; and unstable, which are mixtures that rapidly separate into immiscible layers.

Emulsion stability generally accompanied by a marked increase in *viscosity* and elasticity over that of the parent oil that significantly changes behavior. Coupled with the increased volume due to the introduction of brine, emulsion formation has a large effect on the choice of countermeasures employed to combat a spill.

Engineered bioremediation a type of remediation that increases the growth and degradative activity of microorganisms by using engineered systems that supply nutrients, electron acceptors, and/or other growth-stimulating materials.

Enhanced bioremediation a process that involves the addition of microorganisms (e.g., fungi, bacteria, and other microbes) or nutrients (e.g., oxygen, nitrates) to the subsurface environment to accelerate the natural biodegradation process.

Enzyme any of a group of catalytic proteins that are produced by cells and that mediate or promote the chemical processes of life without themselves being altered or destroyed.

EPA Environmental Protection Agency.

Equipment blank a sample of analyte-free media that has been used to rinse the sampling equipment. It is collected after completion of decontamination and prior to sampling. This blank is useful in documenting and controlling the preparation of the sampling and laboratory equipment.

Ex situ bioremediation a process that involves removing the contaminated soil or water to another location before treatment.

Extract the portion of a sample preferentially dissolved by the solvent and recovered by physically separating the solvent.

Fabric filters filters made from fabric materials and used for removing particulate matter from gas streams (see Baghouse).

FCC fluid catalytic cracking.

FCCU fluid catalytic cracking unit.

Feedstock petroleum as it is fed to the refinery; a refinery product that is used as the raw material for another process; the term is also generally applied to raw materials used in other industrial processes.

Fermentation the process whereby microorganisms use an organic compound as both electron donor and electron acceptor, converting the compound to fermentation products such as organic acids, alcohols, hydrogen, and carbon dioxide.

Filtration the use of an impassable barrier to collect solids but which allows liquids to pass.

Fingerprint a chromatographic signature of relative intensities used in oil–oil or oil–source rock correlations. Mass chromatograms of *steranes* or *terpanes* are examples of *fingerprints* that can be used for qualitative or quantitative comparison of oils.

Fingerprint analysis a direct injection GC/FID analysis in which the detector output—the chromatogram—is compared to chromatograms of reference materials as an aid to product identification.

Fire point the lowest temperature at which, under specified conditions in standardized apparatus, a petroleum product vaporizes sufficiently rapidly to form above its surface an air–vapor mixture that burns continuously when ignited by a small flame.

Fischer–Tropsch process a process for synthesizing hydrocarbons and oxygenated chemicals from a mixture of hydrogen and carbon monoxide.

Fixed bed a stationary bed (of catalyst) to accomplish a process (see Fluid bed).

Flame ionization detector a detector for a gas chromatograph that measures any *detector (FID)* thing that can burn.

Flammability range the range of temperature over which a chemical is flammable.

Flammable a substance that will burn readily.

Flammable liquid a liquid having a flash point below 37.8°C (100°F).

Flammable solid a solid that can ignite from friction or from heat remaining from its manufacture, or which may cause a serious hazard if ignited.

Flash point The temperature at which the vapor over a liquid will ignite when exposed to an ignition source. A liquid is considered to be flammable if its *flash point* is less than 60°C. *Flash point* is an extremely important factor in relation to the safety of spill cleanup operations. Gasoline and other light fuels can ignite under most ambient conditions and therefore are a serious hazard when spilled. Many freshly spilled crude oils also have low *flash points* until the lighter components have evaporated or dispersed.

Flue gas gas from the combustion of fuel, the heating value of which has been substantially spent and which is, therefore, discarded to the flue or stack.

Fluid catalytic cracking cracking in the presence of a fluidized bed of catalyst.

Fluid coking a continuous fluidized solids process that cracks feed thermally over heated coke particles in a reactor vessel to gas, liquid products, and coke.

Fly ash particulate matter produced from mineral matter in coal that is converted during combustion to finely divided inorganic material and which emerges from the combustor in the gases.

Fraction one of the portions of a chemical mixture separated by chemical or physical means from the remainder.

Fractional composition the composition of petroleum as determined by fractionation (separation) methods.

Fractional distillation the separation of the components of a liquid mixture by vaporizing and collecting the fractions, or cuts, which condense in different temperature ranges.

Fractionating column a column arranged to separate various fractions of petroleum by a single distillation and which may be tapped at different points along its length to separate various fractions in the order of their boiling points.

Fractionation the separation of petroleum into the constituent fractions using solvent or adsorbent methods; chemical agents such as sulfuric acid may also be used.

Fuel oil a general term applied to oil used for the production of power or heat. In a more restricted sense, it is applied to any petroleum product that is used as boiler fuel or in industrial furnaces. These oils are normally *residues*, but blends of distillates and *residues* are also used as fuel oil. The wider term *liquid fuel* is sometimes used, but the term *fuel oil* is preferred; also called heating oil; see also No. 1 to No. 4 Fuel oils.

Fuller's earth clay that has high adsorptive capacity for removing color from oils; Attapulgus clay is a widely used fuller's earth.

Functional group the portion of a molecule that is characteristic of a family of compounds and determines the properties of these compounds.

Furnace oil a distillate fuel primarily intended for use in domestic heating equipment.

Gas chromatography (GC) a separation technique involving passage of a gaseous moving phase through a column containing a fixed liquid phase; it is used principally as a quantitative analytical technique for compounds that are volatile or can be converted to volatile forms.

Gas oil a petroleum distillate with a viscosity and *distillation range* intermediate between those of *kerosene* and *light lubricating oil*.

Gaseous nutrient injection a process in which nutrients are fed to contaminated groundwater and soil via wells to encourage and feed naturally occurring microorganisms—the most common added gas is air. In the presence of sufficient oxygen, microorganisms convert many organic contaminants to carbon dioxide, water, and microbial cell mass. In the absence of oxygen, organic contaminants are metabolized to methane, limited amounts of carbon dioxide, and trace amounts of hydrogen gas. Another gas that is added is methane. It enhances degradation by co-metabolism in which as bacteria consume the methane, they produce enzymes that react with the organic contaminant and degrade it to harmless minerals.

Gaseous pollutants gases released into the atmosphere that act as primary or secondary pollutants.

Gasoline fuel for the internal combustion engine that is commonly, but improperly, referred to simply as *gas*; the terms *petrol* and *benzine* are commonly used in some countries.

Gasoline blending components naphtha fractions that will be used for blending or compounding into finished aviation or motor gasoline (e.g., straight-run gasoline, alkylate, reformate, benzene, toluene, and xylene); excludes oxygenates (alcohols, ethers), butane, and pentanes plus.

GC-MS Gas chromatography–mass spectrometry.

GC-TPH GC-detectable total petroleum hydrocarbons, that is the sum of all GC-resolved and unresolved hydrocarbons. The resolvable hydrocarbons appear as peaks, and the unresolvable hydrocarbons appear as the area between the lower baseline and the curve defining the base of resolvable peaks.

Gel permeation chromatography (GPC) method used to measure the molecular weight distribution for synthetic polyethylene waxes. Though it cannot match the resolution available through gas chromatography, gel permeation chromatography can discern molecular weights and molecular weight distribution for products that have high molecular weight too high for gas chromatography.

Gravimetric analysis a technique of quantitative analytical chemistry in which a desired constituent is efficiently recovered and weighed.

Grease a semisolid or solid lubricant consisting of a stabilized mixture of mineral, fatty, or synthetic oil with soaps, metal salts, or other thickeners.

Greenhouse effect warming of the earth due to entrapment of the sun's energy by the atmosphere.

Greenhouse gases gases that contribute to the greenhouse effect.

HAP(s) hazardous air pollutant(s).

Headspace the vapor space above a sample into which volatile molecules evaporate. Certain methods sample this vapor.

Heating oil see Fuel oil.

Heavy ends the highest boiling portion of a petroleum fraction; see also Light ends.

Heavy fuel oil fuel oil having a high density and viscosity; generally residual fuel oil such as No. 5 and No 6. fuel oil (q.v.)

Heavy oil petroleum having an API gravity of less than 20°.

Heavy petroleum see Heavy oil.

Heteroatom compounds chemical compounds that contain nitrogen and/or oxygen and/or sulfur and/or metals bound within their molecular structure(s).

HF alkylation an alkylation process whereby olefins (C_3, C_4, C_5) are combined with *iso*-butane in the presence of hydrofluoric acid catalyst.

Hopane a pentacyclic *hydrocarbon* of the *triterpane* group believed to be derived primarily from bacterio-hopanoids in bacterial membranes.

Hydraulic fluid a fluid supplied for use in hydraulic systems. Low viscosity and low *pour point* are desirable characteristics. Hydraulic fluids may be of petroleum or nonpetroleum origin.

Hydrocarbon one of a very large and diverse group of chemical compounds composed only of carbon and hydrogen; the largest source of hydrocarbons is petroleum crude oil; the principal constituents of crude oils and refined petroleum products; a molecule that consists *only* of hydrogen and carbon atoms.

Hydrocracking a catalytic high-pressure high-temperature process for the conversion of petroleum feedstocks in the presence of fresh and recycled hydrogen; carbon–carbon bonds are cleaved in addition to the removal of heteroatomic species.

Hydrocracking catalyst a catalyst used for hydrocracking that typically contains separate hydrogenation and cracking functions.

Hydrotreating the removal of heteroatomic (nitrogen, oxygen, and sulfur) species by treatment of a feedstock or product at relatively low temperatures in the presence of hydrogen.

Ignitability characteristic of liquids whose vapors are likely to ignite in the presence of ignition source; also characteristic of nonliquids that may catch fire from friction or contact with water and that burn vigorously.

Illuminating oil oil used for lighting purposes.

Immunoassay portable tests that take advantage of an interaction between an antibody and a specific analyte. Immunoassay tests are semiquantitative and usually rely on color changes of varying intensities to indicate relative concentrations.

In situ bioremediation a process that treats the contaminated water or soil where it was found.

Infrared spectroscopy an analytical technique that quantifies the vibration (stretching and bending) that occurs when a molecule absorbs (heat) energy in the infrared region of the electromagnetic spectrum.

Inoculum a small amount of material (either liquid or solid) containing bacteria removed from a culture in order to start a new culture.

Inorganic pertaining to, or composed of, chemical compounds that are not organic, that is, contain no carbon–hydrogen bonds; examples include chemicals with no carbon and those with carbon in non-hydrogen-linked forms.

Interfacial tension the net energy per unit area at the interface of two substances, such as oil and water or oil and air. The air/liquid interfacial tension is often referred to as surface tension. The SI units for *interfacial tension* are milli-Newtons per meter (mN/m). The higher the *interfacial tension*, the less attractive the two surfaces are to each other and the more size of the interface will be minimized. Low surface tensions can drive the spreading of one fluid on another. The surface tension of oil, together its viscosity, affects the rate at which spilled oil will spread over a water surface or into the ground.

Internal Standard (IS) a pure analyte added to a sample extract in a known amount, which is used to measure the relative responses of other analytes and surrogates that are components of the same solution. The *internal standard* must be an analyte that is not a sample component.

Intrinsic bioremediation a type of bioremediation that manages the innate capabilities of naturally occurring microbes to degrade contaminants without taking any engineering steps to enhance the process.

Isomerization the conversion of a *normal* (straight-chain) paraffin hydrocarbon into an *iso* (branched-chain) paraffin hydrocarbon having the same atomic composition.

Jet fuel fuel meeting the required properties for use in jet engines and aircraft turbine engines.

Kauri butanol number A measurement of solvent strength for hydrocarbon solvents; the higher the kauri-butanol (KB) value, the stronger the solvency; the test method (ASTM D1133) is based on the principle that kauri resin is readily soluble in butyl alcohol but not in hydrocarbon solvents, and the resin solution will tolerate only a certain amount of dilution and is reflected as a cloudiness when the resin starts to come out of solution; solvents such as toluene can be added in a greater amount (and thus have a higher KB value) than weaker solvents like hexane.

Kerosene (kerosine) a fraction of petroleum that was initially sought as an illuminant in lamps; a precursor to diesel fuel with a *distillation* range generally falls within the limits of 150 and 300°C; main uses are as a jet engine fuel, an illuminant, for heating purposes, and as a fuel for certain types of internal combustion engines.

K-factor see Characterization factor.

LAER lowest achievable emission rate; the required emission rate in non-attainment permits.

Light ends the lower-boiling components of a mixture of hydrocarbons; see also Heavy ends, Light hydrocarbons.

Light hydrocarbons hydrocarbons with molecular weights less than that of heptane (C_7H_{16}).

Light oil the products distilled or processed from crude oil up to, but not including, the first lubricating-oil distillate.

Light petroleum petroleum having an API gravity greater than 20°.

Ligroine (Ligroin) a saturated petroleum naphtha boiling in the range of 20–135° C (68–275° F) and suitable for general use as a solvent; also called benzine or petroleum ether.

Liquefied petroleum gas propane, butane, or mixtures thereof, gaseous at atmospheric temperature and pressure, held in the liquid state by pressure to facilitate storage, transport, and handling.

Liquid chromatography a chromatographic technique that employs a liquid mobile phase.

Liquid/liquid extraction an extraction technique in which one liquid is shaken with or contacted by an extraction solvent to transfer molecules of interest into the solvent phase.

Lubricants substances used to reduce friction between bearing surfaces or as process materials either incorporated into other materials used as processing aids in the manufacture of other products, or used as carriers of other materials. Petroleum lubricants may be produced either from distillates or residues. Lubricants include all grades of lubricating oils from spindle oil to cylinder oil and those used in greases.

MACT maximum achievable control technology. Applies to major sources of hazardous air pollutants.

Major source a source that has a potential to emit for a regulated pollutant that is at or greater than an emission threshold set by regulations.

Maltenes that fraction of petroleum that is soluble in, for example, pentane or heptane; deasphaltened oil (*q.v.*); also the term arbitrarily assigned to the pentane-soluble portion of petroleum that is relatively high boiling (>300°C, 760 mm) (see also Petrolenes).

Mass spectrometer an analytical technique that *fractures* organic compounds into characteristic "fragments" based on functional groups that have a specific mass-to-charge ratio.

MCL maximum contaminant level as dictated by regulations.

MDL See Method detection limit.

Metabolism the physical and chemical processes by which foodstuffs are synthesized into complex elements, complex substances are transformed into simple ones, and energy is made available for use by an organism; thus, all biochemical reactions of a cell or tissue, both synthetic and degradative, are included.

Metabolize a product of metabolism.

Method Detection Limit the smallest quantity or concentration of a substance that the instrument can measure.

Microbe the shortened term for microorganism.

Microcrystalline wax wax extracted from certain petroleum residua and having a finer and less apparent crystalline structure than paraffin wax.

Microorganism an organism of microscopic size that is capable of growth and reproduction through biodegradation of food sources, which can include hazardous contaminants; microscopic organisms including bacteria, yeasts, filamentous fungi, algae, and protozoa.

Middle distillate one of the distillates obtained between *kerosene* and *lubricating oil* fractions in the refining processes. These include *light fuel oils* and *diesel fuels*.

Mineral hydrocarbons petroleum hydrocarbons, considered *mineral* because they come from the earth rather than from plants or animals.

Mineral oil the older term for petroleum; the term was introduced in the nineteenth century as a means of differentiating petroleum (rock oil) from whale oil or oil from plants that, at the time, were the predominant illuminant for oil lamps.

Mineralization the biological process of complete breakdown of organic compounds, whereby organic materials are converted to inorganic products (e.g., the conversion of hydrocarbons to carbon dioxide and water).

Mobile phase in chromatography, the phase (gaseous or liquid) responsible for moving an introduced sample through a porous medium to separate components of interest.

MSDS Material safety data sheet.

MTBE (Methyl Tertiary Butyl Ether) a fuel additive that has been used in the United States since 1979. Its use began as a replacement for lead in gasoline because of health hazards associated with lead. MTBE has distinctive physical properties that result in it being highly soluble, persistent in the environment, and able to migrate through the ground. Environmental regulations have required the monitoring and cleanup of MTBE at petroleum-contaminated sites since February 1990; the program continues to monitor studies focusing on the potential health effects of MTBE and other fuel additives.

NAAQS National Ambient Air Quality Standards; standards exist for the pollutants known as the criteria air pollutants: nitrogen oxides (NO_x), sulfur oxides (SO_x), lead, ozone, particulate matter less than 10 micrometers in diameter, and carbon monoxide (CO).

Naphtha a generic term applied to refined, partly refined, or unrefined petroleum products and liquid products of natural gas, the majority of which distills below 240°C (464°F); the volatile fraction of petroleum that is used as a solvent or as a precursor to gasoline.

Naphthenes cycloparaffins.

Natural gas a gaseous mixture of hydrocarbon compounds, the primary one being methane.

Natural gas liquids the hydrocarbons in natural gas that are separated from the gas as liquids through the process of absorption, condensation, adsorption, or other methods in gas processing or cycling plants. Generally, such liquids consist of propane and heavier hydrocarbons and are commonly referred to as lease condensate, natural gasoline, and liquefied petroleum gases. Natural gas liquids include natural gas plant liquids (primarily ethane, propane, butane, and isobutane) and lease condensate (primarily pentanes produced from natural gas at field facilities).

NESHAP National Emissions Standards for Hazardous Air Pollutants; emission standards for specific source categories that emit or have the potential to emit one or more hazardous air pollutants; the standards are modeled on the best practices and most effective emission reduction methodologies in use at the affected facilities.

Non-attainment area a geographical area that does not meet NAAQS for criteria air pollutants (See also Attainment area).

NOx oxides of nitrogen.

No. 1 Fuel oil very similar to kerosene (q.v.) and is used in burners where vaporization before burning is usually required and a clean flame is specified.

No. 2 Fuel oil also called domestic heating oil; has properties similar to diesel fuel and heavy jet fuel; used in burners where complete vaporization is not required before burning.

No. 4 Fuel oil a light industrial heating oil and is used where preheating is not required for handling or burning; there are two grades of No. 4 fuel oil, differing in safety (flash point) and flow (viscosity) properties.

No. 5 Fuel oil a heavy industrial fuel oil that requires preheating before burning.

No. 6 Fuel oil a heavy fuel oil and is more commonly known as Bunker C oil when it is used to fuel ocean-going vessels; preheating is always required for burning this oil.

Nuclear magnetic resonance analysis (NMR) used to measure properties such as amount and type of branching; also used for both characterization and quality control of petroleum products.

Olefin synonymous with *alkene*.

Oleophilic oil seeking or oil loving (e.g., nutrients that stick to or dissolve in oil).

Organic chemical compounds based on carbon that also contain hydrogen, with or without oxygen, nitrogen, and other elements.

Organic liquid nutrient injection an enhanced bioremediation process in which an organic liquid, which can be naturally degraded and fermented in the subsurface to result in the generation of hydrogen. The most commonly added for enhanced anaerobic bioremediation include lactate, molasses, hydrogen release compounds (HRCs⁻), and vegetable oils.

Oxidation the transfer of electrons away from a compound, such as an organic contaminant; the coupling of oxidation to reduction (see below) usually supplies energy that microorganisms use for growth and reproduction. Often (but not always), oxidation results in the addition of an oxygen atom and/or the loss of a hydrogen atom.

Oxygen enhancement with hydrogen peroxide an alternative process to pumping oxygen gas into groundwater involves injecting a dilute solution of hydrogen peroxide. Its chemical formula is H_2O_2, and it easily releases the extra oxygen atom to form water and free oxygen. This circulates through the contaminated groundwater zone to enhance the rate of aerobic biodegradation of organic contaminants by naturally occurring microbes. A solid peroxide product [e.g., oxygen-releasing compound (ORC⁻)] can also be used to increase the rate of biodegradation.

Oxygenated gasoline gasoline with added ethers or alcohols, formulated according to the Federal Clean Air Act to reduce carbon monoxide emissions during winter months.

Oxygenates substances that, when added to gasoline, increase the amount of oxygen in that gasoline blend. Fuel ethanol, methyl tertiary butyl ether (MTBE), ethyl tertiary butyl ether (ETBE), and methanol are common oxygenates.

Nitrate enhancement a process in which a solution of nitrate is sometimes added to groundwater to enhance anaerobic biodegradation.

PAHs polycyclic aromatic hydrocarbons. Alkylated *PAHs* are *alkyl group* derivatives of the parent *PAHs*. The five target alkylated *PAHs* referred to in this report are the alkylated naphthalene, phenanthrene, dibenzothiophene, fluorene, and chrysene series.

Paraffin (alkane) one of a series of saturated aliphatic hydrocarbons, the lowest numbers of which are methane, ethane, and propane. The higher homologues are solid waxes.

Paraffin wax the colorless, translucent, highly crystalline material obtained from the light lubricating fractions of paraffinic crude oils (wax distillates).

Particulate matter particles in the atmosphere or on a gas stream that may be organic or inorganic and originate from a wide variety of sources and processes.

Partition ratios, K the ratio of total analytical concentration of a solute in the stationary phase, CS, to its concentration in the mobile phase, CM.

Partitioning in chromatography, the physical act of a solute having different affinities for the stationary and mobile phases.

Penetration a physical phenomenon that measures the depth in tenths of a millimeter that a needle of certain configuration under a given weight penetrates the surface of a wax or asphalt at a given temperature. A series of penetrations measured at different temperatures, rather than a single temperature, is preferred. Penetration of softer waxes and petrolatum may be measured using a cone (*cone penetration*) rather than a needle.

Petrol a term commonly used in some countries for gasoline.

Petrolatum a semisolid product, ranging from white to yellow in color, produced during refining of residual stocks; see Petroleum jelly.

Petrolenes the term applied to that part of the pentane-soluble or heptane-soluble material that is low boiling

(<300°C, <570°F, 760 mm) and can be distilled without thermal decomposition (see also Maltenes).

Petroleum (crude oil) naturally occurring mixture that consists of many types of hydrocarbons, but also containing sulfur, nitrogen, or oxygen derivatives. Petroleum may be of paraffinic, asphaltic, or mixed base, depending on the presence of *paraffin* wax and *bitumen* in the *residue* after atmospheric distillation. Petroleum composition varies according to the geological strata of its origin.

Petroleum refinery see Refinery.

Petroleum refining a complex sequence of events that result in the production of a variety of products.

Photoionization a gas chromatographic detection system that utilizes a *detector (PID)* ultraviolet lamp as an ionization source for analyte detection. It is usually used as a selective detector by changing the photon energy of the ionization source.

Phytodegradation the process in which some plant species can metabolize VOC contaminants. The resulting metabolic products include trichloroethanol, trichloroacetic acid, and dichloracetic acid. Mineralization products are probably incorporated into insoluble products such as components of plant cell walls.

Phytovolatilization the process in which VOCs are taken up by plants and discharged into the atmosphere during transpiration.

Polar compound an organic compound with distinct regions of positive and negative charge. *Polar compounds* include alcohols, such as sterols, and some *aromatics*, such as monoaromatic steroids. Because of their polarity, these compounds are more soluble in polar solvents, including water, compared to nonpolar compounds of similar molecular structure.

Polycyclic aromatic hydrocarbons (PAHs) polycyclic aromatic hydrocarbons are a suite of compounds comprised of two or more condensed aromatic rings. They are found in many petroleum mixtures, and they are predominantly introduced to the environment through natural and anthropogenic combustion processes.

Polynuclear aromatic compound an aromatic compound having two or more fused benzene rings, for example, naphthalene, phenanthrene.

PONA analysis a method of analysis for paraffins (P), olefins (O), naphthenes (N), and aromatics (A).

Porphyrins organometallic constituents of petroleum that contain vanadium or nickel; the degradation products of chlorophyll that became included in the protopetroleum.

Positive bias a result that is incorrect and too high.

Pour point the lowest temperature at which oil will appear to flow under ambient pressure over a period of 5 seconds; the lowest temperature at which oil will pour or onds; the lowest temperature at which oil will pour or flow when it is chilled without disturbance under definite conditions. The *pour point* of crude oils generally varies from -60°C to 30°C. Lighter oils with low *viscosities* generally have lower *pour points*.

Primary substrates The electron donor and electron acceptor that are essential to ensure the growth of microorganisms. These compounds can be viewed as analogous to the food and oxygen that are required for human growth and reproduction.

Propagule any part of a plant (e.g., bud) that facilitates dispersal of the species and from which a new plant may form.

Propane deasphalting solvent deasphalting using propane as the solvent.

Propane dewaxing a process for dewaxing lubricating oils in which propane serves as the solvent.

PSD prevention of significant deterioration.

PTE potential to emit; the maximum capacity of a source to emit a pollutant, given its physical or operation design, and considering certain controls and limitations.

Purge and trap a chromatographic sample introduction technique in volatile components that are purged from a liquid medium by bubbling gas through it. The components are then concentrated by "trapping" them on a short intermediate column, which is subsequently heated to drive the components on to the analytical column for separation.

Purge gas typically helium or nitrogen, used to remove analytes from the sample matrix in purge/trap extractions.

RACT Reasonably Available Control Technology standards; implemented in areas of non-attainment to reduce emissions of volatile organic compounds and nitrogen oxides.

Recycling the use or reuse of chemical waste as an effective substitute for a commercial product or as an ingredient or feedstock in an industrial process.

Reduced crude a residual product remaining after the removal, by distillation or other means, of an appreciable quantity of the more volatile components of crude oil.

Reduction the transfer of electrons to a compound, such as oxygen, that occurs when another compound is oxidized.

Reductive dehalogenation a variation on biodegradation in which microbially catalyzed reactions cause the replacement of a halogen atom on an organic compound with a hydrogen atom. The reactions result in the net addition of two electrons to the organic compound.

Refinery a series of integrated unit processes by which petroleum can be converted to a slate of useful (salable) products.

Refinery gas a gas (or a gaseous mixture) produced as a result of refining operations.

Refining the process(es) by which petroleum is distilled and/or converted by application of physical and chemical processes to form a variety of products.

Reformate the liquid product of a reforming process.

Reforming the conversion of hydrocarbons with low octane numbers (q.v.) into hydrocarbons having higher octane numbers; for example, the conversion of a n-paraffin into an iso-paraffin.

Reformulated gasoline (RFG) gasoline designed to mitigate smog production and to improve air quality by limiting the emission levels of certain chemical compounds such as benzene and other aromatic derivatives; often contains oxygenates.

Residual fuel oil obtained by blending the residual product(s) from various refining processes with suitable diluent(s) (usually middle distillates) to obtain the required fuel oil grades; a general classification for the heavier fuel oils, known as No. 5 and No. 6 fuel oils, that remain after the distillate fuel oils and lighter hydrocarbons are distilled away in refinery operations. No. 6 fuel oil includes Bunker C fuel oil and is used for the production of electric power, space heating, vessel bunkering, and various industrial purposes.

Residual oil see Residuum.

Residuum (resid; _pl:_. residua) the residue obtained from petroleum after nondestructive distillation has removed all the volatile materials from crude oil, for example, an atmospheric (345°C, 650°F+) residuum.

Resins the name given to a large group of _polar compounds_ in oil. These include hetero-substituted _aromatics_, acids, ketones, alcohols, and monoaromatic steroids. Because of their polarity, these compounds are more soluble in _polar_ solvents, including water, than the nonpolar compounds, such as _waxes_ and _aromatics_, of similar molecular weight. They are largely responsible for oil _adhesion_.

Retention time the time it takes for an eluate to move through a chromatographic system and reach the detector. Retention times are reproducible and can therefore be compared to a standard for analyte identification.

Rhizodegradation the process whereby plants modify the environment of the root zone soil by releasing root exudates and secondary plant metabolites. Root exudates are typically photosynthetic carbon, low-molecular-weight molecules, and high-molecular-weight organic acids. This complex mixture modifies and promotes the development of a microbial community in the rhizosphere. These secondary metabolites have a potential role in the development of naturally occurring contaminant-degrading enzymes.

Rhizosphere the soil environment encompassing the root zone of the plant.

RRF relative response factor.

Saponification number the number of milligrams of potassium hydroxide that reacts with 1 g of sample under elevated temperatures and indicates the amount of free carboxylic acid plus any ester materials that may be saponified. Both the acid number and the saponification number are generally provided to give an indication of the acid content of petroleum and petroleum products.

SARA analysis a method of analysis for saturates, aromatics, resins, and asphaltene constituents.

SARA separation see SARA analysis.

Saturated hydrocarbon a saturated carbon–hydrogen compound with all carbon bonds filled; that is, there are no double or triple bonds, as in olefins or acetylenes.

Saturates paraffins and cycloparaffins (naphthenes).

Saybolt Furol viscosity the time, in seconds (Saybolt Furol Seconds, SFS), for 60 ml of fluid to flow through a capillary tube in a Saybolt Furol viscometer at specified temperatures between 70 and 210° F; the method is appropriate for high-viscosity oils such as transmission, gear, and heavy fuel oils.

Saybolt Universal viscosity the time, in seconds (Saybolt Universal Seconds, SUS), for 60 ml of fluid to flow through a capillary tube in a Saybolt Universal viscometer at a given temperature.

Scale wax a semi-refined paraffin wax with an oil content of 1–3% w/w.

Separatory funnel glassware shaped like a funnel with a stoppered rounded top and a valve at the tapered bottom, used for liquid/liquid separations.

SIM (Selecting Ion Monitoring) Mass spectrometric monitoring of a specific mass/charge (m/z) ratio. The _SIM_ mode offers better sensitivity than can be obtained using the full scan mode.

Slack wax generic term for the mixture of wax and oil recovered in a dewaxing process; may contain 2–35% w/w oil.

Sludge a semisolid to solid product that results from the storage instability and/or the thermal instability of petroleum and petroleum products.

Soap an emulsifying agent made from sodium or potassium salts of fatty acids.

Solubility the amount of a substance (solute) that dissolves in a given amount of another substance (solvent). Particularly relevant to oil spill cleanup is the measure of how much and the composition of oil which will dissolve in the water column. This is important as the soluble fractions of the oil are often toxic to aquatic life, especially at high concentrations. The _solubility_ of oil in water is very low, generally less than 1 part per million (ppm).

Soluble capable of being dissolved in a solvent.

Solvent a liquid in which certain kinds of molecules dissolve. While they typically are liquids with low boiling points, they may include high-boiling liquids, supercritical fluids, or gases.

Solvent extraction a process for separating liquids by mixing the stream with a solvent that is immiscible with part of the waste but that will extract certain components of the waste stream.

Sonication a physical technique employing ultrasound to intensely vibrate a sample media in extracting solvent and to maximize solvent/analyte interactions.

Sour crude oil crude oil containing an abnormally large amount of sulfur compounds; see also Sweet crude oil.

SO$_x$ oxides of sulfur.

Soxhlet extraction an extraction technique for solids in which the sample is repeatedly contacted with solvent over several hours, increasing extraction efficiency.

Specific gravity the mass (or weight) of a unit volume of any substance at a specified temperature compared to the mass of an equal volume of pure water at a standard temperature; see also Density.

Spent catalyst catalyst that has lost much of its activity due to the deposition of coke and metals.

Stabilization the removal of volatile constituents from a higher-boiling fraction or product (q.v. stripping); the production of a product that, to all intents and purposes, does not undergo any further reaction when exposed to the air.

Stationary phase in chromatography, the porous solid or liquid phase through which an introduced sample passes. The different affinities the stationary phase has for a sample allow the components in the sample to be separated, or resolved.

Sulfonic acids acids obtained by petroleum or a petroleum product with strong sulfuric acid.

Sulfuric acid alkylation an alkylation process in which olefins (C_3, C_4, and C_5) combine with *iso*-butane in the presence of a catalyst (sulfuric acid) to form branched-chain hydrocarbons used especially in gasoline blending stock.

Supercritical fluid an extraction method where the extraction fluid is present at a pressure and temperature above its critical point.

Surface-active agent a compound that reduces the surface tension of liquids or reduces interfacial tension between two liquids or a liquid and a solid; also known as surfactant, wetting agent, or detergent.

Sustainable enhancement an intervention action that continues until such time that the enhancement is no longer required to reduce contaminant concentrations or fluxes.

Steranes a class of tetracyclic, saturated biomarkers constructed from six isoprene subunits (~C_{30}). *Steranes* are derived from sterols, which are important membrane and hormone components in eukaryotic organisms. Most commonly used *steranes* are in the range of C_{26}–C_{30} and are detected using m/z 217 mass chromatograms.

Surrogate analyte a pure analyte that is extremely unlikely to be found in any sample, which is added to a sample aliquot in a known amount and is measured with the same procedures used to measure other components. The purpose of a *surrogate analyte* is to monitor the method performance with each sample.

SW-846 an EPA multivolume publication entitled *Test Methods for Evaluating Solid Waste, Physical/Chemical Methods*; the official compendium of analytical and sampling methods that have been evaluated and approved for use in complying with the RCRA regulations and that functions primarily as a guidance document setting forth acceptable, although not required, methods for the regulated and regulatory communities to use in responding to RCRA-related sampling and analysis requirements. SW-846 changes over time as new information and data are developed.

TAME (Tertiary amyl methyl ether (CH_3)$_2$(C_2H_5) $COCH_3$ an oxygenate blend stock formed by catalytic etherification of iso-amylene with methanol.

Target analyte target analytes are compounds that are required analytes in US EPA analytical methods. BTEX and PAHs are examples of petroleum-related compounds that are target analytes in US EPA Methods.

Terpanes a class of branched, cyclic alkane biomarkers including *hopanes* and tricyclic compounds. They are commonly monitored using m/z 191 mass chromatograms.

Thermal cracking a process that decomposes, rearranges, or combines hydrocarbon molecules by the application of heat, without the aid of catalysts.

Thin layer chromatography (TLC) a chromatographic technique employing a porous medium of glass coated with a stationary phase. An extract is spotted near the bottom of the medium and placed in a chamber with solvent (mobile phase). The solvent moves up the medium and separates the components of the extract, based on affinities for the medium and solvent.

Total acid number (TAN) the number of milligrams of potassium hydroxide necessary to neutralize 1 gram of sample and indicates the amount of free carboxylic acid present.

Total aromatics the sum of all resolved and unresolved aromatic hydrocarbons including the total of BTEX and other alkyl benzene compounds, total five target alkylated PAH homologues, and other EPA-priority PAHs.

Total five alkylated PAH homologues the sum of the five target PAHs (naphthalene, phenanthrene, dibenzothiophene, fluorene, chrysene) and their alkylated (C_1–C_4) homologues, as determined by GCMS. These five target alkylated PAH homologous series are oil-characteristic aromatic compounds.

Total n-alkanes the sum of all resolved *n-alkanes* (from C_8 to C_{40} plus pristane and phytane).

Total petroleum hydrocarbons (TPH) the family of several hundred chemical compounds that originally come from petroleum.

Total saturates the sum of all resolved and unresolved aliphatic hydrocarbons including the total n-alkanes, branched alkanes, and cyclic saturates.

TPH E gas chromatographic test for TPH extractable organic compounds.

TPH V gas chromatographic test for TPH volatile organic compounds.

TPH-D(DRO) gas chromatographic test for TPH diesel-range organics.

TPH-G(GRO) gas chromatographic test for TPH gasoline-range organics.

Trace element those elements that occur at very low levels in a given system.

Triterpanes a class of cyclic saturated *biomarkers* constructed from six isoprene subunits. Cyclic *terpane* compounds containing two, four, and six isoprene subunits are called monoterpane (C_{10}), diterpane (C_{20}), and *triterpane* (C_{30}), respectively.

UCM GC-unresolved complex mixture of hydrocarbons. The UCM appears as the "envelope" or hump area between the solvent baseline and the curve defining the base of resolvable peaks.

Ultimate analysis elemental composition.

Unresolved complex the thousands of compounds that a gas chromatograph *mixture (UCM)* is unable to fully separate.

Upgrading the conversion of petroleum to value-added salable products.

Vacuum distillation distillation (*q.v.*) under reduced pressure.

Vacuum residuum a residuum (*q.v.*) obtained by distillation of a crude oil under vacuum (reduced pressure); the portion of petroleum that boils above a selected temperature such as 510°C (950°F) or 565°C (1050°F).

Vapor pressure a measure of how oil partitions between the liquid and gas phases, or the partial pressure of a vapor above a liquid oil at a fixed temperature.

Visbreaking a process for reducing the viscosity of heavy feedstocks by controlled thermal decomposition.

Viscosity the resistance of a fluid to shear, movement, or flow. The viscosity of an oil is a function of its composition. In general, the greater the fraction of *saturates* and *aromatics* and the lower the amount of *asphaltene constituents* and *resin constituents*, the lower the viscosity. As oil weathers, the evaporation of the lighter components leads to increased viscosity. Viscosity also increases with decreased temperature, and decreases with increased temperature.

Viscosity a measure of the ability of a liquid to flow or a measure of its resistance to flow; the force required to move a plane surface of area 1 meter^2 (square meter) over another parallel plane surface 1 meter away at a rate of 1 meter/sec when both surfaces are immersed in the fluid.

VGC (viscosity-gravity constant) an index of the chemical composition of crude oil defined by the general relation between specific gravity, sg, at 60° F and Saybolt Universal viscosity, SUV, at 100° F.

VI (Viscosity index) an arbitrary scale used to show the magnitude of viscosity changes in lubricating oils with changes in temperature.

Viscosity-gravity constant see VGC.

Viscosity index see VI.

VOC (VOCs) volatile organic compound(s); volatile organic compounds are regulated because they are precursors to ozone; carbon-containing gases and vapors from incomplete gasoline combustion and from the evaporation of solvents.

Volatile readily dissipating by evaporation.

Volatile compounds a relative term that may mean (i) any compound that will purge, (ii) any compound that will elute before the solvent peak (usually those < C6), or (iii) any compound that will not evaporate during a solvent removal step.

Volatile organic compounds (VOC) organic compounds with high *vapor pressures* at normal temperatures. *VOCs* include light *saturates* and *aromatics*, such as pentane, hexane, *BTEX*, and other lighter substituted benzene compounds, which can make up to a few percentage of the total mass of some crude oils.

Watson characterization factor see Characterization factor.

Wax wax of petroleum origin consists primarily of normal paraffins; wax of plant origin consists of esters of unsaturated fatty acids.

Waxes Waxes are predominately straight-chain *saturates* with melting points above 20°C (generally, the *n*-alkanes C_{18} and higher molecular weight).

Weathered crude oil crude oil that, due to natural causes during storage and handling, has lost an appreciable quantity of its more volatile components; also indicates uptake of oxygen.

Weathering processes related to the physical and chemical actions of air, water, and organisms after oil spill. The major weathering processes include evaporation, dissolution, dispersion, photochemical oxidation, water-in-oil *emulsification*, microbial degradation, and adsorption onto suspended particulate materials, interaction with mineral fines, sinking, sedimentation, and formation of tar balls.

Wobbe Index (or Wobbe Number) the calorific value of a gas divided by the specific gravity.

Zeolite a crystalline aluminosilicate used as a catalyst and having a particular chemical and physical structure.

INDEX

Handbook of Petroleum Product Analysis, Second Edition. James G. Speight.
© 2015 John Wiley & Sons, Inc. Published 2015 by John Wiley & Sons, Inc.

CHEMICAL ANALYSIS

A SERIES OF MONOGRAPHS ON ANALYTICAL CHEMISTRY AND ITS APPLICATIONS

Series Editor
MARK F. VITHA